Interactomics and Systems Biology

Series Editor
Benoit Coulombe
Montreal, Canada

More information about this series at http://www.springer.com/series/8866

Walid A. Houry
Editor

The Molecular Chaperones Interaction Networks in Protein Folding and Degradation

Springer

Editor
Walid A. Houry
Department of Biochemistry
University of Toronto
Toronto
Ontario
Canada

ISBN 978-1-4939-1129-5 ISBN 978-1-4939-1130-1 (eBook)
DOI 10.1007/978-1-4939-1130-1
Springer New York Heidelberg Dordrecht London

Library of Congress Control Number: 2014940931

Printed on acid-free paper

Springer is part of Springer Science+Business Media (www.springer.com)

Preface

Molecular chaperones are a fundamental group of proteins that are key components of protein quality machinery in the cell. This machinery insures that the folding process of any newly synthesized polypeptide chain results in the formation of a properly folded protein and that the folded protein is maintained in an active conformation throughout its functional lifetime. Molecular chaperones have been shown to play essential roles in cell viability under both normal and stress conditions. Chaperones can also assist in the unfolding and degradation of misfolded or regulatory proteins and in the disaggregation of preformed protein aggregates. Chaperones are involved in other cellular functions including protein translocation across membranes, vesicle fusion events, and protein secretion. As a result, there is a growing interest in the role of molecular chaperones in several diseases including cancers.

In recent years, tremendous advances have been made in our understanding of the biology, biochemistry, and biophysics of function of molecular chaperones. In addition, recent technical developments in the fields of proteomics and genomics allowed us to obtain a global view of the chaperone interaction networks.

This book aims at providing a comprehensive analysis of the function of the diverse systems of molecular chaperones and their role in cell stress responses from a global network perspective. The book is divided into eight main parts that highlight network-based analyses of different chaperone systems. The parts trace the role of chaperones at different stages of the protein's lifetime in the cell from its initial translation on the ribosome to its eventual degradation.

In the Part I, a global view of the chaperone network in yeast is presented. Several high-throughput studies were combined with novel computational approaches to eventually provide an in silico assessment of the workload on chaperones in the cell. This allowed for the determination of the flux of substrates through different chaperone systems. Such an analysis brings us closer to understanding the rules that govern cellular protein homeostasis.

In the Part II, a detailed analysis is provided of ribosome-associated chaperones that assist in the folding of newly synthesized nascent chains as they emerge out of the ribosome. These chaperones include trigger factor in bacteria and NAC and RAC chaperone systems in eukaryotes. Trigger factor has been found to have a larger substrate repertoire compared to NAC and RAC, which seem to be more selective.

In Part III, a comprehensive overview is given of the Hsp70 and Hsp40 chaperones. Hsp70, Hsp40, and nucleotide exchange factors form what is known as the Hsp70 chaperone system that typically engages with proteins in their extended state and promotes their folding to the native state. This system is highly adaptable and can form varied interaction networks. As a result, this system functions in many cellular pathways in addition to protein folding including protein transport, iron-sulfur cluster assembly, and amyloid formation. The Hsp70 network has been found to belong to the scale free category with a limited number of nodes having a large number of links.

In Part IV, the Hsp90 interaction networks are discussed and explained. The importance of the Hsp90 system stems from the finding that this chaperone is a novel anticancer drug target with several Hsp90 inhibitors currently in clinical trials. Hence, many academic groups and biotech companies are very keen to understand Hsp90 mechanism of function and its interaction network. Several high-throughput proteomics methods have been used in baker's yeast, mammalian cells, and other model organisms such as *Candida albicans* to specifically elucidate this network and how the network changes under different stress conditions and between species. These efforts have also been driven by the development of first- and second-generation Hsp90 inhibitors to be used to combat cancer as well as fungal infections.

In Part V, the interaction network of p23 is discussed. p23 is a highly conserved molecular chaperone that also functions as an Hsp90 cofactor. p23 is involved in a variety of cellular pathways including chromatin remodeling, protein folding, and protein transport. By applying both genetic and physical high-throughput methods, a large number (about 350) of p23 interactors have been elucidated clearly supporting a broad function of this chaperone beyond its role with Hsp90.

In Part VI, the role of chaperones and their interaction networks in different cellular compartments is addressed. In the endoplasmic reticulum (ER), the entry, oxidative folding, and conformational quality control of proteins is maintained by many chaperones and their cofactors. Furthermore, proteins that fail to fold properly are targeted for degradation via the ER-associated degradation (ERAD) pathway that extracts proteins from the ER for eventual degradation by the proteasome in the cytoplasm. Chaperones also play critical roles in the import of proteins into the mitochondria and in maintaining mitochondrial protein homeostasis. The Hsp70, Hsp40, and AAA+ chaperones together with several ATP-dependent proteases are critical in these activities. The network analyses of both ER and mitochondrial chaperones provide important insights into the crosstalk between different chaperone systems that eventually determines protein fate.

In Part VII, a global view is provided of protein degradation from the perspective of the proteasome and ubiquitylation. Many specialized chaperones are required to bring together and properly assemble the ~33 subunits that make up the proteasome. Furthermore, the application of classical proteomic approaches as well as new systems-wide methodologies allowed for the identification of numerous ubiquitylated proteins and ubiquitylation sites. These studies provided a global understanding of the ubiquitin-proteasome system and what is termed the ubiquitin code. This is particularly important since the impairment of this system has been associated with many neurodegenerative diseases.

In the last part of this book, Part VIII, chaperone and protease networks in model bacteria and parasites, namely *Escherichia coli*, *Mycobacteria*, *Plasmodium*, and *Leishmania*, are discussed. The chapters in this part highlight several novel aspects as relates to the methodologies used and the biological insights gained. These include the application of a novel high-throughput proteomic method to study the role of chaperones in the prevention of protein aggregation in *E. coli*, the discovery of a new prokaryotic ubiquitin-like protein (Pup) modification system to target proteins for degradation in *Mycobacteria*, and the determination of the role of chaperones and their interaction networks in cell cycle progression in *Plasmodium* and *Leishmania.*

In conclusion, the results and insights presented in this book are the products of the tremendous efforts of many groups whose research is aimed at answering the fundamental basic question of how molecular chaperones and proteases maintain protein homeostasis in the cell. Answering such a question at the molecular level will not be an easy task and will require the use of multiple different approaches. The application of high-throughput methods is but one such approach, albeit a very powerful one.

April 2014 Walid A. Houry, PhD

Contents

Contributors

Razvan F. Albu Department of Biochemistry and Molecular Biology, Centre for High-throughput Biology, University of British Columbia, Vancouver, BC, Canada

Júlio César Borges Department of Chemistry and Molecular Physics, Institute of Chemistry of São Carlos, University of São Paulo, São Carlos, SP, Brazil

Michael Bruderek Institut für Biochemie und Molekularbiologie (IBMB), Friedrich-Wilhelms-Universität Bonn, Bonn, Germany

Gabriela Chiosis Department of Molecular Pharmacology and Chemistry, Memorial Sloan-Kettering Cancer Center, New York, NY, USA

Leah E. Cowen Department of Molecular Genetics, University of Toronto, Toronto, ON, Canada

Elizabeth A. Craig Department of Biochemistry, University of Wisconsin–Madison, Madison, WI, USA

Elke Deuerling Department of Biology, University of Konstanz, Konstanz, Germany

Stephanie Diezmann Department of Biology and Biochemistry, University of Bath, Bath, UK

Pablo C. Echeverria Department of Cell Biology, University of Geneva, Geneva 4, Switzerland

Frank J. Echtenkamp Department of Cell and Developmental Biology, University of Illinois, Urbana, IL, USA

Nancy N. Fang Department of Biochemistry and Molecular Biology, Centre for High-throughput Biology, University of British Columbia, Vancouver, BC, Canada

Brian C. Freeman Department of Cell and Developmental Biology, University of Illinois, Urbana, IL, USA

Veronica M. Garcia Department of Microbiology and Molecular Genetics, UTHealth Medical School, Houston, TX, USA

Kalle Gehring Department of Biochemistry, McGill University, Montreal, QC, Canada

Daniel N. Hebert Department of Biochemistry and Molecular Biology, Program in Molecular and Cellular Biology, University of Massachusetts, Amherst, MA, USA

Walid A. Houry Department of Biochemistry, University of Toronto, Toronto, ON, Canada

Simon J. Hubbard Faculty of Life Sciences, University of Manchester, Manchester, UK

Gregor Jansen Department of Biochemistry, McGill University, Montreal, QC, Canada

Guennadi Kozlov Department of Biochemistry, McGill University, Montreal, QC, Canada

Andrew R. Kusmierczyk Department of Biology, Indiana University-Purdue University Indianapolis, Indianapolis, IN, USA

Juerg Laederach Institute of Molecular Biology & Biophysics, ETH Zurich, Zurich, Switzerland

Lydia Lamriben Department of Biochemistry and Molecular Biology, Program in Molecular and Cellular Biology, University of Massachusetts, Amherst, MA, USA

Craig Lawless Faculty of Life Sciences, University of Manchester, Manchester, UK

Julia Leodolter Institute of Molecular Biology & Biophysics, ETH Zurich, Zurich, Switzerland

Pekka Maattanen Department of Biochemistry, McGill University, Montreal, QC, Canada

Department of Cell Biology, SickKids Peter Gilgan Centre for Research and Learning, Toronto, ON, Canada

Jaroslaw Marszalek Laboratory of Evolutionary Biochemistry, Intercollegiate Faculty of Biotechnology, University of Gdansk, Gdansk, Poland

Thibault Mayor Department of Biochemistry and Molecular Biology, Centre for High-throughput Biology, University of British Columbia, Vancouver, BC, Canada

Kevin A. Morano Department of Microbiology and Molecular Genetics, UTHealth Medical School, Houston, TX, USA

Stefan O. Ochiana Department of Molecular Pharmacology and Chemistry, Memorial Sloan-Kettering Cancer Center, New York, NY, USA

Didier Picard Department of Cell Biology, University of Geneva, Geneva 4, Switzerland

Carlos H. I. Ramos Department of Organic Chemistry, Institute of Chemistry, University of Campinas—UNICAMP, Campinas, SP, Brazil

Kamran Rizzolo Department of Biochemistry, University of Toronto, Toronto, ON, Canada

Cornelia Rüb Institut für Biochemie und Molekularbiologie (IBMB), Friedrich-Wilhelms-Universität Bonn, Bonn, Germany

Annika Scior Department of Biology, University of Konstanz, Konstanz, Germany

Thiago Vargas Seraphim Department of Chemistry and Molecular Physics, Institute of Chemistry of São Carlos, University of São Paulo, São Carlos, SP, Brazil

Johan C. Sunryd Department of Biochemistry and Molecular Biology, Program in Molecular and Cellular Biology, University of Massachusetts, Amherst, MA, USA

Hideki Taguchi Department of Biomolecular Engineering, Graduate School of Biosciences and Biotechnology, Tokyo Institute of Technology, Yokohama, Kanagawa, Japan

Tony Taldone Department of Molecular Pharmacology and Chemistry, Memorial Sloan-Kettering Cancer Center, New York, NY, USA

Abla Tannous Department of Biochemistry and Molecular Biology, Program in Molecular and Cellular Biology, University of Massachusetts, Amherst, MA, USA

David Y. Thomas Department of Biochemistry, McGill University, Montreal, QC, Canada

Elisabeth R. M. Tillier Department of Medical Biophysics, University of Toronto, Toronto, ON, Canada

Wolfgang Voos Institut für Biochemie und Molekularbiologie (IBMB), Friedrich-Wilhelms-Universität Bonn, Bonn, Germany

Jannis Warweg Institute of Molecular Biology & Biophysics, ETH Zurich, Zurich, Switzerland

Eilika Weber-Ban Institute of Molecular Biology & Biophysics, ETH Zurich, Zurich, Switzerland

Philip Wong Department of Medical Biophysics, University of Toronto, Toronto, ON, Canada

Part I
A Global View of the Chaperone Network

Chapter 1
Analysis of Chaperone Network Throughput

Craig Lawless and Simon J. Hubbard

Abstract The regulation of protein folding is an important aspect of systems biology that is often overlooked in the modern age of post-genomics. Although the transcriptome and proteome can now be relatively easily quantified, the protein complement in a cell must also be properly folded and delivered to the cognate site of action in order to carry out its function. To understand how a eukaryotic cell can accomplish this task requires an understanding of how the cell's chaperone complement acts to mediate the folding of its substrates. In this chapter, we examine and combine the data available from recent landmark studies to measure the chaperone interactome (the "chaperome") in the model eukaryote Baker's yeast with recent attempts to quantify the levels of yeast proteins. This computational analysis leads to an independent in silico assessment of the workload placed upon the chaperones in a cell, and shows there is a direct relationship between chaperone abundance and properties of their targets. By further considering protein turnover data, we are able to consider the folding flux passing through individual chaperones and chaperone groups, enabling a revaluation of the workload placed upon them, which we estimate exceeds 60% of the cell's protein complement. We also cluster chaperones into coherent groups based on a filtered set of targets. These clusters reproduce some well-known features of the chaperone classes, as well as showing biases in subcellular location of the chaperone targets by factoring in the flux. These integrated analyses show how systems approaches can shed light on proteostasis defined by throughput in the chaperone network.

1 Introduction

Quantitative proteomics is one of the most rapidly advancing fields in the post genomic era. Arguably, it has lagged behind the field of transcriptomics, via microarrays or more recently next generation sequencing (ribonucleic acid, RNA-seq),

S. J. Hubbard (✉) · C. Lawless
Faculty of Life Sciences, University of Manchester, Manchester, UK
e-mail: simon.hubbard@manchester.ac.uk

C. Lawless
e-mail: craig.lawless@manchester.ac.uk

W. A. Houry (ed.), *The Molecular Chaperones Interaction Networks in Protein Folding and Degradation,* Interactomics and Systems Biology, DOI 10.1007/978-1-4939-1130-1_1,

as it has struggled to describe the complement of proteins expressed in a cell in a "genome-wide" fashion. Also, and more importantly, mass spectrometry based proteomics is not inherently quantitative, whereas transcriptomics is. This is a crucial issue for the biologist wishing to understand a process or pathway at the systems level, since this usually requires knowledge of the levels of some of the components, either to parameterize models or to assess the merits of predictions which point to quantitative changes. Moreover, proteins are usually the primary molecules responsible for the delivery of biological function, and the existence of a transcript is not necessarily a guarantee of the presence of a folded, functional protein. Hence, there are many good reasons to want to study the proteome in a comprehensive and quantitative fashion, despite the challenges in doing so.

Most modern proteomics relies on mass spectrometry as the underpinning analytical technology. Advances in instrumentation, chromatography, and allied informatics support the identification and measurement of abundance of an increasingly larger fraction of the protein complement derived from a tissue or cellular context. Quantitation is usually achieved as a relative measure by a variety of techniques that use labelled or label-free approaches to estimate changes in the analytical signal between two samples. These quantitative measurements can then aide our understanding of the mechanisms by which gene products are regulated and organized in order to elicit their cellular effects. Such data is important for generating systems-level models of an organism or a functional pathway, which encompass the interactions of the genome, transcriptome, proteome, and metabolome to describe cellular regulation and reactions to environmental stress. It is this interdisciplinary "systems biology" approach that has allowed the expansion of the relatively simple "central dogma" termed by Crick where "DNA makes RNA makes protein," to complex systems level models that incorporate multiple isoforms and interactions.

Recent studies of the model organism *Saccharomyces cerevisiae*, Baker's yeast, have exemplified this paradigm, providing data for the entire transcriptome [1], interactome [2, 3], translational control rates [4], protein localisation [5], and protein turnover rates [6, 7]. In addition, yeast has also been chosen for several proteome-wide quantitative studies [8–10]. Integration of such data has been used to build a metabolic model in yeast, using glycolysis as proof of principle [11, 12]. Advances have also been made for mammalian systems where transcription and translation data has been integrated with both RNA and protein turnover to develop whole genome models [13], which support global estimates of the relative importance of translational control in regulating gene expression.

Although excellent progress has been made, systems biology still faces many challenges when trying to integrate the vast information that is required to build suitable models to adequately define cellular regulation. Indeed, although we can now measure protein levels relatively routinely, the final and proper function of a protein (or protein complex) requires that it has the correct structure and is localized to the correct part of the cell. The subset of proteins assigned this responsibility are the chaperones, which facilitate correct protein folding, help recognize mis-folding, and prevent protein aggregation within the cell. In eukaryotes, chaperones act in a translation-coupled mechanism and recognize nascent polypeptide chains, thereby ensuring correct de novo folding as translation progresses [14]. Once translated and correctly folded, these proteins are then passed through the general chaperone network and are transported to their correct subcellular localization [15]. Given their critical role in the general maintenance of proteostatsis, many chaperones are heavily involved in stress responses, such as the aptly named heat shock proteins (HSPs), due to their thermo-reactive response [16].

In yeast, there are 63 proteins commonly characterized as chaperones, which have been the subject of a variety of studies [14, 17, 18]. More recently, the focus has moved to a more global view of chaperone function. Houry and colleagues have undertaken a comprehensive affinity purification analysis coupled with mass spectrometry in an attempt to define protein-protein interactions (PPIs) between yeast chaperones and their protein targets [19]. This has expanded on previous network-based analyses that have revealed two distinct networks; de novo protein folding and stress response [14]. Further work on the chaperone interaction network from Gong et al. [19] focused on clustering chaperones based on their target interactions, uncovering 10 chaperone modules that share common features among chaperone targets such as evolutionary rates and expression levels [20].

However, as already suggested, one dimension that must also be considered is quantitation; individual proteins are present in different concentrations in the cell, and therefore the workload placed on individual chaperones will vary depending on the abundance of the substrate/targets. Hence, one should also factor in the target workload (or flux) of each chaperone (or chaperone complex) in terms of protein abundance, or more formally, the total synthesis rate of each protein which can be estimated from protein turnover rates at steady state. To this end, we extend upon previous work that has attempted to characterize the yeast proteome using quantitative approaches, by characterizing the level of the chaperones using the QconCAT approach [21] and integrating other publicly available quantitative datasets, including protein turnover [6]. The QconCAT approach uses stable isotopic heavy-labelled peptide surrogates which are used as an internal standard for specific yeast proteins to provide absolute quantitation via mass spectrometry (see next section for more details). We show that there is a correlation between the abundance of chaperones and the target "folding" workload based on the number of targets and the target abundance. This is expanded further to estimate folding flux through each chaperone (and chaperone complex) using protein turnover rates. We describe these findings in the context of biological annotation which include subcellular localization and protein essentiality.

2 Chaperone Quantification

As part of a larger project to quantify the proteome of *S. cerevisiae* using QconCAT technology [21, 22], three QconCATs were designed to quantify the 63 known chaperones in yeast. Each QconCAT was constructed as a concatamer of heavy-labelled surrogate peptides where by each chaperone protein is represented by two of these target Q-peptides. These recombinant proteins were expressed in *E. coli* grown in media containing stable isotope-labelled arginine and lysine to produce a heavy-labelled QconCAT. The "heavy" QconCAT was used as a reference and spiked-in in known amounts to the yeast protein sample to enable peptide quantification using a selective reaction monitoring (SRM) targeted approach. Each peptide was targeted by three transitions for both the light (target) peptide and the heavy (reference) peptide. Quantification values were calculated from the ratio between the heavy and light extracted ion chromatograms (XICs), averaged across four biological replicates to obtain peptide copies per cell (cpc). Peptide ratios were acquired using the mProphet pipeline [23], which provides false discovery rate (FDR) estimates for the sets of transitions used for each peptide and therefore an estimate of reliability. Final protein abundances were taken as either the average of the peptide quantification values if they were in agreement, the minimum of the two peptides if the higher abundant peptide contained tryptophan, or the maximum peptide value for the remaining cases. The protein quantification values were then assigned into one of three classes; type A proteins where acceptable data above noise for both the target and reference peptides is obtained, type B where only the reference peptide gave acceptable data, and type C where neither gave acceptable data. In the case of type B proteins, we are still able to provide an upper limit of quantification as we know the concentration of the heavy-labelled peptide surrogate and the minimum spike-in level where it is detected in the XIC.

Using our targeted quantification approach, we obtained abundance values for 51 of the 63 chaperones in yeast [24], over a dynamic range covering three orders of magnitude, from 250 to 440,000 cpc as shown in Fig. 1.1. This list contains some chaperones that have eluded previous epitope-tagging approaches, including those that are part of the TRiC/CCT complex, and some proteins that have escaped both label-mediated and label-free MS-based studies. We ascribe our improved ability to quantify these particular proteins to the targeted nature of the QconCAT-SRM approach, which not only eliminates ambiguity by selecting unique peptides and associated productions but is credited with the greatest sensitivity in targeting low abundance proteins [25]. While many proteins are open to quantitation via label-free means to a low level, we believe that targeted approaches are still the gold standard for accurate absolute quantitation of proteins, and in our hands are able to measure down to around 100 cpc with coefficient of variation values (CVs) across replicates routinely less than 20 %.

3 Chaperone Target Quantitation

At present, a complete set of absolute quantitation values for yeast proteins derived via SRM- targeted mass spectrometry is not available. However there are a wide range of proteome-wide quantification data sets in yeast derived from other approaches, catalogued in the protein abundance database PaxDb [26]. We have used this useful data resource for our studies. PaxDb supplies all protein abundance

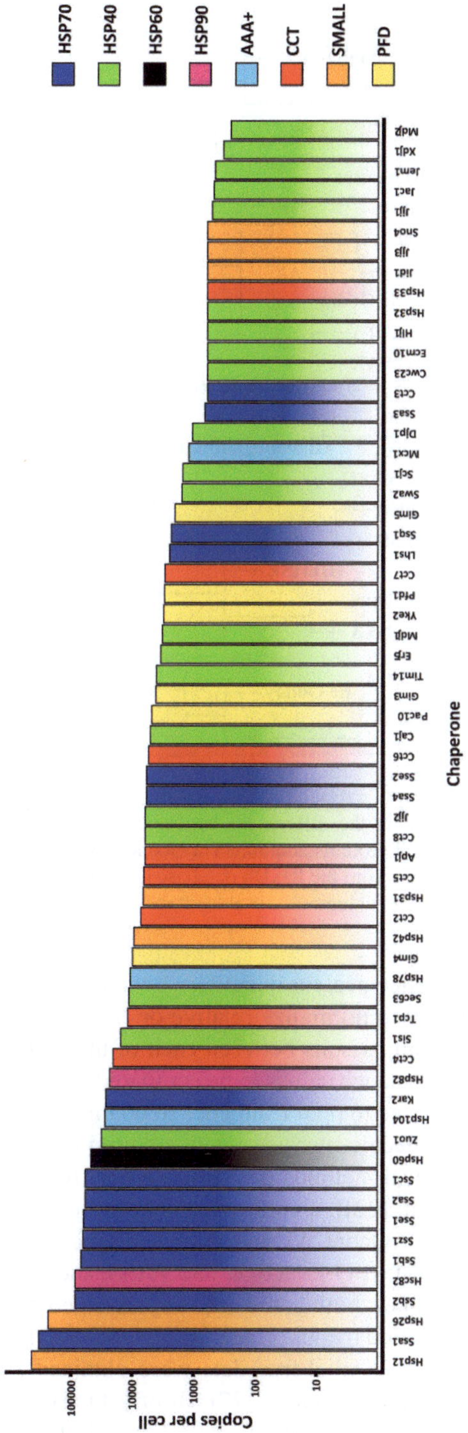

Fig. 1.1 Protein abundance estimates of yeast chaperones in yeast measured using the QconCAT-SRM method. Abundance values range from ~250 copies per cell (*cpc*) to >400,000 cpc. Bars are colored according to the chaperone class they belong to, shown in the accompanying key

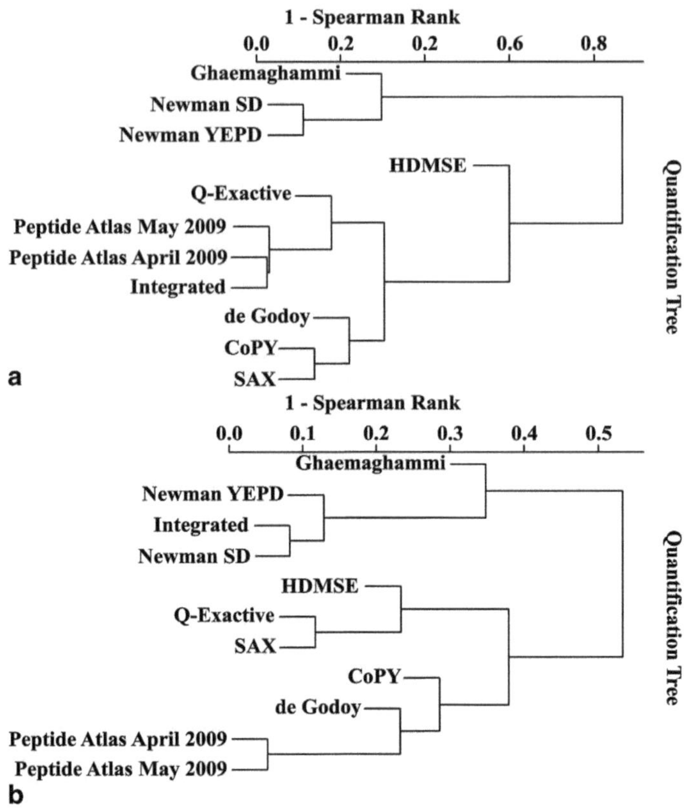

Fig. 1.2 **a** Dendrograms highlighting the differences between protein quantitation datasets for chaperones only and **b** for the 1000 proteins acquired via QconCAT-SRM on an internal project to date. External datasets were taken from the PaxDb database [26], where all protein values were normalized to parts per million (*ppm*) values

values in consistent units. Since protein quantitative values from different studies vary in nature, it is necessary to normalize relative or absolute values to parts per million (ppm), permitting comparisons between each of the methods. In order for our chaperone QconCAT abundance values to be compared to these other approaches they also need to be converted to ppm from cpc. To do this, we made the assumption of 60 million protein molecules per yeast cell and used this value to convert to ppm. While different analytical methods demonstrate positive correlations one to another (c.f. Brownridge et al. [22]), a comparison of all the quantitative methods with each other including our QconCAT approach and label-free acquisitions for the same yeast samples (HDMS^E, Q-Exactive, and SAX), shows considerable discrepancy between the techniques; both in the case of just the chaperones (Fig. 1.2a) and the ~ 1000 proteins quantified by our project using QconCATs so far (Fig. 1.2b).

The HDMS^E and Q-Exactive label-free data were obtained using a Top3 approach, which is described in more detail in [27]. Briefly, the HDMS^E was acquired on a Waters Synapt^{TM} G2, processed using Protein Lynx Global Server v2.5 using the Hi^3 approach [28], and the Q-Exactive was acquired on a Thermo Scientific Q-Exactive^{TM} instrument, processed using Progenesis (Nonlinear Dynamics) and Progenesis PostProcessor [29]. The SAX method was obtained by fractionating the sample using Off-Gel fractionation followed by label-free data acquisition for each fraction on a Thermo Scientific Q-Exactive and processed using Progenesis (Nonlinear Dynamics) and Progenesis PostProcessor [29]. We are indebted to our colleagues Dean Hammond, Philip Brownridge, and Rob Beynon at the University of Liverpool, UK for these data sets.

The most striking outcome of the clustering is the separation between mass spectrometry-based approaches and epitope-tagging-based approaches (Ghaemaghammi and Newman datasets), which co-cluster in unique clades. This is perhaps not surprising given the common analytical technique employed in the mass spectrometry grouping, which operates in a fundamentally different way to quantitative tagging methods that exploit antibody technology. Both have their advantages and disadvantages, but it is worth noting that the mass spectrometry community have recently thrown down the gauntlet to the antibody-based methods [30]. Within the mass spectrometry-based clade, we note there is an apparent separation of the label-free methods from the label-mediated approaches. These features are not perfectly conserved when considering the chaperone-only subset (Fig. 1.2a) compared to ~ 1000 proteins (Fig. 1.2b), but overall, it is noticeable that the major difference between all of the result sets appears to be methodological and not biological. This is exemplified by the GFP-tagging (green fluorescent protein) approach by Newman and colleagues [10] where despite being grown in both yeast extract peptone dextrose (YEPD) and sucrose deficient media (YMD)u, the two datasets cluster very closely together, both for the chaperone and 1000 protein subsets. This is in contrast to our label-free data acquisition on biologically identical yeast samples, but undertaken by three different label-free methods using different mass spectrometers (HDMSE, Q-Exactive, and SAX on Figs. 1.2a and 1.2b); these are not so highly correlated, particularly for the chaperone subset. These dendograms suggest that there is less inherent variance in biological samples than that currently obtained by comparison of two alternate quantitative methods. The absolute data derived from the QconCAT approach produces different abundance values to both the label-free approaches, adding further evidence to this, though interestingly, it clusters closely to another label-mediated data set using stable isotope labeling by amino acids in cell culture (SILAC) [8].

Our experience when comparing these datasets suggests that methodological variance contributes more strongly than biological variance to the protein abundance, and we urge caution in attempting to merge or analyze quantitative data derived from multiple techniques, or even on different instruments. Although this appears to be a somewhat disappointing outcome, it should be noted that the studies considered here have not used the same yeast strain cultured in identical conditions, and despite this, there is still a good correlation between even the most disparate of results. Nevertheless, these findings provide further evidence that there are still large variations between quantitative methodologies despite recent advancements in the field.

To examine this further, we extracted all the unique protein quantitation yeast datasets from PaxDb. These included both tagging based datasets from Ghaemmaghami

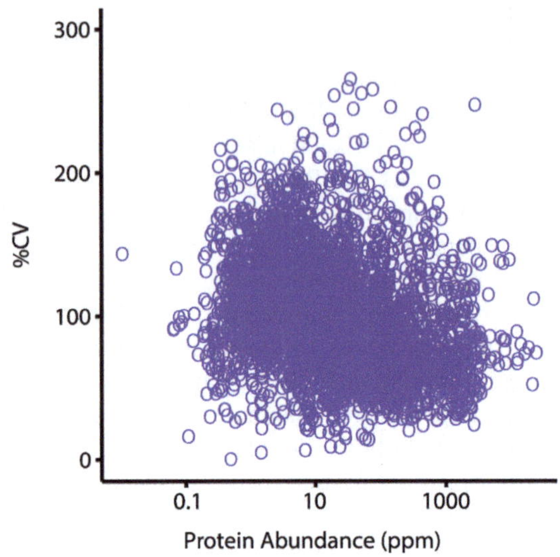

Fig. 1.3 Scatter plot of the coefficient of variances (%CV) of all yeast protein abundance measurements, in parts per million, from QconCAT-SRM, HDMS^E, Q-Exactive, SAX, and all PaxDb datasets plotted against a representative protein abundance. The average (mean) %CV across all datasets is 97%. As can be seen, there is a modest, but significant, negative correlation ($R^2=0.11$, $P<2.2e^{-16}$) between protein abundance and %CV across independent determinations of protein abundance. This suggests there is a weak negative relationship between protein abundance and the ability of independent methods to quantify its abundance as there appears to be greater variance for low abundance proteins, as would be expected

et al. [9] and Newman et al. [10], the MS-based datasets from de Godoy et al. [8], Lu et al. [31], global proteome machine (GPM) [32], Peptide Atlas [33], and the PaxDb yeast-integrated dataset [26]. All datasets were collated by common protein in order to estimate the variation by method. We plot this data in Fig. 1.3 which shows the estimated CV percentage (CV%) plotted against protein abundance in ppm taken from the integrated PaxDB dataset, which is the most comprehensive, as it is a weighted aggregate dataset of the other quantitative datasets [26]. This striking plot shows that generally the CV% exceeds 100% of the protein ppm value; equivalent to a 2x fold change in the protein abundance estimate. This might seem like a large value, but this represents a slightly artificial calculation using protein quantitative data spanning almost 10 years of research, albeit on a well-characterized model eukaryote. It suggests that using such a range of methods one can expect to estimate protein abundance fold changes to within two-fold accuracy. This contrasts with the typical 10–25% CV cited by targeted proteomics studies using SRM approaches [22, 24, 34, 35] which appear more robust. Regardless, we suggest that quantitative proteomics still has work to do to define a definitive absolute quantitative dataset for the complete yeast proteome.

A closer look at specific chaperones and chaperone complexes across the methods highlights some of their shortcomings. Taking the chaperonin-containing TCP1 (CCT) complex as an example, we note that this set of chaperones is covered poorly by the epitope-tagging methods. In fact, CCT4 is the only protein of the eight members of this complex quantified in any of these tagging datasets. Conversely, the mass spectrometry based methods fared much better and consistently acquired abundances for seven or all eight of the CCT chaperones. This outcome is not surprising when taken in context with the structure and function of the CCT chaperones. The CCT proteins form an 8 protein heteromeric ring structure; all in a one-to-one stoichiometry [36]. It seems reasonable to assume that this structure could be disturbed or disrupted by epitope-tagging methods, when an additional protein is tagged on to any one of the CCT chaperones. Indeed, given that the CCT chaperone complex mediates the folding of a large majority of essential proteins with the cell [37], one would expect that any perturbations of this folding mechanism could lead to incorrect folding of these essential proteins and as a result be lethal to the cell. Interestingly, the CV calculated across all the quantitative methods is lowest for the CCT chaperones (44% CV, compared to 84–177% for other classes) when compared to the chaperone groups defined in [19], which is unsurprising given the one-to-one structural stoichiometry.

4 A Correlation Between Chaperone and Substrate Abundance

Chaperones operate by interacting with substrate proteins to ensure correct folding during protein synthesis, to stabilize protein structure during environmental stress, and to mediate their transport to their correct locations within the cell [38]. To identify chaperone substrate interactions, Gong and colleagues [19] undertook a proteome-wide affinity purification study in *S. cerevisiae* and produced a chaperone-interaction map containing 4340 candidate chaperone substrates. However, it is widely known that tandem affinity experiments have a tendency to contain false-positives due to nonspecific binding or common contaminants. Indeed, a variety of sophisticated informatics approaches have been generated to help deal with this [39–41], as well as a dedicated database for biologists to filter their data [42]. Here, to circumvent this, we used a simple consistency based approach and constructed a "high-quality" interaction dataset using three public PPI repositories (STRING [43], BioGRID [44], and MIPS [45]). All chaperone interactor pairs were obtained from each repository (in yeast), retaining only those pairings where the reciprocal (substrate-chaperone) interaction was also observed. For this study, we also excluded chaperone-chaperone interactions, focusing solely on chaperone-target interactions. Additionally, for STRING PPI pairs, a confidence score cutoff of 0.7 was used, retaining only those with superior scores. The three PPI datasets were then combined into a "high-quality" dataset where each interaction had to be present in at least two of the three datasets. This combined dataset contained 60 of the 63 known chaperones, 1711 chaperone substrates, 3649 interactions, and contained all but one of the reciprocal interactions identified by Gong et al. [19]. This qualitative interaction dataset, however, lacks protein abundances required to understand

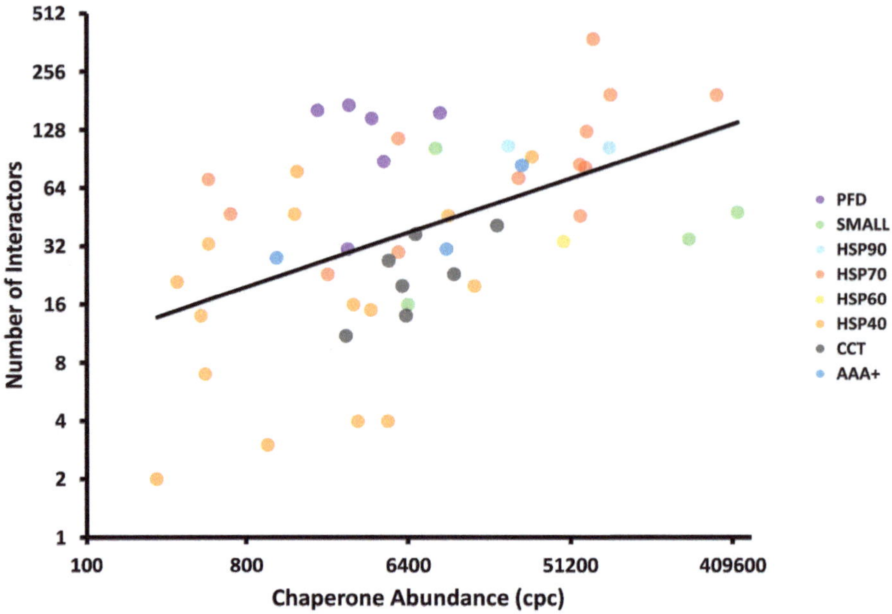

Fig. 1.4 Scatter plot of the number of chaperone interactors against chaperone abundance as measured by QconCAT-SRM mass spectrometry in our (CoPY) lab. There is a good linear correlation between the number of interactors and chaperone abundance ($R^2=0.25$, $P<0.0002$), suggesting that chaperones with more clients are themselves more abundant in steady state conditions. The chaperones have been colored according to their chaperone class (as shown in the key) to visualize the spread of chaperone classes across the chaperone-interaction landscape. (Adapted from Brownridge [24])

the quantitative workload of each chaperone (and chaperone complex) and only provides an account of the different substrates that each chaperone interacts with.

In order to better characterize the workload placed on individual chaperones and chaperone classes, the abundances of their substrate protein targets in the cell needs to be taken into account. In the first instance, we examined chaperone workload with a simple measure, by considering the relationship between the abundance of the chaperone itself compared to the number of different substrate interactions it has, under the assumed logic that the more abundant chaperones would have a higher number of interactors. No prior correlation has been observed by Gong and colleagues [19]. As we reported previously [24], using our filtered interaction dataset we find a significant correlation using the QconCAT chaperone abundance data, as shown in Fig. 1.4, resulting in a Spearman Rank (R_{sp}) correlation of 0.49 ($P<0.00002$). This correlation is also apparent across other quantitative datasets available from PaxDB, with R_{sp} ranging 0.08–0.54 (data not shown), and a significant and positive correlation is present regardless of whether we filter the interactome set or the quantitation data set selected from PaxDB. This supports the general hypothesis that those chaperones responsible for mediating the most folding in the cell are themselves generally highly abundant.

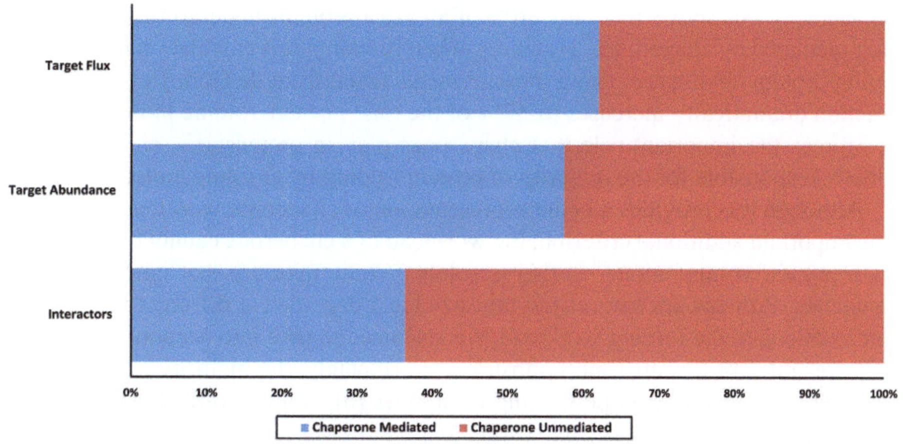

Fig. 1.5 Stacked bar charts showing the proportion of proteome folding mediated by chaperones, as determined by three different measures; the total number of interactors, the aggregated target abundance, and the aggregated target flux (molecules per minute) over the whole proteome. The latter is more representative of the true workload placed on chaperones in the cell, and is the largest relative fraction

Figure 1.4 also shows how breaking the chaperones into classes, as reported by Gong et al. [19], reveals general trends within the classes. For example, HSP40 chaperones are observed here to be low-abundance proteins with fewer interactions than many others. This is interesting, as they are generally considered as co-chaperones of the more abundant HSP70 chaperones, which bind to Hsp70 partners via the ribosome-associated complex (RAC), mediating the folding of the majority of nascent peptides, and thereby regulating the various roles of their Hsp70 partners [15, 38, 46]. Indeed, we observe that the HSP70 class (and similarly SMALL class) chaperones are more abundant and have more interactions, which ties in with their known promiscuity and substantive role [38].

Although looking at the number of different substrate interactions of each chaperone yields valuable information regarding their diversity and specificities, it does not provide a full picture of the true workload placed on any given chaperone. In order to better characterize chaperone workload, we need to have a measure of the total amount or "volume" of protein, in terms of total cpc, that is being mediated by a chaperone. The volume of protein (V_c) of a chaperone (c) can easily be estimated from the total abundance (CPC_n) of all n substrates for c, as shown below.

$$V_c = \sum_{1}^{n} cpc_n$$

Applying this methodology to consider the total cellular workload fundamentally changes the proportion of proteins mediated by chaperones when taking substrate volume and not just the number of substrates into account (Fig. 1.5). When considering just the number of interactors (taken from the unfiltered list from Gong and colleagues), this shows that only 36 % of all yeast proteins undergo folding regulation mediated by chaperones. However, when rescaling this calculation by incorporating protein abundance, using measurements taken from de Godoy et al. [8], this fraction dramatically increases to 57 % of the total protein volume in the cell. This reinforces the important role that chaperones play in proteostasis, since they are clearly responsible for the majority of protein folding by absolute molecular count.

Although this provides a better representation of chaperone workload, it ignores one important additional criterion; the workload of a chaperone cannot be expressed by a protein volume alone, as this is solely a static measure that ignores protein dynamics. Proteins are naturally synthesized and degraded at different rates. In order to calculate the folding workload, we also need to take into account the protein turnover rates of a chaperone's substrates. Fortunately, the majority of these have been measured using epitope-tagging techniques [6], and we were able to use their protein degradation rates (k_{deg}) to estimate protein synthesis rates (k_{syn}) for the substrates of each chaperone.

$$k_{syn} = cpc_n \times k_{deg}$$

The details are described more fully elsewhere [24], but briefly, this calculation presumes the cell is at steady state where protein synthesis and degradation rates are equal $\left(\frac{dCPC_n}{dt} = 0\right)$. From this, we can estimate the rate of synthesis of individual proteins and presume that this flux in molecules per unit time is the responsibility of individual chaperones. This allows us to estimate the total mediation flux on a per chaperone basis, which in turn can then be used to estimate the total workload, or folding flux (F_c), of a chaperone (c) in terms of molecules per minute per cell.

$$F_c = \sum_1^n k_{syn}$$

To circumvent the issue of substrates being mediated by multiple chaperones, the flux values were divided equally pro rata among each chaperone. Any missing values were replaced by the geometric mean across the entire turnover dataset; this was to ensure the mean k_{deg} of the dataset remained unchanged. To calculate the individual flux values F_c, we used data from the de Godoy estimation of yeast protein abundance taken from PaxDb [26], as opposed to our own QconCAT-SRM quantitation, since this currently does not fully cover the yeast proteome.

We recognize that cellular growth rates have not been formally taken into account in our model, primarily because the protein abundances and turnover rates used here have been obtained using different experimental yeast studies where strains and growth rates are not directly matched. However, to consider growth rate formally would simply add an error constant to the calculated k_{deg} values that is equivalent to the dilution rate of cells grown under controlled system (e.g., chemostat culture). Therefore, we believe the current calculations are still representative of the relative split of flux across the chaperone complement in yeast and despite the

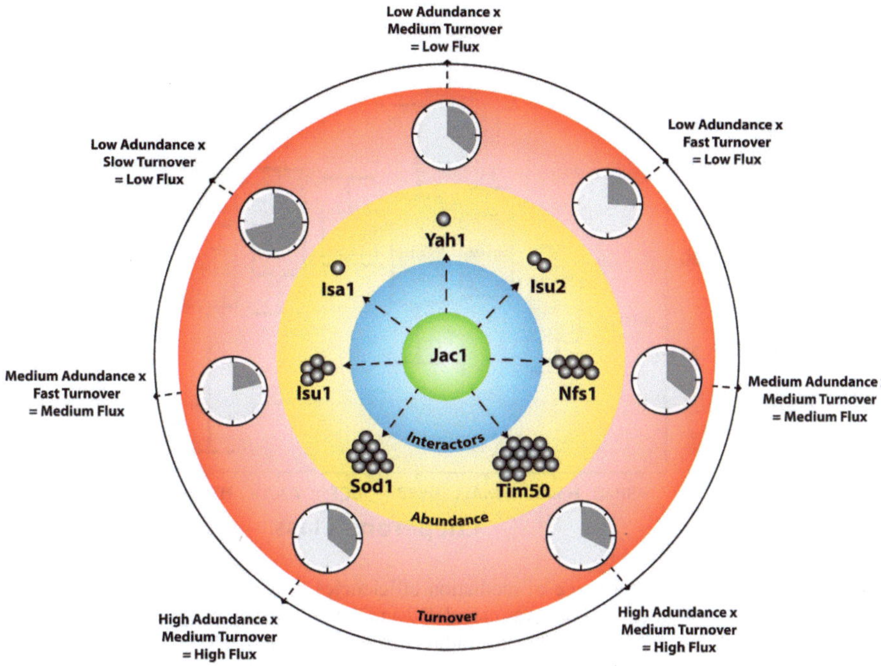

Fig. 1.6 Different levels of chaperone flux, using *Jac1* as an example. The concentric circles show an increasingly representative view of the true workload placed on a chaperone, which takes into account the number of chaperone interactors, the abundance of each target, and the turnover of the targets. Only by taking all three into account can the workload of the chaperone be properly estimated

assumptions made, these flux values are the most accurate proteome-wide estimates currently available.

Before considering the results of these calculations, it is worth putting this into a simple theoretical context describing how proteostasis can be maintained in the cell. This is schematized in Fig. 1.6, which shows different scenarios for regulating protein levels in the cell for a single example chaperone, *Jac1*. This chaperone has seven interactors in our filtered dataset, shown in the yellow concentric circle. The grey spheres broadly correspond to the protein abundance of each substrate with *Tim50* being the highest. As can be seen in the red concentric circle, this protein also apparently has one of the shorter half-lives (shown by the stopwatch graphic), and hence, there is a particularly high-folding flux required. *Tim50* is essential for protein translocation across the mitochondrial inner membrane and its loss can lead to programmed cell death [47, 48]. In contrast, *Isa1*, involved in iron-sulphur assembly displays a low flux, being a low-abundance protein with a relatively longer half-life, and unlike *Tim50* is not essential. The three "levels" of control (chaperone interactions, substrate abundance, and substrate turnover) all relate to the overall folding workload on the cell, and we argue that it is the integrated view on the outside of the circles in Fig. 1.6 which best represents the potential impact on chaperone function and workload.

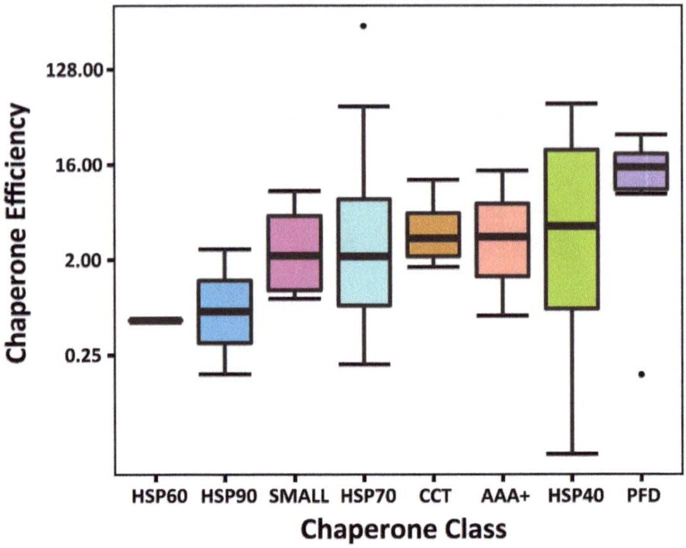

Fig. 1.7 Boxplot showing the range of mediation efficiencies of chaperones in each chaperone class. The efficiency of all chaperones was calculated and the grouped by chaperone class to show the distribution differences by class. The pre-foldin (*PFD*) class is shown to be generally the most efficient with a compact distribution. (Adapted from Brownridge [24])

When applying this theory to estimate substrate flux values (F_c) for each chaperone, we observe that the correlation with chaperone protein abundance generally increases; with R_{sp} ranging from 0.26 to 0.69 across the quantitative datasets [24]. As indicated by the strong Spearman correlation values, the more abundant proteins typically have a larger mediation workload in the cell, i.e., they fold/translocate more substrates per minute within the cell, as one might expect. Introducing the concept of chaperone "efficiency" for chaperone-substrate pairs and calculating a median efficiency for each chaperone class allows us to examine this data further. We estimate efficiency by dividing the substrate flux of a chaperone (F_c) by the abundance of the chaperone itself, reasoning that this effectively describes the number of substrate molecules per minute mediated by a single chaperone molecule. The ranges of efficiencies within each chaperone class are shown as a boxplot in Fig. 1.7. We see a broad range of values, but this suggests that most individual chaperone protein molecules are able to mediate the folding of more than one substrate molecule per minute. The prefoldin class appears to be particularly effective, acting as a major player in transporting proteins to the CCT complex, whose primary substrates are tubulin and actin. Here, the efficiency exceeds 10 molecules per chaperone per minute. The Hsp60 chaperones, which facilitate the transport and import of mitochondrial proteins, appear the least efficient by these criteria, perhaps because they are limited by mitochondrial import processes that do not want to be overloaded. Despite this apparently slow rate, Hsp60 is an essential gene [49].

5 Chaperone Clustering and Annotation

As calculated above, the substrate flux of the chaperones provides a good estimate of the workload diversity placed on them. To extend this analysis further, it is interesting to see how this workload is dispersed within the cell, in terms of function and localization of chaperone targets. It is already well recognized that many chaperones operate on substrates in, or destined for, different subcellular locations and we wished to examine this trend using the flux data. To do this, we used the high-quality interaction dataset to compare substrate specificities across the chaperones, by first grouping the chaperones in to common clusters based on their substrate profiles. The chaperones were assigned a vector specifying each substrate (of the total substrates within the interaction dataset) as either a target (1), or not a target (0), for that chaperone. The chaperone vectors were then submitted to a standard hierarchical clustering algorithm based on the binary distance between the vectors. We then manually assigned the chaperones to nine distinct clusters using a single threshold value, as shown in Fig. 1.8. This produces a set of clusters which closely resemble the chaperone family classifications as described in Gong et al. [19] although it is worth noting that these were arrived at completely independently, based only on chaperone target presence/absence. The clustering procedure carried out here is considerably simpler than that undertaken by Bogumil and colleagues [20], though we filter the interactome data prior to clustering, which derived a similar set of chaperone modules containing both chaperones and their substrate targets for common properties. Their modules were found to be enriched for common gene expression levels and evolutionary rates. Our simple procedure also generally recapitulates these well-known chaperone families and known relationships. For example, the CCT complex members (an octamer of subunits) cluster together and form a single clade, as do the prefoldins. We also observed clustering of coupled chaperone sets, such as the Hsp70 and Hsp40 groups which function as cognate partners. For example, the red clade in the bottom right of the wheel contains the Ssb1 and Ssz1 (two Hsp70s) and Zuo1 (a cognate Hsp40) that are tightly coupled to the ribosome. These clades form coherent groups of chaperones/co-chaperones that can be considered to be acting on common substrate groups.

Another noteworthy feature is the "cluster" of singletons. These are formed as a group that do not share particularly tight clustering with any other chaperones, mostly possessing relatively few targets. They are predominantly Hsp40s, which is not unexpected given their regulatory role in mediating Hsp70 function. Hence, one would expect them to have relatively few direct substrates and our filtering steps when applied to the entire Gong data set would be expected to remove many of the weaker, indirect interactions. Similarly, our procedures currently do not include chaperone-chaperone interactions. These findings are also supported by Fig. 1.4, which highlights the reduced numbers of chaperone clients for most Hsp40s.

We next considered the subcellular location of the chaperones targets, which were assigned to subcellular compartments using the Yeast Gene Ontology Slim from the *Saccharomyces cerevisiae* database (SGD) [50]. This allowed us to

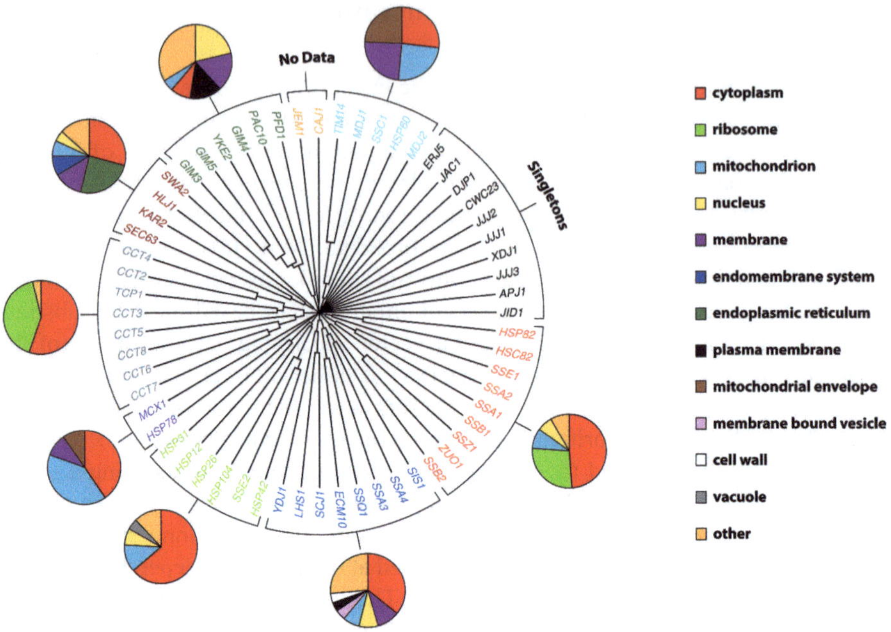

Fig. 1.8 Dendrogram wheel showing chaperones clustered by common target interactors. The chaperones were clustered using hierarchical clustering of the binary distance between their targets. Without any a priori information the clades formed are similar to the chaperone classes defined by Gong et al. [19], such as the eight CCT chaperones highlighted in *grey*. The heat shock protein (*HSP*)40 and HSP70 proteins generally cluster together (*dark blue, light blue,* and *maroon*) supporting the understanding that HSP40 are co-chaperones to HSP70 chaperones. In addition to the dendrogram wheel, each cluster is associated with a pie chart representing the subcellular localization of the cluster targets, weighted by the protein flux (as opposed to counts or protein abundance). These uncover diverse subcellular roles of chaperone mediation by cluster, such as the large ribosome compartment of the *grey* (CCT) and *red* (largely HSP70) clusters suggesting that these chaperones are heavily involved in mediation of proteins required for translation

estimate the spread of chaperone workload across subcellular locations based on the k_{syn} values for folding flux estimated previously. For any substrates that were assigned to more than one cellular localization, the k_{syn} values were split pro rata between the parent chaperones. Although previous analyses had not observed any strong trends in terms of subcellular bias for given chaperones or classes [19], we reasoned this might be different when considering the total folding flux rather than simply counting the number of different targets.

Figure 1.8 shows the target workload/flux for eight of the nine clusters and illustrates not only the differing roles between the clusters but also how the workload is distributed within a cluster. Indeed, there are several notable features of Fig. 1.8. As expected, some of the clusters show strong biases towards substrates destined for that location, such as the top blue cluster containing *Hsp60*, as well as other well-known mitochondrially active chaperones. We also see other strong biases, including the CCT complex which folds cytosolic actins and tubulins, and the *Sec63* cluster that are coupled to endoplasmic reticulum (ER)-associated folding and import.

6 Summary

The work showcased in this chapter highlights how systems biology approaches can shed light on the mechanisms by which the cell maintains proteostasis, and manages the complement of chaperones to ensure correct protein folding on a cellular basis. This is only possible through a lot of hard work and diligence on the part of many groups, to determine the chaperone interactome network via a variety of techniques, coupled to the determination of extensive abundance and turnover data sets. The latter is particularly important, since it represents a tractable way to measure protein-folding rates under steady state conditions and allow us to infer rates of protein synthesis. In other words, we consider this to be the folding flux or workload placed on chaperones within the network, which provides a more holistic understanding of the demand placed on these components than simply counting the number of different protein clients each chaperone has.

The clustered groups of chaperones also show important features that tally with expectation in the literature. We have probably only scratched the surface of the computational analyses that can be performed. To extend this further requires the construction of integrated models of proteostasis that can predict how the cell deals with these processes under perturbation, such as a stress (e.g., heat shock or oxidative stress). This is a current focus of research in our laboratory, coupled to quantitation of the chaperome players via mass spectrometry. We believe that the recent advances made in characterizing the chaperome network structure, coupled to the increasing capacity and decreasing cost of quantitative proteomics, will make this tractable and lead to a much better understanding of how cells manage their protein complement in changing environments.

Acknowledgements We would like to thank Robert Beynon, Philip Brownridge, Dean Hammond, and Stephen Holman from the University of Liverpool for providing the quantitative datasets used in the analysis. We would also like to acknowledge the Biotechnology and Biological Sciences Research Council (BBSRC) for the support of the CoPY project, via grants BB/G009112/1, and BB/G009058/1.

References

1. Nagalakshmi U, Wang Z, Waern K, Shou C, Raha D, Gerstein M, Snyder M (2008) The transcriptional landscape of the yeast genome defined by RNA sequencing. Science 320(5881):1344–1349. doi:10.1126/science.1158441
2. Gavin AC, Aloy P, Grandi P, Krause R, Boesche M, Marzioch M, Rau C, Jensen LJ, Bastuck S, Dumpelfeld B, Edelmann A, Heurtier MA, Hoffman V, Hoefert C, Klein K, Hudak M, Michon AM, Schelder M, Schirle M, Remor M, Rudi T, Hooper S, Bauer A, Bouwmeester T, Casari G, Drewes G, Neubauer G, Rick JM, Kuster B, Bork P, Russell RB, Superti-Furga G (2006) Proteome survey reveals modularity of the yeast cell machinery. Nature 440(7084):631–636. doi:10.1038/nature04532
3. Krogan NJ, Cagney G, Yu H, Zhong G, Guo X, Ignatchenko A, Li J, Pu S, Datta N, Tikuisis AP, Punna T, Peregrin-Alvarez JM, Shales M, Zhang X, Davey M, Robinson MD, Paccanaro

A, Bray JE, Sheung A, Beattie B, Richards DP, Canadien V, Lalev A, Mena F, Wong P, Starostine A, Canete MM, Vlasblom J, Wu S, Orsi C, Collins SR, Chandran S, Haw R, Rilstone JJ, Gandi K, Thompson NJ, Musso G, St Onge P, Ghanny S, Lam MH, Butland G, Altaf-Ul AM, Kanaya S, Shilatifard A, O'Shea E, Weissman JS, Ingles CJ, Hughes TR, Parkinson J, Gerstein M, Wodak SJ, Emili A, Greenblatt JF (2006) Global landscape of protein complexes in the yeast *Saccharomyces cerevisiae*. Nature 440(7084):637–643. doi:10.1038/nature04670

4. Vogel C, Silva GM, Marcotte EM (2011) Protein expression regulation under oxidative stress. Mol Cell Proteomics 10(12):M111.009217. doi:10.1074/mcp.M111.009217
5. Huh WK, Falvo JV, Gerke LC, Carroll AS, Howson RW, Weissman JS, O'Shea EK (2003) Global analysis of protein localization in budding yeast. Nature 425(6959):686–691. doi:10.1038/nature02026
6. Belle A, Tanay A, Bitincka L, Shamir R, O'Shea EK (2006) Quantification of protein half-lives in the budding yeast proteome. Proc Natl Acad Sci U S A 103(35):13004–13009. doi:10.1073/pnas.0605420103
7. Helbig AO, Daran-Lapujade P, van Maris AJA, de Hulster EAF, de Ridder D, Pronk JT, Heck AJR, Slijper M (2011) The diversity of protein turnover and abundance under nitrogen-limited steady-state conditions in *Saccharomyces cerevisiae*. Mol Biosyst 7(12):3316–3326. doi:10.1039/C1MB05250K
8. de Godoy LM, Olsen JV, Cox J, Nielsen ML, Hubner NC, Frohlich F, Walther TC, Mann M (2008) Comprehensive mass-spectrometry-based proteome quantification of haploid versus diploid yeast. Nature 455(7217):1251–1254. doi:10.1038/nature07341
9. Ghaemmaghami S, Huh WK, Bower K, Howson RW, Belle A, Dephoure N, O'Shea EK, Weissman JS (2003) Global analysis of protein expression in yeast. Nature 425(6959):737–741. doi:10.1038/nature02046
10. Newman JR, Ghaemmaghami S, Ihmels J, Breslow DK, Noble M, DeRisi JL, Weissman JS (2006) Single-cell proteomic analysis of *S. cerevisiae* reveals the architecture of biological noise. Nature 441(7095):840–846. doi:10.1038/nature04785
11. Smallbone K, Messiha HL, Carroll KM, Winder CL, Malys N, Dunn WB, Murabito E, Swainston N, Dada JO, Khan F, Pir P, Simeonidis E, Spasić I, Wishart J, Weichart D, Hayes NW, Jameson D, Broomhead DS, Oliver SG, Gaskell SJ, McCarthy JEG, Paton NW, Westerhoff HV, Kell DB, Mendes P (2013) A model of yeast glycolysis based on a consistent kinetic characterisation of all its enzymes. FEBS Lett 587(17):2832–2841. doi:10.1016/j.febslet.2013.06.043
12. Herrgard MJ, Swainston N, Dobson P, Dunn WB, Arga KY, Arvas M, Bluthgen N, Borger S, Costenoble R, Heinemann M, Hucka M, Le Novere N, Li P, Liebermeister W, Mo ML, Oliveira AP, Petranovic D, Pettifer S, Simeonidis E, Smallbone K, Spasic I, Weichart D, Brent R, Broomhead DS, Westerhoff HV, Kirdar B, Penttila M, Klipp E, Palsson BO, Sauer U, Oliver SG, Mendes P, Nielsen J, Kell DB (2008) A consensus yeast metabolic network reconstruction obtained from a community approach to systems biology. Nat Biotechnol 26(10):1155–1160. doi:10.1038/nbt1492
13. Schwanhausser B, Busse D, Li N, Dittmar G, Schuchhardt J, Wolf J, Chen W, Selbach M (2011) Global quantification of mammalian gene expression control. Nature 473(7347):337–342. doi:10.1038/nature10098
14. Albanèse V, Yam AY-W, Baughman J, Parnot C, Frydman J (2006) Systems analyses reveal two chaperone networks with distinct functions in eukaryotic cells. Cell 124(1):75–88. doi:10.1016/j.cell.2005.11.039
15. Vabulas RM, Raychaudhuri S, Hayer-Hartl M, Hartl FU (2010) Protein folding in the cytoplasm and the heat shock response. Cold Spring Harb Perspect Biol 2(12):a004390. doi:10.1101/cshperspect.a004390
16. Parsell DA, Lindquist S (1993) The function of heat-shock proteins in stress tolerance: degradation and reactivation of damaged proteins. Annu Rev Genet 27(1):437–496. doi:10.1146/annurev.ge.27.120193.002253

17. Trotter EW, Kao CM-F, Berenfeld L, Botstein D, Petsko GA, Gray JV (2002) Misfolded proteins are competent to mediate a subset of the responses to heat shock in *Saccharomyces cerevisiae*. J Biol Chem 277(47):44817–44825. doi:10.1074/jbc.M204686200
18. Siegers K, Bolter B, Schwarz JP, Bottcher UMK, Guha S, Hartl FU (2003) TRiC/CCT cooperates with different upstream chaperones in the folding of distinct protein classes. EMBO J 22(19):5230–5240
19. Gong Y, Kakihara Y, Krogan N, Greenblatt J, Emili A, Zhang Z, Houry WA (2009) An atlas of chaperone-protein interactions in *Saccharomyces cerevisiae*: implications to protein folding pathways in the cell. Mol Syst Biol 5:275. doi:10.1038/msb.2009.26
20. Bogumil D, Landan G, Ilhan J, Dagan T (2012) Chaperones divide yeast proteins into classes of expression level and evolutionary rate. Genome Biol Evol 4(5):618–625. doi:10.1093/gbe/evs025
21. Pratt JM, Simpson DM, Doherty MK, Rivers J, Gaskell SJ, Beynon RJ (2006) Multiplexed absolute quantification for proteomics using concatenated signature peptides encoded by QconCAT genes. Nat Protoc 1(2):1029–1043. doi:10.1038/nprot.2006.129
22. Brownridge P, Holman SW, Gaskell SJ, Grant CM, Harman VM, Hubbard SJ, Lanthaler K, Lawless C, O'Cualain R, Sims P, Watkins R, Beynon RJ (2011) Global absolute quantification of a proteome: challenges in the deployment of a QconCAT strategy. Proteomics 11(15):2957–2970. doi:10.1002/pmic.201100039
23. Reiter L, Rinner O, Picotti P, Huttenhain R, Beck M, Brusniak MY, Hengartner MO, Aebersold R (2011) mProphet: automated data processing and statistical validation for large-scale SRM experiments. Nat Methods 8(5):430–435. doi:10.1038/nmeth.1584
24. Brownridge P, Lawless C, Payapilly AB, Lanthaler K, Holman SW, Harman VM, Grant CM, Beynon RJ, Hubbard SJ (2013) Quantitative analysis of chaperone network throughput in budding yeast. Proteomics 13(8):1276–1291. doi:10.1002/pmic.201200412
25. Picotti P, Rinner O, Stallmach R, Dautel F, Farrah T, Domon B, Wenschuh H, Aebersold R (2010) High-throughput generation of selected reaction-monitoring assays for proteins and proteomes. Nat Methods 7(1):43–46. doi:10.1038/nmeth.1408
26. Wang M, Weiss M, Simonovic M, Haertinger G, Schrimpf SP, Hengartner MO, von Mering C (2012) PaxDb, a database of protein abundance averages across all three domains of life. Mol Cell Proteomics. doi:10.1074/mcp.O111.014704
27. Silva JC, Denny R, Dorschel CA, Gorenstein M, Kass IJ, Li G-Z, McKenna T, Nold MJ, Richardson K, Young P, Geromanos S (2005) Quantitative proteomic analysis by accurate mass retention time pairs. Anal Chem 77(7):2187–2200. doi:10.1021/ac048455k
28. Silva JC, Gorenstein MV, Li GZ, Vissers JP, Geromanos SJ (2006) Absolute quantification of proteins by LCMSE: a virtue of parallel MS acquisition. Mol Cell Proteomics 5(1):144–156. doi:10.1074/mcp.M500230-MCP200
29. Qi D, Brownridge P, Xia D, Mackay K, Gonzalez-Galarza FF, Kenyani J, Harman V, Beynon RJ, Jones AR (2012) A software toolkit and interface for performing stable isotope labeling and Top3 quantification using progenesis LC-MS. Omics 16:489–495. doi:10.1089/omi.2012.0042
30. Aebersold R, Burlingame AL, Bradshaw RA (2013) Western blots versus selected reaction monitoring assays: time to turn the tables? Mol Cell Proteomics 12(9):2381–2382. doi:10.1074/mcp.E113.031658
31. Lu P, Vogel C, Wang R, Yao X, Marcotte EM (2007) Absolute protein expression profiling estimates the relative contributions of transcriptional and translational regulation. Nat Biotech 25(1):117–124. doi:10.1038/nbt1270
32. Craig R, Cortens JP, Beavis RC (2004) Open source system for analyzing, validating, and storing protein identification data. J Proteome Res 3(6):1234–1242. doi:10.1021/pr049882h
33. Desiere F, Deutsch EW, King NL, Nesvizhskii AI, Mallick P, Eng J, Chen S, Eddes J, Loevenich SN, Aebersold R (2006) The PeptideAtlas project. Nucleic Acids Res 34(suppl 1):D655–D658. doi:10.1093/nar/gkj040

34. Anderson L, Hunter CL (2006) Quantitative mass spectrometric multiple reaction monitoring assays for major plasma proteins. Mol Cell Proteomics 5(4):573–588. doi:10.1074/mcp.M500331-MCP200
35. Addona TA, Abbatiello SE, Schilling B, Skates SJ, Mani DR, Bunk DM, Spiegelman CH, Zimmerman LJ, Ham A-JL, Keshishian H, Hall SC, Allen S, Blackman RK, Borchers CH, Buck C, Cardasis HL, Cusack MP, Dodder NG, Gibson BW, Held JM, Hiltke T, Jackson A, Johansen EB, Kinsinger CR, Li J, Mesri M, Neubert TA, Niles RK, Pulsipher TC, Ransohoff D, Rodriguez H, Rudnick PA, Smith D, Tabb DL, Tegeler TJ, Variyath AM, Vega-Montoto LJ, Wahlander A, Waldemarson S, Wang M, Whiteaker JR, Zhao L, Anderson NL, Fisher SJ, Liebler DC, Paulovich AG, Regnier FE, Tempst P, Carr SA (2009) Multi-site assessment of the precision and reproducibility of multiple reaction monitoring-based measurements of proteins in plasma. Nat Biotech 27(7):633–641. doi:10.1038/nbt.1546
36. Dekker C, Roe SM, McCormack EA, Beuron F, Pearl LH, Willison KR (2011) The crystal structure of yeast CCT reveals intrinsic asymmetry of eukaryotic cytosolic chaperonins. EMBO J 30(15):3078–3090. doi:10.1038/emboj.2011.208
37. Giaever G, Chu AM, Ni L, Connelly C, Riles L, Veronneau S, Dow S, Lucau-Danila A, Anderson K, Andre B, Arkin AP, Astromoff A, El-Bakkoury M, Bangham R, Benito R, Brachat S, Campanaro S, Curtiss M, Davis K, Deutschbauer A, Entian KD, Flaherty P, Foury F, Garfinkel DJ, Gerstein M, Gotte D, Guldener U, Hegemann JH, Hempel S, Herman Z, Jaramillo DF, Kelly DE, Kelly SL, Kotter P, LaBonte D, Lamb DC, Lan N, Liang H, Liao H, Liu L, Luo C, Lussier M, Mao R, Menard P, Ooi SL, Revuelta JL, Roberts CJ, Rose M, Ross-Macdonald P, Scherens B, Schimmack G, Shafer B, Shoemaker DD, Sookhai-Mahadeo S, Storms RK, Strathern JN, Valle G, Voet M, Volckaert G, Wang CY, Ward TR, Wilhelmy J, Winzeler EA, Yang Y, Yen G, Youngman E, Yu K, Bussey H, Boeke JD, Snyder M, Philippsen P, Davis RW, Johnston M (2002) Functional profiling of the *Saccharomyces cerevisiae* genome. Nature 418(6896):387–391. doi:10.1038/nature00935
38. Kim YE, Hipp MS, Bracher A, Hayer-Hartl M, Ulrich Hartl F (2013) Molecular chaperone functions in protein folding and proteostasis. Annu Rev Biochem 82(1):323–355. doi:10.1146/annurev-biochem-060208-092442
39. Choi H, Larsen B, Lin Z-Y, Breitkreutz A, Mellacheruvu D, Fermin D, Qin ZS, Tyers M, Gingras A-C, Nesvizhskii AI (2011) SAINT: probabilistic scoring of affinity purification-mass spectrometry data. Nat Methods 8(1):70–73. doi:10.1038/nmeth.1541
40. Sardiu ME, Cai Y, Jin J, Swanson SK, Conaway RC, Conaway JW, Florens L, Washburn MP (2008) Probabilistic assembly of human protein interaction networks from label-free quantitative proteomics. Proc Natl Acad Sci U S A 105(5):1454–1459. doi:10.1073/pnas.0706983105
41. Sowa ME, Bennett EJ, Gygi SP, Harper JW (2009) Defining the human deubiquitinating enzyme interaction landscape. Cell 138(2):389–403. doi:10.1016/j.cell.2009.04.042
42. Mellacheruvu D, Wright Z, Couzens AL, Lambert J-P, St-Denis NA, Li T, Miteva YV, Hauri S, Sardiu ME, Low TY, Halim VA, Bagshaw RD, Hubner NC, al-Hakim A, Bouchard A, Faubert D, Fermin D, Dunham WH, Goudreault M, Lin Z-Y, Badillo BG, Pawson T, Durocher D, Coulombe B, Aebersold R, Superti-Furga G, Colinge J, Heck AJR, Choi H, Gstaiger M, Mohammed S, Cristea IM, Bennett KL, Washburn MP, Raught B, Ewing RM, Gingras A-C, Nesvizhskii AI (2013) The CRAPome: a contaminant repository for affinity purification-mass spectrometry data. Nat Methods 10(8):730–736. doi:10.1038/nmeth.2557
43. Szklarczyk D, Franceschini A, Kuhn M, Simonovic M, Roth A, Minguez P, Doerks T, Stark M, Muller J, Bork P, Jensen LJ, Mering CV (2011) The STRING database in 2011: functional interaction networks of proteins, globally integrated and scored. Nucleic Acids Res 39(suppl 1):D561–D568. doi:10.1093/nar/gkq973
44. Stark C, Breitkreutz B-J, Chatr-aryamontri A, Boucher L, Oughtred R, Livstone MS, Nixon J, Van Auken K, Wang X, Shi X, Reguly T, Rust JM, Winter A, Dolinski K, Tyers M (2010) The BioGRID interaction database: 2011 update. Nucleic Acids Res 39(suppl 1):D698–D704
45. Güldener U, Münsterkötter M, Oesterheld M, Pagel P, Ruepp A, Mewes H-W, Stümpflen V (2006) MPact: the MIPS protein interaction resource on yeast. Nucleic Acids Res 34(suppl 1):D436–D441

46. Kampinga HH, Craig EA (2010) The HSP70 chaperone machinery: J proteins as drivers of functional specificity. Nat Rev Mol Cell Biol 11(8):579–592. doi:10.1038/nrm2941
47. Sugiyama S, Moritoh S, Furukawa Y, Mizuno T, Lim YM, Tsuda L, Nishida Y (2007) Involvement of the mitochondrial protein translocator component tim50 in growth, cell proliferation and the modulation of respiration in *Drosophila*. Genetics 176(2):927–936. doi:10.1534/genetics.107.072074
48. Guo Y, Cheong N, Zhang Z, De Rose R, Deng Y, Farber SA, Fernandes-Alnemri T, Alnemri ES (2004) Tim50, a component of the mitochondrial translocator, regulates mitochondrial integrity and cell death. J Biol Chem 279(23):24813–24825. doi:10.1074/jbc.M402049200
49. Cheng MY, Hartl FU, Martin J, Pollock RA, Kalousek F, Neupert W, Hallberg EM, Hallberg RL, Horwich AL (1989) Mitochondrial heat-shock protein hsp60 is essential for assembly of proteins imported into yeast mitochondria. Nature 337(6208):620–625. doi:10.1038/337620a0
50. Cherry JM, Hong EL, Amundsen C, Balakrishnan R, Binkley G, Chan ET, Christie KR, Costanzo MC, Dwight SS, Engel SR, Fisk DG, Hirschman JE, Hitz BC, Karra K, Krieger CJ, Miyasato SR, Nash RS, Park J, Skrzypek MS, Simison M, Weng S, Wong ED (2012) *Saccharomyces* genome database: the genomics resource of budding yeast. Nucleic Acids Res 40(Database issue):D700–D705. doi:10.1093/nar/gkr1029

Part II
Chaperones at the Ribosome

Chapter 2
Functions of Ribosome-Associated Chaperones and their Interaction Network

Annika Scior and Elke Deuerling

Abstract Cells coordinate chaperones at the exit site of the ribosome. Albeit the types and mechanisms of ribosome-associated chaperones differ in the three kingdoms of life, they all share the ability to protect nascent polypeptides from off pathways such as aggregation and degradation and, at least in some cases, support initial folding steps of newly synthesized proteins. Recent progress was made in understanding the nascent interactome of these ribosome-associated chaperones. While the bacteria-specific chaperone trigger factor (TF) binds to almost every nascent polypeptide made by ribosomes except for membrane proteins, the substrate pool of the two eukaryotic ribosome-associated chaperone systems, nascent polypeptide-associated complex (NAC) and ribosome-associated complex (RAC), is more distinct.

Interestingly, there is culminating evidence that these chaperones also display important functions off the ribosome, e.g., in the biogenesis of ribosomal subunits and in protein aggregation under proteotoxic stress conditions. In this chapter, we will discuss the functions of these chaperones with regard to their broad substrate pools.

1 Introduction: Ribosome-Associated Chaperones in De novo Folding

Directly upon their synthesis by the ribosome, proteins have to fold into their unique three-dimensional structure in order to become biologically active. The folding of proteins is problematic since hydrophobic residues of the unfolded polypeptide chain are accessible, which enhances the probability that the newly synthesized protein follows an unproductive off pathway leading to misfolding and aggregation [1]. As protein misfolding and aggregation represent the hallmarks of several neurodegenerative diseases, it is of particular importance to understand the mechanisms by which proteins acquire and maintain their structure under cellular conditions.

E. Deuerling (✉) · A. Scior
Department of Biology, University of Konstanz, Konstanz, Germany
e-mail: elke.deuerling@uni-konstanz.de

W. A. Houry (ed.), *The Molecular Chaperones Interaction Networks in Protein Folding and Degradation,* Interactomics and Systems Biology, DOI 10.1007/978-1-4939-1130-1_2,

To accomplish folding and prevent off pathways, newly synthesized polypeptides interact with at least one cytosolic chaperone that prevents inappropriate inter- and intramolecular interactions, and thus promotes the folding into the native state. Among these chaperones, the ribosome-associated ones interact with nascent polypeptides while they are still attached to the peptidyl-transferase center of the ribosome (Figs. 2.1 and 2.2a) [2–4].

Binding of chaperones to nascent polypeptides can have several effects. The unfolded polypeptide is protected from being degraded or from incorrect contact with other molecules that may lead to aggregation. Moreover, for polypeptides that start their folding program co-translationally, chaperone binding might support very early folding. Beyond the interaction of newly synthesized proteins with ribosome-attached chaperones, additionally other cytosolic chaperones can act during later stages of translation on elongated nascent chains or after release of the polypeptides from the ribosomes. The latter mainly belong to the heat shock protein (Hsp)70/40 and the Hsp60/10 chaperone families. Examples are the DnaK/DnaJ (Hsp70/Hsp40) and GroEL/GroES (Hsp60/Hsp10) systems in the cytosol of *E. coli* cells [3, 5]. Together with ribosome-asssociated chaperones they form a robust network that promotes the de novo folding of newly synthesized polypeptides and prevents off pathways such as aggregation or degradation early in the life of a new protein [4].

Ribosome-associated chaperones are found in every cell but differ significantly among the different kingdoms of life with regard to their number and type. Whereas prokaryotes have only one ribosome-associated chaperone which is called trigger factor (TF) [6–9], eukaryotic ribosomes coordinate two TF-unrelated chaperone systems at the ribosome (Fig. 2.2a). These systems are the highly conserved nascent polypeptide-associated complex (NAC) and an arrangement of specialized Hsp70 and Hsp40 chaperones that includes a heterodimer called ribosome-associated complex (RAC; Fig. 2.2a) [4, 10]. In yeast, the RAC system has a third ribosome-associated partner, a Hsp70 chaperone called Ssb [11] (Fig. 2.2a).

In the following section, the functions and structures of these ribosome-associated chaperones will be discussed together with their respective substrates.

2 The Prokaryotic Ribosome-Associated Chaperone Trigger Factor

The only known chaperone of bacterial cells with a direct binding site on ribosomes is TF. In *E.coli,* TF consists of 432 amino acids and has a molecular weight of 48 kDa. It is a three-domain protein with an N-terminal ribosome-binding domain, a middle domain that displays peptidyl-prolyl *cis/trans* isomerase (PPIase) activity and a C-terminal domain [7, 12] (Fig. 2.1a). Its crystal structure revealed that TF adopts an extended three-dimensional conformation with an unusual domain arrangement [9]. Although distant in the amino acid sequence, the N- and C-terminal domains are located adjacent to each other in the three-dimensional structure to form a cradle-like structure (Fig. 2.1a). The C-terminal domain of TF is located in

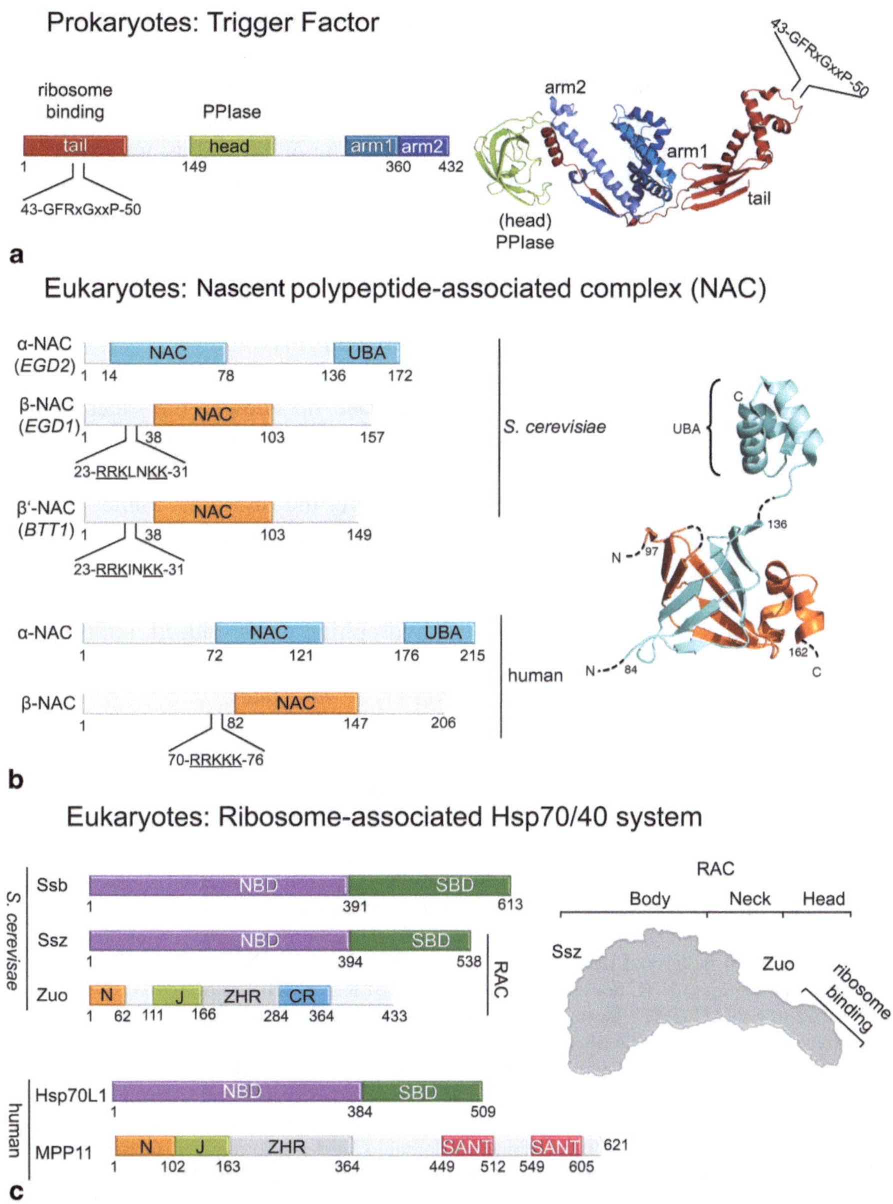

Fig. 2.1 **a** *E. coli* trigger factor. The N-domain (*red*) contains the ribosome-binding motif (40-GFRxGxxP-49). It is located in a loop region between two α-helices, the N-domain is connected to the PPIase domain (*green*) via an extended linker. The C-domain (*blue*) is located in the center of the folded molecule and forms two arm-like protrusions. The N-domain and both arms of the C-domain together form a cavity for nascent polypeptide chains. **b** The nascent polypeptide-associated complex (NAC). The α-subunit of NAC (encoded in yeast by the *EGD2* gene) consists of a NAC-domain in the N-terminal region and an ubiquitin-associated (UBA) domain at its C-terminus. In yeast two alternative β-subunits (β and β′) exist, which are either encoded by the

the center of the molecule and forms two arm-like protrusions [9]. The N-terminal domain contains a signature motif (42-GFRxGxxP-50) that is located in an exposed loop region [9] and binds to the ribosomal protein L23 (Fig. 2.1a). Mutation of either the signature motif or a conserved surface-exposed residue within the ribosomal protein L23 strongly impairs ribosome binding of TF and its activity on nascent polypeptides [13]. The N-terminus is connected via a long linker to the second domain, the PPIase domain, which is located at the opposite end of the molecule [9] (Fig. 2.1a). This domain catalyzes prolyl *cis/trans* isomerization in peptides in vitro and presumably slows down folding processes or acts as auxiliary site to assist the folding of substrate proteins [14–16]. However, the in vivo relevance of this domain and its enzymatic activity is still not well understood [17].

Crystallization of the N-terminal fragment of *E. coli* TF together with the 50S large ribosomal subunit from *Haloarcula marismortui* allowed the superposition of full-length TF with ribosomes and paved the way for understanding how this chaperone functions on ribosomes [9]. TF binds to the ribosomal protein L23 and hunches with its extended cradle-like structure over the ribosomal tunnel exit site (Fig. 2.2a). It was suggested that the cradle-like structure provides a shielded environment for the emerging nascent polypeptides and thus supports co-translational folding events [9]. The interior of the cavity formed by the TF N- and C-terminal domains contains hydrophobic as well as hydrophilic areas. Indeed, multiple potential substrate-binding sites within this cradle have been suggested [18, 19]. This structure might enable TF to bind a large variety of substrates.

The ribosome-binding activity of TF has been extensively characterized. It cycles on and off the ribosomes and binds as a monomer to the translation machinery [14, 20, 21]. TF is present in a two- to threefold molar excess over ribosomes and

---

genes *EGD1* or *BTT1*, respectively. The β-subunits of NAC contain a conserved ribosome-binding motif in the N-terminus and a central NAC domain. Dimerization of the NAC subunits involves the NAC domains of both subunits. α-NAC (*blue*) and β-NAC (*orange*) form a stable heterodimer. The UBA domain is derived from the crystal structure of archaeal NAC (PDB 1TR8) and was modeled on the human NAC domain heterodimer PDB 3LKX, aa 84–136 of α-NAC (NACA) and aa 97–162 of β-NAC (BTF3 isoform). Broken lines indicate unresolved parts of the molecule. **c** The ribosome-asscoiated heat shock protein (Hsp)70/40 system in eukaryotes. The yeast ribosome-associated chaperone system consists of the Hsp70s Ssb, Ssz, and the Hsp40 Zuo. Ssb1 and Ssz contain an N-terminal nucleotide-binding domain (NBD, *violet*) and a C-terminal substrate-binding domain (SBD, *dark green*). The SBD of Ssz is shorter compared to canonical Hsp70s. Like all Hsp40 co-chaperones Zuo contains a J-domain (*light green*) required for stimulation of the adenosine triphosphatase (ATPase) activity of its Hsp70 partner. In addition, Zuo contains a charged region (*blue*) in its C-terminus, which is involved in ribosome binding. The N-terminus (N, *orange*) of Zuo is predicted to be unstructured. Ssz and Zuo form the stable heterodimeric ribosome-associated complex (RAC). A schematic model of RAC based on small-angle X-ray scattering (SAXS) analysis of yeast RAC (on the *right* side, shown in *light grey*) reveals that the complex forms an elongated structure that was divided into body, neck and head. Most of the body is composed of Ssz, whereas Zuo forms the neck and the head. The head region of Zuo was shown to contact the ribosome near the tunnel exit. *Lower panel*: In contrast, in human, no Ssb homolog is found while RAC is conserved. The RAC complex is formed by the Hsp40 MPP11 and Hsp70L1. MPP11 contains two SANT domains (*pink*) at its C-terminus. The functions of the SANT domains are still unknown

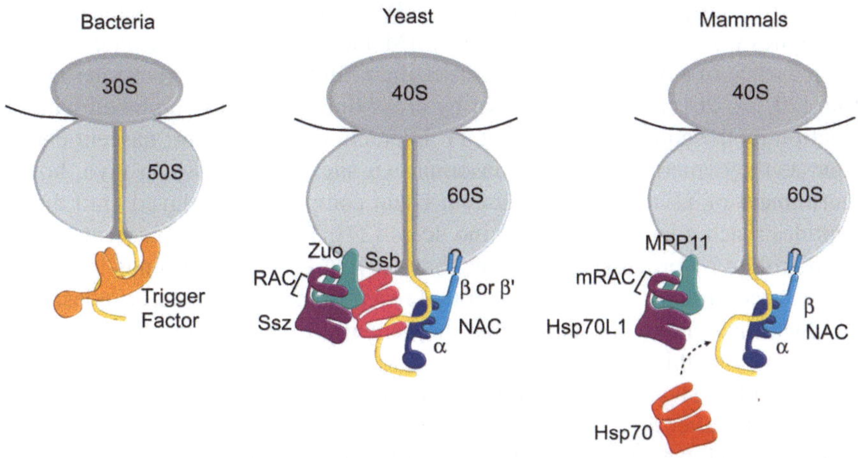

a

	Trigger Factor *E. coli*	Ssb *S. cerevisiae*	NAC (*S. cerevisiae*)	
			α + β	β'
substrate specificity	high hydrophobicity positive charge	high hydrophobicity high intrinsic disorder large size slow synthesis low α–helical content	high hydrophobicity low intrinsic disorder	low hydrophobicity high intrinsic disorder
substrate pool	outer membrane proteins cytosolic proteins	cytosolic proteins nuclear proteins	metabolic enzymes	mitochondrial proteins ribosomal proteins

b

Fig. 2.2 **a** Ribosome-associated chaperones bind to ribosomes (*grey*) in close proximity to the ribosomal tunnel exit and interact with nascent polypeptide chains (*yellow*). In bacteria, only trigger factor (*orange*) binds to the ribosome and interacts with the growing nascent polypeptide. In eukaryotes, two chaperone systems are found in association with the ribosome. On the one hand, the heterodimeric nascent polypeptide-associated complex (NAC) complex that consists of a α- (*dark blue*) and β- (*light blue*) subunit binds to the ribosome via its β-subunit. Additionally, a system consisting of heat shock protein (Hsp)70 and Hsp40 family members binds to ribosomes and interacts with nascent chains. In yeast this system comprises the Hsp70 Ssb (*pink*) as well as the ribosome-associated complex (RAC), which is formed by the Hsp40 Zuo (*green*) and the Hsp70 Ssz (*violet*). Mammalian RAC (mRAC) is formed by the Hsp40 MPP11 and Hsp70L1. As Ssb is restricted to fungi, cytosolic Hsp70s (*red*) are recruited to the nascent chain by mRAC. **b** The table summarizes the co-translational substrate specificities and the substrate pool of trigger factor from *E. coli* as well as Ssb and NAC from *S. cerevisiae*. The substrate specificities of trigger factor were taken from: [36]; the substrate pool from: [27]. The co-translational interactome of Ssb was analyzed in [92] and that of NAC in [79]

thus is bound to most cytosolic ribosomes. Unbound TF diffuses freely in the cytosol, probably as a homodimer. The in vivo relevance of this dimerization is unclear but dimer formation may represent a storage form of this chaperone to prevent its cradle from unspecific associations [22, 23]. Although TF binds to non-translating ribosomes with a K_D of approximately 1 μM and a mean residence time of 10–15 s, the presence of a nascent chain increases the affinity of TF for ribosomes up to 30-fold [20, 24–26]. It could be shown by cross-linking studies combined with cryo-electron microscopy structures that TF interacts with very short nascent chains as soon as they emerge from the ribosomal exit site in vitro [18]. In vivo, however, recruitment of TF to ribosome-nascent chain complexes is delayed until the polypeptides reach a length of 100 amino acids [27]. TF binding to nascent substrates is controlled by processing enzymes that act first on the N-terminus of the newly synthesized protein. It was demonstrated recently that peptide deformylase associates with nascent chains as soon as they emerge from the ribosomal exit tunnel, followed by methionine aminopeptidase [28]. Both enzymes prevent the premature recruitment of TF [27, 29]. Most likely TF can also stay bound to a subset of nascent polypeptides after their release from the ribosome [30].

Initially, TF was discovered as a cytosolic protein involved in the translocation of the outer membrane protein pro-OmpA across the plasma membrane as it has the ability to promote the insertion of chemically denatured pro-OmpA into membrane vesicles [31]. However, TF is not essential for viability of *E. coli* and deletion of its gene *tig* does not cause any growth defect albeit the heat shock response is induced leading to enhanced levels of chaperones and proteases [32, 33]. First evidence that TF fulfills chaperone function in vivo came from the finding that the simultaneous deletion of TF and the cytosolic Hsp70 chaperone DnaK is synthetically lethal at temperatures above 30 °C [32, 34]. Cells with low levels of DnaK and without TF are viable but show slow growth and massive aggregation of several hundreds of cytosolic proteins, especially large-sized proteins and components of protein complexes. Interestingly, TF and DnaK show some overlap in their substrate specificity. Both chaperones interact with peptide segments of an unfolded protein that has a high mean hydrophobicity and an overall positive net charge [35, 36] (Fig. 2.2b). Taken together, these findings suggest that the cytosolic Hsp70 DnaK and the ribosome-associated chaperone TF act on a similar substrate spectrum and thus are able to cooperate in the de novo folding of newly synthesized proteins.

The mechanism by which the non-adenosine triphosphate (non-ATP)-consuming TF promotes de novo folding of newly synthesized proteins is still not fully understood. On the one hand, the cavity formed by the N- and C-terminal domains could provide a folding chamber that protects the emerging nascent chain from unfavorable interactions while translation proceeds. In agreement with these findings it could be demonstrated that TF protects nascent proteins from proteolytic digestion in vitro [37, 38]. Additionally, it was shown recently that TF stimulated native folding in constructs of repeated maltose-binding protein (MBP) domains by protecting partially folded domains from distant interactions that produce stable misfolded states [39]. On the other hand, it was proposed that TF delays the folding of a nascent polypeptide until sufficient sequence information (encoded in the C-terminal area of the synthesized polypeptide chain) is available outside of the ribosome to allow productive folding [35, 38]. Indeed, recent data suggest that TF can unfold preexisting folded states to prevent and revert premature folding, thus limiting the formation of misfolded intermediate states during protein synthesis [8]. In the current model, first the ribosome itself limits the conformational freedom of the newly synthesized polypeptide chains. As translation proceeds and the chain lengthens the influence of the ribosome, in particular on the N-terminal regions, decreases and the risk to form misfolded intermediates increases. Therefore, TF binds to nascent polypeptides of a length of 100 amino acids in repeated binding and release cycles. It limits conformational sampling and folding more efficiently than the ribosome and thereby prevents folding intermediates [8]. TF is not only able to block folding but also unfolds preformed segments in order to give the nascent chain a new opportunity for productive folding. After release from the ribosome, some nascent chains that do not need further assistance acquire their native structure after dissociation of TF. In contrast, other proteins stay associated with TF and are transferred to downstream chaperone systems, in particular to the DnaK/DnaJ/GrpE system or to GroEL/ES.

Very recently, the nascent interactome of ribosome-bound TF was identified using an elegant approach: selective ribosome profiling [27, 40]. For this technique, actively translating ribosomes are isolated; the messenger itochondrial ribonucleic acid (mRNA) bound to them is converted into deoxyribonucleic acid (DNA) and subsequently analyzed by high-throughput sequencing. To gain insights into which nascent polypeptides are bound by TF, only those ribosomes with bound TF were isolated and the mRNA was analyzed. TF was fused with an affinity purification tag, ribosomes were isolated, TF binding to the nascent polypeptides was stabilized by cross-linking, and thus specifically the ribosome-nascent-chain complexes containing TF could be isolated by affinity purification [27].

The data revealed that the medium average length at which TF engaged a polypeptide was 112 amino acids, half of all nascent chains being bound within ±20 amino acids of this position. Nascent chains are generally not engaged by more than one TF molecule. Furthermore, the data revealed that TF interacts with all nascent chains in *E. coli* except for those that localize to the plasma membrane (Fig. 2.2b). Membrane proteins are poorly bound by TF. Interestingly, outer membrane β-barrel proteins (OMPs) represent the strongest TF interactors (Fig. 2.2b). For example, the five best-characterized OMPs (LamB, LptD, OmpA, OmpC, and OmpF) were found among the 25 strongest TF-interacting nascent polypeptides. This finding is in agreement with very early data, as TF was initially discovered by its ability to promote insertion of pro-OmpA into membranes [31]. To prove the in vivo relevance of this finding the authors analyzed the protein content of the outer membrane fraction of wild type *E. coli* cells and compared it with those lacking TF using stable isotope labeling with amino acids in cell culture (SILAC) combined with mass spectrometry. The data revealed that in cells lacking TF, 60% of all detected proteins were found in lower amounts than in the wild type, whereas no proteins were found in higher amounts. In agreement with this result, it was shown that loss of TF causes enhanced sensitivity of *E. coli* to SDS/EDTA (sodium dodecyl sulfate/ethylenediamine tetraacidic acid) treatment or vancomycin, which suggests defects in the biogenesis of outer membrane proteins. In sum, these data clearly suggest that TF not only plays an important role in the folding of cytosolic proteins but is also crucial for the folding-competent conformation of outer membrane proteins during their biogenesis [27].

3 The Nascent Polypeptide-Associated Complex (NAC)

NAC is a heterodimeric complex that associates with ribosomes likely in a 1:1 stoichiometry [41, 42] (Fig. 2.2a). It is highly conserved among eukaryotes and consists of one α- and one β-subunit, referred to as α-NAC and β-NAC (Fig. 2.1b). NAC is not found in bacteria but in archaea, where it is built by an α-NAC homodimer [43, 44].

Both subunits of heterodimeric NAC were shown to contact nascent polypeptide chains [42, 45] while only β-NAC interacts with the ribosome and mediates ribosome binding of the complex [43, 46] (Fig. 2.2a). NAC is present in equimolar concentration relative to ribosomes and binds to them irrespective of their translation status [41]. In yeast, NAC is encoded by three genes. The α-NAC subunit is encoded by *EGD2* while *EGD1* as well as *BTT1* both encode for β-subunits. The gene product of EGD1 is referred to as β-NAC and the one of *BTT1* as β′-NAC (Fig. 2.1b). Both β-subunits are able to form a complex with the α-subunit. However, the β′-NAC version is approximately 100 times less expressed than the β-version [44, 47]. Thus, the most abundant NAC type is formed by α β in yeast cells.

Both subunits of NAC share a homologous structural element, the so-called NAC-domain. These domains are responsible for the dimerization of NAC into the heterodimeric complex (Fig. 2.1b). The NAC-domain is located in the N-terminal part of α-NAC and in the central domain of β-NAC. Crystal structures of parts of the human complex give insights into the molecular interactions governing NAC complex formation [48, 49]. They reveal a handshake interaction of the two NAC-domains consisting of a six-stranded β-barrel that is stabilized by hydrophobic contacts between conserved residues [49, 50]. Additionally, α-NAC contains an ubiquitin-associated (UBA) domain [50] (Fig. 2.1b). UBA domains were described to bind mono- and polyubiquitin but also mediate other protein-protein interactions [51]. They are commonly found in factors dedicated to ubiquitin-dependent protein degradation or in components of signal transduction pathways [51–54]. However, the function of the UBA domain of α-NAC remains elusive.

There is no structure available of NAC in complex with the ribosome and thus its orientation on the ribosome and relative to the exit site for nascent polypeptides is unknown. Different physical contact points for NAC are under debate. Based on cross-linking data various anchor points have been suggested: either Rpl25, the L23 homolog that coordinates TF in bacteria [55] or Rpl31 [56, 57] which is also the main interaction site for RAC (see below). Both Rpl25 and Rpl31 are located directly at the ribosomal exit site and would position NAC in a way that favors its interaction with nascent polypeptides. Previous work suggested that NAC binds to Rpl25 via a conserved motif (RRK-(X)n-KK, amino acids 23–31 located in a loop region between two α-helices in the N-terminus of β-NAC (Fig. 2.1b). Upon mutation of the amino acid residues arginine arginine lysine (RRK) to alanine alanine alanine (AAA) ribosome binding of NAC was abolished both in vivo and in vitro [55].

Despite the evolutionary conservation of NAC in eukaryotes, its in vivo function is still rather obscure. Originally, NAC was discovered in 1994 by Wiedman and coworkers and described as a translocation and sorting factor [42]. Based on in vitro data, they suggested that NAC interacts with ribosomes and nascent polypeptides in order to prevent mislocalization of ribosomes translating cytosolic proteins to the endoplasmic reticulum (ER) [42–46, 57–64]. However, in vivo evidence for such a function is missing till today. Another study reported a correlation between altered NAC levels and neurodegenerative diseases [65]. β-NAC was also described as a target for caspases [66] and an apoptosis-suppressing activity of NAC was reported [67]. Moreover, functions as a transcriptional regulator were described for individual NAC subunits [68–71].

The deletion of the NAC-encoding genes has no effect on the growth or viability of yeast cells. However, β-NAC knockout cells (*egd1Δbtt1Δ*) where only the α-NAC subunit is expressed show a slight growth defect at 37 °C [44]. In contrast, there is an embryonic lethality of NAC mutants in *C. elegans* [67], *D. melanogaster* [72], and mice [73] indicating that this complex fulfills essential functions in higher eukaryotes.

Since NAC binds to ribosomes and interacts with nascent polypeptides, a chaperone-like function of NAC was proposed [4, 74–76]. First evidence for such a role was obtained by a recent study showing that NAC cooperates with the Ssb-RAC chaperone system in yeast [77]. It could be demonstrated that the absence of NAC enhances the sensitivity of *ssbΔ* cells towards drugs that cause misreading during translation or folding stress. Cells lacking Ssb and NAC show aggregation of newly synthesized polypeptides and defects in translation [77]. Another study showed that NAC is a central component of the protein homeostasis network in the metazoan animal model *C. elegans* [78]. Under non-stress conditions NAC is associated with the ribosome where it supports protein folding as well as translation activity. However, NAC seems to fulfill a dual function that becomes evident under stress conditions: Upon proteotoxic stress, NAC interacts with protein aggregates and thereby becomes discharged from its ribosomal housekeeping function. The localization of NAC to protein aggregates induced by heat shock conditions is important for the fast resolubilization of these aggregates after the stress. Likewise, the sequestration of NAC from ribosomes to protein aggregates causes a strong decrease in the cytosolic pool of actively translating polysomes. It is suggested that a decrease of translation is beneficial under stress conditions that cause protein misfolding and aggregation since the influx of newly synthesized and unfolded proteins is reduced. This, in turn, decreases the overall folding load for chaperones and allows the chaperone network to act on the remodeling of existing aggregates. Thus, NAC has crucial housekeeping and stress reducing functions in metazoans. It acts as a modulator of protein synthesis to establish a regulatory feedback mechanism that adjusts translational activity to the cellular protein folding load [78].

A study by Frydman and co-workers recently identified the nascent interactome of NAC in yeast [79]. They expressed individual tandem-affinity-purification (TAP)-tagged NAC subunits to purify the NAC-ribosome-nascent chain complexes (NAC-RNCs) together with the corresponding mRNA. The mRNAs were identified by DNA microarray hybridization. Using this strategy, the authors were able to show that the different subunits of yeast NAC interact with distinct sets of nascent polypeptides, but virtually every nascent protein on ribosomes contacts at least one NAC subunit. In other words, every protein that is translated in a yeast cell interacts co-translationally with NAC. According to these data, β′-NAC associates with ribosomes translating mRNAs that encode proteins with high intrinsic disorder and low hydrophobicity as well as mitochondrial or ribosomal proteins. The substrates of β-NAC and α-NAC showed a large overlap and were more distinct from those of β′-NAC. For example, hydrophobic proteins and metabolic enzymes were highly enriched among their interactors [79] (Fig. 2.2b).

As mentioned above, a role for NAC in co-translational protein targeting has been discussed since decades [42–46, 57–64]. Therefore, the authors also analyzed the interplay between the targeting factor SRP (signal recognition particle) and NAC. Originally, it was proposed that NAC and SRP compete for ribosome binding and that a nascent polypeptide can only interact with one of them at the same time. In contrast, this study indicates that both factors can bind simultaneously to one ribosome and the presence of NAC does not prevent the interaction of nascent chains with SRP in yeast. In the absence of NAC, however, a subset of the nascent secretory proteins was unable to interact efficiently with SRP, whereas on the other hand, increased false contacts between SRP and cytosolic proteins were detected which are no SRP substrates [79].

In addition, NAC plays a role in protecting nascent chains from premature and incorrect co-translational ubiquitination and degradation [80]. Pulse-labeling experiments revealed an increased ubiquitination of nascent chains upon loss of NAC. Thus, NAC protects nascent unfolded polypeptides as they emerge from the ribosome. Many nascent chains shielded by NAC are quite long, suggesting that NAC does not only function by sterically blocking the access of the ubiquitin-proteasome system to the ribosome [80].

4 The Ribosome-Associated Hsp70/Hsp40 System

Besides NAC a second ribosome-associated chaperone system consisting of Hsp70 and Hsp40 chaperones is found in association with eukaryotic ribosomes (Figs. 2.1c and 2.2a). The RAC is the central component of this system and conserved from yeast to mammals [4] (Fig. 2.1c). RAC is a stable heterodimer formed by an Hsp40 chaperone called Zuotin (Zuo) and an Hsp70 called Ssz [10, 11, 81]. In *S. cerevisiae* this system is extended by another ribosome-anchored chaperone, the Hsp70 Ssb (Fig. 2.1c). Ssb binds to the nascent polypeptide, which requires the activity of RAC [11, 82] (Fig. 2.2a). RAC stimulates the ATPase activity of Ssb and thus enhances its affinity for unfolded polypeptides, but RAC seems not to bind to substrate proteins itself (Fig. 2.2a) [83]. Within the RAC complex, Zuo contacts Ssb by its J-domain and in addition anchors RAC at the ribosome [84–86] (Fig. 2.1c). The function of Ssz is less clear. It might fulfill regulatory functions by inducing structural rearrangements within the J-domain of Zuo, which in turn might strengthen the contact to Ssb [87]. Alternatively, Ssz might play a role in the recruitment of substrates to Ssb [84]. However, no experimental data exist so far to support this hypothesis.

In the yeast genome, two genes (*SSB1* and *SSB2*) encode the proteins Ssb1 and Ssb2 that differ only in four amino acids. Therefore, in the following, they are collectively referred to as Ssb. The functional cooperation between the members of the yeast Hsp70/Hsp40-chaperone triad was discovered by genetic experiments. Yeast cells lacking either one or all three proteins of this system show similar phenotypes: sensitivity to high salt concentrations, hypersensitivity to aminoglycosides that increased translational error rates, and cold-sensitivity [11, 82, 86, 88, 89]. The first hint that the RAC-Ssb system might be involved in the folding of nascent chains came from cross-linking experiments showing that Ssb is able to contact short nascent polypeptide chains and this interaction was dependent on RAC [11, 82, 90]. This suggests that Ssb acts as a chaperone for nascent polypeptides. In addition, these data indicate that the chaperone triad must bind to ribosomes in close proximity to the ribosomal tunnel exit site. Moreover, the finding that expression of the prokaryotic ribosome-associated chaperone TF could partially alleviate the aminoglycoside sensitivity of triad-deficient yeast cells indicates overlapping functions of the two chaperone systems from different kingdoms of life [91]. A more recent study shows that the RAC-Ssb and the NAC systems are functionally connected in yeast [77]. The simultaneous deletion of NAC- and Ssb-encoding genes caused conditional loss of cell viability under protein-folding stress conditions. Furthermore, loss of Ssb resulted in the aggregation of newly synthesized polypeptides, ribosomal proteins, as well as several ribosome biogenesis factors. Likewise, the levels of translating ribosomes and 60S ribosomal subunits were decreased in *ssbΔ* cells. These defects were aggravated when NAC was absent in addition to the Ssb chaperone. These findings suggest that not only the folding of cytosolic proteins is affected by these chaperones but also ribosome biogenesis.

Based on the finding that mainly ribosomal proteins are aggregation-prone in the absence of Ssb and on a recent study investigating the nascent interactome of Ssb, it is likely that ribosomal proteins represent important substrates of this chaperone [77, 92].

Ssb represents a member of the Hsp70 chaperone family with a N-terminal nucleotide-binding domain (NBD) and a C-terminal substrate-binding domain (SBD) (Fig. 2.1c). Additionally, it contains a potential nuclear export sequence (NES) at its C-terminus [93]. Ssb is found primarily in the cytosol, upon mutation of the NES however, it strongly accumulates in the nucleus [93]. This suggests that Ssb may rapidly shuttle between the nucleus and cytosol.

Like Ssb, also the Hsp70 Ssz is composed of a NBD and SBD. In contrast to other Hsp70 chaperones, Ssz is not able to hydrolyze ATP and up to now no binding of Ssz to a substrate could be detected [81, 83]. One reason might be its shorter substrate-binding domain (Fig. 2.1c) [90]. Furthermore, it was suggested that amino acid substitutions at three positions in the ATP-binding pocket (compared to other Hsp70s) might be the reason for the inability of Ssz to hydrolyze ATP [84]. As Ssz is continuously in an ATP-bound state its affinity for potential substrates is assumed to be rather low. It was proposed that Ssz could potentially act as a low-affinity holding chaperone that guides the growing nascent chain from the tunnel to Ssb; an interesting aspect that needs to be experimentally tested in the future [84].

The other subunit of RAC, Zuo, consists of an unstructured N-terminal domain that is followed by a J-domain. The J-domain has a conserved HPD (Histidine, Proline, Aspartic acid) motif and is required to stimulate the ATPase activity of Ssb. A characteristic feature of Zuo is a highly positively charged region (charged region) that is located near its C-terminus (Fig. 2.1c) [10, 86, 94]. This region was discussed for a long time to be involved in the ribosome association of Zuo [85, 86].

It was shown that the highly flexible and unstructured N-terminus of Zuo is responsible for the formation of the RAC complex. The first 62 N-terminal amino acids of Zuo contact both domains of Ssz. This interaction is crucial and results in the formation of the unusual, stable chaperone-chaperone complex RAC [87]. Interestingly, upon complex formation the J-domain of Zuo becomes more dynamic. As the J-domain is required for the stimulation of the ATPase activity of Ssb, the complex formation with Ssz might induce a conformation of Zuo that favors the interaction with Ssb and thus might be crucial for the function of RAC as a co-chaperone [87].

Recently, the crystal structure of a truncated RAC version from the thermophilic fungus *Chaetomium thermophilium* could be solved [84]. The structure was modeled into small-angle X-ray scattering (SAXS) densities of RAC bound to the ribosome. The data revealed that RAC forms an elongated complex of 180 Å in length (Fig. 2.1c) that binds to the ribosome near the ribosomal proteins L22 and L31. In combination with cryo-electron microscopy data of ribosome-bound RAC, it was shown that the complex bends over the ribosomal tunnel exit site and contacts the ribosomal extension segment ES27 that stabilizes a distinct conformation of RAC. The elongated RAC structure can be divided into a head, neck and a body region (Fig. 2.1c). Whereas the large body is mainly composed of Ssz, the head is formed by Zuo and mediates the ribosome association (Fig. 2.1c). In agreement with previous data, it could be shown that ribosome association is mediated by a helix that is formed by the C-terminal part of the charged region [84].

While Ssb is found exclusively in fungi, homologs of RAC were identified in mammals as well [95–97] (Fig. 2.1c). This indicates that the presence of Hsp70/Hsp40 chaperones on ribosomes is common in the eukaryotic world (Figs. 2.1c and 2.2a). Supporting this assumption, the human homologs of Ssz (Hsp70L1) and Zuo (MPP11) can partially substitute for the loss of RAC in yeast [95–97]. Additionally, the knockdown of human MPP11 in HeLa cells results in growth defects and sensitivity towards drugs similar to what was observed for yeast cells lacking RAC [96]. These findings indicate that RAC fulfills similar functions in yeast as well as in metazoans. Despite these similarities between yeast and mammalian RAC, there are also large structural and functional differences. Most strikingly, complementation of yeast RAC with the mammalian system is independent of Ssb [97]. It was thus suggested recently that cytosolic Hsp70s, which cannot bind to the ribosome on their own, act as functional partners for RAC in higher eukaryotes (Fig. 2.2a) [96]. In addition, MPP11 has some variation in its domain composition compared to Zuo including two additional SANT-like domains at the C-terminal end of the chaperone (Fig. 2.1c). SANT (Swi3, Ada2, N-Cor, TfIIIB) domains are normally involved in DNA binding and chromatin remodeling, however, their role in MPP11 is unknown.

A recent analysis directly addressed the substrate-specificity of ribosome-bound Ssb [92]. In an elegant approach, a TAP-tagged Ssb version was expressed in yeast and used to pull down associated ribosomes along with the translated mRNA. Subsequently, the mRNAs were subjected to sequence analysis to determine the co-translational interactome of Ssb. The data revealed that Ssb preferentially binds to nascent cytosolic and nuclear proteins, but not secretory polypeptides, which are targeted by SRP (Fig. 2.2b). Moreover, about 80 % of all nascent cytosolic and nuclear proteins interact with Ssb. A common feature among the Ssb substrates is that they are large in size, have a low α-helical content, and an increased content of hydrophobic elements. Additionally, Ssb interacts with many nascent subunits of oligomeric complexes and proteins that are engaged in a large number of interactions (Fig. 2.2b), for example, all subunits of the chaperonin complex TRiC, most subunits of the proteasome, and a large set of ribosomal proteins associated with Ssb during their synthesis [92]. In contrast, the abundance of a protein had no influence on Ssb binding. Thus, Ssb may assist co-translational folding of large proteins with complicated structures or with substrates that require binding partners for their stability. This also agrees with a potential role for Ssb in stabilizing free subunits of oligomeric complexes, which may expose contact sites that are prone to undergo false interactions. In a next step, the authors of the same study analyzed if RAC modulates the substrate specificity of Ssb [92]. To this end, they compared the substrate interactions of Ssb in a wild type yeast strain and in a strain lacking both RAC subunits. They indeed observed an influence of RAC on substrate binding of Ssb. While a subset of Ssb-substrate interactions was unaffected by the loss of RAC, others were significantly altered. For example, the interaction between Ssb and nascent cytosolic proteins was reduced while the interaction with membrane and mitochondrial proteins was enhanced in cells lacking RAC. Thus, RAC influences the co-translational substrate-specificity of Ssb [92].

5 Versatility of Ribosome-Associated Chaperones and their off Ribosomal Functions

The function of ribosome-associated chaperones was initially thought to be limited to assistance in de novo folding of newly synthesized proteins [5]. However, all ribosome-associated chaperones known so far bind to the ribosome only transiently and thus are additionally found in a non-ribosomal state in the cytosol. There is increasing evidence that ribosome-associated chaperones are not in an inactive state when they are off the ribosome, rather they are assumed to display additional functions (Fig. 2.3). In a ribosome-associated state, they all contact the nascent chain and support the de novo folding of newly synthesized proteins by different modes of action. In contrast, the functional spectrum in a non-ribosome-associated state is much more diverse and less well understood.

E. coli TF was found to co-purify with several full-length proteins including many ribosomal proteins, such as ribosomal protein S7 [19]. Interestingly, many

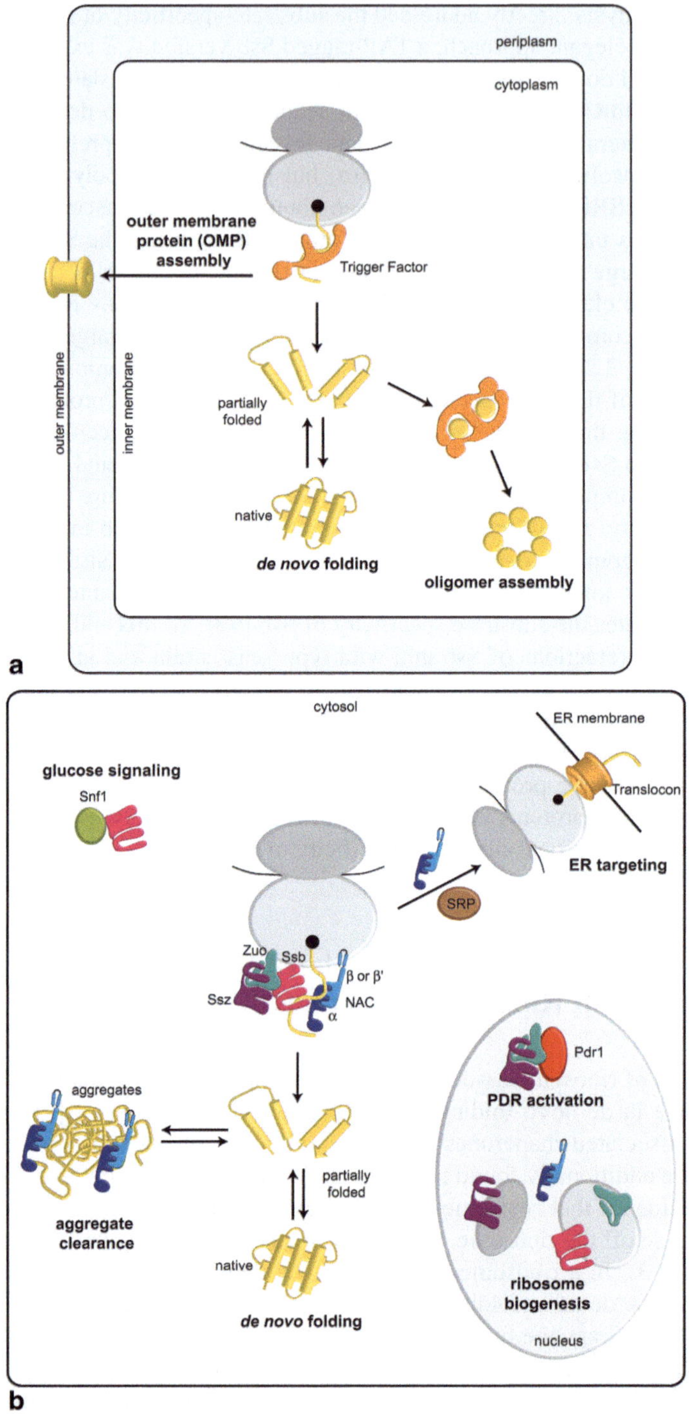

Fig. 2.3 **a** When bound to the ribosome trigger factor (TF) protects cytosolic nascent polypeptides (*yellow*) from aggregation, proteolysis, and premature folding. Thus, TF supports correct de novo protein folding. Thereby, especially outer membrane proteins strongly bind to TF, and TF was

of the interacting proteins were previously found to be aggregation-prone in *E. coli* cells lacking TF and the cytosolic Hsp70 chaperone DnaK [32]. To further investigate the non-ribosomal interaction of TF with substrates, Hendrickson and co-workers co-crystallized TF with the S7 protein, both derived from *Thermotoga maritima* [19]. The structural data revealed that TF binds to almost natively folded S7 proteins in a 2:2 stoichiometry. Through the binding of the TF molecules, large surfaces of the S7 are masked which are normally buried in the interior of the 30S ribosomal subunit. Accordingly, TF might not only co-translationally support protein folding but also function post-translationally in the biogenesis of large macromolecular assemblies such as ribosomes by stabilizing newly synthesized proteins until they become incorporated into complexes [19] (Fig. 2.3a)

Also eukaryotic ribosome-tethered chaperones display off-ribosomal functions. It was shown earlier that depending on the characteristics of the nascent chain between 17.8 and 41.8% of the Ssb molecules, 12.3–16.2% of the RAC molecules, and 29.3–66.4% of all NAC molecules are found in association with actively translating ribosomes in an in vitro translation extract derived from yeast cells while the other molecules diffuse freely [41]. Cells lacking Ssb accumulate insoluble newly synthesized proteins and additionally reveal a strong reduction in translation along with a strong decrease in the levels of ribosomal particles, in particular of the large 60S ribosomal subunit. In line with this finding, Ssb as well as Zuo show genetic interactions with Jjj1, a chaperone specific for ribosome biogenesis [77, 98]. Moreover, Zuo associates with nuclear 60S ribosomal biogenesis intermediates and Zuo (together with Ssb) also participates in the maturation of the 35S ribosomal RNA (rRNA) [98]. Thus, the RAC-Ssb system seems to play an import role in the biogenesis of ribosomal 60S particles (Fig. 2.3b). Whether Ssb and Zuo act independently of each other in ribosome biogenesis or—similar to their activity on ribosomes—always as a RAC-Ssb system is unclear (Fig. 2.3b).

Ribosome biogenesis is a highly complex and energy-consuming process and thus has to be tightly regulated. In eukaryotes, the ribosomal proteins are synthesized in the cytosol and subsequently imported into the nucleus where the assembly of the pre-ribosomal particle takes place. With the help of a variety of ribosome biogenesis factors, the pre-ribosomes undergo several processing steps and are transported out of the nucleus into the cytosol [99]. How might RAC-Ssb act during the biogenesis of new ribosomal particles? One possibility is that the RAC-Ssb system is required for the co-translational folding and prevention of aggregation of

---

shown to facilitate their correct targeting and assembly. In addition, TF is suggested to facilitate the assembly of large protein oligomers and assemblies independent of its function at the ribosome. **b** In yeast, nascent polypeptide-associated complex (NAC) and ribosome-associated complex (RAC)–Ssb bind to ribosomes and interact with nascent polypeptide chains to support de novo folding. NAC is suggested to modulate SRP-dependent co-translational targeting of secretory proteins to the endoplasmic reticulum (ER) translocon. Moreover, NAC interacts with various types of protein aggregates and supports the resolubilization of heat-shock-induced protein aggregates. Non-ribosomal Ssb seems to be involved in glucose signaling by regulating the phosphorylation status of Snf1. Ssb-RAC and NAC have also additional functions in the nucleus. Ssb-RAC and NAC are suggested to be involved in ribosome biogenesis and Zuo or the entire RAC was suggested to interact with the transcription factor Pdr1, which initiates the pleiotropic drug resistance (PDR).

ribosomal proteins and, likewise, Ssb and/or Zuo may additionally assist the transport of the newly synthesized ribosomal proteins into the nucleus where the ribosomal proteins become incorporated into pre-ribosomal particles. In this scenario, Ssb and/or Zuo would stay bound to newly synthesized ribosomal proteins even after their release from the ribosome and accompany them during the transport to the nucleus. Thus, the RAC-Ssb system could act by preventing the aggregation of ribosomal proteins after their synthesis and before they become stabilized by incorporation into a new ribosomal subunit. Alternatively, and not mutually exclusive to this hypothesis, Ssb and Zuo could act by directly promoting ribosome biogenesis in the nucleus, for example, by supporting the processing of rRNA or by modulating the activity of certain ribosome biogenesis factors. The finding that Zuo and Ssb are also detectable in the nucleus albeit only very transiently supports the idea that RAC-Ssb is not only engaged in the co-translational folding of ribosomal proteins in the cytosol but also fulfills functions in the nucleus besides its action on the ribosome [93, 98] (Fig. 2.3b).

Additionally, it was suggested that Ssb regulates the phosphorylation state of the kinase Snf1 and that this function is independent of its ribosome association [100] (Fig. 2.3b). Snf1 is active in a phosphorylated state under non-glucose conditions and drives the expression of glucose-repressed genes. Ssb, however, was shown to be required to keep Snf1 unphosphorylated and inactive when glucose is available. Thus, Ssb might play a role in the appropriate response to changing glucose concentrations [100].

Both RAC subunits where shown to be able to activate the transcription factor Pdr1 resulting in the induction of the pleiotropic drug resistance (PDR) [101–103] (Fig. 2.3b). The PDR is a highly specific transcriptional response that leads to the upregulation of a set of genes involved in the resistance towards several drugs and toxic agents in the cellular environment. For example, Pdr1 target genes include several ATP-binding cassette transporters that extrude xenobiotics from cells, rendering them resistant to a variety of toxic compounds [101]. For activation of Pdr1 in *S. cerevisiae,* the ATPase domain of Ssz or the 13 C-terminal amino acids of Zuo are required and sufficient. Interestingly, when Zuo is bound to the ribosome, its C-terminal 86 residues form a left-handed four helix bundle that buries the residues critical for activation of Pdr1 within its interior. However, the simple dissociation of Zuo1 from the ribosome is not sufficient for Pdr1 transcriptional activity but requires the additional unfolding of the C-terminus of Zuo. How this unfolding is regulated is not clear so far [101]. For Ssz too, it was shown that it has to be free of ribosomes to induce the PDR [101, 103]. These findings imply that the role of the RAC components in the PDR is distinct from their roles as ribosome-associated chaperones (Fig. 2.3b).

A very recent study unraveled a new function of NAC during misfolding and aggregation of cytosolic and nuclear proteins beyond its activity on ribosomes. It was shown that *C. elegans* NAC interacts with different types of amorphous aggregates, like aggregates that accumulate during aging or upon heat shock, as well as with amyloid-like Aβ fibrils [78] (Fig. 2.3b). These findings are in agreement with another study that identified human NAC to associate with artificial β-sheet fibrils

[104]. The physical interaction between NAC and aggregates has also physiological consequences, as disaggregation of heat shock protein aggregates in *C. elegans* is dependent on the presence of NAC [78] (Fig. 2.3b). However, it is not clear how NAC is able to support disaggregation in metazoans. As NAC has no ATPase activity and thus cannot apply mechanical force, its mechanism must be different from the one of classical disaggregases such as the Hsp70-Hsp40-Hsp100 system [105]. One possibility is that NAC in *C. elegans* interacts with non-native proteins during the aggregation process and thereby becomes itself associated with aggregates. NAC associated with aggregates perhaps recruits other chaperones with the capacity to remodel the aggregates, e.g., Hsp70/40 and Hsp110 type chaperones. This is in line with the finding that NAC co-immunoprecipitates with many other chaperones [78]. Thus, NAC is able to dissociate from the ribosome and to re-localize towards misfolded and insoluble proteins. Importantly, the depletion of NAC by RNA interference (RNAi) as well as the re-localization of NAC to protein aggregates upon protein stress conditions is associated with a strong decline of translation in the cytosol in *C. elegans.* This suggests that NAC is also able to modulate ribosome activity. Thus, NAC has a housekeeping function on ribosomes to promote folding and translation while under proteotoxic stress, its action is shifted towards aggregated proteins. This has beneficial consequences: NAC can aid in the re-solubilization of misfolded proteins in cooperation with other molecular chaperones and, likewise, translation and thus the influx of new proteins that need to be assisted by chaperones in their folding program is reduced. The re-localization is reversible, as NAC can re-associate with the ribosome once misfolded and aggregated proteins are eliminated and proteostasis is re-balanced. Based on these observations, it was suggested that the functional depletion of NAC from ribosomes into aggregates is an important regulatory circuit to modulate translation [78, 106].

In sum, there is strong evidence that ribosome-associated chaperones are versatile and vital elements of the chaperone network that interact in vivo with a large variety of substrates to perform more than one function (Fig. 2.3). They co-translationally support de novo folding and also act by unknown mechanisms to regulate ribosome biogenesis and translation which regulates the influx of new proteins into the cellular proteome. In addition, there is a stress-related function at least for NAC. This complex associates with protein aggregates and, as a consequence, either directly or indirectly dims protein synthesis to reduce the new protein load for chaperones and to allow the cells to recover from proteotoxic stress. It will be exciting to illuminate the individual roles, mechanisms, and substrate specificities of these chaperones in their functions on and off the ribosome.

References

1. Hartl FU, Hayer-Hartl M (2009) Converging concepts of protein folding in vitro and in vivo. Nat Struct Mol Biol 16(6):574–581. doi:10.1038/nsmb.1591
2. Hartl FU, Bracher A, Hayer-Hartl M (2011) Molecular chaperones in protein folding and proteostasis. Nature 475(7356):324–332. doi:10.1038/nature10317

3. Bukau B, Deuerling E, Pfund C, Craig EA (2000) Getting newly synthesized proteins into shape. Cell 101(2):119–122. doi:10.1016/S0092-8674(00)80806-5
4. Preissler S, Deuerling E (2012) Ribosome-associated chaperones as key players in proteostasis. Trends Biochem Sci 37(7):274–283. doi:10.1016/j.tibs.2012.03.002
5. Hartl FU, Hayer-Hartl M (2002) Molecular chaperones in the cytosol: from nascent chain to folded protein. Science 295(5561):1852–1858. doi:10.1126/science.1068408
6. Maier T, Ferbitz L, Deuerling E, Ban N (2005) A cradle for new proteins: trigger factor at the ribosome. Curr Opin Struct Biol 15(2):204–212. doi:10.1016/j.sbi.2005.03.005
7. Hoffmann A, Bukau B, Kramer G (2010) Structure and function of the molecular chaperone trigger factor. Biochim Biophys Acta 1803(6):650–661. doi:10.1016/j.bbamcr.2010.01.017
8. Hoffmann A, Becker AH, Zachmann-Brand B, Deuerling E, Bukau B, Kramer G (2012) Concerted action of the ribosome and the associated chaperone trigger factor confines nascent polypeptide folding. Mol Cell 48(1):63–74. doi:10.1016/j.molcel.2012.07.018
9. Ferbitz L, Maier T, Patzelt H, Bukau B, Deuerling E, Ban N (2004) Trigger factor in complex with the ribosome forms a molecular cradle for nascent proteins. Nature 431(7008):590–596. doi:10.1038/nature02899
10. Gautschi M, Lilie H, Funfschilling U, Mun A, Ross S, Lithgow T, Rucknagel P, Rospert S (2001) RAC, a stable ribosome-associated complex in yeast formed by the DnaK-DnaJ homologs Ssz1p and Zuo. Proc Natl Acad Sci U S A 98(7):3762–3767. doi:10.1073/pnas.071057198
11. Gautschi M, Mun A, Ross S, Rospert S (2002) A functional chaperone triad on the yeast ribosome. Proc Natl Acad Sci U S A 99(7):4209–4214. doi:10.1073/pnas.062048599
12. Stoller G, Rucknagel KP, Nierhaus KH, Schmid FX, Fischer G, Rahfeld JU (1995) A ribosome-associated peptidyl-prolyl cis/trans isomerase identified as the trigger factor. EMBO J 14(20):4939–4948
13. Kramer G, Rauch T, Rist W, Vorderwulbecke S, Patzelt H, Schulze-Specking A, Ban N, Deuerling E, Bukau B (2002) L23 protein functions as a chaperone docking site on the ribosome. Nature 419(6903):171–174. doi:10.1038/nature01047
14. Stoller G, Tradler T, Rucknagel KP, Rahfeld JU, Fischer G (1996) An 11.8 kDa proteolytic fragment of the *E. coli* trigger factor represents the domain carrying the peptidyl-prolyl cis/trans isomerase activity. FEBS Lett 384(2):117–122
15. Gupta R, Lakshmipathy SK, Chang HC, Etchells SA, Hartl FU (2010) Trigger factor lacking the PPIase domain can enhance the folding of eukaryotic multi-domain proteins in *Escherichia coli*. FEBS Lett 584(16):3620–3624. doi:10.1016/j.febslet.2010.07.036
16. Liu CP, Zhou QM, Fan DJ, Zhou JM (2010) PPIase domain of trigger factor acts as auxiliary chaperone site to assist the folding of protein substrates bound to the crevice of trigger factor. Int J Biochem Cell Biol 42(6):890–901. doi:10.1016/j.biocel.2010.01.019
17. Kramer G, Patzelt H, Rauch T, Kurz TA, Vorderwulbecke S, Bukau B, Deuerling E (2004) Trigger factor peptidyl-prolyl cis/trans isomerase activity is not essential for the folding of cytosolic proteins in *Escherichia coli*. J Biol Chem 279(14):14165–14170. doi:10.1074/jbc.M313635200
18. Merz F, Boehringer D, Schaffitzel C, Preissler S, Hoffmann A, Maier T, Rutkowska A, Lozza J, Ban N, Bukau B, Deuerling E (2008) Molecular mechanism and structure of Trigger Factor bound to the translating ribosome. Embo J 27(11):1622–1632. doi:10.1038/emboj.2008.89
19. Martinez-Hackert E, Hendrickson WA (2009) Promiscuous substrate recognition in folding and assembly activities of the trigger factor chaperone. Cell 138(5):923–934. doi:10.1016/j.cell.2009.07.044
20. Rutkowska A, Mayer MP, Hoffmann A, Merz F, Zachmann-Brand B, Schaffitzel C, Ban N, Deuerling E, Bukau B (2008) Dynamics of trigger factor interaction with translating ribosomes. J Biol Chem 283(7):4124–4132. doi:10.1074/jbc.M708294200
21. Hesterkamp T, Deuerling E, Bukau B (1997) The amino-terminal 118 amino acids of *Escherichia coli* trigger factor constitute a domain that is necessary and sufficient for binding to ribosomes. J Biol Chem 272(35):21865–21871
22. Patzelt H, Kramer G, Rauch T, Schonfeld HJ, Bukau B, Deuerling E (2002) Three-state equilibrium of *Escherichia coli* trigger factor. Biol Chem 383(10):1611–1619. doi:10.1515/BC.2002.182

23. Zarnt T, Tradler T, Stoller G, Scholz C, Schmid FX, Fischer G (1997) Modular structure of the trigger factor required for high activity in protein folding. J Mol Biol 271(5):827–837. doi:10.1006/jmbi.1997.1206
24. Raine A, Lovmar M, Wikberg J, Ehrenberg M (2006) Trigger factor binding to ribosomes with nascent peptide chains of varying lengths and sequences. J Biol Chem 281(38):28033–28038. doi:10.1074/jbc.M605753200
25. Maier R, Eckert B, Scholz C, Lilie H, Schmid FX (2003) Interaction of trigger factor with the ribosome. J Mol Biol 326(2):585–592
26. Kaiser CM, Chang HC, Agashe VR, Lakshmipathy SK, Etchells SA, Hayer-Hartl M, Hartl FU, Barral JM (2006) Real-time observation of trigger factor function on translating ribosomes. Nature 444(7118):455–460. doi:10.1038/nature05225
27. Oh E, Becker AH, Sandikci A, Huber D, Chaba R, Gloge F, Nichols RJ, Typas A, Gross CA, Kramer G, Weissman JS, Bukau B (2011) Selective ribosome profiling reveals the cotranslational chaperone action of trigger factor in vivo. Cell 147(6):1295–1308. doi:10.1016/j.cell.2011.10.044
28. Kramer G, Boehringer D, Ban N, Bukau B (2009) The ribosome as a platform for co-translational processing, folding and targeting of newly synthesized proteins. Nat Struct Mol Biol 16(6):589–597. doi:10.1038/nsmb.1614
29. Sandikci A, Gloge F, Martinez M, Mayer MP, Wade R, Bukau B, Kramer G (2013) Dynamic enzyme docking to the ribosome coordinates N-terminal processing with polypeptide folding. Nat Struct Mol Biol 20(7):843–850. doi:10.1038/nsmb.2615
30. Lakshmipathy SK, Gupta R, Pinkert S, Etchells SA, Hartl FU (2010) Versatility of trigger factor interactions with ribosome-nascent chain complexes. J Biol Chem 285(36):27911–27923. doi:10.1074/jbc.M110.134163
31. Crooke E, Wickner W (1987) Trigger factor: a soluble protein that folds pro-OmpA into a membrane-assembly-competent form. Proc Natl Acad Sci U S A 84(15):5216–5220
32. Deuerling E, Schulze-Specking A, Tomoyasu T, Mogk A, Bukau B (1999) Trigger factor and DnaK cooperate in folding of newly synthesized proteins. Nature 400(6745):693–696. doi:10.1038/23301
33. Deuerling E, Patzelt H, Vorderwulbecke S, Rauch T, Kramer G, Schaffitzel E, Mogk A, Schulze-Specking A, Langen H, Bukau B (2003) Trigger factor and DnaK possess overlapping substrate pools and binding specificities. Mol Microbiol 47(5):1317–1328
34. Teter SA, Houry WA, Ang D, Tradler T, Rockabrand D, Fischer G, Blum P, Georgopoulos C, Hartl FU (1999) Polypeptide flux through bacterial Hsp70: DnaK cooperates with trigger factor in chaperoning nascent chains. Cell 97(6):755–765
35. Agashe VR, Guha S, Chang HC, Genevaux P, Hayer-Hartl M, Stemp M, Georgopoulos C, Hartl FU, Barral JM (2004) Function of trigger factor and DnaK in multidomain protein folding: increase in yield at the expense of folding speed. Cell 117(2):199–209
36. Patzelt H, Rudiger S, Brehmer D, Kramer G, Vorderwulbecke S, Schaffitzel E, Waitz A, Hesterkamp T, Dong L, Schneider-Mergener J, Bukau B, Deuerling E (2001) Binding specificity of *Escherichia coli* trigger factor. Proc Natl Acad Sci U S A 98(25):14244–14249. doi:10.1073/pnas.261432298
37. Hoffmann A, Merz F, Rutkowska A, Zachmann-Brand B, Deuerling E, Bukau B (2006) Trigger factor forms a protective shield for nascent polypeptides at the ribosome. J Biol Chem 281(10):6539–6545. doi:10.1074/jbc.M512345200
38. Tomic S, Johnson AE, Hartl FU, Etchells SA (2006) Exploring the capacity of trigger factor to function as a shield for ribosome bound polypeptide chains. FEBS Lett 580(1):72–76. doi:10.1016/j.febslet.2005.11.050
39. Mashaghi A, Kramer G, Bechtluft P, Zachmann-Brand B, Driessen AJ, Bukau B, Tans SJ (2013) Reshaping of the conformational search of a protein by the chaperone trigger factor. Nature 500(7460):98–101. doi:10.1038/nature12293
40. Becker AH, Oh E, Weissman JS, Kramer G, Bukau B (2013) Selective ribosome profiling as a tool for studying the interaction of chaperones and targeting factors with nascent polypeptide chains and ribosomes. Nat Protoc 8(11):2212–2239. doi:10.1038/nprot.2013.133

41. Raue U, Oellerer S, Rospert S (2007) Association of protein biogenesis factors at the yeast ribosomal tunnel exit is affected by the translational status and nascent polypeptide sequence. J Biol Chem 282(11):7809–7816. doi:10.1074/jbc.M611436200
42. Wiedmann B, Sakai H, Davis TA, Wiedmann M (1994) A protein complex required for signal-sequence-specific sorting and translocation. Nature 370(6489):434–440. doi:10.1038/370434a0
43. Beatrix B, Sakai H, Wiedmann M (2000) The alpha and beta subunit of the nascent polypeptide-associated complex have distinct functions. J Biol Chem 275(48):37838–37845. doi:10.1074/jbc.M006368200
44. Reimann B, Bradsher J, Franke J, Hartmann E, Wiedmann M, Prehn S, Wiedmann B (1999) Initial characterization of the nascent polypeptide-associated complex in yeast. Yeast 15(5):397–407. doi:10.1002/(SICI)1097-0061(19990330)15:5<397::AID-YEA384>3.0.CO;2-U
45. Wang S, Sakai H, Wiedmann M (1995) NAC covers ribosome-associated nascent chains thereby forming a protective environment for regions of nascent chains just emerging from the peptidyl transferase center. J Cell Biol 130(3):519–528
46. Lauring B, Wang S, Sakai H, Davis TA, Wiedmann B, Kreibich G, Wiedmann M (1995) Nascent-polypeptide-associated complex: a bridge between ribosome and cytosol. Cold Spring Harb Symp Quant Biol 60:47–56
47. Rospert S, Dubaquie Y, Gautschi M (2002) Nascent-polypeptide-associated complex. Cell Mol Life Sci 59(10):1632–1639
48. Liu Y, Hu Y, Li X, Niu L, Teng M (2010) The crystal structure of the human nascent polypeptide-associated complex domain reveals a nucleic acid-binding region on the NACA subunit. BioChemistry 49(13):2890–2896. doi:10.1021/bi902050p
49. Wang L, Zhang W, Zhang XC, Li X, Rao Z (2010) Crystal structures of NAC domains of human nascent polypeptide-associated complex (NAC) and its alphaNAC subunit. Protein Cell 1(4):406–416. doi:10.1007/s13238-010-0049-3
50. Spreter T, Pech M, Beatrix B (2005) The crystal structure of archaeal nascent polypeptide-associated complex (NAC) reveals a unique fold and the presence of a ubiquitin-associated domain. J Biol Chem 280(16):15849–15854. doi:10.1074/jbc.M500160200
51. Hurley JH, Lee S, Prag G (2006) Ubiquitin-binding domains. Biochem J 399(3):361–372. doi:10.1042/BJ20061138
52. Madura K (2002) The ubiquitin-associated (UBA) domain: on the path from prudence to prurience. Cell Cycle 1(4):235–244
53. Searle MS, Garner TP, Strachan J, Long J, Adlington J, Cavey JR, Shaw B, Layfield R (2012) Structural insights into specificity and diversity in mechanisms of ubiquitin recognition by ubiquitin-binding domains. Biochem Soc Trans 40(2):404–408. doi:10.1042/BST20110729
54. Su V, Lau AF (2009) Ubiquitin-like and ubiquitin-associated domain proteins: significance in proteasomal degradation. Cell Mol Life Sci 66(17):2819–2833. doi:10.1007/s00018-009-0048-9
55. Wegrzyn RD, Hofmann D, Merz F, Nikolay R, Rauch T, Graf C, Deuerling E (2006) A conserved motif is prerequisite for the interaction of NAC with ribosomal protein L23 and nascent chains. J Biol Chem 281(5):2847–2857. doi:10.1074/jbc.M511420200
56. Pech M, Spreter T, Beckmann R, Beatrix B (2010) Dual binding mode of the nascent polypeptide-associated complex reveals a novel universal adapter site on the ribosome. J Biol Chem 285(25):19679–19687. doi:10.1074/jbc.M109.092536
57. Zhang Y, Berndt U, Golz H, Tais A, Oellerer S, Wolfle T, Fitzke E, Rospert S (2012) NAC functions as a modulator of SRP during the early steps of protein targeting to the endoplasmic reticulum. Mol Biol Cell 23(16):3027–3040. doi:10.1091/mbc.E12-02-0112
58. Lauring B, Kreibich G, Weidmann M (1995) The intrinsic ability of ribosomes to bind to endoplasmic reticulum membranes is regulated by signal recognition particle and nascent-polypeptide-associated complex. Proc Natl Acad Sci U S A 92(21):9435–9439
59. Lauring B, Sakai H, Kreibich G, Wiedmann M (1995) Nascent polypeptide-associated complex protein prevents mistargeting of nascent chains to the endoplasmic reticulum. Proc Natl Acad Sci U S A 92(12):5411–5415

60. Moller I, Beatrix B, Kreibich G, Sakai H, Lauring B, Wiedmann M (1998) Unregulated exposure of the ribosomal M-site caused by NAC depletion results in delivery of non-secretory polypeptides to the Sec61 complex. FEBS Lett 441(1):1–5
61. Moller I, Jung M, Beatrix B, Levy R, Kreibich G, Zimmermann R, Wiedmann M, Lauring B (1998) A general mechanism for regulation of access to the translocon: competition for a membrane attachment site on ribosomes. Proc Natl Acad Sci U S A 95(23):13425–13430
62. Powers T, Walter P (1996) The nascent polypeptide-associated complex modulates interactions between the signal recognition particle and the ribosome. Curr Biol 6(3):331–338
63. Raden D, Gilmore R (1998) Signal recognition particle-dependent targeting of ribosomes to the rough endoplasmic reticulum in the absence and presence of the nascent polypeptide-associated complex. Mol Biol Cell 9(1):117–130
64. Wickner W (1995) The nascent-polypeptide-associated complex: having a "NAC" for fidelity in translocation. Proc Natl Acad Sci U S A 92(21):9433–9434
65. Kim SH, Shim KS, Lubec G (2002) Human brain nascent polypeptide-associated complex alpha subunit is decreased in patients with Alzheimer' s disease and Down syndrome. J Investig Med 50(4):293–301
66. Thiede B, Dimmler C, Siejak F, Rudel T (2001) Predominant identification of RNA-binding proteins in Fas-induced apoptosis by proteome analysis. J Biol Chem 276(28):26044–26050. doi:10.1074/jbc.M101062200
67. Bloss TA, Witze ES, Rothman JH (2003) Suppression of CED-3-independent apoptosis by mitochondrial betaNAC in *Caenorhabditis elegans*. Nature 424(6952):1066–1071. doi:10.1038/nature01920
68. Quelo I, Hurtubise M, St-Arnaud R (2002) alphaNAC requires an interaction with c-Jun to exert its transcriptional coactivation. Gene Expr 10(5-6):255–262
69. Yotov WV, Moreau A, St-Arnaud R (1998) The alpha chain of the nascent polypeptide-associated complex functions as a transcriptional coactivator. Mol Cell Biol 18(3):1303–1311
70. Yotov WV, St-Arnaud R (1996) Differential splicing-in of a proline-rich exon converts alphaNAC into a muscle-specific transcription factor. Genes Dev 10(14):1763–1772
71. Moreau A, Yotov WV, Glorieux FH, St-Arnaud R (1998) Bone-specific expression of the alpha chain of the nascent polypeptide-associated complex, a coactivator potentiating c-Jun-mediated transcription. Mol Cell Biol 18(3):1312–1321
72. Markesich DC, Gajewski KM, Nazimiec ME, Beckingham K (2000) bicaudal encodes the *Drosophila* beta NAC homolog, a component of the ribosomal translational machinery*. Development 127(3):559–572
73. Deng JM, Behringer RR (1995) An insertional mutation in the BTF3 transcription factor gene leads to an early postimplantation lethality in mice. Transgenic Res 4(4):264–269
74. Deuerling E, Bukau B (2004) Chaperone-assisted folding of newly synthesized proteins in the cytosol. Crit Rev Biochem Mol Biol 39(5-6):261–277. doi:10.1080/10409230490892496
75. Pechmann S, Willmund F, Frydman J (2013) The ribosome as a hub for protein quality control. Mol Cell 49(3):411–421. doi:10.1016/j.molcel.2013.01.020
76. Wegrzyn RD, Deuerling E (2005) Molecular guardians for newborn proteins: ribosome-associated chaperones and their role in protein folding. Cell Mol Life Sci 62(23):2727–2738. doi:10.1007/s00018-005-5292-z
77. Koplin A, Preissler S, Ilina Y, Koch M, Scior A, Erhardt M, Deuerling E (2010) A dual function for chaperones SSB-RAC and the NAC nascent polypeptide-associated complex on ribosomes. J Cell Biol 189(1):57–68. doi:10.1083/jcb.200910074
78. Kirstein-Miles J, Scior A, Deuerling E, Morimoto RI (2013) The nascent polypeptide-associated complex is a key regulator of proteostasis. EMBO J 32(10):1451–1468. doi:10.1038/emboj.2013.87
79. del Alamo M, Hogan DJ, Pechmann S, Albanese V, Brown PO, Frydman J (2011) Defining the specificity of cotranslationally acting chaperones by systematic analysis of mRNAs associated with ribosome-nascent chain complexes. PLoS Biol 9(7):e1001100. doi:10.1371/journal.pbio.1001100
80. Duttler S, Pechmann S, Frydman J (2013) Principles of cotranslational ubiquitination and quality control at the ribosome. Mol Cell 50(3):379–393. doi:10.1016/j.molcel.2013.03.010

81. Huang P, Gautschi M, Walter W, Rospert S, Craig EA (2005) The Hsp70 Ssz1 modulates the function of the ribosome-associated J-protein Zuo1. Nat Struct Mol Biol 12(6):497–504. doi:10.1038/nsmb942
82. Pfund C, Lopez-Hoyo N, Ziegelhoffer T, Schilke BA, Lopez-Buesa P, Walter WA, Wiedmann M, Craig EA (1998) The molecular chaperone Ssb from *Saccharomyces cerevisiae* is a component of the ribosome-nascent chain complex. EMBO J 17(14):3981–3989. doi:10.1093/emboj/17.14.3981
83. Conz C, Otto H, Peisker K, Gautschi M, Wolfle T, Mayer MP, Rospert S (2007) Functional characterization of the atypical Hsp70 subunit of yeast ribosome-associated complex. J Biol Chem 282(47):33977–33984. doi:10.1074/jbc.M706737200
84. Leidig C, Bange G, Kopp J, Amlacher S, Aravind A, Wickles S, Witte G, Hurt E, Beckmann R, Sinning I (2013) Structural characterization of a eukaryotic chaperone-the ribosome-associated complex. Nat Struct Mol Biol 20(1):23–28. doi:10.1038/nsmb.2447
85. Peisker K, Braun D, Wolfle T, Hentschel J, Funfschilling U, Fischer G, Sickmann A, Rospert S (2008) Ribosome-associated complex binds to ribosomes in close proximity of Rpl31 at the exit of the polypeptide tunnel in yeast. Mol Biol Cell 19(12):5279–5288. doi:10.1091/mbc.E08-06-0661
86. Yan W, Schilke B, Pfund C, Walter W, Kim S, Craig EA (1998) Zuo, a ribosome-associated DnaJ molecular chaperone. EMBO J 17(16):4809–4817. doi:10.1093/emboj/17.16.4809
87. Fiaux J, Horst J, Scior A, Preissler S, Koplin A, Bukau B, Deuerling E (2010) Structural analysis of the ribosome-associated complex (RAC) reveals an unusual Hsp70/Hsp40 interaction. J Biol Chem 285(5):3227–3234. doi:10.1074/jbc.M109.075804
88. Kim SY, Craig EA (2005) Broad sensitivity of *Saccharomyces cerevisiae* lacking ribosome-associated chaperone ssb or zuo1 to cations, including aminoglycosides. Eukaryot Cell 4(1):82–89. doi:10.1128/EC.4.1.82-89.2005
89. Nelson RJ, Ziegelhoffer T, Nicolet C, Werner-Washburne M, Craig EA (1992) The translation machinery and 70 kd heat shock protein cooperate in protein synthesis. Cell 71(1):97–105
90. Hundley H, Eisenman H, Walter W, Evans T, Hotokezaka Y, Wiedmann M, Craig E (2002) The in vivo function of the ribosome-associated Hsp70, Ssz1, does not require its putative peptide-binding domain. Proc Natl Acad Sci U S A 99(7):4203–4208. doi:10.1073/pnas.062048399
91. Rauch T, Hundley HA, Pfund C, Wegrzyn RD, Walter W, Kramer G, Kim SY, Craig EA, Deuerling E (2005) Dissecting functional similarities of ribosome-associated chaperones from *Saccharomyces cerevisiae* and *Escherichia coli*. Mol Microbiol 57(2):357–365. doi:10.1111/j.1365-2958.2005.04690.x
92. Willmund F, del Alamo M, Pechmann S, Chen T, Albanese V, Dammer EB, Peng J, Frydman J (2013) The cotranslational function of ribosome-associated Hsp70 in eukaryotic protein homeostasis. Cell 152(1-2):196–209. doi:10.1016/j.cell.2012.12.001
93. Shulga N, James P, Craig EA, Goldfarb DS (1999) A nuclear export signal prevents *Saccharomyces cerevisiae* Hsp70 Ssb1p from stimulating nuclear localization signal-directed nuclear transport. J Biol Chem 274(23):16501–16507
94. Zhang S, Lockshin C, Herbert A, Winter E, Rich A (1992) Zuo, a putative Z-DNA binding protein in *Saccharomyces cerevisiae*. EMBO J 11(10):3787–3796
95. Hundley HA, Walter W, Bairstow S, Craig EA (2005) Human Mpp11 J protein: ribosome-tethered molecular chaperones are ubiquitous. Science 308(5724):1032–1034. doi:10.1126/science.1109247
96. Jaiswal H, Conz C, Otto H, Wolfle T, Fitzke E, Mayer MP, Rospert S (2011) The chaperone network connected to human ribosome-associated complex. Mol Cell Biol 31(6):1160–1173. doi:10.1128/MCB.00986-10
97. Otto H, Conz C, Maier P, Wolfle T, Suzuki CK, Jeno P, Rucknagel P, Stahl J, Rospert S (2005) The chaperones MPP11 and Hsp70L1 form the mammalian ribosome-associated complex. Proc Natl Acad Sci U S A 102(29):10064–10069. doi:10.1073/pnas.0504400102
98. Albanese V, Reissmann S, Frydman J (2010) A ribosome-anchored chaperone network that facilitates eukaryotic ribosome biogenesis. J Cell Biol 189(1):69–81. doi:10.1083/jcb.201001054

99. Henras AK, Soudet J, Gerus M, Lebaron S, Caizergues-Ferrer M, Mougin A, Henry Y (2008) The post-transcriptional steps of eukaryotic ribosome biogenesis. Cell Mol Life Sci 65(15):2334–2359. doi:10.1007/s00018-008-8027-0
100. von Plehwe U, Berndt U, Conz C, Chiabudini M, Fitzke E, Sickmann A, Petersen A, Pfeifer D, Rospert S (2009) The Hsp70 homolog Ssb is essential for glucose sensing via the SNF1 kinase network. Genes Dev 23(17):2102–2115. doi:10.1101/gad.529409
101. Ducett JK, Peterson FC, Hoover LA, Prunuske AJ, Volkman BF, Craig EA (2013) Unfolding of the C-terminal domain of the J-protein Zuo1 releases autoinhibition and activates Pdr1-dependent transcription. J Mol Biol 425(1):19–31. doi:10.1016/j.jmb.2012.09.020
102. Prunuske AJ, Waltner JK, Kuhn P, Gu B, Craig EA (2012) Role for the molecular chaperones Zuo1 and Ssz1 in quorum sensing via activation of the transcription factor Pdr1. Proc Natl Acad Sci U S A 109(2):472–477. doi:10.1073/pnas.1119184109
103. Eisenman HC, Craig EA (2004) Activation of pleiotropic drug resistance by the J-protein and Hsp70-related proteins, Zuo1 and Ssz1. Mol Microbiol 53(1):335–344. doi:10.1111/j.1365-2958.2004.04134.x
104. Olzscha H, Schermann SM, Woerner AC, Pinkert S, Hecht MH, Tartaglia GG, Vendruscolo M, Hayer-Hartl M, Hartl FU, Vabulas RM (2011) Amyloid-like aggregates sequester numerous metastable proteins with essential cellular functions. Cell 144(1):67–78. doi:10.1016/j.cell.2010.11.050
105. Tyedmers J, Mogk A, Bukau B (2010) Cellular strategies for controlling protein aggregation. Nat Rev Mol Cell Biol 11(11):777–788. doi:10.1038/nrm2993
106. Kirstein-Miles J, Morimoto RI (2013) Ribosome-associated chaperones act as proteostasis sentinels. Cell Cycle 12(15):2335–2336. doi:10.4161/cc.25703

Part III
The Hsp70 and Hsp40 Chaperone Networks

Chapter 3
Yeast Hsp70 and J-protein Chaperones: Function and Interaction Network

Elizabeth A. Craig and Jaroslaw Marszalek

Abstract All heat shock protein (Hsp)70 molecular chaperones use the same fundamental biochemical mechanism of action, cycles of interaction with client proteins driven by adenosine triphosphate (ATP) binding and hydrolysis to maintain cellular proteostasis. However, they do not function alone; the action of co-chaperones is required. Together with Hsp70s, the J-proteins and nucleotide exchange factors (NEF) co-chaperones form Hsp70 networks. These networks are complex assembles with the number of components and their mutual interactions varying from species to species. Here, we consider Hsp70 networks functioning in a single cell eukaryote, baker's yeast. By combining network theory perspective with available functional data and evolutionary analysis, we argue that it is possible to make sense of these complexities.

Abbeviations

ER	Endoplasmic reticulum
ERAD	ER-associated degradation
IMS	Intermembrane space domain
mtDNA	Mitochondrial deoxyribonucleic acid (DNA)
mtHsp70	Mitochondrial heat shock protein (Hsp)70
NBD	Nucleotide-binding domain
NEF	Nucleotide exchange factor
PDR	Pleiotropic drug resistance
RAC	Ribosome-associated complex
SBD	Substrate-binding domain
WGD	Whole genome duplication

E. A. Craig (✉)
Department of Biochemistry, University of Wisconsin–Madison, Madison, WI, USA
e-mail: ecraig@wisc.edu

J. Marszalek
Laboratory of Evolutionary Biochemistry, Intercollegiate Faculty of Biotechnology, University of Gdansk, Gdansk, Poland
e-mail: jaroslaw.marszalek@biotech.ug.edu.pl

W. A. Houry (ed.), *The Molecular Chaperones Interaction Networks in Protein Folding and Degradation,* Interactomics and Systems Biology, DOI 10.1007/978-1-4939-1130-1_3,

1 Introduction

Heat shock protein (Hsp)70 molecular chaperones, function in many essential cellular processes. Folding of nascent chains as they emerge from ribosomes, driving protein translocation across membranes, modulating protein-protein interactions by controlling conformational changes, preventing formation of protein aggregates, and facilitating protein degradation are among them [1–3]. By affecting these central processes in all subcellular compartments of eukaryotic cells, Hsp70s play a critical role in managing cellular proteostasis. When functioning in these diverse cellular roles, Hsp70s use the same fundamental biochemical mechanism of action, cycles of interaction with client proteins driven by adenosine triphosphate (ATP) binding and hydrolysis. However, Hsp70s do not function alone. Two obligatory co-chaperones drive this ATP-dependent cycle of interaction: J-proteins and nucleotide exchange factors (NEFs) [4]. The crucial role of J-proteins is stimulation of the ATPase activity of Hsp70, thus stabilizing its interaction with client protein. NEFs facilitate the exchange of the adenosine diphosphate (ADP) formed as the result of J-protein stimulated ATP hydrolysis, thereby allowing client dissociation and the start of a new binding cycle.

As a given Hsp70 typically interacts with several different J-proteins and NEFs, a particular pair of co-chaperones working with a particular Hsp70 is often tailored to work efficiently for a specific task [3]. Together the collection of J-proteins and NEFs functioning with their partner Hsp70 form a collection of highly versatile Hsp70 chaperone machineries. These Hsp70 machines, by interacting with an array of client proteins and with each other, form densely interconnected and highly dynamic functional networks, assisting cellular proteins at every stage of their life cycle from nascent chain folding to proteolytic turnover. In the larger picture, biological networks are defined as the orchestrated activity of many components that interact with each other through pairwise interactions [5]. This definition fits well to the Hsp70 chaperone network, as indeed, its components, individual chaperone proteins and their clients, are involved in a multitude of pairwise interactions [6, 7]. Recently, it has become appreciated that a similar set of rules governs the architecture and functionality of both nonbiological complex systems, such as the Internet, computer chips and society, and biological networks. Thus, tools developed by general network theory can be useful in analyzing and describing biological networks, including those formed by Hsp70 chaperones[5].

In this chapter, we take a combined functional, network and evolutionary perspective to understand how Hsp70 chaperone machines operate in eukaryotic cells. First, we describe the individual components, Hsp70s, J-proteins, and NEFs, focusing on those structural and functional features that allow them to function together as chaperone machines. Next, focusing on the best studied model eukaryotic organism, baker's yeast *Saccharomyces cerevisiae*, we describe the current state of knowledge regarding how these machines work together forming Hsp70 networks, focusing in particular on two cellular compartments, mitochondria, and the cytosol. At the end, we describe the current state of knowledge of the evolution of Hsp70 networks, taking the view that evolutionary thinking is necessary to place Hsp70 networks in a broader biological context by asking where they came from and how their history affected the way they function.

2 Hsp70 Chaperone Machine

Hsp70 Proteins Typically, Hsp70s from distantly related taxons, such as baker's yeast and human, share ~50% amino acid identity [8]. This sequence conservation translates into high structural similarity. Each member of the family has two highly conserved domains, the N-terminal regulatory ATPase domain and the C-terminal client protein-binding domain. The ~44 kDa ATPase domain, often called the nucleotide-binding domain (NBD), consists of two subdomains, forming a cleft in which adenine nucleotide binds. The ~26 kDa client binding domain, often called the substrate-binding domain (SBD), consists of a β-sandwich subdomain (SBDβ), which forms a peptide-binding pocket that can accommodate short, that is ~5 residue, segments of a polypeptide chain that is enriched in hydrophobic amino acids [4]. The binding pocket is covered by a highly mobile lid formed by a portion of the extended α helical subdomain (SBDα). The NBD and SBD are connected by a short highly flexible linker.

Key to Hsp70s' ability to perform their cellular functions is the regulation of the cycles of binding to and release from client proteins (Fig. 3.1). Regulation of interaction with client is determined by the nucleotide-binding status of the NBD. In the ATP-bound state, the SBD is in close proximity to the NBD with the α helical lid held in the open conformation away from the cleft via interaction with the NBD, allowing for fast binding and releasing of a client protein [9, 10]. In the ADP-bound or nucleotide-free state, the two domains do not interact, but tether to each other only by the flexible linker. Unrestrained, the α-helical lid covers the peptide-binding pocket, allowing "trapping" of a client protein [11, 12]. Thus, the productive cycle of interaction of Hsp70 with clients is initiated when Hsp70 is in the ATP-bound, open conformation. However, ATP hydrolysis must occur to stabilize the interaction. But, Hsp70's intrinsic ATPase activity is low and nucleotides typically bind stably. Therefore, Hsp70s function requires co-chaperones: ATPase activity is stimulated by J-proteins, known also as Hsp40s, while dissociation of ADP is stimulated by NEF [4]. Together, these three components form Hsp70 chaperone machines.

2.1 J Proteins

The defining feature of a J-protein and, indeed the only domain required for a protein to be classified as such, is the "J-domain" [13]. It is the J-domain, interacting directly with Hsp70's NBD and interdomain linker that is instrumental in stimulating Hsp70's ATPase activity [4, 14, 15]. Its highly conserved structure consists of four α-helices, with two of them forming an antiparallel coiled-coil connected by a

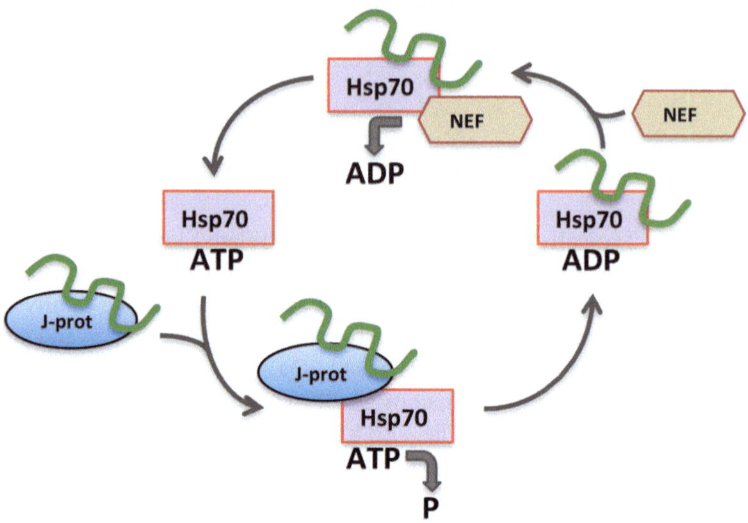

Fig. 3.1 Model of the core heat shock protein (*Hsp*)70 machinery's mode of action. J-protein binds to client protein through its peptide-binding domain and to Hsp70-ATP through its J-domain. The J-domain and client cooperatively stimulate adenosine triphosphate (*ATP*) hydrolysis, stabilizing client interaction. The J-protein dissociates from Hsp70-ADP. A nucleotide exchange factor (*NEF*) binds Hsp70-ADP. Adenosine diphosphate (*ADP*) dissociates from Hsp70, and is replaced by ATP. The client is released because of its low affinity for Hsp70-ATP. If the native state of the client is not attained on release, the J-protein rebinds to its exposed hydrophobic regions and the cycle begins again. Note that all interactions in the functioning ATPase cycle are transient; only if the cycle is blocked are stable protein-protein interactions expected

loop region. This loop invariantly contains a His Pro Asp (HPD) tripeptide, critical for functional interaction with partner Hsp70s [13].

While the J-domain defines a protein as a member of the J-protein class, other domains of J-proteins are very variable, in many cases showing no sequence or structural similarities. This lack of similarity suggests that over time, J-domains have been recruited by unrelated proteins to help accomplish a variety of functions [13]. Some J-proteins have a J-domain localized at the N-terminus followed by various types of substrate-binding domains, which allow them to interact with client on their own and as a consequence to target a client for Hsp70 binding. Cytosolic Ydj1, involved in general protein folding in the cytosol, and mitochondrial Jac1, involved in the biogenesis of *iron-sulfur cluster* (Fe-S) proteins, belong to this group (see Table 3.1 for list of all proteins discussed in the text). On the other hand, a J-domain can be in the middle or at the C-terminus of a J-protein. Often J-proteins with such an architecture lack the ability to bind client, but rather contain domains responsible for their specific cellular localization, e.g., the J-protein Pam18 involved in protein import into mitochondria is localized to the translocon of the inner membrane. However, we emphasize that regardless of their specific domain composition and regardless of the functions associated with these additional domains, all functional J-proteins are able to stimulate ATPase activity via direct interaction of their J-domain with their Hsp70 partner.

Table 3.1 Proteins discussed in text

Protein	Class	Cellular localization	Function/regulation
Ssa1	Hsp70 canonical	Cytosol/nucleus	Folding, protein translocation, stress ↑
Ssa2	Hsp70 canonical	Cytosol	Folding, protein translocation
Ssa3	Hsp70 canonical	Cytosol	Folding, stress ↑, stationary phase ↑
Ssa4	Hsp70 canonical	Cytosol/nucleus	Folding, stress ↑
Ssb1	Hsp70 canonical	Cytosol/ribosome	Folding, nascent polypeptide
Ssb2	Hp70 canonical	Cytosol/ribosome	Folding, nascent polypeptide
Ssc1	Hsp70 canonical	Mitochondrial matrix	Folding, protein import, maintenance of mtDNA, stress ↑
Ssq1	Hsp70 canonical	Mitochondrial matrix	Fe-S biogenesis
Ssc3	Hsp70 canonical	Mitochondrial matrix	Unknown
Kar2	Hsp70 canonical	ER lumen	Folding, protein import, ERAD, stress ↑
Sse1	Hsp70 atypical	Cytosol	NEF for SSA and SSB, stress ↑
Sse2	Hsp70 atypical	Cytosol	NEF for SSA and SSB, stress ↑
Ssz1	Hsp70 atypical	Cytosol/ribosome	Complex with Zuo1 (RAC), SSB's ATPase stimulation, folding, nascent polypeptide, activator TF Pdr1
Lhs1	Hsp70 atypical	ER lumen	NEF for KAR2
Ydj1	J-protein, DnaJ-like	Cytosol/nucleus	Folding, translocation, stress ↑
Apj1	J-protein, DnaJ-like	Cytosol/nucleus	Degradation of sumolyated proteins
Xdj1	J-protein, DnaJ-like	Cytosol/nucleus	Translocation of proteins from cytosol to mitochondria
Mdj1	J-protein, DnaJ-like	Mitochondrial matrix/ nucleoid	Folding, maintenance of mtDNA, stress ↑
Scj1	J-protein, DnaJ-like	ER lumen	Folding, ERAD
Sis1	J-protein	Cytosol/nucleus	(Re)folding, prion maintenance, translation, stress ↑
Djp1	J-protein	Cytosol	Peroxisomal and mitochondrial outer membrane import
Caj1	J-protein	Nucleus	Unknown
Erj5	J-protein	ER	Folding
Jjj1	J-protein	Cytosol/nucleolus	Ribosome biogenesis
Jjj3	J-protein	Cytosol/nucleus	Dipthamide synthesis
Jac1	J-protein	Mitochondrial matrix	Fe-S biogenesis
Cwc23	J-protein	Nucleus	mRNA splicing
Swa2	J-protein	Cytosol	Clathrin uncoating
Jem1	J-protein	ER	Nuclear membrane fusion
Jid1	J-protein	Mitochondria	Unknown
Jjj2	J-protein	Cytosol/nucleus	Unknown
Sec63	J-protein	ER	Protein import
Zuo1	J-protein	Cytosol/ribosome	Complex with Ssz1 (RAC), SSB's ATPase stimulation, folding nascent polypeptide, activation TF Pdr1
Mdj2	J-protein	Mitochondria	Protein import
Pam18	J-protein	Mitochondria	Protein import, in complex with Pam16
Pam16	J-like-protein	Mitochondria	Protein import, in complex with Pam18
Hlj1	J-protein	Cytosol/ER associated	ERAD
Snl1	NEF, Bag domain	Cytosol,	Associated with ER, nucleus, ribosome
Fes1	NEF, armadillo	Cytosol	Folding, protein degradation
Sil1	NEF, armadillo	ER	Protein import, folding

Table 3.1 (continued)

Protein	Class	Cellular localization	Function/regulation
Mge1	NEF, GrpE1-like	Mitochondrial matrix	Folding, protein import, Fe-S biogenesis
Isu1	Fe-S scaffold	Mitochondrial matrix	Fe-S biogenesis, client of Ssq1
Isu2	Fe-S scaffold	Mitochondrial matrix	Fe-S biogenesis, client of Ssq1

ER Endoplasmic reticulum, *Hsp70* Heat shock protein 70, *mtDNA* Mitochondrial deoxyribonucleic acid, *Fe-S* Iron-sulfur cluster, *ERAD* ER-associated degadration, *mRNA* messenger RNA ribonucleic acid, *RAC* Ribosome-associated complex, *ATPase* Adenosine triphosphatase, *NEF* Nucleotide exchange factors, *TF* Transcription factor

Nucleotide Exchange Factors (NEFs) The NEFs, which promote the release of Hsp70-bound ADP, thus facilitating rebinding of ATP and the initiation of a new client-binding cycle, are a heterogeneous group of proteins. Unlike J-proteins, NEFs do not share a common functional domain. Each of the four different types of NEFs identified thus far, bind Hsp70 partner in a different sequence-specific manner. However, they all modulate the same fundamental movement of the ATPase domain, opening the cleft, destabilizing nucleotide binding [16, 17]. Mge1 is a specialized NEF protein, which functions exclusively in mitochondria and is clearly related to the ubiquitous bacterial NEF GrpE. Other cellular compartments contain NEFs, which are unrelated to bacterial GrpE. One class of NEFs functioning both in the cytosol and in the endoplasmic reticulum (ER) has a structural fold called the armadillo repeat domain, which is also present in many other types of proteins [18]. Another class of NEFs uses the so called Bag domain for interaction with the NBD of partner Hsp70s [19]. Bag-type NEFs share a Bag domain, but otherwise differ in their domain composition. There is only one representatives of this class in *S. cerevisiae* [20]. However, in mammals, including humans, Bag proteins have proliferated. The different domains, in addition to the BAG domain in these organisms, allow for BAG proteins to function in a diversity of ways in chaperone networks, beyond nucleotide exchange [21].

The most unusual NEFs are members of the Hsp70 family [22–24]. These are usually larger than typical Hsp70s due to insertions in their sequences. They also diverged functionally from their Hsp70 ancestors. In the process, their ability to perform the ATPase-dependent client-binding/releasing cycle became rather limited, most likely due to the loss of the fine-tuned allosteric communication between their domains. Instead, they function as NEF co-chaperones for partner Hsp70s [23–25]. Yet, they are also able to bind client on their own, thus their role in the Hsp70 reaction cycle could be more complex than just the control of the nucleotide release.

To avoid confusion regarding classification of proteins belonging to Hsp70 family, the ones well established to be able to perform ATPase client-binding/releasing cycle are called canonical Hsp70s. In yeast, this group consists of four subfamilies SSA and SSB in the cytosol, SSC in the mitochondria, and KAR in the ER. Hsp70s, which diverged from canonicals and now function as co-chaperones are called atypical Hsp70s. In yeast, this group consists of three subfamilies SSE and SSZ in the cytosol and LHS in the ER. SSE and LHS function as NEFs while SSZ is a different type of co-chaperone, functioning in complex with a specific J-protein [26].

2.2 Hsp70 Reaction Cycle

As outlined above, the ATPase client-binding/releasing cycle constitutes the basic operational mode of all Hsp70 chaperone machines (Fig. 3.1). Large conformational changes of Hsp70 occur, driven by allosteric communication between the NBD and SBD [27]. The transient interactions of Hsp70 with J-protein and NEF co-chaperones play a pivotal regulatory role in allosteric communication, resulting in stimulation of ATP hydrolysis and ADP release, respectively [16].

The cycle is synchronized and fine tuned in several ways. First, the exposure of the peptide-binding site caused by the docking of the lid on the NBD domain allows for easy access of a client to the cleft [9]. Interaction of client in the cleft results in conformational changes in the ATPase domain that modestly increase ATP hydrolysis. Second, the J-domain, which has a higher affinity for ATP-Hsp70 than ADP-Hsp70, also modestly stimulates the ATPase activity. However, it is the synchronized interaction of J-protein and client that promotes hydrolysis and thus the major conformational changes that lead to SBD dissociation from the NBD and closing of the α-helical lid over the peptide associated with SBDβ [11]. Hydrolysis and lid closure can occur on a *millisecond* timescale, trapping the client.

Since NEFs have higher affinity for ADP compared to ATP-Hsp70, they, unlike J-proteins, bind preferentially to this form, facilitating ADP release [17]. Since cellular ATP concentrations are typically much higher than those of ADP, ATP binding is favored. Both client protein and NEF are released due to their low affinity for ATP-Hsp70. ATP binding is favored, as normally the cellular concentration of ATP is much higher than ADP. Thus, simply release of ADP favors ATP binding and thus initiation of another cycle of client interaction.

Clearly the points of action of J-proteins and NEFs in the Hsp70 reaction cycle can serve as points of regulation. The importance of their function is highlighted not only by their essential nature but also by the effect of altering their cellular concentrations. Overstimulation of Hsp70's ATPase activity might prevent the engagement of clients while an excess of NEF activity may cause a premature release of captured clients. Indeed, in vivo, the J-protein and NEF cofactors are typically present at sub-stoichiometric amounts relative to their partner Hsp70 while overexpression of NEF can have detrimental effects on the Hsp70 machine. For example overexpression of the NEF *SSE1* is detrimental to wild-type yeast, leading to extremely slow growth, even at the optimal temperature of 30 °C [28] while overexpression of NEF Bagl in mammalian cells inhibits the ability of Hsp70 to reactivate heat-inactivated luciferase [29].

3 Hsp70 Chaperone Networks

Components of Hsp70 machines have evolved into large and varied families in virtually all organisms. For example, *S. cerevisiae* encodes 14 Hsp70s, 23 J-proteins, and 5 NEFs, whereas in human cells, 17 Hsp70s cooperate with 50 J-proteins and 11

NEFs [30]. This proliferation is striking, particularly in contrast to genes encoding other molecular chaperones such as the GroEL type. Why? We think the answer lies in the fact that Hsp70 machines are well suited to form complex, highly flexible, and thus, very adaptable networks. In such networks Hsp70 proteins are major players engaged in multitude of pairwise interactions with client proteins, J-proteins, and NEFs. The flexibility of the network is enhanced by the fact that its components are interacting with each other typically act as monomers or homodimers and that those interactions are transient, thus allowing for easy expansion of the network by addition of new partners that emerge as the result of gene duplications. In addition, the adaptability of the network is facilitated by the fact that most protein-protein interactions engage small patches of surface exposed residues thus allowing relatively rapid changes in specificity.

Hsp70 networks function in each major compartment of the eukaryotic cell: cytosol/nucleus, ER, and mitochondrium. While each compartment has its own dedicated Hsp70 network and they are discussed individually below, these networks are also interconnected. For example, Hsp70 machines are involved in the biogenesis of cytosolic ribosomes that synthesize the vast majority of mitochondrial proteins, as well as the translocation of proteins synthesized on these ribosomes across both the ER and mitochondrial membranes [2].

4 Function and Evolution of the Mitochondrial Hsp70 Network

Mitochondria evolved from an α-proteobacterial endosymbiont [31]. Therefore, it is not surprising that many components of the mitochondrial Hsp70 (mtHsp70) network are closely related to those present in currently living bacterial species, including *Escherichia coli*. Of the three Hsp70s, four J-proteins and one NEF in the matrix of *S. cerevisiae* mitochondria, four are orthologous to bacterial proteins. The other four evolved within the eukaryotic lineage, and thus could be considered as mitochondrial innovations [32].

4.1 The Core Mitochondrial (mtHsp70) Machine

The core, that is the most abundant mtHsp70 machine, is composed of the Hsp70 Ssc1/J-protein Mdj1/NEF Mge1 (Fig. 3.2). It is orthologous to well-studied bacterial Hsp70 system, DnaK/DnaJ/GrpE. Not surprisingly, considering the conservation, the yeast systems functions in a similar way to the bacterial one, fully in accordance with the canonical model of Hsp70 function described earlier.

Mdj1 shares structural organization and biochemical properties with DnaJ. DnaJ was the first recognized member of the J-protein class. Its molecular weight inspired the name Hsp40, which is often used interchangeably with "J-protein," regardless of the size of the protein. Mdj1 and DnaJ have the following complex domain com-

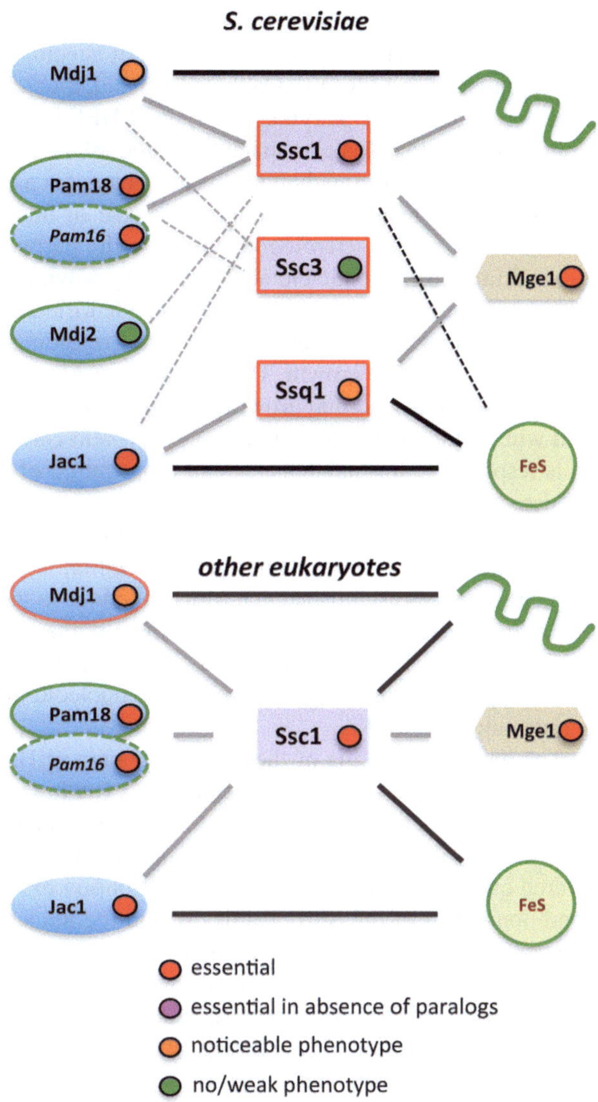

Fig. 3.2 Network of heat shock protein (*Hsp*)70's functional interactions in mitochondria. Binding of clients (*green*) to J-proteins (*blue ovals*) or to Hsp70s (*purple rectangles*) is indicated by *black lines*. Transient interactions between co-chaperones; J-proteins and nucleotide exchange factors (*NEFs*) (*light brown rhomboid*) are indicated by *grey lines*. Solid lines indicate major interactions while *dotted lines* indicate minor interactions. Identical line color around symbols indicates paralogous relationships. The *S. cerevisiae* network (*top panel*) is more complex than the other eukaryote network (*bottom panel*) due to the presence of paralogous Hsp70s. Moreover, due to the specialization of Ssq1 in iron-sulfur cluster (*Fe-S*) biogenesis, the functional module formed by Jac1/Ssq1 and client protein (Fe-S scaffold, *green circle*) is isolated from the major Ssc1 dependent module. Note: only stable interactions between client proteins and Hsp70s or J-proteins can be detected by biochemical methods such as TAP tagged precipitation of protein complexes. Growth phenotype of cells harboring deletion of a given gene is indicated by the color of the dot; red: lethal, pink: lethal with deletion of all other paralogs, orange: noticeable phenotype either alone or in combination with its parlog(s), green: no noticeable phenotype.

position [3]: The N-terminal J-domain is followed by a Gly and Phe-rich region and a C- terminal region composed of two barrel topology domains (CTDI and II), which is known to bind client proteins. CTDI has a hydrophobic peptide-binding pocket. A CysxxCysxGlyxGlyx type zinc finger region protrudes from CTDI and may also be involved in client binding, or may facilitate interaction with Hsp70 [3]. This basic architecture is the same as that of several other J-proteins found in the cytosol/nucleus and ER [33]. On the other hand, though Mge1 is structurally similar to bacterial GrpE, in eukaryotes this type of NEF is found only in mitochondria.

In mitochondria, this core machinery is involved in both de novo protein folding and in protection of mitochondrial proteins against aggregation caused by unfolding initiated by heat stress. Although the role of this machinery has been mainly studied using model heterologous proteins, the native mitochondrial proteins, mitochondrial deoxyribonucleic acid (mtDNA) polymerase (Mip1) and a subunit of mitochondrial ribosome, Var1, were also identified as clients [32, 34, 35]. Consistent with this role in protein folding and refolding, cells lacking Mdj1 exhibit temperature sensitive growth. They are also unable to maintain functional mtDNA. Surprisingly, under normal growth conditions the great majority of Mdj1 is associated with mitochondrial nucleoids, DNA-protein complexes that are the functional units responsible for maintenance and propagation of mtDNA [36, 37]. Biochemical data suggest that Mdj1 is tethered to the nucleoid via DNA binding. Analysis of Mdj1 variants revealed a correlation between nucleoid localization and DNA maintenance activity, suggesting that localization is functionally important. Moreover, this analysis established that Mdj1 functions in the nucleoid as part of the core mtHsp70 machine, since alterations of the J-domain affecting its interaction with Ssc1 resulted in loss of mtDNA [36]. Thus, overall, available data are consistent with a hypothesis that Mdj1 tethered to the nucleoid directs Ssc1 to function in the maintenance of mtDNA.

The molecular details of the role of the core mtHsp70 machine in the maintenance of the mtDNA remain to be elucidated. But it would be consistent with known functions of other Hsp70 machines, if it were involved in facilitating assembly or disassembly of multimeric protein complexes [3]. There is, however, one unanswered question regarding the spatial distribution of Mdj1. If indeed under normal conditions most of the Mdj1 resides within the nucleoid complex, how does such localization affect its function in protein folding? Strikingly, as we discussed above, the two native clients of Mdj1 characterized thus far are residents of the nucleoid complex. mtDNA polymerase is a key component of the nucleoid responsible for mtDNA replication [35], whereas ribosomal protein Var1, which is encoded by mtDNA, is synthesized and folded in the close proximity to the nucleoid [33, 37]. So indeed, core mtHsp70 machinery recruited by Mdj1 to the nucleoid could play an important role in protein folding. Further functional studies are needed however, to verify this hypothesis. Interestingly, although experimental data from functional analyses of Mdj1 is restricted to yeast, both Mdj1 and Ssc1 orthologs have been found in the nucleoid complexes from *Drosophila* and human mitochondria [38], suggesting a universal role of this chaperone system at the nucleoid.

4.2 The Protein Import Motor

Upon conversion of ancestral α-proteobacteria to mitochondria, many of the genes of the bacterial endosymbiont were transferred to the host cell's nuclear genome

[31]. Thus, mitochondria evolved a sophisticated molecular machinery allowing import of nuclear-encoded proteins across the mitochondrial membrane [39, 40]. In the mitochondrial matrix, mtHsp70 was recruited to play a critical role in the translocation process; it became the core of the "import motor," which is directly responsible for the inward movement of polypeptides through the import channel in the inner membrane. The function of the import motor follows the rules for the Hsp70 reaction cycle described earlier (Fig. 3.1). In the process of driving polypeptide import, mtHsp70, Ssc1, binds to the exposed hydrophobic sequences of unfolded polypeptide emerging from the import channel while a J-protein partner stimulates Hsp70's ATPase activity to stabilize its interaction with a polypeptide. The NEF Mge1 triggers nucleotide release, and thus dissociation of the translocating polypeptide from Hsp70. Thus, the Hsp70 machine involved in import consists of the core Hsp70 and NEF components Ssc1 and Mge1. However, Mdj1 does not function in import. Rather, Pam18, which is tethered to the import channel, is the J-protein component of this machinery.

The evolutionary origin of Pam18 remains unclear [41]. It has a unique domain structure, sharing only its J-domain with Mdj1. In contrast to Mdj1, Pam18 is not able to bind client proteins. Thus, its only function during the client interaction cycle of protein import is stimulation of the ATPase activity of Ssc1. The topology of Pam18 maximizes its function in protein import. In addition to the C-terminal J-domain, Pam18 consists of an internal transmembrane domain, which localizes it in the inner mitochondrial membrane. It is tethered to the translocon assembly by a bipartite interaction, one in the matrix and one in the intermembrane space (IMS). The N-terminal IMS domain interacts with a component of the import channel [42]. This domain is not essential, but cells expressing a Pam18 variant lacking the IMS domain exhibit a growth defect under anerobic conditions [43]. The second interaction is functionally more critical: the interaction of the matrix-localized J-domain itself with another motor component, Pam16 [40]. Interestingly, Pam16 has a "J-like" domain that is related by sequence and structure to a typical J-domain, but lacks the HPD motif necessary for stimulation of an Hsp70's ATPase activity. The J-like domain of Pam16 and J-domain of Pam18 interact, forming a stable complex. The interaction of Pam16 with the import machinery is responsible for the proper positioning of Pam18. The obvious sequence and structural similarity of Pam18 and Pam16 implies that the pair likely evolved via gene duplication, and subsequently underwent functional diversification. One paralogous copy maintained the ancestral J-protein function while the other copy lost the ancestral function, but evolved the ability to interact efficiently with the import channel. Thus, Pam16 could be considered a product of an expansion of the Ssc1 chaperone network.

The story of the expansion of mitochondrial J-proteins does not end here, however, as yeast mitochondria contains an additional paralogue of Pam18, Mdj2 [44]. Pam18 and Mdj2 share a high level of sequence identity and a similar domain composition. One important difference, however, is the lack of an IMS domain. While deletion of *PAM18* leads to growth arrest, deletion of *MDJ2* has no obvious phenotypic effects [44]. However, when overexpressed, Mdj2 allows growth of cells lacking Pam18 [45], indicating that it is able to at least substantially functionally replace Pam18 in protein import. An interesting evolutionary twist to this story is that the presence of the IMS domain in orthologs of Pam18 from various fungal species coincides with the presence of *MDJ2* in their genomes, suggesting that the ancestral gene duplication that was responsible for the emergence of the Pam18/Mdj2 paralogous pair was also needed for evolution of the IMS domain to occur [43].

On the surface, the highly evolved mechanism of Pam18 positioning in proximity to the import channel might suggest that one function of Pam18 is the recruitment of Ssc1 into the import motor. However, Ssc1 is positioned at the translocon through its own interaction with another component of the import machinery, Tim44. Thus Ssc1's localization is independent of its interaction with Pam18 [39]. As the Ssc1-Tim44 interaction is not dependent on Ssc1's ability to bind client protein, this Ssc1-Tim44 interaction can be considered a novel biochemical feature that evolved in response to selective pressure for efficient protein import. Thus, two events were involved in expansion of the mtHsp70 network that fosters efficient function in protein import—recruitment of a new type of highly specialized J-protein and the evolution of a new protein-protein interaction between mtHsp70 and the translocon.

4.3 Iron-sulfur Cluster (Fe-S) Biogenesis

Mitochondria are a major metabolic center of eukaryotic cells. Synthesis of Fe-S, prosthetic groups needed for function of many cellular proteins, is one of the critical functions of mitochondria [46]. The mitochondrial pathway of Fe-S biogenesis was inherited from bacteria, including the involvement of Hsp70 chaperone machinery in the process. However, detailed evolutionary and functional analyses revealed that while mitochondria did inherit the J-protein HscB, called Jac1 in *S. cerevisiae*, from bacterial ancestors, it did not inherit the specialized Hsp70 HscA [47]. In most species, including fungi distantly related to *S. cerevisiae*, protozoa, plants, and animals, including humans, the major mtHsp70 functions in the Fe-S biogenesis pathway to transfer Fe-S from the scaffold on which they are built onto recipient proteins.

Available biochemical data indicate that the transfer of a Fe-S requires a typical Hsp70 reaction cycle [48]. The specialized J-protein HscB/Jac1 binds a Fe-S scaffold protein using its unique C-terminal substrate-binding domain. The interaction is highly specific; the Fe-S scaffold protein is the only known client of HscB/Jac1 [31]. Next, HscB/Jac1 interacts with mtHsp70 via the N-terminal J-domain, which is highly similar to the J-domains of other J-proteins. Transfer of client protein to the substrate-binding domain of mtHsp70 and synergistic stimulation of its ATPase activity via J-domain interaction is required for the transfer step. Although the mechanistic details remain unknown, available data suggest that conformational changes induced by mtHsp70 upon ATP hydrolysis affects the stability of the Fe-S binding by the scaffold protein, thus allowing transfer of the cluster onto the recipient protein [48]. Interestingly, due to the low affinity of HscA for nucleotide, no NEF is required in the client interaction cycle. However, in the mitochondrial system, the NEF Mge1 completes the cycle by inducing ADP release and scaffold dissociation from Hsp70.

The system described above functions in most eukaryotes. However, in *S. cerevisiae* and closely related fungal species, the situation is quite different (Fig. 3.2). These organisms harbor an additional Hsp70, called Ssq1, which is specialized in Fe-S biogenesis [32]. Similarly to Ssc1, it functions with Jac1 and Mge1, but in contrast to Ssc1, its client-binding specificity is restricted to the Fe-S scaffold protein, Isu. Evolutionary and biochemical data concerning the origin of Ssq1 are consistent with the hypothesis that Ssq1 is related to Ssc1 via an ancestral gene duplication event [47]. According to this scenario, following the gene duplication, one copy, extant Ssc1, maintained biochemical properties typical for mtHsp70. The other copy diverged functionally to become Ssq1. In post-duplication, species Jac1 coevolved with Ssq1, acquiring structural changes within its J-domain in the process [49]. The altered J-domain became highly specific for Ssq1. Thus, Ssq1 and Jac1 form a highly specialized Hsp70 machine dedicated solely to Fe-S biogenesis. However, the mode of action of this newly evolved machine is not different from the one utilizing multifunctional mtHsp70.

4.4 Atypical Hsp70 of Yeast Mitochondria

Yeast *S. cerevisiae* and a handful of closely related species are unique among eukaryotes, as they all harbor a third mitochondrial Hsp70, Ssc3. Evolutionary analyses revealed that *SSC3* emerged as a result of the whole genome duplication (WGD) [47] that took place in an ancestor of the Saccharomyces lineage [50]. Despite a high level of sequence identity (~80%) the two members of this paralogous pair (Ssc1 and Ssc3) differ significantly in biochemical properties. Ssc3 from *S. cerevisiae* does not bind client peptides with high affinity and its ability to perform the client-binding/release cycle is very limited [51]. Deletion of the *SSC3* gene has no obvious phenotypic consequences. Although its ATPase activity can be stimulated by both Mdj1 and Pam18, and Mge1 is able to release nucleotide bound by Ssc3, it is not know whether Ssc3 forms a functional Hsp70 machine. Thus, the function of Ssc3 currently remains unknown. It is of interest to note that results of evolutionary studies indicate that genes encoding proteins not used in the cell are very quickly lost from yeast genomes [52]. Though, it is possible that Ssc3 has a biological function(s), it is intriguing to consider the possibility that Ssc3 is on its way to becoming the first mitochondrial atypical member of Hsp70 family [26]. Other known and well-characterized atypical Hsp70s function as co-chaperones for Hsp70 machines in cytosol (*SSE1/2*, *SSZ1*) and ER (*LHS1*).

4.5 A Network Theory Perspective on Mitochondrial Hsp70 Machines

According to network theory, each complex system composed of components that interact with each other through pairwise interaction can be studied in an abstract way, ignoring the physical nature of individual components and focusing only on their interactions [5]. At a highly abstract level, the components of the network, whether computer servers or Hsp70s/J-proteins/NEFs, can be reduced to a series of nodes that are connected to each other by links, with each link representing the interaction between two components. Such abstract representations allow the comparison of network properties among unrelated systems, thus enabling the uncovering of common rules that govern their behaviors. One such property is network connectivity. Nodes are linked uniformly in a so-called "random network"; if a network has a limited number of nodes with a large number of links while other nodes are poorly connected, it is referred to as a "scale free network" [5]. An interesting feature of unevenly connected, scale-free networks is the tolerance of local random failures, a property not shared by more evenly connected random networks. This robustness is likely the basis of the tolerance for errors found in many complex systems, ranging from cellular protein networks to the Internet.

Like other biological systems, the *S. cerevisiae* mitochondrial Hsp70 network belongs to the scale free category, with the major mtHsp70 Ssc1 forming a central hub connected to many client proteins, as well as to J-proteins and a NEF (Fig. 3.2). J-proteins vary in the degree of connectivity from very high for Mdj1 to low for Pam18, Mdj2 and Jac1. The J-proteins, however, play an important role in organization of the network topology as they divide the network into a set of separate modules [53] by directing Ssc1 connectivity towards subsets of clients [3]. Mdj1 directs Ssc1 towards those clients with whom it interacts. Pam18 and Mdj2 link Ssc1 with polypeptides crossing the import channel. Jac1 recruits Ssq1 to the Fe-S scaffold protein while at the same time ignoring Ssc1. Thus, it creates its own module, consisting of Jac1/Ssq1/Mge1 interacting with the Fe-S scaffold protein (Fig. 3.2). While modules created by Mdj1 and Pam18 are interconnected as they share the same Ssc1 hub, the one created by Jac1 is isolated from other modules of mitochondrial Hsp70 network.

One may ask how the interconnectivity of modules affects their functionality [53]. This question is especially relevant for Jac1-dependent modules as their interconnectivity varies across species. As we described above, in *S. cerevisiae* and a group of closely related yeast species, Jac1 functions with the specialized mtHsp70 Ssq1 [47]. However, all other eukaryotes do not have Ssq1 as it evolved by a gene duplication at the base of the Saccharomyces lineage. Instead, in these species, Jac1 functions in Fe-S biogenesis with the multifunctional mtHsp70 Ssc1 (Fig. 3.2). While the former Jac1-containing module is isolated or at most weakly interconnected with the major mtHsp70 network, the latter is strongly interconnected, as Jac1 shares its interaction with mtHsp70 with other J-proteins Mdj1 and Pam18.

These differences in interconnectivity between two Jac1-dependent modules could have potential functional consequences [54]. Imagine that due to mutation the affinity of the J-domain of Jac1 for its partner mtHsp70 is reduced to some degree. In the case of the highly interconnected Jac1/mtHsp70 module, this would lead to the situation in which Jac1, which "competes" with other J-proteins for mtHsp70 binding, would not be able to efficiently recruit mtHsp70 for Fe-S biogenesis. In stark contrast, a very similar defect could have much less of a consequence in the case of the Jac1/Ssq1-isolated module. Even if the affinity of Jac1 for Ssq1 is reduced, the fact that Ssq1 is its only partner and thus no other J-protein competes with it for Ssq1 could allow Jac1 to perform its function despite the defect. Thus similar defect could have very different functional consequences, depending on the structural organization of the network, with a high degree of network modularity promoting robustness defined as the ability of the network to function despite genetic and environmental perturbations [54].

Biological networks, similar to human-made networks such as the Internet, expand by duplication of nodes and further re-organization of their connectivity [55]. Functional divergence of specialized Ssq1 following its emergence by duplication of the *mtHsp70* gene, as well as functional diversification of the pair of paralogous J-proteins Pam18 and Mdj2, illustrate that mitochondrial Hsp70 network indeed expanded and diversified by such a mechanism. Moreover, this particular mode of expansion can have profound effect on a network's functionality as it could increase its robustness. Duplicated proteins share with their ancestors an identical or largely overlapping set of interactions, thus creating a redundancy of function [56]. Such redundancy can protect network integrity and functionality in the case that one of the duplicated proteins would malfunction. Indeed, genetic experiments indicate that overexpression of either Ssc1 or Mdj2 can partially compensate for the functional defects of Ssq1 and Pam18, respectively, suggesting that functional redundancy can protect integrity of the mitochondrial Hsp70 network.

5 Function and Evolution of Cytosolic Hsp70/40 Network

The cytosolic network of Hsp70 chaperones is much more complex than its mitochondrial counterpart [2] (Fig. 3.3). The important difference is the greater variety of parts that comprise the cytosol Hsp70 machines compared to the simpler mtHsp70 network. Not only is the number of components involved much larger but also these components are highly diversified, both structurally and functionally. In *S. cerevisiae,* there are two equally abundant classes of canonical Hsp70s, Ssas (4 paralogs) and Ssb (2 paralogs) that constitute major hubs of the cytosolic Hsp70 network. Those hubs interact with 13 J-proteins and 4 NEFs [3]. It is important to remember that fundamentally, despite this overwhelming complexity, available functional data strongly suggest that the mode of function of cytosolic Hsp70s is the same as those observed for mitochondrial and bacterial chaperones. Below we take a functional perspective to describe some of the better characterized Hsp70 machines functioning in the yeast cytosol, beginning by discussing the two Hsp70 subfamilies, the soluble Ssa and ribosome-associated Ssb Hsp70s. In the later parts of this section, we turn to a network perspective and then to a larger evolutionary picture.

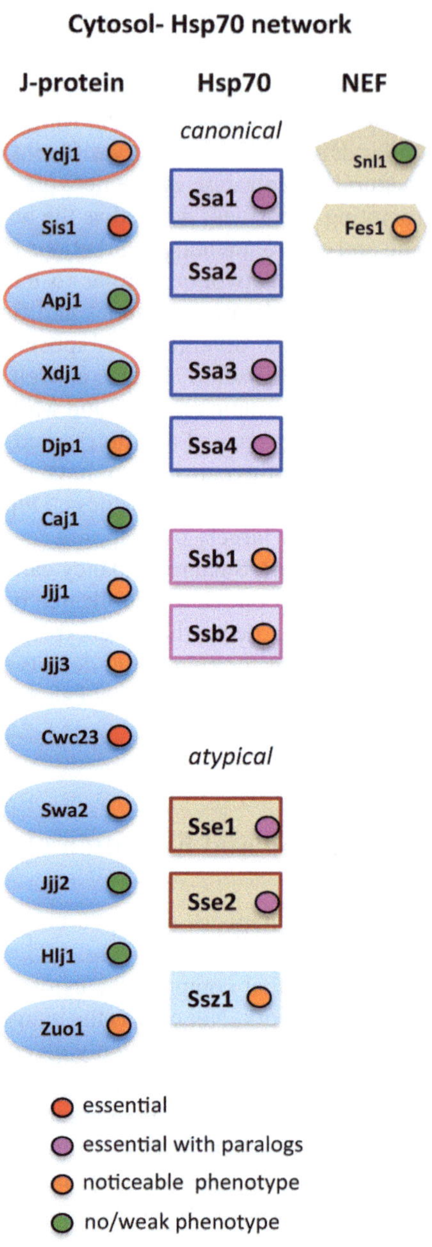

Fig. 3.3 Components of the cytosolic heat shock protein (*Hsp*)70 network in baker's yeast. Shapes indicate homology: J-proteins (*oval*) and Hsp70s (*rectangle*); that two nucleotide exchange factors (*NEFs*) are unrelated is indicated by their different shapes. Identical color of lines around symbols indicates paralogous proteins. Background color of given symbol indicates function in the network; *blue*: stimulation of the adenosine triphosphatase (*ATPase*) activity of partner Hsp70, *light brown*: nucleotide exchange, *pink*: canonical Hsp70 able to perform the ATPase client binding cycle. Note that atypical Hsp70s acting as NEFs are marked by *light brown* background while Ssz1, which forms a stable complex with Zuo1 to form the ribosome-associated complex (*RAC-complex*), is marked by *blue* background as RAC stimulates the ATPase activity of Ssbs. Growth phenotype of cells harboring deletion of a given gene is indicated by the color of the dot; *red*: lethal, *pink*: lethal with deletion of all other paralogs, *orange*: noticeable phenotype either alone or in combination with its parlog(s), *green*: no noticeable phenotype

5.1 Ssa- Core Cytosol Hsp70 Machines

The four members of cytosolic Hsp70 Ssa subfamily of *S. cerevisiae* are closely related, forming two pairs, Ssa1/Ssa2 and Ssa3/Ssa4, whose members share both a high level of sequence identity and similar patterns of expression [57]. Ssa2 is expressed constitutively at a high level; Ssa1 shares its high expression level, but is also induced upon stress. Both Ssa3 and Ssa4 are expressed at low levels under optimal exponential growth condition, but at high levels under stress. *SSA4* is particularly responsive to stresses such as a heat shock; *SSA3* is highly expressed upon entering stationary phase [58]. Functional relationships among the four Ssa Hsp70s are complex. However, it should be noted that each of the four Ssa Hsp70s is a fully functional chaperone able to perform client-binding/release cycles, cooperating with various J-proteins and NEFs. Deletion of all four *SSA* copies is lethal, but expression of any of these genes can substantially rescue growth, providing that its expression level reaches that normally found for the four. However, considering their high expression levels, Ssa1 and Ssa2 can be viewed as the major Ssa players in the cytosolic Hsp70 network. They are involved in a number of facets of proteostasis, including protein folding, protein trafficking across organellar membranes, protein refolding after stresses that induce a loss of structural integrity, maintenance of yeast prions, and directing proteins for degradation [2]. Interestingly, despite their high level of sequence identity, evidence exists that Ssa1/2 have functionally diversified. Ssa2, but not Ssa1, is involved in vacuolar import and degradation of gluconeogenic enzymes after addition of glucose to starved cells [59]. Moreover, Ssa1/2 differ in their involvement in prions propagation, with Ssa2, but not Ssa1, being involved in maintenance of a particular prion, URE3. Strikingly, a single amino acid substitution within the NBD domain (Ala83 in Ssa1 and Gly83 in Ssa2) is solely responsible for this functional difference [60].

Ssa Hsp70s partner with 12 of the 13 cytosolic J-proteins. Three, Ydj1, Xdj1, and Apj1, belong to the classical DnaJ-like family, having extended sequence similarity and the same domain structure as Mdj1 and DnaJ. Ydj1 is the most highly expressed J-protein of the cytosol. Like its mitochondrial and bacterial orthologs, it binds a large variety of client proteins. Recent studies utilizing a high throughput approach estimated that up to 40 % of newly synthesized yeast proteins flux through the Ssa folding system, most of them likely via interaction with Ydj1 [61]. Deletion of Ydj1 is not lethal, yet *YDJ1* deletion cells grow very slowly and are very stress sensitive [62]. Xdj1 and Apj1, whose sequences have significantly diverged from Ydj1, are expressed at low levels [63]. They have also diversified functionally from Ydj1. Xdj1 functions specifically in protein translocation into mitochondria while Apj1 has a specialized role in degradation of sumolyated proteins.

Sis1, the second most abundant cytosolic J-protein that functions with Ssa Hsp70s [64], like Ydj1, interacts with a variety of clients and is thus also involved in the flux of proteins through the Ssa-based folding machine. It shares an overall domain organization with Ydj1, but differs from it in two important regions: the linker between the J-domain and CTDI domain is longer and enriched in Gly and Met residues and CTDI lacks the zinc-binding region. Other important difference from Ydj1 is that a deletion of *SIS1* is lethal, suggesting that Sis1 carries out unique, essential function(s) not compensated by the naturally occurring high concentration of Ydj1 [65]. Known unique functions of Sis1 are its involvement in maintenance of specific yeast prions such as [*RNQ*$^+$] and a role in translation [66, 67]. It is not clear, however, which of Sis1's function(s) determines its essentiality. Yet, functional data indicates that this essential function depends on Sis1 cooperation with Ssa Hsp70s.

Outside their J-domains, other cytosolic J-proteins are at best distantly related to both the DnaJ-type J-proteins and Sis1. Their homology relationships to each other and to Sis1 and Ydj1 remain unclear. Below, we summarize some of the available information regarding the function of 5 other J-proteins, to provide a flavor of the diversity of structure and function of the J-protein partners of Ssa Hsp70s: Jjp1, Jjj1, Jjj3, Swa2, and Cwc23:

1. Djp1: Having an N-terminal J-domain and C-terminal putative client-binding domain, Djp1 has been shown by functional studies to be involved in peroxisomal protein import. In addition, recent results demonstrated its involvement in the import of the mitochondrial outer membrane protein Mim1 [68]. In the absence of Djp1, newly synthesized Mim1 molecules become import incompetent. Therefore, Djp1 could be classified as highly specialized J-protein.
2. Jjj1: The specialized ribosome-associated J-protein, Jjj1, is involved in the biogenesis of the large subunit, specifically facilitating the removal of two ribosome biogenesis factors that travel with the pre60S subunit from the nucleus into the cytosol [69]. This function depends on both Jjj1's cooperation with Hsp70 and another ribosome biogenesis protein, Rei1, with which it interacts via its C-terminal region.
3. Jjj3: Despite its similar name, Jjj3 is very distantly related to Jjj1 and has no functional similarity with it. Jjj3 is a specialized J-protein that plays an essential role in the biosynthesis of diphthamide, an unusual amino acid formed by posttranslational modification of a conserved histidine found in the translation elongation factor, eEF2 [70]. This modified amino acid is the target for ADP-ribosylating diphtheria toxin produced by *Corynebacterium diphtheriae*.
4. Swa2: A complex J-protein, having clathrin-binding domains, a ubiquitin association domain, tetratricopeptide repeat (TPR) motifs, and a C-terminal J domain, Swa2 is involved in the uncoating of clathrin-coated vesicles [71].
5. Cwc23: A specialized J-protein, Cwc23 is required for pre-mitochondrial ribonucleic acid (mRNA) splicing, most likely at the stage of disassembly of the spliceosome [72]. Consistent with its role in this critical biological process, the deletion of *CWC23* is lethal. Yet, although both biochemical and genetic data indicate that the J-domain of Cwc23 functions with Ssa Hsp70, it is not required for it to carry out its role in splicing rather efficiently. Thus, Cwc23 serves as a unique example of a J-protein for which its partnership with Hsp70 plays an auxiliary, rather than a central, role in its essential cellular function.

A comprehensive picture of the functional divergence and convergence among the network of cytosolic J-proteins came from a global analysis asking which of the cytosolic J-proteins are functionally unique [62]. Surprisingly, the severe growth defects caused by the absence of Ydj1 could be substantially alleviated by the expression of J-domain fragments from a variety of J-proteins. Such rescue required neither a J-domain from a particular class of J-proteins nor expression at levels higher than normal for Ydj1. This sufficiency of the J-domains indicates that the core ability of a J-protein, the ability to stimulate the ATPase activity of its partner Hsp70, suffices for J-protein function in many cellular processes that require the Ydj1-containing Hsp70 machine. These results also suggest that, even though Ydj1 interacts with and delivers client proteins to Ssa, this binding activity is not critical for many of its in vivo functions. However, it does not imply that client binding is never critical for the function of Ydj1, and likely other "general" J-proteins as well. Firstly, the J-domain alone does not fully rescue the Δ*ydj1* growth defect. Secondly, it is known that there is some overlap in function of the C-terminal substrate-binding domains of Ydj1 and Sis1, as one or the other, but not both, are required for robust growth of yeast cells [65]. Thus, certain client proteins likely require interaction with a J-protein for efficient presentation to Ssa Hsp70s, but that binding to either Ydj1 or Sis1 suffices in these cases. It is likely that other cytosolic J-proteins also functionally overlap with Ydj1 in vivo, as well. As a counterpoint to the generality of Ydj1 function, a significant degree of specificity was also demonstrated in this global analysis. For example, three deletion strains, Δ*sis1*, Δ*jjj1*, and Δ*jjj3* could only be rescued by expression of the deleted genes, not other J-protein genes, suggesting a high degree of specificity that is largely governed by regions outside the J-domain [62].

5.2 Ssb-Ribosome-Associated Hsp70 Machine

Fungi, including yeast, are unique among eukaryotes as their genomes encode the specialized Hsp70 subfamily, Ssb. In *S. cerevisiae,* the two paralogous proteins, Ssb1 and Ssb2, are 99% identical in sequence and functionally interchangeable [73]. These ribosome-associated Hsp70s function in cotranslational protein folding. At approximately a 1 to 1 stoichiometry with ribosomes, it is perhaps not surprising that their expression level is regulated similarly to that of core ribosomal proteins, and thus in contrast to Ssa Hsp70s, their expression is lowered, not induced, upon a heat shock. Moreover, as overexpression of *SSA* does not complement the lack of *SSB*, their functions are not overlapping.

Like other Hsp70s, Ssb1/2 functions with a J-protein partner. However, its J-protein partner, Zuo1, is unusual. Zuo1 forms a stable heterodimer with the atypical Hsp70 Ssz1. This very stable Zuo1:Ssz1 complex is typically called the "ribosome-associated complex" (RAC) [74]. Like other atypical Hsp70s, Ssz1 is less conserved and has structural differences compared to the canonical Hsp70s, such as Ssa, Ssb, or Ssc. In particular, Ssz1's substrate-binding domain has a shortened C-terminal helical region. There is no evidence that Ssz1 binds client protein, and although it can bind both ATP and ADP, it does not have ATPase activity [75]. Recently published low resolution cryo-electron microscopy data indicates that the heterodimer is an elongated structure, with Zuo1's sequences responsible for ribosome association at one end and Ssz1 and Zuo1's N-terminal Ssz1-interaction domain region at the other [76]. Consistent with their functional cooperation, the deletion of genes encoding Ssb1/2, Ssz1, or Zuo1 causes the same phenotypic effects of slow growth and severe sensitivity to cations [73].

The components of the RAC complex have at least one cellular function in additional to cotranslational protein folding. Zuo1 and Ssz1 activate the transcription factor Pdr1, which is involved in pleiotropic drug resistance (PDR) response and quorum sensing via regulation of expression of transporters of the plasma membrane [77]. However, this function of the RAC complex does not depend on Zuo1's J-domain, as the C-terminal 13 residues of Zuo1 are sufficient to activate Pdr1 [78]. This result serves as another indication that J-proteins can have functions independent of those dependent on their J-domains, and thus cooperation with their Hsp70 partners.

5.3 NEFs of the Cytosolic Hsp70 Network

The NEFs of the yeast cytosol constitute a group of structurally unrelated proteins that facilitate release of ADP from the NBD domain of Hsp70s belonging to both the SSA and SSB subfamilies [16, 17]. Yet, while the bacterial GrpE and mitochondrial Mge1 NEFs were identified long ago, their cytosolic counterparts were discovered more recently, and thus our knowledge about their functional role in the cytosolic Hsp70 network is still rather limited. Cytosolic NEFs belongs to three distinct groups of proteins; Sse1 and Sse2 are members of the Hsp70 family, Snl1 is a member of BAG protein family while Fes1 with its armadillo repeats fold may be a more specialized NEF.

Yeast *S. cerevisaie* harbors two Hsp70-type NEFS, the paralogous Sse1 and Sse2. They belong to the atypical class of Hsp70s also known as Hsp110s, which, compared to canonical Hsp70, contains an insertion in the β-sandwich domain and an extension at the C-terminus. Structural data indicate a stable interaction between the NBD and SBD [79], which resembles the ATP-bound conformation of canonical Hsp70s. Consistent with this structural information, biochemical data suggest that allosteric communication between domains of Sse is limited and that the client-binding/releasing cycle is not fully functional. However, consistent with their ability to function as NEFS, Sse1 in complex with ATP binds a partner canonical Hsp70 [79].

Sse1/2's high level of expression, the lethal phenotype of the double Δ*sse1*Δ*sse2* deletion, and the detection of Sses in complexes with both Ssa and Ssb suggest that Sses are major NEFs for cytosolic Hsp70s [23, 25]. Indeed, high throughput studies utilizing Tandem affinity purification (TAP) tagged Ssa/Ssb/Sse chaperones show that Hsp70/Sse pairs are involved in multiple interactions with cellular proteins [80]. Also consistent with its role as a NEF, accumulation of clients associated with both Ssas and Ssbs was observed in a Δ*sse1* strain [81]. However, the observation that Sse binds client with high affinity and could hold it for a Hsp70 partner, as well as data showing that human Sse ortholog can perform an in vitro refolding reaction, suggest that the functional repertoire of Sses may be larger than previously thought [82, 83]. Their role(s) in the cytosolic chaperone network could extend beyond the release of nucleotide from the Ssa/Ssb Hsp70 machines. For example, they may target partner Hsp70s to specific clients or cellular locations and/or actively participate in refolding of protein aggregates.

Although our knowledge about the role of Sse in the cytosolic network is limited, even less is known about the two other cytosolic NEFs, Snl1 and Fes. Snl1, the only Bag-domain containing protein identified in yeast [21], is unique among NEFs in containing a transmembrane region. This amino terminal segment tethers it to membranes of the nuclear envelope and ER, with the Bag domain, which interacts with Hsp70 and facilitates ADP release, facing towards the cytosol [20]. Intriguingly, recent data indicate that Snl1 binds the 60S ribosomal subunit independently of its transmembrane region and interaction with Hsp70 [84]. The presence of these specific targeting sequences suggests that Snl1 may be a specialized NEF dedicated to recruiting Hsp70 machines to specific sub-cytosolic locations. Although deletion of *SNL1* results in no detectable phenotypes, its evolutionary conservation points to an important, if perhaps subtle or very condition-specific, role.

Unlike Snl1, Fes1 has no identifiable regions involved in functions other than nucleotide release [19, 22]. Structural data available for the human ortholog revealed an all α-helical fold containing four armadillo like repeats [85]. *S. cerevisiae* *Δfes1* cells are moderately impaired in growth at 37 °C, but at optimal temperature, growth is indistinguishable from wt cells. However, recently published data points to a more specialized role of Fes1 in directing Hsp70 bound clients for polyubiquitilation and degradation [86].

In sum, NEFs are the least understood components of cytosolic Hsp70 machines; their functions appear to overlap. However, recent data points to functional diversification, with Sse playing a role in protein folding and refolding while Snl1 and Fes1 having more specialized functions.

5.4 Cytosolic Hsp70 Machines: Network and Evolution Perspective

Despite a larger number of components, we emphasize that the cytosolic Hsp70 network is organized according to the same principles as the mitochondrial Hsp70 network described above. Ssa and Ssb Hsp70s are major hubs linked to clients and co-chaperones. Co-chaperones organize the network into functional modules, either by directing a subset of clients to partner Hsp70s or by recruiting Hsp70s to specific subcellular locations. However, one important difference from the mitochondrial network is the sheer number of the co-chaperones involved: 14 J-proteins and 4 NEFs. However, the most striking feature of the cytosolic network is that its complexity is combined with a high level of redundancy. Both the Ssa and Ssb Hsp70 subfamilies, are encoded by multiple copies of paralogous genes (Fig. 3.3). The same is true for DnaJ like J-proteins, that is Ydj1, Xdj1, and Apj1, as well as for NEFs belonging to the Sse1/2 Hsp70 family. As discussed above, redundancy is a means to increase robustness [56]. Indeed, available functional data are consistent with this hypothesis as they show that the network can sustain functionality when individual redundant elements are missing. For example, a single *SSA* gene is sufficient to maintain network functionality [58]; the same is true for *SSB* and *SSE* genes

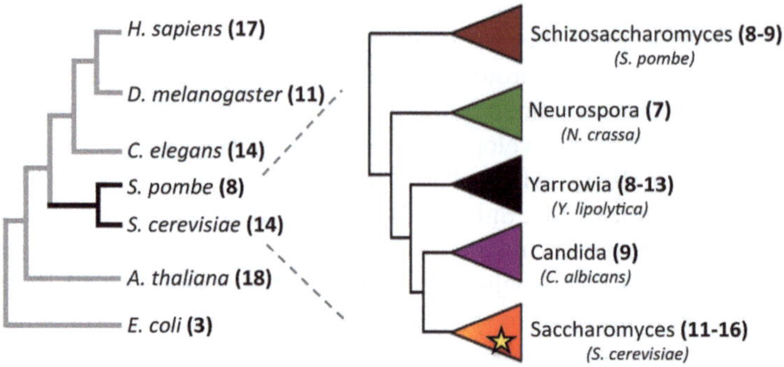

Fig. 3.4 *HSP70* gene copy number across phylogeny. (*Left*) *HSP70* gene copy number in genomes indicated in parentheses. (*Right*) *HSP70* gene copy number in 53 genomes of Ascomycota fungi. Names of major fungal clades are indicated together with most commonly known representative of a given clade. Numbers in brackets indicate the range of HSP70 copy number observed in given clade. *Yellow star* marks the whole genome duplication (*WGD*) that took place in the ancestor of a subset of Saccharomyces species

[2]. Another example of redundancy resulting in the increase of robustness comes from analysis of the yeast cytosol. The absence of the major J-protein Ydj1 can be compensated for, not only by overexpression of many other J-proteins but also by overexpression of the J-domain alone [62].

Yet, the presence of such a high level of redundancy begs for an explanation, especially considering the fact that neither mitochondrial nor bacterial Hsp70 networks exhibit a comparable level of redundancy. One possible explanation is that redundancy was selected for because the cytosol is a cellular center for protein biogenesis and proteostasis, thus an expanded highly robust Hsp70 network is needed for proper function. Indeed, the presence of a similarly expanded and highly redundant cytosolic Hsp70 networks in other eukaryotic organisms, including humans, supports this notion [30]. Yet, rigorous evolutionary analysis of Hsp70 multigene family in fungi showed that the situation is more complex [95]. This analysis revealed that many redundant *Hsp70* gene copies present in *S. cerevisiae* evolved quite recently and are shared with only a handful of closely related species. For example, the evolutionary origin of four copies of *SSA* genes present in *S. cerevisiae* can be traced back to three succeeding gene duplication events. The first took place in a common ancestor of the Candida and Saccharomycetacae clades (Fig. 3.4), approximately 235 million years ago [87] and resulted in formation of the *SSA2/SSA3* gene pair. The second, when both *SSA2* and *SSA3* genes were duplicated, occurred during the WGD event, after which the *SSA3/SSA4* gene pair was retained while the paralogous copy of *SSA2* was lost. The third, most recent, led to formation of *SSA2/SSA1* gene pair by duplication of *SSA2*. This scenario is consistent with the current pattern of sequence similarities among *SSA* genes and with differences in their expression patterns. The origin of *SSB1/2*, *SSE1/2* pairs coincided with the WGD event. Thus, these gene pairs are also relatively recent additions to the yeast cytosolic Hsp70 network.

While S. cerevisiae has 9 *Hsp70* genes in the cytosolic network, other yeast species have significantly fewer: *Kluveromyces lactis*–5, *Candida albicans*–5, *Yarrowia lipolytica*–7, *Neurospora crassa*–4, *Schizosaccharomyces pombe*–5. This taxon-specific copy number variation suggests that cytosolic Hsp70s evolved according to a "birth-death" scenario [87], in which new genes were created by gene duplication. Some were retained in the genome as functional genes, but other genes were inactivated or eliminated from the genome. This phenomenon is also termed "genomic drift" [87], because gene copy number changes randomly from genome to genome.

Assuming that cytosolic Hsp70s indeed evolved according to this mechanism, one might ask why redundant copies were retained, or put another way, why copies were not lost. One possible explanation is selection for robustness, because the presence of duplicated copies would increase robustness of the network [56]. A problem with this explanation is that such robustness would indeed allow for gene loss as the remaining gene copy would be sufficient to maintain network functionality. An alternative explanation is that redundant copies were maintained because they diversified functionally in such a way that functions associated with an ancestral gene were distributed between duplicated copies [88]. Thus, to maintain full functionality of the network, both copies had to be retained. Indeed, available data are consistent in this hypothesis, as in many cases paralogous Hsp70 pairs differ, either in their pattern of expression or in functional attributes (for details, see previous sections on SSAs, SSBs, SSC/SSQ, YDJ1/APJ1). In conclusion, birth-death evolution, combined with retention of functionally diversified genes, could explain network expansion and its apparent redundancy. From a functional perspective, such a network composed of multiple gene copies with partially overlapping functions is able to maintain its integrity despite environmental and genetic perturbations.

6 Hsp70 Network of the ER

While this chapter focuses on the Hsp70 networks of mitochondria and cytosol, we briefly summarize the basics of the Hsp70 network operating in the ER, the initial site of processing and maturation of secretory proteins. The ER of *S. cerevisiae*, like that of the vast majority of other eukaryotes, has a single canonical Hsp70, called Kar2 in yeast and BiP in mammalian cells [2]. Kar2 functions as a hub, interacting with a wide variety of client proteins, through its cooperation with four J-proteins and two NEFs (Fig. 3.5).

Of the four J-proteins in the ER, only one, Scj1, is soluble while the others, Sec63, Jem1, and Erj5, have one or more transmembrane domains with their J-domains facing into the lumen. Scj1 is an ortholog of DnaJ with domain composition similar to that of Ydj1 and Mdj1. Scj1 cooperates with Kar2, not only in protein folding and refolding [89] but also in the ER-associated degradation (ERAD) pathway in which misfolded proteins are translocated back to the cytosol for degradation by the proteasome [90]. Sec63, an essential integral membrane protein that spans the ER

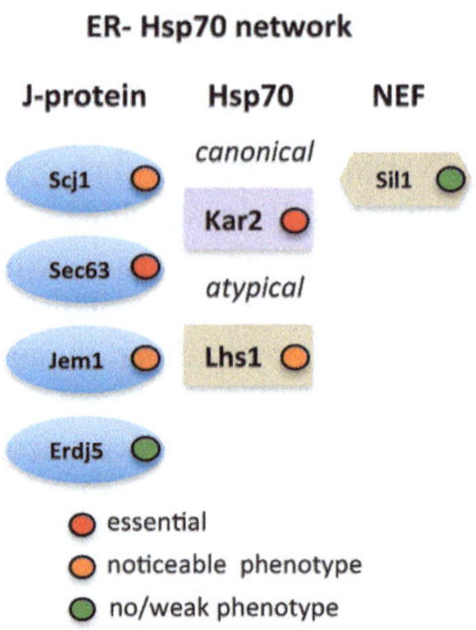

Fig. 3.5 Components of the endoplasmic reticulum (*ER*) heat shock protein (*Hsp*)70 network in baker's yeast. Shapes indicate homology; J-proteins (*oval*), Hsp70s (*rectangle*), and nucleotide exchange factor (*NEF*) (*rhomboid*). Background color of given symbol indicates function in the network; *blue*: stimulation of the Adenosine triphosphatase (*ATPase*) activity of partner Hsp70, *light brown*: nucleotide exchange, *pink*: canonical Hsp70 able to perform the ATPase client binding cycle. Note that Lhs1, which acts as NEF, is marked by *light brown* background. Growth phenotype of cells harboring a deletion of a given gene is indicated by the color of the dot; *red*: lethal, *orange*: noticeable phenotype, *green*: no noticeable phenotype

membrane three times, is localized to the translocon. Like Pam18 of the mitochondria, it plays an important role in protein translocation. It recruits Kar2 to the ER translocon complex and synergistically stimulates its ATPase activity once it binds a nascent chain emerging from the import channel [91]. Both Jem1 and Erj5 have a single membrane-spanning fragment. Δ*jem1* strains are defective in nuclear fusion (karyogamy) during the mating process [92]. Although the molecular details of the role of Jem1 in karyogamy remain unclear, this function requires Jem1 cooperation with Kar2, as the functional J-domain is required. The function of Erj5 function is less clear [93]. However, genetic interaction studies suggest that Erj5 has functional overlap with J-proteins involved in protein folding.

The two NEFs of the ER are structurally and functionally related to counterparts in the cytosol: Lhs1, a major NEF of the lumen of ER, is an atypical Hsp70 similar to Sse NEFS, and Sil1, like Fes1, is an armadillo repeat-containing protein. Individual deletions of *LHS1* and *SIL1* are not lethal, but the double mutation is [94]. This synthetic lethality suggests functional overlap of the two NEFS, as does the fact that the partial defect in protein translocation into the ER caused by the absence of Lhs1 can be compensated by overexpression of Sil1.

In summary, the ER Hsp70 network functions according to the principles that operate in other cellular compartments. Co-chaperones divide the network into functional modules and their overlapping functions provide redundancy, which likely protects network integrity and functionality against environmental and genetic changes.

7 Evolutionary Perspective of the Hsp70 Network in Eukaryotic Cells

When the analysis of the *S. cerevisiae* systems presented in this chapter is challenged with data available for other eukaryotic Hsp70 networks, one fundamental conclusion can be drawn with confidence: All Hsp70 networks are organized and function according to similar principles. However, many of the networks in other eukaryotes are also very complex; the number of components involved is large, making their functional dissection a very difficult task, particularly in organisms for which genetic and cell biological tools are not as highly developed. How can one make sense out of this complexity? One approach is to analyze how genes that form the Hsp70 network are distributed across phylogeny, and in this way determines whether there is a common pattern to their evolutionary conservation and copy number dynamics. One interesting question is whether there are a minimal number of universally conserved genes that constitute the Hsp70 network in a eukaryotic cell? Recently, we began such an analysis by assessing the phylogenetic distribution and evolutionary history of *Hsp70* genes in 53 fungi genomes [95]. The results of our analysis of fungal species, when combined with available published data on other eukaryotes [96–100], are consistent with a scenario in which each eukaryotic genome harbors at least one member of each of six Hsp70 subfamilies present across eukaryotic phylogeny; three canonical Hsp70s (SSA, SSC, KAR) and three atypical Hsp70s (SSE, SSZ, LHS). We note that SSB is a fungal-specific subfamily. Therefore, these six Hsp70 subfamilies constitute a minimal set of *Hsp70* genes required in eukaryotic cells. Indeed, 18 of 57 analyzed fungal species, including *N. crassa,* have only these core six Hsp70s plus the fungal-specific Ssb (Fig. 3.4).

Strikingly, these subfamilies evolved in a taxon-specific manner, such that gene copy number and mode of sequence evolution vary among lineages and sometimes among closely related species [95]. Yet, a common theme emerges when the results of the evolutionary analyses are confronted with functional data: within each Hsp70 subfamily, paralogous genes diversified functionally, regardless of their mode of evolution. Thus, activities that characterize ancestral *Hsp70* genes are distributed among newly formed duplicates. However, as the process of gene duplication is lineage specific, so is the distribution of functions among paralogs. This lineage specificity could have important practical consequences for researchers employing model organisms to study the Hsp70 network. That the manner in which functions associated with a given Hsp70 subfamily are distributed among paralogous genes might vary from model species to model species, and thus needs to be considered when data is interpreted. Yet, in some circumstances, such diversification of function can be exploited. It can allow study of a specific function associated with a specialized Hsp70 without interference from disruption of other activities typically carried out by multifunctional Hsp70s. For example, the presence of the specialized *SSQ1* gene in *S. cerevisiae* has allowed detailed analysis of Hsp70's function in the mitochondrial biogenesis of Fe-S containing proteins where genetic manipulation would have resulted in lethality in taxa without the duplicate gene [32].

Unfortunately, little is known about the evolutionary history and modes of evolution of other genes critical for functioning of the Hsp70 network, those encoding J-proteins and NEFs. Several intriguing questions regarding such genes immediately come to mind. Is there a correlation between a high copy number of *HSP70* genes forming hubs of a given network and the copy number of J-protein or *NEF* genes involved in this network? Put another way, does the complexity of Hsp70s correlate with the complexity of J-proteins and NEFs components? In addition, how did the large number of structurally varied J-proteins and NEFs evolve? Is there a minimal set of genes sufficient to maintain functionality of the Hsp70 network in eukaryotic cells? The large copy number variation and various modes of sequence evolution observed across fungal genomes, combined with the relative ease of gene identification, make fungi an ideal set of organisms to answer these questions.

Acknowledgments Work in the laboratory of E.A. Craig was supported by the National Institutes of Health grants GM31107 and GM27870 (E.A.C.); work in the laboratory of J. Marszalek was supported by Foundation for Polish Science grant TEAM/2009–3/5 and Polish National Science Center grant *Maestro* DEC-2012/06/A/NZ1/000

References

1. Kim YE, Hipp MS, Bracher A, Hayer-Hartl M, Hartl FU (2013) Molecular chaperone functions in protein folding and proteostasis. Annu Rev Biochem 82:323–355
2. Verghese J, Abrams J, Wang Y, Morano KA (2012) Biology of the heat shock response and protein chaperones: budding yeast (*Saccharomyces cerevisiae*) as a model system. Microb Mol Biol Rev 76(2):115–158
3. Kampinga HH, Craig EA (2010) The HSP70 chaperone machinery: J proteins as drivers of functional specificity. Nat Rev Mol Cell Biol 11(8):579–592
4. Mayer MP, Bukau B (2005) Hsp70 chaperones: cellular functions and molecular mechanism. Cell Mol Life Sci 62(6):670–684
5. Barabasi AL, Oltvai ZN (2004) Network biology: understanding the cell's functional organization. Nat Rev Genet 5(2):101–113
6. Gong Y, Kakihara Y, Krogan N, Greenblatt J, Emili A, Zhang Z, Houry WA (2009) An atlas of chaperone-protein interactions in *Saccharomyces cerevisiae*: implications to protein folding pathways in the cell. Mol Syst Biol 5:275
7. Bogumil D, Landan G, Ilhan J, Dagan T (2012) Chaperones divide yeast proteins into classes of expression level and evolutionary rate. Genome Biol Evol 4(5):618–625
8. Gupta RS, Golding GB (1993) Evolution of *HSP70* gene and its implications regarding relationships between archaebacteria, eubacteria, and eukaryotes. J Mol Evol 37(6):573–582
9. Kityk R, Kopp J, Sinning I, Mayer MP (2012) Structure and dynamics of the ATP-bound open conformation of Hsp70 chaperones. Mol Cell 48(6):863–874
10. Qi R, Sarbeng EB, Liu Q, Le KQ, Xu X, Xu H, Yang J, Wong JL, Vorvis C, Hendrickson WA, Zhou L, Liu Q (2013) Allosteric opening of the polypeptide-binding site when an Hsp70 binds ATP. Nat Struct Mol Biol 20(7):900–999
11. Schlecht R, Erbse AH, Bukau B, Mayer MP (2011) Mechanics of Hsp70 chaperones enables differential interaction with client proteins. Nat Struct Mol Biol 18(3):345–351
12. Mapa K, Sikor M, Kudryavtsev V, Waegemann K, Kalinin S, Seidel CA, Neupert W, Lamb DC, Mokranjac D (2010) The conformational dynamics of the mitochondrial Hsp70 chaperone. Mol Cell 38(1):89–100

13. Hennessy F, Nicoll WS, Zimmermann R, Cheetham ME, Blatch GL (2005) Not all J domains are created equal: implications for the specificity of Hsp40-Hsp70 interactions. Protein Sci 14(7):1697–1709
14. Swain JF, Dinler G, Sivendran R, Montgomery DL, Stotz M, Gierasch LM (2007) Hsp70 chaperone ligands control domain association via an allosteric mechanism mediated by the interdomain linker. Mol Cell 26(1):27–39
15. Jiang J, Maes EG, Taylor AB, Wang L, Hinck AP, Lafer EM, Sousa R (2007) Structural basis of J cochaperone binding and regulation of HspMol Cell 28(3):422–433
16. Liu Y, Gierasch LM, Bahar I (2010) Role of Hsp70 ATPase domain intrinsic dynamics and sequence evolution in enabling its functional interactions with NEFs. PLoS Comput Biol 6(9): e1000931
17. Cyr DM (2008) Swapping nucleotides, tuning HspCell 133(6):945–947
18. Kabani M, Beckerich J-M, Brodsky JL (2003) The yeast Sls1p and Fes1p proteins define a new family of Hsp70 nucleotide exchange factors. Curr Genomics 4:465–473
19. Kabani M, Beckerich JM, Brodsky JL (2002) Nucleotide exchange factor for the yeast Hsp70 molecular chaperone Ssa1p. Mol Cell Biol 22(13):4677–4689
20. Sondermann H, Ho AK, Listenberger LL, Siegers K, Moarefi I, Wente SR, Hartl FU, Young JC (2002) Prediction of novel Bag-1 homologs based on structure/function analysis identifies Snl1p as an Hsp70 co-chaperone in *Saccharomyces cerevisiae*. J Biol Chem 277(36):33220–33227
21. Kabbage M, Dickman MB (2008) The BAG proteins: a ubiquitous family of chaperone regulators. Cell Mol Life Sci 65(9):1390–1402
22. Dragovic Z, Shomura Y, Tzvetkov N, Hartl FU, Bracher A (2006) Fes1p acts as a nucleotide exchange factor for the ribosome-associated molecular chaperone Ssb1p. Biol Chem 387(12):1593–1600
23. Raviol H, Sadlish H, Rodriguez F, Mayer MP, Bukau B (2006) Chaperone network in the yeast cytosol: Hsp110 is revealed as an Hsp70 nucleotide exchange factor. EMBO J 25(11):2510–2518
24. Steel GJ, Fullerton DM, Tyson JR, Stirling CJ (2004) Coordinated activation of Hsp70 chaperones. Science 303(5654):98–101
25. Dragovic Z, Broadley SA, Shomura Y, Bracher A, Hartl FU (2006) Molecular chaperones of the Hsp110 family act as nucleotide exchange factors of Hsp70s. EMBO J 25(11):2519–2528
26. Shaner L, Morano KA (2007) All in the family: atypical Hsp70 chaperones are conserved modulators of Hsp70 activity. Cell Stress Chaperones 12(1):1–8
27. Zhuravleva A, Clerico EM, Gierasch LM (2012) An interdomain energetic tug-of-war creates the allosterically active state in Hsp70 molecular chaperones. Cell 151(6):1296–1307
28. Shaner L, Trott A, Goeckeler JL, Brodsky JL, Morano KA (2004) The function of the yeast molecular chaperone Sse1 is mechanistically distinct from the closely related hsp70 family. J Biol Chem 279(21):21992–22001
29. Nollen EA, Brunsting JF, Song J, Kampinga HH, Morimoto RI (2000) Bag1 functions in vivo as a negative regulator of Hsp70 chaperone activity. Mol Cell Biol 20(3):1083–1088
30. Kampinga HH, Hageman J, Vos MJ, Kubota H, Tanguay RM, Bruford EA, Cheetham ME, Chen B, Hightower LE (2009) Guidelines for the nomenclature of the human heat shock proteins. Cell Stress Chaperones 14(1):105–111
31. Gray MW (2012) Mitochondrial evolution. Cold Spring Harb Perspect Biol 4(9):a011403
32. Craig EA, Marszalek J (2011) Hsp70 chaperones. Encyclopedia of life sciences (ELS). Wiley, Chichester
33. Landry SJ (2003) Swivels and stators in the Hsp40-Hsp70 chaperone machine. Structure 11(12):1465–1466
34. Westermann B, Gaume B, Herrmann JM, Neupert W, Schwarz E (1996) Role of the mitochondrial DnaJ homolog Mdj1p as a chaperone for mitochondrially synthesized and imported proteins. Mol Cell Biol 16(12):7063–7071
35. Germaniuk A, Liberek K, Marszalek J (2002) A bichaperone (Hsp70-Hsp78) system restores mitochondrial DNA synthesis following thermal inactivation of Mip1p polymerase. J Biol Chem 277(31):27801–27808

36. Ciesielski GL, Plotka M, Manicki M, Schilke BA, Dutkiewicz R, Sahi C, Marszalek J, Craig EA (2013) Nucleoid localization of Hsp40 Mdj1 is important for its function in maintenance of mitochondrial DNA. Biochim Biophys Acta 1833(10):2233–2243
37. Spelbrink JN (2010) Functional organization of mammalian mitochondrial DNA in nucleoids: history, recent developments, and future challenges. IUBMB Life 62(1):19–32
38. Hayashi M, Imanaka-Yoshida K, Yoshida T, Wood M, Fearns C, Tatake RJ, Lee JD (2006) A crucial role of mitochondrial Hsp40 in preventing dilated cardiomyopathy. Nat Med 12(1):128–132
39. Neupert W, Herrmann JM (2007) Translocation of proteins into mitochondria. Annu Rev Biochem 76:723–749
40. Chacinska A, Koehler CM, Milenkovic D, Lithgow T, Pfanner N (2009) Importing mitochondrial proteins: machineries and mechanisms. Cell 138(4):628–644
41. Clements A, Bursac D, Gatsos X, Perry AJ, Civciristov S, Celik N, Likic VA, Poggio S, Jacobs-Wagner C, Strugnell RA, Lithgow T (2009) The reducible complexity of a mitochondrial molecular machine. Proc Natl Acad Sci U S A 106(37):15791–15795
42. Chacinska A, Lind M, Frazier AE, Dudek J, Meisinger C, Geissler A, Sickmann A, Meyer HE, Truscott KN, Guiard B, Pfanner N, Rehling P (2005) Mitochondrial presequence translocase: switching between TOM tethering and motor recruitment involves Tim21 and TimCell 120(6):817–829
43. Hayashi M, Schilke B, Marszalek J, Williams B, Craig EA (2011) Ancient gene duplication provided a key molecular step for anaerobic growth of baker's yeast. Mol Biol Evol 28(7):2005–2017
44. Westermann B, Neupert W (1997) Mdj2p, a novel DnaJ homolog in the mitochondrial inner membrane of the yeast *Saccharomyces cerevisiae*. J Mol Biol 272(4):477–483
45. Mokranjac D, Sichting M, Neupert W, Hell K (2003) Tim14, a novel key component of the import motor of the TIM23 protein translocase of mitochondria. EMBO J 22(19):4945–4956
46. Lill R, Hoffmann B, Molik S, Pierik AJ, Rietzschel N, Stehling O, Uzarska MA, Webert H, Wilbrecht C, Muhlenhoff U (2012) The role of mitochondria in cellular iron-sulfur protein biogenesis and iron metabolism. Biochim Biophys Acta 1823(9):1491–1508
47. Schilke B, Williams B, Knieszner H, Pukszta S, D'Silva P, Craig EA, Marszalek J (2006) Evolution of mitochondrial chaperones utilized in Fe-S cluster biogenesis. Curr Biol 16(16):1660–1665
48. Vickery LE, Cupp-Vickery JR (2007) Molecular chaperones HscA/Ssq1 and HscB/Jac1 and their roles in iron-sulfur protein maturation. Crit Rev Biochem Mol Biol 42(2):95–111
49. Pukszta S, Schilke B, Dutkiewicz R, Kominek J, Moczulska K, Stepien B, Reitenga KG, Bujnicki JM, Williams B, Craig EA, Marszalek J (2010) Co-evolution-driven switch of J-protein specificity towards an Hsp70 partner. EMBO Rep 11(5):360–365
50. Scannell DR, Wolfe KH (2008) A burst of protein sequence evolution and a prolonged period of asymmetric evolution follow gene duplication in yeast. Genome Res 18(1):137–147
51. Pareek G, Samaddar M, D'Silva P (2011) Primary sequence that determines the functional overlap between mitochondrial heat shock protein 70 Ssc1 and Ssc3 of *Saccharomyces cerevisiae*. J Biol Chem 286(21):19001–19013
52. Hittinger CT, Rokas A, Carroll SB (2004) Parallel inactivation of multiple GAL pathway genes and ecological diversification in yeasts. Proc Natl Acad Sci U S A 101(39):14144–14149
53. Wagner GP, Pavlicev M, Cheverud JM (2007) The road to modularity. Nat Rev Genet 8(12):921–931
54. Espinosa-Soto C, Wagner A (2010) Specialization can drive the evolution of modularity. PLoS Comput Biol 6(3):e1000719
55. Yamada T, Bork P (2009) Evolution of biomolecular networks: lessons from metabolic and protein interactions. Nat Rev Mol Cell Biol 10(11):791–803
56. Wagner A (2008) Gene duplications, robustness and evolutionary innovations. Bioessays 30(4):367–373
57. Boorstein WR, Ziegelhoffer T, Craig EA (1994) Molecular evolution of the HSP70 multigene family. J Mol Evol 38(1):1–17

58. Werner-Washburne M, Stone DE, Craig EA (1987) Complex interactions among members of an essential subfamily of *Hsp70* genes in *Saccharomyces cerevisiae*. Mol Cell Biol 7(7):2568–2577
59. Brown CR, McCann JA, Chiang HL (2000) The heat shock protein Ssa2p is required for import of fructose-1, 6-bisphosphatase into Vid vesicles. J Cell Biol 150(1):65–76
60. Sharma D, Masison DC (2011) Single methyl group determines prion propagation and protein degradation activities of yeast heat shock protein (Hsp)-70 chaperones Ssa1p and Ssa2p. Proc Natl Acad Sci U S A 108(33):13665–13670
61. Brownridge P, Lawless C, Payapilly AB, Lanthaler K, Holman SW, Harman VM, Grant CM, Beynon RJ, Hubbard SJ (2013) Quantitative analysis of chaperone network throughput in budding yeast. Proteomics 13(8):1276–1291
62. Sahi C, Craig EA (2007) Network of general and specialty J protein chaperones of the yeast cytosol. Proc Natl Acad Sci U S A 104(17):7163–7168
63. Sahi C, Kominek J, Ziegelhoffer T, Yu HY, Baranowski M, Marszalek J, Craig EA (2013) Sequential duplications of an ancient member of the DnaJ-family expanded the functional chaperone network in the eukaryotic cytosol. Mol Biol Evol 30(5):985–998
64. Lu Z, Cyr DM (1998) Protein folding activity of Hsp70 is modified differentially by the Hsp40 co-chaperones Sis1 and Ydj1. J Biol Chem 273(43):27824–27830
65. Johnson JL, Craig EA (2001) An essential role for the substrate-binding region of Hsp40s in *Saccharomyces cerevisiae*. J Cell Biol 152(4):851–856
66. Zhong T, Arndt KT (1993) The yeast SIS1 protein, a DnaJ homolog, is required for the initiation of translation. Cell 73(6):1175–1186
67. Higurashi T, Hines JK, Sahi C, Aron R, Craig EA (2008) Specificity of the J-protein Sis1 in the propagation of 3 yeast prions. Proc Natl Acad Sci U S A 105(43):16596–16601
68. Papic D, Elbaz-Alon Y, Koerdt SN, Leopold K, Worm D, Jung M, Schuldiner M, Rapaport D (2013) The role of Djp1 in import of the mitochondrial protein Mim1 demonstrates specificity between a cochaperone and its substrate protein. Mol Cell Biol 33(20):4083–4094
69. Meyer AE, Hoover LA, Craig EA (2010) The cytosolic J-protein, Jjj1, and Rei1 function in the removal of the pre-60 S subunit factor Arx1. J Biol Chem 285(2):961–968
70. Liu S, Milne GT, Kuremsky JG, Fink GR, Leppla SH (2004) Identification of the proteins required for biosynthesis of diphthamide, the target of bacterial ADP-ribosylating toxins on translation elongation factor 2. Mol Cell Biol 24(21):9487–9497
71. Pishvaee B, Costaguta G, Yeung BG, Ryazantsev S, Greener T, Greene LE, Eisenberg E, McCaffery JM, Payne GS (2000) A yeast DNA J protein required for uncoating of clathrin-coated vesicles in vivo. Nature Cell Biol 2(12):958–963
72. Sahi C, Lee T, Inada M, Pleiss JA, Craig EA (2010) Cwc23, an essential J protein critical for pre-mRNA splicing with a dispensable J domain. Mol Cell Biol 30(1):33–42
73. Pfund C, Huang P, Lopez-Hoyo N, Craig EA (2001) Divergent functional properties of the ribosome-associated molecular chaperone Ssb compared with other Hsp70s. Mol Biol Cell 12(12):3773–3782
74. Rakwalska M, Rospert S (2004) The ribosome-bound chaperones RAC and Ssb1/2p are required for accurate translation in *Saccharomyces cerevisiae*. Mol Cell Biol 24(20):9186–9197
75. Huang P, Gautschi M, Walter W, Rospert S, Craig EA (2005) The Hsp70 Ssz1 modulates the function of the ribosome-associated J-protein Zuo1. Nat Struct Mol Biol 12(6):497–504
76. Leidig C, Bange G, Kopp J, Amlacher S, Aravind A, Wickles S, Witte G, Hurt E, Beckmann R, Sinning I (2013) Structural characterization of a eukaryotic chaperone-the ribosome-associated complex. Nat Struct Mol Biol 20(1):23–28
77. Prunuske AJ, Waltner JK, Kuhn P, Gu B, Craig EA (2012) Role for the molecular chaperones Zuo1 and Ssz1 in quorum sensing via activation of the transcription factor Pdr1. Proc Natl Acad Sci U S A 109(2):472–477
78. Ducett JK, Peterson FC, Hoover LA, Prunuske AJ, Volkman BF, Craig EA (2013) Unfolding of the C-terminal domain of the J-protein Zuo1 releases autoinhibition and activates Pdr1-dependent transcription. J Mol Biol 425(1):19–31
79. Liu Q, Hendrickson WA (2007) Insights into Hsp70 chaperone activity from a crystal structure of the yeast Hsp110 Sse1. Cell 131(1):106–120

80. Makhnevych T, Wong P, Pogoutse O, Vizeacoumar FJ, Greenblatt JF, Emili A, Houry WA (2012) Hsp110 is required for spindle length control. J Cell Biol 198(4):623–636
81. Yam AY, Albanese V, Lin HT, Frydman J (2005) Hsp110 cooperates with different cytosolic HSP70 systems in a pathway for de novo folding. J Biol Chem 280(50):41252–41261
82. Xu X, Sarbeng EB, Vorvis C, Kumar DP, Zhou L, Liu Q (2012) Unique peptide substrate binding properties of 110-kDa heat-shock protein (Hsp110) determine its distinct chaperone activity. J Biol Chem 287(8):5661–5672
83. Mattoo RU, Sharma SK, Priya S, Finka A, Goloubinoff P (2013) Hsp110 is a bona fide chaperone using ATP to unfold stable misfolded polypeptides and reciprocally collaborate with hsp70 to solubilize protein aggregates. J Biol Chem 288(29):21399–21411
84. Verghese J, Morano KA (2012) A lysine-rich region within fungal BAG domain-containing proteins mediates a novel association with ribosomes. Eukaryot Cell 11(8):1003–1011
85. Shomura Y, Dragovic Z, Chang HC, Tzvetkov N, Young JC, Brodsky JL, Guerriero V, Hartl FU, Bracher A (2005) Regulation of Hsp70 function by HspBP1: structural analysis reveals an alternate mechanism for Hsp70 nucleotide exchange. Mol Cell 17(3):367–379
86. Gowda NK, Kandasamy G, Froehlich MS, Dohmen RJ, Andreasson C (2013) Hsp70 nucleotide exchange factor Fes1 is essential for ubiquitin-dependent degradation of misfolded cytosolic proteins. Proc Natl Acad Sci U S A 110(15):5975–5980
87. Nei M (2007) The new mutation theory of phenotypic evolution. Proc Natl Acad Sci U S A 104(30):12235–12242
88. Hughes AL (1994) The evolution of functionally novel proteins after gene duplication. Proc Biol Sci 256(1346):119–124
89. Silberstein S, Schlenstedt G, Silver PA, Gilmore R (1998) A role for the DnaJ homologue Scj1p in protein folding in the yeast endoplasmic reticulum. J Cell Biol 143(4):921–933
90. Brodsky JL (2012) Cleaning up: ER-associated degradation to the rescue. Cell 151(6):1163–1167
91. Johnson N, Powis K, High S (2013) Post-translational translocation into the endoplasmic reticulum. Biochim Biophys Acta 1833(11):2403–2409
92. Nishikawa S, Endo T (1997) The yeast JEM1p is a DnaJ-like protein of the endoplasmic reticulum membrane required for nuclear fusion. J Biol Chem 272(20):12889–12892
93. Carla Fama M, Raden D, Zacchi N, Lemos DR, Robinson AS, Silberstein S (2007) The *Saccharomyces cerevisiae* YFR041 C/ERJ5 gene encoding a type I membrane protein with a J domain is required to preserve the folding capacity of the endoplasmic reticulum. Biochim Biophys Acta 1773(2):232–242
94. Hale SJ, Lovell SC, de Keyzer J, Stirling CJ (2010) Interactions between Kar2p and its nucleotide exchange factors Sil1p and Lhs1p are mechanistically distinct. J Biol Chem 285(28):21600–21606
95. Kominek J, Marszalek J, Neuveglise C, Craig EA, Williams BL (2013) The complex evolutionary dynamics of Hsp70s: a genomic and functional perspective. Genome Biol Evol. doi:10.1093/gbe/evt192
96. Nikolaidis N, Nei M (2004) Concerted and nonconcerted evolution of the *Hsp70* gene superfamily in two sibling species of nematodes. Mol Biol Evol 21(3):498–505
97. Wada S, Hamada M, Satoh N (2006) A genomewide analysis of genes for the heat shock protein 70 chaperone system in the ascidian *Ciona intestinalis*. Cell Stress Chaperones 11(1):23–33
98. Brocchieri L, Conway deME, Macario AJ (2008) *Hsp70* genes in the human genome: conservation and differentiation patterns predict a wide array of overlapping and specialized functions. BMC Evol Biol 8:19
99. Louw CA, Ludewig MH, Mayer J, Blatch GL (2010) The Hsp70 chaperones of the Tritryps are characterized by unusual features and novel members. Parasitol Int 59(4):497–505
100. Lin BL, Wang JS, Liu HC, Chen RW, Meyer Y, Barakat A, Delseny M (2001) Genomic analysis of the Hsp70 superfamily in *Arabidopsis thaliana*. Cell Stress Chaperones 6(3):201–208

Chapter 4
The Chaperone Networks: A Heat Shock Protein (Hsp)70 Perspective

Veronica M. Garcia and Kevin A. Morano

Abstract The heat shock protein (Hsp)70 protein chaperone is a ubiquitous and promiscuous simple machine that functions in a wide variety of cellular processes. Its primary role of binding short, exposed hydrophobic stretches in misfolded polypeptides is augmented by the participation of an array of partners that act at multiple levels to govern Hsp70 functionality. Protein folding by Hsp70 is adenosine triphosphate (ATP)-dependent and the state of nucleotide binding is driven by dedicated Hsp70 cofactors. Moreover, these and other associates provide functional specificity by virtue of pathway-specific interactions that both recruit and regulate Hsp70 to provide critical protein remodeling contributions. In this chapter, we break down these interactions into two broad themes: protein biogenesis and maturation, and quality control surveillance and degradation. Key findings that establish the pathway- or process-specific network around Hsp70 supporting each cellular activity are discussed, with special attention paid to protein interactions that dictate the contextual role Hsp70 plays. Understanding these various chaperone networks is a requisite first step in designing pathway-specific pharmacological agents for potential therapeutic intervention in protein misfolding disorders.

1 Introduction

The heat shock protein (Hsp)70 superfamily of molecular chaperones is among the most highly conserved proteins in life. Representatives are found in bacteria, archaea, and eukaryotes, in the latter case including compartment-specific isoforms. Although usually considered as "heat shock proteins," aptly named due to their significant increase in abundance during cytotoxic stress, these chaperones contribute to a plethora of cellular functions such as protein folding, maturation, translocation, refolding, and degradation under essentially all environmental conditions. Indeed, Hsp70s are tightly integrated into the cellular circuitry as powerful modulators of protein structure and dynamics, serving both support and regulatory roles.

K. A. Morano (✉) · V. M. Garcia
Department of Microbiology and Molecular Genetics,
UTHealth Medical School, Houston, TX, USA
e-mail: kevin.a.morano@uth.tmc.edu

W. A. Houry (ed.), *The Molecular Chaperones Interaction Networks in Protein Folding and Degradation,* Interactomics and Systems Biology, DOI 10.1007/978-1-4939-1130-1_4,

The functional diversity of Hsp70-dependent processes is astounding when one considers the extremely high level of sequence and structural conservation across species: This relatively simple protein folding machine has been co-opted to support diverse cellular processes, in diverse subcellular locations, with exceedingly little variability in its fundamental nature. In the most reductionist view, Hsp70 binds and releases hydrophobic segments of polypeptides in an adenosine triphosphate (ATP)-driven cycle. How then is the chaperone targeted to specific cellular processes such as protein translocation or ubiquitin-mediated degradation? How does Hsp70 appear to act like the molecular equivalent of a Swiss Army knife while only possessing a single tool? As we will elaborate in this chapter, the answer appears to be twofold: co-chaperone proteins that both modulate activity and provide some degree of specificity, and the organization of subsets of these partnerships into more complex protein networks dedicated to the cellular process in question. In this respect, these cofactors and interactors provide the informational content required to support the functional complexity inherent in a living cell.

1.1 The Hsp70 Machine

Hsp70 chaperones can be divided into two basic functional units: an amino-terminal nucleotide-binding domain (NBD) and a carboxyl-terminal substrate-binding domain (SBD) [1, 2]. Biochemical and structural studies have lucidated a complex relationship between these two domains based on an allosteric regulatory mechanism. This mechanism communicates nucleotide state in the form of NBD conformational dynamics to differential docking of the SBD to the NBD, through a short extended linker region [3]. The SBD itself consists of a β-sandwich domain with a peptide-binding cleft followed by an α-helical domain that exists in two distinct conformational states. In the so-called open conformation, SBDα is rotated away from SBDβ, resulting in measured low affinity binding of substrate polypeptide to SBDβ. A second “closed” state is achieved in the presence of substrate with SBDαrotated in close proximity to the bound peptide, significantly reducing the off-rate [4]. The NBD regulates SBD conformational gymnastics via an amino acid relay chain based on the presence of ATP, or after hydrolysis, adenosine diphosphate (ADP), which dictate the SBD low- and high-affinity substrate-binding states, respectively [5–11]. In this manner, polypeptides cycle through bound and released states at a rate governed by nucleotide hydrolysis in the NBD. SBDβ specifically recognizes extended hydrophobic regions, whose availability is reduced through iterative rounds of folding as these substrate domains are buried away from hydrophilic solvent. Once no exposed hydrophobic regions remain, the substrate is effectively folded and independent of Hsp70. The basal ATPase level of DnaK is ~0.04 min^{-1} and is accelerated by different partner proteins [12]. Two major classes of co-factors (and in some cases substrate-binding co-chaperones), J proteins and nucleotide exchange factors (NEFs), impact the hydrolysis and nucleotide exchange steps, respectively, and together with Hsp70 form a maximally operational chaperone machine.

1.2 Essential Partners: J-proteins and Nucleotide Exchange Factors

All J-domain proteins are categorized by a 70-amino acid signature region, found in the archetypal *E. coli* protein DnaJ. A J-domain is defined by a highly conserved histidine-proline-aspartate (HPD) tripeptide sequence located between helix II and helix III [13]. While the Hsp40 class of J-protein co-chaperones bears the closest resemblance to DnaJ, J-domains are modular in nature and are found appended to pathway-specific proteins ostensibly in order to recruit Hsp70, as will be discussed below. The HPD motif is thought to allow J-proteins to stimulate the Hsp70 ATPase via direct interaction with key NBD residues, and at least one co-crystal structure supports this contention [14–16]. In addition to stimulating Hsp70 nucleotide hydrolysis, resulting in increased substrate affinity, many Hsp40s also possess a substrate-binding domain with a similar preference for exposed hydrophobic peptides [17–19]. In vitro modeling has demonstrated that a tri-protein complex can exist where Hsp40 first binds substrate and then interacts with the Hsp70 NBD. Presumably the Hsp40-bound substrate is then delivered to Hsp70, effectively increasing the chances that a misfolded protein will engage the chaperone machine [20]. At least one of the substrate-binding domains of the two Hsp40s in *Saccharomyces cerevisiae*, *YDJ1* or *SIS1*, is required for cell viability indicating the importance of this role [21].

In contrast to the high sequence and structural conservation of J-domains, at least four NEFs completely unrelated in sequence and structure have been described to date: homologs of the *E. coli* protein GrpE, the Hsp70-binding protein 1 (HspBP1) family, proteins with the Bcl-2 associated athanogene (BAG) domain, and the Hsp110 family of atypical Hsp70s. Co-crystal structures revealed that they bind to the Hsp70 NBD in distinct modalities, causing conformational changes that induce release of ADP and acquisition of fresh ATP, effectively resetting the nucleotide cycle to allow for the next round of substrate dynamics [10, 22–24]. As with the Hsp40 co-chaperones, lack of cytosolic NEF function is lethal in yeast as a double deletion of the two duplicated stress seventy E (*SSE*) (Hsp110) genes is non-viable, demonstrating that NEFs are crucial for Hsp70 function in vivo [25]. Cells lacking stress seventy e1 (Sse1) show defects in both Hsp70- and Hsp90-dependent functions, including mitosis through interactions with the spindle, and cell integrity through interactions with the mitogen activated protein (MAP) kinase signaling pathway [26, 27]. It is not yet understood why NEF activity has independently evolved multiple times. One possibility lies in the fact that in addition to their nucleotide exchange function, the Hsp110 family also possesses passive chaperone activity and may contribute to Hsp70 substrate recruitment [27–33].

1.3 Hsp70 Functional Plasticity via Protein Chaperone Networks

Hsp70 client recognition is by default nonspecific and likely modulated by collaborating co-chaperones and co-factors, which are in many cases structurally divergent,

Table 4.1 Major chaperones and cofactors

Class	Bacterial	Yeast	Human
Hsp70			
Cytosolic	DnaK	Ssa, Ssb	Hsc70, Hsp70
mitochondria	HscA	Ssc1, Ssq1	mtHsp70
ER		Kar2	BiP
Hsp40			
Cytosolic	DnaJ	Ydj1, Sis1	
mitochondria	n/a	Mdj1, Jac1	Tid-1
ER		Scj1, Jem1	ERdj3
ribosome biogenesis		Jjj1	
RAC	n/a	Ssz1/Zuo1	Mpp11, Hsp70L1
Hsp70 NEF, cytosolic			
Hsp110	n/a	Sse1/2	Hsp105, Apg1
BAG	n/a	Snl1	BAG1–6
HspBP1	n/a	Fes1	HspBP1
GrpE	GrpE	Mge1 (mitochondrial)	GrpEL1 (mitochondrial)
Hsp70 NEF, ER			
Hsp110	n/a	Lhs1	Grp170
HspBP1	n/a	Sil1	BAP
Hsp90			
Cytosolic	HtpG	Hsp/c82	Hsp90
ER	n/a	n/a	GRP94
Hsp90/Hsp70 organizing protein	n/a	Sti1	HOP
Cdc37	n/a	Cdc37	CDC37
C-term Hsp70 interacting protein proteins	n/a	n/a	CHIP
Hsp70 interacting protein	n/a	n/a	HIP
Chaperonin	GroEL/ES	TRiC	TRiC
sHSP	n/a	Hsp26, Hsp42	HSPB1–10
Hsp100			
cytosolic	ClpB	Hsp104	n/a
mitochondrial	ClpB	Hsp78	CLPB

and temporally, developmentally, and spatially regulated. These features make co-chaperones ideal "adaptors" to couple the protein remodeling power of Hsp70 to specific cellular processes. Mapping interactions between Hsp70 and other proteins is complicated by the fact that such candidates may be binding partners or folding substrates of the chaperone, akin to deciphering protein kinase networks where novel interactors may be regulators or phosphorylation targets. In this chapter, we will focus on cellular factors with which Hsp70 interacts to target the chaperone to diverse processes, with the goal of identifying functional protein networks that dictate this behavior. Although much of these relationships have been elucidated in model systems such as budding yeast, orthologous associations in many cases have been defined in metazoan cells and these parallels will be discussed where appropriate (Table 4.1).

2 Protein Biogenesis: Biosynthetic Networks

2.1 Protein Synthesis at the Ribosome

During protein synthesis, nascent polypeptide chains exiting the ribosome run the risk of misfolding or aggregating due to their largely unfolded nature. Large and complex polypeptides frequently cannot assume the appropriate tertiary structure until required domains have exited the ribosome and acquired significant secondary structure. There are approximately 10^6–10^7 active ribosomes in a eukaryotic cell at any given time [34]. With so much nascent peptide biomass, cells have evolved mechanisms to ensure that protein production is both efficient and productive. To prevent misfolding of nascent chains, cells use chaperone-mediated co-translational systems where newly synthesized polypeptides can fold in a protective environment [35]. Prokaryotes and eukaryotes both possess ribosome-bound chaperone systems that promote folding of nascent chains prior to opportunities for misfolding and subsequent targeting for degradation.

In bacteria, trigger factor (TF) is a ribosome-bound chaperone that delivers nascent chains to the cytosolic Hsp70/Hsp40 system, represented by DnaK/DnaJ, for proper folding (Fig. 4.1) [36–39]. Fungi, on the other hand, utilize two distinct ribosome-bound protein quality control systems [40]. The first is the ribosome-associated complex (RAC) which is composed of an atypical Hsp70, Ssz1, and a J-domain protein, Zuo1 [41, 42]. The ATP-independent nascent polypeptide-associated complex (NAC) is the second system, which in contrast is not composed of molecular chaperones. Two subunits, denoted α-NAC and β-NAC, heterodimerize, with β-NAC associating directly with the ribosome near the polypeptide exit tunnel. NAC likely competes with other nascent chain interacting complexes including the signal recognition particle (SRP), for binding to the ribosome, generating specificity. Both NAC and RAC are structurally and functionally conserved in yeast and higher eukaryotes [40, 43, 44].

In addition to the ribosome associated complexes, Hsp70s bind nascent chains during translation to promote folding, delivery to downstream chaperones, and translocation [45]. Co-translational Hsp70 function has been studied best in the yeast *Saccharomyces cerevisiae* where there are two different ribosome associated Hsp70 proteins. As previously mentioned, Ssz1 is a component of RAC. The other representative is the functionally redundant pair of nearly identical Ssb1 and Ssb2 isoforms (referred to as Ssb) [42, 46]. The primary function of Ssb is to protect newly synthesized polypeptides as it is almost exclusively found associated with nascent chains. The other yeast cytosolic Hsp70s are the four homologous Ssa proteins and they do not typically associate with ribosome-bound polypeptides. Although Ssa and Ssb seem to have distinct and non-overlapping functions, Ssa can switch to a co-translational function if Ssb is not available [42, 47, 48]. Ssb interaction with the ribosome is largely mediated by RAC (Ssz1 and Zuo1) which binds in very close proximity to the ribosomal exit tunnel. Ssz1 is an atypical Hsp70 lacking important catalytic residues and therefore does not cycle through nucleotide

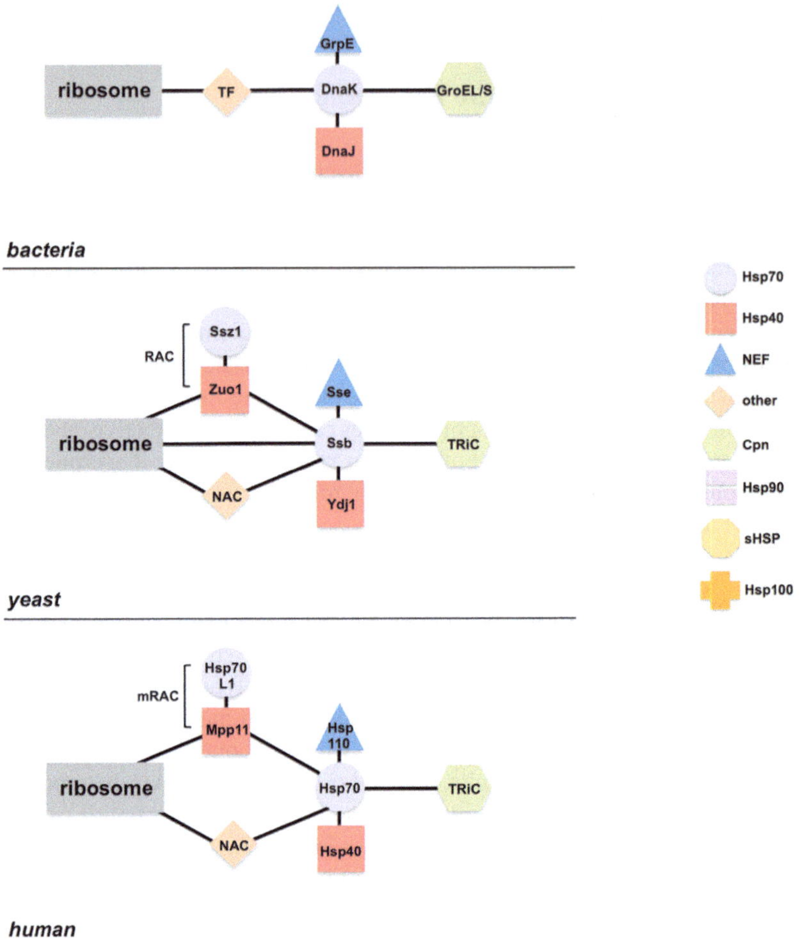

Fig. 4.1 Comparison of biosynthetic Hsp70 network interactions at the ribosome. Key proteins interacting with Hsp70 are shown using *colored* geometric shapes as identified in the key at *right*. Chaperone families are indicated by shape and color while genus-specific members of each family are specified by name. *Black* connecting lines indicate functional connections. Cpn; chaperonin.

binding and hydrolysis phases, leaving it perpetually in an ATP-locked state [49]. Due to this permanent conformation, Ssz1binds the ribosome but does not itself interact with substrate. Ssz1 may be present to help Zuo1 maintain a competent structure, regulate Ssb ribosomal binding, or to act like the bacterial TF to recruit Hsp70 to ribosome-bound nascent chains [50, 51]. While Ssz1 is not required for Ssb to bind nascent polypeptides in vivo, it cooperates with Zuo1 to stimulate Ssb ATPase activity and may thus influence cycling rates and polypeptide dynamics [42]. Ribosomal interactions with RAC are conferred exclusively by Zuo1 [49, 52]. As a J-domain protein, Zuo1 stimulates the ATPase activity of Ssb, but it also possesses a charged region that allows RAC to bind the ribosomal protein Rpl31 located at the ribosomal tunnel exit [40, 49, 53]. To act as a co-chaperone for Ssb, Zuo1 must stably interact with Ssz1 [41] via its N-terminus [50]. As neither protein directly binds substrate, the role of Ssz1 as a "facilitator" of Zuo1 J-domain stimulation of Ssb is highly unusual for an Hsp70 protein [42].

Recent proteomic studies conducted in yeast revealed that the Ssb-RAC network interacts with more than 3200 non-chaperone substrates or approximately 70 % of newly translated polypeptides [54, 55]. Of those 70 %, 45 % were highly enriched for cytosolic and nuclear-localized proteins, in contrast to the population of secretory and membrane proteins recognized by the signal recognition particle (SRP) system. Ssb itself does not interact with every nascent chain but preferentially associates with substrates that have certain qualities. Bioinformatic analysis of biophysical characteristics of these substrates identified a number of determinant factors, including aggregation prone domains, hydrophobic stretches, intrinsically disordered regions, and overall polypeptide length [55]. Based on its association with the ribosome and RAC, Ssb is recruited to interact mostly with cytoplasmic and nuclear proteins whose slow translation rates and other intrinsic properties make timely and proper folding a challenge [55]. The mammalian RAC complex is homologous to that of the *S. cerevisiae* RAC complex as it consists of Mpp11, a J-protein and Zuo1 ortholog, and the atypical Hsp70, Hsp70L1 [43, 56]. Although the mRAC proteins are slightly structurally divergent, they functionally complement yeast RAC deletions, assuring similar if not identical biochemical behavior [44]. One important difference in the two systems is that the "dedicated" Ssb-RAC pairing only exists in fungi, whereas mRAC recruits cytosolic Hsp70 (analogous to yeast Ssa) to nascent chains on the ribosome [43, 44]. Therefore, cytosolic de novo protein folding in mammals is carried out by the same Hsp70s that participate in all other aspects of cytosolic proteome maintenance.

Ssb has been shown to dissociate from the ribosomal complex while remaining substrate-bound, when it subsequently interacts with the Hsp110 nucleotide exchange factor (NEF) Sse1 to induce nucleotide cycling [42]. A recent report demonstrated that Hsp110 interacts with Ssb bound to nascent polypeptides but only after Ssb has moved away from the ribosome [55]. Interestingly Sse1 displays a higher binding affinity in vivo for Ssb relative to its homolog Sse2, presumably due to differences in the respective NBDs [57]. It remains unclear whether Hsp110 itself productively binds substrate in vivo, although many studies have demonstrated Hsp110 chaperoning activity uncoupled from Hsp70 in vitro [28, 31–33, 58]. It is therefore possible that Sse1/Hsp110 may also interact with unfolded nascent chains while simultaneously catalyzing nucleotide exchange on Ssb. Recently the yeast BAG homolog Snl1 was demonstrated to associate with the ribosome as well as with Hsp70 in a non-competitive fashion [59]. The binding events are dependent on distinct regions within the Snl1 sequence, indicating that Snl1 has the ability to simultaneously bind Hsp70s and the ribosome, possibly aiding in de novo folding. Although more studies are needed, these data suggest that Snl1 may function like RAC to recruit Hsp70 to the ribosome. Moreover, in both *S. cerevisiae* and the distantly related yeast *Candida albicans*, Snl1 is a transmembrane-anchored protein localized to the endoplasmic reticulum (ER) membrane, allowing speculation that the NEF could complement the role of SRP in recruiting a sub-population of assembled ribosomes to the ER surface to promote translocation or insertion. Most proteins destined for the ER are inserted co-translationally via recruitment of the previously mentioned SRP [60, 61]. By stopping elongation SRP binding effectively prevents a folding problem from existing in the cytoplasm for these polypeptides [62]. However, a small subset of ER proteins (i.e., yeast mating pheromone α-factor precursor) do not engage the SRP system and are instead synthesized on soluble ribosomes, released into the cytosol, and bound by cytosolic Hsp70. In this capacity, the chaperone stabilizes the unfolded conformation, staging it for transfer through the ER translocon. A similar interaction is thought to occur for mitochondrial import where precursors displaying a transit peptide are stabilized by Hsp70 prior to transport across the outer membrane. Experiments in yeast using a strain lacking three of the four Ssa genes and bearing a temperature sensitive allele of *SSA1* have been instrumental in uncovering these relationships, allowing investigators to conditionally eliminate cytosolic Hsp70 function and follow substrate maturation kinetics in real time [63].

2.2 Ribosomal Assembly

In addition to interacting with ribosomes to aid de novo folding, Hsp70s (Ssa and Ssb), RAC, NAC, and the J-protein Jjj1 have all been implicated in assembly of ribosomes themselves [64]. Ribosome assembly is a complicated process that involves coordinating transcription, folding and maturation of ribosomal ribonucleic acid (rRNA) and ribosomal proteins. Genetic ablation of the Ssb/RAC system results in aggregation of ribosomal biogenesis factors, ribosomal proteins, and concomitant reduction in the abundance of active ribosomes. Ribosomal biogenesis factors such as Dbp8, Tif6, Nog1, and Rlp24 along with ribosomal protein Rpl35 have been detected in the insoluble aggregates that form in cells lacking the Ssb/RAC and NAC systems [50]. The Ssa/Jjj1 and the Ssb/RAC complexes are both needed for efficient ribosome production. Ssb/RAC along with Jjj1 is required to aid in processing of 27S and 35S rRNA and of mature 60S subunits [64, 65]. Albanese et. al. conducted studies using a truncated *JJJ1* allele Jjj1-ΔC that was solely nuclear-localized which demonstrated that overexpression of the mutant protein can complement the growth defect of cells lacking Ssb or Zuo1. The Jjj1 mutant also restored the normal polysome profile. These results indicate that the functional overlap between Ssa/Jjj1 and Ssb/RAC occurs in the nuclear/nucleoloar steps of ribosome biogenesis. Additionally, Jjj1 accelerates ATPase activity of nucleolar Ssa and is required for assembly and maturation of the 60S subunit through removal of Arx1, an export adaptor, from the pre-60S subunit [50, 65–67]. Mitochondrial Hsp70 (mtHSP70), represented in yeast by Scc1, also acts to assemble mitochondrial ribosomes. Var1, one of the two ribosomal proteins encoded in the mitochondria, requires mtHSP70 to remain in a soluble state for assembly into the small ribosomal subunit [68].

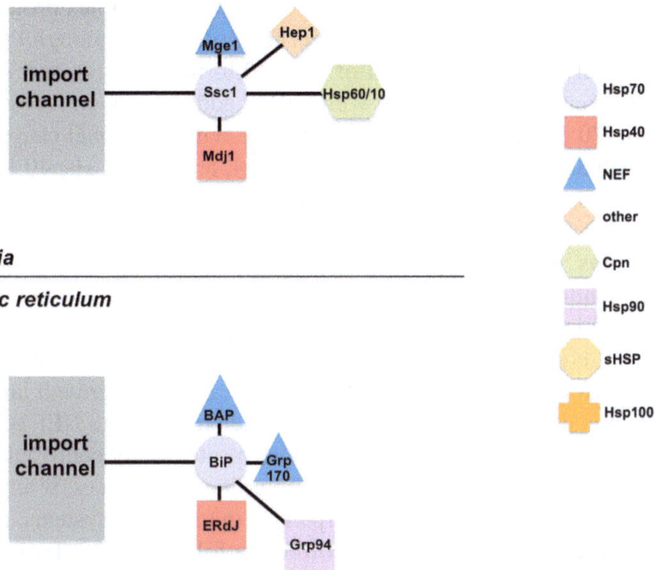

Fig. 4.2 Organellar post-translocation Hsp70 chaperone networks. Interacting proteins are identified as described in Fig. 4.1

Work in bacteria has extended the parallels between prokaryotic and eukaryotic ribosome-associated chaperones. DnaK and DnaJ together promote the conversion of a small subunit assembly intermediate to native 30S subunits, and cells expressing temperature sensitive DnaK mutants at non-permissive temperatures are also found to accumulate ribosomal particles [69–71]. Bacterial TF acts on ribosomal protein S7 to prevent its aggregation in vivo [72]. Additionally, TF also binds to other ribosomal proteins, suggesting that TF has more functions related to the ribosome than solely chaperoning nascent polypeptides [39, 64].

2.3 Organelle Protein Folding

The yeast ER contains two members of the Hsp70 superfamily, Kar2 and Lhs1, represented in mammals by the archetypal BiP and less well-understood Grp170, respectively (Fig. 4.2) [73]. These ER Hsp70s are required for folding of lumenal polypeptides, most notably after passage through the translocon in an unfolded state [74, 75]. Interestingly, these two Hsp70s are in a sense co-chaperones for each other: When the Hsp70 (Kar2/BiP) and the atypical Hsp70 (Lhs1/Grp170) associate, each induces the ATPase activity of the other via mutual acceleration of nucleotide exchange [76, 77]. This is most apparent for BiP/Kar2 with its higher ATPase rate [75]. Because of this *quid pro quo* interaction and their unique ability to associate in a heterodimeric complex, it is very likely that the two Hsp70s are coordinating binding and release of polypeptides in order to fold complex proteins [78]. In mammalian cells, Grp170 and BiP play a critical role in immunoglobulin (IgG) folding and assembly. The contribution of Grp170 to this process, specifically whether it operates exclusively in tandem with BiP or additionally on its own, is unclear [79, 80]. Grp170 and BiP can further coordinate with additional chaperones in a supercomplex that includes the ER Hsp90 Grp94, and an ER Hsp40 ERdj3, to regulate maturation of protein clients [73]. Another lumenal NEF, BiP associated protein (BAP) (Sil1 in yeast) interacts with BiP/Kar2 to enhance folding of newly translocated chains. Interestingly, mutant alleles of BAP are causally linked to Marinesco-Sjøgren syndrome, an early-onset disease marked by pronounced myopathy, ataxia, intellectual impairment and skeletal defects [30, 81]. Although rare, this is the first significant human disease directly linked to dysfunction of a protein chaperone cofactor. Yeast cells lacking either *LHS1* or *SIL1* display growth and functional phenotypes but are viable while simultaneous deletion of both NEF genes results in cell death [82]. These results imply that either protein may contribute to essential BiP functions, and it is not clear how NEF-binding preferences, if any, are established. BiP and its cofactors, along with Grp94, in mammalian systems, also network with other ER proteins responsible for correcting folding errors. Two unique features of ER resident and secreted proteins are the presence of disulfide bonds and glycosylation. In some cases, these modifications are required for protein maturation and acquisition of a stable fold, and improper disulfide bond formation or delays in glycosylation are recognized by dedicated machineries. A family of thioredoxin-related enzymes termed protein disulfide isomerases (PDI) recognize mis- or unfolded substrates and resolve improper disulfide bonds. Although these important enzymes interact with substrate mostly independently of Hsp70, mammalian ERdj5 is both a PDI and Hsp40 chaperone and therefore partners directly with BiP [83, 84].

The mitochondrial matrix contains its own dedicated chaperone network responsible for the folding of proteins synthesized on mitochondrial ribosomes as well as those posttranslationally imported from nuclear-encoded genes. Mammalian mtHsp70 is also known as GRP75 and "mortalin," and it is functionally analogous to the well-studied yeast Ssc1 protein [85, 86]. As with other Hsp70 machineries, dedicated Hsp40s human Tid-1/yeast Mdj1 and yeast Jac1 both associate with client substrates and accelerate mtHsp70 nucleotide hydrolysis [87, 88]. A single NEF, GrpEL1/Mge1, has been described thus far, and in keeping with the predicted bacterial origin of mitochondria, is highly homologous to bacterial GrpE [89]. Ssq1, a yeast mitochondrial Hsp70, specifically functions in the maturation of iron-sulfur proteins, an essential process in mitochondria. Unlike multifunctional Hsp70s that are promiscuous in the variety of hydrophobic substrates they bind, Ssq1/HscA (yeast/bacterial) selectively recognizes and interacts with the conserved LPPVK peptide loop of Isu1/IscU [90]. This interaction is regulated by the mitochondrial Hsp40 co-chaperone, Jac1, which delivers clients to and stimulates the ATPase activity of Ssq1 [91]. The monothiol glutaredoxin Grx5, associates with ADP-bound Ssq1 to complete maturation of mitochondrial Fe/S proteins [92]. Bacteria exhibit a homologous interaction where the J-protien, HscB, cooperates with HscA to assemble Fe-S centers in *E. coli* [93].

2.4 Regulated Maturation of Hsp90 Client Proteins

The Hsp90 molecular chaperone system is a complex, multistage machine for the regulation and final maturation of select substrate, or "client" proteins. Cytosolic Hsp70 is an integral part of this machine. While the process varies for different clients, most are likely initially recognized by Hsp70 in conjunction with its J-domain co-chaperones. Recognition likely occurs early during synthesis as little evidence exists for Hsp90 participating in refolding of damaged client proteins. Clients bound to Hsp70 are subsequently transferred to Hsp90 in a manner assisted by the Hsp70-Hsp90 organizing protein (Hop) [15, 94]. Hop achieves this coordination through two separate tetratricopeptide repeat (TPR) domains that specifically bind Hsp70 (TPR1, TPR3) or Hsp90 (TPR2a) [94, 95]. Once this complex is assembled and the substrate has been delivered to Hsp90, Hsp70 and Hop dissociate via competition with other TPR-containing subunits. Hsp90 is itself an ATPase whose cycling governs conformational dynamics and is in turn governed by co-chaperone binding. For example, Hop binding to Hsp90 blocks its ATPase activity, effectively stalling the system by stabilizing the "open" Hsp90 conformation to promote client transfer and binding. Association with the Hsp70 sub-complex (70/40/NEF) is critical for most clients to ultimately achieve Hsp90 binding, suggesting that Hsp70 is capable of recognizing substrates in a format that Hsp90 is not. Exceptions are protein kinases, well-documented clients of Hsp90, which appear to rely on the conserved protein Cdc37 to perform the same role. Indeed, like the Hop/Hsp70 complex, Cdc37 binding inhibits Hsp90 ATPase activity, providing a parallel mechanism for client transfer [96].

3 Protein Quality Control: Damage and Degradation Networks

In the preceding sections, we considered Hsp70 roles in protein biogenesis, with the implicit assumption that these activities are occurring under favorable conditions. We now turn our attention to situations in which a cell experiences proteotoxic stress resulting in protein misfolding, damage, and aggregation. These insults cause an imbalance in proteostasis that is recognized and dealt with by molecular chaperones in a process termed protein quality control. This surveillance system has three responsibilities: (1) to shield hydrophobic, aggregation-prone regions of misfolded or otherwise damaged proteins, (2) to promote refolding of these same substrates, or (3) to target terminally damaged proteins for degradation. While the former role is essentially an extension of the same activities chaperones, especially Hsp70, engage in under normal conditions, the latter two outcomes require the surveillance system to perform molecular triage and make the "decision" to fold or degrade a given substrate. Under optimal conditions, cells express the requisite amounts of chaperone machinery involved in de novo folding, maturation, translocation, and proteolysis.

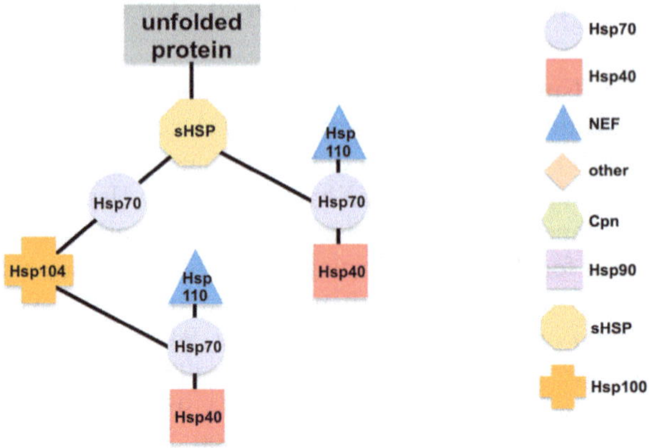

Fig. 4.3 Multiple Hsp70 interactions govern recognition and refolding of damaged proteins. Interacting proteins are identified as described in Fig. 4.1

When a cell is undergoing stress the expression of many of these components is greatly increased at the transcriptional level via dedicated systems. For example, the presence of misfolded proteins in the ER lumen activates the multi-pronged unfolded protein response (UPR) while misfolded cytoplasmic/nuclear proteins instead activate the heat shock response (HSR) [97]. In both cases these pathways guard against the respective chaperone networks becoming overwhelmed, allowing the buildup of potentially toxic aggregates [98].

3.1 Recognition of Damaged Proteins

Hsp70 is generally classified as a major component of the HSR. Although Hsp70 is among the most abundant cellular proteins, a subset of Hsp70 genes are primarily expressed during heat stress as well as other stresses that cause protein misfolding. These distinct gene products are delineated as constitutive or inducible Hsp70s, respectively. In yeast, *SSA3* and *SSA4* are examples of Hsp70s whose expression is tightly repressed during standard growth conditions but highly induced during times of stress [99, 100]. Ydj1 and Sis1 (J-proteins) collaborate with the cytosolic Ssa proteins to prevent aggregation [101]. *SIS1* is essential at all temperatures, whereas *ydj1Δ* cells are strongly temperature sensitive, presumably due to an inability to enhance Ssa functions [17]. Additionally, Ydj1 is approximately four times more efficient at catalyzing Ssa-mediated refolding of the model protein firefly luciferase than Sis1 in vitro, providing a rationale for why *ydj1Δ* cells exhibit a phenotype in the presence of Sis1 [101].

Although the ubiquitous Hsp70 machine (Hsp70 along with an Hsp40 and an NEF) can effectively prevent aggregation and target small aggregates for repair, large aggregate re-solubilization requires the cooperation of the ATP-dependent Hsp100 family of proteins (Fig. 4.3) [102]. While all eukaryotes harbor Hsp100 proteins in the mitochondrial matrix, only plant and fungal cells appear to possess a cytosolic homolog of this family to provide compartment-specific protection by interacting with resident Hsp70s (see further discussion on mammalian disaggregase activity below) [103]. *E. coli* ClpB is part of the Clp family that includes ClpP, ClpA, and ClpX that associate to sequentially unfold (ClpA, ClpX) and/or degrade (ClpP) aggregates. In contrast, the highly homologous ClpB and eukaryotic Hsp100s do not act on protein aggregates by promoting their eradication, acting instead in a non-destructive manner to re-solubilize aggregated proteins for delivery to protein-folding chaperones [98, 104]. Hsp104 is the yeast Hsp100 in the cytosol and while not essential, is highly induced upon heat stress and required for tolerance to extreme heat. Hsp100s are members of the ATPases associated with diverse activities (AAA+) superfamily, with two NBD per monomer that come into contact when assembled into the functional homo-hexameric form. To achieve disaggregation and reactivation of misfolded proteins, Hsp100s act in concert with Hsp70 and Hsp40, which are thought to "accept" the now-unfolded polypeptide extracted from the aggregate and promote its refolding [105, 106]. In mitochondria, Hsp100 homologs (Hsp78 in yeast) bind misfolded mitochondrial matrix proteins to stabilize them and prevent aggregation, thus overlapping in part with the chaperone functions of mtHsp70. As with cytosolic Hsp104, Hsp78 is required for mitochondrial thermotolerance as it is crucial to maintain respiration and the genome integrity of the mitochondrion [103].

A distinguishing feature of ClpB and Hsp104 is a coiled-coil middle-domain (M-domain,) consisting of four α-helices, which is necessary for disaggregation activity and appears to contact Hsp70 directly through interactions in helix two [107–110]. Recent work indicates that Hsp70 acts as an activator of the Hsp104 motor by promoting coordination of the Hsp104 subunits [106]. This makes for an interesting relationship between the two proteins where Hsp70 likely recognizes misfolded protein segments in an aggregate, and recruits Hsp104 to the aggregate while simultaneously activating the M-domain "switch" to extract the polypeptide in an ATP-dependent manner [102, 111]. Hsp104 then in turn delivers the unfolded substrate to Hsp70 for refolding. Hsp70 therefore acts at both ends of a linear pathway that requires a protein with disaggregase activity in between [112]. Fascinatingly, higher eukaryotes lack cytosolic Hsp100 chaperones, prompting two competing theories that either they do not need them, or that some other protein or protein complex plays an orthologous role. Very recently, a possible candidate machinery has been proposed. In the metazoan, *Caenorhabditis elegans*, Hsp110 is required to empower Hsp70 and Hsp40 to efficiently disaggregate and refold aggregated proteins [113]. This same activity is not observed with Hsp110, Hsp70, or Hsp40 alone suggesting that together the triad forms a powerful remodeling machine capable of dissaggregation. It is not clear if the peptide-binding activity of Hsp110 is also required, but as mentioned earlier, Hsp110 is the only NEF that exhibits peptide binding and the only one shown to hyperactivate the Hsp70/Hsp40 dimer from a foldase into a disaggregase [33, 105].

In addition to the Hsp100s, another set of eukaryotic proteins collaborates with Hsp70 to manage proteostasis [114]. The small heat shock proteins (sHSP), yeast Hsp26 and Hsp42, act as energy-independent "holdases" through binding of misfolded polypeptides to prevent aggregation [115–117]. There are ten members of the human sHSP family, and their genes are upregulated during stress and provide cytoprotection by binding unfolded substrate until it can be transferred to Hsp70 [118]. sHSPs protect substrate through formation of high molecular weight homo- or hetero-oligomers of up to 50 subunits [119]. Hsp42 was shown to prevent citrate synthase, insulin, and rhodanese aggregation in vitro in a concentration-dependent manner [115]. Because each subunit has the capacity to bind a substrate molecule, the total number of unfolded substrates in the sHSP complex can be quite high. Consistent with the formation of such super-structures, when exposed to heat shock Hsp26 becomes insoluble along with the unfolded substrate and only returns to a soluble state after substrate processing by Hsp104 [117]. In support of a critical role in substrate re-solubilization, yeast cells lacking Hsp26 cannot disaggregate heat denatured luciferase [116]. Substrates bound to small HSPs are only released through the interaction of Hsp70 in an energy-dependent manner [119]. In a study by Mogk et al., heat dentured malate dehydrogenase formed aggregates that decreased in size as concentration of Hsp26 exposure increased [120]. Additionally, this study demonstrated that DnaK, DnaJ, and GrpE can refold sHSP-bound substrate aggregates that are small, whereas the addition of ClpB allowed resolubilization of even the largest aggregates that occurred with lower concentrations of Hsp26. This indicates that Hsp26 acts to reduce aggregate size during heat stress, possibly reducing insoluble aggregates in the process by making aggregated substrates more accessible to Hsp70 and Hsp100.

In recent years it has been appreciated that misfolded proteins may be concentrated into targeted "depots" during processing. In mammalian cells, these structures are termed aggresomes and their behavior has been well documented. For example, protein stress exerted on human astrocytes results in increased expression of Hsp70 and Hsp25 and the formation of aggresomes that include Hsp25 [121]. Removal of the proteotoxic stress allowed for the re-solublization of the aggresome-bound proteins. Recently, two distinct "bodies" have been described in yeast, the insoluble protein deposits (IPOD) and the juxtanuclear quality control (JUNQ) compartment [122]. Btn2, a trafficking protein, can associate with Hsp42 or the Hsp40 Sis1 to sort misfolded proteins to protein deposits, including JUNQ [123]. Research is currently being conducted to try and understand the spatial and temporal regulations of these inclusion bodies with the small HSPs and Hsp104. Questions still remain about how and why the cell may be forming these protein deposits and what the roles of Hsp70 may be in governing substrate flux through these compartments.

3.2 Control of Substrate Ubiquitination and Degradation

Biological systems are inherently conservative with their resources and protein synthesis requires significant energy input. It is therefore logical that the protein quality control machinery would prioritize repair of proteins where energy has already been expended [124]. Although molecular chaperones are typically thought of as protein folding machines that correct damage, it is becoming clear that they also play important roles in deciding when a substrate is fatally flawed and must

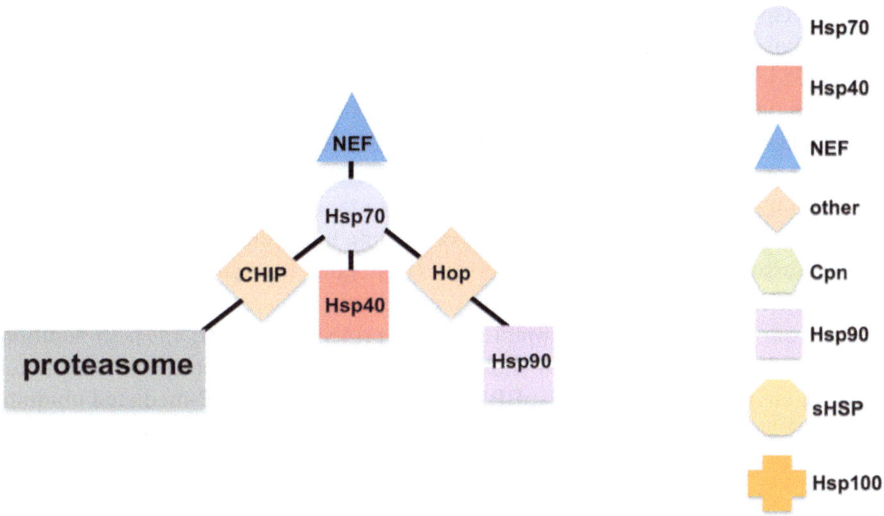

Fig. 4.4 Conversion of Hsp70 into a pro-degradation machine via network interactions. Interacting proteins are identified as described in Fig. 4.1

be targeted for degradation. By virtue of its ability to iteratively bind and release substrate, Hsp70 is in a prime position to "evaluate" the folding status of a given substrate once during each nucleotide cycle. Hsp70s located in the cytoplasm influence the degradation of nuclear and cytosolic proteins while the ER-associated degradation (ERAD) pathway relies on lumenal BiP to perform the same role, in concert with additional chaperones and co-chaperones. Both of these processes ultimately utilize the ubiquitin-proteosome system (UPS) to degrade aberrant proteins, necessitating that an early, if not the first, committed step be ubiquitination of the terminally misfolded substrate. Ubiquitin conjugating enzymes (E2) and ubiquitin ligases (E3) cooperatively recognize and target defective proteins to the 26S proteasome through ubiquitin modification [125–128]. How are substrates engaged in a futile cycle of folding by Hsp70 targeted to the UPS? The cytosolic C-terminus of Hsc70 interacting protein (CHIP) was first identified as a binding partner of Hsp70 in mammalian cells (Fig. 4.4). Hsp70 association is mediated primarily through a tetratricopeptide repeat (TPR) at the N-terminus of the protein [129]. CHIP binding to Hsp70 negatively regulates Hsp40-induced ATPase activity, effectively stalling the protein folding cycle [130, 131]. CHIP also contains an E3 ubiquitin ligase domain that subsequently ubiquitinates Hsp70-bound substrate. Binding of CHIP in the presence of unfolded substrate therefore switches Hsp70 from a machine dedicated to folding to one primarily functioning in degradation [132, 133]. CHIP-Hsp70 interactions are in turn dramatically affected by other co-chaperones. By virtue of their similar TPR domains, Hop and CHIP compete for binding to Hsp70 [134]. Hop is discussed earlier in this chapter as a protein that aids in maturation of Hsp90 clients. Competition between Hop and CHIP may therefore be a determining factor in protein triage for clients failing to be properly matured by Hsp90. Factors dictating binding of CHIP versus Hop for a given client are currently unknown.

The Hsp40 Hsj1 contains two ubiquitin interaction motifs (UIM) that allow it to function as a chaperone substrate escort during proteosomal targeting in neuronal cells [135]. Its combined abilities of recognizing ubiquitinated substrates and loading them onto Hsp70 allow neuronal cells to prevent cytotoxic protein aggregation by enhanced delivery of substrates to the degradation pathway. DnaJB1 is another Hsp40 that, like its yeast homolog Sis1, is necessary for proteosomal degradation, although it remains unclear whether its role is to facilitate ubiquitination of chaperone substrates or to deliver misfolded proteins into the nucleus [122, 136]. The NEF Bag-1 possesses an ubiquitin-like domain (UBL) that mediates binding to the 26S proteasome, indicating that Bag-1 may coordinate transfer of ubiquitinated substrate from the Hsp70-CHIP complex to the proteasome [130, 137]. In addition, the related Bag-6 has also been shown to facilitate Hsp70/proteasome interaction [138]. Conversely, the NEFs HspBP1 and BAG-2 inhibit CHIP-mediated ubiquitination of Hsp70 substrates when bound to the Hsp70 ATPase domain [125]. Fes1, the HspBP1 homolog in yeast, was recently shown to promote ubiquitin-dependent degradation of misfolded protein in the cytosol [139]. Although this activity is in stark contrast to HspBP1, it is important to note that yeast does not possess CHIP (see below). In this instance, Fes1 may be acting solely as an Hsp70 NEF to promote substrate release rather than actively promoting degradation.

The less well-understood Hsp70 interacting protein HIP is a modestly abundant protein but has not been detected in an Hsp70 complex along with CHIP [136]. Although a negative result should not be over-interpreted, this implies that steric considerations may prohibit CHIP and HIP from binding Hsp70 simultaneously even though they each bind on different domains of the Hsp70 protein [140]. By preventing formation of the CHIP-Hsp70 complex, HIP therefore promotes the Hsp70 folding function. HIP preferentially binds to the ADP bound state of Hsp70 and competes for Hsp70 NBD binding with the NEF BAG-1 [141]. This interaction additionally prevents exchange of nucleotide in the Hsp70 NBD thus, keeping Hsp70 in a high affinity substrate-binding state. As a consequence, dwell time of the substrate with the chaperone is increased, allowing greater folding opportunities.

Interestingly, CHIP is absent from the yeast *Saccharomyces cerevisiae*, yet Hsp70 and NEFs have still been implicated in substrate ubiquitination and degradation. Instead, the E3 Ubr1 appears to be recruited to Hsp70/substrate complexes in the cytoplasm via unknown targeting mechanisms to initiate the degradation process. Ubr1 lacks any known Hsp70- or substrate-binding features, yet is required for ubiquitination of many Hsp70 and Hsp90 clients. The predominantly nuclear E3 ligase San1 contains a disordered region that may interact in a non-specific fashion with misfolded substrates, thereby mimicking the role of the Hsp70 SBD. Indeed in vitro Hsp70 and the NEF Sse1 antagonize substrate ubiquitination by San1. Interactions between Hsp70s, NEFs, and dedicated or recruited E3 ligases therefore profoundly influence protein triage decisions and client fate. Substrates that fold quickly and efficiently do not have a high risk of being prematurely degraded. It is the substrates that take a long time to fold and remain in the Hsp70 folding cycle for a prolonged period of time that are at risk for being targeted for degradation by collaborating E3 ligases [134].

While substrates targeted to the proteasome are tagged through typical lysine-48 ubiquitin chains, chaperones are often targeted to the proteasome along with the substrate without inducing degradation of the chaperones. This is accomplished through ubiquitination at alternative lysine sites. For example, BAG-1 is ubiquitinated with lysine 27 linkages [142]. Interestingly, Hsp70 is also ubiquitinated at alternative residues only when Hsp70 must be recruited to the 26S proteasome but it itself is not the degradation target. When Hsp70 is targeted for degradation, for example when cells are recovering after heat shock and attempting to restore chaperone levels back to normal, it is ubiquitinated at the typical lysine residues, implying the existence of unknown factors that govern lysine selectivity on a substrate-specific basis [143].

The ERAD system, which clears the eukaryotic secretory pathway of defective proteins, likewise has an effective surveillance machinery to sort through the folding intermediates that it encounters in the ER from the proteins that are fatally flawed. Retrotranslocated ER proteins are degraded in the cytoplasm through the pathways involving Hsp70s that have already been discussed, with the addition of dedicated membrane-associated E3 ligases localized near translocons. As introduced earlier in this chapter, Kar2, the yeast homolog of human BiP, and the major ER Hsp70 in yeast, in cooperation with Yos9 and Hrd3, plays a major role in identifying terminally misfolded proteins and delivering them to the ERAD pathway [144, 145]. An additional protein called EDEM (ER degradation-enhancing α-mannosidase like protein) participates in recognition of improperly glycosylated proteins, directing them to the retrotranslocon. In mammals, ERdj5 uses its J domain to interact with BiP in an ATP-dependent manner. ERdj5 then functions as a PDI and reduces the disulfide bonds from terminally misfolded ER proteins followed by inducing the ATP hydrolysis activity of BiP [83]. The function of these proteolysis inhibitors is necessary to ensure that slow-folding proteins reach their final configuration without being prematurely degraded. For example, the CFTR ion channel is a slow-folding protein. Because of this, only 30% of CFTR produced in the ER reaches its final destination at the plasma membrane. HspBP1 and Bag-2 can inhibit CHIP-mediated degradation of CFTR. Additionally, these NEFs compete for the ATPase docking site on Hsp70 with the degradation-promoting BAG-1. These antagonistic interactions allow for at least a portion of CFTR to properly mature and be delivered. These factors become critical with certain known mutations in CFTR linked to the disease cystic fibrosis, which results from little to no CFTR at the cell surface due to enhanced degradation of folding-challenged variants.

4 Conclusions

The ubiquitous Hsp70 protein chaperone forms the hub of multiple protein biogenesis and protein quality control networks. As described in this chapter, the chaperone itself while a powerful remodeler of protein structure, lacks innate specificity. It is therefore the interacting partners that recruit it to support numerous cellular processes, and these associations may further govern substrate interactions and decisions regarding client fate. Despite the significant progress made in delineating these networks, specifically at the level of partner discovery and proteome-scale interaction analysis, much is still unclear. What substrate characteristics dictate Hsp70 involvement in folding at the ribosome? How does Hsp70, in concert with Hsp40 co-chaperones and NEFs, identify hopeless folding cases to target for ubiquitination in the cytoplasm or ER? On the network level, similar outstanding questions remain unanswered. With the plethora of J-proteins available to recruit Hsp70 to distinct cellular processes, what is the distribution of association with respect to total Hsp70 abundance? Are complexes formed simply as a function of relative expression levels or do regulatory factors play a role? Recent studies have found phosphorylation events influencing co-chaperone binding to Hsp70 and Hsp90, suggesting another layer of informational complexity that may establish differential binding within networks [146, 147].

Identification of chaperone network interactions is a crucial first step toward therapeutic intervention to treat diseases of protein misfolding, ostensibly the driving force behind years of research in the field. Because the processes of protein folding and regulated degradation are so fundamental to life, small perturbations in chaperone behavior targeted to rescue a specific misfolded protein of interest may have unintended cytotoxic effects. Knowledge of the precise set of interactions for any given chaperone or cofactor at the subcellular level will help minimize off-pathway negative effects by informed selection of ideal targets. For example, future work may select cytosolic chaperone-assisted degradation as a druggable point by targeting CHIP, which based on network analysis would be predicted to have little effect on nascent protein folding. Conversely, treatments to reduce global protein synthesis by a few percent may be sufficient to reduce secretory load in the ER to allow higher yield of a poorly folding protein. Efforts to identify novel small molecule "proteostasis regulators" that enhance or inhibit chaperone abundance or activity are therefore widespread in both academic and industry laboratories. The products of this research are likely to be clinically useful compounds that ameliorate or at a minimum delay the onset or severity of protein misfolding diseases, increasing what has been termed human "healthspan"—the promise of enhancing and extending the quality of disease-free life.

References

1. Zhu X, Zhao X, Burkholder WF, Gragerov A, Ogata CM, Gottesman ME, Hendrickson WA (1996) Structural analysis of substrate binding by the molecular chaperone DnaK. Science 272(5268):1606–1614
2. Saibil H (2013) Chaperone machines for protein folding, unfolding and disaggregation. Nat Rev Cancer 13(10):630–642. doi:10.1038/nrm3658
3. Vogel M, Mayer MP, Bukau B (2006) Allosteric regulation of Hsp70 chaperones involves a conserved interdomain linker. J Biol Chem 281(50):38705–38711. doi:10.1074/jbc.M609020200
4. Popp S, Packschies L, Radzwill N, Vogel KP, Steinhoff HJ, Reinstein J (2005) Structural dynamics of the DnaK-peptide complex. J Mol Biol 347(5):1039–1052. doi:10.1016/j.jmb.2005.02.026

5. Flaherty KM, DeLuca-Flaherty C, McKay DB (1990) Three-dimensional structure of the ATPase fragment of a 70 K heat-shock cognate protein. Nature 346(6285):623–628
6. Takeda S, McKay DB (1996) Kinetics of peptide binding to the bovine 70 kDa heat shock cognate protein, a molecular chaperone. BioChemistry 35(14):4636–4644. doi:10.1021/bi952903o
7. Theyssen H, Schuster HP, Packschies L, Bukau B, Reinstein J (1996) The second step of ATP binding to DnaK induces peptide release. J Mol Biol 263(5):657–670. doi:10.1006/jmbi.1996.0606
8. Schlecht R, Erbse AH, Bukau B, Mayer MP (2011) Mechanics of Hsp70 chaperones enables differential interaction with client proteins. Nat Struct Mol Biol 18(3):345–351. doi:10.1038/nsmb.2006
9. Kityk R, Kopp J, Sinning I, Mayer MP (2012) Structure and dynamics of the ATP-bound open conformation of Hsp70 chaperones. Mol Cell 48(6):863–874. doi:10.1016/j.molcel.2012.09.023
10. Wu CC, Naveen V, Chien CH, Chang YW, Hsiao CD (2012) Crystal structure of DnaK protein complexed with nucleotide exchange factor GrpE in DnaK chaperone system: insight into intermolecular communication. J Biol Chem 287(25):21461–21470. doi:10.1074/jbc.M112.344358
11. Zhuravleva A, Clerico EM, Gierasch LM (2012) An interdomain energetic tug-of-war creates the allosterically active state in Hsp70 molecular chaperones. Cell 151(6):1296–1307. doi:10.1016/j.cell.2012.11.002
12. Davis JE, Voisine C, Craig EA (1999) Intragenic suppressors of Hsp70 mutants: interplay between the ATPase- and peptide-binding domains. Proc Natl Acad Sci U S A 96(16):9269–9276
13. Craig EA, Huang P, Aron R, Andrew A (2006) The diverse roles of J-proteins, the obligate Hsp70 co-chaperone. Rev Physiol Biochem Pharmacol 156:1–21
14. Jiang J, Maes EG, Taylor AB, Wang L, Hinck AP, Lafer EM, Sousa R (2007) Structural basis of J cochaperone binding and regulation of Hsp70. Mol Cell 28(3):422–433. doi:10.1016/j.molcel.2007.08.022
15. Johnson BD, Schumacher RJ, Ross ED, Toft DO (1998) Hop modulates Hsp70/Hsp90 interactions in protein folding. J Biol Chem 273(6):3679–3686
16. Tsai J, Douglas MG (1996) A conserved HPD sequence of the J-domain is necessary for YDJ1 stimulation of Hsp70 ATPase activity at a site distinct from substrate binding. J Biol Chem 271(16):9347–9354
17. Cyr DM (1995) Cooperation of the molecular chaperone Ydj1 with specific Hsp70 homologs to suppress protein aggregation. FEBS Lett 359(2–3):129–132
18. Li J, Qian X, Sha B (2003) The crystal structure of the yeast Hsp40 Ydj1 complexed with its peptide substrate. Structure 11(12):1475–1483
19. Wall D, Zylicz M, Georgopoulos C (1995) The conserved G/F motif of the DnaJ chaperone is necessary for the activation of the substrate binding properties of the DnaK chaperone. J Biol Chem 270(5):2139–2144
20. Kampinga HH, Craig EA (2010) The HSP70 chaperone machinery: J proteins as drivers of functional specificity. Nat Rev Mol Cell Biol 11 (8):579–592. doi:nrm2941 [pii], 10.1038/nrm2941
21. Johnson JL, Craig EA (2001) An essential role for the substrate-binding region of Hsp40 s in *Saccharomyces cerevisiae*. J Cell Biol 152(4):851–856
22. Liu Q, Hendrickson WA (2007) Insights into Hsp70 chaperone activity from a crystal structure of the yeast Hsp110 Sse1. Cell 131(1):106–120. doi:10.1016/j.cell.2007.08.039
23. Policr S, Dragovic Z, Hartl FU, Bracher A (2008) Structural basis for the cooperation of Hsp70 and Hsp110 chaperones in protein folding. Cell 133(6):1068–1079. doi:10.1016/j.cell.2008.05.022
24. Schuermann JP, Jiang J, Cuellar J, Llorca O, Wang L, Gimenez LE, Jin S, Taylor AB, Demeler B, Morano KA, Hart PJ, Valpuesta JM, Lafer EM, Sousa R (2008) Structure of the

Hsp110:Hsc70 nucleotide exchange machine. Mol Cell 31(2):232–243. doi:10.1016/j.molcel.2008.05.006
25. Trott A, Shaner L, Morano KA (2005) The molecular chaperone Sse1 and the growth control protein kinase Sch9 collaborate to regulate protein kinase A activity in *Saccharomyces cerevisiae*. Genetics 170(3):1009–1021. doi:10.1534/genetics.105.043109
26. Makhnevych T, Wong P, Pogoutse O, Vizeacoumar FJ, Greenblatt JF, Emili A, Houry WA (2012) Hsp110 is required for spindle length control. J Cell Biol 198(4):623–636. doi:10.1083/jcb.201111105
27. Shaner L, Gibney PA, Morano KA (2008) The Hsp110 protein chaperone Sse1 is required for yeast cell wall integrity and morphogenesis. Curr Genet 54(1):1–11. doi:10.1007/s00294-008-0193-y
28. Oh HJ, Easton D, Murawski M, Kaneko Y, Subjeck JR (1999) The chaperoning activity of hsp110. Identification of functional domains by use of targeted deletions. J Biol Chem 274(22):15712–15718
29. Park J, Easton DP, Chen X, MacDonald IJ, Wang XY, Subjeck JR (2003) The chaperoning properties of mouse grp170, a member of the third family of hsp70 related proteins. BioChemistry 42(50):14893–14902. doi:10.1021/bi030122e
30. Anttonen AK, Mahjneh I, Hamalainen RH, Lagier-Tourenne C, Kopra O, Waris L, Anttonen M, Joensuu T, Kalimo H, Paetau A, Tranebjaerg L, Chaigne D, Koenig M, Eeg-Olofsson O, Udd B, Somer M, Somer H, Lehesjoki AE (2005) The gene disrupted in Marinesco-Sjogren syndrome encodes SIL1, an HSPA5 cochaperone. Nat Genet 37(12):1309–1311
31. Goeckeler JL, Petruso AP, Aguirre J, Clement CC, Chiosis G, Brodsky JL (2008) The yeast Hsp110, Sse1p, exhibits high-affinity peptide binding. FEBS Lett 582(16):2393–2396. doi:S0014-5793(08)00477-8 [pii], 10.1016/j.febslet.2008.05.047
32. Polier S, Hartl FU, Bracher A (2010) Interaction of the Hsp110 molecular chaperones from S. cerevisiae with substrate protein. J Mol Biol 401(5):696–707. doi:10.1016/j.jmb.2010.07.004
33. Xu X, Sarbeng EB, Vorvis C, Kumar DP, Zhou L, Liu Q (2012) Unique peptide substrate binding properties of 110-kDa heat-shock protein (Hsp110) determine its distinct chaperone activity. J Biol Chem 287(8):5661–5672. doi:10.1074/jbc.M111.275057
34. Pechmann S, Willmund F, Frydman J (2013) The ribosome as a hub for protein quality control. Mol Cell 49(3):411–421. doi:10.1016/j.molcel.2013.01.020
35. Duttler S, Pechmann S, Frydman J (2013) Principles of cotranslational ubiquitination and quality control at the ribosome. Mol Cell 50(3):379–393. doi:10.1016/j.molcel.2013.03.010
36. Deuerling E, Schulze-Specking A, Tomoyasu T, Mogk A, Bukau B (1999) Trigger factor and DnaK cooperate in folding of newly synthesized proteins. Nature 400(6745):693–696. doi:10.1038/23301
37. Teter SA, Houry WA, Ang D, Tradler T, Rockabrand D, Fischer G, Blum P, Georgopoulos C, Hartl FU (1999) Polypeptide flux through bacterial Hsp70: DnaK cooperates with trigger factor in chaperoning nascent chains. Cell 97(6):755–765
38. Deuerling E, Patzelt H, Vorderwulbecke S, Rauch T, Kramer G, Schaffitzel E, Mogk A, Schulze-Specking A, Langen H, Bukau B (2003) Trigger factor and DnaK possess overlapping substrate pools and binding specificities. Mol Microbiol 47(5):1317–1328
39. Rutkowska A, Mayer MP, Hoffmann A, Merz F, Zachmann-Brand B, Schaffitzel C, Ban N, Deuerling E, Bukau B (2008) Dynamics of trigger factor interaction with translating ribosomes. J Biol Chem 283(7):4124–4132. doi:10.1074/jbc.M708294200
40. Preissler S, Deuerling E (2012) Ribosome-associated chaperones as key players in proteostasis. Trends Biochem Sci 37(7):274–283. doi:10.1016/j.tibs.2012.03.002
41. Huang P, Gautschi M, Walter W, Rospert S, Craig EA (2005) The Hsp70 Ssz1 modulates the function of the ribosome-associated J-protein Zuo1. Nat Struct Mol Biol 12(6):497–504
42. Yam AY, Albanese V, Lin HT, Frydman J (2005) Hsp110 cooperates with different cytosolic HSP70 systems in a pathway for de novo folding. J Biol Chem 280(50):41252–41261. doi:10.1074/jbc.M503615200
43. Hundley HA, Walter W, Bairstow S, Craig EA (2005) Human Mpp11 J protein: ribosome-tethered molecular chaperones are ubiquitous. Science 308(5724):1032–1034

44. Jaiswal H, Conz C, Otto H, Wolfle T, Fitzke E, Mayer MP, Rospert S (2011) The chaperone network connected to human ribosome-associated complex. Mol Cell Biol 31(6):1160–1173. doi:10.1128/MCB.00986-10
45. Broadley SA, Hartl FU (2009) The role of molecular chaperones in human misfolding diseases. FEBS Lett 583(16):2647–2653. doi:10.1016/j.febslet.2009.04.029
46. Pfund C, Lopez-Hoyo N, Ziegelhoffer T, Schilke BA, Lopez-Buesa P, Walter WA, Wiedmann M, Craig EA (1998) The molecular chaperone Ssb from *Saccharomyces cerevisiae* is a component of the ribosome-nascent chain complex. EMBO J 17(14):3981–3989. doi:10.1093/emboj/17.14.3981
47. James P, Pfund C, Craig EA (1997) Functional specificity among Hsp70 molecular chaperones. Science 275(5298):387–389
48. Peisker K, Chiabudini M, Rospert S (2010) The ribosome-bound Hsp70 homolog Ssb of *Saccharomyces cerevisiae*. Biochim Biophys Acta 1803(6):662–672. doi:10.1016/j.bbamcr.2010.03.005
49. Leidig C, Bange G, Kopp J, Amlacher S, Aravind A, Wickles S, Witte G, Hurt E, Beckmann R, Sinning I (2013) Structural characterization of a eukaryotic chaperone-the ribosome-associated complex. Nat Struct Mol Biol 20(1):23–28. doi:10.1038/nsmb.2447
50. Fiaux J, Horst J, Scior A, Preissler S, Koplin A, Bukau B, Deuerling E (2010) Structural analysis of the ribosome-associated complex (RAC) reveals an unusual Hsp70/Hsp40 interaction. J Biol Chem 285(5):3227–3234. doi:M109.075804 [pii], 10.1074/jbc.M109.075804
51. Calloni G, Chen T, Schermann SM, Chang HC, Genevaux P, Agostini F, Tartaglia GG, Hayer-Hartl M, Hartl FU (2012) DnaK functions as a central hub in the E. coli chaperone network. Cell Rep 1 (3):251–264. doi:10.1016/j.celrep. 2011.12.007
52. Yan W, Schilke B, Pfund C, Walter W, Kim S, Craig EA (1998) Zuotin, a ribosome-associated DnaJ molecular chaperone. EMBO J 17(16):4809–4817. doi:10.1093/emboj/17.16.4809
53. Peisker K, Braun D, Wolfle T, Hentschel J, Funfschilling U, Fischer G, Sickmann A, Rospert S (2008) Ribosome-associated complex binds to ribosomes in close proximity of Rpl31 at the exit of the polypeptide tunnel in yeast. Mol Biol Cell 19(12):5279–5288. doi:10.1091/mbc.E08-06-0661
54. Gong Y, Kakihara Y, Krogan N, Greenblatt J, Emili A, Zhang Z, Houry WA (2009) An atlas of chaperone-protein interactions in *Saccharomyces cerevisiae*: implications to protein folding pathways in the cell. Mol Syst Biol 5:275. doi:msb200926 [pii], 10.1038/msb.2009.26
55. Willmund F, del Alamo M, Pechmann S, Chen T, Albanese V, Dammer EB, Peng J, Frydman J (2013) The cotranslational function of ribosome-associated Hsp70 in eukaryotic protein homeostasis. Cell 152(1-2):196–209. doi:10.1016/j.cell.2012.12.001
56. Otto H, Conz C, Maier P, Wolfle T, Suzuki CK, Jeno P, Rucknagel P, Stahl J, Rospert S (2005) The chaperones MPP11 and Hsp70L1 form the mammalian ribosome-associated complex. Proc Natl Acad Sci U S A 102(29):10064–10069. doi:10.1073/pnas.0504400102
57. Shaner L, Sousa R, Morano KA (2006) Characterization of Hsp70 binding and nucleotide exchange by the yeast Hsp110 chaperone Sse1. BioChemistry 45(50):15075–15084. doi:10.1021/bi061279k
58. Oh HJ, Chen X, Subjeck JR (1997) Hsp110 protects heat-denatured proteins and confers cellular thermoresistance. J Biol Chem 272(50):31636–31640
59. Verghese J, Morano KA (2012) A lysine-rich region within fungal BAG domain-containing proteins mediates a novel association with ribosomes. Eukaryot Cell 11(8):1003–1011. doi:10.1128/EC.00146-12
60. Kurzchalia TV, Wiedmann M, Girshovich AS, Bochkareva ES, Bielka H, Rapoport TA (1986) The signal sequence of nascent preprolactin interacts with the 54 K polypeptide of the signal recognition particle. Nature 320(6063):634–636. doi:10.1038/320634a0
61. Halic M, Becker T, Pool MR, Spahn CM, Grassucci RA, Frank J, Beckmann R (2004) Structure of the signal recognition particle interacting with the elongation-arrested ribosome. Nature 427(6977):808–814. doi:10.1038/nature02342

62. Walter P, Blobel G (1981) Translocation of proteins across the endoplasmic reticulum III. Signal recognition protein (SRP) causes signal sequence-dependent and site-specific arrest of chain elongation that is released by microsomal membranes. J Cell Biol 91(2 Pt 1):557–561
63. Becker J, Walter W, Yan W, Craig EA (1996) Functional interaction of cytosolic hsp70 and a DnaJ-related protein, Ydj1p, in protein translocation in vivo. Mol Cell Biol 16(8):4378–4386
64. Karbstein K (2010) Chaperoning ribosome assembly. J Cell Biol 189(1):11–12. doi:10.1083/jcb.201002102
65. Albanese V, Reissmann S, Frydman J (2010) A ribosome-anchored chaperone network that facilitates eukaryotic ribosome biogenesis. J Cell Biol 189(1):69–81. doi:10.1083/jcb.201001054
66. Meyer AE, Hung NJ, Yang P, Johnson AW, Craig EA (2007) The specialized cytosolic J-protein, Jjj1, functions in 60 S ribosomal subunit biogenesis. Proc Natl Acad Sci U S A 104(5):1558–1563. doi:10.1073/pnas.0610704104
67. Meyer AE, Hoover LA, Craig EA (2010) The cytosolic J-protein, Jjj1, and Rei1 function in the removal of the pre-60 S subunit factor Arx1. J Biol Chem 285(2):961–968. doi:10.1074/jbc.M109.038349
68. Herrmann JM, Stuart RA, Craig EA, Neupert W (1994) Mitochondrial heat shock protein 70, a molecular chaperone for proteins encoded by mitochondrial DNA. J Cell Biol 127(4):893–902
69. Maki JA, Southworth DR, Culver GM (2003) Demonstration of the role of the DnaK chaperone system in assembly of 30 S ribosomal subunits using a purified in vitro system. RNA 9(12):1418–1421
70. Maki JA, Schnobrich DJ, Culver GM (2002) The DnaK chaperone system facilitates 30S ribosomal subunit assembly. Mol Cell 10(1):129–138
71. Alix JH, Guerin MF (1993) Mutant DnaK chaperones cause ribosome assembly defects in *Escherichia coli*. Proc Natl Acad Sci U S A 90(20):9725–9729
72. Martinez-Hackert E, Hendrickson WA (2009) Promiscuous substrate recognition in folding and assembly activities of the trigger factor chaperone. Cell 138(5):923–934. doi:10.1016/j.cell.2009.07.044
73. Meunier L, Usherwood YK, Chung KT, Hendershot LM (2002) A subset of chaperones and folding enzymes form multiprotein complexes in endoplasmic reticulum to bind nascent proteins. Mol Biol Cell 13(12):4456–4469. doi:10.1091/mbc.E02-05-0311
74. Brodsky JL, Werner ED, Dubas ME, Goeckeler JL, Kruse KB, McCracken AA (1999) The requirement for molecular chaperones during endoplasmic reticulum-associated protein degradation demonstrates that protein export and import are mechanistically distinct. J Biol Chem 274(6):3453–3460
75. Steel GJ, Fullerton DM, Tyson JR, Stirling CJ (2004) Coordinated activation of Hsp70 chaperones. Science 303(5654):98–101. doi:10.1126/science.1092287
76. Weitzmann A, Baldes C, Dudek J, Zimmermann R (2007) The heat shock protein 70 molecular chaperone network in the pancreatic endoplasmic reticulum—a quantitative approach. FEBS J 274(19):5175–5187. doi:10.1111/j.1742-4658.2007.06039.x
77. Hale SJ, Lovell SC, de Keyzer J, Stirling CJ (2010) Interactions between Kar2p and its nucleotide exchange factors Sil1p and Lhs1p are mechanistically distinct. J Biol Chem 285(28):21600–21606. doi:10.1074/jbc.M110.111211
78. Simons JF, Ferro-Novick S, Rose MD, Helenius A (1995) BiP/Kar2p serves as a molecular chaperone during carboxypeptidase Y folding in yeast. J Cell Biol 130(1):41–49
79. Lin HY, Masso-Welch P, Di YP, Cai JW, Shen JW, Subjeck JR (1993) The 170-kDa glucose-regulated stress protein is an endoplasmic reticulum protein that binds immunoglobulin. Mol Biol Cell 4(11):1109–1119
80. Craven RA, Egerton M, Stirling CJ (1996) A novel Hsp70 of the yeast ER lumen is required for the efficient translocation of a number of protein precursors. Embo J 15(11):2640–2650
81. Senderek J, Krieger M, Stendel C, Bergmann C, Moser M, Breitbach-Faller N, Rudnik-Schoneborn S, Blaschek A, Wolf NI, Harting I, North K, Smith J, Muntoni F, Brockington M, Quijano-Roy S, Renault F, Herrmann R, Hendershot LM, Schroder JM, Lochmuller

H, Topaloglu H, Voit T, Weis J, Ebinger F, Zerres K (2005) Mutations in SIL1 cause Marinesco-Sjogren syndrome, a cerebellar ataxia with cataract and myopathy. Nat Genet 37(12):1312–1314. doi:10.1038/ng1678

82. Tyson JR, Stirling CJ (2000) LHS1 and SIL1 provide a lumenal function that is essential for protein translocation into the endoplasmic reticulum. Embo J 19(23):6440–6452. doi:10.1093/emboj/19.23.6440
83. Ushioda R, Hoseki J, Araki K, Jansen G, Thomas DY, Nagata K (2008) ERdj5 is required as a disulfide reductase for degradation of misfolded proteins in the ER. Science 321(5888):569–572. doi:10.1126/science.1159293
84. Hagiwara M, Maegawa K, Suzuki M, Ushioda R, Araki K, Matsumoto Y, Hoseki J, Nagata K, Inaba K (2011) Structural basis of an ERAD pathway mediated by the ER-resident protein disulfide reductase ERdj5. Mol Cell 41(4):432–444. doi:10.1016/j.molcel.2011.01.021
85. Mizzen LA, Chang C, Garrels JI, Welch WJ (1989) Identification, characterization, and purification of two mammalian stress proteins present in mitochondria, grp 75, a member of the hsp 70 family and hsp 58, a homolog of the bacterial groEL protein. J Biol Chem 264(34):20664–20675
86. Bhattacharyya T, Karnezis AN, Murphy SP, Hoang T, Freeman BC, Phillips B, Morimoto RI (1995) Cloning and subcellular localization of human mitochondrial hsp70. J Biol Chem 270(4):1705–1710
87. Syken J, De-Medina T, Munger K (1999) TID1, a human homolog of the Drosophila tumor suppressor l(2)tid, encodes two mitochondrial modulators of apoptosis with opposing functions. Proc Natl Acad Sci U S A 96(15):8499–8504
88. Voos W, Rottgers K (2002) Molecular chaperones as essential mediators of mitochondrial biogenesis. Biochim Biophys Acta 1592(1):51–62
89. Voos W, Gambill BD, Laloraya S, Ang D, Craig EA, Pfanner N (1994) Mitochondrial GrpE is present in a complex with hsp70 and preproteins in transit across membranes. Mol Cell Biol 14(10):6627–6634
90. Andrew AJ, Dutkiewicz R, Knieszner H, Craig EA, Marszalek J (2006) Characterization of the interaction between the J-protein Jac1p and the scaffold for Fe-S cluster biogenesis, Isu1p. J Biol Chem 281(21):14580–14587. doi:10.1074/jbc.M600842200
91. Dutkiewicz R, Schilke B, Knieszner H, Walter W, Craig EA, Marszalek J (2003) Ssq1, a mitochondrial Hsp70 involved in iron-sulfur (Fe/S) center biogenesis. Similarities to and differences from its bacterial counterpart. J Biol Chem 278(32):29719–29727
92. Uzarska MA, Dutkiewicz R, Freibert SA, Lill R, Muhlenhoff U (2013) The mitochondrial Hsp70 chaperone Ssq1 facilitates Fe/S cluster transfer from Isu1 to Grx5 by complex formation. Mol Biol Cell 24(12):1830–1841. doi:10.1091/mbc.E12-09-0644
93. Bonomi F, Iametti S, Morleo A, Ta D, Vickery LE (2008) Studies on the mechanism of catalysis of iron-sulfur cluster transfer from IscU[2Fe2S] by HscA/HscB chaperones. BioChemistry 47(48):12795–12801. doi:10.1021/bi801565j
94. Chen S, Smith DF (1998) Hop as an adaptor in the heat shock protein 70 (Hsp70) and hsp90 chaperone machinery. J Biol Chem 273(52):35194–35200
95. Smith DF (2004) Tetratricopeptide repeat cochaperones in steroid receptor complexes. Cell Stress Chaperones 9(2):109–121
96. Siligardi G, Panaretou B, Meyer P, Singh S, Woolfson DN, Piper PW, Pearl LH, Prodromou C (2002) Regulation of Hsp90 ATPase activity by the co-chaperone Cdc37p/p50cdc37. J Biol Chem 277(23):20151–20159. doi:10.1074/jbc.M201287200
97. Heldens L, Hensen SM, Onnekink C, van Genesen ST, Dirks RP, Lubsen NH (2011) An atypical unfolded protein response in heat shocked cells. PLoS ONE 6(8):e23512. doi:10.1371/journal.pone.0023512
98. Hodson S, Marshall JJ, Burston SG (2012) Mapping the road to recovery: the ClpB/Hsp104 molecular chaperone. J Struct Biol 179(2):161–171. doi:10.1016/j.jsb.2012.05.015
99. Werner-Washburne M, Becker J, Kosic-Smithers J, Craig EA (1989) Yeast Hsp70 RNA levels vary in response to the physiological status of the cell. J Bacteriol 171(5):2680–2688

100. Boorstein WR, Craig EA (1990) Structure and regulation of the SSA4 HSP70 gene of *Saccharomyces cerevisiae*. J Biol Chem 265(31):18912–18921
101. Lu Z, Cyr DM (1998) Protein folding activity of Hsp70 is modified differentially by the hsp40 co-chaperones Sis1 and Ydj1. J Biol Chem 273(43):27824–27830
102. Winkler J, Tyedmers J, Bukau B, Mogk A (2012) Hsp70 targets Hsp100 chaperones to substrates for protein disaggregation and prion fragmentation. J Cell Biol 198(3):387–404. doi:10.1083/jcb.201201074
103. Schmitt M, Neupert W, Langer T (1996) The molecular chaperone Hsp78 confers compartment-specific thermotolerance to mitochondria. J Cell Biol 134(6):1375–1386
104. Lum R, Tkach JM, Vierling E, Glover JR (2004) Evidence for an unfolding/threading mechanism for protein disaggregation by *Saccharomyces cerevisiae* Hsp104. J Biol Chem 279(28):29139–29146. doi:10.1074/jbc.M403777200
105. Shorter J (2011) The mammalian disaggregase machinery: Hsp110 synergizes with Hsp70 and Hsp40 to catalyze protein disaggregation and reactivation in a cell-free system. PLoS ONE 6(10):e26319. doi:10.1371/journal.pone.0026319
106. Lee J, Kim JH, Biter AB, Sielaff B, Lee S, Tsai FT (2013) Heat shock protein (Hsp) 70 is an activator of the Hsp104 motor. Proc Natl Acad Sci U S A 110(21):8513–8518. doi:10.1073/pnas.1217988110
107. Haslberger T, Weibezahn J, Zahn R, Lee S, Tsai FT, Bukau B, Mogk A (2007) M domains couple the ClpB threading motor with the DnaK chaperone activity. Mol Cell 25(2):247–260. doi:10.1016/j.molcel.2006.11.008
108. Desantis ME, Shorter J (2012) The elusive middle domain of Hsp104 and ClpB: location and function. Biochim Biophys Acta 1823(1):29–39. doi:10.1016/j.bbamcr.2011.07.014
109. Miot M, Reidy M, Doyle SM, Hoskins JR, Johnston DM, Genest O, Vitery MC, Masison DC, Wickner S (2011) Species-specific collaboration of heat shock proteins (Hsp) 70 and 100 in thermotolerance and protein disaggregation. Proc Natl Acad Sci U S A 108(17):6915–6920. doi:10.1073/pnas.1102828108
110. Seyffer F, Kummer E, Oguchi Y, Winkler J, Kumar M, Zahn R, Sourjik V, Bukau B, Mogk A (2012) Hsp70 proteins bind Hsp100 regulatory M domains to activate AAA+ disaggregase at aggregate surfaces. Nat Struct Mol Biol 19(12):1347–1355. doi:10.1038/nsmb.2442
111. Rosenzweig R, Moradi S, Zarrine-Afsar A, Glover JR, Kay LE (2013) Unraveling the mechanism of protein disaggregation through a ClpB-DnaK interaction. Science 339(6123):1080–1083. doi:10.1126/science.1233066
112. Zietkiewicz S, Krzewska J, Liberek K (2004) Successive and synergistic action of the Hsp70 and Hsp100 chaperones in protein disaggregation. J Biol Chem 279(43):44376–44383. doi:10.1074/jbc.M402405200
113. Rampelt H, Kirstein-Miles J, Nillegoda NB, Chi K, Scholz SR, Morimoto RI, Bukau B (2012) Metazoan Hsp70 machines use Hsp110 to power protein disaggregation. EMBO J 31(21):4221–4235. doi:10.1038/emboj.2012.264
114. Jakob U, Gaestel M, Engel K, Buchner J (1993) Small heat shock proteins are molecular chaperones. J Biol Chem 268(3):1517–1520
115. Haslbeck M, Braun N, Stromer T, Richter B, Model N, Weinkauf S, Buchner J (2004) Hsp42 is the general small heat shock protein in the cytosol of *Saccharomyces cerevisiae*. EMBO J 23(3):638–649. doi:10.1038/sj.emboj.7600080, 7600080 [pii]
116. Cashikar AG, Duennwald M, Lindquist SL (2005) A chaperone pathway in protein disaggregation. Hsp26 alters the nature of protein aggregates to facilitate reactivation by Hsp104. J Biol Chem 280 (25):23869–23875. doi:M502854200 [pii], 10.1074/jbc.M502854200
117. Specht S, Miller SB, Mogk A, Bukau B (2011) Hsp42 is required for sequestration of protein aggregates into deposition sites in *Saccharomyces cerevisiae*. J Cell Biol 195(4):617–629. doi:10.1083/jcb.201106037
118. Lee GJ, Vierling E (2000) A small heat shock protein cooperates with heat shock protein 70 systems to reactivate a heat-denatured protein. Plant Physiol 122(1):189–198

119. Garrido C, Paul C, Seigneuric R, Kampinga HH (2012) The small heat shock proteins family: the long forgotten chaperones. Int J Biochem Cell Biol 44(10):1588–1592. doi:10.1016/j.biocel.2012.02.022
120. Mogk A, Schlieker C, Friedrich KL, Schonfeld HJ, Vierling E, Bukau B (2003) Refolding of substrates bound to small Hsps relies on a disaggregation reaction mediated most efficiently by ClpB/DnaK. J Biol Chem 278(33):31033–31042. doi:10.1074/jbc.M303587200
121. Goldbaum O, Riedel M, Stahnke T, Richter-Landsberg C (2009) The small heat shock protein HSP25 protects astrocytes against stress induced by proteasomal inhibition. Glia 57(14):1566–1577. doi:10.1002/glia.20870
122. Shiber A, Breuer W, Brandeis M, Ravid T (2013) Ubiquitin conjugation triggers misfolded protein sequestration into quality control foci when Hsp70 chaperone levels are limiting. Mol Biol Cell 24(13):2076–2087. doi:10.1091/mbc.E13-01-0010
123. Malinovska L, Kroschwald S, Munder MC, Richter D, Alberti S (2012) Molecular chaperones and stress-inducible protein-sorting factors coordinate the spatiotemporal distribution of protein aggregates. Mol Biol Cell 23(16):3041–3056. doi:10.1091/mbc.E12-03-0194
124. Dokladny K, Zuhl MN, Mandell M, Bhattacharya D, Schneider S, Deretic V, Moseley PL (2013) Regulatory coordination between two major intracellular homeostatic systems: heat shock response and autophagy. J Biol Chem 288(21):14959–14972. doi:10.1074/jbc.M113.462408
125. Arndt V, Rogon C, Hohfeld J (2007) To be, or not to be–molecular chaperones in protein degradation. Cell Mol Life Sci 64(19-20):2525–2541. doi:10.1007/s00018-007-7188-6
126. Kravtsova-Ivantsiv Y, Ciechanover A (2012) Non-canonical ubiquitin-based signals for proteasomal degradation. J Cell Sci 125(Pt 3):539–548. doi:10.1242/jcs.093567
127. Voges D, Zwickl P, Baumeister W (1999) The 26 S proteasome: a molecular machine designed for controlled proteolysis. Annu Rev Biochem 68:1015–1068. doi:10.1146/annurev.biochem.68:1.1015
128. Guerriero CJ, Weiberth KF, Brodsky JL (2013) Hsp70 targets a cytoplasmic quality control substrate to the San1p ubiquitin ligase. J Biol Chem 288(25):18506–18520. doi:10.1074/jbc.M113.475905
129. Ballinger CA, Connell P, Wu Y, Hu Z, Thompson LJ, Yin LY, Patterson C (1999) Identification of CHIP, a novel tetratricopeptide repeat-containing protein that interacts with heat shock proteins and negatively regulates chaperone functions. Mol Cell Biol 19(6):4535–4545
130. Buchberger A, Bukau B, Sommer T (2010) Protein quality control in the cytosol and the endoplasmic reticulum: brothers in arms. Mol Cell 40(2):238–252. doi:10.1016/j.molcel.2010.10.001
131. Murata S, Minami Y, Minami M, Chiba T, Tanaka K (2001) CHIP is a chaperone-dependent E3 ligase that ubiquitylates unfolded protein. EMBO Rep 2(12):1133–1138. doi:10.1093/embo-reports/kve246
132. Meacham GC, Patterson C, Zhang W, Younger JM, Cyr DM (2001) The Hsc70 co-chaperone CHIP targets immature CFTR for proteasomal degradation. Nat Cell Biol 3(1):100–105. doi:10.1038/35050509
133. Demand J, Alberti S, Patterson C, Hohfeld J (2001) Cooperation of a ubiquitin domain protein and an E3 ubiquitin ligase during chaperone/proteasome coupling. Curr Biol 11(20):1569–1577
134. Stankiewicz M, Nikolay R, Rybin V, Mayer MP (2010) CHIP participates in protein triage decisions by preferentially ubiquitinating Hsp70-bound substrates. FEBS J 277(16):3353–3367. doi:10.1111/j.1742-4658.2010.07737.x
135. Westhoff B, Chapple JP, van der Spuy J, Hohfeld J, Cheetham ME (2005) HSJ1 is a neuronal shuttling factor for the sorting of chaperone clients to the proteasome. Curr Biol 15(11):1058–1064. doi:10.1016/j.cub.2005.04.058
136. Park SH, Kukushkin Y, Gupta R, Chen T, Konagai A, Hipp MS, Hayer-Hartl M, Hartl FU (2013) PolyQ proteins interfere with nuclear degradation of cytosolic proteins by sequestering the Sis1p chaperone. Cell 154(1):134–145. doi:10.1016/j.cell.2013.06.003

137. Luders J, Demand J, Hohfeld J (2000) The ubiquitin-related BAG-1 provides a link between the molecular chaperones Hsc70/Hsp70 and the proteasome. J Biol Chem 275(7):4613–4617
138. Kikukawa Y, Minami R, Shimada M, Kobayashi M, Tanaka K, Yokosawa H, Kawahara H (2005) Unique proteasome subunit Xrpn10c is a specific receptor for the antiapoptotic ubiquitin-like protein Scythe. FEBS J 272(24):6373–6386. doi:10.1111/j.1742-4658.2005.05032.x
139. Gowda NK, Kandasamy G, Froehlich MS, Dohmen RJ, Andreasson C (2013) Hsp70 nucleotide exchange factor Fes1 is essential for ubiquitin-dependent degradation of misfolded cytosolic proteins. Proc Natl Acad Sci U S A 110(15):5975–5980. doi:10.1073/pnas.1216778110
140. Arndt V, Daniel C, Nastainczyk W, Alberti S, Hohfeld J (2005) BAG-2 acts as an inhibitor of the chaperone-associated ubiquitin ligase CHIP. Mol Biol Cell 16(12):5891–5900. doi:10.1091/mbc.E05-07-0660
141. Gebauer M, Zeiner M, Gehring U (1997) Proteins interacting with the molecular chaperone hsp70/hsc70: physical associations and effects on refolding activity. FEBS Lett 417(1):109–113
142. Qian SB, McDonough H, Boellmann F, Cyr DM, Patterson C (2006) CHIP-mediated stress recovery by sequential ubiquitination of substrates and Hsp70. Nature 440(7083):551–555. doi:10.1038/nature04600
143. Alberti S, Demand J, Esser C, Emmerich N, Schild H, Hohfeld J (2002) Ubiquitylation of BAG-1 suggests a novel regulatory mechanism during the sorting of chaperone substrates to the proteasome. J Biol Chem 277(48):45920–45927. doi:10.1074/jbc.M204196200
144. Denic V, Quan EM, Weissman JS (2006) A luminal surveillance complex that selects misfolded glycoproteins for ER-associated degradation. Cell 126(2):349–359. doi:10.1016/j.cell.2006.05.045
145. Kabani M, Kelley SS, Morrow MW, Montgomery DL, Sivendran R, Rose MD, Gierasch LM, Brodsky JL (2003) Dependence of endoplasmic reticulum-associated degradation on the peptide binding domain and concentration of BiP. Mol Biol Cell 14(8):3437–3448. doi:10.1091/mbc.E02-12-0847
146. Soroka J, Wandinger SK, Mausbacher N, Schreiber T, Richter K, Daub H, Buchner J (2012) Conformational switching of the molecular chaperone Hsp90 via regulated phosphorylation. Mol Cell 45(4):517–528. doi:10.1016/j.molcel.2011.12.031
147. Truman AW, Kristjansdottir K, Wolfgeher D, Hasin N, Polier S, Zhang H, Perrett S, Prodromou C, Jones GW, Kron SJ (2012) CDK-dependent Hsp70 Phosphorylation controls G1 cyclin abundance and cell-cycle progression. Cell 151(6):1308–1318. doi:10.1016/j.cell.2012.10.051

Part IV
The Hsp90 Chaperone Network

Chapter 5
The Interaction Network of the Hsp90 Molecular Chaperone

Kamran Rizzolo, Philip Wong, Elisabeth R. M. Tillier and Walid A. Houry

Abstract Heat shock protein 90 (Hsp 90) is a highly abundant and critical molecular chaperone that plays key roles in cellular quality control systems. The Hsp90 mechanism of function has been the subject of extensive investigation by many groups using traditional biochemical approaches as well as high-throughput methods, however, the Hsp90 functional cycle still remains enigmatic. A complicating factor in understanding Hsp90 function is the presence of many cofactors and co-chaperones that assist in Hsp90 chaperoning activity. The widely used model organism, *Saccharomyces cerevisiae* (budding yeast), contains two highly conserved members of the Hsp90 family, Hsp82 and Hsc82, and has been used as a model organism for mapping Hsp90 interactors. High-throughput proteomic studies on the yeast Hsp90 provided a wealth of information from a global perspective. More recently, such studies in mammalian cells have led to a better understanding of Hsp90 function. Here, we discuss the Hsp90 functions and highlight the most recent efforts leading to the construction of the Hsp90 interaction networks.

1 The Hsp90 Chaperone

The heat shock protein 90 (Hsp 90) is a member of a widespread family of molecular chaperones found in both eukaryotes and prokaryotes. Hsp90s are highly abundant within the cell accounting for about 1–2 % of total protein [1, 2]. Bacterial Hsp90s are typically non-essential, however, a functional Hsp90 is required for cell viability in eukaryotes [3, 4]. In mammalian cells, Hsp90 is predominantly localized to the cytoplasm and the nucleus but is also present in several organelles such as the mitochondria and endoplasmic reticulum [5].

Hsp90 functions as a homo-dimeric protein, with each monomer consisting of three domains: the N-terminal domain (N-domain) involved in adenosine triphosphate (ATP) binding and hydrolysis, a middle domain involved in substrate binding

W. A. Houry (✉) · K. Rizzolo
Department of Biochemistry, University of Toronto, Toronto, ON, Canada
e-mail: walid.houry@utoronto.ca

P. Wong · E. R. M. Tillier
Department of Medical Biophysics, University of Toronto, Toronto, ON, Canada

W. A. Houry (ed.), *The Molecular Chaperones Interaction Networks in Protein Folding and Degradation,* Interactomics and Systems Biology, DOI 10.1007/978-1-4939-1130-1_5,

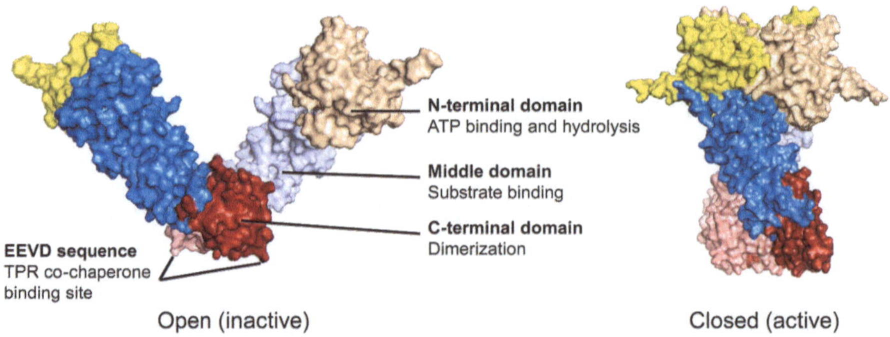

Fig. 5.1 Heat shock protein 90 (Hsp90) structure and conformational states. The Hsp90 homodimer with three domains in each monomer is shown: The N-terminal domain (N-domain in *yellow* and wheat) is the nucleotide binding domain (NBD) and is primarily responsible for adenosine triphosphate (ATP) binding and hydrolysis, the middle-domain (M-domain in *blue* and *sky blue*) has been implicated in client binding while the C-terminal domain (C-domain in *red* and salmon) is responsible for dimerization and contains the tetratricopeptide repeat (TPR) co-chaperone binding motif sequence EEVD at the very C-terminus. The structure on the *left* shows the inactive open form of *Escherichia coli* Hsp90 (PDB entry 2IOQ) [93]. The structure on the *right* shows the closed active form of *S. cerevisiae* Hsp90 (PDB entry 2CG9) [43].

(M-domain), and a C-terminal domain necessary for dimerization (Fig. 5.1) [6, 7]. Hsp90 has hundreds of substrate proteins, which are referred to as clients, belonging to different protein families but have no obvious sequence, structural, or functional similarities [8, 9]. It has been shown that yeast Hsp90 interacts genetically or physically with ~10% of the yeast proteome [8]. Also, it is speculated that Hsp90 in mammalian cells may assist in the folding and maturation of up to 10% of cytosolic proteins at some stage in their life cycle [2]. The majority of Hsp90 client proteins play essential roles in many biological processes and, therefore, Hsp90 is a master regulator of very diverse cellular activities including: signal transduction, vesicle transport, cell cycle regulation, telomere maintenance, transcription regulation, steroid signaling, immune response, viral infections, and cancer development [10, 11]. Hsp90 is currently considered as a validated anticancer drug target [12].

One key difference between Hsp90 and other chaperones is that Hsp90 can interact with a great number of co-chaperones and, in turn, these associations are critical for its activity [9]. The co-chaperones regulate Hsp90 ATPase activity, influence the chaperone's affinity for client proteins, or cause the chaperone to be targeted to a particular cellular pathway. Traditional biochemical and genetic approaches to characterize Hsp90 clients, co-chaperones, and other cofactors have been challenging primarily due to the low affinity of the chaperone-substrate interactions and the pleiotropic functions of this chaperone [1, 2, 13]. Biochemical and genetic studies suggest that Hsp90 is not a chaperone that interacts with unfolded or newly translated proteins, rather it seems to act on proteins that are significantly folded. Hsp90 most likely plays a role in remodeling protein conformation or in protein stabilization [6, 14].

The model organism *Saccharomyces cerevisiae* has served as an ideal system in providing a global cellular view of Hsp90 functions along with the discovery of novel interactors. There are two isoforms of Hsp90 in yeast: one that is constitutively expressed (Hsc82) and another one that is induced under heat shock conditions (Hsp82) [15]. The two proteins are 97% identical at the protein sequence level [3]. In this review, we provide an overview of the Hsp90 functions and discuss studies involving Hsp90 chaperone network characterization including methods used and possible future directions.

2 Hsp90 Co-chaperones

Given the breadth of Hsp90 involvement in various cellular pathways, an efficient control mechanism is needed to ensure proper function and avoid deleterious effects that may occur when the activity of this abundant chaperone is not regulated. The regulation of Hsp90 function is performed by numerous co-chaperones that have co-evolved with Hsp90 [16]. The repertoire of co-chaperones varies greatly between eukaryotes and may well be dependent on the presence of specific client proteins [9, 17]. More than 20 co-chaperones are known to be involved in affecting the ATP-dependent functional cycle of Hsp90 and the recruitment of client proteins (Fig. 5.2). By taking into account the effect on Hsp90 and its clients' maturation, co-chaperones can be divided into three groups: client recruiters, Hsp90 remodelers, and late-acting co-chaperones. These groups have overlapping members [9].

Many co-chaperones of Hsp90 contain the tetratricopeptide repeat (TPR) domains that bind the C-terminal Glu-Glu-Val-Asp (EEVD) sequence present in eukaryotic cytoplasmic Hsp90s [9]. These TPR domains are typically composed of tandem helical repeats of seven antiparallel α-helices that form a cleft capable of binding the EEVD C-terminal residues of Hsp90 and also Hsp70 [18]. Some important TPR domain-containing co-chaperones of Hsp90 are listed in Table 5.1. Heat shock organizing protein (Hop; yeast Sti1) has three TPR domains and forms a ternary complex with Hsp90 and Hsp70 [19, 20]. Protein phosphatase PP5 (yeast Ppt1) is involved in the maturation of the glucocorticoid receptor [21, 22]. Yeast Tah1 [8, 23] and mammalian RPAP3 [24] form a ternary complex with Hsp90 and another protein termed Pih1 involved in the assembly of box C/D small nucleolar ribonucleoprotein complexes (snoRNPs). Myosin-folding factor Unc45 (yeast She4) is involved in myosin fiber assembly [25, 26]. Members of the peptidylprolyl isomerase (PPIase) family Cyp40 (yeast Cpr6/Cpr7) [17, 27] are involved in the maturation of certain clients. Tom70 is involved in mitochondrial import [28]. Sgt1 is a subunit of the core kinetochore and the Skp1-Cul1-F-box (SCF) ubiquitin ligase complex [29]. However, it should be noted that unlike other TPR-domain-containing co-chaperones of Hsp90, Sgt1 binds Hsp90 through its p23-like cysteine and histidine-rich domain (CHORD) and Sgt1 (CS) domain and not through its TPR domain [30].

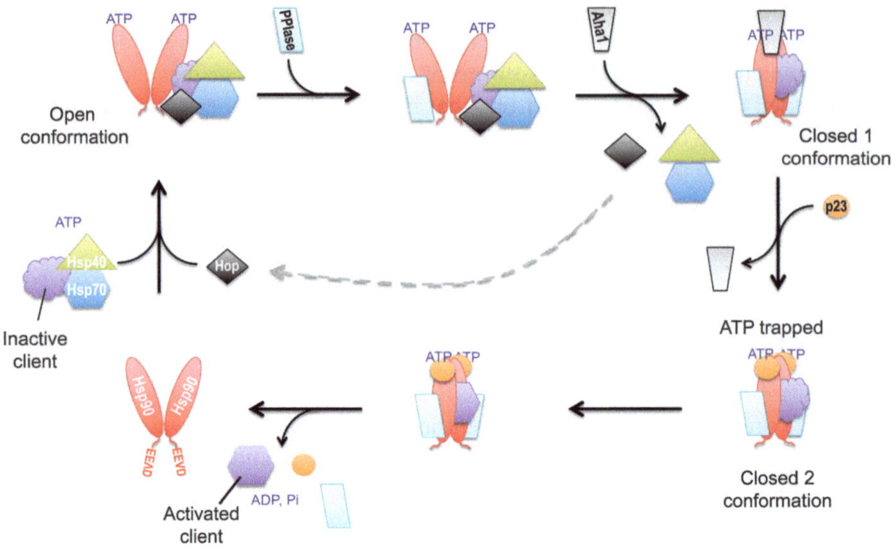

Fig. 5.2 Heat shock protein 90 (Hsp90) cycle and client maturation. Heat shock organizing protein (Hop) binds to the open state of Hsp90 together with the Hsp70-Hsp40-client protein complex. The adenosine triphosphate (ATP) hydrolysis-inhibited state may allow the binding of peptidyl-prolyl isomerase (PPIase) producing an asymmetric Hsp90 protein complex. One or two PPIases might bind the Hsp90 dimer. ATP binding to the complex induces a conformational change causing the transfer of the client from Hsp70 to Hsp90. Co-chaperone Aha1 binds Hsp90 and weakens the Hop-Hsp90 interaction causing the release of Hop from the chaperone. As a result, the Hsp70-Hsp40 complex is also released. Aha1 induces a conformational change in Hsp90 producing the closed state (closed 1). The resulting Hsp90-client complex is stabilized by the binding of the p23 co-chaperone and leads to the release of Aha1 (closed 2). Subsequently, ATP hydrolysis generates a conformational change that releases the mature client protein, the PPIase, and p23. The EEVD sequence at the C-terminus of Hsp90 is shown.

Other less-characterized TPR-domain-containing co-chaperones of Hsp90 (Table 5.1) include the essential yeast protein Cns1 whose activity is poorly understood but might overlap with that of Cpr7 [31]. Ttc4 has been postulated to link Hsp90 chaperone activity with deoxyribonucleic acid (DNA) replication [32]. The J-domain containing protein TPR2 is proposed to play a role in the Hsp90-dependent chaperoning of the progesterone receptor [33, 34].

There are three main co-chaperones of Hsp90 that do not contain a TPR domain (Table 5.1). Aha1 enhances the Hsp90 ATPase activity and acts to generally enhance Hsp90's chaperoning function [35–37]. Aha1 forms an intricate complex with Hsp90: the Aha1 N-terminal domain binds to Hsp90's middle domain while the C-terminal domain of Aha1 interacts with the N-terminal domain of the chaperone [36, 38–40]. The close paralog of Aha1, Hch1, also enhances Hsp90 ATPase activity. However, despite having more than 50% amino acid similarity with Aha1, Hch1 has been shown to regulate Hsp90 function in a way distinct from that of Aha1 and independent from Hch1's effect on Hsp90 ATPase activity [41]. p23 is an abundant co-chaperone of Hsp90. It forms co-complexes with the chaperone

Table 5.1 Hsp90 co-chaperones

Co-chaperones		Function
Yeast	Mammals	
TPR co-chaperones		
Sti1	Hop	Scaffold protein for Hsp90/Hsp70, involved in maturation of client proteins, inhibits Hsp90 ATPase
Ppt	PP5	Phosphatase
Tah1	RPAP3	Forms complex with Pih1 and Hsp90
She4	Unc45	Involved in Myosin fiber assembly
Cpr6/Cpr7	Cyp40	Peptidy-prolyl-isomerase, involved in maturation of client proteins
Tom70	Tom70	Involved in Mitochondrial protein import
Sgt1	Sgt1	Involved in kinetochore assembly
Cns1		Not well characterized, essential in yeast
	Ttc4	Acts as a link between Hsp90 and DNA replication
	TPR2	Involved in the maturation of the progesterone receptor
Non-TPR co-chaperones		
Aha1/Hch1	Aha1	Stimulates ATPase activity of Hsp90 and modulates the conformational changes of the chaperone
Sba1	p23	Inhibits Hsp90 ATPase activity, has independent chaperone activity
Cdc37	Cdc37	Inhibits Hsp90 ATPase activity, targets Hsp90 to kinases

TPR Tetratricopeptide, *ATP* Adenosine triphosphate, *Hop* Heat shock organizing protein, *Hsp* Heat shock protein, *DNA* Deoxyribonucleic acid

and client proteins, however, it might also function independently [42]. p23 binds the N-terminal domain of Hsp90 and stabilizes the closed ATP-bound state of the chaperone (Fig. 5.2) [43]. Finally, the cell division cycle 37 protein (Cdc37) acts as the co-chaperone of Hsp90 for kinase client proteins [44]. Cdc37 binds both the N-terminal and middle domains of the chaperone and restricts Hsp90 conformational mobility [45–47]. As a result, Cdc37 stabilizes the client-loading phase of the chaperone cycle prior to the dimerization of the Hsp90 N-terminal domains and inhibits the chaperone's ATPase [9] (Fig. 5.2).

The Hsp90 functional cycle is not fully elucidated and might not be strictly regulated by ATP binding and hydrolysis [48, 49]. It is well established that the Hsp90 cycle also involves the Hsp70 and Hsp40 chaperones (Fig. 5.2). Hsp70 has an N-terminal nucleotide-binding domain (NBD) followed by a substrate-binding domain (SBD) and a C-terminal region. Hsp40 has a J-domain that binds to the Hsp70 NBD and enhances the ATPase activity of Hsp70 [50]. In a simplistic model of the Hsp90 cycle (Fig. 5.2), substrates initially bound by Hsp70-Hsp40 are transferred to Hsp90 through the bridging action of Hop that interacts with both Hsp90 and Hsp70. There is a minimal complex consisting of Hsp40, Hsp70, and Hsp90 and including the co-chaperone Hop that has been shown to form with clients such as the hormone receptors [51, 52]. Subsequently, Aha1 and then p23 stabilize the closed ATP-bound and client-loaded state of Hsp90 resulting in the release of Hsp70-Hsp40. Upon ATP hydrolysis by Hsp90, the client protein is released from the chaperone in a mature folded state (Fig. 5.2). Certain PPIases also act during this cycle on the Hsp90-containing complexes (Fig. 5.2). It should be noted that Hsp90 has very weak ATPase activity that is regulated by the different cofactors.

3 Coevolution of Hsp90 and Its Co-chaperones

Coevolution of protein-protein interactions have been used to study the interaction of Hsp90 with its co-chaperones (Table 5.1). Coevolution analysis highlights important genetic factors responsible for the robustness of biological systems. Furthermore, NxN sequence comparisons of homologous proteins produce distance matrices that allow the construction of phylogenetic trees [53]. Alignment of different sequences from many homologs enables statistical inference of coevolution between residue positions within the same protein family or between protein families. When homologous residues are aligned, mutations at the homologous site that coincide with mutations at other homologous positions tend to be easier to detect.

Travers and Fares [54] investigated the presence of coevolving amino acid residues between Hsp70 and Hop, as well as, between Hsp90 and Hop. They explored 25 organisms for Hop-Hsp70 and 27 organisms for Hop-Hsp90 inter-protein coevolution. They then used the coevolution analysis using protein sequences (CAPS) method [55] to identify coevolving groups. Such a computational method can suggest related constraints over large evolutionary distances between proteins and is an alternate way to determine biological associations between them. For Hsp70, the authors found a total of 30 amino acid residues coevolving with Hop. The majority of these residues lie within the NBD or the C-terminal region (50 % of the total) with only four residues in the SBD. For Hsp90, 51 amino acid residues where observed to coevolve with Hop. The majority of the residues were located within the N-terminal and middle domains. They also found great conservation and overlap between the sites in Hop and the various domains in Hsp90 suggesting inter-domain communication as essential for Hsp90 function.

Furthermore, the Hsp90-Hsp70-Hop system was found to involve complex interactions at the intra-molecular level: A residue change in one domain was found to have effects on domains from other proteins even if they do not directly interact with each other. For example, the first TPR domain of Hop, which binds to Hsp70, was found to have four residues that coevolve with Hsp90, two of which also coevolve with Hsp70. These results demonstrate coevolution at the systems level in the Hsp90 network and further highlight the functional dependence between its components.

4 Posttranslational Modifications of Hsp90

Hsp90 activity is also regulated by posttranslational modifications such as phosphorylation, acetylation, nitrosylation, and SUMOylation (conjugation with small ubiquitin-like modifier) [9, 17, 56]. This confers an additional layer of modulation of Hsp90 activity by the extra- and intracellular environment in order to guarantee a fast and efficient response to specific cellular requirements [17, 57]. These modifications greatly affect Hsp90 ATPase activity, conformational changes, dimerization, co-chaperone binding, and, as a result, impact client maturation.

Hsp90 is a phosphoprotein whose levels of phosphorylation vary greatly depending on its cellular environment. CK2 is a serine-threonine acidophilic kinase whose activity is highly dependent on Hsp90 function. In yeast, residue T22 in the N-domain of Hsp90 is phosphorylated by CK2, which then modulates the ATPase activity of the chaperone. Mutations of this residue have been shown to affect Hsp90 chaperoning of kinase and non-kinase clients [58, 59]. These mutants significantly affected the interaction of Hsp90 with its co-chaperone Aha1. Interestingly, overexpression of Aha1 can compensate for the various chaperoning defects caused by mutating T22 in Hsp90 [58].

Yeast Swe1 (also known as Wee1) tyrosine kinase is an Hsp90 client that regulates G2/M cell cycle transition [57]. Phosphorylation of Hsp90 in S-phase by this enzyme causes the chaperone to be translocated from the nucleus to the cytoplasm where it is tagged for degradation by proteasomes as a way to eliminate this particular population of Hsp90 in the cell [60]. In addition, deletion of the *SWE1* gene causes increased sensitivity towards Hsp90 inhibiting drugs.

Acetylation of Hsp90 was also shown to regulate Hsp90 functions in client maturation and co-chaperone binding [61]. Nitrosylation (the attachment of nitric oxide, NO, to the thiol side chain of cysteine) of Hsp90 has been found to occur at the C-terminal domain residue C597 of Hsp90α in endothelial cells [62, 63]. This modification inhibits Hsp90 ATPase and chaperone activities. Hsp90 also undergoes asymmetric SUMOylation at a conserved lysine in the N-terminal domain (K178 in yeast Hsp82, K191 in human Hsp90). This modification facilitates the interaction of Aha1 with the chaperone, as well as, the binding of Hsp90 inhibitors [56].

In summary, these studies highlight the importance of various posttranslational modifications of Hsp90 on the ATPase activity, maturation of clients, co-chaperone interactions, conformational changes, and Hsp90 stability.

5 Examples of Hsp90 Clients

In order to understand the diversity in functions of the Hsp90 system, it is crucial to identify the chaperone's client proteins, and more importantly discover the link between them. Despite the increasing list of clients, very little is actually known about client recognition and binding sites. These clients are frequently intrinsically unstable proteins and bind to Hsp90 in a partially unfolded (or folded) state [9]. They are very diverse in structure and function suggesting that Hsp90 interacts with them using a general principal of recognition, probably modulated by the co-chaperones, as opposed to binding specifically to defined motifs.

The instability of Hsp90 clients makes Hsp90-client interaction and structural studies very difficult. Many in vivo and in vitro studies have approached this challenge by combining multiple biochemical and cell biological methods. For example,

Table 5.2 Examples of Hsp90 clients in yeast

Client protein	Description	Client class	Co-chaperones involved	Reference
Ste11	MEK kinase involved in pheromone response	Protein kinase	Cdc37	[66]
Gcn2	Stimulates translation of the yeast transcription factor Gcn4 upon amino acid starvation.	Transcription factor	Cdc37, Sti1, Sba1	[68]
Swe1	Protein kinase that regulates the G2/M transition	Protein kinase	Sba1, Cpr6/Cpr7	[71]
Slt2	Serine/threonine MAP kinase	Protein kinase		[72]
Hap1	Heme-responsive zinc finger transcription factor	Transcription factor	Hsp40 (Ydj1), Hsp70 (Ssa1–4)	[74]
Mal63	MAL activator, a DNA-binding transcription activator	Transcription factor	Cpr7	[76]

DNA Deoxyribonucleic acid, *Hsp* Heat shock protein

Genest et al. [64] developed a screen to identify *E. coli* mutants defective in Hsp90 function. They found that mutations in the M and C domains of Hsp90 affect chaperone activity and client binding. Interestingly, after making the same mutations in yeast Hsp90 (Hsp82), the authors found that mutant Hsp82 was less affected in client-binding capacity. This suggested that Hsp90 client-binding capabilities have diverged through evolution. Table 5.2 lists a few examples of Hsp90 clients that have been studied in some detail.

The first yeast Hsp90 client identified was Ste11. It is a MEK kinase involved in pheromone response and in pseudohyphal/invasive growth pathways in yeast [65]. The Hsp90 co-chaperone Cdc37 is required for Ste11 maturation as it is the co-chaperone that generally interacts with Hsp90 protein kinase clients (Tables 5.1 and 5.2) [66]. A comprehensive study on the folding of Ste11 indicated that, in addition to Cdc37, the Hsp90 co-chaperones Sti1, Cpr7, and Cns1 are also involved in Ste11 maturation [67].

The protein kinase Gcn2 (Table 5.2) phosphorylates the α-subunit of translation initiation factor eIF2 (Sui2) in response to starvation. Phosphorylated eIF2α then becomes an inhibitor of eIF2β thus decreasing the rate of translational initiation by interfering with the eIF2 recycling pathway. Hsp90 binds and modulates the function of Gcn2, which requires co-chaperones Cdc37, Sti1, and Sba1 (Tables 5.1 and 5.2) [68].

Yeast gene *SWE1* (also known as *WEE1*) encodes a protein kinase involved in regulating cell cycle transition from G2 to M phase as mentioned above [69]. Phosphorylated Swe1 inhibits the kinase activity of the main cell cycle cyclin-dependent kinase Cdc28 by phosphorylating the conserved residue Tyr19 [70]. When tested in the fission yeast strain *Schizosaccharomyces pombe*, overexpressed Swe1 triggered cell cycle arrest if Hsp90 was defective [71]. Swe1 was shown to co-immunoprecipitate with Hsp90 and the co-chaperones Sba1 and Cpr6/Cpr7 (Table 5.2).

The mitogen-activated protein kinase, MAP kinase, Slt2 (Table 5.2) was identified as a strong interactor of the yeast Hsp90 mutant, Hsp82 (E33A), in a two-hybrid screen [72]. Detailed analysis indicated that Hsp90 binds exclusively to the dually phosphorylated version Thr190/Tyr192 of the stress-activated form of Slt2. This was found when mutations in these two sites abolished interaction with Hsp90. Furthermore, in vivo analysis indicated that the Slt2 MAP kinase cascade requires the proper function of Hsp90.

Transcription factor Hap1 is another client of Hsp90 (Table 5.2). It modulates the expression of oxygen-dependent genes such as *CYC1* and *CYC7* in yeast [73]. The activity of Hap1 is regulated by heme binding. In the absence of heme, Hap1 forms a high-molecular-weight complex (HMC) that has low affinity for DNA. This HMC includes the molecular chaperones Hsp90, Hsp70, and Hsp40 [74]. Hsp70 plays a major role in Hap1 repression in the absence of heme while Hsp90 activates Hap1 when heme is present [75].

Mal63 is a transcription activator of genes involved in maltose metabolism including maltose permease and maltase. Yeast cells expressing a temperature-sensitive Hsp90 allele or lacking the co-chaperone Cpr7 were shown to down regulate Mal63 when grown in maltose limiting medium [76]. This suggests that Mal63 is a substrate of Hsp90 (Table 5.2).

6 Building the Global Hsp90 Interaction Network—Yeast

Hsp90 is involved in a very broad variety of processes with many of its functions still to be identified. However, it should be noted that despite the fact that Hsp90 is highly abundant under normal growth conditions and that its levels are elevated under environmental stress conditions, its functions are not related to promoting de novo protein folding or protein disaggregation but rather in the maturation or activation of proteins that are in a near-native state [77, 78]. In this sense, building a comprehensive Hsp90 chaperone network allows for a global view of this chaperone, its co-chaperones, and its clients.

In an application of an integrative proteomic approach, Zhao et al. [8] combined interaction data obtained from four different approaches. In the first approach, the authors carried out genome-wide systematic two-hybrid screens (2H screen) using ordered strain arrays to identify protein-protein interactions with full-length and different domains of Hsp90. In the second approach, they used the yeast library in which each open reading frame (ORF) is tagged at the C-terminus with a tandem affinity purification tag (TAP-tag) to carry out pulldowns of tagged ORFs and identify proteins interacting with Hsp90 using mass spectrometry (TAP screen). In the third approach, they screened for synthetic lethal interactions between a specific mutant allele of Hsp90 and members of a panel of about 4700 single yeast gene deletion strains (synthetic genetic array, SGA, screen). In the fourth approach, they screened the deletion mutant strains for differential hypersensitivity to the Hsp90 inhibitor geldanamycin (GS screen). These efforts identified 627 putative Hsp90 interactors, representing about 10 % of the yeast proteome (Fig. 5.3a).

From these studies, Zhao et al. [8] were able to elucidate new links between Hsp90 and transcriptional regulation, cell cycle, DNA processing, and cellular transport, among others. Furthermore, they were able to distinguish 84 Hsp90 clients identified in the two-hybrid and TAP screens that were also found to physically interact with 1 of 15 Hsp90 cofactors (Fig. 5.3b). Hence, this provided a first clue as to how client interaction with Hsp90 is stratified by the chaperone cofactors.

To highlight the strength of such a high-throughput analysis, in this study, the authors also identified new cofactors for Hsp90, namely Tah1 (Table 5.1) and its interactor Pih1. Tah1 and Pih1 were found to associate with the essential helicases Rvb1 and Rvb2 to form what was termed by the authors the R2TP complex. Hsp90-R2TP was subsequently shown to be involved in the assembly of box C/D small nucleolar ribonucleoprotein particles (snoRNPs) [79] required for pre-ribosomal ribonucleic acid (pre-rRNA) modification, phosphatidylinositol-3 kinase-related protein kinase (PIKK) signaling complexes [80], the telomerase reverse transcriptase (TERT) core complex [81], and RNA polymerase II [82–84]. Hence, the Zhao et al. study [8] allowed the characterization of new cellular pathways regulated by Hsp90 and its cofactors.

In another growth-based genetic interaction study, McClellan et al. [1] performed a genome-wide screen for negative genetic interactions between Hsp90 and target genes by examining the fitness of Hsp90-inhibited yeast cells, using the Hsp90 inhibitor macbecin II, versus Hsp90-non-inhibited yeast cells when these target genes were deleted. Furthermore, in order to gain insights into Hsp90 function in normal versus stress conditions, such experiments were conducted at 30 °C (normal yeast growth conditions) and 37 °C (heat stress conditions). The top 5 % most growth-inhibited strains were selected as candidates for further analysis (Fig. 5.4).

By linking the identified genetic interaction pairs with both physical and genetic interaction pairs from the BioGRID database [85], the authors were able to construct the Hsp90 interaction network at the two temperatures. Under normal conditions in yeast, Hsp90 was found to be involved in protein secretion and trafficking. However, under heat shock conditions, this chaperone was found to be required for normal progression of the cell cycle, meiosis, and cytokinesis (Fig. 5.4). Further analysis using GO annotations [86] allowed the discovery of the modular nature by which the Hsp90 chaperone system operates. The two main groups that were found to form the Hsp90 network were (1) cellular trafficking and transport, and (2) regulation of the cell cycle. In a follow-up study, it was found that a total of 113 ORFs and 417 interactions are part of an Hsp90 functional network involved in the secretory pathway [1]. This network verified the notion that Hsp90 acts both in the exocytic and endocytic parts of the secretory pathway.

The Zhao et al. [8] and the McCellan et al. [1] analyses demonstrate the level of complexity of the Hsp90 network acting in many cellular pathways. The yeast Hsp90 network offers a significant level of completeness. This, therefore, becomes

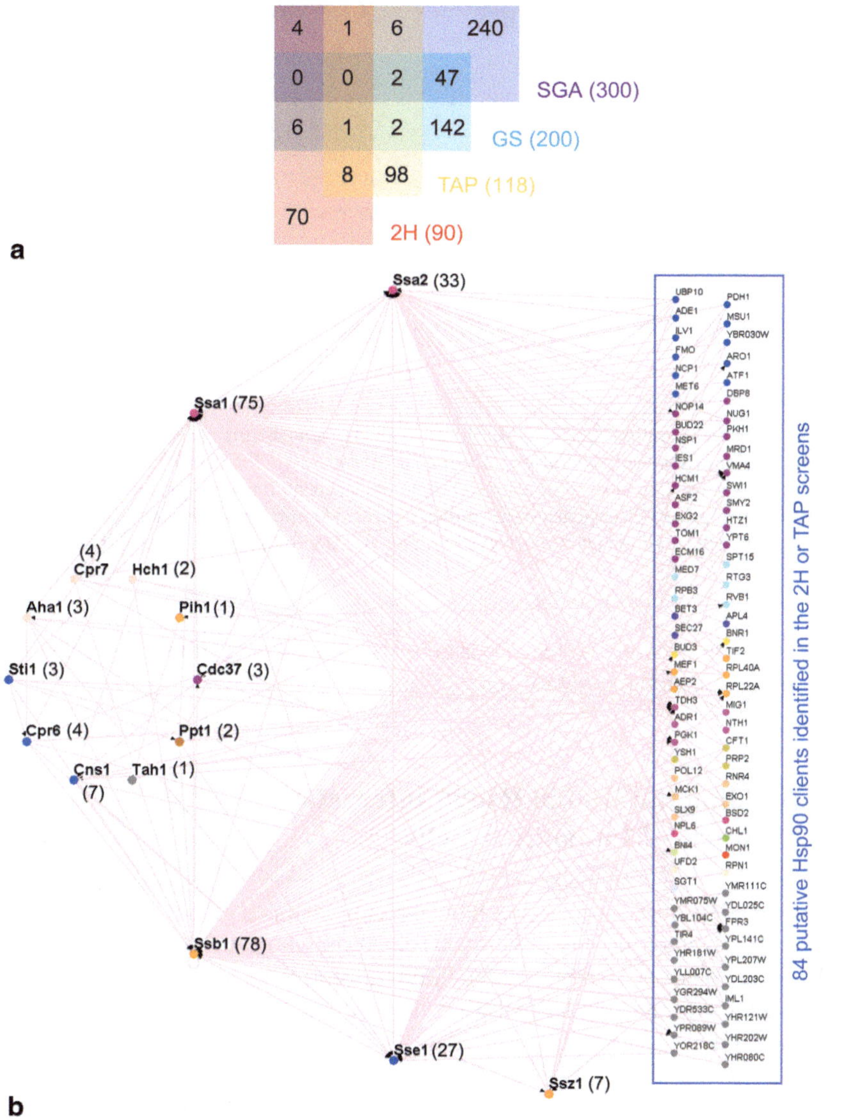

Fig. 5.3 Hsp90 interaction data from Zhao et al. [8]. **a** Venn diagram showing the overlap among proteins that were found to physically or genetically interact with Hsp90 using 2H, tandem affinity purification (TAP), synthetic genetic array (SGA), and geldanamycin screen (GS) screens. The total number of interactions for each method is indicated in brackets. **b** Interaction network of heat shock protein 90 (Hsp90) and its co-chaperones. Eighty-four putative clients identified in the 2H or TAP screens for Hsp90 were found to also interact with at least one of the 15 Hsp90 co-chaperones in TAP-tag pulldowns. The putative Hsp90 clients are grouped according to GO terms and are placed in the *blue rectangle*. *Lines* refer to TAP-based interactions. The numbers in brackets refer to the number of putative clients in the rectangle that interact with a given co-chaperone. (Modified from Zhao et al. [8]).

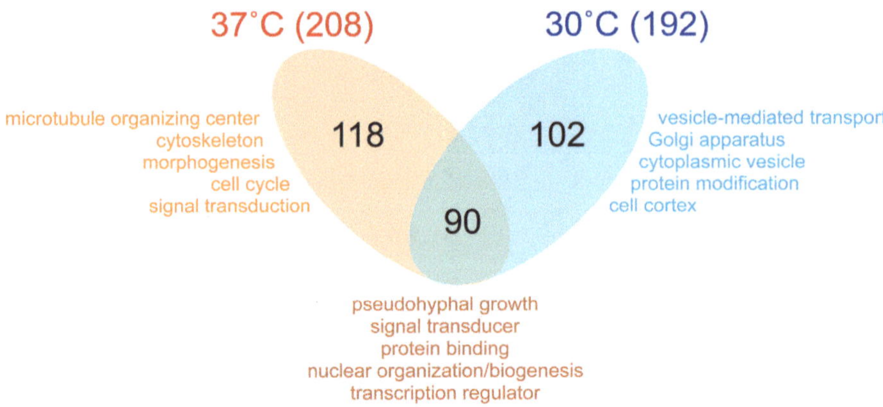

Fig. 5.4 Hsp90 interaction data from McClellan et al. [1]. Venn diagram showing the overlap among the Hsp90 chemical-genetic interactors identified at 30 and 37 °C. Total number of genes is indicated in brackets. The top five most significant GO slim compartment, function, and process categories are shown from data sets where genes were similarly enriched (*overlapping region*) and independently enriched for each condition (*non-overlapping regions*) (Data from McClellan et al. [1]).

an exceptional resource to tease apart the Hsp90 networks from many other organisms through comparative biology.

7 Building the Global Hsp90 Interaction Network—Mammalian Cells

In mammalian cells, most studies have typically mapped global cellular changes in protein levels upon treatment with an Hsp90 inhibitor. Although, such studies do not provide data for Hsp90-mediated protein-protein interactions, they do provide a global view of the cellular pathways that are directly or indirectly modulated by the chaperone.

By inhibiting Hsp90 ATPase activity using geldanamycin, Wu et al. [87] were able to identify protein level changes using different mass spectrometry-based methods including stable isotope labeling with amino acids in cell culture (SILAC) approach. They used four cancer cell lines K562 (blood), Colo205 (colon), Cal27 (head and neck), and MDAMB231 (breast) and identified about 1600 proteins that showed changes in levels upon drug treatment (Fig. 5.5a). They found that the rate of Hsp90 inhibition-induced protein downregulation correlates with protein half-life and that protein kinases have significantly shorter half-lives than other proteins.

In a similar study, Sharma et al. [88] used the Hsp90 inhibitor 17-(dimethylaminoethylamino)-17-demethoxygeldanamycin (17-DMAG) on HeLa cells followed by SILAC analysis. They observed activation of the heat shock response and found that protein kinases and proteins involved in the DNA damage response were particularly affected by the inhibitor (Fig. 5.5b).

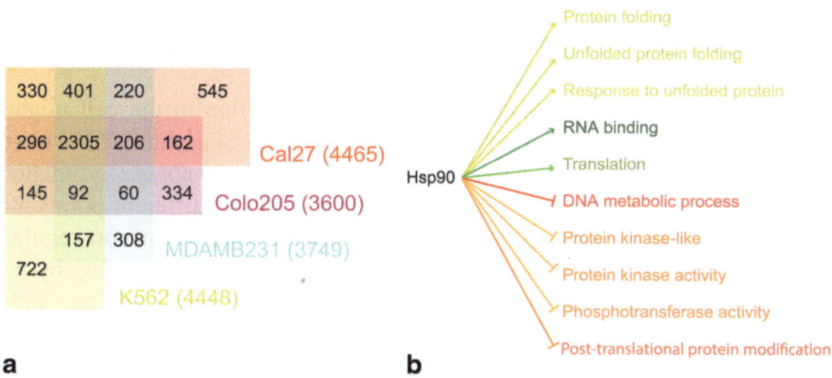

Fig. 5.5 Protein level changes upon heat shock protein 90 (Hsp90) inhibition in mammalian cells [87, 88]. **a** Summary of the number of proteins identified by Wu et al. [87] to significantly change levels upon Hsp90 inhibition in four human cancer cell lines: Cal27 (*head and neck*), Colo205 (colon), MDAMB231 (breast), and K562 (blood). Total numbers of proteins are indicated in brackets. **b** Statistically significant annotated terms (KEGG, GO, PFAM, and SCOP) found for proteins whose levels significantly changed upon treatment of HeLa cells with the Hsp90 inhibitor 17-DMAG ((dimethylaminoethylamino)-17-demethoxygeldanamycin) as determined by Sharma et al. [88]. Upregulated groups are shown in *green*, downregulated groups are in *red*. Groups that correspond to the same pathway are color-coded in different shades of *green* or *red*.

By performing a quantitative high-throughput analysis, Taipale et al. [89] were able to measure the interaction of Hsp90 and the co-chaperone Cdc37 with human protein kinases, transcription factors and ubiquitin ligases *in vivo*. They found interactions between Hsp90 and almost 400 client proteins with kinases making up the largest proportion of interactors. Cdc37 was found to serve as an adaptor to provide specificity for Hsp90 interaction with kinases. They were able to show that Hsp90 clients within the kinase family can be differentiated from non-clients on the basis of their intrinsic stability. They propose that Hsp90 kinase clients are generally unstable in their fully folded state and this makes them more prone to recognition by Hsp90 and Cdc37. Transcription factors were found to be chaperoned by Hsp90 in a similar way as kinases but represent only a small fraction of Hsp90 clients.

All studies clearly suggest that kinases are the major group of Hsp90 clients in mammalian cells and that they are destabilized upon Hsp90 inhibition.

8 Building the Co-chaperone Interaction Network—The Case of Sba1

Building the interaction network for Hsp90 co-chaperones has so far only been done for yeast Sba1 (yeast ortholog of mammalian p23, Table 5.1). Echtenkamp et al. [42] carried out an SGA analysis of *SBA1* yeast deletion strain against a deletion library to identify parallel or compensatory pathways affected by Sba1. They found that Sba1 genetic interactors were enriched in signal transduction, protein catabolism, cell budding, DNA metabolism, cellular respiration, and vesicle-mediated transport. Sba1 was also found to serve as a negative regulator of vesicle transport.

Interestingly, the data suggest that Sba1's main function in the cell might be independent of Hsp90 since only ~25% of the 348 total Sba1 interactors overlap with known Hsp90 interactors. Nevertheless, Sba1 and Hsp90 appear to function in the same biological pathways but interacting with clients at different points along those pathways. For example, Sba1 seems to modulate protein transport regulating protein mannosylation in the Golgi, whereas Hsp82 affects that pathway by interacting with vesicle-tethering complexes [1, 42]. In a similar scenario, the authors show that both Hsp82 and Sba1 act on cell mobility by interacting with different clients. By intersecting at individual proteins, protein complexes, and protein pathways, Sba1 likely established a relationship with Hsp90 and cellular processes that systematically expand the reach of the Hsp90 system.

9 Hsp90 Interaction Networks in Pathogens

Hsp90 has gained much popularity as a target for treating various infectious diseases in addition to cancer. It is involved in many important cellular processes of intracellular protozoans and other important human pathogens [90, 91]. Building a chaperone network to understand the biology of these pathogens is a powerful tool for drug discovery efforts [10].

In *Candida albicans* (the most prevalent human fungal pathogen), Hsp90 regulates not only drug resistance but also morphogenesis and virulence, making it an ideal organism to study Hsp90 functions in pathogenesis [91]. Diezmann et al. [91] performed a chemical-genetic screen using geldanamycin as an Hsp90 inhibitor on mutant strains from a homozygous transposon insertion library. This was done under a variety of stress conditions, such as: NaCl (osmotic stress), tunicamycin (causes the unfolded protein response in the endoplasmic reticulum), azole fluconazole (targets the cell membrane), chinocandin caspofungin (targets the cell wall), and growth at 37 or 41 °C (heat stress). A network was built with Hsp90 targets as nodes linked to the associated genetic interaction condition. In addition to genetic interactivity, the dependence of certain target genes on Hsp90 was determined by measuring protein or transcript levels upon Hsp90 depletion. The authors found that the networks obtained under different conditions are quite different with little overlap.

A comparison of the genetic interactions between *C. albicans* and *S. cerevisiae* revealed that certain interactions are conserved between the two fungi and that the conservation was dependent on the experimental conditions employed. The authors found that, for interactors in *C. albicans* that have homologs in *S. cerevisiae*, only 17% of *C. albicans* Hsp90 interactors were conserved in *S. cerevisiae*. This indicates that there has been considerable rewiring of the Hsp90 network over evolution. However, although the inferred interactions were different between the two organisms, they both retained similar proportions of their genome associated with Hsp90.

In the malarial parasite *Plasmodium falciparum*, a chaperone network was inferred using computational approaches [92]. This parasite undergoes frequent episodes of heat shock during febrile periods, which is a clinical hallmark of malaria. It is therefore reasonable to assume that chaperones play an important role in the parasite's adaptation. Hsp90 has been shown to be essential for parasite viability. The analysis was performed by interolog mapping of a human network derived from both the human protein reference database (HPRD) and from predicted malaria-human orthologs using BLASTP. Interologs are protein-protein interactions that are conserved across species. This network was then combined with protein-protein interactions from a yeast two-hybrid screen for malarial proteins to construct the chaperone network for this parasite. The network allowed the authors to perform various putative predictions on chaperone function during parasite development. For example, they were able to find an Hsp40 molecule that acts in concert with Hsp90 in membrane protein trafficking.

The networks described for *C. albicans* and *P. falciparum* highlight the importance of Hsp90 in pathogen drug resistance, morphogenesis, stage development, and virulence. More importantly, they provide clues towards selecting Hsp90 interactors that may serve as putative drug targets. These analyses reveal again the extraordinary versatility of Hsp90 activities and its involvement in multiple cellular pathways. They also demonstrate that there is still much to be discovered about this essential chaperone system.

10 Concluding Remarks

Hsp90 is a central molecular chaperone for maintaining protein homeostasis. Its multiple co-chaperones and its functional interaction with several other chaperones makes mapping the interaction network of Hsp90 a requirement for understanding its cellular activities. Currently, the Hsp90 interaction network has been extensively studied only in yeast and future studies have to concentrate on mapping the network in different mammalian cell lines under normal and disease states. This is especially important due to the fact that Hsp90 is an anticancer drug target. Furthermore, mapping the Hsp90 network in different pathogens will help in drug discovery efforts targeting these pathogens by targeting Hsp90. Multiple proteomic approaches have to be employed to elucidate such networks. The techniques currently being employed are still under development and will require further improvement to be able to be implemented by the many groups working on Hsp90. Gathering many types of high-throughput data simultaneously (genetic and physical interactions; localization of small molecules, RNA, lipids, proteins) could be a challenging task but is absolutely required to allow for the generation of an integrated network that more clearly reflects the plethora of functions of this chaperone.

Acknowledgments This work was funded by a Canadian Institutes of Health Research grant (MOP-93778) to WAH and a Natural Sciences and Engineering Research Council of Canada (RGPIN 283337-11) grant to ERMT.

References

1. McClellan AJ, Xia Y, Deutschbauer AM, Davis RW, Gerstein M, Frydman J (2007) Diverse cellular functions of the Hsp90 molecular chaperone uncovered using systems approaches. Cell 131(1):121–135. doi:10.1016/j.cell.2007.07.036
2. Echeverria PC, Bernthaler A, Dupuis P, Mayer B, Picard D (2011) An interaction network predicted from public data as a discovery tool: application to the Hsp90 molecular chaperone machine. PLoS ONE 6(10):e26044. doi:10.1371/journal.pone.0026044
3. Borkovich KA, Farrelly FW, Finkelstein DB, Taulien J, Lindquist S (1989) hsp82 is an essential protein that is required in higher concentrations for growth of cells at higher temperatures. Mol Cell Biol 9(9):3919–3930
4. Versteeg S, Mogk A, Schumann W (1999) The Bacillus subtilis htpG gene is not involved in thermal stress management. Mol Gen Genet MGG 261(3):582–588. doi:10.1007/s004380051004
5. Langer T, Rosmus S, Fasold H (2003) Intracellular localization of the 90 kDA heat shock protein (HSP90alpha) determined by expression of a EGFP-HSP90alpha-fusion protein in unstressed and heat stressed 3T3 cells. Cell Biol Int 27(1):47–52
6. Pearl LH, Prodromou C (2006) Structure and mechanism of the Hsp90 molecular chaperone machinery. Annu Rev Biochem 75(1):271–294. doi:10.1146/annurev.biochem.75.103004.142738
7. Krukenberg KA, Street TO, Lavery LA, Agard DA (2011) Conformational dynamics of the molecular chaperone Hsp90. Q Rev Biophys 44(2):229–255. doi:10.1017/S0033583510000314
8. Zhao R, Davey M, Hsu YC, Kaplanek P, Tong A, Parsons AB, Krogan N, Cagney G, Mai D, Greenblatt J, Boone C, Emili A, Houry WA (2005) Navigating the chaperone network: an integrative map of physical and genetic interactions mediated by the hsp90 chaperone. Cell 120(5):715–727. doi:10.1016/j.cell.2004.12.024
9. Rohl A, Rohrberg J, Buchner J (2013) The chaperone Hsp90: changing partners for demanding clients. Trends Biochem Sci 38(5):253–262. doi:10.1016/j.tibs.2013.02.003
10. Echeverria PC, Figueras MJ, Vogler M, Kriehuber T, de Miguel N, Deng B, Dalmasso MC, Matthews DE, Matrajt M, Haslbeck M, Buchner J, Angel SO (2010) The Hsp90 co-chaperone p23 of *Toxoplasma gondii*: Identification, functional analysis and dynamic interactome determination. Mol Biochem Parasitol 172(2):129–140. doi:10.1016/j.molbiopara.2010.04.004
11. Makhnevych T, Houry WA (2012) The role of Hsp90 in protein complex assembly. Biochim Biophys Acta 1823(3):674–682. doi:10.1016/j.bbamcr.2011.09.001
12. Neckers L, Workman P (2012) Hsp90 molecular chaperone inhibitors: are we there yet? Clin Cancer Res 18(1):64–76. doi:10.1158/1078-0432.CCR-11-1000
13. Samant RS, Clarke PA, Workman P (2012) The expanding proteome of the molecular chaperone HSP90. Cell Cycle 11(7):1301–1308. doi:10.4161/cc.19722
14. Wandinger SK, Richter K, Buchner J (2008) The Hsp90 chaperone machinery. J Biol Chem 283(27):18473–18477. doi:10.1074/jbc.R800007200
15. Csermely P, Schnaider T, Soti C, Prohaszka Z, Nardai G (1998) The 90-kDa molecular chaperone family: structure, function, and clinical applications: a comprehensive review. Pharmacol Ther 79(2):129–168. doi:10.1016/S0163-7258(98)00013-8
16. Johnson JL, Brown C (2009) Plasticity of the Hsp90 chaperone machine in divergent eukaryotic organisms. Cell Stress Chaperones 14(1):83–94. doi:10.1007/s12192-008-0058-9
17. Li J, Soroka J, Buchner J (2012) The Hsp90 chaperone machinery: conformational dynamics and regulation by co-chaperones. Biochim Biophys Acta 1823(3):624–635. doi:10.1016/j.bbamcr.2011.09.003
18. Smith DF (2004) Tetratricopeptide repeat cochaperones in steroid receptor complexes. Cell Stress Chaperones 9(2):109–121. doi:10.1379/CSC-31.1
19. Chang HC, Lindquist S (1994) Conservation of Hsp90 macromolecular complexes in *Saccharomyces cerevisiae*. J Biol Chem 269(40):24983–24988
20. Johnson BD, Schumacher RJ, Ross ED, Toft DO (1998) Hop modulates Hsp70/Hsp90 interactions in protein folding. J Biol Chem 273(6):3679–3686. doi:10.1074/jbc.273.6.3679

21. Silverstein AM, Galigniana MD, Chen MS, Owens-Grillo JK, Chinkers M, Pratt WB (1997) Protein phosphatase 5 is a major component of glucocorticoid receptor.hsp90 complexes with properties of an FK506-binding immunophilin. J Biol Chem 272(26):16224–16230
22. Wandinger SK, Suhre MH, Wegele H, Buchner J (2006) The phosphatase Ppt1 is a dedicated regulator of the molecular chaperone Hsp90. EMBO J 25(2):367–376. doi:10.1038/sj.emboj.7600930
23. Jimenez B, Ugwu F, Zhao R, Orti L, Makhnevych T, Pineda-Lucena A, Houry WA (2012) Structure of minimal tetratricopeptide repeat domain protein Tah1 reveals mechanism of its interaction with Pih1 and Hsp90. J Biol Chem 287(8):5698–5709. doi:10.1074/jbc.M111.287458
24. Boulon S, Marmier-Gourrier N, Pradet-Balade B, Wurth L, Verheggen C, Jady BE, Rothe B, Pescia C, Robert MC, Kiss T, Bardoni B, Krol A, Branlant C, Allmang C, Bertrand E, Charpentier B (2008) The Hsp90 chaperone controls the biogenesis of L7Ae RNPs through conserved machinery. J Cell Biol 180(3):579–595. doi:10.1083/jcb.200708110
25. Wesche S, Arnold M, Jansen RP (2003) The UCS domain protein She4p binds to myosin motor domains and is essential for class I and class V myosin function. Curr Biol 13(9):715–724
26. Etard C, Behra M, Fischer N, Hutcheson D, Geisler R, Strahle U (2007) The UCS factor Steif/Unc-45b interacts with the heat shock protein Hsp90a during myofibrillogenesis. Dev Biol 308(1):133–143. doi:10.1016/j.ydbio.2007.05.014
27. Mayr C, Richter K, Lilie H, Buchner J (2000) Cpr6 and Cpr7, two closely related Hsp90-associated immunophilins from *Saccharomyces cerevisiae*, differ in their functional properties. J Biol Chem 275(44):34140–34146. doi:10.1074/jbc.M005251200
28. Bhangoo MK, Tzankov S, Fan AC, Dejgaard K, Thomas DY, Young JC (2007) Multiple 40-kDa heat-shock protein chaperones function in Tom70-dependent mitochondrial import. Mol Biol Cell 18(9):3414–3428. doi:10.1091/mbc.E07-01-0088
29. Lingelbach LB, Kaplan KB (2004) The interaction between Sgt1p and Skp1p is regulated by HSP90 chaperones and is required for proper CBF3 assembly. Mol Cell Biol 24(20):8938–8950. doi:10.1128/MCB.24.20.8938-8950.2004
30. Lee YT, Jacob J, Michowski W, Nowotny M, Kuznicki J, Chazin WJ (2004) Human Sgt1 binds HSP90 through the CHORD-Sgt1 domain and not the tetratricopeptide repeat domain. J Biol Chem 279(16):16511–16517. doi:10.1074/jbc.M400215200
31. Tesic M, Marsh JA, Cullinan SB, Gaber RF (2003) Functional interactions between Hsp90 and the co-chaperones Cns1 and Cpr7 in *Saccharomyces cerevisiae*. J Biol Chem 278(35):32692–32701. doi:10.1074/jbc.M304315200
32. Crevel G, Bennett D, Cotterill S (2008) The human TPR protein TTC4 is a putative Hsp90 co-chaperone which interacts with CDC6 and shows alterations in transformed cells. PLoS ONE 3(3):e0001737. doi:10.1371/journal.pone.0001737
33. Brychzy A, Rein T, Winklhofer KF, Hartl FU, Young JC, Obermann WM (2003) Cofactor Tpr2 combines two TPR domains and a J domain to regulate the Hsp70/Hsp90 chaperone system. EMBO J 22(14):3613–3623. doi:10.1093/emboj/cdg362
34. Moffatt NS, Bruinsma E, Uhl C, Obermann WM, Toft D (2008) Role of the cochaperone Tpr2 in Hsp90 chaperoning. Biochemistry (Mosc) 47(31):8203–8213. doi:10.1021/bi800770g
35. Panaretou B, Siligardi G, Meyer P, Maloney A, Sullivan JK, Singh S, Millson SH, Clarke PA, Naaby-Hansen S, Stein R, Cramer R, Mollapour M, Workman P, Piper PW, Pearl LH, Prodromou C (2002) Activation of the ATPase activity of hsp90 by the stress-regulated cochaperone aha1. Mol Cell 10(6):1307–1318
36. Lotz GP, Lin H, Harst A, Obermann WM (2003) Aha1 binds to the middle domain of Hsp90, contributes to client protein activation, and stimulates the ATPase activity of the molecular chaperone. J Biol Chem 278(19):17228–17235. doi:10.1074/jbc.M212761200
37. Li J, Richter K, Reinstein J, Buchner J (2013) Integration of the accelerator Aha1 in the Hsp90 co-chaperone cycle. Nat Struct Mol Biol 20(3):326–331. doi:10.1038/nsmb.2502
38. Meyer P, Prodromou C, Liao C, Hu B, Mark Roe S, Vaughan CK, Vlasic I, Panaretou B, Piper PW, Pearl LH (2004) Structural basis for recruitment of the ATPase activator Aha1 to the Hsp90 chaperone machinery. EMBO J 23(3):511–519. doi:10.1038/sj.emboj.7600060

39. Koulov AV, LaPointe P, Lu B, Razvi A, Coppinger J, Dong MQ, Matteson J, Laister R, Arrowsmith C, Yates JR 3rd, Balch WE (2010) Biological and structural basis for Aha1 regulation of Hsp90 ATPase activity in maintaining proteostasis in the human disease cystic fibrosis. Mol Biol Cell 21(6):871–884. doi:10.1091/mbc.E09-12-1017
40. Retzlaff M, Hagn F, Mitschke L, Hessling M, Gugel F, Kessler H, Richter K, Buchner J (2010) Asymmetric activation of the hsp90 dimer by its cochaperone aha1. Mol Cell 37(3):344–354. doi:10.1016/j.molcel.2010.01.006
41. Armstrong H, Wolmarans A, Mercier R, Mai B, LaPointe P (2012) The co-chaperone Hch1 regulates Hsp90 function differently than its homologue Aha1 and confers sensitivity to yeast to the Hsp90 inhibitor NVP-AUY922. PLoS ONE 7(11):e49322. doi:10.1371/journal.pone.0049322
42. Echtenkamp FJ, Zelin E, Oxelmark E, Woo JI, Andrews BJ, Garabedian M, Freeman BC (2011) Global functional map of the p23 molecular chaperone reveals an extensive cellular network. Mol Cell 43(2):229–241. doi:10.1016/j.molcel.2011.05.029
43. Ali MM, Roe SM, Vaughan CK, Meyer P, Panaretou B, Piper PW, Prodromou C, Pearl LH (2006) Crystal structure of an Hsp90-nucleotide-p23/Sba1 closed chaperone complex. Nature 440(7087):1013–1017. doi:10.1038/nature04716
44. Mandal AK, Lee P, Chen JA, Nillegoda N, Heller A, DiStasio S, Oen H, Victor J, Nair DM, Brodsky JL, Caplan AJ (2007) Cdc37 has distinct roles in protein kinase quality control that protect nascent chains from degradation and promote posttranslational maturation. J Cell Biol 176(3):319–328. doi:10.1083/jcb.200604106
45. Vaughan CK, Gohlke U, Sobott F, Good VM, Ali MM, Prodromou C, Robinson CV, Saibil HR, Pearl LH (2006) Structure of an Hsp90-Cdc37-Cdk4 complex. Mol Cell 23(5):697–707. doi:10.1016/j.molcel.2006.07.016
46. Sreeramulu S, Jonker HR, Langer T, Richter C, Lancaster CR, Schwalbe H (2009) The human Cdc37.Hsp90 complex studied by heteronuclear NMR spectroscopy. J Biol Chem 284(6):3885–3896. doi:10.1074/jbc.M806715200
47. Eckl JM, Rutz DA, Haslbeck V, Zierer BK, Reinstein J, Richter K (2013) Cdc37 (cell division cycle 37) restricts Hsp90 (heat shock protein 90) motility by interaction with N-terminal and middle domain binding sites. J Biol Chem 288(22):16032–16042. doi:10.1074/jbc.M112.439257
48. Hessling M, Richter K, Buchner J (2009) Dissection of the ATP-induced conformational cycle of the molecular chaperone Hsp90. Nat Struct Mol Biol 16(3):287–293. doi:10.1038/nsmb.1565
49. Mickler M, Hessling M, Ratzke C, Buchner J, Hugel T (2009) The large conformational changes of Hsp90 are only weakly coupled to ATP hydrolysis. Nat Struct Mol Biol 16(3):281–286. doi:10.1038/nsmb.1557
50. Kampinga HH, Craig EA (2010) The HSP70 chaperone machinery: J proteins as drivers of functional specificity. Nat Rev Mol Cell Biol 11(8):579–592. doi:10.1038/nrm2941
51. Dittmar KD, Demady DR, Stancato LF, Krishna P, Pratt WB (1997) Folding of the glucocorticoid receptor by the heat shock protein (hsp) 90-based chaperone machinery—the role of p23 is to stabilize receptor-hsp90 heterocomplexes formed by hsp90-p60-hsp70. J Biol Chem 272 (34):21213–21220. doi:DOI 10.1074/jbc.272.34.21213
52. Kosano H, Stensgard B, Charlesworth MC, McMahon N, Toft D (1998) The assembly of progesterone Receptor-hsp90 complexes using purified proteins. J Biol Chem 273 (49):32973–32979. doi:DOI 10.1074/jbc.273.49.32973
53. Korber BT, Farber RM, Wolpert DH, Lapedes AS (1993) Covariation of mutations in the V3 loop of human immunodeficiency virus type 1 envelope protein: an information theoretic analysis. Proc Natl Acad Sci U S A 90(15):7176–7180
54. Travers SA, Fares MA (2007) Functional coevolutionary networks of the Hsp70-Hop-Hsp90 system revealed through computational analyses. Mol Biol Evol 24(4):1032–1044. doi:10.1093/molbev/msm022
55. Fares MA, McNally D (2006) CAPS: coevolution analysis using protein sequences. Bioinformatics 22(22):2821–2822. doi:10.1093/bioinformatics/btl493

56. Mollapour M, Bourboulia D, Beebe K, Woodford MR, Polier S, Hoang A, Chelluri R, Li Y, Guo A, Lee MJ, Fotooh-Abadi E, Khan S, Prince T, Miyajima N, Yoshida S, Tsutsumi S, Xu W, Panaretou B, Stetler-Stevenson WG, Bratslavsky G, Trepel JB, Prodromou C, Neckers L (2014) Asymmetric Hsp90 N domain SUMOylation recruits Aha1 and ATP-competitive inhibitors. Mol Cell 53(2):317–329. doi:10.1016/j.molcel.2013.12.007
57. Mollapour M, Neckers L (2012) Post-translational modifications of Hsp90 and their contributions to chaperone regulation. Biochim Biophys Acta 1823(3):648–655. doi:10.1016/j.bbamcr.2011.07.018
58. Mollapour M, Tsutsumi S, Truman AW, Xu W, Vaughan CK, Beebe K, Konstantinova A, Vourganti S, Panaretou B, Piper PW, Trepel JB, Prodromou C, Pearl LH, Neckers L (2011) Threonine 22 phosphorylation attenuates Hsp90 interaction with cochaperones and affects its chaperone activity. Mol Cell 41(6):672–681. doi:10.1016/j.molcel.2011.02.011
59. Mollapour M, Tsutsumi S, Kim YS, Trepel J, Neckers L (2011) Casein kinase 2 phosphorylation of Hsp90 threonine 22 modulates chaperone function and drug sensitivity. Oncotarget 2(5):407–417
60. Mollapour M, Tsutsumi S, Donnelly AC, Beebe K, Tokita MJ, Lee MJ, Lee S, Morra G, Bourboulia D, Scroggins BT, Colombo G, Blagg BS, Panaretou B, Stetler-Stevenson WG, Trepel JB, Piper PW, Prodromou C, Pearl LH, Neckers L (2010) Swe1Wee1-dependent tyrosine phosphorylation of Hsp90 regulates distinct facets of chaperone function. Mol Cell 37(3):333–343. doi:10.1016/j.molcel.2010.01.005
61. Scroggins BT, Robzyk K, Wang D, Marcu MG, Tsutsumi S, Beebe K, Cotter RJ, Felts S, Toft D, Karnitz L, Rosen N, Neckers L (2007) An acetylation site in the middle domain of Hsp90 regulates chaperone function. Mol Cell 25(1):151–159. doi:10.1016/j.molcel.2006.12.008
62. Martinez-Ruiz A, Villanueva L, Gonzalez deOC, Lopez-Ferrer D, Higueras MA, Tarin C, Rodriguez-Crespo I, Vazquez J, Lamas S (2005) S-nitrosylation of Hsp90 promotes the inhibition of its ATPase and endothelial nitric oxide synthase regulatory activities. Proc Natl Acad Sci U S A 102(24):8525–8530. doi:10.1073/pnas.0407294102
63. Retzlaff M, Stahl M, Eberl HC, Lagleder S, Beck J, Kessler H, Buchner J (2009) Hsp90 is regulated by a switch point in the C-terminal domain. EMBO Rep 10(10):1147–1153. doi:10.1038/embor.2009.153
64. Genest O, Reidy M, Street TO, Hoskins JR, Camberg JL, Agard DA, Masison DC, Wickner S (2013) Uncovering a region of heat shock protein 90 important for client binding in *E. coli* and chaperone function in yeast. Mol Cell 49(3):464–473. doi:10.1016/j.molcel.2012.11.017
65. Louvion JF, Abbas-Terki T, Picard D (1998) Hsp90 is required for pheromone signaling in yeast. Mol Biol Cell 9(11):3071–3083. doi:10.1091/mbc.9.11.3071
66. Abbas-Terki T, Donze O, Picard D (2000) The molecular chaperone Cdc37 is required for Ste11 function and pheromone-induced cell cycle arrest. FEBS Lett 467(1):111–116
67. Lee P, Shabbir A, Cardozo C, Caplan AJ (2004) Sti1 and Cdc37 can stabilize Hsp90 in chaperone complexes with a protein kinase. Mol Biol Cell 15(4):1785–1792. doi:10.1091/mbc.E03-07-0480
68. Donze O, Picard D (1999) Hsp90 binds and regulates Gcn2, the ligand-inducible kinase of the alpha subunit of eukaryotic translation initiation factor 2 [corrected]. Mol Cell Biol 19(12):8422–8432
69. Aligue R, Akhavan-Niak H, Russell P (1994) A role for Hsp90 in cell cycle control: Wee1 tyrosine kinase activity requires interaction with Hsp90. EMBO J 13(24):6099–6106
70. Booher RN, Deshaies RJ, Kirschner MW (1993) Properties of *Saccharomyces cerevisiae* wee1 and its differential regulation of p34CDC28 in response to G1 and G2 cyclins. EMBO J 12(9):3417–3426
71. Goes FS, Martin J (2001) Hsp90 chaperone complexes are required for the activity and stability of yeast protein kinases Mik1, Wee1 and Swe1. Eur J Biochem 268(8):2281–2289. doi:10.1046/j.1432-1327.2001.02105.x
72. Millson SH, Truman AW, King V, Prodromou C, Pearl LH, Piper PW (2005) A two-hybrid screen of the yeast proteome for Hsp90 interactors uncovers a novel Hsp90 chaperone re-

quirement in the activity of a stress-activated mitogen-activated protein kinase, Slt2p (Mpk1p). Eukaryot Cell 4(5):849–860. doi:10.1128/EC.4.5.849-860.2005
73. Creusot F, Creusot F, Verdière J, Verdière J, Gaisne M, Gaisne M (1988) CYP1 (HAP1) regulator of oxygen-dependent gene expression in yeast: I. Overall organization of the protein sequence displays several novel structural domains. J Mol Biol 204(2):263–276
74. Hon T, Lee HC, Hach A, Johnson JL, Craig EA, Erdjument-Bromage H, Tempst P, Zhang L (2001) The Hsp70-Ydj1 molecular chaperone represses the activity of the heme activator protein Hap1 in the absence of heme. Mol Cell Biol 21(23):7923–7932. doi:10.1128/MCB.21.23.7923-7932.2001
75. Hon T, Lee HC, Hu Z, Iyer VR, Zhang L (2005) The heme activator protein Hap1 represses transcription by a heme-independent mechanism in *Saccharomyces cerevisiae*. Genetics 169(3):1343–1352. doi:10.1534/genetics.104.037143
76. Bali M, Zhang B, Morano KA, Michels CA (2003) The Hsp90 molecular chaperone complex regulates maltose induction and stability of the Saccharomyces MAL gene transcription activator Mal63p. J Biol Chem 278(48):47441–47448. doi:10.1074/jbc.M309536200
77. Nathan DF, Vos MH, Lindquist S (1997) In vivo functions of the *Saccharomyces cerevisiae* Hsp90 chaperone. Proc Natl Acad Sci U S A 94(24):12949–12956
78. Doyle SM, Genest O, Wickner S (2013) Protein rescue from aggregates by powerful molecular chaperone machines. Nat Rev Mol Cell Biol 14(10):617–629. doi:10.1038/nrm3660
79. Zhao R, Kakihara Y, Gribun A, Huen J, Yang G, Khanna M, Costanzo M, Brost RL, Boone C, Hughes TR, Yip CM, Houry WA (2008) Molecular chaperone Hsp90 stabilizes Pih1/Nop17 to maintain R2TP complex activity that regulates snoRNA accumulation. J Cell Biol 180(3):563–578. doi:10.1083/jcb.200709061
80. Horejsi Z, Takai H, Adelman CA, Collis SJ, Flynn H, Maslen S, Skehel JM, de Lange T, Boulton SJ (2010) CK2 phospho-dependent binding of R2TP complex to TEL2 is essential for mTOR and SMG1 stability. Mol Cell 39(6):839–850. doi:10.1016/j.molcel.2010.08.037
81. Venteicher AS, Meng Z, Mason PJ, Veenstra TD, Artandi SE (2008) Identification of ATPases pontin and reptin as telomerase components essential for holoenzyme assembly. Cell 132(6):945–957. doi:10.1016/j.cell.2008.01.019
82. Boulon S, Pradet-Balade B, Verheggen C, Molle D, Boireau S, Georgieva M, Azzag K, Robert MC, Ahmad Y, Neel H, Lamond AI, Bertrand E (2010) HSP90 and its R2TP/Prefoldin-like cochaperone are involved in the cytoplasmic assembly of RNA polymerase II. Mol Cell 39(6):912–924. doi:10.1016/j.molcel.2010.08.023
83. Gomez-Navarro N, Peiro-Chova L, Rodriguez-Navarro S, Polaina J, Estruch F (2013) Rtp1p is a karyopherin-like protein required for RNA polymerase II biogenesis. Mol Cell Biol 33(9):1756–1767. doi:10.1128/MCB.01449-12
84. Mita P, Savas JN, Ha S, Djouder N, Yates JR 3rd, Logan SK (2013) Analysis of URI nuclear interaction with RPB5 and components of the R2TP/prefoldin-like complex. PLoS ONE 8(5):e63879. doi:10.1371/journal.pone.0063879
85. Chatr-Aryamontri A, Breitkreutz BJ, Heinicke S, Boucher L, Winter A, Stark C, Nixon J, Ramage L, Kolas N, O'Donnell L, Reguly T, Breitkreutz A, Sellam A, Chen D, Chang C, Rust J, Livstone M, Oughtred R, Dolinski K, Tyers M (2013) The BioGRID interaction database: 2013 update. Nucleic Acids Res 41(Database issue):D816–D823. doi:10.1093/nar/gks1158
86. Gene Ontology C, Blake JA, Dolan M, Drabkin H, Hill DP, Li N, Sitnikov D, Bridges S, Burgess S, Buza T, McCarthy F, Peddinti D, Pillai L, Carbon S, Dietze H, Ireland A, Lewis SE, Mungall CJ, Gaudet P, Chrisholm RL, Fey P, Kibbe WA, Basu S, Siegele DA, McIntosh BK, Renfro DP, Zweifel AE, Hu JC, Brown NH, Tweedie S, Alam-Faruque Y, Apweiler R, Auchinchloss A, Axelsen K, Bely B, Blatter M, Bonilla C, Bouguerleret L, Boutet E, Breuza L, Bridge A, Chan WM, Chavali G, Coudert E, Dimmer E, Estreicher A, Famiglietti L, Feuermann M, Gos A, Gruaz-Gumowski N, Hieta R, Hinz C, Hulo C, Huntley R, James J, Jungo F, Keller G, Laiho K, Legge D, Lemercier P, Lieberherr D, Magrane M, Martin MJ, Masson P, Mutowo-Muellenet P, O'Donovan C, Pedruzzi I, Pichler K, Poggioli D, Porras Millan P, Poux S, Rivoire C, Roechert B, Sawford T, Schneider M, Stutz A, Sundaram S, Tognolli M, Xenarios I, Foulgar R, Lomax J, Roncaglia P, Khodiyar VK, Lovering RC, Talmud PJ, Chi-

bucos M, Giglio MG, Chang H, Hunter S, McAnulla C, Mitchell A, Sangrador A, Stephan R, Harris MA, Oliver SG, Rutherford K, Wood V, Bahler J, Lock A, Kersey PJ, McDowall DM, Staines DM, Dwinell M, Shimoyama M, Laulederkind S, Hayman T, Wang S, Petri V, Lowry T, D'Eustachio P, Matthews L, Balakrishnan R, Binkley G, Cherry JM, Costanzo MC, Dwight SS, Engel SR, Fisk DG, Hitz BC, Hong EL, Karra K, Miyasato SR, Nash RS, Park J, Skrzypek MS, Weng S, Wong ED, Berardini TZ, Huala E, Mi H, Thomas PD, Chan J, Kishore R, Sternberg P, Van Auken K, Howe D, Westerfield M (2013) Gene ontology annotations and resources. Nucleic Acids Res 41(Database issue):D530–D535. doi:10.1093/nar/gks1050

87. Wu Z, Moghaddas Gholami A, Kuster B (2012) Systematic identification of the HSP90 candidate regulated proteome. Mol Cell Proteomics 11(6):M111 016675. doi:10.1074/mcp.M111.016675
88. Sharma K, Vabulas RM, Macek B, Pinkert S, Cox J, Mann M, Hartl FU (2012) Quantitative proteomics reveals that Hsp90 inhibition preferentially targets kinases and the DNA damage response. Mol Cell Proteomics 11(3):M111 014654. doi:10.1074/mcp.M111.014654
89. Taipale M, Krykbaeva I, Koeva M, Kayatekin C, Westover KD, Karras GI, Lindquist S (2012) Quantitative analysis of HSP90-client interactions reveals principles of substrate recognition. Cell 150(5):987-1001. doi:10.1016/j.cell.2012.06.047
90. Folgueira C, Requena JM (2007) A postgenomic view of the heat shock proteins in kinetoplastids. FEMS Microbiol Rev 31(4):359–377. doi:10.1111/j.1574-6976.2007.00069.x
91. Diezmann S, Michaut M, Shapiro RS, Bader GD, Cowen LE (2012) Mapping the Hsp90 genetic interaction network in *Candida albicans* reveals environmental contingency and rewired circuitry. PLoS Genet 8(3):e1002562. doi:10.1371/journal.pgen.1002562
92. Pavithra SR, Kumar R, Tatu U (2007) Systems analysis of chaperone networks in the malarial parasite *Plasmodium falciparum*. PLoS Comput Biol 3(9):1701–1715. doi:10.1371/journal.pcbi.0030168
93. Shiau AK, Harris SF, Southworth DR, Agard DA (2006) Structural analysis of *E. coli* hsp90 reveals dramatic nucleotide-dependent conformational rearrangements. Cell 127(2):329–340. doi:10.1016/j.cell.2006.09.027

Chapter 6
A Global View of the Proteome Perturbations by Hsp90 Inhibitors

Pablo C. Echeverria and Didier Picard

Abstract Heat shock protein 90 (Hsp90) is a highly efficient molecular chaperone and a major hub in the protein network that maintains cellular homeostasis and function. The qualitative and quantitative changes and rewiring of this protein network in tumor cells make them vastly dependent on Hsp90, which therefore becomes a key target to fight cancer. The inhibition of Hsp90 creates a profound transformation in the cell proteome. In this chapter, we review and analyze the most recent efforts that take advantage of the druggability of Hsp90 in order to understand the global changes at the proteome level that this inhibition produces. The considerable impact that the targeting of Hsp90 has on the structure of these protein networks is also discussed.

1 Introduction

Heat shock protein 90 (Hsp90) was initially described in the early 1980s as an interactor of the tyrosine kinase pp60src of the Rous sarcoma virus, as abundant in infected and non-infected cells and as a protein induced by cellular stress [1, 2]. These studies kicked off the idea that it may be interesting to explore further the relationships between the cellular response to cellular stress such as heat shock and the neoplastic transformation of cells by pp60src. More than 30 years later, we know that Hsp90 is not just an interactor of a tumorigenic protein. Hsp90 is a key proteostatic factor in the cell, playing essential roles in health and disease. Hsp90 interacts with hundreds or thousands of proteins ("clients") to assist them in their transition to thermodynamically stable and ultimately active conformations [3–5]. Estimated to be the most abundant protein in the cell, accounting for 1–2 % of the total protein levels [6, 7], Hsp90 may deal with 1–10 % of all cellular proteins. This massive "clientele" makes Hsp90 a strategic hub that is attractive to target in order

P. C. Echeverria (✉) · D. Picard
Department of Cell Biology, University of Geneva,
30 Quai Ernest-Ansermet, Sciences III, 1211 Geneva 4, Switzerland
e-mail: Pablo.Echeverria@unige.ch

D. Picard
e-mail: didier.picard@unige.ch

W. A. Houry (ed.), *The Molecular Chaperones Interaction Networks in Protein Folding and Degradation,* Interactomics and Systems Biology, DOI 10.1007/978-1-4939-1130-1_6,

to control cellular processes gone astray (e.g., cancer). Moreover, it is important to understand how and to which extent its perturbations affect the cellular proteome. In this chapter, we review the latest studies that have addressed these questions by taking advantage of the druggability of the protein using specific inhibitors.

2 Hsp90, a Multitasking Assistant of Proteins

Hsp90 is a molecular chaperone, which is highly abundant in bacterial and eukaryotic cells under physiological conditions, but which can also be modestly further induced by cellular stress. The Hsp90 family in mammalian cells is composed of four isoforms: Hsp90α (the more inducible form) and Hsp90β (the more constitutive form) are cytosoli cisoforms; the 94 kDa glucose-regulated protein (GRP94) is localized in the endoplasmic reticulum [8] and tumor necrosis factor receptor-associated protein 1 (TRAP1) resides in the mitochondrial matrix [9]. The Hsp90 molecular chaperone machine of the cytosol is constituted of Hsp90 molecules that act concurrently with a group of co-chaperones that modulate their client recognition and chaperone functions. Hsp90 contains an N-terminal conserved adenosine triphosphatase (ATPase) domain, whose adenosine triphosphate (ATP) binding and hydrolysis drives a complex cycle [10, 11]. For many years, the main strategy to block Hsp90 has been to employ competitive inhibitors that target this N-terminal ATP-binding pocket [12, 13].

In contrast to other molecular chaperones, which mainly recognize unfolded polypeptides or protein aggregates, Hsp90 seems to identify proteins that are already partially folded but still in thermodynamically unstable conformations [4, 5]. In addition, Hsp90 assists some of its substrates in other ways (e.g., in nucleo-cytoplasmic trafficking, complex assembly, as a scaffold protein, recycling from chromatin complexes). It is therefore not surprising that the inhibition of Hsp90 leads to the depletion of (or impaired accumulation of) many of its clients. This inhibition of Hsp90 function alters chaperone complex composition and promotes the association with ubiquitin ligases, leading to client ubiquitination and degradation [14–18]. Consequently, several essential cellular activities such as development, transcription, cell cycle, intracellular signaling, apoptosis, protein degradation, and innate and adaptive immunity are affected by Hsp90 inhibition [5, 19, 20, 21–23]. However, what are the global changes at the proteome level that occur upon Hsp90 inhibition? What are the functions affected in different cell types? What are the differences in response to Hsp90 inhibition in cancer cells compared to normal cells? Some of these questions are starting to be answered and they will be discussed in the following sections. It has been argued that Hsp90 is highly exploited by cancer cells to support client oncoproteins, which may require assistance because of overexpression and/or mutation [24]. The increased reliance on Hsp90 may explain the experimentally observed higher sensitivity of cancer cells to Hsp90 inhibitors. The increased sensitivity may also reflect the fact that the commonly observed aneuploidy of cancer cells causes a proteotoxic imbalance [25, 26]. Surprisingly, despite a strong interest in understanding this differential sensitivity, a systematic and proteome-wide study with genetically really comparable and matched normal and cancer cells has not yet been undertaken.

3 Perturbing the Hsp90-Dependent Proteome

Many efforts aimed at understanding the functions of Hsp90 have involved inhibiting, blocking, or overexpressing Hsp90 itself, and analyzing the general phenotypic effects, the effects on specific cellular processes, or client proteins. It is not the goal here to review all this information, which is extensively summarized on the regularly updated Picard lab web page (http://www.picard.ch/downloads/Hsp90facts.pdf). Our goal is to focus on the consequences of Hsp90 inhibition at the proteomic level. We will review what can be learned from these recent studies, notably by integrating all of them in a comprehensive analysis.

3.1 Early Approaches

It has been generally accepted that the net abundances of Hsp90 clients decrease upon Hsp90 inhibition due to proteasomal degradation [27, 28]. But, it was not until a few years ago that the global impact of this inhibition on the cell was explored. The first effort to characterize the global effects of Hsp90 inhibition on mammalian cells monitored the changes in the mRNA expression profile of human colon cancer cell lines treated with the Hsp90 inhibitor 17-allylamino-17-demethoxygeldanamycin (17-AAG), a derivative of the prototypical Hsp90 inhibitor geldanamycin (GA) [29]. Apart from the expected destabilization of Hsp90 clients, this work highlighted the global heat shock factor 1 (HSF1)-dependent upregulation of several heat shock proteins and molecular chaperones, which is due to the disruption of the Hsp90-mediated inhibition of this transcription factor [29, 30]. Another study by Maloney et al. [31], which examined the global changes in mRNA and protein levels induced by treatment of human ovarian cancer cells with 17-AAG, showed that around 3 % of all examined transcripts and 4 % of the detectable proteins were altered by the drug. A similar proteomic approach with human retinal pigment epithelial cells yielded a similar number of proteins altered in their accumulation by 17-AAG [32]. These early studies presented the idea that the sole inhibition of Hsp90 has pleiotropic effects not only affecting the stability of proteins but also affecting gene expression levels. Maloney et al. also found that the global levels of protein acetylation were reduced in treated cells, most likely due to the reduced levels of histone acetyltransferase-1 (HAT-1) protein [31], even in the presence of the histone deacetylase inhibitor tricostatin A (TSA). They proposed a therapeutic synergy between 17-AAG and TSA that comes about from combining the inhibition of Hsp90 with the anti-proliferative properties of TSA without the histone hyperacetylation that is normally induced by this drug [31]. This is an example that highlights the importance of these global analysis strategies in the quest for novel cancer therapeutics. The latter study used two-dimensional polyacrylamide gel electrophoresis (2D-PAGE) followed by protein identification through mass spectrometry. In another study, the Hsp90-dependent proteome was analyzed using liquid chromatography tandem mass spectrometry (LC-MS/MS) in combination with cleavable isotope-coded affinity tag (cICATTM) labeling of GA-treated anaplastic lymphoma kinase (ALK)-positive human anaplastic large-cell lymphoma cells [33]. Here, they found that 119 out of 176 cICAT-labeled peptides differentially accumulated 12 h after GA treatment, being depleted by almost 60% [33]. A similar study using isobaric tag for relative and absolute quantitation (iTRAQ)-labeled peptides investigated proteomic changes in human pancreatic cancer cell lines treated with the GA analog IPI-504 [34]. It found 20 downregulated and 42 upregulated proteins. Even though these global profiling reports were useful and provided interesting information, the small number of detected proteins did not fulfill the ambitious goals of the Hsp90 research community of revealing the full Hsp90-dependent proteome of a given cell type. Too many well-known Hsp90 clients (e.g., protein kinases) were missed, probably due to their low levels in the cell and the poor sensitivity of these methods. In order to tackle this issue, more sensitive techniques and more efficient pipelines were needed.

3.2 Quantitative Proteomic Approaches with Stable Isotope Labeling with Amino Acids in Cell Culture (SILAC)

A quantum leap has recently been made with several large scale studies that have tried to identify and to quantify the portion of the proteome that is perturbed when Hsp90 is inhibited, employing standard SILAC (stSILAC) [35] technology for quantitative proteomics [36–42]. These surveys comprise the assessment of the components of the inhibitor-bound Hsp90 complexes [37] or quantitative and kinase-targeted chemo-proteomic profiles [38, 40]. In addition, these studies covered the changing phosphoprotein profile of the proteome or the dynamic (over time) fluctuations in abundances during short- (6 h) and long-term (20 h) Hsp90 inhibition [39, 41].

The rest of the review will mainly focus on the comparative analysis of the studies performed by the three independent groups that conducted global quantitative proteomics primarily using SILAC approaches in combination with other methods. Their goal was to characterize the full dynamic range of changes in abundances over time after Hsp90 inhibition with the additional exploration in some cases of the kinome, the phosphoproteome, or of the effects of different drugs [39–42]. Despite the use of several different cell lines, inhibitors, drug concentrations, and incubation times, a comparative analysis of these datasets reveals several common features. Compiling the Log2 normalized H/L ratios from the datasets of the Sharma et al., Wu et al., and Fierro et al. papers [39–41], it is possible to extract a common group of 1223 detected proteins. As a first approach to analyze these combined data, we used a principal component analysis (PCA) as a dimension-reduction tool to reduce this large set of variables (five different cell lines) to a smaller set that still contains most of the information (Fig. 6.1a). The PCA highlights that the different cell lines respond to Hsp90 inhibitors differently with HeLa cells being the most divergent one

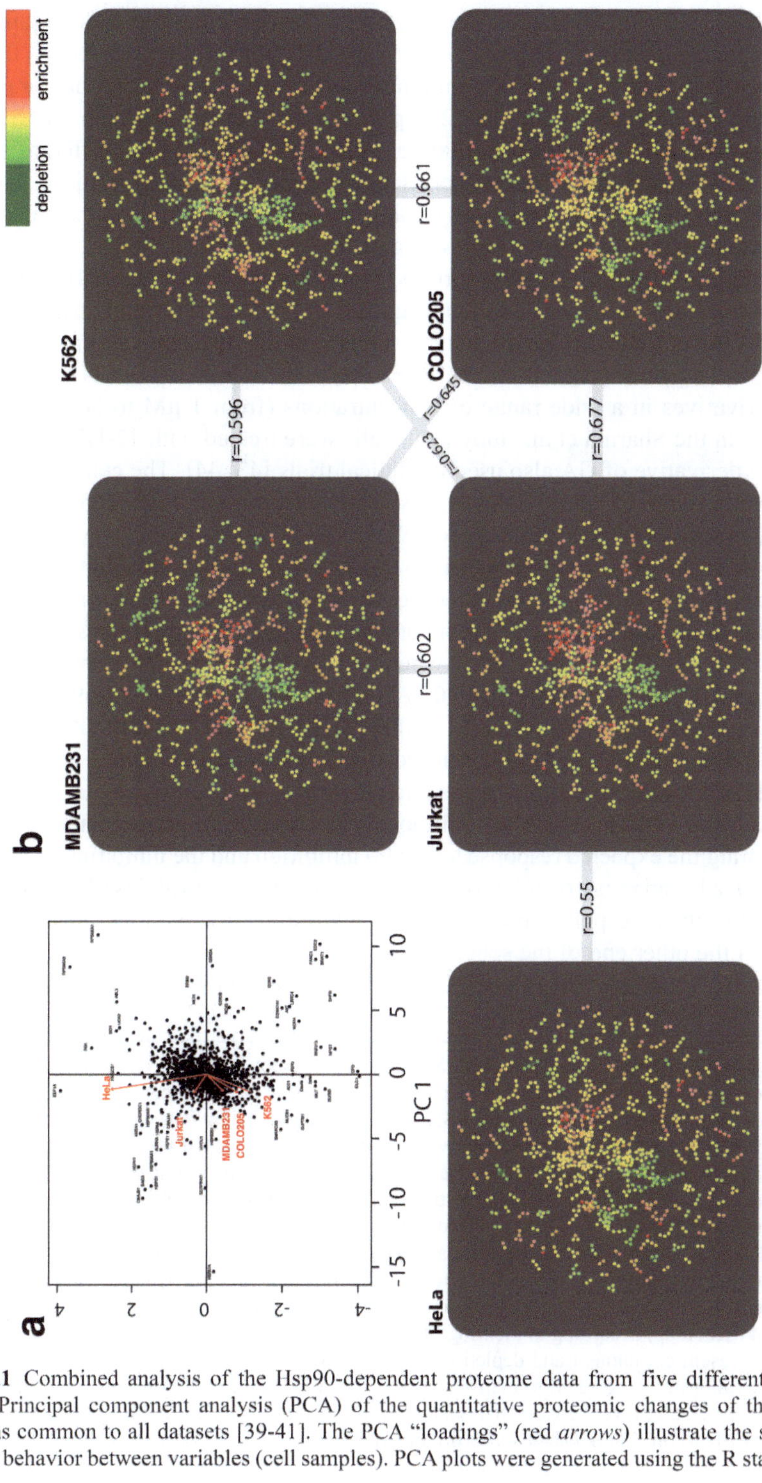

Fig. 6.1 Combined analysis of the Hsp90-dependent proteome data from five different datasets. **a** Principal component analysis (PCA) of the quantitative proteomic changes of the 1223 proteins common to all datasets [39-41]. The PCA "loadings" (red *arrows*) illustrate the similar/distant behavior between variables (cell samples). PCA plots were generated using the R statistical

(Fig. 6.1a). Although there are a large number of proteins that respond the same way (Fig. 6.1a; center of the scatterplot), there are also many outliers (proteins) that can define both the extremes of the common response or the particular behavior of each cell line. We next assembled the 1223 proteins of all five datasets in a correlation network, according to their quantitative proteomic data along the different experiments, by computing the Pearson correlation coefficient for all conditions. We used a cutoff of 0.97 in this analysis, and so, any correlation above this threshold value is again as an edge between two nodes (Fig. 6.1b). Out of which, 842 proteins showed a correlation (referred to here as group C proteins; shown in Fig. 6.1b) while 381 did not have any correlation partner. Figure 6.1b quantitatively and qualitatively illustrates the correlations between the responses of the different cell lines.

Depending on the specific experiment, cells were treated with GA [40, 41] or GA derivatives in a wide range of concentrations (from 1 μM to 50 μM) for 6 h to 24 h. In the Sharma et al. study [39], cells were treated with 17-DMAG, a more soluble derivative of GA also used in clinical trials [43, 44]. The cell lines were an interesting range of cancer lines representing solid tumors (Colo205, colon; MD-AMB231, breast [40]; HeLa, cervix [39]) to blood tumors (K562, erythroleukemia type [40] and Jurkat, T cell leukemia [41]). Remarkably, despite all the differences, one can identify a consistent signature of depleted and enriched proteins that is common to all cases (Fig. 6.2; bottom and top panels). Eleven proteins were significantly enriched (fold change$\geq$1.3) when cells were treated with Hsp90 inhibitors. They are: BAG3, DNAJB1, HSP90AA1 (Hsp90α), HSP90B1 (Grp94), HSPA1A (Hsp70), HSPA5, HSPB1 (Hsp27), HSPH1 (Hsp105), HYOU1, PDIA4, and SERPINH1 (Fig. 6.2). According to a survey of HSF1-dependent genes performed by Page et al. [45], almost all the genes encoding these proteins (with the exception of HYOU1 and PDIA4, which are still poorly characterized) are regulated by HSF1, confirming the expected response to Hsp90 inhibition and the inhibition of HSF1 by Hsp90 in a broad range of cell types. All these upregulated proteins play an essential role in the stress response, protein folding, cytoprotection, and resistance to apoptosis. At the other end of the spectrum, 11 proteins were systematically depleted in all studies. These are: BRAT1, CDC2, CDK5, FANCI, GSK3A, GSK3B, LRRC47,

package Modern Applied Statistics with S, MASS (R Project; http://www.r-project.org/). **b** Protein abundance correlation map of the Hsp90-dependent proteome. Proteins showing a significant change in their relative abundance by stSILAC in the five different datasets after treatment with Hsp90 inhibitors were organized in a network-based map using the software Cytoscape [62]. Correlation networks were computed with the Cytoscape plug-in ExpressionCorrelation (http://www.baderlab.org/Software/ExpressionCorrelation), which facilitates assembly of a network; nodes with similar "behavior" (Log2 normalized H/L ratios) were connected and clustered on the basis of the Pearson correlation coefficient for all pairwise comparisons. A cutoff of 0.97 was used and any correlations above this threshold value are displayed as an edge (*line*) between two nodes (proteins). Note that only group C proteins are shown. Nodes are colored with a *red-yellow-green* gradient (heat map) according to their log2 fold change values where *red-to-yellow* and *yellow-to-green* represent enrichment and depletion, respectively (see inset for color gradient). Each inset corresponding to the correlation map for each cell line is connected with links representing the Pearson correlation level between datasets. *stSILAC* Standard stable isotope labeling with amino acids in cell culture, *Hsp90* Heat shock protein 90

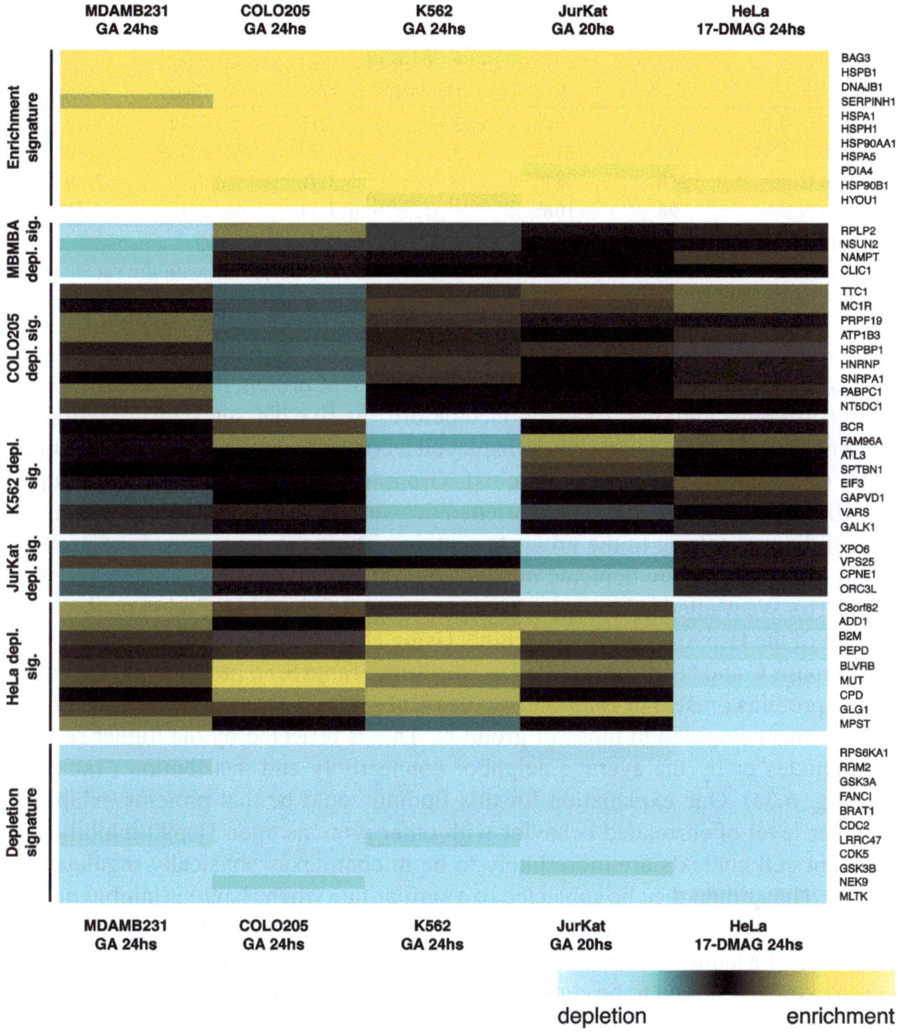

Fig. 6.2 Proteomic signatures of Hsp90 inhibition. Heat map depicting the proteins that are up- or downregulated by at least 1.3-fold and clustered by common enrichment and depletion signatures for the five datasets (*top* and *bottom* panels) or distinctive depletion signature for each cell line (intermediate panels). To generate the heat map, data were processed and clustered using the R statistical package MASS (R Project; http://www.r-project.org/). *Hsp90* Heat shock protein 90

MLTK, NEK9, RPS6KA1, and RRM2 (Fig. 6.2). According to our public database of the Hsp90 interactome, Hsp90Int (http://www.picard.ch/Hsp90Int) [3], CDK5, CDC2, GSK3A, GSK3B, RPS6KA1, and NEK9 interact with the Hsp90 chaperone machine. The signature of downregulated proteins contains protein kinases that are important for the progression of the cell cycle, and proteins involved in deoxyribonucleic acid (DNA) repair and synthesis. Interestingly, BRAT1, a protein proposed and validated as a novel Hsp90 client protein in the Fierro et al. study [41] is part

Table 6.1 Comparison of the topologies of the correlation (C) and non-correlation networks (N) with Hsp90Int

	# Nodes	# Nodes in Hsp90Int	# Edges in Hsp90Int	% Nodes in Hsp90Int	Ratio edges/ nodes	Avg. degree	Avg. neighbors
Group C	842	216	900	25.65	4.17	7.39	7.42
Group N	381	98	217	25.72	2.21	4.48	3.39
Random C	–	216	381[a]	–	1.76[a]	3.2[a]	2.79[a]
Random N	–	98	168[a]	–	1.71[a]	3.43[a]	2.71[a]

Hsp90Int: http://www.picard.ch/Hsp90Int

[a] Average out of five different random networks generated from Hsp90Int with the same size of groups C or N

of this signature, confirming that it is an Hsp90 client protein that is widely spread across different cell types. As mentioned before, the five datasets had 1223 proteins in common, of which 842 proteins displayed a correlation (group C), and 381 none at all (referred to as group N proteins). Group C consists of proteins that, upon Hsp90 inhibition, are depleted or enriched in a similar fashion as a cluster in different cell lines, contrary to the group N proteins. There are some groups of proteins (from group C) that are depleted in some cell lines and enriched in others (explore Fig. 6.1b). We wanted to know the common impact of Hsp90 inhibition on the interactome of the Hsp90 chaperone machine. Using Hsp90Int [3], we found that 25 % of both groups C and N are represented in Hsp90Int (Table 6.1). Intriguingly, as judged by the proteins present in Hsp90Int, proteins of group C have more protein-protein interactions compared to those of group N. This is reflected by the higher ratio of edges/nodes or by the average neighbor connectivity and distribution (Table 6.1 and Fig. 6.3a). One explanation for this finding could be that proteins exhibiting a higher level of correlated behavior with other proteins upon Hsp90 inhibition in different cell contexts are more likely to be in complexes physically regulated by Hsp90. They might then be subjected to a similar fate when Hsp90 is inhibited. This is supported by the fact that in the Wu et al. [40] paper; using the Comprehensive Resource of Mammalian protein complex database [46], members of some protein complexes were found to be coregulated by Hsp90 inhibition. In contrast, the group C proteins that are not present in Hsp90Int could be proteins that are indirectly (not physically) regulated by Hsp90, for example by genuine Hsp90 clients.

In addition to the signature that is common to all five datasets, there are also specific signatures for each cell type. These consist of proteins that are depleted upon Hsp90 inhibition (Fig. 6.2). This set of proteins reflects the cellular processes that are significantly affected by Hsp90 inhibition in a cell type-specific fashion (e.g., kinases, metabolic proteins, processing and modification of DNA and RNA). These proteins are poorly represented in Hsp90Int, which might be due to their cell-specific nature. Clearly, their cell specificity makes them very interesting new Hsp90 client candidates.

The general message about the global functions of Hsp90 from these proteomic studies is in accordance with what was already suspected from previous biochemical studies with individual proteins: depletion of functional groups of proteins involved in cellular homeostasis and in cancer development, and the enrichment of proteins

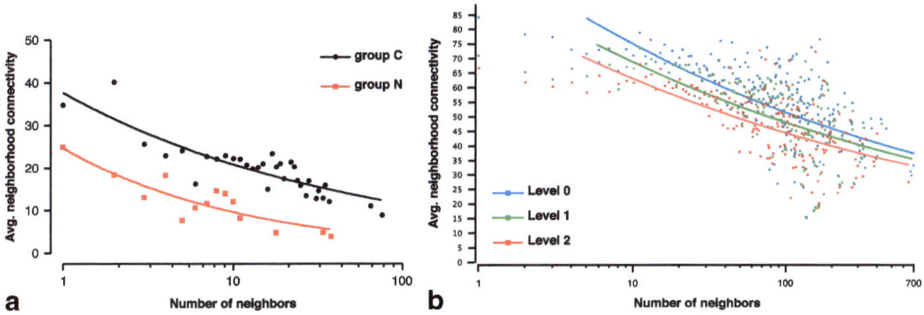

Fig. 6.3 Topology features of Hsp90-dependent networks. The neighborhood connectivity of a node *n* is the average connectivity of all neighbors of *n* [63]. The neighborhood connectivity distribution gives the average of the neighborhood connectivities of all nodes *n* with *k* neighbors for $k=0,1,\ldots$. **a** Neighborhood connectivity distribution of the correlated (group C) versus non-correlated (group N) proteins in the Hsp90Int network [3]. Each group of data points is fitted to a power-law. Topological parameters were computed with the Cytoscape plug-in NetworkAnalyzer (http://med.bioinf.mpi-inf.mpg.de/netanalyzer). **b** Neighborhood connectivity distribution of the human interactome without Hsp90 inhibition (Level 0), post inhibition with just the functional removal of Hsp90α and β (Level 1), and following the additional depletion of Hsp90 clients (Level 2). Each group of data points is fitted to a power-law. *Hsp90* Heat shock protein 90

involved in the stress response due the activation of HSF1. It is also possible to discover new and distinctive features from these studies. In Sharma et al. [40], the treatment of HeLa cells with 17-DMAG led to more downregulation because of the compromised chaperoning activity of Hsp90 than upregulation because of HSF1 activation. In their functional analysis, these authors determined which gene ontology (GO) groups deviated from the normal distribution. Linking these significant groups to Kyoto Encyclopedia of Genes and Genomes (KEGG) pathways, they found an enrichment of the pathways relating to the proteasome together with protein folding, unfolded protein binding, and response to unfolded protein [40]. In contrast, a significantly depleted group was constituted of proteins in the excision repair pathway (“cellular response to DNA damage,” “DNA modification,” “response to DNA damage stimulus,” and “regulation of transcription”). As expected, protein kinases were highly represented and downregulated (especially the protein tyrosine kinase subset). A special feature of this study was the analysis of the Hsp90-dependent phosphoproteome, which is significantly downregulated. More than half of the downregulated phosphorylation events contain “proline-directed” motifs, probably due to the decrease of cyclin-dependent kinases and mitogen-activated protein kinase (MAPK) signaling. As a novelty, the paper highlighted the downregulation of the sphingolipid metabolism, a pathway that has been implicated in cancer [47]. Another remarkable finding of this report is the fact that larger proteins are more prompt to be degraded than smaller ones; this is reflected by the increase in the median molecular weight of downregulated proteins. One of the ultimate goals of using Hsp90 inhibitors is to block the development of cancer processes. In this context, Sharma et al. [40] discussed some positive effects of treating Hela cells with 17-DMAG. These are the following: Hsp90 inhibition downregulates ROR2

(Wnt receptor), a protein overexpressed in oral and renal cancer, and in osteosarcoma [48, 49], and upregulates ADAMTS1, a protein that has been shown to inhibit angiogenesis by interacting with vascular endothelial growth factor A [50]. Some undesired effects of Hsp90 inhibition were also exposed in this paper. It reported that 17-DMAG causes the downregulation of LRIG1, a proposed tumor suppressor and feedback attenuator of signaling by receptor tyrosine kinases. Intriguingly, the downregulation of LRIG1 had been observed in several cancers and cancer lines [51–53]. In addition, the tumor-associated antigen L6 (TM4SF1), a key regulator of pathological angiogenesis involved in cancer invasion and metastasis [54], is also upregulated.

The study reported by Wu et al. covered a broader range of cell lines and its results also agree with those of the previously discussed study in that after 24 h of Hsp90 inhibition; the whole proteome exhibits a higher level of downregulated (52–65 %) than upregulated proteins (36–48 %) [40]. Using an affinity precipitation of kinases with kinobeads, they could observe that this downregulation was more dramatic for protein kinases reaching 89–92 % depending on the cell type [40]. While the general functional analysis in this report matches that of the previous one, a cell line-specific examination gives interesting information about the different responses elicited upon Hsp90 inhibition in blood cancer versus solid tumors. For instance, the GO term oxygen transport is only significantly present in the dataset obtained with K562 cells (blood). Clathrin-mediated endocytosis is a feature of many receptor-regulated processes, and it is increased in all solid cancer lines but not in the blood cell line. Peroxisome proliferator activated receptor α/retinoid X receptor α (PPARα/RXRα) activation has been implicated in several cancers [55, 56], and these proteins are increased in all cell lines but those of the colon. Ephrin receptor signaling is strongly downregulated in all three solid tumor lines but not in the leukemia cell line. The downregulation of mammalian target of rapamycin (mTOR) signaling is much more pronounced in the colon cancer line Colo205 and the breast cancer line MDAMB231 than in the other two cell lines examined in this study [40]. The full analysis of all these three major global proteomic studies clearly shows that the responses of proteins to Hsp90 inhibition do include a cell-type specific component (Figs. 6.1 and 6.2).

As we mentioned before, it seems that the proteins of the same complex are doomed to have a similar fate when Hsp90 is inhibited. Examples in the Wu et al. paper are a nice validation of that. The core members of the kinase maturation complex 1 (Hsp90α and β, CDC37, Hsp70) and the Arp2/3 complex (a cellular complex responsible for actin filament nucleation and branching) are all upregulated. The components of the DNA-PK-Ku-eIF2-NF90-NF45 complex (DNA repair), the Nop56p-associated pre-ribosomal ribonucleoprotein complex (ribosome biogenesis), the spliceosome (mRNA splicing), and the MNK1-eIF4F complex (protein translation initiation) are all downregulated [40]. An interesting analysis for this kind of global studies of the Wu et al. paper was inspired after the detection, in a time course performed by immunoblotting for ten proteins, that some proteins were much more rapidly removed from cells upon drug treatment than others. These different kinetics of loss of proteins might be related to the individual protein turnover rates. To understand this, the authors determined protein turnover by pulsed SILAC of untreated cells at 24 h. When comparing the turnover rates of kinases versus the extent of downregulation by Hsp90 inhibition (24 h pulsed SILAC versus 24 h GA treatment), they found that these values correlate for many proteins, suggesting that even when proteins depend on Hsp90, their removal from cells primarily depends on their intrinsic turnover rates. For example, the protein kinase DDR1a has a half-life of 7 h, which is quite similar to the time required to deplete 50 % of DDR1 from cells upon GA treatment. At the same time, few proteins are degraded faster than their intrinsic half-lives impose, suggesting that they may be in immature complexes that are highly dependent on Hsp90, and that they are quickly degraded when they are abandoned on their own. Conversely, other proteins show a delayed degradation, presumably because of the induced synthesis of some regulatory partner elicited by Hsp90 inhibition.

As a first step towards a comparative proteomic profiling of different Hsp90 inhibitors, Wu et al. analyzed the quantitative differences in a subpopulation of the proteome in kinases in cells treated with the Hsp90 inhibitors GA or PU-H71. Under these conditions, they could observe a high degree of correlation in the response to both drugs, suggesting that the mechanism by which these drugs inhibit Hsp90 and the consequences of this inhibition are similar, at least for kinases.

To prove that the proteomic profiling data and subsequent bioinformatic analysis can provide a wealth of meaningful information that can be biologically interpreted to fight cancer more effectively, Wu et al. considered combinatorial drug strategies. Since mTOR signaling in Cal27 cells was strongly inhibited by Hsp90 inhibition and these cells are dependent on the epidermal growth factor (EGF) receptor for growth, they concluded that it may be worth combining GA with lapatinib (a highly selective EGF receptor inhibitor) or testing PF-04691502 (a potent dual PI3K/mTOR inhibitor) by itself. Indeed, the combination of GA and lapatinib was more effective than either drug alone, and PF-04691502 efficiently killed Cal27 cells [40]. This paper also discusses the potentially undesired effects of treating cancer with Hsp90 inhibitors. The treatment with GA could lead to the upregulation of 15 kinases, including AURA, PERK, and AXL, and the downregulation of p53 in MDA-MB-231 cells. It is also expected that the upregulation of the chaperone machinery itself may delay cancer cell death and thus increase the required drug doses [57, 58].

3.3 Novel Quantitative Modeling and Network-Based Approaches

The most recently published investigation using SILAC [41] includes some novel experimental strategies. These authors studied the dynamic changes in the Hsp90-dependent proteome using a stSILAC approach, but by comparing two time points (6 h and 20 h) after the treatment of Jurkat cells with GA. Among the short-term effects, it is worth mentioning the upsurge of proteins involved in cell survival pathways (stress response), a transient general inhibition of protein synthesis, and a rapid decrease in some protein kinases and the T-cell receptor signaling pathway. As long-term effects, they observed a decrease in the levels of RNA polymerase II in proteins related with the regulation of splicing, the biosynthesis of macromolecules, RNA transport, translation initiation and elongation, rRNA processing, ribonucleoprotein complexes, and ribosome biogenesis. Other groups of proteins that are only depleted at later times include proteins involved in cell-cycle progression and proteins linked to kinetochore and condensed chromosome functions (DNA maintenance). In other studies, these had been linked with aneuploidy and overduplication of functional centrosomes [59]. After 20 h of treatment with GA, there is also a depletion of proteins implicated in nucleo-cytoplasmic transport.

In addition, they applied pulse-chase SILAC (pcSILAC), a novel labelling scheme, which is an improvement of the pulsed-SILAC method [60]. With this pcSILAC [42] methodology, one can compare the decay rate constants and synthesis rates for two biological conditions and model their relative contributions to the changes in net protein levels. Specifically, the sequential metabolic labeling was used to differentiate the effects on preexistent versus newly synthesized proteins. The fluctuations observed in the relative levels of preexisting proteins in the two conditions (GA versus untreated) was used to calculate decay rates. The population of proteins synthesized upon GA treatment was monitored and used to estimate net de novo synthesis rates. The main message of Fierro et al. [41] is that the key role played by Hsp90 in gene expression imposes a global reduction in net de novo protein synthesis, which appears to be greater in magnitude than the concomitantly observed global increase in protein decay. Globally, the median of total protein half-life decreased from 55.9 h in the control to 32.0 h in the GA-treated cells. Surprisingly, for molecular chaperones and co-chaperones, the increase in synthesis is accompanied by a shorter half-life. Proteins in charge of DNA repair and the cell cycle showed a greater decrease in their synthesis (most of them below average) than a decrease in half-life.

These investigations yielded a slew of other interesting insights. GA treatment induces an arrest of the cell cycle. A flow cytometry analysis showed an enrichment of cells blocked either in the G1 or in the G2/M phases. Under the specific experimental conditions, there was no strong induction of apoptosis. By integrating the quantitative proteomic information in maps of protein-protein interactions, a network analysis could be performed. Several highly interconnected subgraphs and functional modules were extracted from Hsp90Int [3] to display dynamic changes in protein complexes present in the stSILAC dataset. Remarkably, while several cell cycle-regulated Hsp90 clients are, as expected, degraded upon Hsp90 inhibition, their ubiquitination machinery eventually succumbs as well. As partially shown by Wu et al., Fierro et al. further emphasized that the Hsp90 chaperone machine [3] is considerably remodeled in response to GA. In addition to the previous reports, Fierro et al. specifically evaluated the oncoproteins and tumor suppressors that were significantly affected by Hsp90 inhibition. They found that among the proteins that are degraded and induced, there were critical areas of the cancer-related proteome that one may not want to change when treating cancer. At least in Jurkat cells, GA provokes the depletion of a whole panel of tumor suppressors. Conversely, oncoproteins such as several members of the Ras superfamily of GTPases together with Rho proteins and regulators of the cell cycle, migration, metastasis, and angiogenesis are significantly upregulated.

3.4 Hsp90 Wires the Cellular Protein Networks

Molecular chaperones in general are sensitive hubs of the proteostasis network. Hsp90 in particular is a key hub of the human interactome. As of today, if one retrieves a curated human protein-protein interactome, it contains a network of almost 12,000 proteins with more than 70,000 reported interactions. Right after drugging Hsp90, this interactome "loses" only two members (i.e., Hsp90α and Hsp90β), but the number of interactions falls by at least 1200, with a huge impact on the topology of the network (Fig. 6.3b). Sometime later, the inhibition produces a significant ($\geq$1.3-fold) depletion of approximately 500 proteins, and as a consequence, there is a drop of around 8000 extra interactions in the human interactome. In undirected networks, a connected component is a subgraph where any two nodes are connected to each other by paths (a string of connected nodes), and which is not connected to any other nodes in the supergraph (the whole network). In the human interactome, the inhibition of Hsp90 provokes a twofold increase in the number of these isolated subgraphs (from 170 to 342), which is a sign of an increased fragmentation of the network. The analysis of topological parameters shows that the transition from basal state (Level 0) to Hsp90 inhibition (Level 1) and later client depletion (Level 2) produces a perturbation in the human interactome that reduces the number of physically connected neighbors that a protein possesses with the additional reduction in the connectivity of its neighbors to other proteins (Fig. 6.3b). As a result, the human interactome becomes laxer and presumably less efficient in communicating changes and stimuli to maintain cellular homeostasis.

4 Discussion

It was clearly demonstrated in all the discussed global proteomic analyses that the inhibition of Hsp90 globally produces more downregulation than upregulation. Fierro et al. [41] shed more light on the underlying mechanisms by discriminating the decay and synthesis rates that contribute to the net levels of proteins upon exposure of cells to Hsp90 inhibitors. This recalls that an observed downregulation of a particular protein is not necessarily due to degradation. It could be due to a mixture of varying levels of decay and synthesis. The same holds true for the upregulation of specific proteins. It remains to be understood if the large number of proteins with correlated behaviors in response to Hsp90 inhibition actually have similar decay and/or synthesis rates or only a similar change in net abundance. These studies also expose the fact that inhibiting Hsp90 is not exclusively beneficial in the context of cancer therapy. The induction of the stress response [61], the upregulation of some oncogenes, and the downregulation of some tumor suppressors could lead to undesired side effects and compromise the therapeutic benefits of Hsp90 inhibition. A detailed knowledge of how the human proteome is changed and shaped by targeting Hsp90 could set the stage for the pharmacoproteomic profiling of responses to sets of different inhibitors and of cancer types with different molecular backgrounds. These drug profiling analyses will be essential for the development of tailored Hsp90 inhibitors for the treatment of different tumor types and for the use in personalized medicine. Curiously, the differential perturbation that Hsp90 inhibition elicits in a normal cell compared to a cancer cell remains largely unexplored.

Acknowledgments The authors would like to thank Manfredo Quadroni (University of Lausanne, Lausanne, Switzerland) for providing the common quantitative proteomic data for all the datasets analyzed in this chapter, and Tai Wang (University of Geneva, Picard Lab) for helping in the preparation of Figs. 6.1 and 6.2.

References

1. Oppermann H, Levinson W, Bishop JM (1981) A cellular protein that associates with the transforming protein of Rous sarcoma virus is also a heat-shock protein. Proc Natl Acad Sci U S A 78:1067–1071
2. Brugge JS, Erikson E, Erikson RL (1981) The specific interaction of the Rous sarcoma virus transforming protein, pp60src, with two cellular proteins. Cell 25:363–372
3. Echeverría PC, Bernthaler A, Dupuis P, Mayer B, Picard D (2011) An interaction network predicted from public data as a discovery tool: application to the Hsp90 molecular chaperone machine. PLoS ONE 6:e26044
4. Jakob U, Lilie H, Meyer I, Buchner J (1995) Transient interaction of hsp90 with early unfolding intermediates of citrate synthase - implications for heat shock in vivo. J Biol Chem 270:7288–7294
5. Taipale M, Krykbaeva I, Koeva M, Kayatekin C, Westover KD, Karras GI, Lindquist S (2012) Quantitative analysis of HSP90-client interactions reveals principles of substrate recognition. Cell 150:987–1001
6. Borkovich KA, Farrelly FW, Finkelstein DB, Taulien J, Lindquist S (1989) Hsp82 is an essential protein that is required in higher concentrations for growth of cells at higher temperatures. Mol Cell Biol 9:3919–3930
7. Finka A, Goloubinoff P (2013) Proteomic data from human cell cultures refine mechanisms of chaperone-mediated protein homeostasis. Cell Stress Chaperones 18:591–605
8. Sorger PK, Pelham HR (1987) The glucose-regulated protein grp94 is related to heat shock protein hsp90. J Mol Biol 194:341–344
9. Felts SJ, Owen BA, Nguyen P, Trepel J, Donner DB, Toft DO (2000) The hsp90-related protein TRAP1 is a mitochondrial protein with distinct functional properties. J Biol Chem 275:3305–3312
10. Richter K, Muschler P, Hainzl O, Buchner J (2001) Coordinated ATP hydrolysis by the Hsp90 dimer. J Biol Chem 276:33689–33696
11. Hessling M, Richter K, Buchner J (2009) Dissection of the ATP-induced conformational cycle of the molecular chaperone Hsp90. Nat Struct Mol Biol 16:287–293

12. Massey AJ (2010) ATPases as drug targets: insights from heat shock proteins 70 and 90. J Med Chem 53:7280–7286
13. Wang T, Echeverria PC, Picard D (2013) Overview of molecular chaperones in health and disease. Inhibitors of molecular chaperones as therapeutic agents; Machajewski TD, Gao Z, Eds.; RSC Publishing: Cambridge
14. McClellan AJ, Frydman J (2001) Molecular chaperones and the art of recognizing a lost cause. Nat Cell Biol 3:E51–E53
15. Meacham GC, Patterson C, Zhang W, Younger JM, Cyr DM (2001) The Hsc70 co-chaperone CHIP targets immature CFTR for proteasomal degradation. Nat Cell Biol 3:100–105
16. Murata S, Chiba T, Tanaka K (2003) CHIP: a quality-control E3 ligase collaborating with molecular chaperones. Int J Biochem Cell Biol 35:572–578
17. Whitesell L, Mimnaugh EG, De Costa B, Myers CE, Neckers LM (1994) Inhibition of heat shock protein HSP90-pp60v-src heteroprotein complex formation by benzoquinone ansamycins: essential role for stress proteins in oncogenic transformation. Proc Natl Acad Sci U S A 91:8324–8328
18. Mimnaugh EG, Chavany C, Neckers L (1996) Polyubiquitination and proteasomal degradation of the p185c-erbB-2 receptor protein-tyrosine kinase induced by geldanamycin. J Biol Chem 271:22796–22801
19. Yeyati PL, Bancewicz RM, Maule J, van Heyningen V (2007) Hsp90 selectively modulates phenotype in vertebrate development. PLoS Genet 3:e43
20. Sawarkar R, Sievers C, Paro R (2012) Hsp90 globally targets paused RNA polymerase to regulate gene expression in response to environmental stimuli. Cell 149:807–818
21. Bishop SC, Burlison JA, Blagg BS (2007) Hsp90: a novel target for the disruption of multiple signaling cascades. Curr Cancer Drug Targets 7:369–388
22. Sato N, Yamamoto T, Sekine Y, Yumioka T, Junicho A, Fuse H, Matsuda T (2003) Involvement of heat-shock protein 90 in the interleukin-6-mediated signaling pathway through STAT3. Biochem Biophys Res Commun 300:847–852
23. Kunisawa J, Shastri N (2006) Hsp90α chaperones large C-terminally extended proteolytic intermediates in the MHC class I antigen processing pathway. Immunity 24:523–534
24. Whitesell L, Lindquist SL (2005) HSP90 and the chaperoning of cancer. Nat Rev Cancer 5:761–772
25. Tang YC, Williams BR, Siegel JJ, Amon A (2011) Identification of aneuploidy-selective antiproliferation compounds. Cell 144:499–512
26. Oromendia AB, Dodgson SE, Amon A (2012) Aneuploidy causes proteotoxic stress in yeast. Genes Dev 26:2696–2708
27. da Silva VC, Ramos CH (2012) The network interaction of the human cytosolic 90 kDa heat shock protein Hsp90: a target for cancer therapeutics. J Proteomics 75:2790–2802
28. Taipale M, Jarosz DF, Lindquist S (2010) HSP90 at the hub of protein homeostasis: emerging mechanistic insights. Nat Rev Mol Cell Biol 11:515–528
29. Clarke PA, Hostein I, Banerji U, Stefano FD, Maloney A, Walton M, Judson I, Workman P (2000) Gene expression profiling of human colon cancer cells following inhibition of signal transduction by 17-allylamino-17-demethoxygeldanamycin, an inhibitor of the hsp90 molecular chaperone. Oncogene 19:4125–4133
30. Zou J, Guo Y, Guettouche T, Smith DF, Voellmy R (1998) Repression of heat shock transcription factor HSF1 activation by HSP90 (HSP90 complex) that forms a stress-sensitive complex with HSF1. Cell 94:471–480
31. Maloney A, Clarke PA, Naaby-Hansen S, Stein R, Koopman JO, Akpan A, Yang A, Zvelebil M, Cramer R, Stimson L, Aherne W, Banerji U, Judson I, Sharp S, Powers M, deBilly E, Salmons J, Walton M, Burlingame A, Waterfield M, Workman P (2007) Gene and protein expression profiling of human ovarian cancer cells treated with the heat shock protein 90 inhibitor 17-allylamino-17-demethoxygeldanamycin. Cancer Res 67:3239–3253
32. Yao JQ, Liu QH, Chen X, Yang Q, Xu ZY, Hu F, Wang L, Li JM (2010) Hsp90 inhibitor 17-allylamino-17-demethoxygeldanamycin inhibits the proliferation of ARPE-19 cells. J Biomed Sci 17:30

33. Schumacher JA, Crockett DK, Elenitoba-Johnson KS, Lim MS (2007) Proteome-wide changes induced by the Hsp90 inhibitor, geldanamycin in anaplastic large cell lymphoma cells. Proteomics 7:2603–2616
34. Song D, Chaerkady R, Tan AC, Garcia-Garcia E, Nalli A, Suarez-Gauthier A, Lopez-Rios F, Zhang XF, Solomon A, Tong J, Read M, Fritz C, Jimeno A, Pandey A, Hidalgo M (2008) Antitumor activity and molecular effects of the novel heat shock protein 90 inhibitor, IPI-504, in pancreatic cancer. Mol Cancer Ther 7:3275–3284
35. Ong SE, Blagoev B, Kratchmarova I, Kristensen DB, Steen H, Pandey A, Mann M (2002) Stable isotope labeling by amino acids in cell culture, SILAC, as a simple and accurate approach to expression proteomics. Mol Cell Proteomics 1:376–386
36. Caldas-Lopes E, Cerchietti L, Ahn JH, Clement CC, Robles AI, Rodina A, Moulick K, Taldone T, Gozman A, Guo Y, Wu N, de Stanchina E, White J, Gross SS, Ma Y, Varticovski L, Melnick A, Chiosis G (2009) Hsp90 inhibitor PU-H71, a multimodal inhibitor of malignancy, induces complete responses in triple-negative breast cancer models. Proc Natl Acad Sci U S A 106:8368–8373
37. Moulick K, Ahn JH, Zong H, Rodina A, Cerchietti L, Gomes Dagama EM, Caldas-Lopes E, Beebe K, Perna F, Hatzi K, Vu LP, Zhao X, Zatorska D, Taldone T, Smith-Jones P, Alpaugh M, Gross SS, Pillarsetty N, Ku T, Lewis JS, Larson SM, Levine R, Erdjument-Bromage H, Guzman ML, Nimer SD, Melnick A, Neckers L, Chiosis G (2011) Affinity-based proteomics reveal cancer-specific networks coordinated by Hsp90. Nat Chem Biol 7:818–826
38. Haupt A, Joberty G, Bantscheff M, Frohlich H, Stehr H, Schweiger MR, Fischer A, Kerick M, Boerno ST, Dahl A, Lappe M, Lehrach H, Gonzalez C, Drewes G, Lange BM (2012) Hsp90 inhibition differentially destabilises MAP kinase and TGF-β signalling components in cancer cells revealed by kinase-targeted chemoproteomics. BMC Cancer 12:38
39. Sharma K, Vabulas RM, Macek B, Pinkert S, Cox J, Mann M, Hartl FU (2012) Quantitative proteomics reveals that Hsp90 inhibition preferentially targets kinases and the DNA damage response. Mol Cell Proteomics 11:M111.014654
40. Wu Z, Moghaddas Gholami A, Kuster B (2012) Systematic identification of the Hsp90 candidate regulated proteome. Mol Cell Proteomics 11:M111.016675
41. Fierro-Monti I, Echeverria P, Racle J, Hernandez C, Picard D, Quadroni M (2013) Dynamic impacts of the inhibition of the molecular chaperone Hsp90 on the T-cell proteome have implications for anti-cancer therapy. PLoS ONE 8:e80425
42. Fierro-Monti I, Racle J, Hernandez C, Waridel P, Hatzimanikatis V, Quadroni M (2013) A novel pulse-chase SILAC strategy measures changes in protein decay and synthesis rates induced by perturbation of proteostasis with an Hsp90 inhibitor. PLoS ONE 8:e80423
43. Jez JM, Chen JC, Rastelli G, Stroud RM, Santi DV (2003) Crystal structure and molecular modeling of 17-DMAG in complex with human Hsp90. Chem Biol 10:361–368
44. Pacey S, Wilson RH, Walton M, Eatock MM, Hardcastle A, Zetterlund A, Arkenau HT, Moreno-Farre J, Banerji U, Roels B, Peachey H, Aherne W, de Bono JS, Raynaud F, Workman P, Judson I (2011) A phase I study of the heat shock protein 90 inhibitor alvespimycin (17-DMAG) given intravenously to patients with advanced solid tumors. Clin Cancer Res 17:1561–1570
45. Page TJ, Sikder D, Yang L, Pluta L, Wolfinger RD, Kodadek T, Thomas RS (2006) Genome-wide analysis of human HSF1 signaling reveals a transcriptional program linked to cellular adaptation and survival. Mol Biosyst 2:627–639
46. Ruepp A, Brauner B, Dunger-Kaltenbach I, Frishman G, Montrone C, Stransky M, Waegele B, Schmidt T, Doudieu ON, Stumpflen V, Mewes HW (2008) CORUM: the comprehensive resource of mammalian protein complexes. Nucleic Acids Res 36:D646–D650
47. Ponnusamy S, Meyers-Needham M, Senkal CE, Saddoughi SA, Sentelle D, Selvam SP, Salas A, Ogretmen B (2010) Sphingolipids and cancer: ceramide and sphingosine-1-phosphate in the regulation of cell death and drug resistance. Future Oncol 6:1603–1624
48. Wright TM, Brannon AR, Gordan JD, Mikels AJ, Mitchell C, Chen S, Espinosa I, van de Rijn M, Pruthi R, Wallen E, Edwards L, Nusse R, Rathmell WK (2009) Ror2, a developmentally regulated kinase, promotes tumor growth potential in renal cell carcinoma. Oncogene 28:2513–2523

49. Enomoto M, Hayakawa S, Itsukushima S, Ren DY, Matsuo M, Tamada K, Oneyama C, Okada M, Takumi T, Nishita M, Minami Y (2009) Autonomous regulation of osteosarcoma cell invasiveness by Wnt5a/Ror2 signaling. Oncogene 28:3197–3208
50. Luque A, Carpizo DR, Iruela-Arispe ML (2003) ADAMTS1/METH1 inhibits endothelial cell proliferation by direct binding and sequestration of VEGF165. J Biol Chem 278:23656–23665
51. Gur G, Rubin C, Katz M, Amit I, Citri A, Nilsson J, Amariglio N, Henriksson R, Rechavi G, Hedman H, Wides R, Yarden Y (2004) LRIG1 restricts growth factor signaling by enhancing receptor ubiquitylation and degradation. EMBO J 23:3270–3281
52. Laederich MB, Funes-Duran M, Yen L, Ingalla E, Wu X, Carraway KL 3rd, Sweeney C (2004) The leucine-rich repeat protein LRIG1 is a negative regulator of ErbB family receptor tyrosine kinases. J Biol Chem 279:47050–47056
53. Segatto O, Anastasi S, Alema S (2011) Regulation of epidermal growth factor receptor signalling by inducible feedback inhibitors. J Cell Sci 124:1785–1793
54. Kao YR, Shih JY, Wen WC, Ko YP, Chen BM, Chan YL, Chu YW, Yang PC, Wu CW, Roffler SR (2003) Tumor-associated antigen L6 and the invasion of human lung cancer cells. Clin Cancer Res 9:2807–2816
55. Nagao Y, French BA, Cai Y, French SW, Wan YJ (1998) Inhibition of PPAR alpha/RXR alpha-mediated direct hyperplasia pathways during griseofulvin-induced hepatocarcinogenesis. J Cell Biochem 69:189–200
56. Li MY, Yuan H, Ma LT, Kong AW, Hsin MK, Yip JH, Underwood MJ, Chen GG (2010) Roles of peroxisome proliferator-activated receptor-alpha and -gamma in the development of non-small cell lung cancer. Am J Respir Cell Mol Biol 43:674–683
57. Guo F, Rocha K, Bali P, Pranpat M, Fiskus W, Boyapalle S, Kumaraswamy S, Balasis M, Greedy B, Armitage ES, Lawrence N, Bhalla K (2005) Abrogation of heat shock protein 70 induction as a strategy to increase antileukemia activity of heat shock protein 90 inhibitor 17-allylamino-demethoxy geldanamycin. Cancer Res 65:10536–10544
58. Goloudina AR, Demidov ON, Garrido C (2012) Inhibition of HSP70: a challenging anti-cancer strategy. Cancer Lett 325:117–124
59. Chen G, Bradford WD, Seidel CW, Li R (2012) Hsp90 stress potentiates rapid cellular adaptation through induction of aneuploidy. Nature 482:246–250
60. Schwanhausser B, Gossen M, Dittmar G, Selbach M (2009) Global analysis of cellular protein translation by pulsed SILAC. Proteomics 9:205–209
61. Duerfeldt AS, Blagg BS (2010) Hsp90 inhibition: elimination of shock and stress. Bioorg Med Chem Lett 20:4983–4987
62. Cline MS, Smoot M, Cerami E, Kuchinsky A, Landys N, Workman C, Christmas R, Avila-Campilo I, Creech M, Gross B, Hanspers K, Isserlin R, Kelley R, Killcoyne S, Lotia S, Maere S, Morris J, Ono K, Pavlovic V, Pico AR, Vailaya A, Wang PL, Adler A, Conklin BR, Hood L, Kuiper M, Sander C, Schmulevich I, Schwikowski B, Warner GJ, Ideker T, Bader GD (2007) Integration of biological networks and gene expression data using Cytoscape. Nat Protoc 2:2366–2382
63. Maslov S, Sneppen K (2002) Specificity and stability in topology of protein networks. Science 296:910–913

Chapter 7
Designing Drugs Against Hsp90 for Cancer Therapy

Stefan O. Ochiana, Tony Taldone and Gabriela Chiosis

Abstract Heat shock protein 90 (Hsp90) is a molecular chaperone exploited by cancer cells to aid the function of numerous oncoproteins. The recognition of Hsp90 as a critical facilitator for oncogene addiction and survival of the cancer cell has opened a promising new niche for cancer treatment. The serendipitous discovery that the broad spectrum anticancer activity of the natural products geldanamycin (GM) and radicicol (RD) was a result of inhibition of Hsp90 resulted in the development of improved derivatives of these natural products. One of these was 17-allylaminogeldanamycin (17-AAG), a closely related analog of GM, and was in fact the first Hsp90 inhibitor to enter the clinic. However, GM and its analogs suffer from poor "drug-like" properties and this served as a strong impetus for the development of novel synthetic Hsp90 inhibitors. These efforts resulted in the development of numerous potent synthetic small molecule inhibitors with significant scaffold diversity as well as superior pharmacokinetic and toxicity profile to have entered clinical trials. This review highlights the drug discovery efforts pertaining to the development of the first and second-generation Hsp90 inhibitors, and also gleans over their individual promise as clinical agents for anticancer therapy.

1 Introduction

Heat shock proteins were first reported in the early 1960s when they were serendipitously discovered by the Italian scientist Ferruccio Ritossa [1] who observed that cells could mount very strong transcriptional activity upon exposure to heat. Subsequent work undertaken by Tissieres et al. led to the discovery of heat shock proteins by means of electrophoretically studying the newly synthesized proteins

S. O. Ochiana (✉) · T. Taldone · G. Chiosis
Department of Molecular Pharmacology and Chemistry, Memorial Sloan-Kettering Cancer Center, New York, NY, USA
e-mail: ochianas@mskcc.org

T. Taldone
e-mail: taldonet@mskcc.org

G. Chiosis
e-mail: chiosisg@mskcc.org

W. A. Houry (ed.), *The Molecular Chaperones Interaction Networks in Protein Folding and Degradation,* Interactomics and Systems Biology, DOI 10.1007/978-1-4939-1130-1_7,

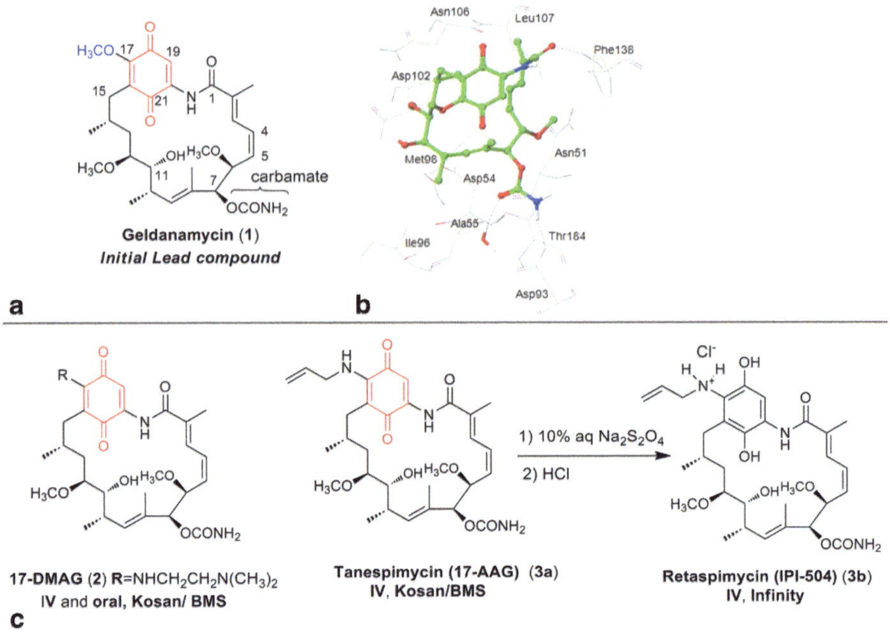

Fig. 7.1 **a** Structure of Geldanamycin, the first reported Hsp90 inhibitor. **b** Structure of human Hsp90α bound to (**1**; PDB ID: 1YET), H-bonds are shown by *dotted lines*. **c** Ansamycin inhibitors of Hsp90; in *red* is shown the *benzoquinone* (i.e., *quinone ring*) moiety and in *blue* the methoxy group at C17

in heat-shocked *Drosophila* larvae [2]. Over the course of a few decades, various research groups have shown that the vast majority of these proteins are molecular chaperones that aid the correct folding of proteins and guard the proteome from the threat of misfolding and aggregation [3, 4]. A noteworthy member of this family is heat shock protein 90 (Hsp90), a chaperone that employs a complex cycle of client protein binding and hydrolysis of adenosine triphosphate (ATP) to regulate both the stability and activity of its client proteins, including kinases, steroid hormone receptors, and transcription factors [5, 6]. Furthermore, when the Hsp90 chaperone cycle is inhibited, its client proteins become destabilized, are ubiquitinated, and consequently degraded by the proteasome [7–9]. The knowledge that many Hsp90 client proteins (e.g., human epidermal growth factor receptor 2, HER2; AKT; met proto-oncogene (MET); epidermal growth factor receptor, EGFR; estrogen receptor (ER); Raf-1) are known to play significant roles in the development and promotion of cancer has sparked a strong interest in this chaperone as a viable target for cancer therapy [10–12].

As is often the case in drug discovery, natural products played an important role in the discovery of Hsp90 inhibitors. In fact, the very concept of inhibition of Hsp90 as a viable anticancer strategy was demonstrated with two seminal natural products, geldanamycin (GM; Fig. 7.1a) and radicicol (RD; Fig. 7.2a). GM is a benzoquinone ansamycin which was first isolated in 1970 from a fermentation broth

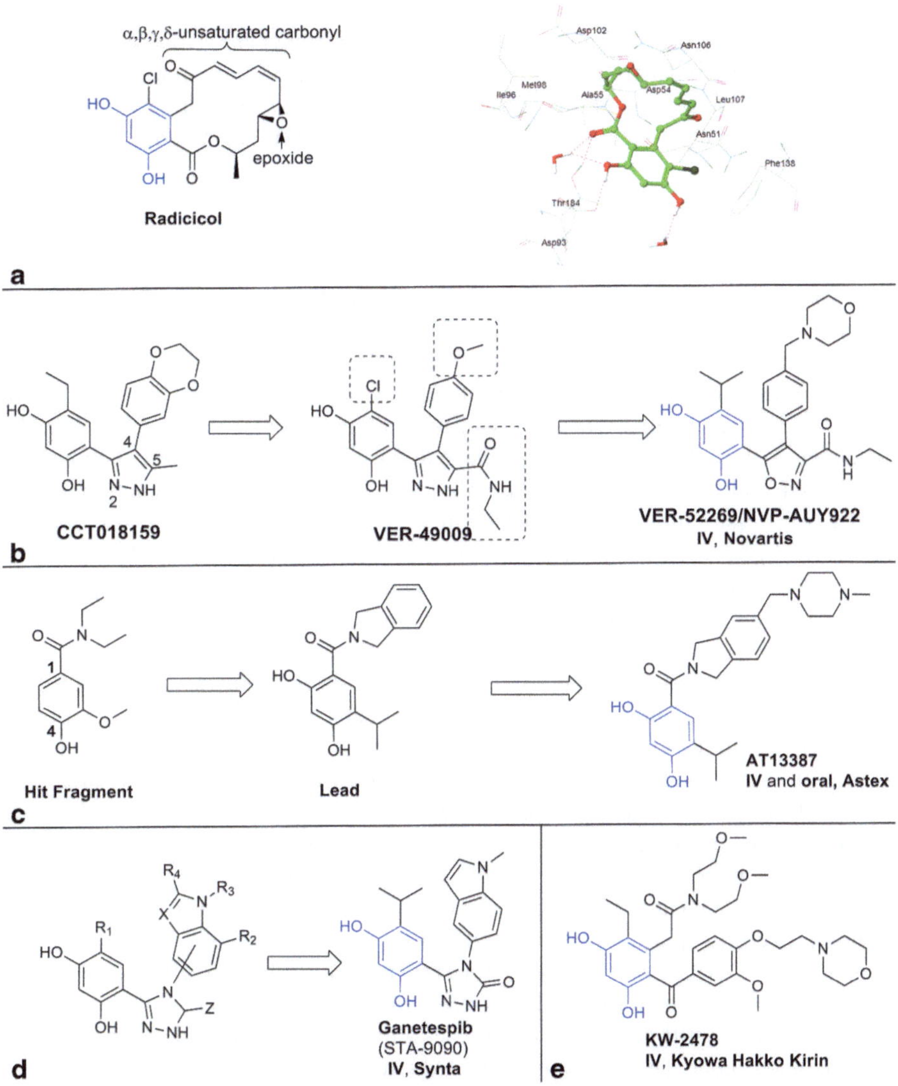

Fig. 7.2 **a** Resorcinol inhibitor of Hsp90, radicicol (in *blue* is shown the resorcinol moiety), and the structure of human Hsp90α bound to RD (Radicicol) (PDB ID: 4EGK), H-bonds are shown by *dotted lines*. **b** Scaffold modifications of CCT018159 leading first to VER-49009, and then to the clinical candidate NVP-AUY922. **c** From fragment to lead to clinical candidate AT13387. **d** Regions explored leading to Ganetespib. **e** KW2478 a resorcinol type inhibitor of Hsp90

of *Streptomyces hygroscopicus* [13]. Though GM was originally pursued as an antibiotic, its anticancer properties were discovered later from a phenotypic screen of compounds able to reverse the phenotype of *v-src* transformed cells [14], and these screening efforts also identified other structurally related benzoquinone ansamycins including herbimycin and macbecin. A landmark publication by Whitesell et al. in

1994 showed that GM bound to Hsp90 [15]. Subsequently, it was demonstrated that GM binds to the N-terminal nucleotide-binding pocket [16, 17] and inhibits the adenosine triphosphatase (ATPase) activity of Hsp90 [18]. The binding of GM to Hsp90 disrupts formation of the "closed" Hsp90 complex, inhibiting progression of the chaperone cycle, resulting in client protein degradation by the ubiquitin proteasome system [7, 19].

The crystal structures of the natural products GM and RD with the Hsp90 N-terminal domain provide essential aspects of their nucleotide mimicry [20]. In the case of GM, the crystal structure showed that the macrocyclic ansa ring and pendant carbamate are directed towards the bottom of the binding pocket (Fig. 7.1b), whereas the benzoquinone ring (depicted in red; Fig. 7.1a) was oriented towards the top of the pocket. On the other hand, the orientation of RD is opposite to that of GM since the resorcinol ring (depicted in blue, Fig. 7.2a) directed towards the bottom of the pocket and the macrocycle, with its conjugated bond system and epoxide, is facing the top of the pocket. The adenosine triphosphate/adenosine diphosphate (ATP/ADP) adenine's NH_2 group is mimicked by the carbamate functionality of GM and the resorcinol moiety of RD, which make direct and indirect (water mediated) hydrogen interactions with the protein (Figs. 7.1b and 7.2a). Additionally, hydrophobic interactions are also observed in this pocket for both natural products. As this chapter will show, the motifs of GM and RD are an occurring theme in many Hsp90 inhibitors that have reached the clinic.

Initially, there was considerable hesitancy towards targeting a housekeeping protein that is abundantly expressed in normal cells due to the perceived risk that inhibition might lead to undesirable toxicity. However, numerous groups were undeterred by this, buoyed in the knowledge that GM had previously been shown to be selectively toxic towards cancer cells, though the reasons for the observed therapeutic index for Hsp90 inhibitors was not yet known. The first major insight into the reasons underlying this was presented in a seminal paper by Kamal et al. which showed that cancer cells are composed of a distinct pool of Hsp90 compared to normal cells [21]. This work showed that the Hsp90 inhibitor 17-allylaminogeldanamycin (17-AAG; Fig. 7.1c) bound Hsp90 in tumor cells with a 100-fold greater affinity than Hsp90 in normal cells. Furthermore, Hsp90 in cancer cells was shown to be entirely in multi-chaperone complexes with co-chaperones (i.e., Hsp70/Hsp90 organizing protein, HOP, p23, Hsp70, etc.) with high ATPase activity, whereas Hsp90 in normal cells was determined to be in a latent uncomplexed state. This work provided a solid rationale for the observed selectivity of GM for cancer cells over normal cells, and served as a powerful impetus for the development of Hsp90 inhibitors by multiple institutions throughout the world. Later on, Moulick et al. revised this view and showed that Hsp90 forms biochemically distinct complexes in malignant cells where a major fraction of cancer cell Hsp90 retained "housekeeping" chaperone functions similar to normal cells, whereas a functionally distinct Hsp90 pool enriched or expanded in cancer cells and referred to as "oncogenic Hsp90" executed functions necessary to maintain the malignant phenotype [22]. This work also indicated that Hsp90 inhibitors have distinct selectivity profiles for the two Hsp90 species, providing perhaps a rationale for the non-overlapping therapeutic index and tumor retention profiles noted for the several Hsp90 chemotypes currently in clinical evaluation.

While the early stages in the development of Hsp90 inhibitors was undertaken by only a select group of research institutions and small biotechnology companies, our increased understanding of tumor Hsp90 biology has changed this landscape dramatically. Hsp90 has currently emerged as one of the most highly pursued drug targets for the treatment of cancer with 17 agents already in various stages of clinical trials [23]. Hsp90 is a very attractive target for a number of reasons. Hsp90 regulates multiple oncogenic signaling pathways and its inhibition enables these pathways to be targeted simultaneously [8, 24]. Additionally, it provides a potential advantage over other anticancer strategies (i.e., kinase inhibitors) in that it may be less susceptible to resistance mechanisms [25]. For this reason, the use of Hsp90 inhibitor can be envisioned not only as a monotherapy but also as combination therapy to combat the developing of resistance observed with current anti-cancer drugs. Finally, as we shall show, from a drug discovery perspective Hsp90 has turned out to be a highly druggable target.

It is significant that each of the compounds currently in the clinic target the N-terminal ATP pocket of Hsp90 [11, 23, 26]. There is also future promise for other clinical agents which disrupt Hsp90 chaperone cycle by alternative mechanisms such as targeting co-chaperone-Hsp90 interactions (e.g., Cdc37, HOP), targeting client-Hsp90 interactions or developing inhibitors that bind to the Hsp90 C-terminal ATP-binding site [12, 27, 28]. To date, there are no approved Hsp90 targeted agents, but the fact that clinical activity has been obtained with multiple drugs in various tumors is promising to change this status quo.

The great amount of work that has been undertaken on developing Hsp90 inhibitors is evinced by the sheer number of compounds reported in the literature to modulate Hsp90 function through a variety of mechanisms and constitute a broad array of structural classes [11, 12, 23, 29, 30]. Indeed, the complex and dynamic nature of Hsp90 function presents multiple opportunities for the design of drugs that target it and is a major reason why so many compounds have been identified. The dynamic nature of Hsp90 is exemplified by the significant structural changes it undergoes throughout the various stages of the chaperone cycle that ultimately result in client protein folding. What is more, it is becoming increasingly clear that even if these clinical agents share a similar binding site, their precise mode of binding may be different resulting in wide-ranging effects on biology [22, 31]. Conformational dynamism of Hsp90 enables an inhibitor to potentially target multiple diverse conformations and it is becoming apparent that the ability of an inhibitor to sample a variety of conformations can have dramatic effects on the observed biology. While a better understanding of the observed biological effects to the precise binding mode of inhibitor is clearly necessary, it is obviously clear that the plasticity of Hsp90 has not prevented the discovery and development of inhibitors using standard approaches in drug discovery.

Herein, we would like to review the strategies used to design some of the major classes of Hsp90 inhibitors that are being evaluated in clinical trials. In this chapter, we focus on the medicinal chemistry efforts pertaining to select scaffolds and describe progress from initial discovery to their advancement into clinical agents. While various methods have been used to discover the initial lead molecules (i.e., phenotypic screening, structure-based drug design (SBDD) [32], biochemical and cell-based screening, and fragment-based drug discovery (FBDD) [33]), in each case standard medicinal chemistry concepts were utilized to optimize the initial leads into clinical candidates. This was done with strong absorption, distribution, metabolism excretion, and toxicity (ADMET) considerations and with an increasing focus on tumor Pharmacokinetics (PK) properties. The focus on tumor PK is derived from the peculiar PK profile that Hsp90 inhibitors display, being selectively retained in tumor rather than normal tissues, often at pharmacologically relevant concentrations for an extended period of time. This effect was first noted by Vilenchik et al. [34]. This profile augurs well for development of these compounds into drugs as one can envision a molecule which is rapidly distributed to the tumor tissues where the drug is retained in sufficient concentration to cause marked anticancer activity and at the same time is rapidly cleared from non-cancerous tissues, thus decreasing the probability of observing toxic side effects.

2 Hsp90 Chemotypes: From Discovery to Development

2.1 Ansamycin Natural Product Class as the First Hsp90 Inhibitors

The discovery of lead agents for Hsp90 inhibition was initially paved by the thorough investigation of the ansamycin natural product class (Fig. 7.1). This was also the first class to have been evaluated in clinical trials. GM demonstrated both in vitro and in vivo antitumor activity in preclinical models, but it never reached clinical trials as a result of liver toxicity as well as poor solubility and metabolic and chemical instability [35, 36]. The hepatotoxicity observed with GM was attributed to the chemical instability of the quinone ring, which is known to be reactive towards nucleophiles that are typically present in biological molecules [37, 38], and is exacerbated by the C-17 methoxy group (Fig. 7.1a, methoxy group shown in blue, quinone shown in red). However, analogs of GM that contain alkylamino groups instead of the C-17 methoxy group retain their biological activity and show decreased hepatotoxicity [39]. Reflective of its high reactivity, the methoxy group of GM is easily substituted by reaction with amines and at least two significant derivatives have been prepared by this approach. Substitution with an allylamino group resulted in 17-allylamino-17-demethoxygeldanamycin (17-AAG; Fig. 7.1c), a compound that shared important biological features with GM but displayed a more favorable toxicity profile [40]. 17-AAG was the first Hsp90 inhibitor to be evaluated in clinical trials in 1999, and subsequent published clinical data for 17-AAG demonstrated in tumors molecular signatures (i.e., c-Raf-1 and CDK4 depletion and Hsp70 induction) indicative of target inhibition [41]. This information combined with the knowledge that clinical activity has been seen for 17-AAG in melanoma, multiple myeloma, prostate, and breast cancer [8] provided proof of concept for Hsp90 inhibition in patients. The further advancement of this compound was impeded as a result of its poor pharmaceutical and toxicity profile. For example, because of its low solubility, 17-AAG was administered intravenously using a dimethyl sulfoxide (DMSO) and egg phospholipid vehicle, which, at doses higher than 100 mg/m^2 was itself toxic (i.e., bad odor, nausea, anorexia), and these vehicle-related toxicities, attributed to the DMSO formulation, were as major as the ones resulting from the drug itself [23]. Additionally, 17-AAG is a substrate of the P-glycoprotein-mediated efflux and the polymorphic cytochrome P450 CYP3A4 [42]. Finally, the delayed hepatotoxicity observed for 17-AAG in the clinic with the twice a week continuous dosing schedule [43] further supported ongoing endeavors to develop non-ansamycin Hsp90 inhibitors with an improved toxicity profile.

Kosan Biosciences initiated a program aimed at finding potent, water soluble inhibitors of Hsp90 based on the GM scaffold, and thus synthesized more than 60 17-alkylamino-17-demethoxygeldanamycin analogs [11, 39]. Their efforts resulted in 17-(2-dimethylaminoethyl)-amino-17-demethoxygeldanamycin (17-DMAG; Fig. 7.1c), an analog which contains the ionizable N,N-dimethylethylamine group instead of the methoxy group at C-17. The ionizable amino group results in an increased solubility profile as well as higher oral bioavailability and equal if not better antitumor activity than 17-AAG [44]. Thus, in 2004, 17-DMAG was advanced into clinical trials where it was evaluated both orally and as an intravenous (IV) agent. Clinical benefit was reported in 42 % (i.e., 8 out of 19) of patients with refractory HER2 + metastatic breast cancer and in patients with refractory ovarian cancer who were progressing on standard chemotherapy, and also promising results were seen in patients with chemotherapy refractory acute myelogenous leukemia since 3 out of 17 patients showed complete response to the Hsp90 therapy [26]. Nonetheless, in 2008, the clinical development of 17-DMAG was discontinued owing to unacceptable toxicity, which was target unrelated [26].

Infinity Pharmaceuticals attempted to address the pharmaceutical deficiencies characteristic to the ansamycin inhibitors through IPI-504 (Retaspimycin), the hydroquinone hydrochloride salt of 17-AAG (Fig. 7.1c). This was obtained by reducing 17-AAG to the corresponding hydroquinone using $Na_2S_2O_4$ which was subsequently treated with hydrochloric acid (HCl; Fig. 7.1c). The hydrochloride salt was chemically stable to oxidation and improved the aqueous solubility > 4000-fold when compared to 17-AAG [45, 46]. Moreover, the hydroquinone compound (IPI-504) was shown to be the more potent Hsp90 inhibitor when compared to the quinone (17-AAG) and this was suggested to be due to the various hydrogen bonding now possible with the hydroquinone derivative [45, 46]. In the clinic, IPI-504 has shown promising activity in patients with non-small cell lung cancer (NSCLC) and oncogenic anaplastic lymphoma kinase (ALK) gene rearrangements [47]. In a Phase II trial, 28 % of patients with NSCLC were able to see a stabilization in their condition as well as tumor reduction [29]. In the clinic, the evaluation of this agent against gastrointestinal stromal tumor (GIST) was terminated due to unexpectedly high hepatic toxicity. The exact mechanism of the liver toxicity is still elusive but it may perhaps relate to the potential to form a quinone through oxidation in vivo. The quinone moiety can be seen as a liability since it has potential for adduct formation and redox metabolism, and this most likely contributes to the liver toxicity [48]. On the whole, IPI-504 is a clinical agent with improved PK and toxicity properties and this profile makes it the most promising candidate from the ansamycin class.

Overall, the natural product-based Hsp90 inhibitors have pharmaceutical limitations related to their ADME properties. The pioneering work with GM served as a proof-of-concept for this cancer target. However, the combined knowledge that GM and related derivatives show dose-dependent liver toxicity, and that RD derivatives suffer from severe toxicities in rats and mice [49] opened the landscape to the discovery and development of synthetic inhibitors. These are compounds based on novel chemical scaffolds that sought to overcome the limitations observed with GM, RD, and their derivatives.

2.2 *Resorcinol Derivatives*

A number of compounds which have been advanced into the clinic contain a resorcinol moiety, and in this regard can be related to RD (Fig. 7.2a). RD is a resorcylic acid lactone first isolated in 1953 from *Monosporium bonorden* [50] and later shown to be a potent Hsp90 inhibitor where it affects key components of the molecular chaperone in a manner similar to GM [51]. The resorcinol core of RD binds deep in the ATP-binding pocket and provides key interactions which are essential for its potent inhibitory activity [20, 51]. Although RD has potent activity in vitro, this did not translate to in vivo activity [52]. The in vivo inactivity is attributed to the highly reactive functionalities in its structure such as α, β, γ, δ-unsaturated carbonyl group and an epoxide (Fig. 7.2a) which are readily metabolized [53, 54]. These liabilities were partially addressed through synthetic modifications. In one approach, the electrophilic epoxide was replaced with a relatively unreactive cyclopropyl group [53], which resulted in an analog that retained potency when tested in the MCF-7 breast cancer cell line with an IC_{50} of 54 nM vs. 23 nM for RD [53] and also showed activity in vivo in a MX-1 adenocarcinoma mouse model [55]. In another approach, the biologically unstable carbonyl was converted to oxime derivatives [52, 56]. One of the most potent of these oxime derivatives was shown to have significant antitumor activity in MX-1 and A431 epidermoid carcinoma xenograft models [57]. Despite these results, any attempts, this far, to directly modify RD have failed to deliver clinically viable compounds. Nonetheless, the resorcinol core of RD is an essential motif distinguishing several small molecule Hsp90 clinical agents which are described below.

2.2.1 NVP-AUY922/VER52296

A high-throughput screen of approximately 56,000 compounds was initiated to identify novel small molecule inhibitors of Hsp90. The assay used in this screen measured inhibition of yeast Hsp90 ATPase activity using the malachite green detection of inorganic phosphate [58] and identified the resorcinolic pyrazole compound CCT018159 as the most potent hit (Fig. 7.2b). Crystallographic data revealed that CCT018159 bound to the ATP pocket in the N-terminal domain of yeast Hsp90 in a manner similar to RD [59]. The two phenolic hydroxyls and the adjacent pyrazole nitrogen of CCT018159 form a network of hydrogen bonds, similar to RD, with Asp79, Gly83, and the hydroxyl side chain of Thr171, and water molecules found at the base of the pocket. From the crystal structure, it was deduced that the 5-methyl group (Fig. 7.2b, CCT018159) was only 4 Å away from the carbonyl oxygen of Gly97 and this hydrogen bond acceptor could potentially be exploited to increase activity by the installation of a donor group. Replacement of the C5-methyl group by an amide group allowed the formation of a hydrogen bond with Gly97 as predicted by modeling, and this is supported by the increase in potency against the target and functional assays resulting in compound VER-49009 (Fig. 7.2b) which showed inhibition of cell proliferation comparable to the clinical agent 17-AAG [60].

Further medicinal chemistry efforts were focused on varying the pyrazole, introduction of a solubilizing group on the 4-arylpyrazole substituents, and also optimization of the 5'-substituent on the resorcinol ring [61]. Replacement of the more polar pyrazole with the isoxazole was well tolerated and bound Hsp90 with similar affinity, and also showed that isoxazole derivatives were more potent in cell growth inhibition assays [61]. Prior knowledge gained from other work [60, 62–64] suggested the importance of a solubilizing group on the 4-aryl ring, and thus their subsequent efforts focused on substitutions on the para-position of the aromatic ring. The morpholine moiety was found to be an optimal choice, one that also had additional hydrophobic interactions with Thr109 and Gly135 [61]. From the 4,5-diarylisoxazole series, VER-52296/NVP-AUY922 (Fig. 7.2b) emerged as a potent Hsp90 inhibitor which also showed antitumor activity in colon and breast cancer xenograft models [61, 65]. This lead candidate subsequently advanced to clinical trials where it is evaluated not only as a single agent but also as a combination therapy in various cancers. The first clinical results with NVP-AUY922 in patients with advanced solid tumors were disclosed in mid 2013 and showed that the drug was well tolerated, though among other common toxicities (e.g., diarrhea, nausea, fatigue) visual side effects were noticed at higher doses in some patients, but these visual symptoms were typically reversed upon interruption or discontinuation of the treatment [66]. The drug also provided evidence of Hsp90 inhibition in both peripheral blood mononuclear cells (Hsp70 induction) and tumor (reduction in levels of AKT and decrease of metabolic activity by ^{18}F-FDG PET) [66]. Following these results, Phase II single-agent and combination studies are in progress in patients with HER2+ breast, gastric and non-small cell lung cancers [66].

2.2.2 AT-13387

Scientists at Astex Pharmaceuticals used a FBDD approach to identify AT-13387. They used nuclear magnetic resonance (NMR) and high throughput X-ray crystallography to discover fragments with Hsp90-binding affinity. These efforts resulted in the discovery of two fragments, one of which was a phenolic chemotype (Fig. 7.2c, **Hit Fragment**) with affinity in the high micromolar range as determined by isothermal titration calorimetry [67].

The follow-up on this hit fragment used crystallographic information which indicated that the methoxy group could be substituted to fill a proximal lipophilic pocket [67]. Analogs substituted with chloro, ethyl, isopropyl, and *tert*-butyl were initially prepared, and it was shown that a 100-fold improvement in potency was obtained for isopropyl and *tert*-butyl analogs, which are able to fill the lipophilic pocket. Then, two analogs containing the hydroxyl (OH) at the 2-position of the phenol instead of the 4-OH were pursued given the knowledge that 2-OH group in RD forms a direct hydrogen bond to Asp93 (Fig. 7.2c, **Hit Fragment**). However, these analogs were tenfold less potent than the 4-OH substituted compounds, and this implied that the 4-OH group must be an important contributor for affinity in the series. Their next synthetic iteration was aimed at modifying the diethylamide in the original hit fragment, and guided by X-ray crystallography, they pursued tertiary amides substitutions. The isoindoline moiety emerged as the best choice, one that provided a significant improvement in affinity of several-hundred-fold. Further modification of the phenol into a resorcinol resulted in a compound (Fig. 7.2c, **Lead**) with subnanomolar affinity, very good ligand efficiency and cell activity [67]. This initial lead was further modified first by the introduction of halogens at the 4, 5, and 7 positions of the isoindoline which were all tolerated. The subsequent structure-activity relationship (SAR) was primarily focused on substitutions at the 5-position of the isoindoline ring system shown to be best overall for activity, though the 4-position was also explored, with a focus on basic and other solubilizing functionalities. Among the substitutions explored at the 5-position were N-methylpiperazine, N,N-dimethylethanolamine, 2-methoxyethoxy, and 4-morpholinylmethyl, and all these compounds had good enzyme and cell activity [68]. Additionally, analogs containing N-methylpiperazine or N,N-dimethylethanolamine had high tumor retention and efficacy in HCT116 human colon carcinoma xenografts when compared to their original lead [68]. In order to increase the pKa of the basic center and also lower lipophilicity, a methylene linker was introduced at the 5-position linking the isoindoline with N-methylpiperazine or N,N-dimethyl. A compound preclinical selection process then ensued where the key parameters evaluated were in vivo target modulation, predicted human dose, selectivity, solubility, stability, formulation, and ease of synthesis [68]. From this triage process emerged the lead candidate AT-13387 (Fig. 1.2c) [69]. Same study also showed that the suppression of client proteins can last up to 72 h in NCI-H1975 xenograft tumors in mice when given as a single dose. The drug has subsequently advanced to the clinic where it is evaluated in patients with metastatic solid tumors.

2.2.3 STA-9090 (Ganetespib)

Ganetespib (Fig. 7.2d) is a novel resorcinol-containing triazole compound that is being developed by Synta Pharmaceuticals. The medicinal chemistry efforts resulting in the identification of STA-9090 have not yet been described. However, the patent literature does provide the main regions explored for the resorcinol-containing triazole series, and these are shown in Fig. 7.2d (i.e., R_1, R_2, R_3, R_4, Z, and X). Briefly, the Hsp90 inhibitory compounds disclosed explored at R_1, R_2=H, OH, SH, alkyl, etc.; R_3=H, alkyl, alkenyl, cycloalkyl, heteroaryl, etc.; R_4=H, OH, alkyl, alkenyl, cycloalkyl, etc.; Z=OH, SH, NH_2; and X=N, alkyl, H, OH, etc. [70]. Ganetespib is among the Hsp90 triazole-containing inhibitors disclosed in the patent. This compound has a molecular weight (MW=364.4) that is considerably lower than most of the newer second generation Hsp90 inhibitors and is relatively hydrophobic (cLogP=3.3) [71]. X-ray studies of STA-9090 bound to Hsp90 not only confirmed the critical hydrogen interactions of the resorcinol hydroxyl group with Asp93 and the carbonyl of triazolone with Lys58 but also showed that the hydrazinecarboxamide moiety of triazolone makes a distinctive hydrogen bond with Gly97 that is not observed with ansamycin analogs [71]. Additionally, there are water bridge bonds between the 4-hydroxyl of resorcinol and Leu48 and Ser52 that are vital for binding efficiency [71].

STA-9090 was shown to have potent in vitro cytotoxicity when tested in various solid and hematologic tumor cell lines [71]. Additionally, ganetespib demonstrated antitumor efficacy in solid and hematologic xenograft models. Ganetespib subsequently advanced to clinical trials both as single agent and in combination therapy for a variety of cancers. In the clinic, ganetespib showed manageable side effects [72] and also clinical activity, for example, in previously heavily pretreated patients with NSCLCs [73]. This drug candidate is the most advanced Hsp90 inhibitor currently in the clinic where it is undergoing evaluation in several Phase II trials as well as a Phase III clinical trial for various human cancers.

2.2.4 KW-2478

The focus of Kyowa Hakko Kirin Pharma was initially on the development of oxime derivatives of RD, but their progress into the clinic was discontinued due to severe toxicities seen in rats and mice. As a result Kyowa Hakko Kirin Pharma decided to shift its focus towards the development of new small molecule Hsp90 inhibitors with an improved safety profile [49]. To achieve their goal, a unique screening system was established which consisted of a binding assay where Hsp90 was fixed onto plates, and to these wells, the compounds to be screened were added. The plates were washed a few times, after which biotin-labeled RD was added and poly-HRP streptavidin solution was added to verify the competitive inhibition with RD [49]. These efforts resulted in the identification of a number of leads containing a resorcinol moiety. Although detailed SAR leading to the identification of KW-2478 (Fig. 7.2e) has yet to be disclosed, lead optimization utilized X-ray crystallography, cell-based assays, and in vivo models. KW-2478-inhibited biotin-RD binding to Hsp90α with an IC_{50} value of 3.8 nM [74] and was found to have good solubility in saline (>30 mg/mL), no hepatotoxicity in animal studies, good PK profile, and also no metabolism by CYP3A4 enzyme [49]. Hence, it was subsequently advanced into Phase I clinical trials in patients with multiple myeloma and non-Hodgkin's lymphoma where no dose limiting toxicities (DLTs) or retinal toxicity was noticed at doses up to 176 mg/m[2] [49]. This clinical agent is also currently being evaluated as a combination therapy with bortezomib in a Phase I/II trial in patients with multiple myeloma.

2.3 Purine and Purine-Like Analogs

Structure-based drug design of Hsp90 inhibitors was made possible with the advent of crystal structures of GM [16, 17], RD [20], and ADP [16] bound to the N-terminal nucleotide-binding domain (NBD) of Hsp90. These structures show the ATP-binding pocket to have a distinctive Bergerat fold that is observed only in the GHKL ATPase protein family [75, 76]. The unique "bent" conformation adopted by ATP when bound to Hsp90 was incorporated into the design of the synthetic class of inhibitors. Chiosis et al. utilized co-crystal structures of Hsp90 with its ligands (i.e., GM, RD, and ADP) to rationally design the first synthetic Hsp90 inhibitor, purine-based PU3 (Fig. 7.3a) [77]. PU3 was shown to inhibit the binding of purified Hsp90 to immobilized GM with an $EC_{50} = 15-20$ μM. Furthermore, PU3 was the first synthetic molecule to show similar phenotypic properties to GM, including degradation of HER2 and inhibition of cancer cell growth [77]. In the design of PU3, the purine core of the endogenous ligands were linked to an aryl moiety via the 8-position through a methylene linker (i.e., Purine-CH_2-Aryl, Fig. 7.3a). This structural arrangement allows PU3 to adopt a bent shape even in the unbound state. Although previous crystal structures aided in the design of this class, they could not predict the precise mode of binding for these molecules. Interestingly, a crystal structure of PU3 bound to the NBD of Hsp90α showed that it induces a conformational shift that exposes a pocket not observed in any of the previous structures [78].

As discussed below, the essential motif exemplified by PU3, purine-linker-aryl, would be maintained for essentially all of the inhibitors of this class, with only slight variations. The SAR that was constructed on the purine scaffold can be divided into four parts: (1) the purine ring, (2) the linker, (3) the 8-aryl ring, and (4) the 9-alkyl substituent. The purine core is not very amenable to structural modifications given its overall tight fit in the pocket and important binding interactions made by the exocyclic amino group. The same holds true for the linker which is critically important for imparting an acceptable angle between the purine and aryl rings. The best results in advancing this class of compounds into clinical candidates came through modifications of the 8-aryl ring and the N9-alkyl substituent. Affinity has generally been improved through modifications of the 8-aryl ring that impart strong binding interactions within the newly created binding pocket. Additionally, PK properties have been improved largely through modifications to the 9-substituent which is oriented towards the solvent exposed region. This strategy was indeed successful as changes made to the aryl fragment resulted in compounds that went from micromolar potency (e.g., PU3, bound Hsp90 with moderate activity and showed biological activity in the 50 μM concentration range or PU24FCl which exhibited biological effects in the 2–6 μM range) through to the nanomolar level (e.g., PU-

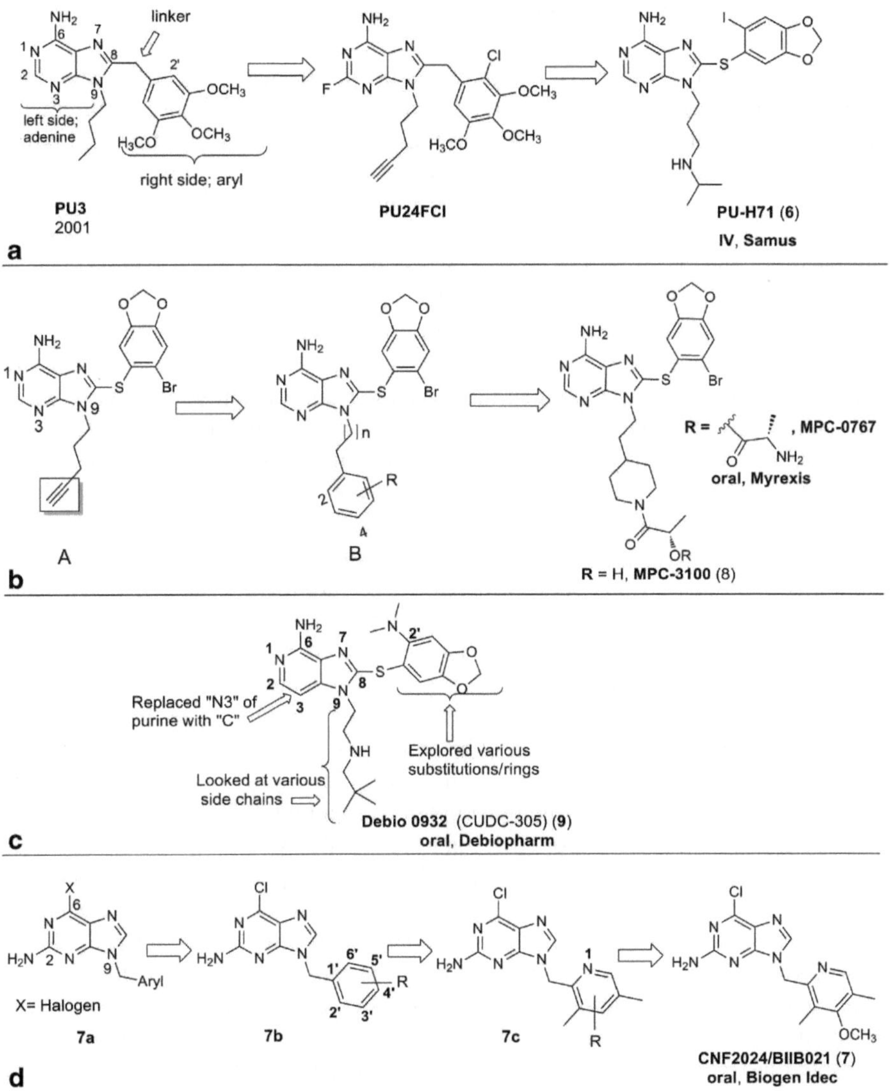

Fig. 7.3 **a** PU-H71, from discovery to clinic. **b** Development of MPC-3100 and MPC-0767 from the initial lead A, following extensive SAR. **c** Imidazopyridine 9 (purine-like) SAR regions. **d** BIIB021, from lead to clinic

H71) [79], and this is schematically depicted in Fig. 7.3a. The initial lead PU3 was subsequently picked up by numerous research groups and optimized via various strategies to lead to potent drug candidates that display a much improved PK and toxicity profile compared to the ansamycin analogs. Combined these efforts have resulted in the advancement to clinical trials of PU-H71, MPC-3100, CNF2024/BIIB021, and Debio 0932 (CUDC-305; Fig. 7.3).

2.3.1 Early SAR of the Purine-Scaffold Class

The discovery of PU3 triggered an interest in the purine-scaffold class demonstrated by early publications aimed at studying the main pharmacophore requirements for this class of inhibitor [78, 80, 81]. In the first reported SAR, extensive knowledge was obtained from a careful study of the almost 70 derivatives synthesized [80]. These analogs were evaluated for Hsp90 binding via a solid phase competition assay for their ability to compete with immobilized GM and their cellular activity was tested by the ability of the drugs to arrest the growth of MCF-7 breast cancer cells and to induce the degradation of HER2 tyrosine kinase [80]. The advent of multiple co-crystal structures of a series of purine-based inhibitors bound to Hsp90 [81] have allowed to explain the earlier observed SAR and subsequent ones. For example, it was previously noticed that when the butyl chain of PU3 is replaced with pent-4-ynyl the activity increases, and this was attributed to an increase interaction between the alkyne group and hydrophobic lid of the pocket. The crystal structure also helped to explain why functionalities on the 9N moiety could not have any substituent on C1 where a steric clash with Leu107 would occur. In addition, at position 2 of the purine only fluorine was tolerated which increased potency and water solubility in this series whereas larger groups like cyano, vinyl, iodo, methoxy, ethoxy, or amino decreased activity [80]. Modification of the methylene linker with -O-, -NH-, $-CH_2CH_2-$, $-OCH_2-$ were each detrimental to activity, which, hinted at the importance of the dihedral angle between the purine and the aryl moiety. With regards to the 8-aryl moiety, the addition of chlorine to the 2'-position increases activity while the addition of bromine was found to have little to no effect compared to PU3. The combination of all the favorable functionalities provided an additive effect and resulted in the most potent compound, PU24FCl, which had an $IC_{50}=1-2$ μM in degradation of HER2 in MCF-7cells and was shown to be 30-fold more potent than PU3 (Fig. 7.3a). Notably, PU24FCl displayed antiproliferative effects against various cancer cell lines with $IC_{50}=2-7$ μM and caused the degradation of Hsp90 client proteins at similar concentrations [34]. Importantly, PU24FCl showed 10- to 50-fold higher affinity for tumor vs. normal tissue Hsp90. This compound also showed antitumor effects in vivo when administered at 200 mg/Kg ip on alternate days in mice bearing MCF-7 breast cancer xenografted tumors where it resulted in significant depletion of HER2, AKT, and Raf-1 in tumors and also a 72% decrease in tumor burden vs. untreated control mice [34]. Various groups took on the prospect of further developing the purine class of inhibitors by performing extensive SAR, and included the testing strategy in the preclinical development of their Hsp90 inhibitors. These combined efforts have culminated in four compounds being evaluated in the clinic for cancer.

2.3.2 PU-H71

PU-H71 was discovered by researchers at Memorial Sloan-Kettering Cancer Center through the further development of PU24FCl (Fig. 7.3a). These efforts were greatly assisted by improvements in binding assays designed to measure affinity to Hsp90. Development of fluorescent polarization (FP) assays enabled for the accurate evaluation of the potency of compounds in a more high-throughput manner. These assays were first developed for purified protein and then later for cancer cell lysates which more accurately define the affinity of the compound for the oncogenic Hsp90 complex [82–84]. A significant development within the purine-scaffold series came from the finding that the sulfur linker showed activity similar to the methylene [79, 85]. This modification enabled for more extensive SAR evaluation of the 8-aryl moiety through the advent of suitable chemical methods as well as the broader availability of aryl iodides required in the synthesis of the requisite 8-arylsulfanyl adenine derivatives, since the rather limited availability of aryl acetic acids needed for the synthesis of methylene-linked compounds would have limited development [80, 83, 86]. Through these efforts, it was found that the 2-halo-4,5-methylenedioxy series was the best aryl moiety to fit into the Hsp90 hydrophobic pocket [79, 83]. As can be seen from PU-H71, CUDC305, and MPC-3100, the methylenedioxy moiety is indeed a recurring theme in 3 of the 4 clinical candidates within the series.

Orientation of the N9-substituent towards solvent-exposed region makes this part of the molecule an ideal place to introduce moieties that would improve PK properties. Thus, the pent-4-ynyl previously determined to be favored by the Hsp90 pocket [80] was subsequently replaced with ionizable amino groups that enabled the formation of salt, and hence dramatically enhanced solubility while retaining biological activity of these compounds. PU-H71, a potent water soluble compound with an $IC_{50} \sim 50$ nM in cellular models of cancer, contains a 3-isopropylamino-propyl group at N9, sulfur as the linker, and 2-iodo-4,5-methylenedioxy for the aryl group [79]. This inhibitor was shown to have potent activity in various preclinical models of triple-negative breast cancer (TNBC) [87], small-cell lung carcinoma (SCLC) [88], hepatocellular carcinoma [89], diffuse large B-cell lymphomas [90], and myeloproliferative disorders [91]. In addition, PU-H71 was also shown to have extended tumor retention and prolonged PD effects; one such example is the retention of this compound at pharmacological doses and the associated suppression of Hsp90 oncoclients such as AKT for over 48 h in TNBC tumors. Drug concentrations in normal tissues and plasma declined quickly being almost undetectable by 6 h [87]. This drug candidate has subsequently advanced to Phase I clinical trials where it is being evaluated in patients with solid tumors, lymphoma, and myeloproliferative disorders [23]. The first results from the two Phase I clinical trials with PU-H71 are anticipated in mid 2014.

2.3.3 MPC-3100

MPC-3100 was discovered by scientists at Myrexis Inc. who decided to pursue their Hsp90 discovery program using the intrinsic activity of the purine scaffold as a starting point [92]. Their first SAR efforts entailed the replacement of the propargyl group of **A**, a derivative previously reported by the Chiosis group [79] (Fig. 7.3b), with the phenyl bioisostere linked to the N9 via alkyl chains of different lengths

(Fig. 7.3b; **B**). The two carbon linker was determined to be optimal and various substitutions on the aromatic ring (i.e., at position 2, R=F, Cl, Br, compound **B**; Fig. 7.3b) were probed for best activity using FP. Unfortunately, the limiting factor of these analogs was their decreased metabolic stability, therefore, further structural changes to the pendant aryl group were explored. Introduction of heteroaromatic ring systems (e.g., pyridine, pyrole, imidazole, etc.) or replacement of the phenyl ring with the thiophene bioisostere generally resulted in a significant loss of potency [92]. Since these modifications were not successful and some were even detrimental, other strategies were pursued. To this end, their first attempt was substitution of the phenylethyl with an aminomethyl- cyclopropyl group, still this choice did not bear fruit as the potency was not improved and more this analog suffered from high clearance rate and very low oral bioavailability [92]. Thus, their subsequent endeavors were towards improving oral bioavailability while preserving potency, and they opted for the introduction of the piperidine moiety which is frequently used by medicinal chemists as it provides an expedient platform for drug discovery. Next, variations were made to the piperidine N-substituent in order to improve both potency as well as ADME properties consisting of N-alkyl (e.g., *i*-Pr, *i*-Bu, CH_2CF_3, CH_2-cyclo-hexyl) and N-acylpiperidine analogs. From these efforts, it was discovered that the optimal piperidine N-substituents were N-hydroxyacyl moieties which provided both potency and metabolic stability.

Further extensive SAR studies were carried out for the 8-aryl region in an effort to replace the 6-bromo-1,3-benzodioxole moiety with other functionalities such as dihydrobenzofuran, benzofuran, benzodioxane, benzoxazine. However, none had more favorable PK profiles than the 1,3-benzodioxole analogs [92]. It should also be noted that these derivatives containing a bromine on the 2′-position of the 8-aryl group are less potent than their iodine derivatives at the same position. Overall, the SAR of the pendant N-9 piperidine moiety and of the 1,3-benzodioxole ring, complemented by in vitro profiles and PK properties led to the selection of MPC-3100 (Fig. 7.3b) as the most suitable drug for advancement. MPC-3100 was administered orally and demonstrated significant antitumor activity in the NCI-N87 gastric cancer xenograft model [92]. MPC-3100 has recently completed Phase I clinical trials in refractory or relapsed cancer. This study revealed the safety and tolerability when administered orally of MPC-3100 at doses lower than 600 mg/day with the most common side effects reported having gastrointestinal origin (i.e., diarrhea, nausea, and vomiting) [92].

As a result of the relatively poor solubility and bioavailability of MPC-3100, Myrexis developed MPC-0767 (Fig. 7.3b) [93], a novel L-alanine ester pro-drug of MPC-3100 which has completed all requirements for Investigational New Drug (IND) filing. Nonetheless, in February 2012, the development of activities in the preclinical and clinical oncology programs of Myrexis were suspended [94], and currently, there are no ongoing clinical trials indicative, perhaps of the uncertain future of these two potential drugs.

2.3.4 Debio 0932 (CUDC-305)

The structure of CUDC-305, though technically an imidazopyridine, bears strong resemblance to other members of the purine class (Fig. 7.3c). To date, the only report regarding the design and synthesis of CUDC-305 was provided in a poster presented at the American Association for Cancer Research (AACR) 102nd Annual Meeting that took place in 2011 at Orlando, Florida [95]. Upon analysis of the existing SAR for the purine class [79, 81, 83], efforts at optimization were focused primarily on the regions indicated by arrows in Fig. 7.3c. The impetus to develop the imidazopyridine scaffold was likely the result of available crystal structures of purine-scaffold compounds which showed that the purine N3 interaction with Hsp90 is less critical for activity. Hence, their research team concluded that replacement of nitrogen with carbon should not result in diminished potency and could provide novelty and potentially unique properties [95]. Their SAR for the solvent exposed region examined various side chains such as pentynyl, butynyl, hexynyl, pentyl, replacement of terminal alkynyl with cyano, ether, and amide chains, as well as amino containing chains. This screen provided as the best side chain the neopentylamine with an ethyl linker connected to N9. The 8-aryl region was also subject to SAR exploration, and again the methylenedioxy analogs, similar to PU-H71 and MPC-3100, was determined to be best. Finally, they also examined substituents at position 2′ of the phenyl ring and found the dimethylamine moiety to give a favorable Hsp90 inhibitor [95]. Their most potent analog was CUDC-305 ($IC_{50}=100$ nM) [95]. This compound, later renamed Debio 0932, is orally bioavailable, blood-brain barrier-permeable, and can achieve therapeutic levels in brain tissue (suggesting perhaps its potential role to treat brain tumors). Additionally, in tumor tissues the drug displays high concentrations and prolonged half-life ($t_{1/2}=20.4$ h) [95, 96]. It was subsequently advanced to the clinic where Debio 0932 went through the dose escalation portion of a Phase I study where it was evaluated for safety and tolerability at various doses given orally in patients with advanced solid tumors or lymphoma [97]. The results of the Phase Ib study are expected in the near term, and currently the drug is the subject of a Phase I/II clinical study in patients with advanced lung cancer [97].

2.3.5 BIIB021

BIIB021 (Fig. 7.3d) was discovered by scientists at Conforma Therapeutics [98] and is unique from the other compounds in this class in that the aryl moiety is attached to the N9 position of the purine instead of the C8 position (Fig. 7.3d; **7a**). They showed that potent inhibition of Hsp90 was possible when the NH_2 group was also moved from the 6- to the 2-position. These changes maintained the optimal six bond distance between the NH_2 and the aryl group required for potent binding by the pharmacophore (purine-linker-aryl). Their first efforts were on exploring the effect of the halogen (i.e., Cl, Br, and I) at the 2′-position of the benzyl ring of compound **7b** (Fig. 7.3d), and their expectation to see an increase in potency was confirmed. Then, the importance of the chlorine atom at the C6 position of the purine ring was studied. Attempts to modify the chlorine with other functionalities (NH_2, OCH_3, OH, SH, H, or CH_3) resulted in a decrease of activity suggesting that the halogen is important not only for hydrophobic contacts but also for its electron withdrawing characteristics [98]. Their next optimization steps were aimed at improving the aqueous solubility and to achieve this they replaced the phenyl ring with a pyridine. The subsequent SAR on the 9-heteroaryl moiety (Fig. 7.3d; **7c**) determined 3,5-dimethyl-4-methoxypyridine as ideal, and from this emerged the potent inhibitor 6-chloro-9-((4-methoxy-3,5-dimethylpyridin-2-yl)methyl)-9H-purin-2-amine (Fig. 7.3d; **BIIB021**) which had a better profile based on the in vitro potency, pharmaceutical properties, and in vivo oral efficacy. The antitumor activity of BIIB021 was demonstrated in a number of human tumor xenograft models where the drug was given orally on both daily and intermittent dosing schedules [99]. BIIB021 was advanced in the clinic by Biogen Idec where it has been evaluated in several Phase I and II clinical trials in patients with gastrointestinal stromal tumors (GIST) and hormone receptor positive metastatic breast cancer [11]. Additionally, a Phase II study of BIIB021 in patients with GIST refractory to imatinib and sunitinib was published in 2013 and showed that treatment with BIIB021 led to metabolic responses in >20 % of patients with no significant hepatotoxicity when compared to IPI-504 [100]. However, the further advancement of BIIB021 in the clinic by Biogen Idec seems unlikely given the desire of the company to shift away from the oncology development program [101], and currently the company is seeking to out-license BIIB021.

2.4 Other Chemotypes

2.4.1 SNX-5422/PF-04929113, a Dihydroindazolone Derivative

SNX-5422 (Fig. 7.4a) is based on a unique dihydroindazolone scaffold and was discovered by scientists at Serenex. They used a chemoproteomics-based strategy in screening a focused chemical library against a diverse set of potential targets [102]. In this unbiased approach both the hit compound and target were identified concomitantly, so that the screen enabled for the identification of the most optimal starting point at the level of both biology and chemistry. This was achieved using an ATP-affinity column to which ATP-binding proteins were first loaded, and this was then challenged with a diverse set of 8000 compounds for their ability to displace unspecified proteins from this column. Some of the compound-protein interactions found from this screen involved Hsp90. The decision to pursue Hsp90 stemmed from the potential of it as a target and by the novelty and diversity of the chemical leads identified by the screen.

The researchers at Serenex were inspired to pursue a synthetically tractable benzamide hit (Fig. 7.4a; **compound A**) which conferred structural novelty, chemi-

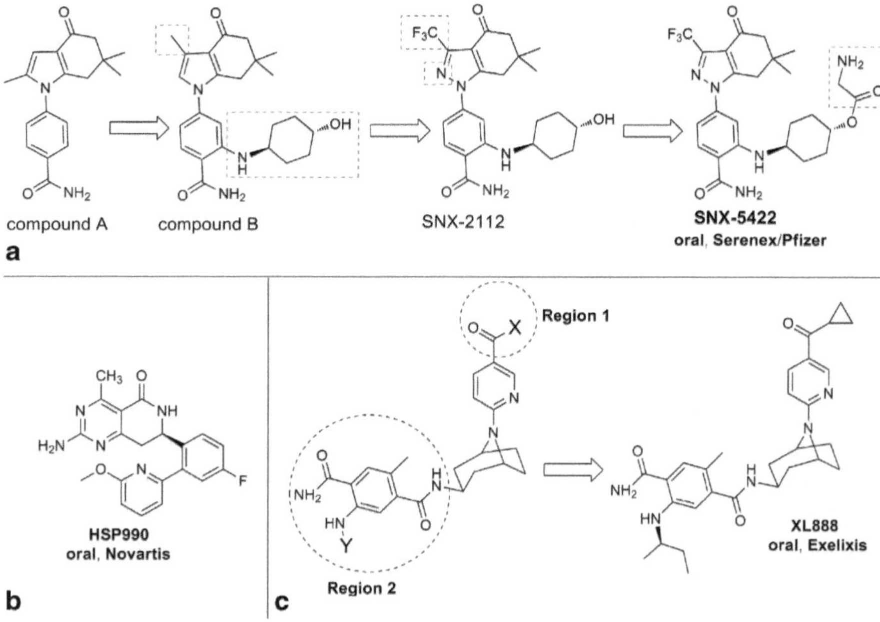

Fig. 7.4 **a** Discovery of SNX-5422, from hit to clinical candidate. **b** Clinical candidate HSP990. **c** Important SAR optimization regions leading to XL888

cal tractability, high selectivity for Hsp90, low molecular weight, and heteroatom count [102]. The screen was further validated by X-ray crystallography analysis. Their follow-up synthetic design complemented by structural information resulted in **compound B** (Fig. 7.4a) which met their requirements of strong affinity to Hsp90 and cell activity. The Serenex efforts extended further since precautions were taken to replace the possible metabolic liability of the methyl at C2 of the pyrrole moiety of **B** with a trifluoromethyl, a group known to enhance metabolic stability. Additionally, the pyrrole moiety of **compound B** was substituted with a pyrazole designed to confer an increase in polarity and solubility [103]. These changes resulted in the new lead candidate SNX-2112 which exceeded the potency of 17-AAG [102]. The last optimization step was focused on improving the oral bioavailability of SNX-2112, and this was achieved via the introduction of the glycine ester prodrug moiety to give their clinical candidate SNX-5422 (Fig. 7.4a) [102]. The prodrug is rapidly hydrolyzed to the parent SNX-2112 in vivo following oral dosing given that the levels of SNX-5422 were below the lower limit of quantitation at all time points. Additionally, SNX-5422 showed significant antitumor activity in different human tumor xenograft models in mice when administered orally [102].

In 2007, Serenex began a Phase I trial of SNX-5422 in patients with solid tumor and lymphoma. Later in 2008, Pfizer Inc. had reached an agreement to acquire Serenex [23] and thus to continue development of SNX-5422. However, the advancement of SNX-5422 in the clinic was halted in 2010 by Pfizer as a result of reports of ocular toxicity in a Phase I study [104]. At that time, Pfizer decided that comprehensive ocular assessments are needed for the drug before any further development. The drug was subsequently picked up by Esanex Inc. which is presently evaluating the drug in the clinic in patients with resistant lung adenocarcinoma and selected HER2+ cancers.

2.4.2 HSP990 (NVP-HSP990)

HSP990 (Fig. 7.4b) is a potent and selective Hsp90 inhibitor that was discovered by Novartis. The discovery process for this compound relied on high throughput screening which was complemented by structure based lead optimization [105], though the detailed SAR is yet to be published. Potency against the various Hsp90 isoforms was evaluated in a binding assay using biotinylated-GM and in ATPase activity assays [105]. The 2-amino-4-methyl-7,8-dihydropyrido[4,3-d]pyrimidin-5(6H)-one scaffold is structurally diverse from the other Hsp90 inhibitors, and crystallographic work shows that NVP-HSP990 binds to the N-terminal ATP-binding domain of Hsp90 [105]. Though the crystal structure is yet to be published, it is likely that the 2-NH_2 group of NVP-HSP990 interacts with Asp93 similarly to NH_2 of purine and the resorcinol moiety.

This drug candidate was shown to have potent antiproliferative activity in various cell lines as well as efficacy in vivo when tested in the GLT-16 xenograft model [105]. NVP-HSP990 was under investigation as an oral agent in two Phase I trials in patients with advanced solid tumors. However, the further development of NVP-HSP990 has been stopped due to its failure to achieve clinically meaningful responses at the maximum tolerated dose [106].

2.4.3 XL888, A 2-amino Terephthalamide Derivative

XL888 (Fig. 7.4c) is a novel tropane-derived inhibitor of Hsp90 discovered by Exelixis. The strategy used to obtain some hits involved an HTS campaign of 4.1 million compounds from the company's in-house chemical library [107]. The percentage displacement of biotinylated-GM from the Hsp90 chaperone complex was the method used to screen compounds and to identify initial leads. The most promising scaffold was a 3-amidotropane which provided sub-micromolar IC_{50}s in the Hsp90 displacement assay, and this was subsequently optimized through a structure-guided approach. Some problems encountered during optimization related to high molecular weights (>650 amu) and polar surface areas (>135 Å^2) which were outside of the desired range for orally bioavailable drugs [107]. This observation was further confirmed by rat PK which indicated high clearance and no calculable oral bioavailability. SAR efforts were focused on reducing the polarity and molecular weight of the nicotinamide portion (Region 1; Fig. 7.4c) of the scaffold. Thus, the amide functionality of the nicotinamide portion was replaced with ketone derivatives (Region 1; Fig. 7.4c) which had lower molecular weights (<520 amu) and PSA (< 120 Å^2) and improved rat PK when compared to the amides. There was little tolerability for ketone steric bulk (i.e., progressing from methyl ketone to ethyl, isopropyl, or cyclopropylketone) since Hsp90 inhibition decreased with increasing size, but noticeably the cyclopropylketone derivative showed encouraging results in the cell assay measuring HER2 degradation in NCI-N87 cells [107]. Next, efforts were directed towards introducing substituents on the benzamide 5-position (Region 2; Fig. 7.4c) portion of the molecule. A screen of a varied set of small alkyl amines was performed and these showed similar activity in the biochemical assay, still they did show variation in the NCI-N87 HER2 degradation assay. The most potent compound in cells was the R-*sec*-butyl compound (Fig. 7.4c; XL888) with an $IC_{50}=56$ nM and an attractive rat PK. As a whole, the extensive medicinal chemistry efforts coupled with biochemical and X-ray crystallographic methods were successful in the optimization process of the 3-amidotropane hit leading to XL888. The efficacy experiments performed in NCI-87 mouse xenograft model showed tumor regression, and further preclinical investigations coupled with more toxicity assessment resulted in the selection and advancement of XL888 into clinical trials [107, 108].

The first Phase I safety study of XL888 was terminated by the sponsor Exelixis [23] with no particular reason being disclosed. Another Phase I study of XL888 in combination with vemurafenib for patients with unresectable Serine/threonine-protein kinase B-raf (BRAF) mutated stage III/IV melanoma is underway [108, 109]. This study is supported by a report showing that signaling proteins involved in intrinsic and acquired resistance to BRAF inhibitors are clients of Hsp90 and inhibiting Hsp90 restored sensitivity to vemurafenib [108].

3 Future Drug-Discovery Efforts

3.1 Selective Inhibition of Hsp90 Isoforms

Hsp90 has four different paralogs, Hsp90α and Hsp90β in the cytoplasm, Grp94 in the endoplasmatic reticulum (ER), and tumor necrosis factor receptor-associated protein 1 (Trap-1) in the mitochondria. The Hsp90 clinical candidates described in this chapter are reported to be pan-Hsp90 inhibitors. Nevertheless, there is a growing interest in the development of isoform-selective inhibitors. A better understanding of the biological activity of each isoform would potentially allow the development of better and more selective therapeutics. The advent of new tools to study the contribution of the four Hsp90 paralogs [110, 111] has already set the ground for the design of paralog-selective inhibitors which will not only provide a better understanding of the role of each isoform in disease, but could potentially improve the toxicity profile of the pan-inhibitors. There are currently significant efforts underway and some of these directed towards Trap-1 and Grp94 selective inhibitors are discussed below.

3.1.1 Targeting Trap-1 for Cancer Therapy

Recent studies have demonstrated that Trap-1 levels are highly expressed in mitochondria of tumor cells, but are either very low or undetectable in normal tissues [112, 113]. As a result, selective targeting of Trap-1 has become of interest as a potential cancer-specific target. The first reported class of small molecule to selectively target the Hsp90 network in tumor mitochondria is represented by Gamitrinibs (GA mitochondrial matrix inhibitors) [112, 114]. Gamitrinibs are analogs of GM containing a mitochondrial targeting moiety connected through a linker at the C17 position. The gamitrinibs are distinct from current Hsp90 inhibitors since they display a "mitochondriotoxic" mechanism of action, resulting in very fast and complete tumor cell killing by apoptosis [114]. This property combined with the information that gamitrinibs have no effect on overall Hsp90 homeostasis outside of mitochondria make these anticancer agents novel and potentially attractive for human testing [114]. Extensive preclinical evaluation of gamitrinib for prostate cancer revealed that it induces prostate cancer cell death, and this is achieved mechanistically via acute mitochondrial dysfunction including loss of membrane potential and release of cytochrome c [115]. Overall, it has been demonstrated in vivo that gamitrinibs show preclinical efficacy combined with a good safety profile in models of drug-resistant and bone metastatic prostate cancer [115].

3.1.2 Targeting Grp94 for Cancer Therapy

Glucose-regulated protein 94 (Grp94) expression is associated with advanced stage and low survival in various cancers and correlates well with cancer growth and metastasis [116–119]. Most cancer-related studies on this chaperone looked at the immunogenic activity of Grp94-peptide complexes [117] and its role in the regulation of EGFR and HER2 in the ER, secretion of IGF-I and -II, and the regulation of Toll-like receptors (i.e., TLR1, TLR2, TLR4, and TLR9) and integrins [116–118, 120]. This was complemented by another report which used library screening (i.e., screened > 130 purine-scaffold compounds) in a fluorescence polarization assay to test the compounds for all four Hsp90 paralogs, and were able to find derivatives that had selectivity for Grp94 [111]. The library screen was further analyzed by structural and computational analysis which indicated the presence of a new allosteric pocket where these Grp94 selective compounds insert. These paralog selective inhibitors also inhibited Grp94-mediated cellular events such as IGF-II secretion and TLR9 trafficking [111]. Importantly, Hsp70 induction, which is a hallmark of cytosolic Hsp90 inhibition, was not observed. Additionally, this work also provided proof for the role of Grp94 in preserving the architecture of high-density HER2 formations at plasma membrane, specifically in cancer cells where HER2 has to channel the amplified signal through the receptor [111]. The new mechanistic insights reveal the importance of Grp94 inhibition in certain breast cancers (i.e., HER2 overexpressing), and the newly uncovered allosteric-binding site in Grp94 is an avenue to be explored by future drug discovery efforts.

3.2 *Alternative Modes of Modulating Hsp90 Activity in Cancer*

As already mentioned, all of the Hsp90 inhibitors that have entered clinical trials target the N-terminal ATP pocket of Hsp90. One drawback of these agents is that they induce a heat shock response which in turn increases the cellular levels of prosurvival chaperones (i.e., Hsp27 and Hsp70). Therefore, alternative approaches to inhibit Hsp90 function without inducing a heat shock response would be highly desirable. Efforts in this regard are increasingly being pursued and include developing inhibitors that bind to the C-terminal domain, molecules that target the binding of Hsp90 to either co-chaperones, or to client protein. Though none of these approaches have yet provided a clinical candidate they offer a potential for further future drug development that may result in one.

3.2.1 Inhibitors that Target the C-terminus of Hsp90

Novobiocin (NB), a coumarin antibiotic, was the first compound shown to bind weakly to the C-terminal nucleotide-binding site (~700 μM in SKBr3 cells) [27] and was shown to affect the association of Hsp90 with its co-chaperones Hsc70 and p23 [121, 122]. Subsequent work aimed at improving the observed low affinity of NB for Hsp90 has resulted in more potent analogs such as KU174 (K_d=94 μM) which has a 12-fold higher affinity than NB (K_d=1.1 mM) when analyzed by surface plasmon resonance (SPR) [123]. Importantly, KU174 does not induce a heat shock response and displays antitumor activity in prostate cancer cells [123, 124]. However, evaluation of KU174 in mouse was prevented by the extensive metabolism and clearance observed when pilot PK studies were performed with this species [123].

Epigallocatechin-3-gallate (EGCG) has also been shown to inhibit Hsp90 by binding to the C-terminus [125]. Recently, a report on SAR study was disclosed [126] and the best analog displayed anti-proliferative activity against the MCF-7 breast cancer cell line with an IC_{50}=4 μM, an 18-fold improvement over EGCG (MCF-7 IC_{50}=74 μM) [126]. Other C-terminus inhibitors reported in the literature are cisplatin [127], taxol [128], and withaferin A [129]. It is interesting to speculate how much of the anticancer activity of cisplatin and taxol, two popular clinical agents, is as a result of affecting Hsp90 function. Despite an undeniable interest in Hsp90 inhibitors for cancers, the further advancement of C-terminal inhibitors in the clinic is sluggish perhaps due to their observed weak biological activity and potential pleiotropic mechanisms of action.

3.2.2 Targeting Co-chaperone–Hsp90 Interactions

Hsp90 is assisted in its chaperone activity by a cohort of co-chaperones (e.g., Hsp70, Cdc37, Aha1, HOP, etc.) which modulate Hsp90 ATPase and determine the rate of chaperone cycling [10]. Therefore, Hsp90 activity can be indirectly modulated by targeting co-chaperone function, and some initial efforts towards targeting co-chaperone-Hsp90 interactions are indeed described in the literature [12].

3.2.3 Targeting Client Protein Binding to Hsp90

Strategies which target the binding of client proteins to Hsp90 offer an exciting opportunity to develop novel anticancer drugs. Developing such molecules which target-specific protein-protein interactions is notoriously difficult; however, this has already been attempted with some success for the client protein survivin. Survivin is an inhibitor of apoptosis protein that is overexpressed in almost all human tumors and has critical roles in tumor cell proliferation and cell viability [130, 131]. Thus targeting the survivin-Hsp90 complex may be a suitable strategy for cancer therapy. To this end, a peptide sequence of survivin called shepherdin was designed via structure-based mimicry as a high affinity ($K_d \sim 80$ nM) inhibitor of the survivin-Hsp90 interaction, but subsequent data suggested that shepherdin could perhaps function as a more global antagonist of Hsp90 activity [132]. Later work screened a database of nonpeptidic structures using combined structure and dynamics based computational design, and this resulted in the identification of the small compound 5-aminoimidazole-4-carboxamide-1-β-D-ribofuranoside (AICAR) which binds to the Hsp90 N-domain, in vivo destabilizes numerous Hsp90 client proteins (counting survivin), and also displays selective anticancer activity in several tumor cell lines [131]. Overall, the experimental structural data (i.e., docking studies, molecular dynamics, and NMR) for these inhibitors provide insights for the future design of better compounds.

4 Conclusions

Undoubtedly, there has been considerable progress in the development of Hsp90 inhibitors over the past decade. While the first generation of compounds, natural products GM and RD, served as proof of concept for Hsp90 as an anticancer target, their further advancement in the clinic has met many obstacles as a result of their poor drug like properties. The challenge to develop better therapeutics that would overcome most of the limitations observed with the first generation drugs resulted in the second generation synthetic Hsp90 inhibitors. This chapter summarized the main medicinal chemistry SAR and drug design strategies performed by various research groups for a number of candidates being evaluated in clinical trials. All the clinical candidates are indeed a testimony to the arduous efforts of both academia and industry towards the development of better Hsp90 therapeutics for the treatment of cancer. Furthermore, herein we also touched upon other chemotypes with alternative binding modes that are currently in preclinical development.

To date, no Hsp90 inhibitor has yet been approved by the Food and Drug Administration (FDA), a reality which may reflect our little understanding about the best use of these compounds in the clinic, and also our limited knowledge in selecting those patients who will likely benefit most from Hsp90 inhibitor therapy. This may soon change as results from the numerous clinical trials become available and are carefully scrutinized. To conclude, this chapter highlighted potent and pharmaceutically promising Hsp90 clinical agents that have populated the field, some of which are credibly poised to ultimately reach market approval in the future. Understandably, progress in this field will be aided by a better comprehension of the interaction of the inhibitor with the target thus leading to better clinical response.

References

1. Ritossa F (1962) A new puffing pattern induced by temperature shock and DNP in drosophila. Cell Mol Life Sci 18(12):571–573
2. Tissieres A, Mitchell HK, Tracy UM (1974) Protein synthesis in salivary glands of *Drosophila melanogaster*: relation to chromosome puffs. J Mol Biol 84(3):389–398
3. Whitesell L, Lindquist SL (2005) HSP90 and the chaperoning of cancer. Nat Rev Cancer 5(10):761–772. doi:10.1038/nrc1716
4. Wegele H, Muller L, Buchner J (2004) Hsp70 and Hsp90–a relay team for protein folding. Rev Physiol Biochem Pharmacol 151:1–44. doi:10.1007/s10254-003-0021-1
5. Pearl LH, Prodromou C (2006) Structure and mechanism of the Hsp90 molecular chaperone machinery. Annu Rev Biochem 75:271–294. doi:10.1146/annurev.biochem.75.103004.142738
6. Wandinger SK, Richter K, Buchner J (2008) The Hsp90 chaperone machinery. J Biol Chem 283(27):18473–18477. doi:10.1074/jbc.R800007200
7. Zuehlke A, Johnson JL (2010) Hsp90 and co-chaperones twist the functions of diverse client proteins. Biopolymers 93 (3):211–217. doi:10.1002/bip.21292
8. Workman P, Burrows F, Neckers L, Rosen N (2007) Drugging the cancer chaperone HSP90: combinatorial therapeutic exploitation of oncogene addiction and tumor stress. Ann N Y Acad Sci 1113:202–216. doi: 10.1196/annals.1391.012
9. Neckers L (2002) Hsp90 inhibitors as novel cancer chemotherapeutic agents. Trends Mol Med 8(4 Suppl):S55–S61
10. Trepel J, Mollapour M, Giaccone G, Neckers L (2010) Targeting the dynamic HSP90 complex in cancer. Nat Rev Cancer 10(8):537–549. doi:10.1038/nrc2887
11. Porter JR, Fritz CC, Depew KM (2010) Discovery and development of Hsp90 inhibitors: a promising pathway for cancer therapy. Curr Opin Chem Biol 14(3):412–420. doi:10.1016/j.cbpa.2010.03.019
12. Patel HJ, Modi S, Chiosis G, Taldone T (2011) Advances in the discovery and development of heat-shock protein 90 inhibitors for cancer treatment. Expert Opin Drug Discov 6(5):559–587. doi:10.1517/17460441.2011.563296
13. DeBoer C, Meulman PA, Wnuk RJ, Peterson DH (1970) Geldanamycin, a new antibiotic. J Antibiot (Tokyo) 23(9):442–447
14. Uehara Y, Hori M, Takeuchi T, Umezawa H (1986) Phenotypic change from transformed to normal induced by benzoquinonoid ansamycins accompanies inactivation of p60src in rat kidney cells infected with Rous sarcoma virus. Mol Cell Biol 6(6):2198–2206
15. Whitesell L, Mimnaugh EG, De Costa B, Myers CE, Neckers LM (1994) Inhibition of heat shock protein HSP90-pp60v-src heteroprotein complex formation by benzoquinone ansamycins: essential role for stress proteins in oncogenic transformation. Proc Natl Acad Sci U S A 91(18):8324–8328
16. Prodromou C, Roe SM, O'Brien R, Ladbury JE, Piper PW, Pearl LH (1997) Identification and structural characterization of the ATP/ADP-binding site in the Hsp90 molecular chaperone. Cell 90(1):65–75

17. Stebbins CE, Russo AA, Schneider C, Rosen N, Hartl FU, Pavletich NP (1997) Crystal structure of an Hsp90-geldanamycin complex: targeting of a protein chaperone by an antitumor agent. Cell 89(2):239–250. doi:S0092-8674(00)80203-2
18. Grenert JP, Sullivan WP, Fadden P, Haystead TA, Clark J, Mimnaugh E, Krutzsch H, Ochel HJ, Schulte TW, Sausville E, Neckers LM, Toft DO (1997) The amino-terminal domain of heat shock protein 90 (hsp90) that binds geldanamycin is an ATP/ADP switch domain that regulates hsp90 conformation. J Biol Chem 272(38):23843–23850
19. Pratt WB (1998) The hsp90-based chaperone system: involvement in signal transduction from a variety of hormone and growth factor receptors. Proc Soc Exp Biol Med 217(4):420–434
20. Roe SM, Prodromou C, O'Brien R, Ladbury JE, Piper PW, Pearl LH (1999) Structural basis for inhibition of the Hsp90 molecular chaperone by the antitumor antibiotics radicicol and geldanamycin. J Med Chem 42(2):260–266. doi:10.1021/jm980403y
21. Kamal A, Thao L, Sensintaffar J, Zhang L, Boehm MF, Fritz LC, Burrows FJ (2003) A high-affinity conformation of Hsp90 confers tumour selectivity on Hsp90 inhibitors. Nature 425(6956):407–410. doi:10.1038/nature01913
22. Moulick K, Ahn JH, Zong H, Rodina A, Cerchietti L, Gomes DGEM, Caldas-Lopes E, Beebe K, Perna F, Hatzi K, Vu LP, Zhao X, Zatorska D, Taldone T, Smith-Jones P, Alpaugh M, Gross SS, Pillarsetty N, Ku T, Lewis JS, Larson SM, Levine R, Erdjument-Bromage H, Guzman ML, Nimer SD, Melnick A, Neckers L, Chiosis G (2011) Affinity-based proteomics reveal cancer-specific networks coordinated by Hsp90. Nat Chem Biol 7(11):818–826. doi:10.1038/nchembio.670
23. Jhaveri K, Taldone T, Modi S, Chiosis G (2012) Advances in the clinical development of heat shock protein 90 (Hsp90) inhibitors in cancers. Biochim Biophys Acta 1823(3):742–755. doi:10.1016/j.bbamcr.2011.10.008
24. Kamal A, Burrows FJ (2004) Hsp90 inhibitors as selective anticancer drugs. Discov Med 4(23):277–280
25. Ardini E, Galvani A (2012) ALK inhibitors: a pharmaceutical perspective. Front Oncol 2:17. doi:10.3389/fonc.2012.00017
26. Kim YS, Alarcon SV, Lee S, Lee MJ, Giaccone G, Neckers L, Trepel JB (2009) Update on Hsp90 inhibitors in clinical trial. Curr Top Med Chem 9(15):1479–1492. doi:CTMC-Abs-031-9-15
27. Donnelly A, Blagg BS (2008) Novobiocin and additional inhibitors of the Hsp90 C-terminal nucleotide-binding pocket. Curr Med Chem 15(26):2702–2717
28. Brandt GE, Blagg BS (2009) Alternate strategies of Hsp90 modulation for the treatment of cancer and other diseases. Curr Topics Med Chem 9(15):1447–1461
29. Gao Z, Garcia-Echeverria C, Jensen MR (2010) Hsp90 inhibitors: clinical development and future opportunities in oncology therapy. Curr Opin Drug Discov Dev 13(2):193–202
30. Janin YL (2010) ATPase inhibitors of heat-shock protein 90, second season. Drug Discov Today 15(9-10):342–353. doi:10.1016/j.drudis.2010.03.002
31. Beebe K, Mollapour M, Scroggins B, Prodromou C, Xu W, Tokita M, Taldone T, Pullen L, Zierer BK, Lee MJ, Trepel J, Buchner J, Bolon D, Chiosis G, Neckers L (2013) Posttranslational modification and conformational state of Heat Shock Protein 90 differentially affect binding of chemically diverse small molecule inhibitors. Oncotarget 4(7):1065–1074
32. Anderson AC (2003) The process of structure-based drug design. Chem Biol 10(9):787–797. doi:10.1016/j.chembiol.20 03.09.002
33. Erlanson DA, McDowell RS, O'Brien T (2004) Fragment-based drug discovery. J Med Chem 47(14):3463–3482. doi:10.1021/jm040031v
34. Vilenchik M, Solit D, Basso A, Huezo H, Lucas B, He H, Rosen N, Spampinato C, Modrich P, Chiosis G (2004) Targeting wide-range oncogenic transformation via PU24FCl, a specific inhibitor of tumor Hsp90. Chem Biol 11(6):787–797. doi:10.1016/j.chembiol.2004.04.008
35. Supko JG, Hickman RL, Grever MR, Malspeis L (1995) Preclinical pharmacologic evaluation of geldanamycin as an antitumor agent. Cancer Chemother Pharmacol 36(4):305–315

36. Neckers L (2006) Chaperoning oncogenes: Hsp90 as a target of geldanamycin. Handb Exp Pharmacol 172:259–277
37. Samuni A, Goldstein S (2012) Redox properties and thiol reactivity of geldanamycin and its analogues in aqueous solutions. J Phys Chem B 116(22):6404–6410. doi:10.1021/jp304206n
38. Samuni Y, Ishii H, Hyodo F, Samuni U, Krishna MC, Goldstein S, Mitchell JB (2010) Reactive oxygen species mediate hepatotoxicity induced by the Hsp90 inhibitor geldanamycin and its analogs. Free Radic Biol Med 48(11):1559–1563. doi:10.1016/j.freeradbiomed.2010.03.001
39. Tian ZQ, Liu Y, Zhang D, Wang Z, Dong SD, Carreras CW, Zhou Y, Rastelli G, Santi DV, Myles DC (2004) Synthesis and biological activities of novel 17-aminogeldanamycin derivatives. Bioorg Med Chem 12(20):5317–5329. doi:10.1016/j.bmc.2004.07.053
40. Schulte TW, Neckers LM (1998) The benzoquinone ansamycin 17-allylamino-17-demethoxygeldanamycin binds to HSP90 and shares important biologic activities with geldanamycin. Cancer Chemother Pharmacol 42(4):273–279
41. Banerji U, O'Donnell A, Scurr M, Pacey S, Stapleton S, Asad Y, Simmons L, Maloney A, Raynaud F, Campbell M, Walton M, Lakhani S, Kaye S, Workman P, Judson I (2005) Phase I pharmacokinetic and pharmacodynamic study of 17-allylamino, 17-demethoxygeldanamycin in patients with advanced malignancies. J Clin Oncol 23(18):4152–4161. doi:10.1200/JCO.2005.00.612
42. Kelland LR, Sharp SY, Rogers PM, Myers TG, Workman P (1999) DT-diaphorase expression and tumor cell sensitivity to 17-allylamino, 17-demethoxygeldanamycin, an inhibitor of heat shock protein 90. J Natl Cancer Inst 91(22):1940–1949
43. Solit DB, Ivy SP, Kopil C, Sikorski R, Morris MJ, Slovin SF, Kelly WK, DeLaCruz A, Curley T, Heller G, Larson S, Schwartz L, Egorin MJ, Rosen N, Scher HI (2007) Phase I trial of 17-allylamino-17-demethoxygeldanamycin in patients with advanced cancer. Clin Cancer Res 13(6):1775–1782. doi:10.1158/1078-0432.CCR-06-1863
44. Hollingshead M, Alley M, Burger AM, Borgel S, Pacula-Cox C, Fiebig HH, Sausville EA (2005) In vivo antitumor efficacy of 17-DMAG (17-dimethylaminoethylamino-17-demethoxygeldanamycin hydrochloride), a water-soluble geldanamycin derivative. Cancer Chemother Pharmacol 56(2):115–125. doi:10.1007/s00280-004-0939-2
45. Ge J, Normant E, Porter JR, Ali JA, Dembski MS, Gao Y, Georges AT, Grenier L, Pak RH, Patterson J, Sydor JR, Tibbitts TT, Tong JK, Adams J, Palombella VJ (2006) Design, synthesis, and biological evaluation of hydroquinone derivatives of 17-amino-17-demethoxygeldanamycin as potent, water-soluble inhibitors of Hsp90. J Med Chem 49(15):4606–4615. doi:10.1021/jm0603116
46. Sydor JR, Normant E, Pien CS, Porter JR, Ge J, Grenier L, Pak RH, Ali JA, Dembski MS, Hudak J, Patterson J, Penders C, Pink M, Read MA, Sang J, Woodward C, Zhang Y, Grayzel DS, Wright J, Barrett JA, Palombella VJ, Adams J, Tong JK (2006) Development of 17-allylamino-17-demethoxygeldanamycin hydroquinone hydrochloride (IPI-504), an anti-cancer agent directed against Hsp90. Proc Natl Acad Sci U S A 103(46):17408–17413. doi:10.1073/pnas.0608372103
47. Sequist LV, Gettinger S, Senzer NN, Martins RG, Janne PA, Lilenbaum R, Gray JE, Iafrate AJ, Katayama R, Hafeez N, Sweeney J, Walker JR, Fritz C, Ross RW, Grayzel D, Engelman JA, Borger DR, Paez G, Natale R (2010) Activity of IPI-504, a novel heat-shock protein 90 inhibitor, in patients with molecularly defined non-small-cell lung cancer. J Clin Oncol 28(33):4953–4960. doi:10.1200/JCO.2010.30.8338
48. Floris G, Debiec-Rychter M, Wozniak A, Stefan C, Normant E, Faa G, Machiels K, Vanleeuw U, Sciot R, Schoffski P (2011) The heat shock protein 90 inhibitor IPI-504 induces KIT degradation, tumor shrinkage, and cell proliferation arrest in xenograft models of gastrointestinal stromal tumors. Mol Cancer Ther 10(10):1897–1908. doi:10.1158/1535-7163.MCT-11-0148
49. Soga S, Akinaga S, Shiotsu Y (2013) Hsp90 inhibitors as anti-cancer agents, from basic discoveries to clinical development. Curr Pharm Des 19(3):366–376
50. Delmotte P, Delmotte-Plaque J (1953) A new antifungal substance of fungal origin. Nature 171(4347):344

51. Schulte TW, Akinaga S, Soga S, Sullivan W, Stensgard B, Toft D, Neckers LM (1998) Antibiotic radicicol binds to the N-terminal domain of Hsp90 and shares important biologic activities with geldanamycin. Cell Stress Chaperones 3(2):100–108
52. Soga S, Neckers LM, Schulte TW, Shiotsu Y, Akasaka K, Narumi H, Agatsuma T, Ikuina Y, Murakata C, Tamaoki T, Akinaga S (1999) KF25706, a novel oxime derivative of radicicol, exhibits in vivo antitumor activity via selective depletion of Hsp90 binding signaling molecules. Cancer Res 59(12):2931–2938
53. Yang ZQ, Geng X, Solit D, Pratilas CA, Rosen N, Danishefsky SJ (2004) New efficient synthesis of resorcinylic macrolides via ynolides: establishment of cycloproparadicicol as synthetically feasible preclinical anticancer agent based on Hsp90 as the target. J Am Chem Soc 126(25):7881–7889. doi:10.1021/ja0484348
54. Soga S, Shiotsu Y, Akinaga S, Sharma SV (2003) Development of radicicol analogues. Curr Cancer Drug Targets 3(5):359–369
55. Blagg BS, Kerr TD (2006) Hsp90 inhibitors: small molecules that transform the Hsp90 protein folding machinery into a catalyst for protein degradation. Med Res Rev 26(3):310–338. doi:10.1002/med.20052
56. Agatsuma T, Ogawa H, Akasaka K, Asai A, Yamashita Y, Mizukami T, Akinaga S, Saitoh Y (2002) Halohydrin and oxime derivatives of radicicol: synthesis and antitumor activities. Bioorg Med Chem 10(11):3445–3454
57. Ikuina Y, Amishiro N, Miyata M, Narumi H, Ogawa H, Akiyama T, Shiotsu Y, Akinaga S, Murakata C (2003) Synthesis and antitumor activity of novel O-carbamoylmethyloxime derivatives of radicicol. J Med Chem 46(12):2534–2541. doi:10.1021/jm030110r
58. Rowlands MG, Newbatt YM, Prodromou C, Pearl LH, Workman P, Aherne W (2004) High-throughput screening assay for inhibitors of heat-shock protein 90 ATPase activity. Anal Biochem 327(2):176–183. doi:10.1016/j.ab.2003.10.038
59. Cheung KM, Matthews TP, James K, Rowlands MG, Boxall KJ, Sharp SY, Maloney A, Roe SM, Prodromou C, Pearl LH, Aherne GW, McDonald E, Workman P (2005) The identification, synthesis, protein crystal structure and in vitro biochemical evaluation of a new 3,4-diarylpyrazole class of Hsp90 inhibitors. Bioorg Med Chem Lett 15(14):3338–3343. doi:10.1016/j.bmcl.2005.05.046
60. Dymock BW, Barril X, Brough PA, Cansfield JE, Massey A, McDonald E, Hubbard RE, Surgenor A, Roughley SD, Webb P, Workman P, Wright L, Drysdale MJ (2005) Novel, potent small-molecule inhibitors of the molecular chaperone Hsp90 discovered through structure-based design. J Med Chem 48(13):4212–4215. doi:10.1021/jm050355z
61. Brough PA, Aherne W, Barril X, Borgognoni J, Boxall K, Cansfield JE, Cheung KM, Collins I, Davies NG, Drysdale MJ, Dymock B, Eccles SA, Finch H, Fink A, Hayes A, Howes R, Hubbard RE, James K, Jordan AM, Lockie A, Martins V, Massey A, Matthews TP, McDonald E, Northfield CJ, Pearl LH, Prodromou C, Ray S, Raynaud FI, Roughley SD, Sharp SY, Surgenor A, Walmsley DL, Webb P, Wood M, Workman P, Wright L (2008) 4,5-diarylisoxazole Hsp90 chaperone inhibitors: potential therapeutic agents for the treatment of cancer. J Med Chem 51(2):196–218. doi:10.1021/jm701018h
62. Brough PA, Barril X, Beswick M, Dymock BW, Drysdale MJ, Wright L, Grant K, Massey A, Surgenor A, Workman P (2005) 3-(5-Chloro-2,4-dihydroxyphenyl)-pyrazole-4-carboxamides as inhibitors of the Hsp90 molecular chaperone. Bioorg Med Chem Lett 15(23):5197–5201. doi:10.1016/j.bmcl.2005.08.091
63. Barril X, Beswick MC, Collier A, Drysdale MJ, Dymock BW, Fink A, Grant K, Howes R, Jordan AM, Massey A, Surgenor A, Wayne J, Workman P, Wright L (2006) 4-Amino derivatives of the Hsp90 inhibitor CCT018159. Bioorg Med Chem Lett 16(9):2543–2548. doi:10.1016/j.bmcl.2006.01.099
64. Sharp SY, Prodromou C, Boxall K, Powers MV, Holmes JL, Box G, Matthews TP, Cheung KM, Kalusa A, James K, Hayes A, Hardcastle A, Dymock B, Brough PA, Barril X, Cansfield JE, Wright L, Surgenor A, Foloppe N, Hubbard RE, Aherne W, Pearl L, Jones K, McDonald E, Raynaud F, Eccles S, Drysdale M, Workman P (2007) Inhibition of the heat shock protein 90 molecular chaperone in vitro and in vivo by novel, synthetic, potent resorcinylic pyrazole/

isoxazole amide analogues. Mol Cancer Ther 6(4):1198–1211. doi:10.1158/1535-7163.MCT-07-0149
65. Jensen MR, Schoepfer J, Radimerski T, Massey A, Guy CT, Brueggen J, Quadt C, Buckler A, Cozens R, Drysdale MJ, Garcia-Echeverria C, Chene P (2008) NVP-AUY922: a small molecule HSP90 inhibitor with potent antitumor activity in preclinical breast cancer models. Breast Cancer Res 10(2):R33. doi:10.1186/bcr1996
66. Sessa C, Shapiro GI, Bhalla KN, Britten C, Jacks KS, Mita M, Papadimitrakopoulou V, Pluard T, Samuel TA, Akimov M, Quadt C, Fernandez-Ibarra C, Lu H, Bailey S, Chica S, Banerji U (2013) First-in-human phase I dose-escalation study of the HSP90 inhibitor AUY922 in patients with advanced solid tumors. Clin Cancer Res 19(13):3671–3680. doi:10.1158/1078-0432.CCR-12-3404
67. Murray CW, Carr MG, Callaghan O, Chessari G, Congreve M, Cowan S, Coyle JE, Downham R, Figueroa E, Frederickson M, Graham B, McMenamin R, O'Brien MA, Patel S, Phillips TR, Williams G, Woodhead AJ, Woolford AJ (2010) Fragment-based drug discovery applied to Hsp90. Discovery of two lead series with high ligand efficiency. J Med Chem 53(16):5942–5955. doi:10.1021/jm100059d
68. Woodhead AJ, Angove H, Carr MG, Chessari G, Congreve M, Coyle JE, Cosme J, Graham B, Day PJ, Downham R, Fazal L, Feltell R, Figueroa E, Frederickson M, Lewis J, McMenamin R, Murray CW, O'Brien MA, Parra L, Patel S, Phillips T, Rees DC, Rich S, Smith DM, Trewartha G, Vinkovic M, Williams B, Woolford AJ (2010) Discovery of (2,4-dihydroxy-5-isopropylphenyl)-[5-(4-methylpiperazin-1-ylmethyl)-1,3-dihydrois oindol-2-yl]methanone (AT13387), a novel inhibitor of the molecular chaperone Hsp90 by fragment based drug design. J Med Chem 53(16):5956–5969. doi:10.1021/jm100060b
69. Graham B, Curry J, Smyth T, Fazal L, Feltell R, Harada I, Coyle J, Williams B, Reule M, Angove H, Cross DM, Lyons J, Wallis NG, Thompson NT (2012) The heat shock protein 90 inhibitor, AT13387, displays a long duration of action in vitro and in vivo in non-small cell lung cancer. Cancer Sci 103(3):522–527. doi:10.1111/j.1349-7006.2011.02191.x
70. El-Hariry I, Proia D, Vukovic V (2013) Treating cancer with heat-shock protein-90 (HSP90) inhibitory compounds such as ganetespib. WO2013006864A2
71. Ying W, Du Z, Sun L, Foley KP, Proia DA, Blackman RK, Zhou D, Inoue T, Tatsuta N, Sang J, Ye S, Acquaviva J, Ogawa LS, Wada Y, Barsoum J, Koya K (2012) Ganetespib, a unique triazolone-containing Hsp90 inhibitor, exhibits potent antitumor activity and a superior safety profile for cancer therapy. Mol Cancer Ther 11(2):475–484. doi:10.1158/1535-7163.MCT-11-0755
72. Goldman JW, Raju RN, Gordon GA, El-Hariry I, Teofilivici F, Vukovic VM, Bradley R, Karol MD, Chen Y, Guo W, Inoue T, Rosen LS (2013) A first in human, safety, pharmacokinetics, and clinical activity phase I study of once weekly administration of the Hsp90 inhibitor ganetespib (STA-9090) in patients with solid malignancies. BMC Cancer 13:152. doi:10.1186/1471-2407-13-152
73. Socinski MA, Goldman J, El-Hariry I, Koczywas M, Vukovic V, Horn L, Paschold E, Salgia R, West H, Sequist LV, Bonomi P, Brahmer J, Chen LC, Sandler A, Belani CP, Webb T, Harper H, Huberman M, Ramalingam S, Wong KK, Teofilovici F, Guo W, Shapiro GI (2013) A multicenter phase II study of ganetespib monotherapy in patients with genotypically defined advanced non-small cell lung cancer. Clin Cancer Res 19(11):3068–3077. doi:10.1158/1078-0432.CCR-12-3381
74. Nakashima T, Ishii T, Tagaya H, Seike T, Nakagawa H, Kanda Y, Akinaga S, Soga S, Shiotsu Y (2010) New molecular and biological mechanism of antitumor activities of KW-2478, a novel nonansamycin heat shock protein 90 inhibitor, in multiple myeloma cells. Clin Cancer Res 16(10):2792–2802. doi:10.1158/1078-0432.CCR-09-3112
75. Dutta R, Inouye M (2000) GHKL, an emergent ATPase/kinase superfamily. Trends Biochem Sci 25(1):24–28
76. Chene P (2002) ATPases as drug targets: learning from their structure. Nat Rev Drug Discov 1(9):665–673. doi:10.1038/nrd894

77. Chiosis G, Timaul MN, Lucas B, Munster PN, Zheng FF, Sepp-Lorenzino L, Rosen N (2001) A small molecule designed to bind to the adenine nucleotide pocket of Hsp90 causes Her2 degradation and the growth arrest and differentiation of breast cancer cells. Chem Biol 8(3):289–299. doi:S1074-5521(01)00015-1
78. Wright L, Barril X, Dymock B, Sheridan L, Surgenor A, Beswick M, Drysdale M, Collier A, Massey A, Davies N, Fink A, Fromont C, Aherne W, Boxall K, Sharp S, Workman P, Hubbard RE (2004) Structure-activity relationships in purine-based inhibitor binding to HSP90 isoforms. Chem Biol 11(6):775–785. doi:10.1016/j.chembiol.2004.03.033
79. He H, Zatorska D, Kim J, Aguirre J, Llauger L, She Y, Wu N, Immormino RM, Gewirth DT, Chiosis G (2006) Identification of potent water soluble purine-scaffold inhibitors of the heat shock protein 90. J Med Chem 49(1):381–390. doi:10.1021/jm0508078
80. Chiosis G, Lucas B, Shtil A, Huezo H, Rosen N (2002) Development of a purine-scaffold novel class of Hsp90 binders that inhibit the proliferation of cancer cells and induce the degradation of Her2 tyrosine kinase. Bioorg Med Chem 10(11):3555–3564
81. Dymock B, Barril X, Beswick M, Collier A, Davies N, Drysdale M, Fink A, Fromont C, Hubbard RE, Massey A, Surgenor A, Wright L (2004) Adenine derived inhibitors of the molecular chaperone HSP90-SAR explained through multiple X-ray structures. Bioorg Med Chem Lett 14(2):325–328
82. Kim J, Felts S, Llauger L, He H, Huezo H, Rosen N, Chiosis G (2004) Development of a fluorescence polarization assay for the molecular chaperone Hsp90. J Biomol Screen 9(5):375–381. doi:10.1177/1087057104265995
83. Llauger L, He H, Kim J, Aguirre J, Rosen N, Peters U, Davies P, Chiosis G (2005) Evaluation of 8-arylsulfanyl, 8-arylsulfoxyl, and 8-arylsulfonyl adenine derivatives as inhibitors of the heat shock protein 90. J Med Chem 48(8):2892–2905. doi:10.1021/jm049012b
84. Llauger-Bufi L, Felts SJ, Huezo H, Rosen N, Chiosis G (2003) Synthesis of novel fluorescent probes for the molecular chaperone Hsp90. Bioorg Med Chem Lett 13(22):3975–3978
85. Biamonte MA, Shi J, Hong K, Hurst DC, Zhang L, Fan J, Busch DJ, Karjian PL, Maldonado AA, Sensintaffar JL, Yang YC, Kamal A, Lough RE, Lundgren K, Burrows FJ, Timony GA, Boehm MF, Kasibhatla SR (2006) Orally active purine-based inhibitors of the heat shock protein 90. J Med Chem 49(2):817–828. doi:10.1021/jm0503087
86. He H, Llauger L, Rosen N, Chiosis G (2004) General method for the synthesis of 8-arylsulfanyl adenine derivatives. J Org Chem 69(9):3230–3232. doi:10.1021/jo049875c
87. Caldas-Lopes E, Cerchietti L, Ahn JH, Clement CC, Robles AI, Rodina A, Moulick K, Taldone T, Gozman A, Guo Y, Wu N, de Stanchina E, White J, Gross SS, Ma Y, Varticovski L, Melnick A, Chiosis G (2009) Hsp90 inhibitor PU-H71, a multimodal inhibitor of malignancy, induces complete responses in triple-negative breast cancer models. Proc Natl Acad Sci U S A 106(20):8368–8373. doi:10.1073/pnas.0903392106
88. Rodina A, Vilenchik M, Moulick K, Aguirre J, Kim J, Chiang A, Litz J, Clement CC, Kang Y, She Y, Wu N, Felts S, Wipf P, Massague J, Jiang X, Brodsky JL, Krystal GW, Chiosis G (2007) Selective compounds define Hsp90 as a major inhibitor of apoptosis in small-cell lung cancer. Nat Chem Biol 3(8):498–507. doi:10.1038/nchembio.2007.10
89. Breinig M, Caldas-Lopes E, Goeppert B, Malz M, Rieker R, Bergmann F, Schirmacher P, Mayer M, Chiosis G, Kern MA (2009) Targeting heat shock protein 90 with non-quinone inhibitors: a novel chemotherapeutic approach in human hepatocellular carcinoma. Hepatology 50 (1):102–112. doi:10.1002/hep. 22912
90. Cerchietti LC, Lopes EC, Yang SN, Hatzi K, Bunting KL, Tsikitas LA, Mallik A, Robles AI, Walling J, Varticovski L, Shaknovich R, Bhalla KN, Chiosis G, Melnick A (2009) A purine scaffold Hsp90 inhibitor destabilizes BCL-6 and has specific antitumor activity in BCL-6-dependent B cell lymphomas. Nat Med 15(12):1369–1376. doi:10.1038/nm.2059
91. Marubayashi S, Koppikar P, Taldone T, Abdel-Wahab O, West N, Bhagwat N, Caldas-Lopes E, Ross KN, Gonen M, Gozman A, Ahn JH, Rodina A, Ouerfelli O, Yang G, Hedvat C, Bradner JE, Chiosis G, Levine RL (2010) HSP90 is a therapeutic target in JAK2-dependent myeloproliferative neoplasms in mice and humans. J Clin Invest 120(10):3578–3593. doi:10.1172/JCI42442

92. Kim SH, Bajji A, Tangallapally R, Markovitz B, Trovato R, Shenderovich M, Baichwal V, Bartel P, Cimbora D, McKinnon R, Robinson R, Papac D, Wettstein D, Carlson R, Yager KM (2012) Discovery of (2 S)-1-[4-(2-{6-amino-8-[(6-bromo-1,3-benzodioxol-5-yl)sulfanyl]-9 H-purin-9-yl}ethyl)piperidin-1-yl]-2-hydroxypropan-1-one (MPC-3100), a purine-based Hsp90 inhibitor. J Med Chem 55(17):7480–7501. doi:10.1021/jm3004619
93. Bajji AC, Tangallapally RP, Kim IC, Kim SH, Parker DP, Trovato R, Papac DI, Yager K, Baichwal VR (2012) Preparation of 6-amino-9 H-purine prodrugs as HSP90 inhibitors for the treatment of cancer. WO2012148550A1
94. 10-Q: MYREXIS, INC (2012). http://www.marketwatch.com/story/10-q-myrexis-inc-2012-11-09. Accessed 8 Oct 2013
95. Cai X, Zhai H-X, Wang J, Samson M, Atoyan R, Forrester J, Qu H, Yin L, Wang D, Zifcak B, DellaRocca S, Xu G-X, Lai C-J, Bao R, Simonin M-P, Keegan M, Pepicelli CV, Changgeng Q (2011) Design and synthesis of imidazopyridine derivatives as novel HSP90 inhibitors for the treatment of cancer. Proceedings: AACR 102nd Annual Meeting 2011, Apr 2–6, Orlando, Florida. doi:10.1158/1538-7445.AM2011-3249
96. Bao R, Lai CJ, Qu H, Wang D, Yin L, Zifcak B, Atoyan R, Wang J, Samson M, Forrester J, DellaRocca S, Xu GX, Tao X, Zhai HX, Cai X, Qian C (2009) CUDC-305, a novel synthetic HSP90 inhibitor with unique pharmacologic properties for cancer therapy. Clin Cancer Res 15(12):4046–4057. doi:10.1158/1078-0432.CCR-09-0152
97. Curis (2013) Debio 0932. http://www.curis.com/pipeline_detail.php?id=15. Accessed 21 Sept 2013
98. Kasibhatla SR, Hong K, Biamonte MA, Busch DJ, Karjian PL, Sensintaffar JL, Kamal A, Lough RE, Brekken J, Lundgren K, Grecko R, Timony GA, Ran Y, Mansfield R, Fritz LC, Ulm E, Burrows FJ, Boehm MF (2007) Rationally designed high-affinity 2-amino-6-halopurine heat shock protein 90 inhibitors that exhibit potent antitumor activity. J Med Chem 50(12):2767–2778. doi:10.1021/jm050752+
99. Lundgren K, Zhang H, Brekken J, Huser N, Powell RE, Timple N, Busch DJ, Neely L, Sensintaffar JL, Yang YC, McKenzie A, Friedman J, Scannevin R, Kamal A, Hong K, Kasibhatla SR, Boehm MF, Burrows FJ (2009) BIIB021, an orally available, fully synthetic small-molecule inhibitor of the heat shock protein Hsp90. Mol Cancer Ther 8(4):921–929. doi:10.1158/1535-7163.MCT-08-0758
100. Dickson MA, Okuno SH, Keohan ML, Maki RG, D'Adamo DR, Akhurst TJ, Antonescu CR, Schwartz GK (2013) Phase II study of the HSP90-inhibitor BIIB021 in gastrointestinal stromal tumors. Ann Oncol 24(1):252–257. doi:10.1093/annonc/mds275
101. Mitchell P (2011) Biogen Idec restructures, sharpens neurology focus. Nat Biotechnol 29(1):7–8. doi:10.1038/nbt0111-7
102. Fadden P, Huang KH, Veal JM, Steed PM, Barabasz AF, Foley B, Hu M, Partridge JM, Rice J, Scott A, Dubois LG, Freed TA, Silinski MA, Barta TE, Hughes PF, Ommen A, Ma W, Smith ED, Spangenberg AW, Eaves J, Hanson GJ, Hinkley L, Jenks M, Lewis M, Otto J, Pronk GJ, Verleysen K, Haystead TA, Hall SE (2010) Application of chemoproteomics to drug discovery: identification of a clinical candidate targeting hsp90. Chem Biol 17(7):686–694. doi:10.1016/j.chembiol.2010.04.015
103. Huang KH, Veal JM, Fadden RP, Rice JW, Eaves J, Strachan JP, Barabasz AF, Foley BE, Barta TE, Ma W, Silinski MA, Hu M, Partridge JM, Scott A, DuBois LG, Freed T, Steed PM, Ommen AJ, Smith ED, Hughes PF, Woodward AR, Hanson GJ, McCall WS, Markworth CJ, Hinkley L, Jenks M, Geng L, Lewis M, Otto J, Pronk B, Verleysen K, Hall SE (2009) Discovery of novel 2-aminobenzamide inhibitors of heat shock protein 90 as potent, selective and orally active antitumor agents. J Med Chem 52(14):4288–4305. doi:10.1021/jm900230j
104. Rajan A, Kelly RJ, Trepel JB, Kim YS, Alarcon SV, Kummar S, Gutierrez M, Crandon S, Zein WM, Jain L, Mannargudi B, Figg WD, Houk BE, Shnaidman M, Brega N, Giaccone G (2011) A phase I study of PF-04929113 (SNX-5422), an orally bioavailable heat shock protein 90 inhibitor, in patients with refractory solid tumor malignancies and lymphomas. Clin Cancer Res 17(21):6831–6839. doi:10.1158/1078-0432.CCR-11-0821

105. Menezes DL, Taverna P, Jensen MR, Abrams T, Stuart D, Yu GK, Duhl D, Machajewski T, Sellers WR, Pryer NK, Gao Z (2012) The novel oral Hsp90 inhibitor NVP-HSP990 exhibits potent and broad-spectrum antitumor activities in vitro and in vivo. Mol Cancer Ther 11(3):730–739. doi:10.1158/1535-7163.MCT-11-0667
106. Vernalis (2012) Interim results for the six months ended 30 June 2012. http://www.vernalis.com/media-centre/latest-releases/2012-releases/642-interim-results-for-the-six-months-ended-30-june-2012. Accessed 21 Sept 2013
107. Bussenius J, Blazey CM, Aay N, Anand NK, Arcalas A, Baik T, Bowles OJ, Buhr CA, Costanzo S, Curtis JK, DeFina SC, Dubenko L, Heuer TS, Huang P, Jaeger C, Joshi A, Kennedy AR, Kim AI, Lara K, Lee J, Li J, Lougheed JC, Ma S, Malek S, Manalo JC, Martini JF, McGrath G, Nicoll M, Nuss JM, Pack M, Peto CJ, Tsang TH, Wang L, Womble SW, Yakes M, Zhang W, Rice KD (2012) Discovery of XL888: a novel tropane-derived small molecule inhibitor of HSP90. Bioorg Med Chem Lett 22(17):5396–5404. doi:10.1016/j.bmcl.2012.07.052
108. Paraiso KH, Haarberg HE, Wood E, Rebecca VW, Chen YA, Xiang Y, Ribas A, Lo RS, Weber JS, Sondak VK, John JK, Sarnaik AA, Koomen JM, Smalley KS (2012) The HSP90 inhibitor XL888 overcomes BRAF inhibitor resistance mediated through diverse mechanisms. Clin Cancer Res 18(9):2502–2514. doi:10.1158/1078-0432.CCR-11-2612
109. Study of XL888 With Vemurafenib for patients with unresectable BRAF mutated stage III/IV melanoma. U.S. National Institutes of Health. http://clinicaltrials.gov/ct2/show/NCT01657591?term=XL888&rank=2. Accessed 5 July 2013
110. Taldone T, Patel PD, Patel M, Patel HJ, Evans CE, Rodina A, Ochiana S, Shah SK, Uddin M, Gewirth D, Chiosis G (2013) Experimental and structural testing module to analyze paralogue-specificity and affinity in the Hsp90 inhibitors series. J Med Chem 56(17):6803–6818. doi:10.1021/jm400619b
111. Patel PD, Yan P, Seidler PM, Patel HJ, Sun W, Yang C, Que NS, Taldone T, Finotti P, Stephani RA, Gewirth DT, Chiosis G (2013) Paralog-selective Hsp90 inhibitors define tumor-specific regulation of HER2. Nat Chem Biol 9(11):677−684.doi:10.1038/nchembio.1335
112. Kang BH, Plescia J, Dohi T, Rosa J, Doxsey SJ, Altieri DC (2007) Regulation of tumor cell mitochondrial homeostasis by an organelle-specific Hsp90 chaperone network. Cell 131(2):257–270. doi:10.1016/j.cell.2007.08.028
113. Altieri DC, Stein GS, Lian JB, Languino LR (2012) TRAP-1, the mitochondrial Hsp90. Biochim Biophys Acta 1823(3):767–773. doi:10.1016/j.bbamcr.2011.08.007
114. Kang BH, Plescia J, Song HY, Meli M, Colombo G, Beebe K, Scroggins B, Neckers L, Altieri DC (2009) Combinatorial drug design targeting multiple cancer signaling networks controlled by mitochondrial Hsp90. J Clin Invest 119(3):454–464. doi:10.1172/JCI37613
115. Kang BH, Siegelin MD, Plescia J, Raskett CM, Garlick DS, Dohi T, Lian JB, Stein GS, Languino LR, Altieri DC (2010) Preclinical characterization of mitochondria-targeted small molecule hsp90 inhibitors, gamitrinibs, in advanced prostate cancer. Clin Cancer Res 16(19):4779–4788. doi:10.1158/1078-0432.CCR-10-1818
116. Marzec M, Eletto D, Argon Y (2012) GRP94: An HSP90-like protein specialized for protein folding and quality control in the endoplasmic reticulum. Biochim Biophys Acta 1823(3):774–787. doi:10.1016/j.bbamcr.2011.10.013
117. Zheng H, Dai J, Stoilova D, Li Z (2001) Cell surface targeting of heat shock protein gp96 induces dendritic cell maturation and antitumor immunity. J Immunol 167(12):6731–6735
118. Ni M, Lee AS (2007) ER chaperones in mammalian development and human diseases. FEBS Lett 581(19):3641–3651. doi:10.1016/j.febslet.2007.04.045
119. Eletto D, Dersh D, Argon Y (2010) GRP94 in ER quality control and stress responses. Semin Cell Dev Biol 21(5):479–485. doi:10.1016/j.semcdb.2010.03.004
120. Duerfeldt AS, Peterson LB, Maynard JC, Ng CL, Eletto D, Ostrovsky O, Shinogle HE, Moore DS, Argon Y, Nicchitta CV, Blagg BS (2012) Development of a Grp94 inhibitor. J Am Chem Soc 134(23):9796–9804. doi:10.1021/ja303477g

121. Marcu MG, Chadli A, Bouhouche I, Catelli M, Neckers LM (2000) The heat shock protein 90 antagonist novobiocin interacts with a previously unrecognized ATP-binding domain in the carboxyl terminus of the chaperone. J Biol Chem 275(47):37181–37186. doi:10.1074/jbc.M003701200
122. Centenera MM, Fitzpatrick AK, Tilley WD, Butler LM (2013) Hsp90: still a viable target in prostate cancer. Biochim Biophys Acta 1835(2):211–218. doi:10.1016/j.bbcan.2012.12.005
123. Eskew JD, Sadikot T, Morales P, Duren A, Dunwiddie I, Swink M, Zhang X, Hembruff S, Donnelly A, Rajewski RA, Blagg BS, Manjarrez JR, Matts RL, Holzbeierlein JM, Vielhauer GA (2011) Development and characterization of a novel C-terminal inhibitor of Hsp90 in androgen dependent and independent prostate cancer cells. BMC Cancer 11:468. doi:10.1186/1471-2407-11-468
124. Matthews SB, Vielhauer GA, Manthe CA, Chaguturu VK, Szabla K, Matts RL, Donnelly AC, Blagg BS, Holzbeierlein JM (2010) Characterization of a novel novobiocin analogue as a putative C-terminal inhibitor of heat shock protein 90 in prostate cancer cells. Prostate 70(1):27–36. doi:10.1002/pros.21035
125. Palermo CM, Westlake CA, Gasiewicz TA (2005) Epigallocatechin gallate inhibits aryl hydrocarbon receptor gene transcription through an indirect mechanism involving binding to a 90 kDa heat shock protein. BioChemistry 44(13):5041–5052. doi:10.1021/bi047433p
126. Khandelwal A, Hall J, Blagg BS (2013) Synthesis and structure-activity relationships of egcg analogues, a recently identified Hsp90 inhibitor. J Org Chem 78(16):7859−7884. doi:10.1021/jo401027r
127. Rosenhagen MC, Soti C, Schmidt U, Wochnik GM, Hartl FU, Holsboer F, Young JC, Rein T (2003) The heat shock protein 90-targeting drug cisplatin selectively inhibits steroid receptor activation. Mol Endocrinol 17(10):1991–2001. doi:10.1210/me.2003-0141
128. Chaudhury S, Welch TR, Blagg BS (2006) Hsp90 as a target for drug development. ChemMedChem 1(12):1331–1340. doi:10.1002/cmdc.200600112
129. Yu Y, Hamza A, Zhang T, Gu M, Zou P, Newman B, Li Y, Gunatilaka AA, Zhan CG, Sun D (2010) Withaferin A targets heat shock protein 90 in pancreatic cancer cells. Biochem Pharmacol 79 (4):542–551. doi:10.1016/j.bcp. 2009.09.017
130. Fortugno P, Beltrami E, Plescia J, Fontana J, Pradhan D, Marchisio PC, Sessa WC, Altieri DC (2003) Regulation of survivin function by Hsp90. Proc Natl Acad Sci U S A 100(24):13791–13796. doi:10.1073/pnas.2434345100
131. Meli M, Pennati M, Curto M, Daidone MG, Plescia J, Toba S, Altieri DC, Zaffaroni N, Colombo G (2006) Small-molecule targeting of heat shock protein 90 chaperone function: rational identification of a new anticancer lead. J Med Chem 49(26):7721–7730. doi:10.1021/jm060836y
132. Plescia J, Salz W, Xia F, Pennati M, Zaffaroni N, Daidone MG, Meli M, Dohi T, Fortugno P, Nefedova Y, Gabrilovich DI, Colombo G, Altieri DC (2005) Rational design of shepherdin, a novel anticancer agent. Cancer Cell 7(5):457–468. doi:10.1016/j.ccr.2005.03.035

Chapter 8
The *Candida albicans* Hsp90 Chaperone Network Is Environmentally Flexible and Evolutionarily Divergent

Stephanie Diezmann and Leah E. Cowen

Abstract The essential molecular chaperone heat shock protein 90 (Hsp90) is a highly conserved protein hub, which modulates the genotype to phenotype translation in animals, plants, and fungi. In doing so, Hsp90 interacts with up to 10% of the eukaryotic cell's proteome. In the leading fungal pathogen of humans, *Candida albicans*, Hsp90 governs virulence, morphogenesis, and drug resistance. While specific Hsp90 clients have been identified and described, a global overview of Hsp90 interactions in this pathogen remained elusive until recently.

Here, we discuss recent advancements in mapping the *C. albicans* Hsp90 chaperone network. We describe the first Hsp90 genetic interaction network in *C. albicans*, discuss its divergence from that of its relative *Saccharomyces cerevisiae* and illustrate how the network informs our understanding of fungal biology, stress responses and virulence. Deciphering the Hsp90 chaperone network holds great promise for the development of suitable measures to combat fungal drug resistance and counter the ever-increasing number of *Candida* infections.

Abbreviations

Hsp90	Heat shock protein 90
GdA	Geldanamycin (Hsp90 inhibitor)
GO	Gene ontology
cAMP	3'-5'-Cyclic adenosine monophosphate
PKA	Protein kinase A
MAPK	Mitogen-activated protein kinase
17-AAG	17-N-allylamino-17-demethoxygeldanamycin
SGA	Synthetic genetic array

L. E. Cowen (✉)
Department of Molecular Genetics, University of Toronto, 1 Kings College Circle, Medical Sciences Building, Room 4268, Toronto, ON M5S 1A8, Canada
e-mail: leah.cowen@utoronto.ca

S. Diezmann
Department of Biology and Biochemistry, University of Bath, Claverton Down, 4 South, Bath BA2 7AY, UK
e-mail: s.diezmann@bath.ac.uk

W. A. Houry (ed.), *The Molecular Chaperones Interaction Networks in Protein Folding and Degradation,* Interactomics and Systems Biology, DOI 10.1007/978-1-4939-1130-1_8,

KDAC	Lysine deacetylase
PPIs	Protein-protein interactions
TAP	Tandem affinity purification
CO2	Carbon dioxide
RPMI	Roswell Park Memorial Institute medium
UV	ultraviolet
Sgt1	Suppressor of G2 allele of skp1
MALDI ToF	Matrix-assisted laser desorption/ionization Time of Flight
SAGA	Spt-Ada-Gcn5-acetyltransferase
UTP-C	U three proteins complex
CUG	Cysteine Uracil Guanine

1 The Commensal Yeast *Candida albicans* Is the Leading Fungal Pathogen of Humans

Candida albicans is a member of the human microbiome, colonizing mucosal surfaces, such as the oral cavity and the gastrointestinal tract in 30–70% of healthy adults [1]. While fungal carriage does not present a problem in most immunocompetent individuals, it can cause serious complications in individuals with compromised immune function. The impact on human health and quality of life is immense. Every year up to 11,200 patients in the USA alone die from *C. albicans* infections of the bloodstream or the organs (candidemia) [2]. Recent estimates suggest that ~400,000 patients suffer from life-threatening *C. albicans* infections worldwide [3]. *C. albicans* is the fourth most common cause of hospital-acquired bloodstream infections [4], with mortality rates of up to 50% and an attributable economic burden of US$ 40,000 per episode and patient due to extended stays at the hospital [1].

Patients at risk of contracting candidemia include those who received solid organ transplants, are cared for at neonatal intensive care units, underwent abdominal surgery, or were diagnosed with solid tumors [4]. As this patient population is growing due to recent advances in medical care, so is the incidence rate of candidemia, which increased between 1979 and 2000 by 207% [5]. This already dire situation is exacerbated by the two main challenges associated with the treatment of *C. albicans* infections. First, due to the close association between *C. albicans* and its human host, the patient acts as a reservoir for endogenous infections. Second, therapeutic strategies are quickly compromised by the rapid emergence of antifungal drug resistance, the occurrence of intrinsically resistant *C. albicans* strains, or severe side effects caused by antifungal drugs. To make matters worse, development of more effective therapeutic strategies is hampered by the paucity of fungal-specific drug targets and *C. albicans*' status as commensal. This emphasizes a pressing need for more efficacious treatment strategies.

Currently, the two most widely utilized drug classes for the treatment of life-threatening candidemia are the azoles and the echinocandins. They affect the fungal cell's outer layers, the cell membrane and the cell wall. The azoles block the production of ergosterol, the fungal analog of human cholesterol, by inhibiting C14α-demethylase, thereby causing a toxic sterol intermediate to accumulate in the cell membrane [6]. This results in increased vulnerability to further membrane damage. The echinocandins are the only new class to enter the clinic in decades and inhibit 1,3-β-D-glucan synthase, which is essential for cell wall production [7]. Importantly, both basal tolerance and resistance to these drugs acquired by diverse mechanisms depend upon cellular stress responses, such as those regulated by the molecular chaperone Hsp90. The combination of antifungal drug treatment with Hsp90 compromise abrogates drug resistance *in vitro*, emphasizing the role of Hsp90 as the emerging fungal Achilles' heel [8].

2 Hsp90 Stabilizes Stress-Responsive Signal Transducers and Governs the Genotype-to-Phenotype Transition

Hsp90 was originally discovered as essential for high temperature growth in the eukaryotic model organism *Saccharomyces cerevisiae.* Unlike other heat shock proteins, Hsp90 is not required for *de novo* protein folding but keeps metastable client proteins poised for activation [9]. It does so through a chaperone cycle, during which adenosine triphosphate (ATP) hydrolysis and associated co-chaperones assist in opening of the Hsp90 homodimer and client release [10]. Protein folding is facilitated by Hsp90's molecular structure [11–13]. The tri-partite composition of an ATP-binding N-terminal domain, a client-binding middle segment, and a C-terminal dimerization domain promote chaperoning [13]. Hsp90 client proteins important for human health and biology include the steroid hormone receptors and oncoproteins, such as p53 [14]. The intricate liaison between Hsp90 and cancer proteins is being exploited for the development of anticancer therapeutics, and there are currently 13 Hsp90 inhibitors in clinical trials for the treatment of diverse cancers [15]. Hsp90 inhibitors commonly used in laboratory settings for the dissection of Hsp90 function include the antibiotics geldanamycin (GdA) and radicicol, which are structurally unrelated and competitively bind Hsp90's ATP-binding pocket, thereby inhibiting Hsp90 function [16].

Hsp90 serves as a central hub in the cellular circuitry by chaperoning diverse regulators of cellular signalling in a manner that is contingent upon environmental cues. Although Hsp90 is one of the most abundant proteins in the cell, environmental perturbation can exhaust the Hsp90 system by increasing the cellular burden of misfolded proteins. Modulation of Hsp90 function by environmental stress can have profound impacts on the genotype-to-phenotype translation on a global level [17]. Hsp90 can function as a capacitor by buffering genetic or epigenetic variation, such that it remains in a phenotypically silent state until Hsp90 function is compromised and new traits are revealed [18]. Perturbations of the Hsp90 chaperone system have been found to reveal new traits in animals, plants, and fungi [17, 19, 20]. Hsp90 can also function as a potentiator of genetic variation by stabilizing mutant regulators and enabling new genetic variants to have immediate phenotypic effects.

In this context, compromise of the Hsp90 chaperone system has been found to mask new traits in fungi as well as mammalian cancer cells [14, 17, 21].

Hsp90 does not operate solitarily. For proper function, it requires more than 20 different co-chaperones that contribute to distinct tasks, such as regulating ATPase activity, targeting specific clients, and assisting with client maturation [12]. The composition of the Hsp90 co-chaperone machinery differs considerably across the tree of life [22] while Hsp90 itself is highly conserved. For example, Hsp90 amino acid sequence identity between humans and *C. albicans* is 62%, and between *C. albicans* and *S. cerevisiae*, it is 84% [23].

3 *C. albicans* Hsp90 Modulates Antifungal Drug Resistance, Morphogenesis, and Virulence in Concert With its Co-chaperone Sgt1 (suppressor of G2 allele of skp1), and its Client Proteins Mkc1 and Calcineurin

Hsp90 operates in concert with its co-chaperone machinery, which comprises 10 co-chaperones in both *C. albicans* and *S. cerevisiae* [24]. In *S. cerevisiae*, the co-chaperone Sgt1 has been demonstrated to function as an adaptor protein [25], as it connects Hsp90 with several of its clients, one of which is the adenylyl cyclase Cyr1 [26]. Cyr1 generates 3′-5′-cyclic adenosine monophosphate (cAMP), whose signalling properties are a key factor in *C. albicans*' yeast-to-filament transition [27]. The morphogenetic transition between yeast and filamentous growth is a temperature-dependent developmental program, and a key virulence trait [28]. The basis of temperature dependence of *C. albicans* morphogenesis remained enigmatic for decades, until it was discovered that Hsp90 represses filamentation, such that inhibition of Hsp90 induces filamentation under conditions that normally favor yeast-form growth [29]. Consistent with the importance of Sgt1 for cAMP-protein kinase A (PKA) signalling in *S. cerevisiae* [30], *C. albicans* Sgt1 was found to be a key regulator of cAMP-PKA signalling and morphogenesis [31]. Genetic depletion of Sgt1 phenocopies Hsp90 compromise and induces filamentation in the absence of other cues, such as elevated temperature, serum, or CO_2 [31]. Co-immunoprecipitation experiments showed that Hsp90 and Sgt1 interact with each other, and that Sgt1 interacts with Cyr1. Consistent with the impact of Hsp90 on antifungal drug resistance, Sgt1 was also found to be a key regulator of basal tolerance and resistance to both the azoles and echinocandins [31].

Hsp90 governs the emergence and maintenance of antifungal drug resistance [8, 32]. Initial studies with *S. cerevisiae* revealed that genetic reduction of Hsp90 levels blocks the rapid evolution of resistance to azoles [21]. Pharmacological inhibition of Hsp90 with GdA or radicicol, structurally distinct Hsp90 inhibitors that competitively bind to the ATP-binding pocket [16], blocks the rapid evolution of azole resistance in *C. albicans*, and abrogates resistance acquired by different mechanisms [21]. Hsp90 also has a profound impact on azole resistance of *C. albicans* that evolved in a human host. Analysis of a series of 17 *C. albicans* isolates from a human immunodeficiency virus (HIV) patient undergoing therapy with the azole fluconazole for two years [33] revealed that inhibition of Hsp90 reduced resistance of all of the isolates, with the largest impact on those isolates recovered earlier during treatment [21]. Targeting Hsp90 also transforms azoles from fungistatic to fungicidal [34] and overcomes the intrinsic resistance of *C. albicans* biofilms to azole treatment [35]. Inhibition of Hsp90 function in *C. albicans* abrogates resistance not only to azoles but also to echinocandins. Genetic or pharmacological compromise of Hsp90 function reduces resistance of laboratory strains and clinical isolates that evolved resistance in the human host [36, 37].

One of the key downstream effectors through which Hsp90 governs resistance to both azoles and echinocandins is calcineurin. Calcineurin is a protein phosphatase that was known to regulate cellular responses to various environmental stressors, including the azoles [38]. Work in *S. cerevisiae* demonstrated that calcineurin physically interacts with Hsp90 [39]. Consistent with an important role for Hsp90 in regulating calcineurin-dependent stress responses essential for azole resistance, inhibition of calcineurin with the structurally distinct molecules cyclosporin A or FK506, phenocopies Hsp90 inhibition and reduces azole resistance [21]. Deletion of the catalytic or regulatory subunits of calcineurin also phenocopies genetic depletion of Hsp90, abrogating resistance to both azoles and echinocandins [21, 36]. Calcineurin was the first Hsp90 client protein identified in *C. albicans* based on the physical interaction between Hsp90 and the catalytic subunit of calcineurin, Cna1, and the destabilization of Cna1 upon depletion of Hsp90 [36].

Although calcineurin is a key mediator of Hsp90-dependent drug resistance, additional targets have recently been implicated in *C. albicans*. Another client protein that is important for resistance to both azoles and echinocandins is Mkc1, the terminal mitogen-activated protein kinase (MAPK) of the Pkc1 cell wall integrity pathway. Depletion of Hsp90 leads to destabilization of Mkc1 [40], consistent with the interaction of Hsp90 with the homologous MAPK, Slt2, in *S. cerevisiae* [41]. Beyond Mkc1, targeting Pkc1 with a specific inhibitor phenocopies inhibition of Hsp90 or calcineurin, suggesting that Pkc1 may be an additional client protein in *C. albicans*, as is the case in mammalian cells [42].

Given Hsp90's key role in fungal drug resistance, morphogenesis, and virulence, targeting Hsp90 with one of the inhibitors currently in clinical trials as an anticancer agent in combination with an antifungal has emerged as a promising approach for treating candidemia. This prospect has been explored in multiple models of infection. In larvae of the Greater Wax Moth *Galleria mellonella,* combination therapy with an Hsp90 inhibitor structurally related to GdA, 17-N-allylamino-17-demethoxygeldanamycin (17-AAG), rescues infections with *C. albicans* that are otherwise lethal and recalcitrant to treatment with azoles alone [34]. In a rat central venous catheter model of *C. albicans* biofilm infection, 17-AAG transforms azoles from ineffective to highly efficacious, such that the combination sterilizes an infected catheter [35]. While there was no toxicity in the biofilm model where the drug delivery and fungal infection are localized, systemic inhibition of Hsp90 with 17-AAG in the context of a disseminated infection in a murine model of candidemia proved toxic, likely due to inhibition of host Hsp90. This emphasizes the need for Hsp90 inhibitors that can selectively target the pathogen or for alternate strategies to inhibit the Hsp90 chaperone machinery. Recent work has established a powerful approach to map the *C. albicans* Hsp90 chaperone network, revealing novel effectors upstream and downstream of Hsp90 and new candidate therapeutic targets for combatting fungal infectious disease.

4 The *C. albicans* Hsp90 Genetic Interaction Network Reveals Environmental Contingency and Evolutionary Rewiring

Genetic interaction networks are important illustrations of which genes are involved in similar biological processes or pathways, permitting gene function annotation, and providing functional insights into protein-protein complexes [43]. These interaction profiles typically emerge from synthetic lethality screens. Synthetic lethality describes the situation in which single mutants are viable while concurrent perturbation of two genes causes lethality, which is indicative of a genetic interaction [44]. To obtain genome-wide genetic interaction data, genomic tools such as gene deletion libraries [45], synthetic genetic array technology (SGA) [43], or ribonucleic acid (RNA) interference [46], are an absolute necessity. Until recently, such tools were restricted to model eukaryotic systems such as *S. cerevisiae* and the nematode *Caenorhabditis elegans*.

Proteomic and genomic analyses in the model yeast *S. cerevisiae* revealed that Hsp90 interacts with approximately 10% of the proteome. Many interactors are signal transducers such as kinases and transcription factors [47] that are involved in different stress response pathways, such as those governing high osmolarity growth (Hog1) and cell wall disturbances (Mkc1/Slt2) [41]. Detection of genetic and physical interactions was facilitated by the extensive genomic and proteomic tool set available for *S. cerevisiae*. Hsp90 genetic interactions have been identified by an SGA screen for mutants with a synthetic lethal or synthetic sick phenotype when combined with a temperature-sensitive hsp82 allele as the only source of Hsp90, as well as by screens of gene deletion mutant libraries for hypersensitivity to Hsp90 inhibitors [47–49]. Physical interactions have been uncovered by MALDI-ToF (Matrix-assisted laser desorption/ionization Time of Flight) mass spectrometric analyses following tandem affinity purification of Hsp82-tandem affinity purification (TAP), as well as by yeast two-hybrid assays employing Hsp82 as bait [41, 47]. These studies provide a foundation for analyses of the Hsp90 chaperone network in *C. albicans*, contingent on the availability of genomic resources such as mutant libraries.

In 2002, the first large-scale set of *C. albicans* mutants was made available by Prof. Aaron Mitchell [50], and is referred to here as the Mitchell Library. This library of homozygous transposon insertion mutants was created by random insertions of a transposon followed by selection for homozygosity. It quickly became a vital tool for the *C. albicans* community and expedited mapping of the *C. albicans* Hsp90 genetic interaction network. Over the years, additional mutant strains have been generated and made publically available. The library now comprises ~1,200 mutant strains representing 704 genes, which cover about 10% of the *C. albicans* genome. More than half the genes are represented by two mutants, about a third by one mutant, and very few genes are represented by 3–6 mutants. A genome-scale library with 5,400 bar-coded heterozygous gene deletion mutants and 1,152 mutants with promoter replacement of the intact allele with the tetracycline-repressible promoter have only very recently been made publically available [51, 52], providing unprecedented opportunities to expand the Hsp90 genetic interaction network.

We employed the most recent version of the Mitchell Library to map the *C. albicans* Hsp90 genetic interaction network and to explore the impact of environmental stress on network dynamics [53]. To do so, we optimized six environmental conditions identified as general or specific stresses, which were selected to exemplify conditions similar to those the pathogen may encounter during the course of infection. As a general stressor, we chose high temperature growth, which mimics a febrile host situation (41 °C). To determine if stressors that affect a wide array of cellular processes elicit similar Hsp90 genetic interaction profiles, we compared high temperature growth to osmotic stress. This is a particularly well-characterized stressor, here exerted by NaCl. To test specific stressors, we turned to drugs that target cellular compartments in a well-defined manner. The nucleoside inhibitor tunicamycin exerts stress on the endoplasmatic reticulum by inhibiting the first step of glycoprotein synthesis, thus inducing the unfolded protein response. The azole fluconazole inhibits ergosterol biosynthesis and exerts stress on the cell membrane while the echinocandin caspofungin targets cell wall synthesis and exerts cell wall stress. All assays were carried out in RPMI (Roswell Park Memorial Institute), a cell culture medium that is formulated to approximate human physiological conditions and is the standard used for clinical analysis of antifungal drug susceptibility.

We screened the Mitchell Library, which is arrayed in 96-well format, for growth inhibition in the presence of GdA in six different environmental conditions (standard growth at 37 °C, two general stresses, and three specific stresses [53]). Hypersensitivity to Hsp90 inhibition is indicative of an Hsp90 genetic interactor. The screen revealed a total of 226 Hsp90 genetic interactors (Table 8.1), including the only two Hsp90 interactors that were previously described in *C. albicans*. Identification of the two reported client proteins, Mkc1 [40] and Cna1 [36], lent support to the validity of our approach. This *C. albicans* Hsp90 network also includes known *S. cerevisiae* Hsp90 interactors, such as Hog1 and Mkc1/Slt2 [41].

A subsequent network analysis using Cytoscape [54] revealed relationships between interactors and the environmental conditions they were identified in [53]. The network structure was found to be highly dependent on environmental conditions (Fig. 8.1), such that the vast majority of the interactors were relevant for growth under only one or two conditions. Only four interactors were identified in five or six screens. The number of interactors differed dramatically between screens (Table 8.1). The cell wall stressor caspofungin elicited a total of 73 interactions, only 21 of which were shared with other conditions. The dependence of cell wall integrity on Hsp90 function suggests a broader role for the chaperone in cell wall regulation than previously appreciated.

Table 8.1 GO category enrichment in C. albicans Hsp90 genetic interaction screens. (Reproduced from [53])

Stress condition	Screen	Total number of interactions	Number of unique interactions	GO category enrichment (*P* value)
No stress	37 °C	69	27	
General stress	41 °C	64	32	Phosphorylation (0.005) Protein amino acid phosphorylation (0.005) Phosphorus metabolic process (0.024) Phosphate metabolic process (0.024) Regulation of cellular process (0.007) Regulation of biological process (0.03)
	0.55 M NaCl	27	5	SAGA complex (0.033) SAGA-like and SAGA-type complex (0.033) CK2 complex (0.033) UTP-C complex (0.033) Protein complex (0.015) Macromolecular complex (0.036) Posttranslational modification (0.043) Nucleolar part (0.033) Drug response (0.001) Chemical stimulus response (0.034) Stimulus response (0.039)
Specific stress	1 μg/ml Tunicamycin (T)	57	29	Cell cortex (0.04) Protein complex (0) Protein amino acid phosphorylation (0) Macromolecule modification (0) Protein modification process (0) Macromolecular complex (0) Macromolecule metabolic process (0) Cellular macromolecule metabolic process (0) Phosphorylation (0) Phosphorus metabolic process (0.005) Phosphate metabolic process (0.005) Cellular protein metabolic process (0) Protein metabolic process (0) Posttranslational protein modification (0) Drug response (0.001) Stimulus response (0.005) Chemical stimulus response (0.021)
	0.1 μg/ml Caspofungin (C)	73	52	
	0.1 μg/ml Fluconazole (F)	32	12	

GO gene ontology

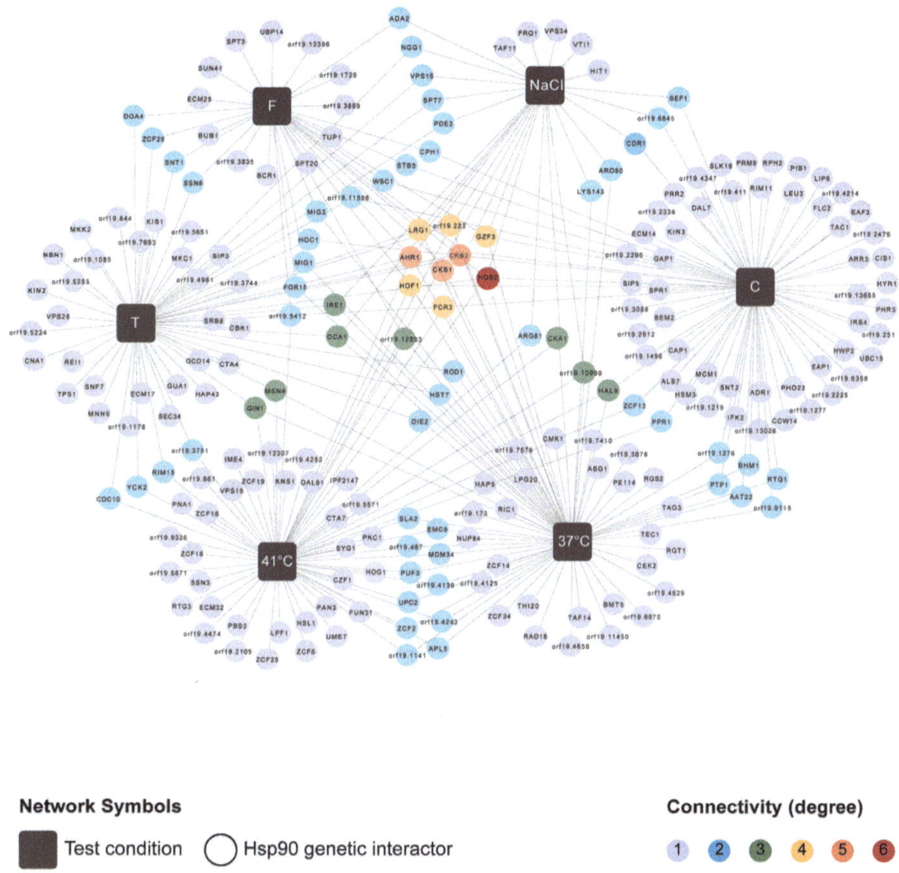

Fig. 8.1 Global *C. albicans* Hsp90 genetic interaction network. The Cytoscape network analysis of 226 genetic interactors (*filled circles*) revealed an environmentally contingent network structure. Interactors were identified in six different environmental conditions (*black squares*), including standard growth (37 °C), general stress conditions (41 °C, NaCl), and the specific stressors tunicamycin (T), caspofungin (C), and fluconazole (F). Each interactor is color-coded depending on its degree of connectivity (frequency of detection in six screens), which ranges from 1 (*grey*) to 6 (*red*). The edges connecting interactors indicate the screen in which the interactor had originally been identified. Thus, an interactor that has been identified more than once is connected by multiple edges to different conditions. (Figure reproduced from [53])

Next, we screened the set of Hsp90 genetic interactors for enrichment of gene ontology (GO) gene function categories relative to the set represented by the Mitchell Library [53]. Hsp90 genetic interactors were enriched for macromolecular complexes, protein complexes, protein modification processes, biopolymer modification, and posttranslation protein modifications. We then analyzed GO category enrichment for each environmental condition separately (Table 8.1). Significant enrichment could be detected in three environmental conditions, namely both general stresses and tunicamycin. Interestingly, patterns of GO category enrichment differed between the general stresses. Phosphate-based protein modifications were more relevant during high temperature growth while various complexes were identified as important for growth during osmotic stress. Protein phosphorylation and phosphate/phosorphorus metabolic processes were particularly important for growth at 41 °C. During osmotic stress, the SAGA (Spt-Ada-Gcn5-acetyltransferase) complex, which regulates transcription by RNA polymerase II [55], the CK2 complex, a promiscuous kinase complex with some 300 targets in the cell [56], and the UTP-C (U three proteins complex) complex, which comprises the four CK2 subunits as well as two more proteins (Rrp7 and Utp22) that form a subcomplex of the 90S preribosome [57], were relevant for growth during osmotic stress. This conforms with previous insights gleaned from Hsp90 clients in *S. cerevisiae*, many of which are signal transducers [47], and the finding of distinct interactors to be relevant during different conditions [48].

We found that analysis of Hsp90 genetic interactions in multiple conditions is crucial for a comprehensive appreciation of client proteins. For example, we identified Hog1 as a *C. albicans* Hsp90 interactor only in our 41 °C screen [53]. Depletion of Hsp90 reduced Hog1 transcript and protein levels, and blocked Hog1 activation, confirming its position as a conserved Hsp90 client protein. *S. cerevisiae* Hog1 [41] and the mammalian Hog1 homolog p38 [58] are both established Hsp90 clients. Thus, analysis of genetic interactors in multiple environmental conditions was instrumental for the identification of client proteins in *C. albicans*.

The extensive degree of environmental contingency of the *C. albicans* Hsp90 chaperone network revealed important properties that enabled prediction of whether interactors function upstream or downstream of Hsp90 [53]. Our finding that only four of the 226 interactors were important for growth in five or six conditions led to the model that interactors important for growth in many conditions (defined here as high-connectivity), operate upstream of Hsp90 affecting its expression or function. In contrast, those interactors that were only important for growth in one or two conditions (low-connectivity) function downstream of Hsp90 to regulate a specific subset of Hsp90-dependent cellular processes. The simple expectation is that for downstream effectors, Hsp90 would impact on their stability or activation while Hsp90 would have little impact on upstream effectors. We tested this model by analyzing the effect of Hsp90 depletion on the transcript and protein levels of three high-connectivity interactors (Hos2, Ckb1, and Ckb2) and six low-connectivity interactors (Hog1, Cka1, Cka2, Mkk2, Cdr1, and Cmk1). As predicted, Hsp90 depletion had a significantly greater effect on levels of the low-connectivity interactors than the high-connectivity interactors ($P < 0.0430$). Thus, our chemical genomic approach to identify Hsp90 interactors in multiple environmental conditions provided a powerful strategy to predict functional relationships within the Hsp90 chaperone network.

We aimed to further investigate how the four high-connectivity interactors might affect Hsp90 function. First, we focused on Ckb1 and Ckb2, the regulatory subunits of the protein kinase CK2 complex. We found this complex enriched during osmotic stress, one of the general stressors (Table 8.1). Previous research in *S. cerevisiae* demonstrated that CK2 phosphorylates Hsp90 residue T22, which regulates high temperature growth [59]. Thus, we monitored levels of phosphorylation of Hsp90 and its kinase-targeting co-chaperone Cdc37 in CK2 subunit deletion mutants [53]. As expected, deletion of *CKA1* or *CKA2*, encoding the low-connectivity catalytic subunits, had negligible impact on phosphorylation of Hsp90 or Cdc37. Strikingly, deletion of *CKB1* reduced phosphorylation of Hsp90 and Cdc37 while deletion of *CKB2* abolished Cdc37 phosphorylation. This provides a compelling example of high connectivity Hsp90 interactors functioning upstream of Hsp90.

To determine the phenotypic consequences of aberrant Hsp90 and Cdc37 phosphorylation, we measured Hog1 protein levels and activation during standard growth and oxidative stress, and further assessed the ability of CK2 deletion mutants to grow in sorbitol, which exerts osmotic stress [53]. During oxidative stress, Hog1 protein levels are reduced in *ckb1*Δ/Δ and *ckb2*Δ/Δ mutants but Hog1 activation, as measured by phosphorylation, remains unaffected. Given that Hsp90 is required for Hog1 stability and phosphorylation, this suggests that blocking CK2-mediated phosphorylation of Hsp90 and/or Cdc37 reduces but does not abrogate the function of chaperoning Hog1. The phenotypic assay revealed that the CK2 regulatory subunit deletion mutants were hypersensitive to sorbitol, as is the case with the *hog1Δ/Δ* mutant, confirming the importance of Ckb1 and Ckb2 for Hog1-mediated growth in osmotic stress. This provides the first example of regulation of Hsp90 function by posttranslational modification in *C. albicans*, and implicates protein kinase CK2 as an evolutionarily conserved regulator of Hsp90 function.

Next, we sought to explore the function of another high-connectivity interactor that was poised to modulate *HSP90* expression, the zinc finger transcription factor Ahr1. This *Candida*-specific transcription factor has been found to influence morphology [60] and high-temperature growth [61], such that deletion of *AHR1* causes aberrant morphologies and reduced biofilm formation [60], as well as reduces the ability to growth at 37 °C and 42 °C [61]. Here, we predicted that Ahr1 regulates *HSP90* expression and that the morphological alterations of the *ahr1Δ/Δ* mutant might reflect reduced *HSP90* levels, given that reduction of Hsp90 function or levels induces filamentation [29]. These predictions were confirmed by qRT-PCR and microscopic analyses of the *ahr1Δ/Δ* mutant grown at 30 °C in rich medium that favors yeast growth [53]. Hsp90 levels are significantly reduced in the *ahr1Δ/Δ* mutant, and the strain is filamentous, consistent with reduced Hsp90 function. Thus, Ahr1 is a novel regulator of Hsp90 expression, emphasizing the power of the Hsp90 genetic network for identifying upstream effectors of this global cellular regulator.

Next, we focussed on the lysine deacetylase Hos2, a high connectivity Hsp90 interactor important for growth in all six environmental conditions. Hos2 is part of the Set3C lysine deacetylase (KDAC) complex. Together with Set3, Snt1, and Sif2, Hos2 forms the core of this complex [62] and is required for lysine deacetylation of H3 and H4 histones, thereby ensuring efficient transcription [63]. Hos2 protein levels are not reduced upon Hsp90 depletion, suggesting that Hos2 may function upstream of Hsp90, as with other high-connectivity interactors [53]. Deletion of *HOS2* causes colony morphology to appear wrinkly at 37 °C due to filamentous growth [64], consistent with a modest reduction of Hsp90 function. Further, the broad spectrum lysine deacetylase inhibitor trichostatin A has recently been demonstrated to phenocopy Hsp90 inhibitors, blocking the emergence of azole resistance and abrogating resistance that evolved in a human host [65]. In *S. cerevisiae*, Hsp90 function is regulated by acetylation, such that inhibition of KDACs compromises Hsp90 function, blocks Hsp90 interaction with key client proteins for drug resistance, and alters client protein stability and activation [65]. The key KDACs in *S. cerevisiae* are Rpd3 and Hda1 while the relevant players in *C. albicans* remain to be identified. Our network analysis suggests that Hos2 may well be a key regulator of Hsp90 function in this pathogen.

In addition to illuminating novel functional insights into *C. albicans* Hsp90 and its targets, this network provided the first opportunity for a comparative analysis of a chaperone network. Until recently, comparative analyses were limited by the lack of large-scale interaction data in species other than *S. cerevisiae* [47, 48] or by the extensive protein interaction network divergence from other eukaryotes, such as with *Plasmodium falciparum* [66]. Thus, we seized this opportunity and compared *C. albicans* Hsp90 genetic interactors to those identified in *S. cerevisiae* [47].

These yeast species diverged from their most recent common ancestor approximately 200 million years ago [67]. During this time, they adapted to diverse environmental niches. *C. albicans* is a commensal of warm-blooded animals, mostly humans [68] while *S. cerevisiae* is a saprophyte living on fruit and in the soil. These environments exert distinct selective pressures, such as temperature and glucose content. Given that genetic variation between strains of *S. cerevisiae* can alter the dependence of diverse traits on Hsp90 [17], it is thus conceivable that the Hsp90 network has diverged considerably between *S. cerevisiae* and *C. albicans*. Although a small but significant ($P=0.004$) overlap in Hsp90 genetic interactors remained between the species, the network underwent extensive rewiring over evolutionary time [53]. Less than 20% of the interactors are conserved, despite very similar Hsp90 chaperone machineries in the two species [22].

A total of 428 *C. albicans* genes present in the Mitchell Library have homologs in *S. cerevisiae* [53], 59 of those are Hsp90 genetic interactors in *S. cerevisiae* as well (Fig. 8.2). Of these 59, 29 were shared between both species and 30 were unique to *S. cerevisiae*, indicating extensive rewiring. Conserved interactors are distributed throughout the network. While there is no detectable pattern regarding conservation of high- or low-connectivity interactors, there is some degree of conservation detectable within different environmental conditions. For example, a quarter of Hsp90 interactors relevant for growth during exposure to the cell membrane stressor fluconazole are conserved, but less than 10% of those important for growth during cell wall stress exerted by caspofungin are conserved. While the extent of interactor conservation may be low, response profiles are conserved. Both species maintain a large number of their interactors when temperatures shifts between standard growth (30 °C for *S. cerevisiae* [48] and 37 °C for *C. albicans*) and high-temperature growth (37 °C for *S. cerevisiae* [48] and 41 °C for *C. albicans*; Fig. 8.2 inset). Even though the specific interactions may change, similar proportions of the genome remain associated during high-temperature

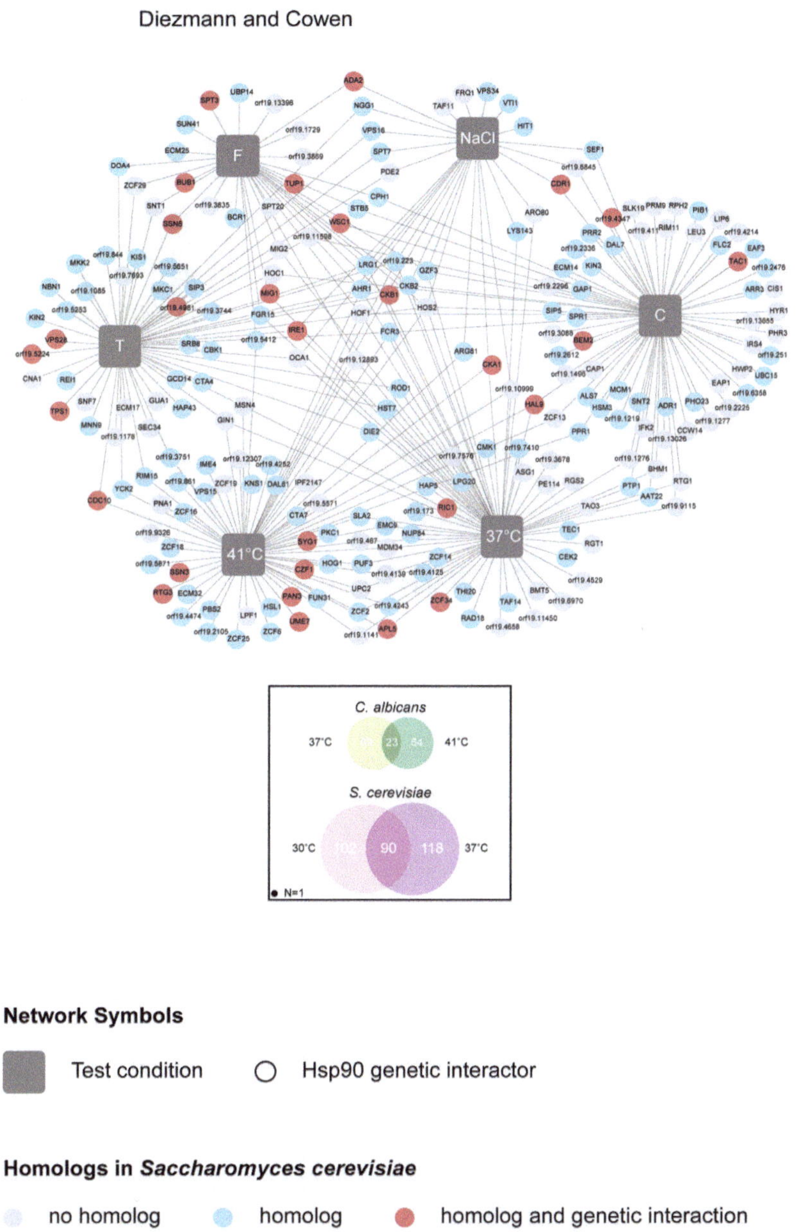

Fig. 8.2 The *C. albicans* Hsp90 genetic interaction network has been substantially rewired. The *C. albicans* Hsp90 genetic interactor set was compared to that previously identified for *S. cerevisiae*, revealing that less than 20 % of Hsp90 genetic interactions are conserved. Interactors with a homolog in *S. cerevisiae* (*blue*) and those that are also Hsp90 genetic interactors in the model yeast (*red*) are mapped onto the global network. The Hsp90 genetic interaction network has been largely rewired since both yeasts diverged from their most common ancestor. (Figure reproduced from [53])

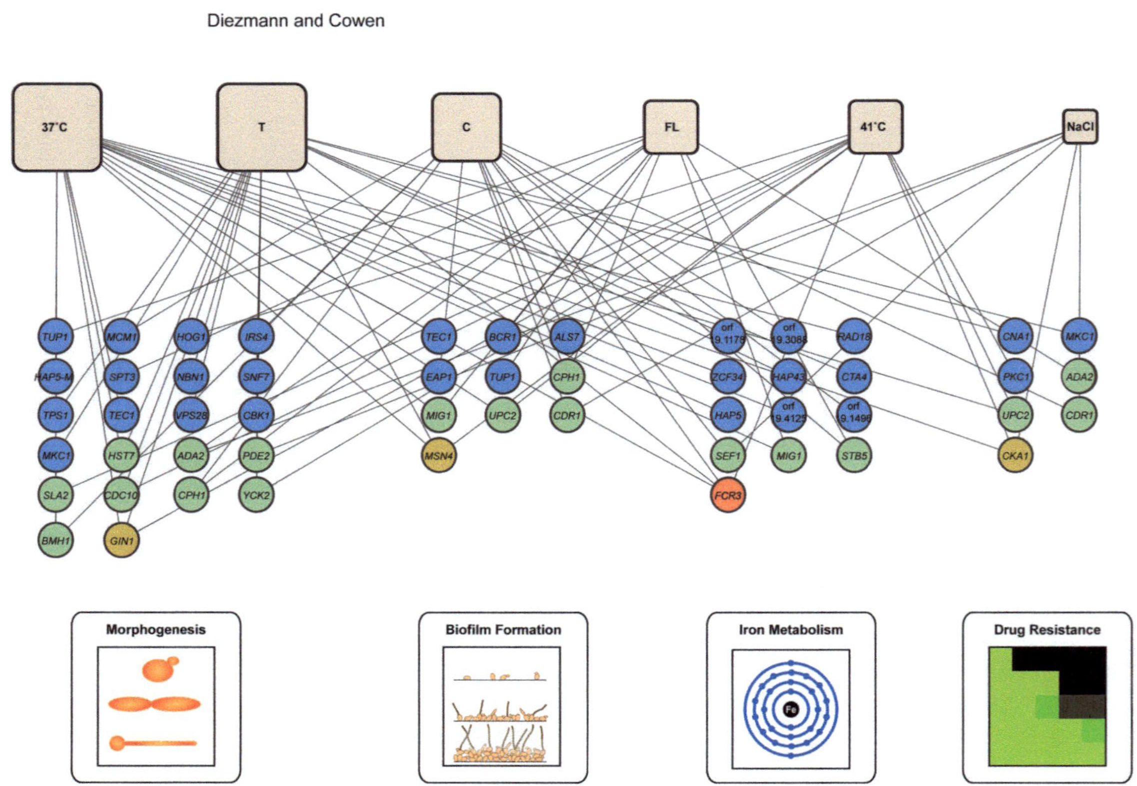

Fig. 8.3 *C. albicans* Hsp90 genetic interactors relevant for virulence phenotypes. Fifty-two of the 226 genetic interactors have GO annotations for phenotypes important in *C. albicans* virulence. This subset of interactors is grouped here by phenotype (morphogenesis, biofilm formation, iron metabolism, and drug resistance) and connected by edges to the environmental condition they were identified in. Environmental condition box size is relative to the number of edges connected to this condition. All interactors highlighted here are low-connectivity, except for *FCR3*, which is connected to four conditions. This emphasizes the high degree of specificity through which Hsp90 fine-tunes complex processes such as biofilm formation or drug resistance

growth. The mechanisms underpinning Hsp90 chaperone network dynamics and evolution remains largely unexplored.

Taken together, the initial work towards elucidating the *C. albicans* Hsp90 genetic interaction network has imparted a suite of critical insights regarding Hsp90 biology in this leading fungal pathogen of humans: (i) High connectivity Hsp90 genetic interactors are poised to regulate Hsp90 function or expression; (ii) Hsp90 genetic interaction profiles are highly contingent on the environment; and (iii) The network has been substantially rewired over the last 200 million years in response to niche-specific selective pressures. This work has also revealed connections of a multitude of Hsp90 genetic interactors with virulence phenotypes; 52 of the 226 genetic interactors have GO annotations related to traits important for *C. albicans* virulence, including morphogenesis, biofilm formation, iron metabolism, and drug resistance (Fig. 8.3).

5 Beyond the Current State of the *C. albicans* Hsp90 Chaperone Network

To date, 10 % of the *C. albicans* genome have been screened for Hsp90 genetic interactors. Since the first report of the Mitchell Library in 2002 and its extension over the next few years, other labs have created *C. albicans* mutant libraries and Merck has released its genome-scale mutant libraries. Currently, there are four additional collections available. Two of the libraries were generated by Prof. Alexander Johnson's laboratory: One library consists of 143 homozygous transcription factor gene deletion mutants [61], and the other consists of ~3,000 homozygous gene deletion mutants representing a total of 674 genes [69]. Screens of these libraries have revealed novel insights into *C. albicans* regulatory circuitries, and identified interesting instances of uncoupling of the important connections between morphology and virulence [61, 69]. Notably, the mutants in the Noble Library are marked with a set of distinct bar codes that enables pooled assays and increases speed of discovery dramatically [69]. Most recently, Merck has made available its gene replacement and conditional expression (GRACE™) [51] and double bar-coded (DBC) [52] heterozygous mutant libraries to the academic community. These resources provide unprecedented opportunity to expand the *C. albicans* Hsp90 chaperone network.

Despite the creation of five libraries in different strain backgrounds using different strategies and selectable markers, comprehensive screens have the potential to provide important insights and hold a number of advantages. Screening for genetic interactions in different genetic backgrounds could elucidate epistatic regulation of Hsp90 interactions. One gene may interact with Hsp90 only if there are other genes present in a specific background that facilitate interaction. Furthermore, comparing transposon insertions, which only affect a specific region of a gene, with a clean gene deletion may yield insights as to which protein domain or gene region are relevant for the interaction. Additional insight may be leveraged by comparison of homozygous deletion mutants with heterozygotes, and by analysis of essential genes using a gene depletion system.

In addition to developing mutant libraries that could be exploited to study Hsp90 interactions, the *C. albicans* community has created a suite of tools that facilitate genetic analyses and the detection of protein-protein interactions (PPIs) in this finicky yeast. There are three main challenges associated with conducting molecular biological work in *C. albicans*. First, *C. albicans* employs an alternative codon usage. This yeast translates the CUG (cysteine-uracil-guanine) codon into serine instead of leucine [70], which necessitates codon-by-codon alterations of constructs that were previously used successfully in *S. cerevisiae*. Second, *C. albicans* is almost exclusively found in the diploid state in nature despite recent advances in the construction of stable haploids that are competent for mating [71]. The lack of a complete meiotic cycle hampers construction of double mutants for SGA and monitoring segregation patterns following sporulation, as can be done with ease in *S. cerevisiae*. Third, *C. albicans* does not maintain plasmids, which can render the development of some proteomic tools and the study of essential genes more challenging. Despite these hurdles, significant progress has been made in adapting the yeast two-hybrid assay [72], tandem-affinity purification [73], and the usage of an expanded genetic code for *C. albicans* [74].

The yeast two-hybrid assay uncovers PPIs via expression of reporter genes. To do so, the protein of interest, such as Hsp90, can be fused to the DNA-binding domain of a transcription factor and any other gene is fused to the activation domain of that transcription factor. Consequently, an interaction between the two test proteins reconstitutes a functional transcription factor that in turn activates expression of a reporter gene. Stynen et al. [72] constructed a yeast two-hybrid system that is compatible for use in *C. albicans*. This includes selection of a functional DNA-binding domain and activation domain, as well as an effective reporter gene. Following optimization of this system, it was used to verify previously identified PPIs and uncovered novel PPIs in the MAPK pathway regulating filamentous growth [72].

An alternative strategy for identifying PPIs is based on purification of an epitope-tagged version of the protein of interest followed by mass spectrometry of the proteins that co-purify. TAP has been used extensively in *S. cerevisiae* [75], including for the identification of Hsp90 physical interactions [47]. TAP was optimized for *C. albicans* with two different epitope tags. First, Kaneko et al. [73] applied this method to study members of the *C. albicans* septin complex by constructing a tagging vector and 6xHis-FLAG tagged strains. Second, Blackwell and Brown [76] optimized the most widely used TAP tag, which comprises a proximal calmodulin-binding peptide, a cleavage site for the tobacco etch virus protease, and a distal repeated protein A IgG-binding domain. Proteomic approaches for identifying Hsp90 interactors can be challenging due to the transient nature of many Hsp90-client interactions, but remain a promising strategy to expand the chaperone network in *C. albicans*.

The expanded genetic code is a standard tool for the study of PPIs in model organisms, such as *Escherichia coli*, *S. cerevisiae*, and *C. elegans*. Palzer et al. [74]

optimized this approach for use in *C. albicans*. This required codon optimization of the *E. coli* orthogonal transfer RNA (tRNA) and tRNA synthetase for the unnatural amino acid p-azido-L-phenylalanine. This amino acid can be photoactivated by ultraviolet (UV) light, which leads to the establishment of covalent bonds between molecules in close vicinity. This system is particularly amenable to investigations of weak or transient interactions in living cells. Covalent bonds are subsequently stabilized by UV cross-linking and interaction partners can be identified using mass spectrometry [74].

With these new tools in place, it should soon be possible to map the Hsp90 physical interaction network and expand the Hsp90 genetic interaction network. Further resolution of the *C. albicans* Hsp90 chaperone network is particularly relevant in the context of identifying urgently needed regulators of Hsp90 function that are more divergent between pathogen and host than Hsp90, facilitating selective targeting of the pathogen. Furthermore, understanding the broad spectrum of genetic and physical interactors will reveal the mechanisms by which Hsp90 governs fungal morphogenesis, drug resistance, and virulence.

References

1. Zaoutis TE, Argon J, Chu J, Berlin JA, Walsh TJ, Feudtner C (2005) The epidemiology and attributable outcomes of candidemia in adults and children hospitalized in the United States: a propensity analysis. Clin Infect Dis 41:1232–1239
2. Zilberberg MD, Shorr AF, Kollef MH (2008) Secular trends in candidemia-related hospitalization in the United States, 2000–2005. Infect Control Hosp Epidemiol 29:978–980
3. Brown GD, Denning DW, Gow NAR, Levitz SM, Netea MG, White TC (2012) Hidden killers: human fungal infections. Sci Transl Med 4:165rv13
4. Pfaller MA, Diekema DJ (2010) Epidemiology of invasive mycoses in North America. Crit Rev Microbiol 36:1–53
5. Martin GS, Mannino DM, Eaton S, Moss M (2003) The epidemiology of sepsis in the United States from 1979 through 2000. N Engl J Med 348:1546–1554
6. Lupetti A, Danesi R, Campa M, Del Tacca M, Kelly S (2002) Molecular basis of resistance to azole antifungals. Trends Mol Med 8:76–81
7. Turner MS, Drew RH, Perfect JR (2006) Emerging echinocandins for treatment of invasive fungal infections. Expert Opin Emerg Drugs 11:231–250
8. Cowen LE (2013) The fungal Achilles' heel: targeting Hsp90 to cripple fungal pathogens. Curr Opin Microbiol 16:377–384
9. Nathan DF, Vos MH, Lindquist S (1997) In vivo functions of the *Saccharomyces cerevisiae* Hsp90 chaperone. PNAS 94:12949–12956
10. Young JC, Moarefi I, Hartl FU (2001) Hsp90: a specialized but essential protein-folding tool. J Cell Biol 154:267–273
11. Ali MMU, Roe SM, Vaughan CK, Meyer P, Panaretou B, Piper PW et al (2006) Crystal structure of an Hsp90-nucleotide-p23/Sba1 closed chaperone complex. Nature 440:1013–1017
12. Li J, Buchner J (2013) Structure, function and regulation of the Hsp90 machinery. Biomed J 36:106–117
13. Pearl LH, Prodromou C (2006) Structure and mechanism of the Hsp90 molecular chaperone machinery. Annu Rev Biochem 75:271–294
14. Whitesell L, Lindquist SL (2005) HSP90 and the chaperoning of cancer. Nat Rev Cancer 5:761–772

15. Sidera K, Patsavoudi E (20134) Hsp90 inhibitors: current development and potential in cancer therapy. Recent Pat Anticancer Drug Discov; 9(1):1–20
16. Roe SM, Prodromou C, O'Brien R, Ladbury JE, Piper PW, Pearl LH (1999) Structural basis for inhibition of the Hsp90 molecular chaperone by the antitumor antibiotics radicicol and geldanamycin. J Med Chem 42:260–266
17. Jarosz DF, Lindquist S (2010) Hsp90 and environmental stress transform the adaptive value of natural genetic variation. Science 330:1820–1824
18. Wong KSK, Houry WA (2006) Hsp90 at the crossroads of genetics and epigenetics. Cell Res 16:742–749
19. Queitsch C, Sangster TA, Lindquist S (2002) Hsp90 as a capacitor of phenotypic variation. Nature 417:618–624
20. Rutherford SL, Lindquist S (1998) Hsp90 as a capacitor for morphological evolution. Nature 396:336–342
21. Cowen LE, Lindquist S (2005) Hsp90 potentiates the rapid evolution of new traits: drug resistance in diverse fungi. Science 309:2185–2189
22. Johnson JL, Brown C (2009) Plasticity of the Hsp90 chaperone machine in divergent eukaryotic organisms. Cell Stress Chaperon 14:83–94
23. Swoboda RK, Bertram G, Budge S, Gooday GW, Gow NA, Brown AJ (1995) Structure and regulation of the *HSP90* gene from the pathogenic fungus *Candida albicans*. Infect Immun 63:4506–4514
24. Leach MD, Klipp E, Cowen LE, Brown AJP (2012) Fungal Hsp90: a biological transistor that tunes cellular outputs to thermal inputs. Nat Rev Micro 10:693–704
25. Bansal PK, Abdulle R, Kitagawa K (2004) Sgt1 associates with Hsp90: an initial step of assembly of the core kinetochore complex. Mol Cell Biol 24:8069–8079.
26. Dubacq C, Guerois R, Courbeyrette R, Kitagawa K, Mann C (2002) Sgt1p contributes to cyclic AMP pathway activity and physically interacts with the adenylyl cyclase Cyr1p/Cdc35p in budding yeast. Eukaryot Cell 1:568–582
27. Rocha CR, Schröppel K, Harcus D, Marcil A, Dignard D, Taylor BN et al (2001) Signaling through adenylyl cyclase is essential for hyphal growth and virulence in the pathogenic fungus *Candida albicans*. Mol Biol Cell 12:3631–3643
28. Shapiro RS, Robbins N, Cowen LE (2011) Regulatory circuitry governing fungal development, drug resistance, and disease. Microbiol Mol Biol Rev 75:213–267
29. Shapiro RS, Uppuluri P, Zaas AK, Collins C, Senn H, Perfect JR et al (2009) Hsp90 orchestrates temperature-dependent *Candida albicans* morphogenesis via Ras1-PKA signaling. Curr Biol 19:621–629
30. Flom GA, Langner E, Johnson JL (2012) Identification of an Hsp90 mutation that selectively disrupts cAMP/PKA signaling in *Saccharomyces cerevisiae*. Curr Biol 58:149–163
31. Shapiro RS, Zaas AK, Betancourt-Quiroz M, Perfect JR, Cowen LE (2012) The Hsp90 co-chaperone Sgt1 governs *Candida albicans* morphogenesis and drug resistance. PLoS ONE 7:e44734
32. Cowen LE (2008) The evolution of fungal drug resistance: modulating the trajectory from genotype to phenotype. Nat Rev Micro 6:187–198
33. White TC (1997) The presence of an R467K amino acid substitution and loss of allelic variation correlate with an azole-resistant lanosterol 14alpha demethylase in *Candida albicans*. Antimicrob Agents Chemother 1997;41:1488−1494
34. Cowen LE, Singh SD, Köhler JR, Collins C, Zaas AK, Schell WA et al (2009) Harnessing Hsp90 function as a powerful, broadly effective therapeutic strategy for fungal infectious disease. PNAS 106:2818–2823
35. Robbins N, Uppuluri P, Nett J, Rajendran R, Ramage G, Lopez-Ribot JL et al (2011) Hsp90 governs dispersion and drug resistance of fungal biofilms. PLoS Pathog 7:e1002257
36. Singh SD, Robbins N, Zaas AK, Schell WA, Perfect JR, Cowen LE (2009) Hsp90 governs echinocandin resistance in the pathogenic yeast *Candida albicans* via calcineurin. PLoS Pathog 5:e1000532

37. Singh-Babak SD, Babak T, Diezmann S, Hill JA, Xie JL, Chen Y-L et al (2012) Global analysis of the evolution and mechanism of echinocandin resistance in *Candida glabrata*. PLoS Pathog 8:e1002718
38. Cruz MC, Goldstein AL, Blankenship JR, Del Poeta M, Davis D, Cardenas ME et al (2002) Calcineurin is essential for survival during membrane stress in *Candida albicans*. EMBO J 21:546–559
39. Imai J, Yahara I (2000) Role of *HSP90* in salt stress tolerance via stabilization and regulation of calcineurin. Mol Cell Biol 20:9262–9270
40. LaFayette SL, Collins C, Zaas AK, Schell WA, Betancourt-Quiroz M, Gunatilaka AAL et al (2010) PKC signaling regulates drug resistance of the fungal pathogen *Candida albicans* via circuitry comprised of Mkc1, calcineurin, and Hsp90. PLoS Pathog 6:e1001069
41. Millson SH, Truman AW, King V, Prodromou C, Pearl LH, Piper PW (2005) A two-hybrid screen of the yeast proteome for Hsp90 interactors uncovers a novel Hsp90 chaperone requirement in the activity of a stress-activated mitogen-activated protein kinase, Slt2p (Mpk1p). Eukaryot Cell 4:849–860
42. Gould CM, Kannan N, Taylor SS, Newton AC (2008) The chaperones Hsp90 and Cdc37 mediate the maturation and stabilization of protein kinase c through a conserved PXXP motif in the c-terminal tail. J Biol Chem 284:4921–4935
43. Tong AHY, Lesage G, Bader GD, Ding H, Xu H, Xin X et al (2004) Global mapping of the yeast genetic interaction network. Science 303:808–813
44. Dobzhansky T. (1946) Genetics of natural populations. xiii. Recombination and variability in populations of *Drosophila pseudoobscura*. Genetics 31:269–290
45. Giaever G, Chu AM, Ni L, Connelly C, Riles L, Veronneau S et al (2002) Functional profiling of the *Saccharomyces cerevisiae* genome. Nature 418:387–391
46. Tischler J, Lehner B, Fraser AG (2008) Evolutionary plasticity of genetic interaction networks. Nat Genet 40:390–391
47. Zhao R, Davey M, Hsu Y-C, Kaplanek P, Tong A, Parsons AB et al (2005) Navigating the chaperone network: an integrative map of physical and genetic interactions mediated by the Hsp90 chaperone. Cell 120:715–727
48. McClellan AJ, Xia Y, Deutschbauer AM, Davis RW, Gerstein M, Frydman J (2007) Diverse cellular functions of the Hsp90 molecular chaperone uncovered using systems approaches. Cell 131:121–135
49. Franzosa EA, Albanèse V, Frydman J, Xia Y, McClellan AJ (2011) Heterozygous yeast deletion collection screens reveal essential targets of Hsp90. PLoS ONE 6:e28211
50. Davis DA, Bruno VM, Loza L, Filler SG (2002) Mitchell AP. *Candida albicans* Mds3p, a conserved regulator of pH responses and virulence identified through insertional mutagenesis. Genetics 162:1573–1581
51. Roemer T, Jiang B, Davison J, Ketela T, Veillette K, Breton A et al (2003) Large-scale essential gene identification in *Candida albicans* and applications to antifungal drug discovery. Mol Microbiol 50:167–181
52. Xu D, Jiang B, Ketela T, Lemieux S, Veillette K, Martel N et al (2007) Genome-wide fitness test and mechanism-of-action studies of inhibitory compounds in *Candida albicans*. PLoS Pathog 3:e92
53. Diezmann S, Michaut M, Shapiro RS, Bader GD, Cowen LE (2012) Mapping the Hsp90 genetic interaction network in *Candida albicans* reveals environmental contingency and rewired circuitry. PLoS Genet 8:e1002562
54. Shannon P, Markiel A, Ozier O, Baliga NS, Wang JT, Ramage D et al (2003) Cytoscape: a software environment for integrated models of biomolecular interaction networks. Genome Res 13:2498–2504
55. Wu P-YJ, Ruhlmann C, Winston F, Schultz P (2004) Molecular architecture of the *S. cerevisiae* SAGA complex. Mol Cell 15:199–208
56. Pinna LA (2002) Protein kinase CK2: a challenge to canons. J Cell Sci 115:3873–3878

57. Perez-Fernandez J, Roman A, Las RDeJ, Bustelo XR, Dosil M (2007) The 90S preribosome is a multimodular structure that is assembled through a hierarchical mechanism. Mol Cell Biol 27:5414–5429
58. Ota A, Zhang J, Ping P, Han J, Wang Y (2010) Specific regulation of noncanonical p38 activation by Hsp90-Cdc37 chaperone complex in cardiomyocyte. Circ Res 106:1404–1412
59. Mollapour M, Tsutsumi S, Truman AW, Xu W, Vaughan CK, Beebe K et al (2011) Threonine 22 phosphorylation attenuates Hsp90 interaction with cochaperones and affects its chaperone activity. Mol Cell 41:672–681
60. Askew C, Sellam A, Epp E, Mallick J, Hogues H, Mullick A et al (2010) The zinc cluster transcription factor Ahr1p directs Mcm1p regulation of *Candida albicans* adhesion. Mol Microbiol 79:940–953
61. Homann OR, Dea J, Noble SM, Johnson AD (2009) A phenotypic profile of the *Candida albicans* regulatory network. PLoS Genet 5:e1000783
62. Pijnappel WW, Schaft D, Roguev A, Shevchenko A, Tekotte H, Wilm M et al (2001) The *S. cerevisiae* SET3 complex includes two histone deacetylases, Hos2 and Hst1, and is a meiotic-specific repressor of the sporulation gene program. Gene Dev 15:2991–3004
63. Wang A, Kurdistani SK, Grunstein M (2002) Requirement of Hos2 histone deacetylase for gene activity in yeast. Science 298:1412–1414
64. Hnisz D, Majer O, Frohner IE, Komnenovic V, Kuchler K (2010) The Set3/Hos2 histone deacetylase complex attenuates cAMP/PKA signaling to regulate morphogenesis and virulence of *Candida albicans*. PLoS Pathog 6:e1000889
65. Robbins N, Leach MD, Cowen LE (2012) Lysine deacetylases Hda1 and Rpd3 regulate Hsp90 function thereby governing fungal drug resistance. Cell Reports 2:878–888
66. Suthram S, Sittler T, Ideker T (2005) The plasmodium protein network diverges from those of other eukaryotes. Nature 438:108–112
67. Taylor JW, Berbee ML (2006) Dating divergences in the fungal tree of life: review and new analyses. Mycologia 98:838–849
68. Odds FC. (1988) *Candida* and Candidosis. Elsevier Science Health Science Division, 468 p.
69. Noble SM, French S, Kohn LA, Chen V, Johnson AD (2010) Systematic screens of a *Candida albicans* homozygous deletion library decouple morphogenetic switching and pathogenicity. Nat Genet 42:590–598
70. Santos MA, Tuite MF (1995) The CUG codon is decoded in vivo as serine and not leucine in *Candida albicans*. Nucleic Acids Res 23:1481–1486
71. Hickman MA, Zeng G, Forche A, Hirakawa MP, Abbey D, Harrison BD et al (2013) The “obligate diploid” *Candida albicans* forms mating-competent haploids. Nature 494:55–59
72. Stynen B, Van Dijck P, Tournu HA (2010) CUG codon adapted two-hybrid system for the pathogenic fungus *Candida albicans*. Nucleic Acids Res 38:e184–e184
73. Kaneko A, Umeyama T, Hanaoka N, Monk BC, Uehara Y, Niimi M (2004) Tandem affinity purification of the *Candida albicans* septin protein complex. Yeast 21:1025–1033
74. Palzer S, Bantel Y, Kazenwadel F, Berg M, Rupp S, Sohn K (2013) An expanded genetic code in *Candida albicans* to study protein-protein interactions in vivo. Eukaryot Cell 12:816–827.
75. Gavin A-C, Bösche M, Krause R, Grandi P, Marzioch M, Bauer A et al (2002) Functional organization of the yeast proteome by systematic analysis of protein complexes. Nature 415:141–147
76. Blackwell C, Brown JD (2009) The application of tandem-affinity purification to *Candida albicans*. Methods Mol Biol 499:133–148

Part V
The p23 Chaperone Network

Chapter 9
Emergence and Characterization of the p23 Molecular Chaperone

Frank J. Echtenkamp and Brian C. Freeman

Abstract The p23 molecular chaperone is an abundant eukaryotic protein that shares genetic and physical connections to a variety of proteins working in diverse biological processes including protein transport, ribosome biogenesis, transcription activation, and chromatin remodeling. Here, we describe the initial founding of the p23 chaperone along with the early discoveries related to p23's function, its relationship to small heat shock proteins and other CHORD and Sgt1 (CS)-domain-containing factors, p23's biological relevancies in different cellular pathways, and how this chaperone has been exploited by both parasites and viruses. Finally, we discuss the correlation between p23 levels and cancerous cell growth.

1 Introduction

The promiscuous members of the heterogeneous molecular chaperone family manage the conformational status of cellular proteins [1]. Typically, chaperone function is considered through the study of the classic heat shock proteins (HSPs) in preventing the aggregation of nascent or metastable (inherent and stress-induced) polypeptides, thereby allowing low-energy states to be achieved [2]. Yet, seemingly native proteins also require assistance under normal physiological conditions. Most proteins are not single-use factors. Rather, they perform multiple activities by interacting with various partners and/or ligands (i.e., coenzymes). By adopting distinct physical forms, cofactor binding and release is coordinated [3, 4]. While discrete polypeptide shapes can dictate partner selection, the mechanisms triggering the transitions between physical states are not understood. We believe that molecular chaperones facilitate the interchange between structural forms thereby enhancing coenzyme exchange. To illustrate our point, we will describe the p23 molecular chaperone including its apparent rise from the classic small heat shock proteins (sHSPs) and its functional interaction network adapted for manipulating

B. C. Freeman (✉) · F. J. Echtenkamp
Department of Cell and Developmental Biology, University of Illinois,
Urbana-Champaign, Urbana, IL, USA
e-mail: bfreeman@life.illinois.edu

W. A. Houry (ed.), *The Molecular Chaperones Interaction Networks in Protein Folding and Degradation,* Interactomics and Systems Biology, DOI 10.1007/978-1-4939-1130-1_9,

the conformations of native or near native client proteins to promote an effective and dynamic protein environment.

Historically, molecular chaperone activities have been aligned with the process of nascent polypeptide folding where the complexity of a linear chain necessitates assistance to avoid aggregating within the context of a cytosol [2]. For example, a single polypeptide consisting of 149 amino acids can have 4^{149} to 9^{149} different conformations in solution assuming two rotational bonds per residue [5]. However, the likelihood that an elongated polypeptide is exposed to the cytosol is rather limited. When a nascent chain emerges from a ribosome, it is met by a legion of molecular chaperones including Hsp70, TRiC, Hsp90, along with their respective cochaperones. Although the absolute contributions made by each type of chaperone to protein folding is still being resolved, chaperones minimally shield neighboring chains to avoid nonspecific aggregation through the exposed hydrophobic side groups. Significantly, collapse of the chaperone system managing the early nascent chains leads to cell death [6].

The folding process initiates as a naïve polypeptide exits the ribosome with the chain reaching a biologically active, low-energy state by proceeding down a folding landscape pervaded with nuances, which creates a variety of potential folding routes [7]. Regardless of the precise pathway traversed, each polypeptide will encounter unproductive intermediates trapping the molecule in an ineffective state. Fortunately, molecular chaperones circumvent these wasteful intermediaries by overcoming the energy barriers thwarting the capacity of a polypeptide to reach its native state (Fig. 9.1a). The effectiveness of the chaperone system is enhanced by a differential use of the molecular chaperones along the folding process. Early off-pathway species are managed by the Hsp70s, which bind to ribosome-associated nascent chains [8–10]. Presumably, binding by Hsp70 restricts the conformation of the chain and biases the folding towards on-pathway steps. Late maturation events are mediated by the chaperonin (e.g., GroEL, TRiC) and Hsp90 paralogs [2]. For example, chaperonins coordinate the assembly of multi-subunit structures whereas Hsp90 maintains metastable factors in a soluble state until activated (e.g., ligand binding by a steroid hormone receptor, SHR).

In the event that the chaperones fail to guide a naïve polypeptide to a productive structure, the network is designed to deliver the castoffs to the proteolysis machine for disposal [11]. The relationship between the chaperone and proteolysis systems is further exploited to remove damaged proteins occurring under both normal and stressful (e.g., heat shock) circumstances. To effectively handle the aggregation prone polypeptides an additional class of molecular chaperones, the sHSPs, is employed [12]. In general, the expression of sHSPs is highly stress-inducible. Yet, several sHSP derivatives, including p23, have emerged for prominent use during normal growth (discussed below). Overall, molecular chaperones collectively monitor the cellular proteome and maintain homeostasis by minimizing unproductive conformational protein species under all physiological conditions.

In addition to the classic folding activities, molecular chaperones contribute to more subtle protein conformational changes. Many, if not all, cellular proteins utilize distinct structural forms to accomplish their cellular tasks. To partition the

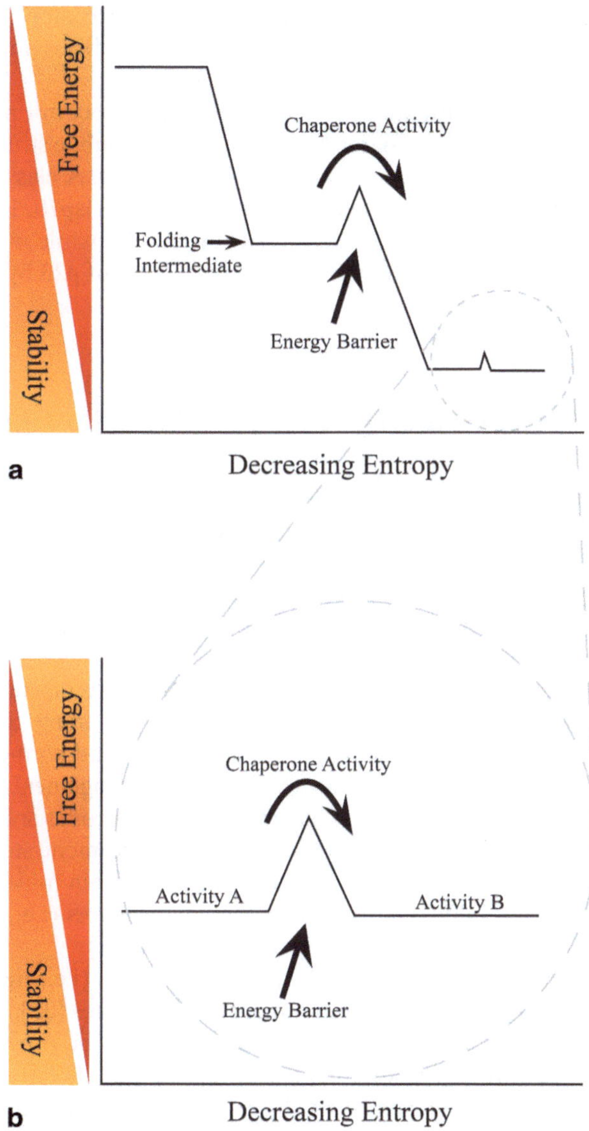

Fig. 9.1 Molecular chaperones mediate structural changes throughout the lifetime of a polypeptide. Molecular chaperones facilitate protein conformational changes necessary for the folding of nascent polypeptides (**a**) and for coenzyme interactions separating functional activities of a client protein (**b**)

different configurations relatively small energy barriers can be used to bias residence in one low-energy state or another (Fig. 9.1b). Hence, the ability of a protein to perform different activities is distinguished. For example, Clathrin heavy and light chains form a triskelion-shaped structure that oligomerizes into a polyhedral lattice around a membrane vesicle [13]. While cooperative interactions foster assembly of the Clathrin triskelion lattice, an incumbent energy barrier prevents the cage from spontaneously dissociating [14]. The clathrin-coated vesicles are an effective means to transport molecules between intracellular compartments or even to other cells [14]. Upon connection to an acceptor membrane, Hsp70 triggers a conformational change in clathrin promoting disassembly of the coat and allowing the release of the vesicle's contents [15, 16].

While contributions to the protein quality control process are well established, how molecular chaperones manage the more subtle structural fluctuations that drive coenzyme interactions is less well understood. A classic example of client dependence for coenzyme binding involves SHRs, Hsp70, Hsp90, and a plethora of co-chaperones. As p23 was originally discovered while investigating Hsp90-mediated SHR-biogenesis, we will start our discussion of p23 by describing its contributions to SHR biology.

2 The Founding of the p23 Molecular Chaperone

SHRs are the protein-binding partners for glucocorticoid, progestin, androgen, and estradiol hormones [17]. SHRs are a subfamily within the intracellular hormone receptor (IR) super family, which also contains binders for retinoids and thyroids along with numerous orphan receptors [18]. In brief, IRs convert hormonal signals into gene regulatory programs important for diverse physiological functions including embryonic development, cell differentiation, inflammation, and memory [19–22]. In addition, IRs are one of the most therapeutically targeted protein families being relevant to numerous diseases including cancer and diabetes [23–27]. Central to SHR function are the abilities to bind hormone, execute ligand-induced allostery, and undergo further structural rearrangements to release the hormone [28, 29]. It is well established that molecular chaperones are required for ligand binding, yet the influence of chaperones in the later hormone steps is not as well understood [30]. Nevertheless, it is now evident that certain molecular chaperones, including p23, have additional roles with the holoreceptors (i.e., receptor-ligand complexes) downstream of the hormone-binding event [31].

Prior to hormone association, aporeceptors (i.e., ligand free) are in heterotypic assemblies with numerous molecular chaperones [32]. The p23 chaperone was discovered as a component of the apo form of the progesterone receptor (PR) [33, 34]. Follow-up studies demonstrated that p23 directly binds to Hsp90 through an adenosine triphosphate (ATP)-dependent mechanism [35]. Importantly, both PR and the glucocorticoid receptor (GR) rely on the weakly associated p23 chaperone for proper long-term maintenance of hormone-binding activity [34, 36]. However, the initial formation of a SHR hormone-binding state only requires Hsp70, Hsp90/Hsp70 organizing protein (HOP), and Hsp90 [37]. Extensive additional research identified Hsp70, Hsp40, HOP, Hsp90, and p23 as the minimum chaperones sufficient for both establishing and maintaining the hormone-binding competence of a SHR in vitro [38, 39]. In a biochemical context, Hsp70 acts at an early stage of folding and then the SHR is transferred to Hsp90 in a HOP-dependent manner.

In the typical model, the maturation process continues with p23 interacting with the complex by binding to the $(Hsp90\text{-}ATP)_2$:SHR assembly where it stabilizes the closed conformation of the Hsp90 dimer [38, 40, 41]. Initial work suggested that the chaperones could associate, albeit ineffectually, with an aporeceptor independent of ATP but later studies showed the structure to be ATP-dependent [34, 42, 43]. Suggestively, p23 joins the complex to effectively prolong the Hsp90-client association and increase the time period an SHR is maintained in a hormone-binding competent state [40, 44]. Besides SHRs, the tyrosine kinase feline sarcoma (FES), heat shock factor 1 (HSF1), and aryl hydrocarbon receptor (AhR) form similar complexes with the chaperones suggesting that numerous proteins rely on this system [45].

The p23 molecular chaperone is a highly conserved protein being present from yeast to human cells with a broad tissue distribution in mammals [34]. At least in plants, p23 supports thermal stress tolerance [46]. Although p23 is nonessential in yeast, it is essential in vertebrates [47–51]. p23 is perhaps best known as an Hsp90 cochaperone where it binds to the amino-terminus of Hsp90 in a structure dependent manner [52, 53]. Hsp90 works as a highly flexible dimer anchored through carboxyl-termini oligomerization motifs [54]. ATP binding by Hsp90 induces a transition from an open conformation to a closed state by promoting further cross subunit interactions at the amino-termini [53]. Once in proximity the amino-ends form a p23-binding site [56]. Correspondingly, p23 association stabilizes the Hsp90 amino-terminal interaction, which functionally results in a ~2-fold reduction in Hsp90's ATPase rate [57]. Whether p23's relevance in the SHR maturation pathway is its ability to prolong the closed conformation of Hsp90, to impair Hsp90's ATPase, or to bind directly to the SHR will require more work to resolve. Likely, it is a combination of these benefits.

Besides being an Hsp90 cochaperone, p23 has an intrinsic chaperone function. Notably, p23 prevents the in vitro aggregation of denatured model substrates at a level comparable to Hsp90 or Hsp70 [58, 59]. Intriguingly, however, the mechanisms used by p23 and the classic HSPs to prevent protein aggregation are different. Hsp90 and Hsp70 both mediate a collapse of non-native polypeptides while keeping the species in a soluble form but p23 maintains non-native proteins in a soluble, extended state [59]. If p23 has a distinct structural consequence on a client, how did this chaperone evolve?

3 The Relationship Between p23, Small Heat Shock Proteins, and Other CS-domain Factors

The p23 molecular chaperone family is expressed in most eukaryotes being represented from yeast to humans (29 % identity with 54 % conservation); however, a prokaryotic counterpart is yet to be described. In general, p23 proteins are acidic (pI ~ 4) and typically range in size from 18 to 25 kDa. The signature motif of the family is a central WPRLTKE sequence, which has been preserved in human, mouse, chicken, budding yeast, fission yeast with elements of the motif remaining in round worms and other species. In fact, the WPRLTKE motif was key for identifying the budding yeast homolog Sba1 (sensitive to benzoquinoid ansamycins 1) since it shares only 24 % amino acid identity with the human homolog [47, 48].

However, the potential origin of the p23 homologs did not become apparent until a structure of p23 was solved. The atomic structure of human p23 missing the 35 carboxyl-terminal residues, which are essential for chaperone activity, revealed a compact antiparallel β-sandwich fold containing eight β-strands [60, 61]. A comparison of the p23 structure yielded the *Methanococcus jannaschii* sHSPs Hsp16.5 in which 73 α-carbon atoms of the amino acid backbone displayed a root mean square deviation of 1.27 Å despite only ~10% sequence identity [61]. This fold had been previously coined HSP20 or the α-crystallin domain (ACD) [62].

SHSPs are ubiquitously expressed across all three domains of life [63]. The sHSPs are of variable sequence and size (12–42 kDa range) except for a carboxyl-terminal ~90 residue ACD. The sHSPs are ATP-independent molecular chaperones that form polydispersed assemblies ranging from 2–32 subunits that minimally support thermotolerance [64]. The chaperone active oligomeric state of sHSPs is of debate. One proposed mechanism involves the larger structures (i.e., 12–32 subunit complexes) breaking down into active dimers in response to unfolded protein substrates [65]. For example, during heat shock conditions, the yeast Hsp26 is expressed and transitions from a large inactive oligomer (24–26 subunits) to a dimer displaying chaperone activity as unfolded proteins appear [65]. Yet, another yeast sHSP Hsp42 is constitutively expressed assembling into large (24–26 subunits) chaperone-active structures under all growth conditions [66, 67]. Together, Hsp26 and Hsp42 can suppress the aggregation of approximately one third of all yeast proteins [67]. Given their capacity, sHSPs have been referred to as molecular sponges capable of broadly reducing or preventing irreversible protein aggregation [12]. Basically, sHSPs have evolved as a cellular mechanism for functionally interacting with a wide-range of non-native proteins.

The unifying physical characteristic of all sHSPs is the presence of an ACD flanked by variable amino- and carboxyl-terminal extensions [63, 68]. The end regions contribute to functions specific to each sHSP and also are used for self-association [69]. Besides the highly oligomeric sHSPs, the α-crystallin-like domains are found in a number of monomeric proteins including p23, suppressor of the G_2 allele of *SKP1* (Sgt1), nuclear distribution gene C (NudC), melusin, and b5+b5R flavoheme cytochrome NAD(P)H oxidoreductase type B [70]. To distinguish between these classes of proteins, the domain in sHSPs is typically referred to as an ACD while the structure is marked as a CHORD and Sgt1 (CS)-like domain in the other proteins. No matter the nomenclature, the structure is typically a compact antiparallel β-sandwich fold consisting of seven β-strands [70] (Fig. 9.2a). Interestingly, p23 contains an eighth β-strand in its CS domain that is critical for modulating the ATPase activity of Hsp90 [53] (Fig. 9.2b). Yet, the eighth β-strand is not essential for a CS protein to interact with Hsp90 since Sgt1, which has a seven β-strand CS domain, associates with Hsp90 [71–73]. However, the mode of assembly is distinct as Sgt1 binds to Hsp90 independent of ATP and Sgt1 has no apparent effect on Hsp90's ATPase activity [74].

Nevertheless, the presence of a common fold opens the possibility that CS and ACD proteins are distantly related [70]. The melusin and cytochrome reductase type B proteins have diversified the use of a CS domain, as these proteins

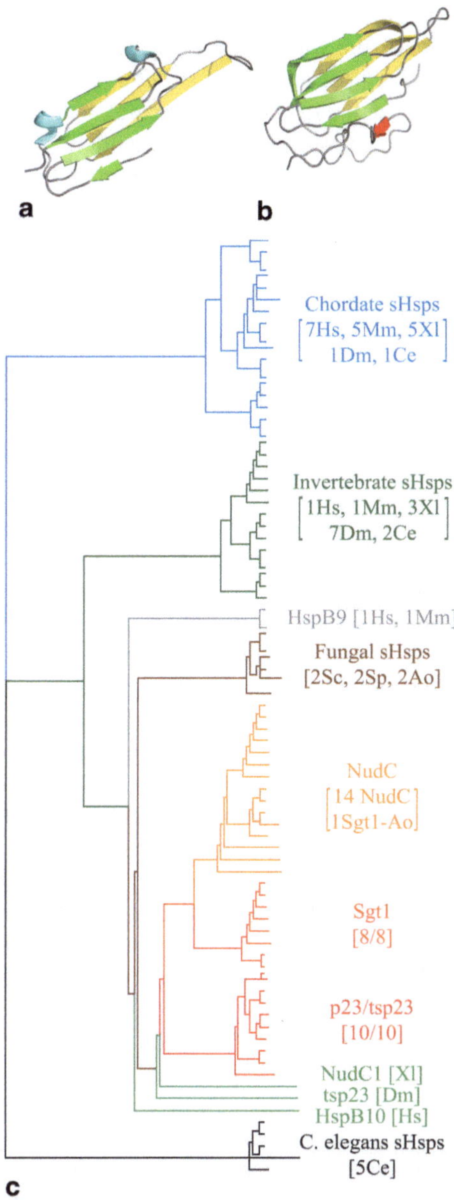

Fig. 9.2 Phylogenetic map and structural comparison of CS/ACD-containing proteins. Representative structures of human p23 (accession code 1EJF) (**a**) and the human small heat shock protein (sHSP) alpha B crystallin (accession code 2WJ7) (**b**) were obtained from the protein data bank (http://www.rcsb.org) and modified with Pymol Version 1.5.0.4 (Molecular Graphics System, Schrödinger). Homologous secondary structures are highlighted in green (β-stand Face A) and yellow (β-stand Face B) with the p23-specific eighthβ-strand depicted in red. Protein sequences of sHSP, NudC, Sgt1, and p23 family members from *Homo sapiens* (Hs), *Mus musculus* (Mm), *Xenopus laevis* (Xl), *Drosophila melanogaster* (Dm), *Caenorhabditis elegans* (Ce), *Saccharomyces cerevisiae* (Sc), *Schizosaccharomyces pombe* (Sp), and *Aspergillus oryzae* (Ao) were aligned and mapped using the ClustalW multiple alignment program (http://www.bioinformatics.nl/tools/clustalw.html). The protein clusters, based on relative homology levels, were color coded and the species along with number of homologs within each group are indicated (**c**)

are important for mammalian cell focal adhesions and protection against oxidative stress, respectively [75–78]. Intriguingly, three of the CS domain protein groups have retained chaperone activity—p23, NudC, and Sgt1. A phylogenetic analysis of the full-length sequences of these CS proteins along with numerous sHSPs illustrates the apparent relationships (Fig. 9.2c). While the CS/ACD domains drive the arrangement of the homology tree, the protein regions outside of this domain demark the various subgroups.

A comparison of the structural organization of the p23s to the sHSPs highlights the absence of an amino-terminal extension and a significant expansion of p23's carboxyl-terminus [61]. As mentioned previously, the CS domain of p23 has been lengthened by an additional β-strand in order to gain a regulatory capacity with Hsp90 [70]. Importantly, a further extension of the carboxyl-terminus is critical for the chaperone activity of p23 used both to prevent the aggregation of model substrates in vitro and to affect clients in vivo [61, 79–81]. On the other hand, the NudC and Sgt1 families can have extensions on either side of the CS domain.

NudC was first identified as a regulator of nuclear movement in the asexual reproductive cycle of the filamentous fungus *Emericella nidulans* [82]. Metazoans express three NudC paralogues (NudC, NudCL, and NudCL2) with a fourth distantly related isoform, NudCD1 (CML66), being a tumor antigen [83]. The NudC proteins are involved in numerous events including nuclear migration, cell cycle progression, cell proliferation, dynein- and kinesin1-mediated transport, and have been implicated in a variety of malignancies [84–86]. All four NudC isoforms have a carboxyl-terminal CS domain, yet only NudC and NudCL display in vitro chaperone activity [83]. Both NudC and NudCL have amino-terminal extensions with a coiled-coil site along with an unstructured region—truncations comprising just the amino-termini and CS domains are sufficient for in vitro chaperone activity [83]. Comparable to p23, NudC can dampen the ATPase activity of Hsp90 [87]. Given the different extensions on p23 and NudC, it seems that CS domains provide an adaptable module to generate a protein capable of managing labile polypeptides and also interacting with Hsp90.

Sgt1 is conserved among eukaryotes being identified as a suppressor of the G_2 allele of SKP1 where it fosters assembly of yeast kinetochores and is a subunit of the SCF (Skp1-Cul1-F-box) ubiquitin ligase complex [88]. Further work demonstrated that Sgt1 supports innate immunity pathways by stabilizing resistance (R) and nucleotide-binding domain and leucine-rich repeat containing (NLR) proteins [89–91]. Sgt1s are extended on both sides of the CS domain with an amino-terminal tetratricopeptide repeat domain (TPR) and a carboxyl-terminal SGT1-specific (SGS) domain [89]. The CS fold is used to associate with Hsp90 while Hsp70 binds to the SGS motif [71, 92]. The ability of Sgt1 to interact with both Hsp90 and Hsp70 suggests that it may serve a role similar to the well-established cochaperone HOP but for specializing the chaperone system for R and NLR proteins [93, 94]. Unlike other cochaperones, Sgt1 does not affect the ATPase activity of Hsp90 [95]. However, similar to p23 and NudC, Sgt1 suppresses the aggregation of a model non-native protein in vitro by working as a molecular chaperone [96].

Among the CS protein family is a poorly understood p23 paralog tsp23 (transcript similar to p23), also referred to as "putative protein PTGES3 Lisoform 5," with 44% sequence identity to p23 [97]. The tsp23 protein has the WPRLTKE signature motif and a carboxyl-terminal extension with a length intermediate between human and yeast p23. Tsp23 is a poorly studied protein though it is known to stimulate certain SHRs in a manner similar to p23 yet it does not inhibit other

SHRs that are p23-repressive [97]. Interestingly, tsp23 is only expressed in cardiac and skeletal muscle whereas p23 is broadly expressed in tissues except for cardiac and skeletal muscles [34, 97]. In addition, tsp23 can be found in skeletal muscle as a splice variant fused to the alanyl-transfer ribonucleic acid (tRNA) synthetase AlaXp to form $p23^{H}$AlaXp [98]. However, the fusion protein is inactive and has no known function although it was suggested that $p23^{H}$AlaXp might link the Hsp90 machine to the prevention of alanine-tRNAs misactivation [98].

In sum, the CS/ACD domain provides a versatile module to cultivate a protein family with diverse cellular functions using a few basic properties—chaperone activity, partner interaction site (e.g., Hsp90), and self-association capacity. While no single member seems to use all three assets (e.g., p23 is a monomeric chaperone that can bind to Hsp90 yet Hsp26 is an autonomous, oligomeric chaperone), the flexibility of this relatively simple domain appears to condone a fitness advantage. The benefit of the CS/ACD domain is obvious for eukaryotes where the variety of sHSPs significantly increases and the domain is commandeered to develop numerous non-HSPs including p23. Perhaps the broad binding capacity of the ancient sHSPs, which is used to prevent general protein aggregation, provided a readily exploitable module for widely interacting with cellular proteins under normal physiological conditions. At least for p23, we suspect this is the case since the p23 molecular chaperone has a role in a broad array of cellular processes important for mediating cofactor binding and release.

4 Cellular Processes Governed by p23 Molecular Chaperones

The exploration of molecular chaperone biology is particularly demanding since these proteins tend to be highly abundant and promiscuous [2]. Further complicating matters is the capacity of chaperones to associate with both coregulators (i.e., cochaperones) and clients (i.e., protein substrates) and therefore discrimination between these two classes is a matter of concern. Often the cellular roles of a protein, including the elucidation of previously unknown functions, are gained by identifying other proteins capable of interacting with the target. Unfortunately, the fundamental properties of molecular chaperones challenge this proposition at least when using classic assays. For example, a 2-hybird screen incorporating both yeast Hsp90 homologs as baits did not identify any strong interactors and found only four hits (two known cochaperones and two open reading frames (ORFs) with unknown functions) under mild selection conditions [99]. Similar 2-hybrid results were obtained using four different cochaperones, including yeast p23, as bait [99]. However, if the Hsp90 mutant E33A was the bait, then 177 interactors were identified [100]. Presumably, the propensity of the E33A mutant to form the closed conformation of Hsp90, which might trap clients and certain cochaperones, accounts for the increase in hits. Together, these studies support the notion that the transient nature of chaperone-binding limits the ability to detect interacting proteins.

As a comprehensive and reliable interaction network is highly beneficial for comprehending the cellular roles of a given protein, interest in constructing system-wide maps of the molecular chaperones persisted especially in light of their increased medicinal targeting [101–104]. With the recent rise in the types of available high-throughput approaches, it is now feasible to generate a variety of networks for a target protein that are based upon physical, genetic, or functional relationships [105, 106]. The selection of readily accessible large-scale methods along with their declining costs makes it possible to chart the global connections for most, if not all, cellular proteins. Application of tactics such as synthetic genetic array (SGA) analysis, small-molecule chemical screens, protein microarrays, affinity purification/mass spectrometry methods have led to the development of extensive networks for numerous proteins including molecular chaperones [107–112]. For instance, application of such techniques raised the number of known yeast Hsp90 interactors to ~1000 [113, 114]. Given that budding yeast only has ~6000 ORFs, the two studies showed that Hsp90 has a significant relationship with the yeast proteome.

With the application of both high-throughput genetic (SGA) and physical (protein microarrays) tactics to the yeast p23 chaperone, the first functional interaction network for an Hsp90 cochaperone was established [115]. Prior to this work 34 proteins had a known association with p23. After the study, 348 unique p23 interactors had been revealed including 234 new genetic hits and 80 direct binders. A bioinformatic analysis of the compiled p23-interactors using the gene ontology (GO) Slim program freely available at the *Saccharomyces* genome database (SGD) suggests p23 functions within numerous biological processes and is associated with all compartments of the cell (Fig. 9.3a). Such a broad distribution is supported by indirect immunofluorescent images using mouse embryonic fibroblasts and antibodies directed against vertebrate p23 (Fig. 9.3b) [115]. Of note, if the localization pattern of p23 is examined using fluorescent protein fusion technology, the relative level of p23 nuclear occupancy is reduced [116, 117]. As fusion tags on either terminus of p23 abrogate its chaperone activity [118], the altered pattern likely reflects this functional change. The broad cellular localization along with the apparent wide distribution of interactors indicates that p23 might be a general molecular chaperone with a bias for nuclear pathways.

As p23 is typically considered an Hsp90 cochaperone, where it is used to modulate Hsp90's ATPase and possibly direct client selection for Hsp90, it is reasonable to expect a significant overlap between the interactors for p23 and Hsp90. Yet, only ~25% of the p23 hits were found within the Hsp90 network suggesting p23 also functions autonomously of Hsp90 [115]. As the potential for independent activities of several Hsp90 cochaperones has been recently described [119], we will refrain from an elaborate discussion of this topic. Rather, we will delve into less explored areas of the p23 network.

with their predicted cellular locations were determined using the Gene Ontology (GO) Slim Mapper program (**a**). The cellular distribution of mouse p23 was determined by indirect immunofluorescence using wild type mouse embryonic fibroblasts and a monoclonal antibody (JJ3) that recognizes vertebrate p23 [34] (**b**)

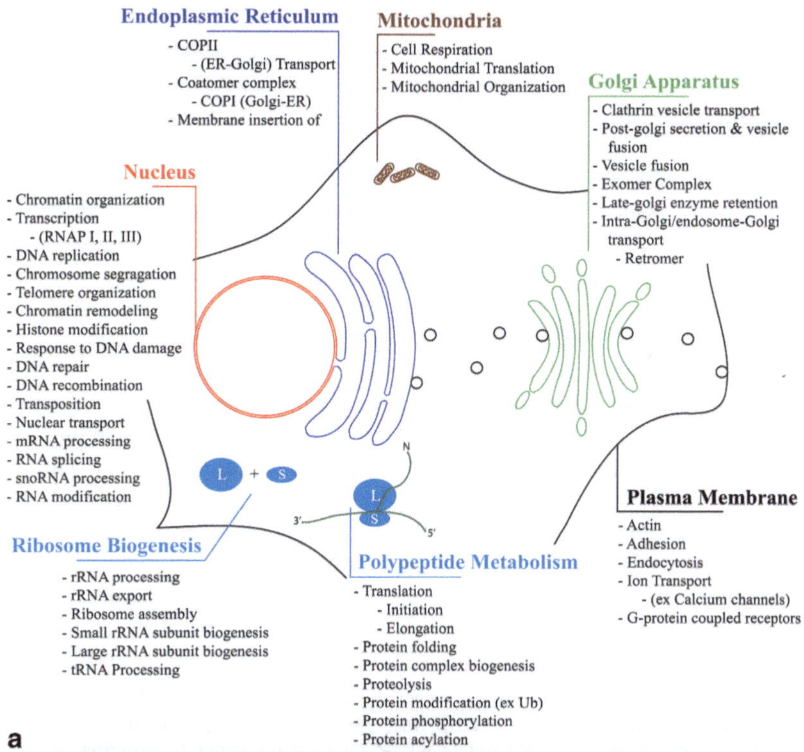

a

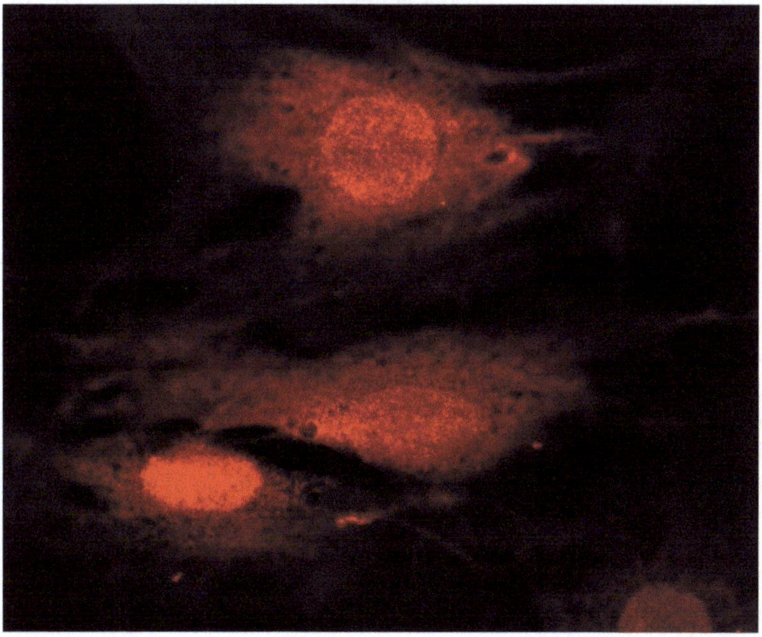

b

Fig. 9.3 The p23 chaperone is a ubiquitous protein with an interaction network encompassing a variety of biological processes. The expanse of the p23 network is illustrated using select, established p23-interactors [115]. The biological processes associated with various p23 targets along

The most evident biological process noted in the p23 GO Slim analysis was a link to intracellular protein transport underscored by a prevalence of Golgi apparatus associated factors [115]. Yet a rudimentary check of p23's localization pattern does not highlight the Golgi organelle, as p23 seems to be uniformly distributed within the cytosol (Fig. 9.3b). However, comparisons with Golgi and endoplasmic reticulum (ER) markers suggest that p23 preferentially associates with the Golgi. Briefly, p23 displayed a ~98% coincidence with a Golgi marker but only a ~50% overlap with an ER indicator [115]. While the seemingly substantial phase with the ER might normally tempt the conclusion that p23 has an ER interaction, the high abundance and broad distribution of many molecular chaperones, including p23, warrants a more significant overlap to conclude a working relationship.

Significantly, functional studies identified one of the p23 affects in the protein transport process. In both yeast and mammalian backgrounds the efficiency of transport-associated activities (e.g., secretion and processing) were increased in p23 null cells indicating that p23 serves as a negative regulator [115]. Through an epistasis approach one p23 contribution to the transport pathway was discovered—modulation of cargo mannosylation within the Golgi. However, the mannosylation step only accounts for 2 of the 31 transportation-related hits within the p23 network.

Interestingly, normal yeast are fairly unresponsive to the Golgi inhibitor Brefeldin A, yet *p23Δ* cells are sensitive and p23 overexpression is protective [115, 120]. Brefeldin A is a heterocyclic lactone that inhibits coat protein I (COPI)-mediated retrograde transport from the Golgi to the ER [121]. The intracellular transport of vesicles is a complicated process with numerous steps reliant on dynamic changes in protein conformations and in the composition of complexes [122]. To briefly illustrate we will focus on the final step of the pathway when a vesicle merges with its target membrane [123]. Fusion occurs through the pairing of soluble N-ethylmaleimide-sensitive factor (NSF) attachment protein receptor (SNARE) proteins when a vesicle-SNARE (v-SNARE) docks with a target-SNARE (t-SNARE) protein to form a labile trans-SNARE complex [122]. Following membrane fusion, the resultant protein structure (the cis-SNARE complex) must be "primed" and then disassembled minimally by the actions α-SNAP and NSF homologs [124].

While vesicular trafficking requires numerous conformational changes to proceed, the direct involvement of p23 has not been established. The Brefeldin A phenotypes implicate a p23 involvement in COPI retrograde transport delivering components from the Golgi back to the ER. Intriguingly, p23 shares a genetic connection with *SEC28* [115]. Sec28 (epsilon-COP) stabilizes the cytosolic coatomer complexes that are required for the formation of COPI-coated vesicles [125]. An additional p23-link to vesicular transport is made with Vam7, which stabilizes a t-SNARE complex required for the docking and/or fusion of multiple transport intermediates destined for the vacuole [115, 126, 127]. Besides Sec28 and Vam7, p23 has 27 other interactors linked to the protein transport pathway whose functional connection to p23 is yet to be defined. Moreover, there are over 300 hits within the p23 map yet to be explored. Hence, significant additional work is required to fully appreciate the utility of p23.

An unexpected outcome found within the p23 network was a preponderance of nuclear hits—almost one third of the 348 interactors are considered nuclear proteins [115]. At the time, the predominant model indicated that the Hsp90 molecular chaperone system was a strict cytoplasmic machine [128]. This erroneous theory was proposed to rationalize the cytosolic localization of certain unactivated SHRs [129]. Now, p23 and Hsp90 are known to have a variety of nucleoplasmic roles in diverse activities including transcription activation and telomere deoxyribonucleic acid (DNA) maintenance [119, 130]. A survey of the p23 network reveals a wide-range of nuclear processes including ribosome biogenesis, chromatin structure, transcription events, RNA processing, DNA replication, telomere maintenance, and DNA repair [115]. The simplest explanation for the numerous connections is that p23 serves as a general chaperone overseeing nuclear pathways. But why would there be a need for chaperone activity in the nucleus?

5 The Functional Requisite of p23 in the Nucleoplasm

A dynamic protein environment along the genome is essential to appropriately respond to the numerous and diverse conditions that challenge cells [131]. For example, fluctuations in carbon sources or exposure to stress conditions prompt rapid changes in chromatin status and gene expression patterns. Coordinating the activities of these pathways are various transcription factors along with other DNA-binding proteins—for simplicity we will refer to this group as TFs. The TFs use cooperative interactions to bind select DNA sites and recruit various protein machines (e.g., chromatin remodelers, histone acetylases, RNA polymerases) to perform the requisite work (i.e., remodel chromatin or transcribe DNA into RNA). While the cooperativity drives precise and rapid assembly, the integral stability of these structures interferes with the timing of biological systems [132]. Therefore, TF-nucleated DNA complexes must be actively and persistently disassembled to function on a useful time scale. Fortunately, the abundant and promiscuous nature of the p23 molecular chaperone places it in a position to foster a dynamic genomic environment that is sufficiently pliable to meet the requirements of homeostasis.

With the application of modern cell biology techniques it was realized that nuclear factors are highly motile [133]. In a seminal study, Phair and Misteli [134] used green fluorescent protein (GFP) fusions to show that diverse nuclear proteins, regardless of functional purpose, move with near diffusion constants in live cells. Of note, the rapid motilities of the fusions were not restricted by chemical or thermal energy. Yet, the candidates moved slower than GFP alone suggesting that each momentarily interacts with partners, thereby marginally reducing their movements. If a factor is recognizing a partner, why isn't there a more pronounced change in mobility especially when bound to chromatin? While it is clear that dynamic action is a fundamental property of nuclear proteins [135–141], the mechanisms driving the rapid actions are not understood.

To limit the timing of protein interactions and to accelerate pathways, we believe that molecular chaperones, especially p23, create a highly dynamic nuclear environment [142]. The chaperones contribute by circumventing otherwise long-lived, unproductive structures. Significantly, molecular chaperones do not interfere with the formation of transient, functional complexes since chaperone-client-binding events are typically fleeting. Of note, chaperones work over a broad temperature range and do not require chemical energy to functionally impact a client [143, 144]. Hence, chaperones can account for the observed dynamic protein behavior that persists following chemical and thermal energy fluctuations in vivo [134].

Perhaps fittingly, an early activity revealed for any molecular chaperone was the ability to dissociate a protein-DNA complex. While identifying the host proteins supporting the life cycle of the bacterial phage λ, both DnaK and DnaJ were discovered [145]. DnaK and DnaJ are the *Escherichia coli* homologs of Hsp70 and Hsp40. Further work demonstrated that DnaK and DnaJ dissociate the λP protein from preprimosomal structures leading to firing of origins of λ DNA replication [146]. Subsequent studies led to the suggestion that DnaK, in conjunction with DnaJ, creates a feedback reaction by cyclically building and destroying the λ preprimosomal DNA replication complexes [147]. Hence, one of the first discovered functions for a molecular chaperone was the assembly and disassembly of protein-DNA complexes.

The original report suggesting a role for the Hsp90 chaperone system in DNA-associated events was the finding that Hsp90 fosters the DNA-binding activity of the myogenic transcription factor MyoD in vitro [148]. Further biochemical work showed that Hsp90 might serve as a general chaperone for basic helix-loop-helix (bHLH) transcription factors by manipulating the conformation of preformed dimeric bHLH transcription factors into a DNA-binding competent state [149]. Numerous other in vitro studies indicate that Hsp90 facilitates the DNA-binding activity of diverse transcription factors including p53, hypoxia-inducible factor 1 (HIF-1), and AhR [150–153]. Significantly, in vivo data support the idea that Hsp90 promotes the DNA-binding activity of a transcription factor as it engages a gene promoter [154].

Early p23 studies showed a nucleoplasmic presence of this chaperone as it was found in nuclear extracts prepared from *Mus musculus*, *Xenopus laevis,* and Schizosaccharomyces pombe cells [155–157]. In murine erythroleukemia (MEL) cells p23 directly interacted with DNA cystosine-5 methyltransferase independent of Hsp90 [155]. In fission yeast the ubiquitous cellular distribution of p23 minimally was required for suppression of the Wee1 kinase [157]. The frog work provided the first indication that p23 might affect the DNA-binding activity of a transcription factor, as injection of anti-p23 antibodies into oocytes prevented attenuation of stress-induced HSF1 DNA-binding activity [156]. Interestingly, additional research demonstrated that overexpression or deletion of p23 in budding yeast results in chromosome instability [158]. While a mechanism is yet to be discovered, the phenomena suggest p23 has a broad role along the genome.

Subsequent work to these studies showed that p23 overexpression altered ligand efficacy of several intracellular receptors [97]. Typically, molecular chaperones modulate ligand potency of SHRs [41]. In brief, ligand efficacy gauges the transcription activation potential whereas potency is an indirect measure of hormone-binding activity. Follow-up work using purified thyroid hormone receptor (TR) and p23 showed that the change in efficacy correlated with an ability of p23 to dissociate TR from its cognate DNA element [97]. Additional work indicated that p23 could alter the transcriptional output of non-IR factors including activation protein 1 (AP1), nuclear factor κB (NF-κB), and vitamin D receptor (VDR) [159, 160].

The breadth of pathways regulated by p23 at the level of DNA binding has been expanded to include proteins involved in telomere biogenesis (Telomerase), DNA repair (RAD51), and DNA replication (CDC6) [79, 81, 161, 162]. The most notable addition to our understanding of p23's range with DNA-binding proteins was recently shown using a high-throughput tactic (DNase-Seq) for comprehensively mapping protein-DNA interactions along a genome. Basically, the preferential cleavage of naked DNA by the DNase I nuclease is exploited along with massive parallel DNA sequencing (Seq). In brief, nuclei are lightly treated with DNase I and the fragmentation pattern of the mildly digested chromatin reveals both areas of open chromatin (i.e., DNase I hypersensitive sites (DHSs)) and TF interactions (bound TFs protect DNA from digestion) [163]. The identities of the bound TFs are made by analogy with their known consensus binding elements. Application of DNase-Seq to both parental and *p23Δ* yeast revealed changes in open chromatin with the number of DHSs declining upon loss of p23 yet the average size of the prevailing DHSs doubling [81]. Significantly, a dramatic increase in the extent of DNA-bound transcription factors occurs within the affected DHSs since p23 is required to disassemble the otherwise stable protein-DNA structures. Of the 178 TFs followed in yeast using this approach, 110 are associated with DHSs and of those proteins 71 are impacted in *p23Δ* yeast. Hence, p23 modulates ~70% of the yeast TFs with known DNA-binding motifs.

Even the seemingly simple feat of a transcription factor finding its appropriate DNA-binding element requires dynamic action. A classic conundrum within the transcription field has been the observation that many DNA motifs are not utilized even when they are near consensus sequences with close proximity to transcription start sites [164–168]. Although many reasons have been postulated to account for this phenomenon including incomplete activation of a TF or a need to dimerize with a partner, these explanations do not account for the broad absence of binding. For example, the human genome contains thousands of potential glucocorticoid-binding elements, yet only a few hundred are apparently utilized in any given cell type [169]. While many regulatory features contribute to these patterns, including the state of the chromatin encompassing each site [170], the ability of the p23 molecular chaperone to broadly impede DNA-binding activities indicates that numerous transcription factors are actively removed from their motifs. Supporting this conclusion is the finding that many transcription factors inappropriately occupy their cognate sites in the absence of the p23 global regulator [81]. Hence, a prominent means to prevent spurious transcription factor-DNA interactions is the active removal of the DNA-binding proteins by p23.

Intriguingly, however, p23 is not necessarily a negative regulator of DNA-binding events. Studies focusing on either estrogen receptor beta (ERβ) or androgen receptor (AR) demonstrate that the gene promoter occupancies of these SHRs could increase following an elevation in p23 levels [171, 172]. In addition, p23 overexpression enhances both basal and ligand-induced peroxisome proliferator-activated receptor (PPAR) transcription [173]. Whether these effects are through a direct change in the DNA-binding function of a receptor or through an indirect mechanism (e.g., changes to open chromatin exposing additional receptor response elements) is yet to be revealed.

We believe a broad use of the p23's nuclear activities is to foster the dynamics of protein DNA complexes. Prior studies have shown that the Hsp90 and p23 chaperones facilitate DNA binding and release for a variety of proteins including transcription factors and telomerase [130]. Certainly there is a need for rapid action by DNA-binding proteins as most genome-related pathways utilize multiple steps to function. For instance, both transcription initiation and DNA repair pathways employ a series of histone acetyltransferase- and chromatin remodeling-complexes to clear the chromatin and expose the DNA for RNA polymerase or DNA repair enzymes, respectively [174]. Other events at or near DNA including DNA replication and telomere maintenance also utilize multi-step pathways to function [175, 176]. In our model, DNA-associated cellular processes move forward using high affinity interactions between low abundant proteins, which are specific for a particular path, and transitions between the diverse structures are mediated by transient, low affinity interactions with the highly abundant p23 molecular chaperone.

6 Hijacking of p23 by Viruses and Parasites

One indicator of a protein's central importance is the usurping of that factor by a virus or parasite to maintain a pest's life cycle. An early observation on p23 demonstrated that the Hepatitis B virus, a hepadnavirus that can cause liver cancer, packages the host's p23 and Hsp90 chaperones into the viral nucleocapsids along with its RNA genome [177]. Both p23 and Hsp90 bind independently to the polymerase, which is essential to convert the RNA strand into DNA upon entry into a host cell [177].

Nearly one third of the human population is chronically infected with *Toxoplasma gondii* [178]. During infection *T. gondii* undergoes a transition from a rapidly replicating, infectious tachyzoite stage to a slower dividing encysted bradyzoite phase. The stage conversion is critical to the pathogenesis and longevity of infection. Interestingly, p23 transitions from a solely cytosolic protein in tachyzoite to being evenly distributed in the nucleoplasm and cytosol in bradyzoite period [179]. Hsp90's pattern mirrors that of p23's. Of note, the altered localization of p23 and Hsp90, as the cells transition to a more quiescent growth phase, parallels observations in budding yeast [180]. Unfortunately, the functional purpose of the redistribution in either *T. gondii* or *Saccharomyces cerevisiae* is yet to be resolved.

In *Plasmodium falciparum*, which causes a severe form of malaria, the p23 homolog also undergoes changes during the parasites life cycle. In the early ring and trophozoite stages (actively dividing within blood cells), p23 is expressed but its levels decline at the beginning of schizogony (vector production for entering the mosquito) [181]. While the relevance of the change is not known, the results parallel fission yeast data showing a decline in p23 levels as yeast enter stationary phase [157]. Comparable to the initial studies on mammalian p23, *P. falciparum* p23 does not form a stable interaction with *P. falciparum* Hsp90 in cell extracts [36, 182]. Yet, the purified proteins can assemble and p23 does suppress Hsp90's ATPase in vitro [182]. As *P. falciparum* p23 has independent chaperone activity [182], it too may function autonomously of Hsp90 in live parasites.

7 Cancer Relevance of p23

At the clinical level, cancer includes a diverse set of illnesses that share a common feature of inappropriate cell growth and at the molecular level an accumulation of unacceptable genetic alterations [183, 184]. Almost certainly cancerous growth results from the buildup of DNA mutations that eventually initiate and drive the transformation process. In addition to genomic alterations, significant changes in gene expression and epigenome programs often are observed along with increased activities in many enzymes—telomerase being a notable example [185–188].

Although the p23 molecular chaperone has not been shown to be a causative agent for any cancer, increased expression of p23 correlates with cellular transformation, increased cancer cell invasiveness, and reduced prognosis for cancer patients [162, 189–192]. Significantly, p23 influences genome stability and regulates numerous key factors associated with cancer including nuclear hormone receptors and telomerase [79, 81, 97, 158, 159, 161, 162]. In addition, p23 has been linked to endoplasmic reticulum-stress induced apoptosis [193–196]. While a causative association has not been made, mitochondrial RNA (miRNA) mediated upregulation of p23 can act as an anti-apoptotic factor in childhood acute lymphoblastic leukemia [197].

A promising new report identifies Gedunin as a potential small molecule inhibitor of p23 [198]. Although Gedunin had been suggested to inactivate Hsp90 [199, 200], the Patwardhan study reveals p23 as the primary target. Significantly, Gedunin is a more potent growth inhibitor of cancerous lines relative to non-transformed cells. Besides the benefits of directly inhibiting p23, targeting p23 might enhance the effectiveness of Hsp90 inhibitors since p23 has a protective effect against Hsp90 drugs [80]. Hence, strategies targeting p23 have significant therapeutic promise.

8 Summary

Homeostasis requires cellular processes to function efficiently and selectively within the crowded milieu of the cell [142]. While cooperative interactions drive rapid and precise assembly, the inherent stability of such organized structures interferes with the timing of biological systems. Further complicating performance is the nature of the cell interior, it is densely packed and often contains multiple binding partners for each protein—both features increase non-productive or off-pathway interactions. These variables present great challenges for achieving homeostasis especially in the midst of fluctuating internal and external stimuli that must be monitored constantly to appropriately initiate, continue or halt cellular pathways [31]. Hence, biological complexes must be actively and persistently disassembled in order to work on a useful time scale. We suggest that the apparent broad binding specificity and energy-independent chaperone activity provided by the CS/ACD domain has been exploited to allow p23 to govern the kinetic-behavior of the heterologous proteins within a cell [139]. In our view, p23 resolves inherently stable cooperative-complexes into dynamic machinery capable of rapid action that enables efficient and timely biological pathways [142].

References

1. Ellis JR (2006) Molecular chaperones: assisting assembly in addition to folding. Trends Biochem Sci 31:395–401
2. Hartl FU, Bracher A, Hayer-Hartl M (2011) Molecular chaperones in protein folding and proteostasis. Nature 475:324–332
3. Henzler-Wildman K, Kern D (2007) Dynamic personalities of proteins. Nature 450:964–972
4. Mitternacht S, Berezovsky IN (2011) Coherent conformational degrees of freedom as a structural basis for allosteric communication. PLoS Comput Biol 7:1–12
5. Anfinsen CB (1973) Principles that govern the folding of protein chains. Science 181:223–230
6. Deuerling E, Schulze-Specking A, Tomoyasu T et al (1999) Trigger factor and DnaK cooperate in folding of newly synthesized proteins. Nature 400:693–696
7. Chan HS, Dill KA (1998) Protein Folding in the landscape perspective: chevron plots and non-arrhenius kinetics. Proteins 30:2–33
8. Beckmann RP, Mizzen LA, Welch WJ (1990) Interaction of Hsp 70 with newly synthesized proteins: implications for protein folding and assembly. Science 248:850–854
9. Pfund C, Lopez-Hoyo N, Ziegelhoffer T (1998) The molecular chaperone Ssb from *Saccharomyces cerevisiae* is a component of the ribosome-nascent chain complex. EMBO J 17:3981–3989
10. Gautschi M, Lilie H, Fünfschilling U (2001) RAC, a stable ribosome-associated complex in yeast formed by the DnaK-DnaJ homologs Ssz1p and zuotin. Proc Natl Acad Sci U S A 98:3762–2767
11. McClellan AJ, Tam S, Kaganovich D (2005) Protein quality control: chaperones culling corrupt conformations. Nat Cell Biol 7:736–741
12. Eyles SJ, Gierasch LM (2010) Nature's molecular sponges: Small heat shock proteins grow into their chaperones roles. Proc Natl Acad Sci U S A 107:2727–2728
13. Pearse BMF (1975) Coated vesicles from pig brain: purification and biochemical characterization. J Mol Biol 97:93–98
14. Kirchhausen T (2000) Clathrin. Annu Rev Biochem 69:699–727
15. Braell WA, Schlossman DM, Schmid SL et al (1984) Dissociation of clathrin coats coupled to the hydrolysis of ATP: role of an uncoating ATPase. J Cell Biol 99:734–741
16. Ungewickell E, Ungewickell H, Holstein SEH et al (1995) Role of auxilin in uncoating clathrin-coated vesicles. Nature 378:632–635
17. Tsai MJ, O'Malley BW (1994) Molecular mechanisms of action of steroid/thyroid receptors superfamily members. Annu Rev Biochem 63:451–486

18. Mangelsdorf DJ, Thummel C, Beato et al (1995) The nuclear receptor superfamily: the second decade. Cell 83:835–839
19. Jeong Y, Mangelsdorf DJ (2009) Nuclear receptor regulation of stemness and stem cell differentiation. Exp Mol Med 41:525–537
20. McEwan IJ (2009) Nuclear receptors: one big family. In: McEwan IJ (ed) Methods in molecular biology: the nuclear receptor superfamily, vol 505. Humana, Totowa, pp 3–18
21. Stanišić V, Lonard DM, O'Malley BW (2010) Modulation of steroid hormone receptor activity. In: Martini L (ed) Progress in brain research, vol 181. Elsevier, Amsterdam, pp 153–176
22. Kadmiel M, Cidlowski JA (2013) Glucocorticoid receptor signaling in health and disease. Trends Pharmocol Sci 34:518–530
23. Stahn C, Löwenberg M, Hommes DW, et al (2007) Molecular mechanisms of glucocorticoid action and selective glucocorticoid receptor agonists. Mol Cell Endocrinol 275:71–78
24. Stingl J (2011) Estrogen and progesterone in normal mammary gland development and in cancer hormones. Cancer 2:85–90
25. Ahmadian M, Suh JM, Hah N et al (2013) PPARγ signaling and metabolism: the good, the bad and the future. Nat Med 19:557–566
26. Saylor PJ (2013) The androgen receptor remains front and centre. Nat Rev Clin Oncol 10:126–128
27. Zennaro MC, Rickard AJ, Boulkroun S (2013) Genetics of mineralocorticoid excess: an update for clinicians. Eur J Endocrinol 169:R15–R25
28. Renaud JP, Moras D (2000) Structural studies on nuclear receptors. Cell Mol Life Sci 57:1748–1769
29. Bain DL, Heneghan AF, Connaghan-Jones KD (2007) Nuclear receptor structure implications for function. Annu Rev Physiol 69:201–220
30. Grad I, Picard D (2007) The glucocorticoid responses are shaped by molecular chaperones. Mol Cell Endocrinol 275:2–12
31. Freeman BC, Yamamoto KR (2001) Continuous recycling: a mechanism for modulatory signal transduction. Trends Biochem Sci 26:285–290
32. Pratt WB, Toft DO (1997) Steroid receptor interactions with heat shock protein and immunophilin chaperones. Endocr Rev 18:306–360
33. Smith DF, Faber LE, Toft DO (1990) Purification of unactivated progesterone receptor and identification of novel receptor-associated proteins. J Biol Chem 265:3996–4003
34. Johnson JL, Beito TG, Krco CJ et al (1994) Characterization of a novel 23-kilodalton protein of unactive progesterone receptor complexes. Mol Cell Biol 14:1956–1963
35. Johnson JL, Toft DO (1994) A novel chaperone complex for steroid receptors involving heat shock proteins, immunophilins, and p23. J Biol Chem 269:24989–24993
36. Hutchison KA, Stancat LF, Owens-Grillo JK (1995) The 23-kDa acidic protein in reticulocyte lysate is the weakly bound component of the hsp foldosome that is required for assembly of the glucocorticoid receptor into a functional heterocomplex with Hsp90. J Biol Chem 270:18841–18847
37. Dittmar KD, Pratt WB (1997) Folding of the glucocorticoid receptor by the Reconstituted hsp90-based chaperone machinery: the initial hsp90-p60-hsp70-dependent step is sufficient for creating the steroid binding conformation. J Biol Chem 272:13047–13054
38. Dittmar KD, Hutchison KA, Owens-Grillo JK et al (1996) Reconstitution of the steroid receptor-Hsp90 heterocomplex assembly system of rabbit reticulocyte lysate. J Biol Chem 271:12833–12839
39. Kosano H, Stensgard B, Charlesworth MC et al (1998) The assembly of progesterone receptor-Hsp90 complexes using purified proteins. J Biol Chem 273:32973–32979
40. McLaughlin SH, Sobott F, Yao Z et al (2006) The co-chaperone p23 arrests the Hsp90 ATPase cycle to trap client proteins. J Mol Biol 356:746–758
41. Smith DF, Toft DO (2008) The intersection of steroid receptors with molecular chaperones: observations and questions. Mol Endocrinol 22:2229–2240
42. Smith DF, Schowalter DB, Kost SL et al (1990) Reconstitution of progesterone receptor with heat shock proteins. Mol Endocrinol 4:1704–1711

43. Smith DF, Toft DO (1992) Composition, assembly and activation of the avian progesterone receptor. J Steroid Biochem 41:201–207
44. Morishima Y, Kanelakis KC, Murphy PJM et al (2003) The Hsp90 Cochaperone p23 is the limiting component of the multiprotein Hsp90/Hsp70-based chaperone system *in Vivo* where it acts to stabilize the client protein-Hsp90 complex. J Biol Chem 278:48754–48763
45. Nair SC, Toran EJ, Rimerman RA et al (1996) A pathway of multi-chaperone interactions common to diverse regulatory proteins: estrogen receptor, Fes tyrosine kinase, heat shock transcription factor Hsf1, and the aryl hydrocarbon receptor. Cell Stress Chaperones 1:237–250
46. Cha JY, Ermawati N, Jung MH et al (2009) Characterization of orchardgrass p23, a flowering plant Hsp90 cohort protein. Cell Stress Chaperones 14:233–243
47. Bohen SP (1998) Genetic and biochemical analysis of p23 and ansamycin antibiotics in the function of Hsp90-dependent signaling proteins. Mol Cell Biol 18:3330–3339
48. Fang Y, Fliss AE, Rao J et al (1998) SBA1 encodes a yeast Hsp90 cochaperone that is homologous to vertebrate p23 proteins. Mol Cell Biol 18:3727–3734
49. Grad I, McKee TA, Ludwig SM et al (2006) The Hsp90 cochaperone p23 is essential for perinatal survival. Mol Cell Biol 26:8976–8983
50. Lovgren AK, Kovarova M, Koller BH (2007) cPGES/p23 is required for glucocorticoid receptor function and embryonic growth but not prostaglandin E_2 synthesis. Mol Cell Biol 27:4416–4430
51. Nakatani Y, Hokonohara Y, Kakuta S et al (2007) Knockout mice lacking the cPGES/p23, a constitutively expressed PGE_2 synthetic enzyme are peri-natally lethal. Biochem Biophys Res Commun 362:387–392
52. Prodromou C, Panaretou B, Chohan S et al (2000) The ATPase cycle of Hsp90 drives a molecular 'clamp' via transient dimerization of the N-terminal domains. EMBO J 19:4383–4392
53. Ali MMU, Roe SM, Vaughan CK et al (2006) Crystal structure of an Hsp90-nucleotide-p23/Sba1 closed chaperone complex. Nature 440:1013–1017
54. Minami Y, Kimura Y, Kawasaki H et al (1994) The carboxy-terminal region of mammalian Hsp90 is required for its dimerization and function in vivo. Mol Cell Biol 14:1459–1464
55. Sullivan WP, Owen BAL, Toft DO (2002) The Influence of ATP and p23 on the conformation of hsp90. J Biol Chem 277:45942–45948
56. Zhu S, Tytgat J (2004) Evolutionary epitopes of Hsp90 and p23: implications for their interaction. FASEB J 18:940–947
57. McLaughlin SH, Smith HW, Jackson SE (2002) Stimulation of the weak ATPase activity of human Hsp90 by a client protein. J Mol Biol 315:787–798
58. Böse S, Weikl T, Bügl H, et al (1996) Chaperone function of Hsp90-associated proteins. Science 274:1715–1717
59. Freeman BC, Toft DO, Morimoto RI (1996) Molecular chaperone machines: chaperone activities of the cyclophilin Cyp-40 and the steroid aporeceptor-associated protein p23. Science 274:1718–1720
60. Weikl T, Abelmann K, Buchner J (1999) An unstructured C-terminal region of the Hsp90 cochaperone p23 is important for its chaperone function. J Mol Biol 293:685691
61. Weaver AJ, Sullivan WP, Felts SJ et al (2000) Crystal structure and activity of human p23, a heat shock protein 90 co-chaperone. J Biol Chem 275:23045–23052
62. Kim KK, Kim R, Kim SH (1998) Crystal structure of a small heat-shock protein. Nature 394:595–599
63. Poulain P, Gelly JC, Flatters D (2010) Detection and architecture of small heat shock protein monomers. PLoS ONE 5:1–10
64. Haslbeck M, Franzmann T, Weinfurtner D et al (2005) Some like it hot: the structure and function of small heat-shock proteins. Nat Struct Mol Biol 12:842–846
65. Haslbeck M, Walke S, Stromer T et al (1999) Hsp26: a temperature-regulated chaperone. EMBO J 18:6744–6751
66. Wotton D, Freeman K, Shore D (1996) Multimerization of Hsp42, a novel heat shock protein of *Saccharomyces cerevisiae*, is dependent on a conserved carboxyl-terminal sequence. J Biol Chem 271:2717–2723

67. Haslbeck M, Braun N, Stromer et al (2004) Hsp42 is the general small heat shock protein in the cytosol of *Saccharomyces cerevisiae*. EMBO J 23:638–649
68. Kriehuber T, Rattei T, Weinmaier T et al (2010) Independent evolution of the core domain and its flanking sequences in small heat shock proteins. FASEB J 24:3633–3642
69. Chen J, Feige MJ, Franzmann TM et al (2010) Regions outside the α-crystallin domain of the small heat shock protein Hsp26 are required for its dimerization. J Mol Biol 398:122–131
70. Garcia-Ranea JA, Mirey G, Camonis J et al (2002) p23 and HSP20/α-crystallin proteins define a conserved sequence domain present in other eukaryotic protein families. FEBS Lett 529:162–167
71. Takahashi A, Casais C, Ichimura K et al (2003) Hsp90 interacts with RAR1 and SGT1 and is essential for RPS2-mediated disease resistance in *Arabidopsis*. Proc Natl Acad Sci U S A 100:11777–11782
72. Liu Y, Burch-Smith T, Schiff M et al (2004) Molecular chaperone Hsp90 associates with resistance protein N and its signaling proteins SGT1 and Rar1 to modulate an innate immune response in plants. J Biol Chem 279:2101–2108
73. Hubert DA, Tornero P, Belkhadir Y et al (2003) Cytosolic Hsp90 associates with and modulates the *Arabidopsis* RPM1 disease resistance protein. EMBO J 22:5679–5689
74. Lee YT, Jacob J, Michowski W et al (2004) Human Sgt1 binds Hsp90 through the CHORD-Sgt1 domain and not the tetratricopeptide repeat domain. J Biol Chem 279:16511–16517
75. Sbroggiò M, Ferretti R, Percivalle E et al (2008) The mammalian CHORD-containing protein melusin is a stress response protein interacting with Hsp90 and Sgt1. FEBS Lett 582:1788–1794
76. Deng B, Parthasarathy S, Wang W et al (2010) Study of the individual cytochrome b_5 and cytochrome b_5 reductase domains of Ncb5Or reveals a unique heme pocket and a possible role of the CS domain. J Biol Chem 285:30181–30191
77. Sbroggiò M, Bertero A, Velasco S et al (2011) ERK1/2 activation in heart is controlled by melusin focal adhesion kinase and the scaffold protein IQGAP1. J Cell Sci 124:3515–3524
78. Kálmán FS, Lizák B, Nagy SK, et al (2013) Natural mutations lead to enhanced proteasomal degradation of human Ncb5or, a novel flavoheme reductase. Biochimie 95:1403–1410
79. Toogun OA, Zeiger W, Freeman BC (2007) The p23 molecular chaperone promotes functional telomerase complexes through DNA dissociation. Proc Natl Acad Sci USA 104:5765–5770
80. Forafonov F, Toogun OA, Grad I et al (2008) p23/Sba1p Protects against Hsp90 inhibitors independently of its intrinsic chaperone activity. Mol Cell Biol 28:3446–3456
81. Zelin E, Zhang Y, Toogun OA et al (2012) The p23 molecular chaperone and GCN5 acetylase jointly modulate protein-DNA dynamics and open chromatin status. Mol Cell 48:1–12
82. Osmani AH, Osmani SA, Morris NR (1990) The molecular cloning and identification of a gene product specifically required for nuclear movement in *Aspergillus nidulans*. J Cell Biol 111:543–551
83. Zheng M, Cierpicki T, Burdette AJ et al (2011) Structural features and chaperone activity of the NudC protein family. J Mol Biol 409:722–741
84. Gocke CD, Reaman GH, Stine C et al (2000) The nuclear migration gene NudC and human hematopoiesis. Leuk Lymphoma 39:447–454
85. Riera J, Lazo PS (2009) The mammalian *NudC*-like genes: a family with functions other than regulating nuclear distribution. Cell Mol Life Sci 66:2383–2390
86. Toba S, Hirotsune S (2012) A unique role of dynein and nud family proteins in corticogenesis. Neuropathology 32:432–439
87. Zhu XJ, Liu X, Jin Q et al (2010) The L279P mutation of nuclear distribution gene C (NudC) influences its chaperone activity and lissencephaly protein 1 (Lis1) stability. J Biol Chem 285:29903–29910
88. Kitagawa K, Skowyra D, Elledge SJ et al (1999) *SGT1* encodes an essential component of the yeast kinetochore assembly pathway and a novel subunit of the SCF ubiquitin ligase complex. Mol Cell 4:21–33
89. Azevedo C, Sadanandom A, Kitagawa K et al (2002) The RAR1 interactor SGT1, an essential component of *R* gene-triggered disease resistance. Science 295:2073–2076

90. da Silva Correia JM, Leonard N et al (2007) SGT1 is essential for Nod1 activation. Proc Natl Acad Sci U S A 104:6764–6769
91. Mayor A, Martinon F, De Smedt T et al (2007) A crucial function of SGT1 and Hsp90 in inflammasome activity links mammalian and plant innate immune responses. Nat Immunol 8:497–503
92. Spiechowicz M, Zylicz A, Bieganowski P et al (2007) Hsp70 is a new target of Sgt1—an interaction modulated by S100A6. Biochem Biophys Res Commun 357:1148–1153
93. Botër M, Amigues B, Peart J et al (2007) Structural and functional analysis of SGT1 reveals that its interaction with Hsp90 is required for the accumulation of Rx, an R protein involved in plant immunity. Plant Cell 19:3791–3804
94. Stuttmann J, Parker JE, Noël LD (2008) Staying in the fold. Plant Signal Behav 3:283–285
95. Catlett MG, Kaplan KB (2006) Sgt1p is a unique co-chaperone that acts as a client adaptor to link Hsp90 to Skp1p. J Biol Chem 281:33739–33748
96. Żabka M, Leśniak W, Prus W et al (2008) Sgt1 has co-chaperone properties and is up-regulated by heat shock. Biochem Biophys Res Commun 370:179–183
97. Freeman BC, Felts SJ, Toft DA et al (2000) The p23 molecular chaperones act at a late step in intracellular receptor action to differentially affect ligand efficacies. Gene Dev 14:422–434
98. Nawaz MH, Merriman E, Yang XL et al (2011) p23^{H} implicated as *cis/trans* regulator of AlaXp-directed editing for mammalian cell homeostasis. Proc Natl Acad Sci U S A 108:2723–2328
99. Millson SH, Truman AW, Wolfram F et al (2004) Investigating the protein-protein interactions of the yeast Hsp90 chaperone system by two-hybrid analysis: potential uses and limitations of this approach. Cell Stress Chaperones 9:359–368
100. Millson SH, Truman AW, King V et al (2005) A two-hybrid screen of the yeast proteome for Hsp90 interactors uncovers a novel Hsp90 chaperone requirement in the activity of a stress-activated mitogen-activated protein kinase, Slt2p (Mpk1p). Eukaryot Cell 4:849–860
101. Kalia SK, Kalia LV, McLean PJ (2010) Molecular chaperones as rational drug targets for parkinson's disease therapeutics. CNS Neurol Disord Drug Targets 9:741–753
102. Hartson SD, Matts RL (2012) Approaches for defining the Hsp90-dependent proteome. BBA-Mol Cell Res 1823:656–667
103. Neckers L, Workman P (2012) Hsp90 molecular chaperone inhibitors: are we there yet? Clin Cancer Res 18:64–76
104. Patki JM, Pawar SS (2013) Hsp90: chaperone-me-not. Pathol Oncol Res 19:631–640
105. Joyce AR, Palsson B (2006) The model organism as a system integrating 'omics' data sets. Nat Rev Mol Cell Biol 7:198–210
106. Fraser JS, Gross JD, Krogan NJ (2013) From systems to structure: bridging networks and mechanism. Mol Cell 49:222–231
107. Tong AHY, Evangelista M, Parsons AB et al (2001) Systematic genetic analysis with ordered arrays of yeast deletion mutants. Science 294:2364–2368
108. Ho Y, Gruhler A, Hellbut A et al (2002) Systematic identification of protein complexes in *Saccharomyces cerevisiae* by mass spectrometry. Nature 415:180–183
109. Kreutzberger J (2006) Protein microarrays: a chance to study microorganisms? Appl Microbiol Biotechnol 70:383–390
110. Kawasumi M, Nghiem P (2007) Chemical genetics: elucidating biological systems with small-molecule compounds. J Invest Dermatol 127:1577–1584
111. Gong Y, Zhang Z, Houry WA (2011) Bioinformatic approach to identify chaperone pathway relationship from large-scale interaction networks. In: Calderwood SK, Prince TL (eds) Molecular chaperones: methods and protocols, methods in molecular biology. Springer, New York, pp 189–203
112. Enserink JM (2012) Chemical genetics: budding yeast as a platform for drug discovery and mapping of genetic pathways. Molecules 17:9258–9273

113. Zhao R, Davey M, Hsu YC et al (2005) Navigating the chaperone network: an integrative map of physical and genetic interactions mediated by the Hsp90 chaperone. Cell 120:715–727
114. McClellan AJ, Xia Y, Deutschbauer AM, Davis RW et al (2007) Diverse cellular functions of the Hsp90 molecular chaperone uncovered using systems approaches. Cell 131:121–135
115. Echtenkamp FJ, Zelin E, Oxelmark E et al (2011) Global functional map of the p23 molecular chaperone reveals an extensive cellular network. Mol Cell 43:229–241
116. Knoblauch R, Garabedian MJ (1999) Role for Hsp90-associated cochaperone p23 in estrogen receptor signal transduction. Mol Cell Biol 19:3748–3759
117. Picard D, Suslova E, Briand PA (2006) 2-color photobleaching experiments reveal distinct intracellular dynamics of two components of the Hsp90 complex. Exp Cell Res 312:3949–3958
118. Freeman BC, Unpublished Data
119. Echtenkamp FJ, Freeman BC (2012) Expanding the cellular molecular chaperone network through the ubiquitous cochaperones. BBA-Mol Cell Res 1823:668–673
120. Shah N, Klausner RD (1993) Brefeldin a reversibly inhibits secretion in *Saccaromyces cerevisiae*. J Biol Chem 268:5345–5348
121. Jackson CL, Casanova JE (2000) Turning on ARF: the Sec7 family of guanine-nucleotide exchange factors. Trends Cell Biol 10:60–67
122. Jahn R, Scheller RH (2006) SNAREs-engines for membrane fusion. Nat Rev Mol Cell Biol 7:631–643
123. Cai H, Reinisch K, Ferro-Novick S (2007) Coats, tethers, rabs, and SNAREs work together to mediate the intracellular destination of a transport vesicle. Dev Cell 12:671–682
124. Chen YA, Scheller RH (2001) SNARE-mediated membrane fusion. Nat Rev Mol Cell Biol 2:98–106
125. Duden R, Kajikawa L, Wuestehube L et al (1998) ε-COP is a structural component of coatomerthat functions to stabilize α-COP. EMBO J 17:985–995
126. Sato TK, Darsow T, Emr S (1998) Vam7p a SNAP-25-like molecule, and Vam3p, a syntaxin homolog, function together in yeast vacuolar protein trafficking. Mol Cell Biol 18:5308–5319
127. Ungermann C, Wickner W (1998) Vam7p, a vacuolar SNAP-25 homolog, is required for SNARE complex integrity and vacuole docking and fusion. EMBO J 17:3269–3276
128. Pratt WB (1993) The role of heat shock proteins in regulating the function, folding, and trafficking of the glucocorticoid receptor. J Biol Chem 268:21455–21458
129. Sanchez ER, Toft DO, Schlesinger MJ et al (1985) Evidence that the 90-kDa phosphoprotein associated with the untransformed L-cell glucocorticoid receptor is a muring heat shock protein. J Biol Chem 260:12398–12401
130. DeZwaan DC, Freeman BC (2010) Hsp90 manages the ends. Trends Biochem Sci 35:384–391
131. Misteli T (2001) Protein dynamics: implications for nuclear architecture and gene expression. Science 291:843–847
132. Stasevich TJ, McNally JG (2011) Assembly of the transcription machinery: ordered and stable, random and dynamic, or both? Chromosoma 120:533–545
133. Dundr M, Misteli T (2001) Functional architecture in the cell nucleus. Biochem J 356:297–310
134. Phair RD, Misteli T (2000) High mobility of proteins in the mammalian cell nucleus. Nature 404:604–609
135. Houtsmuller AB, Rademakers S, Nigg AL (1999) Action of DNA repair endonuclease ERCC1/XPF in living cells. Science 284:958–961
136. McNally JG, Müller WG, Walker D (2000) The glucocorticoid receptor: rapid exchange with regulatory sites in living cells. Science 287:1262–1265
137. Phair RD, Scaffidi P, Elbi C et al (2004) Global nature of dynamic protein-chromatin interactions in vivo: three-dimensional genome scanning and dynamic interaction networks of chromatin proteins. Mol Cell Biol 24:6393–6402

138. Gorski SA, Snyder SK, John S et al (2008) Modulation of RNA polymerase assembly dynamics in transcriptional regulation. Mol Cell 30:486–497
139. DeZwaan DC, Freeman BC (2008) Hsp90 the rosetta stone for cellular protein dynamics? Cell Cycle 7:1006–1012
140. Hager GL, McNally JG, Misteli T (2009) Transcription dynamics. Mol Cell 35:741–753
141. Hübner MR, Spector DL (2010) Chromatin dynamics. Ann Rev Biophys 39:471–489
142. Echtenkamp FJ, Freeman BC (2013) Molecular chaperone-mediated nuclear protein dynamics. Curr. Protein Pept. Sci., 15, 216–224
143. Freeman BC, Morimoto RI (1996) The human cytosolic molecular chaperones Hsp90, Hsp70 (hsc70) and hdj-1 have distinct roles in recognition of a non-native protein and protein refolding. EMBO J 15:2969–2979
144. Hayer-Harl MK, Weber F, Hartl FU (1996) Mechanism of chaperonin action: GroES binding and release can drive GroEL-mediated protein folding in the absence of ATP hydrolysis. EMBO J 15:6111–6121
145. Yochem J, Uchida H, Sunshine M et al (1978) Genetic analysis of two genes, *dnaJ* and *dnaK*, necessary for *Escherichia coli* and bacteriophage lambda DNA replication. Mol Genet Genomics 164:9–14
146. LeBowitz JH, Zylicz M, Georgopoulos C et al (1985) Initiation of DNA replication on single-stranded DNA templates catalyzed by purified replication proteins of bacteriophage λ and *Escherichia coli*. Proc Natl Acad Sci U S A 82:3988–3992
147. Liberek K, Georgopoulos C, Zylicz M (1988) Role of the *Escherichia coli* DnaK and DnaJ heat shock proteins in the initiation of bacteriophage λ DNA replication. Proc Natl Acad Sci U S A 85:6632–6636
148. Shaknovich R, Shue G, Kohtz S (1992) Conformational activation of a basic helix-loop-helix protein (MyoD1) by the C-terminal region of murine Hsp90 (HSP84). Mol Cell Biol 12:5059–5068
149. Shue G, Kohtz DS (1994) Structural and functional aspects of basic helix-loop-helix protein folding by heat-shock protein 90. J Biol Chem 269:2707–2711
150. Antonsson C, Whitelaw ML, McGuire J et al (1995) Distinct roles of the molecular chaperone hsp90 in modulating dioxin receptor function via the basic helix-loop-helix and PAS domains. Mol Cell Biol 15:756–765
151. Hur E, Kim HH, Choi SM et al (2002) Reduction of hypoxia-induced transcriptino through the repression of hypoxia-inducible factor-1α/Aryl hydrocarbon receptor nuclear translocator DNA binding by the 90-kDa heat-shock protein inhibitor radicicol. Mol Pharmacol 62:975–982
152. Müller L, Schaupp A, Walerych D et al (2004) Hsp90 regulates the activity of wild type p53 under physiological and elevated temperatures. J Biol Chem 279:48846–48854
153. Walerych D, Kudla G, Gutkowska M et al (2004) Hsp90 chaperones wild-type p53 tumor suppressor protein. J Biol Chem 279:48836–48845
154. Stavreva DA, Müller WG, Hager GL, et al (2004) Rapid glucocorticoid receptor exchange at a promoter is coupled to transcription and regulated by chaperones and proteasomes. Mol Cell Biol 24:2682–2697
155. Zhang X, Verdine GL (1996) Mammalian DNA cytosine-5 methyltransferase interacts with p23 protein. FEBS Lett 392:178–183
156. Bharadwaj S, Ali A, Ovsenek N (1999) Multiple component of the HSP90 chaperone complex function in regulation of heat shock factor 1 in vivo. Mol Cell Biol 19:8033–8041
157. Muñoz MJ, Bejarano ER, Daga RR et al (1999) The identification of Wos2, a p23 homologue that interacts with Wee1 and Cdc2 in the mitotic control of fission yeast. Genetics 153:1561–1572
158. Ouspenski II, Elledge SJ, Brinkley BR (1999) New yeast genes important for chromosome integrity and segregation identified by dosage effects on genome stability. Nucleic Acids Res 27:3001–3008
159. Freeman BC, Yamamoto KR (2002) Disassembly of transcriptional regulatory complexes by molecular chaperones. Science 296:2232–2235

160. Bikle D, Teichert A, Hawker N et al (2007) Sequential regulation of keratinocyte differentiation by 1,25(OH)$_2$D$_3$, VDR, and its coregulators. J Steroid Biochem 103:396–404
161. Holt SE, Aisner DL, Baur J et al (1999) Functional requirement of p23 and Hsp90 in telomerase complexes. Gene Dev 13:817–826
162. Forsythe HL, Jarvis JL, Turner JW et al (2001) Stable association of hsp90 and p23, but Not hsp70, with active human telomerase. J Biol Chem 276:15571–15574
163. Hesselberth JR, Chen X, Zhang Z et al (2009) Global mapping of protein-DNA interactions *in vivo* by digital genomic footprinting. Nat Methods 6:283–289
164. Carr A, Biggin MD (1999) A comparison of *in vivo* and *in vitro* DNA-binding specificities suggest a new model for homeoprotein DNA binding in *Drosophila* embryos. EMBO J 18:1598–1608
165. Iyer VR, Horak CE, Scafe CS et al (2001) Genomic binding sites of the yeast cell-cycle transcription factors SBF and MBF. Nature 409:533–538
166. Yang A, Zhu Z, Kapranov P et al (2006) Relationships between p63 binding DNA sequence, transcription activity, and biological function in human cells. Mol Cell 24:593–602
167. Joseph R, Orlov YL, Huss M et al (2010) Integrative model of genomic factors for determining binding site selection by estrogen receptor-α. Mol Syst Biol 456:1–13
168. Kaplan T, Li XY, Sabo PJ et al (2011) Quantitative models of the mechanisms that control genome-wide patterns of transcription factor binding during early *Drosophila* development. PLoS Genet 7:1–15
169. Wang JC, Derynck MK, Nonaka DF et al (2004) Chromatin immunoprecipitation (ChIP) scanning identifies primary glucocorticoid receptor target genes. Proc Natl Acad Sci U S A 101:15603–15608
170. John S, Sabo PJ, Thurman RE et al (2011) Chromatin accessibility pre-determines glucocorticoid receptor binding patterns. Nat Genet 43:264–270
171. Reebye V, Cano LQ, Lavery DN et al (2012) Role of the HSP90-associated cochaperone p23 in enhancing activity of the androgen receptor and significance for prostate cancer. Mol Endocrinol 26:1694–1706
172. Simpson NE, Gertz J, Imberg K et al (2012) Research resource: enhanced genome-wide occupancy of estrogen receptor α by the cochaperone p23 in breast cancer cells. Mol Endocrinol 26:194–202
173. Sumanasekera WK, Tien ES, Davis JW et al (2003) Heat shock protein-90 (Hsp90) acts as a repressor of peroxisome proliferator-activated receptor-α (PPARα) and PPARβ activity. Biochemistry 42:10726–10735
174. Lee KK, Workman JL (2007) Histone acetyltransferase complexes: one size doesn't fit all. Nat Rev Mol Cell Biol 8:284–295
175. Bell SP, Dutta A (2002) DNA replication in eukaryotic cells. Annu Rev Biochem 71:333–374
176. Gilson E, Géli V (2007) How telomeres are replicated. Nat Rev Mol Cell Biol 8:825–838
177. Hu J, Toft DO, Seeger C (1997) Hepadnavirus assembly and reverse transcription require a multi-component chaperone complex which is incorporated into nucleocapsids. EMBO J 16:59–68
178. Halonen SK, Weiss LM (2013) Toxoplasmosis. In: Garcia HH, Tanowitz HB, Del BOH (eds) Handbook of clinical neurology, 3rd series. Elsevier, Amsterdam, pp 125–145
179. Echeverria PC, Figueras MJ, Vogler M et al (2010) The Hsp90 co-chaperone p23 of *Toxoplasma gondii*: identification, functional analysis and dynamic interactome determination. Mol Biochem Parasit 172:129–140
180. Tapia H, Morano K (2010) Hsp90 nuclear accumulation in quiescence is linked to chaperone function and spore development in yeast. Mol Biol Cell 21:63–72
181. Wiser MF (2003) A *Plasmodium* homologue of cochaperone p23 and its differential expression during the replicative cycle of the malaria parasite. Parasitol Res 90:166–170
182. Chua CS, Low H, Goo KS et al (2010) Characterization of *Plasmodium falciparum* co-chaperone p23: its intrinsic chaperone activity and interaction with Hsp90. Cell Mol Life Sci 67:1675–1686

183. Luo J, Isaacs WB, Trent JM et al (2003) Looking beyond morphology: cancer gene expression profiling using DNA microarrays. Cancer Invest 21:937–949
184. Storchova Z, Pellman D (2004) From polyploidy to aneuplody, genome instability and cancer. Nat Rev Mol Cell Biol 5:45–54
185. Hermeking H (2003) Serial analysis of gene expression and cancer. Curr Opin Oncol 15:44–49
186. Meeker AK, De Marzo A (2004) Recent advances in telomere biology: implications for human cancer. Curr Opin Oncol 16:32–38
187. van't Veer LJ, Bernards R (2008) Enabling personalized cancer medicine through analysis of gene-expression patterns. Nature 452:564–570
188. Chi P, Allis CD, Wang GG (2010) Covalent histone modification-miswritten, misinterpreted and mis-erased in human cancers. Nat Rev Cancer 10:457–469
189. Boltze C, Schneider-Stock R, Roessner A et al (2003) Function of HSP90 and p23 in the telomerase complex of thyroid tumors. Pathol Res Pract 199:573–579
190. Mollerup J, Krogh TN, Nielsen PF et al (2003) Properties of the co-chaperone protein p23 erroneously attributed to ALG-2 (apoptosis-linked gene 2). FEBS Lett 555:478–482
191. Elmore LW, Forsythe R, Forsythe H et al (2008) Overexpression of telomerase-associated chaperone proteins in prostatic intraepithelial neoplasia and carcinomas. Oncol Rep 20:613–617
192. Simpson NE, Lambert WM, Watkins R et al (2010) High levels of Hsp90 cochaperone p23 promote tumor progression and poor prognosis in breast cancer by increasing lymph node metastases and drug resistance. Cancer Res 70:8446–8456
193. Rao RV, Niazi K, Mollahan P et al (2006) Coupling endoplasmic reticulum stress to the cell-death program: a novel HSP90-independent role for the small chaperone protein p23. Cell Death Differ 13:415–425
194. Chinta SJ, Rane A, Poksay KS et al (2008) Coupling endoplasmic reticulum stress to the cell death program in dopaminergic cells: effect of paraquat. Neuromol Med 10:333–342
195. Radanyi C, Le Bras G, Bouclier C et al (2009) Tosylcyclonovobiocic acids promote cleavage of the hsp90-associated cochaperone p23. Biochem Bioph Res Co 379:514–518
196. Poksay KS, Banwait S, Crippen D et al (2012) The small chaperone protein p23 and its cleaved product p19 in cellular stress. J Mol Neurosci 46:303–314
197. Liu X, Zou L, Zhu L et al (2012) miRNA mediated up-regulation of cochaperone p23 acts as an anti-apoptotic factor in childhood acute lymphoblastic leukemia. Leuk Res 36:1098–1104
198. Patwardhan CA, Fauq A, Peterson LB et al (2013) Gedunin inactivates the co-chaperone p23 protein causing cancer cell death by apoptosis. J Biol Chem 288:7313–7325
199. Hieronymus H, Lamb J, Ross KN et al (2006) Gene expression signature-based chemical genomic prediction identifies a novel class of HSP90 pathway modulators. Cancer Cell 10:321–330
200. Brandt GEL, Schmidt MD, Prisinzano TE et al (2008) Gedunin, a novel Hsp90 inhibitor: semisynthesis of derivatives and preliminary structure-activity relationships. J Med Chem 51:6495–6502

Part VI
Chaperone Networks in Organelles

Chapter 10
Chaperones in the Endoplasmic Reticulum (ER): Function and Interaction Network

Pekka Maattanen, Gregor Jansen, Guennadi Kozlov, Kalle Gehring and David Y. Thomas

Abstract The directional entry, oxidative folding, and quality control of proteins that enter the secretory pathway is mediated by chaperones and foldases in and adjacent to the endoplasmic reticulum (ER). Properly folded and assembled proteins continue along the secretory pathway while proteins that ultimately fail quality control are targeted to the proteasome by removal from the ER in a process called ER-associated degradation (ERAD). The protein folding machineries in the ER interact with each other to form functional complexes. Studies have revealed that abundant chaperones and foldases serve multiple functions in the ER through membership in diverse complexes that can target their activities to substrates at different stages of maturation. These findings are providing insight into how ER complexes combine various functions together to engage substrates and determine their fates.

P. Maattanen (✉) · G. Jansen · G. Kozlov · K. Gehring · D. Y. Thomas
Department of Biochemistry, McGill University, Montreal, QC, Canada
e-mail: pekka.maattanen@sickkids.ca

P. Maattanen
Department of Cell Biology, SickKids Peter Gilgan Centre for Research and Learning, Toronto, ON, Canada

G. Kozlov. K. Gehring
Groupe de Recherche axé sur la Structure des Protéines, McGill University, Montreal, QC, Canada

G. Kozlov
e-mail: guennadi.kozlov@mcgill.ca

K. Gehring
e-mail: kalle.gehring@mcgill.ca

G. Jansen
e-mail: gregor.jansen@mcgill.ca

D. Y. Thomas
e-mail: david.thomas@mcgill.ca

W. A. Houry (ed.), *The Molecular Chaperones Interaction Networks in Protein Folding and Degradation,* Interactomics and Systems Biology, DOI 10.1007/978-1-4939-1130-1_10,

Abbreviations

AAT	α-1 antitrypsin
AGR	Anterior gradient
AP	Affinity purification
APEX	Ascorbate peroxidase
ATP	Adenosine triphosphate
BAP	BiP-associated protein
CFTR	Cystic Fibrosis Transmembrane Conductance Regulator
CGHC	Cysteine-Glycine-Histidine-Cysteine
Co-IP	Co-immunoprecipitation
COP	Coat protein
CPHC	Cysteine-Proline-Histidine-Cysteine
CSMC	Cysteine-Serine-Methionine-Cysteine
EDEM	ER degradation enhancing mannosidase-like protein
EGF	Epidermal growth factor
ENaC	Epithelial sodium channel
ER	Endoplasmic reticulum
ER-MYTHS	Endoplasmic reticulum-membrane yeast two hybrid system
ERAD	Endoplasmic reticulum-associated degradation
FAD	Flavine adenine dinucleotide
FKPB	FK506 binding protein
GFP	Green fluorescent protein
Glc	Glucose
GlcNAc	N-acetylglucosamine
GPI	Glycophosphatidylinositol
HPD	Histidine-proline-aspartate
Hsp	Heat shock protein
ITC	Isothermal titration calorimetry
LDL	Low density lipoprotein
Man	Mannose
MAP	Membrane Yeast Two Hybrid System, Affinity Purification, NMR
MHC	Major Histocompatability Complex
MS	Mass spectrometry
NEF	Nucleotide exchange factor
NHK	Null Hong Kong
NMR	Nuclear magnetic resonance
NOX	Nicotinamide adenine dinucleotide oxidase
OST	Oligosaccharyltransferase
PCA	Protein complementation assay
PDB	Protein Data Bank
PDI	Protein disulfide isomerase
PDIr	Protein Disulfide Isomerase-related protein
PERK	PKR-like endoplasmic reticulum kinase
PPI	Peptidyl-prolyl *cis-trans* isomerase
PPI	Protein-protein Interaction
PRIME	Probe incorporation mediated by enzymes

Prx	Peroxiredoxin
RAMP	Ribosome associated membrane protein
shRNA	Short hairpin ribonucleic acid (RNA)
SIL 1	Suppressor of the Ire1/Lhs1 double mutant 1
SRP	Signal recognition particle
TRAP	Translocon-associated protein
UGGT	UDP-glucose:glycoprotein-glucosyltransferase
UPR	Unfolded protein response
Y2H	Yeast two-hybrid

1 Introduction

Recent studies have uncovered many novel interactions between proteins within the endoplasmic reticulum (ER). These advances have come through application of interaction detection methods in both yeast and mammalian studies. While studies of yeast ER protein-protein interactions (PPIs) provides a rational starting point to understand interactions of homologous mammalian ER proteins, it is not surprising that there are many more specialized mammalian ER chaperones and folding enzymes that do not have specific yeast counterparts, limiting the extent to which comparisons can be made. However, expression of mammalian proteins in yeast systems allows for simpler genetic manipulations and can uncover binary interactions. The yeast two-hybrid (Y2H) methods can be combined with affinity purification and mass spectrometry (AP-MS) methods and nuclear magnetic resonance (NMR) using mammalian proteins, and ideally, short hairpin ribonucleic acid (shRNA) methods to provide broader perspectives and functional insight. We discuss some of the recently identified interactions between the mammalian ER proteins with a focus on proteins involved in initial entry into the ER, protein folding, sorting, and ER-associated degradation (ERAD). Significant advances have been made in understanding ER membrane architecture [1], the proteins, and complexes involved in determining reticular shape [2] and fusion [3] and cytosolic components involved in ERAD [4]. We will focus on luminal chaperone interactions and their functional consequences.

2 Overview

Relatively few PPIs between ER resident chaperones were known until the discovery of the unfolded protein response and the link between ER protein folding and human disease. Historically, George Palade's pioneering work on this organelle demonstrated its role as the entry point for secreted proteins. We aim to trace, from the perspective of ER-protein complexes encountered, the maturation of a newly synthesized secretory protein from nascent chain entry into the ER to exit from the ER as either a correctly folded polypeptide, or a terminally misfolded protein targeted for retrotranslocation and proteasomal degradation. Along the way, we will highlight the latest complexes identified, and describe their functional contributions to the process of protein folding.

It is not currently clear how ER protein folding machineries may be spatially arranged for sequential protein maturation; many of the protein complexes we will discuss act simultaneously, so our journey from translocon to exit sites is heuristic, but we hope this narrative will offer food for thought to develop testable models of ER function. Furthermore, while it is not possible to understand the functions of all identified interactions, and it has been argued that non-functional interactions may be more common than appreciated [5], we will focus on interactions validated through various means that have been implicated both directly and indirectly in ER functions through additional evidences. We will also address less well-characterized interactions with a more philosophical discussion of possible functions for further investigation. We do not attempt to describe in detail all that is known regarding nascent protein entry until exit from the ER, as these have been extensively reviewed elsewhere (see a recent special issue on ER structure and function in Biochemica et Biophysica Acta, Vol. 1833 Issue 11), but rather highlight the novel interactions recently identified and discuss their place within the context of ER protein folding, quality control, and secretion. Particular emphasis is placed on interactions between chaperones and foldases of different functional annotations. We begin with a discussion of the methods used to identify interactions in the ER.

3 Methods for Studying Interactions

Protein interactions maps for cytosolic and nuclear proteins have been useful for the elucidation of the function of proteins and of protein complexes. These maps have relied on two hybrid systems that establish binary interactions and methods such as tandem affinity purification (TAP)-tagging and mass spectrometric identification of protein complexes. Interaction maps of membrane proteins have been more of a technical challenge, but robust methods are being developed. What is lacking is a toolbox of methods to study interactions of proteins of cellular compartments that compose organelles such as the ER.

Most ER proteins are membrane proteins that require both specific posttranslational modifications and the specific environmental control of the ER to be able to fold and function properly. These specific requirements have hindered many attempts using conventional methods to interrogate the properties of ER-localized proteins and their physical and functional interactions.

Tailoring existing approaches to the necessities of ER proteins has been essential to capture physical associations on the level of complexes and defined binary interactions. Table 10.1 summarizes some of the methods that have been used to identify protein interactions of ER proteins. Traditionally, ER proteins were first defined in complexes using co-immunoprecipitation (IP) with either unfolded substrate baits or functional interaction partners. Binding partners were identified with using available antibodies [6–8], and more recently, in high-throughput studies using affinity puri-

Table 10.1 Methods for studying interactions

ID	Details	References
CoIP/AP	Affinity purification	[6–8]
TAP-MS	Tandem-tag purification and mass spectrometry	[9, 10]
AP-cross	Affinity purification with cross-linker	[4, 6, 11]
Assembly	In vitro complex assembly	[13]
PCA	Protein complementation assay	[12]
Y2H	Yeast two hybrid	[13, 14]
OP-MS	Organelle purification and mass spectrometry	[15, 16]
Enzyme-shRNA	Functional assay and knockdown	[17]
APEX	Affinity purification, mass spectrometry, microscopy	[18]
ER-MAP	Affinity purification, ER-MYTHS, NMR, functional assay	[13]
Integrative mapping	Affinity purification, mass spectrometry, knockdown	[4]
Real time homeostasis	Microscopy, affinity purification	[19]
–	High-resolution electron microscopy, knockdown	[20, 21]
–	In vivo complex tagging	[22, 23]

fication with tandem tags in combination with mass spectrometry [9, 10]. The more static complexes can be readily identified, but the discovery of other more transient ones must be aided by stabilizing agents such as cross-linkers [4, 6, 11]. These newly identified complexes are then dissected into binary interactions using in vitro assembly studies with purified proteins [12], or different protein complementation assay (PCA) and two-hybrid technologies [12–14]. Where applicable, protein complexes are subjected to enzymatic characterization. In the proteomics area of ER research, mass spectrometry coupled to pull-down and knockdown approaches is being used to monitor changes in complex compositions and their functional consequences on ER processes such as ERAD [4]. In contrast to top-down approaches like affinity purification (AP) and Y2H, mass spectrometry of purified whole ER is used in a bottom-up approach to identify novel ER components [15, 16].

More recent developments in technologies for genetic manipulation have opened the research for the in-depth characterization of the mammalian ER function as it is known for model organisms like *Saccharomyces cerevisiae*. Next to transcription activator-like effector nucleases (TALENS) and clustered, regularly interspersed, short palindromic repeats (CRISPR) associated (Cas) manipulations to introduce very specific genetic changes in genomic DNA, shRNA has become a valuable tool in ER-Omics that allows the directed inactivation of single genes in mammalian genomes. These mutations can then be subjected to a multitude of functional tests. Rutkevitch et al. for example have used shRNA technology to study the functional relationship between different protein disulfide isomerases illustrating overlap in substrate specificities [17].

However it has become more and more evident that no single method is sufficient to study the complex organization of the ER, and research groups have now moved on to using combinatorial experimental approaches like ascrobate peroxidase (APEX)-protein fusions for electron microscopy, ER-MAP (ER-Membrane yeast two hybrid system and Affinity Purification, NMR), integrative mapping (ERAD using AP; mass spectronomy, MS; shRNA), and real-time homeostasis (ERdj3 using green fluorescent protein (GFP)-fusion microscopy, AP) [4, 13, 18, 19].

There have been several exciting recent methodological advances that have allowed more detailed study of the secretory pathway and relevant interactions. Coupling high-throughput high resolution light microscopy with electron microscopy and shRNA has allowed unprecedented details to be revealed including network structures [20, 21]. These approaches not only offer promising ways to validate hypotheses from less direct observations (AP-MS and Y2H) but also provide ways to visualize how multi-chaperone complexes may be organized within the ER in a single cell. Similarly, recent advances in nanotechnology have allowed microscopic visualization of molecular assemblies in living cells independent of fluorescence, opening up new avenues to study organelle architecture and organization [22]. Finally, a recently developed methodology that relies on probe incorporation mediated by enzymes (PRIME) has been adapted for the secretory pathway in yeast, and interaction-dependent coumarin probes have been developed to allow specific labeling of proteins of interest [23]. This technology requires genetic addition of a 13-mer peptide acceptor and expression of a ligase that couples the probe specifically to acceptor peptides, and has been applied to extra-cellular protein interactions. While these technologies have not yet been applied to protein-protein interactions in the ER, they provide additional tools to enhance the clarity of previously used combinatorial approaches in the future.

4 Protein Complexes of Nascent Chain Entry

4.1 SRP-Dependent Entry

Perhaps one of the most well studied steps in the process of protein folding and secretion through the ER is the entry of nascent chains into the ER [24–29], but this process has recently been updated with fresh data pertaining to alternative entry mechanisms [30, 31], and is far from completely understood. Nascent chain entry into the mammalian ER occurs primarily through recognition of signal peptides on ribosome-bound proteins destined for secretion by the signal recognition particle (SRP) [28] followed by recruitment of the proteins either co or posttranslationally to the SRP receptor (SR) docked at the translocation channel (translocon). The translocon is comprised of Sec61α, β, and γ subunits, and depending on its association with other proteins, it can function in either co-translational translocation [28], posttranslational translocation [32], and perhaps retro-translocation of proteins for ERAD [33, 34] although other proteins such as Derlins have been implicated and may be more suitable for this process [35] (ERAD will be revisited later in this chapter). The prokaryotic SecY has been posited to have a similar structure to that of eukaryotic Sec61, and structural insights into this pore indicate that during translocation, a small molecule barrier is maintained around the translocating peptide by amino acids lining the pore that form a "gasket-like" seal [36]. When the channel is not translocating peptides, it is plugged on the cytosolic side by a helix of the channel that shifts into a blocking position [36]. In eukaryotes, chaperones including the ER-resident Hsp70-like protein BiP aid the directional translocation of the nascent chain, and further translocation after cleavage of the signal sequence by the translocon-associated signal peptidase [37].

The Sec61 translocon is part of a much larger ribosome-anchored membrane protein (RAMP) complex that resolves by high resolution native electrophoresis into three distinct multicomponent complexes that at their core include the oligosaccharyltransferase complex (OST), glucosidase I, the microtubule tethering protein CLIMP63, and on the luminal side, the J-domain co-chaperone ERdj3 [11]. The larger two complexes contain the kinesin-motor-component-binding protein p180 and Sec61 while the largest complex contains the translocon associated protein (TRAP) complex and Bap31 (apoptotic regulator protein), as determined using proteomics and high-resolution native electrophoresis [11]. Interaction of ERdj3 with the Sec61α subunit of the translocon has been confirmed in separate Co-IP, native gel, and mobility studies [19]. The translocon complex also interacts in a less stable manner with the signal peptidase complex.

Importantly, in addition to ERdj3, there are several other translocon-associated J-domain co-chaperones of BiP that interact with BiP through hydroxyphenylpyruvate dioxygenase (HPD) motifs in their J-domains, and/or through other, still poorly understood mechanisms. These other J-domain containing co-chaperones include ERdj1, ERdj2/Sec63, and the Sec63 interacting protein Sec62. ERdj1 is a membrane protein that recruits BiP to ribosomes and also regulates translation [38, 39]. Sec63 contains three transmembrane domains with a luminal J domain and is found in stoichiometric amounts with the Sec61 alpha subunit in dog pancreatic microsomes [40]. While loss of function mutations of mammalian Sec63 are not lethal (as observed for Sec63p in yeast), they are associated with polycystic liver disease [41]. Sec62, a double-spanning membrane protein, associates with ribosomes and with Sec63, and has been recently shown to play a role in mediating membrane insertion and orientation of moderately hydrophobic signal anchor proteins in the ER [42]. Sec62 is also critical for SRP-independent translocation of short peptides (~ 160 amino acids) into the ER [43]. Although the precise functions of each mammalian translocon-associated co-chaperone are still being worked out, studies in yeast indicate that their ability to recruit BiP defines the specific roles of BiP in translocation [44].

New studies have shed light on the functions of the translocon and associated J-domain co-chaperones. A recent study in human cells examined gene silencing of Sec61α, Sec62, and Sec63 and its effects on growth and trafficking of proteins of different topologies [45]. Sec61α was found to be essential for growth (also lethal as expected), but analysis of time-points prior to cell death showed no defects in tail-anchored protein insertion while signal-peptide-dependent translocation was inhibited [45]. Silencing of Sec62 inhibited posttranslational transport of small presecretory proteins into the ER while silencing of Sec63 only affected a subset of signal-peptide containing precursor proteins, including Prion protein [45]. A subsequent gene silencing and overexpression study also in human cells found that Sec63 has a negative feedback role in multi-spanning membrane protein insertion that is independent of its interaction with Sec62 [46]. This suggested that Sec63 may play a quantity control function in transmembrane protein biosynthesis, and interestingly, this down-regulating function relied on HPD-mediated interaction with BiP. Perhaps the Sec63-BiP interaction functions to slow translocation of polytopic membrane proteins to prevent overload of the possibly complex chaperone-mediated folding of these proteins on both the ER and cytosolic side. The interaction may provide a feedback to slow translation to a rate that complies with chaperone availability. The role for BiP in protein translocation as a molecular ratchet through binding and release [37, 47] may be not only to maintain inward directional flow but also to control the rate, and perhaps in concert with Sec63 to slow it for more complex proteins. The J-domain co-chaperone and BiP interactions located at the translocon may represent one of the first places where chaperone interactions function to assure quality of products in the secretory pathway.

4.2 SRP Independent Entry

While SRP-independent mechanisms for nascent chain entry into the ER have been known to exist for some time [25, 29] there is now strong evidence for some of the machineries involved and possible mechanisms used, particularly in yeast [30–32]. These alternative entry methods utilize cytosolic factors and require chaperoning of the proteins prior to entry, efficient targeting to the ER, and directional entry of the nascent chains (reviewed in [48]). It has been suggested that alternative entry methods function to facilitate specific clients with signal sequences that are not optimal for SRP recruitment (lacking adequate hydrophobicity) and are unable or poorly able to bind the SRP receptor. Alternative entry methods may direct certain nascent chains to particular fates on the luminal side better suited for their particular folding needs; for example near the nucleus [30].

Regardless of entry point, even before the nascent protein fully enters the ER lumen, oxidative protein folding involving interactions with co-chaperones, chaperones, and foldases (some of which are translocon-associated) begins.

5 Early Chaperone Encounters

5.1 BiP and Its J-Domain Co-chaperones

Probably the first chaperone encountered in the ER lumen by most nascent chains is BiP, given its direct role in translocation. BiP is the ER-resident Hsp70 that functions to bind and release substrates through cycles of adenosine triphosphate (ATP) hydrolysis and nucleotide exchange. A family of J-domain co-chaperones (Erdj1–7) stimulate hydrolysis of ATP bound to BiP, and several nucleotide exchange factors (NEFs) including BiP-associated protein (BAP)/SIL1, and glucose regulated protein GRp170 help exchange adenosine diphosphate (ADP) to ATP, crucial for substrate release. Importantly, a rapid regulatory modification of BiP has been uncovered where ADP-ribosylation of BiP by an unknown ADP-ribosyltransferase acts to destabilize substrate binding, acting as a rapid brake to inhibit BiP-substrate interactions [49]. This rapid regulation of the substrate binding of BiP could be crucial to prevent aggregation of substrates at lower concentrations, and also allow more rapid folding without overzealous chaperone binding of less abundant substrates. ADP-ribosylation (of Arg470 and Arg492) of BiP can be rapidly reversed upon increased protein load [49], presumably by an as yet unidentified ADP-ribosylhydrolase. The future identification of the ER located ADP-ribosyltransferase and ribosylhydrolase will provide further insight into rapid temporal regulation of chaperone networks in the ER, which may be critical for optimal protein folding quality control in the ER.

As described above, ERdj1 and Erdj2/Sec63 are closely associated with the translocon, and, along with ERdj3, serve to recruit BiP to nascent chains as they emerge. Available evidence suggests that the other J-domain co-chaperones ERdj4, ERdj5, and ERdj6/DnajC3 (p58 IPK), and ERdj7 likely function as specialized adapter proteins to target BiP to specific substrates and/or functions in the ER. While ERdj3 can recruit BiP to newly synthesized proteins through direct interaction with unfolded protein regions [50], ERdj5 functions as a reductase to prepare substrates for ERAD in concert with ER degradation enhancing mannosidase like protein (EDEM)1 and BiP [51], and as a specific chaperone/foldase for the low-density lipoprotein (LDL) receptor [52]. The crystal structure and in vitro functional analysis of ERdj5 [53] revealed how it interacts with EDEM and BiP, and utilizes two highly reductive thioredoxin domains to facilitate reduction of disulfide bonds in proteins destined for ERAD. Structural insight into DnaJc3/ERdj6 (P58/IPK), that is strongly induced by the unfolded protein response, revealed that the putative nascent chain or unfolded protein-binding site is situated 100 Å away from the HPD motif that interacts with BiP, suggesting that handoff of substrates to BiP may involve structural re-arrangements of the J-domain [54].

It is also possible that co-chaperones with hydrophobic substrate-binding sites provide a distant alternative "safe-holding" site for exposed hydrophobic regions of substrate proteins in between cycles of binding and release by BiP. It is important to note that for at least one J-domain co-chaperone (ERdj3), and likely others, release of ERdj3 from substrates is in turn regulated by BiP, and only occurs in the presence of ATP [55]. This suggests that ERdjs could be interacting with clients and chaperoning them until they recruit BiP to engage in protein binding and release cycles that result in back and forth handoff of the protein until correctly folded. This relationship is more than a simple binding and release, as the J-domain co-chaperone ERdj3 can help open the lid of BiP to facilitate peptide binding in a nucleotide dependent manner [56]. The different domain architectures of the other co-chaperones might provide versatile adapters to facilitate productive binding and release of substrates with different folds.

Importantly, of the seven J-domain containing co-chaperones in the ER, 4 are induced by the unfolded protein response (UPR; ERdj3, ERdj4, ERdj5, and DNAJc3/ERdj6), suggesting that these have important roles in coping with unfolded protein stress. The function of ERdj7 is not yet known, but its unresponsiveness to the UPR suggests it may be involved in constitutive processes like translocation or

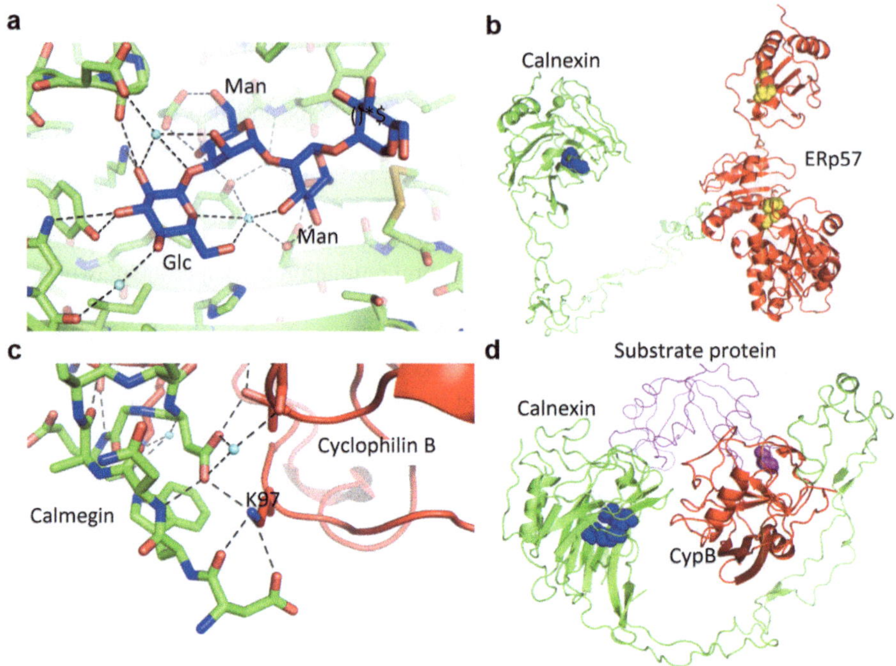

Fig. 10.1 Structural basis of ER interactions. **a** Structure of the tetrasaccharide Glc_1Man_3 (*blue*) bound to calreticulin (PDB code 3O0W). **b** Model of the complex between calnexin (PDB code 1JHN) and ERp57 (*red*; PDB code 3F8U). The glycan-binding site in calnexin is marked *blue* and the catalytic cysteines in ERp57 are highlighted in *yellow*. **c** Detailed view of hydrogen bonds and electrostatic interactions at the interface between the lectin chaperone, calmegin and cyclophilin B (PDB code 3ICI). K97 of cyclophilin B plays a key role in orchestrating the interaction. **d** Model of calnexin (*green*) and cyclophilin B (*red*; PDB code 3ICI) with a substrate protein (RNase B; *magenta*; PDB code 1Z6 S). The substrate is positioned so that proline 93 occupies the cyclophilin B active site.

membrane insertion as observed for ERdj1 and ERdj2/Sec63. While the specific functions of each ERdj co-chaperone are not fully understood, the emerging pattern is that they specify spatial targeting of BiP to specific ER functions through interactions with BiP and the machineries involved in these functions.

Recent insight into J-domain interactions with other ER luminal proteins has provided additional clues regarding their targeting to specific ER processes (Fig. 10.1 and Table 10.1). These interactions were identified mostly using an ER-specific membrane yeast two-hybrid system (ER-MYTHS) [13]. Interactions between the J-domain co-chaperones tested (ERdj3, ERdj4, ERdj5, DnaJc3) and BiP were confirmed, in addition to several novel interactions. Surprisingly, ERdj3, ERdj4, and ERdj5 all exhibited binary interactions with the protein disulfide isomerase (PDI) family member P5 that is known to interact with BiP. P5 is comprised of two active site containing thioredoxin domains and a single inactive thioredoxin-like domain that resembles the b domain of PDI [57]. The b domain of PDI is not sufficient for binding peptides alone [58]. Similarly, the inactive b-like domain of P5 is unable to bind peptides directly (our unpublished observations). Perhaps these three J-domain co-chaperones act to recruit substrates to both BiP and P5, providing recruitment services to BiP through either or both proteins. Interaction of both ERdj3 and ERdj4 with the FK506 binding protein (FKBP) family member FKBP60, an ER-resident peptidyl-prolyl *cis-trans* isomerase (PPI) that is poorly characterized but contains four PPI domains [59] is a novel way to recruit PPI activity to nascent substrates. Interestingly, ERdj4 was found to interact with calnexin, and ERdj3 was found to interact with UDP-glucose:glycoprotein glucosyltransferase (UGGT), providing connections between the BiP and lectin chaperone systems (to be discussed below). Interaction between ERdj3 and UGGT, a protein strongly implicated in misfolding recognition and quality control of N-glycosylated proteins, suggests a previously unappreciated cooperation between the lectin and BiP chaperone systems. While ERdj3 has been most associated with folding of nascent chains, both ERdj3 and ERdj4 have been recently implicated in BiP-independent selection of substrates for ERAD [60], perhaps analogous to misfolding recognition by UGGT.

5.2 *Nucleotide Exchange Factor Interactions of BiP*

The NEFs BAP and GRp170 are important for release of substrates from BiP. GRp170 is unique in that it is also a large Hsp70 with capacity to bind unfolded/misfolded proteins. Recent studies have shown that GRp170 remains bound to unfolded substrates after BiP release [61], and functions in ERAD of the unglycosylated form of the epithelial sodium channel (ENaC) [62]. It is interesting to note that GRp170 interacts with ERdj5 [13], a J-domain co-chaperone with a key role in reducing disulfides of misfolded proteins prior to ERAD. GRp170 might be a glycan-independent chaperone of misfolded proteins, taking the place of EDEM in the well-known EDEM-ERdj5-BiP ERAD complex [51], but this hypothesis requires further investigation.

5.3 *Early Lectin Chaperone Interaction Networks*

As nascent polypeptide chains enter the ER, a large proportion of them are glycosylated on Asn-X-S/T consensus sequences with Glc3Man9GlcNAc2 N-glycans by the translocon-associated oligosaccharyltransferase (OST). Aside from providing an additional mechanism for directional peptide sequestration in the ER, the glycan acts directly to increase hydrophilicity of the protein and indirectly as a ligand for recruitment of the nascent chain to the membrane anchored lectin chaperone calnexin and its soluble counterpart calreticulin. The terminal two glucose moieties must first be trimmed by glucosidase I (translocon associated), and glucosidase II, to generate the Glc1Man9GlcNAc2 N-glycan recognized by calnexin/calreticulin. The structural determinants of glycan recognition by calreticulin have been determined, and are illustrated in Fig. 10.1a [63]. To efficiently capture nascent glycoproteins as they enter the ER, calnexin associates directly with the ribosome-translocon complex through its cytosolic tail [64]. The lectins recruit the protein disulfide isomerase ERp57 [65, 66] (Fig. 10.1b), or the peptidyl prolyl cis-trans isomerase cyclophilin B [13, 67] (Fig. 10.1c) to the nascent chain to aid the folding process (Fig. 10.1d). In yeast, Mpd1p (a yeast PDI family protein) interacts with Cne1p (a yeast homolog of calnexin) [68].

A significant body of research has been dedicated to understanding the specific roles of the ERp57-calnexin/calreticulin interaction in the folding of glycoproteins [69, 70] and also in the MHC Class I peptide loading complex with tapasin [71, 72]. Structural insights, NMR, mutagenesis, and isothermal titration calorimetry (ITC)-binding studies have defined the molecular details of the interaction between the extended proline-rich (P) domain of calnexin, and the noncatalytic bb' domains of ERp57 [66]. The molecular details have allowed theoretical modeling of the interaction, revealing a complex that positions the thiol active sites of ERp57 in proximity to a glycoprotein substrate bound to the lectin-binding site of calnexin [69] (Fig. 10.1d). Specific abrogation of the ERp57/calnexin interaction through the point mutation (R282A) of ERp57 has made it possible to ascertain the importance of this interaction for recruitment of ERp57 to specific glycoprotein substrates or complexes containing calnexin/calreticulin [73]. The ERp57-calnexin/calreticulin interaction is crucial for recruitment of ERp57 to the majority of its glycoprotein substrates, but is not required for interaction with tapasin or the primary oxidant Ero1 [73].

The ERp57/calnexin or calreticulin and cyclophilin B/calnexin or calreticulin complexes serve as examples of how ER complexes can be best defined through detailed molecular characterization that subsequently provides the means to define interaction importance *in vivo* using targeted genetic manipulations. In the case of the cyclophilin B/calnexin or calreticulin interactions, the specific functional consequences remain to be elucidated, but genetic diseases linked to mutations of cyclophilin B are providing clues [74] (discussed in a subsequent section below).

Interaction mapping in the ER has revealed much more interconnections between different functional categories of chaperones and foldases than previously anticipated, and the lectin chaperones are no exception (Fig. 10.2). Besides the surprising interactions of both calnexin and calreticulin with cyclophilin B, novel disulfide isomerase interactions have been identified for both lectin chaperones. Calnexin interacts with ERp29 [75] while calreticulin interacts with PDIr [13, 76, 77]. ERp29 does not contain a thiol-reactive active site, and is comprised of a helical D domain similar to Drosophila Wind protein [78], and a thioredoxin-like b domain. Its function appears to be in chaperoning a broad range of secreted and ER-resident proteins [79] including thyroglobulin, and PERK [75]. More recently, ERp29 has been implicated in trafficking of wild-type and F508del CFTR. ERp29 expression increases significantly in response to low concentrations of 4-phenylbutyrate (4PBA) that do not induce UPR, suggesting that 4PBA induces ERp29 expression through an unknown mechanism independent of UPR [80]. Increased levels of ERp29 enhance trafficking to the cell surface of both the mutant and wild-type CFTR channels in CF epithelial cells [80].

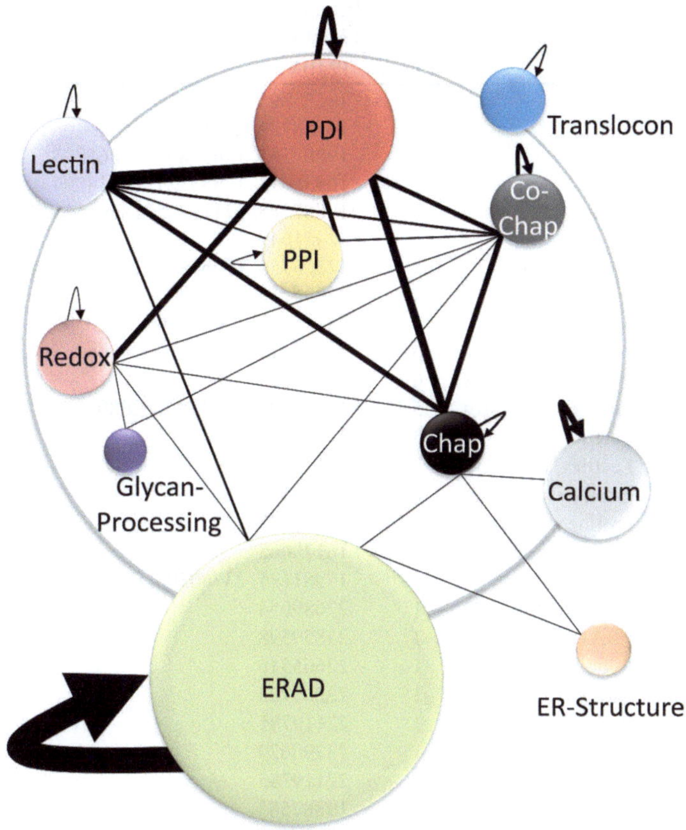

Fig. 10.2 Physical interactions of functional complexes in the ER. Represented are functional complexes that have been shown to associate directly. The size of the spheres correlates to the number of members for each functional complex found to be involved in direct interactions (as provided in Table 10.2). The weight of the edges indicates the number of literature-documented interactions between different pairs of functional complexes, or interactions of a functional complex with itself.

This pivotal role for ERp29 as a chaperone may also be related to its emerging function as a tumor suppressor [81]. PDIr is less well-studied, but its domain architecture is unique among ~20 ER-resident PDI family proteins [57], containing three slightly different thiol-reactive active sites. The active sites are found in three adjacent C-terminal thioredoxin domains with a single N-terminal non-catalytic thioredoxin-like domain mediating interactions with ERp72 and Calreticulin [77]. A PDIr-ERp72 complex would provide a large number of thiol reactive active sites in a single unit (six), and it is unclear what advantage this might provide. PDIr contains three distinct active sites CSMC, CGHC, and CPHC, and its thiol oxidase activity is minimal compared to other human PDIs [82] suggesting that it may play a more specialized role in isomerization. Perhaps calreticulin associates with both a potent oxidase (ERp57), and a more specialized isomerase (PDIr) to recruit slightly different activities to N-glycoproteins during different stages of protein folding.

Table 10.2 Functional connections between different classes of ER proteins, indicating the number of individual connections, referenced by PubMed-IDs

Class I	Class II	Count	PMID
Lectin	PDI	17	18653895, 22665516, 15865205, 10436013, 14988724, 23614004
Chaperone	PDI	10	19887585, 18653895, 15865205, 22665516, 10436013, 14988724, 23614004
PDI	Redox	8	19887585, 22119785, 22665516, 23979138, 22451649
Chaperone	Co-chaperone	7	12356756, 17567950, 18923428, 22267725, 22665516, 18653895
Chaperone	Lectin	7	22119785, 22665516, 12610305
Co-chaperone	PDI	6	22665516
PDI	Proline isomerases	6	12204109, 22665516
Chaperone	Proline isomerases	5	14960307, 18946027, 20801878
Co-chaperone	Lectin	3	22665516
ERAD	Lectin	3	22119785
Co-chaperone	Proline isomerases	2	22665516
Lectin	Proline isomerases	2	20801878
Redox	Glycan processing	1	16129668
Chaperone	Calcium	1	17981125, 23760505
Chaperone	ER structure	1	22689054
Chaperone	ERAD	1	23859598
Chaperone	Redox	1	22665516
Co-chaperone	ERAD	1	22267725
Co-chaperone	Redox	1	22119785
ERAD	ER structure	1	23790629
ERAD	Redox	1	22119785
Lectin	Redox	1	19887585
Co-chaperone	Glycan processing	1	22665516

ER Endoplasmic reticulum, *PDI* Protein disulfide isomerase, *ERAD* ER-associated degradation

An unexpected role for calnexin in regulating the transcriptional response to epidermal growth factor (EGF) receptor stimulation and apoptosis has been recently uncovered. Calnexin is highly abundant and has been found to associate with areas of the ER involved in mitochondrial contacts. The cytosolic domain of calnexin is cleaved by caspase-8 in response to EGF signaling, and this cytosolic peptide of calnexin translocates to the nucleus where it binds a protein inhibitor of activated STAT3 (PIAS3) [83]. Competitive binding of the STAT3 inhibitor enhances the STAT3-mediated response to EGF, linking calnexin to this critical apoptotic signal [83]. However, more than this, calnexin responds to ER stress by rendering its cytosolic peptide uncleavable by an unknown mechanism dependent on the glycan-binding luminal domain of calnexin. Previous studies have shown that calnexin can promote apoptosis through caspase-8 mediated cleavage of Bap31 in response to prolonged ER stress [84]. Inhibition of apoptosis through caspase-8 mediated cleavage of the calnexin cytosolic domain [83, 85] seems to indicate that calnexin can act as a switch to either promote or inhibit apoptosis. The interactions and regulation of calnexin at ER-mitochondrial membrane interaction sites will be a subject of intense interest, relating ER stress, oncology, and potentially providing additional insight into the established role of ERp57 in cancer [86–88].

6 Protein Disulfide Isomerase Interactions

A critical role in protein folding in the ER is performed by protein disulfide isomerases (PDIs) that introduce disulfide bonds between cysteine residues (oxidation), re-arrange already formed disulfide bonds (isomerization), and break incorrectly formed disulfide bonds (reduction). These functions are performed by thiol reactive di-cysteine motifs (CXXC, commonly CGHC) that are found in the active domains of PDIs. There are 20 different PDI-like proteins in the ER, and the majority of them contain thiol-reactive active sites. The versatility of PDIs to perform reactions in opposite directions has led to much interest into their mechanisms of action, interactions, and regulation, particularly focused on the most abundant and prolific family member (PDI). Several key aspects of PDI activity have been worked out. For example, in reduction and isomerization reactions, the more C-terminal cysteine within the active site of PDI provides an "escape-pathway" [89]. This property has been used to trap substrates with second cysteine mutants of PDI family proteins [90], and has provided some insight into their substrate preferences.

The ER is, on average, more oxidizing than the cytosol, but recent evidence has called into question the long-held redox status of glutathione in the ER, and its capacity to oxidize PDIs based on their specific standard redox equilibria [91]. The biochemical standard reduction potential in millivolts ($E^{0\prime}$) is a measure of the tendency to gain or lose electrons at pH 7.0 in the presence of another electron acceptor or donor. $E^{0\prime}$ is measured in relation to glutathione. A lower $E^{0\prime}$ indicates that a particular species is more likely to be oxidized (or donate electrons) than a species with a higher $E^{0\prime}$. For example, human PDI has an $E^{0\prime}$ of approximately −180 mV [92] while *E.coli* thioredoxin has an $E^{0\prime}$ of −270 mV [93]. Thus thioredoxin is an efficient reductase of denatured proteins, ($E^{0\prime}$ of approximately −220 mV) [93] while PDI is an efficient oxidase of denatured proteins. A redox sensitive GFP revealed that the ER is not as oxidizing as suggested, having an $E^{0\prime}$of −208 +/− 4 mV at pH 7.0 [91]. This less oxidizing environment introduces a challenge for PDI family members because their reducing potentials vary between −219 mV (for the very reducing ERdj5) [51] to −157 mV (for the more oxidizing a' domain of ERp57) [94]. Thus, for the more oxidizing PDIs, specific sources of oxidative equivalents that can accept electrons from these PDIs are required to facilitate their functions in oxidative protein folding. Such systems of electron transport (described below) are stabilized by local interactions and structural properties of the proteins involved [95]. One potential benefit of a less oxidizing ER is that PDIs should have more targeted control over the disulfide bonding process, with less rapid formation of potentially incorrect disulfides spontaneously upon ER entry.

6.1 PDI Interactions with Electron Acceptors and Donors

Acting upstream of PDI-mediated oxidation of cysteine thiols is ER oxidoreductin-1 (Ero1) [96] that is a flavin adenine dinucleotide (FAD)-binding protein that oxidizes PDI [89, 97], allowing it in turn to oxidize protein substrates. Studies in yeast have uncovered that the oxidative function of Ero1p is modulated tightly by Pdi1p [98], and similar regulation occurs with the mammalian proteins [99]. In mammals there are two isoforms of Ero1 (α and β), and Ero1α is the most abundant and ubiquitous isoform, and appears to function most prominently as a source of oxidative equivalents [100], while Ero1β has a more specialized role in the pancreas [101]. In S. *cerevisiae* and C. *elegans*, Ero1p is essential for viability, but neither Ero1α or β are essential for viability of mice [101], indicating that in mammals additional sources of disulfide bond equivalents are also important. These alternative sources include but may not be limited to glutathione and glutathione peroxidases, peroxiredoxin IV, vitamin K epoxide reductase, sulfhydryl oxidases, and the selenoprotein Sep15. Particularly in mammals, low molecular weight oxidants may contribute quite significantly to oxidative protein folding, and this area requires more research [102]. Ero1α is exquisitely redox sensitive [99, 100], and studies of the functional relationship between mammalian Ero1α and PDI have revealed that PDI regulates Ero1α oxidative function through rearrangement of a regulatory disulfide between C94 and C131 (inactive) or C94 and C99 (active) [103–105]. The functions and interactions of the alternative sources of oxidative equivalents are less well-characterized, but interactions with substrates and PDI family members [106] and other chaperones [13] are providing clues.

Recent work has determined a hierarchy of interaction of several of the most abundant PDIs with Ero1α [107] and other sources of oxidative equivalents [106]. While many of the soluble oxidoreductases in the ER (PDI, ERp57, ERp72, ERp46, and P5) are capable of complexing with Ero1α in isolation, PDI appears to be the most engaged with Ero1α, and can also act as a dispenser of oxidative equivalents to less oxidizing PDI family proteins [107]. Furthermore, the two separate redox-active sites of PDI have distinctive roles where the a domain accepts electrons from reduced glutathione while the a' domain accepts electrons from other PDIs [107]. These findings hint at an electron transport system where PDI and Ero1α act as a regulatory hub, and the biochemical redox potentials of the other PDI family members provide a variety of oxidative potentials to substrates with varying needs. This model is consistent with an earlier notion that the many different protein disulfide isomerases may provide slightly different redox capabilities to suit different substrates to which they are recruited [69], and that these are tightly regulated through the redox interactions of the PDIs with Ero1α [99].

6.2 Complexes for Safe Disposal of Electrons from PDIs

Importantly, as Ero1α oxidizes PDIs it accepts electrons and must dispose of these safely [108–110]. Electrons are transferred to molecular oxygen, forming hydrogen peroxide that must be neutralized by ER-resident peroxiredoxin 4 (PrxIV) [111, 112] or glutaredoxins (GPx7, GPx8) [110], or properly directed for use in other functions such as ROS signaling [113]. Importantly, other proteins including nicotinamide adenine dinucleotide oxidase (NOX) family members that can reside in the ER (NOX2 and 4) also produce peroxides and are involved in signaling the UPR, calcium release regulation, and interactions with PDIs [114]. The protein complexes involved in these processes are not currently well characterized. However, in the case of PrxIV, which requires a disulfide in its active site to be reduced prior to reaction with H_2O_2, studies have shown it cooperates with PDI family members (PDI, ERp46, P5, ERp57, and ERp18) that reduce this disulfide bond, promoting PrxIV activity [115]. Other investigations have revealed a substrate-like specificity of ERp46 for PrxIV [90], and a physical interaction between ERp72 and PrxIV [13]. Redox active proteins in the ER exhibit close functional and physical interactions, and future studies will reveal more clearly how tightly knit these systems are.

6.3 Diverse Structures, Functions, and Interactions of PDIs

The PDI family of proteins is one of the largest families of ER-resident proteins, and not surprisingly, the members are varied in structure [57], function, and interactions. An important longstanding question has been: Why are there so many PDI family proteins in the ER? There appears to be at least two main categories of PDIs—those that interact directly with substrates (e.g., PDI can bind substrates directly), and those that interact with substrates through partner proteins (e.g., ERp57 interacts with calnexin or calreticulin to recruit glycoprotein substrates). This does not mean that a particular PDI that is capable of interacting with substrates on its own works independently of other ER resident proteins, but simply that it does not require those interactions for substrate binding. The specific roles of the many different PDIs is becoming more and more clear through studies looking at both how each one interacts with specific sources of oxidative equivalents [106], and by how each interacts with other ER resident chaperones and foldases [13].

While it is expected that redox active proteins involved in electron transfer reactions in the ER should be physically associated, additional insight into the varied functions of the different PDIs can be gained from understanding their unique interactions with proteins of different functional annotations. Guilt by association reveals a different but perhaps complementary perspective on the unique redox potential hypothesis for PDIs. This perspective suggests that the specific interactions of each PDI are critical for its unique substrate-specific functions.

The reason for so many different PDIs in the ER is a question that is gradually being answered. On the one hand, redundancy of PDIs allows the ER to adapt when overloaded with a particular substrate, and thus proteins with overlapping functions would be beneficial. Gene silencing studies of four abundant PDIs (PDI, ERp57, ERp72, and P5) and the resultant impact on the folding of a panel of five secreted proteins revealed that, at least for the specific substrates chosen, PDI is most important [17]. However, the other PDIs were able to compensate for PDI loss quite effectively although a loss of both PDI and ERp57 impeded oxidative protein folding efficiency most significantly [17]. Given the many different client proteins passing through the ER, it is difficult to generalize these results more broadly, but the relative abundance of PDI and ERp57 supports their dominant roles in disulfide bond formation for non-glycosylated and glycosylated substrates, respectively. The specialized functions of other PDI family proteins continue to be uncovered. For example, the PDI family member anterior gradient-2 (AGR2) functions as a shuttling factor for mucins [116], and is implicated as an oncogene in a variety of cancers [117]. The role of AGR2 in cancer has prompted research into its physical and functional interactions, and the AAA+ tumor suppressor Reptin was found to bind its substrate-binding loop region in an ATP-regulated fashion [118]. It is alarming how many PDI family members have been implicated in cancer using unbiased genomic screens [119], hinting at additional unexpected and specialized interactions for this family to be uncovered in the future.

6.4 PDI and PPI Partnerships

While it is difficult to surmise the implications of all of the known PDI interactions, a trend of partnering with peptidyl-prolyl cis-trans isomerases is clear (Table 10.2). While disulfide and prolyl isomerases carry out very different biochemical functions, the utility of joining them together in a functional unit can be specifically appreciated for substrates such as Immunoglobulin G (IgG) where rate-determining steps for disulfide bond formation require prolyl-cis-trans isomerization [120, 121]. ERp72, PDI, and P5 can partner with cyclophilin B, and the assembly of C_H1 and C_L fragments of IgG in vitro is enhanced in the presence of both ERp72 and cyclophilin B [13]. This physical and functional cooperation appears to be replicated in several other PDI-PPI pairs including ERp29-FKBP23, ERp19-FKBP65, and ERp57-FKBP13. The functional implications of these other partnerships remain unexplored but results with cyclophilin B suggest that efficiency of protein folding could be enhanced by concerted action of both PDIs and PPIs on the same substrate.

6.5 Other PDI Interactions

The PDI family contains members involved in diverse ER functions. As the chaperone systems devoted to protein retention, secretion, quality control, and ERAD are being identified, more PDIs are finding their places involved in one or more of these processes. In addition, unexpected connections between PDI family proteins are being revealed, such as interaction between ERp57 and ERp27 (a non-redox active 2-domain PDI) [122]. This interaction may serve as an alternative means to recruit substrates to ERp57 via the recently defined chaperone-like binding activity of ERp27 [123]. ERp44 has been found to cycle between the ER and Golgi, engaging in a pH-dependent retrieval of incompletely disulfide bonded substrates like IgM

[124]. PDIs are also clearly involved in ERAD, with diverse substrate-binding capabilities likely contributing to ERAD of diverse substrates [125], and documented reduction of disulfides in terminally misfolded proteins by highly reductive ERdj5 [51]. The targeting of PDIs to ERAD is in at least some cases clearly determined through interactions with ERAD components, and it is likely that there are additional interactions between the different PDIs and ERAD machineries.

7 Interactions between Functionally Distinct ER Proteins

While it is to be expected that proteins of different functions will interact, the identification of these interactions provides a wealth of insight into how machineries assemble and are organized within the cell. Many such interactions have been recently identified in the ER, and here we attempt to highlight some of these novel complexes encountered by nascent chains, with a view towards their functional consequences.

Aside from its interactions with protein disulfide isomerases, the prolyl isomerase cyclophilin B interacts with GRp94, BiP, and calnexin/calreticulin [13]. These interactions suggest that cyclophilin B is a very versatile foldase, capable of being recruited to multiple chaperone complexes to aid the folding of proteins that may benefit from prolyl isomerization. The importance of cyclophilin B recruitment to specific chaperone complexes is supported by recent genetic evidence where a mutation within the polybasic region of cyclophilin B (G6R), known to be important for chaperone binding from NMR and crystallography studies [13, 67] leads to hyperelastosis cutis in inbred quarter horses [74]. Collagen assembly is impaired in fibroblasts from the horses, implying that in order for cyclophilin B to carry out its critical function for assembly of the collagen triple-helix, it must be properly targeted to chaperone complexes [74].

BiP and cyclophilin B are perhaps two of the best examples of proteins that function in very diverse capacities in the ER, depending on the complexes that they associate with. For example, BiP not only functions to bind and aid in the directional entry of nascent chains at the translocon when recruited by specific J-domain co-chaperones like ERdj1, 2, and 3 (described earlier) but also plays a crucial role in unfolding and disposal of terminally misfolded proteins as part of a complex with ERdj5 and EDEMs [51]. Similarly, cyclophilin B not only has roles in anterograde folding in the aforementioned chaperone complexes [74, 126] but also plays critical roles in ERAD of soluble proteins [127]. While the different chaperones and foldases in the ER may appear to have prominent roles in specific complexes, it is important to keep in mind that entirely new and diverse functions for these same chaperones may be uncovered through their associations with other complexes, particularly if they are relatively abundant in the ER.

This emerging trend of chaperone sharing between functional complexes can be seen in Fig. 10.2. The number of interactions between proteins in the ER that carry out diverse functions are illustrated by the thickness of the edges linking different functional annotations (nodes). While we do not specifically describe each chaperone protein that finds itself playing multiple parts in the ER, it is clear that multi-function chaperones are common. Table 10.2 summarizes the number of interactions between different functional classes. Generally, as might be expected, the most abundant ER-resident proteins are found to interact with the largest variety of functional categories. The most connected functional classes are the lectins and PDIs followed by chaperones and PDIs. Less abundant ER proteins likely fill specialized roles in specific complexes that are spatially arranged in the ER in defined microdomains or subcompartments. These complexes help polarize the ER into regions specialized for entry, oxidative folding functions (like the Ero1 electron transport system previously described), quality control, retention, and ERAD. It appears that the functions of many proteins within the ER are more precisely defined by the complexes that they belong to than by their individual annotated functions. With further investigation, involvement of different chaperone proteins in diverse complexes and functions will be uncovered. These trends speak to the complex interplay between ER resident proteins and their versatility, and further agree with data from real time imaging of abundant chaperones under different conditions in living cells [19]. The most abundant chaperones diffuse freely throughout the ER, and upon increases in protein load, their localization to specific regions (likely complexes engaged with substrates involved in folding or ERAD) is very rapid. This intrinsic buffering capacity also provides feedback through, for example, lack of BiP availability, as a mechanism for initiating the unfolded protein response [128].

8 Cargo Sorting and Protein Retention Complexes

As proteins fold in the ER, their subsequent fate must be determined based on their fidelity. Proteins that are incompletely or improperly folded or assembled must be retained until defects are corrected, or the terminally defective protein is earmarked for ERAD. This is not a trivial process, and there appear to be glycan-dependent and independent sorting and retention mechanisms. Glycan independent mechanisms include interactions with BiP. There are also retrieval mechanisms that recapture proteins that have erroneously exited the ER in vesicles headed for the Golgi.

8.1 The UGGT Cycling Complex

UDP-glucose:glycoprotein glucosyltransferase (UGGT) functions to recognize and reglucosylate incompletely folded proteins on their high mannose N-linked glycans so that they can be recycled for productive folding attempts in association with calnexin/calreticulin. How exactly UGGT recognizes misfolding remains a mystery, but the mechanism appears to involve conformational changes in UGGT and binding through hydrophobic surfaces [129]. The recognition mechanism is distinct from active binding and release cycles typical of most chaperones. There are two isoforms of UGGT in mammals (UGGT1 and UGGT2), and in *C.elegans* UGGT2 appears to have a distinct but critical function with weak or non-existent glucosylation activity (although the lack of observed activity may be due to low expression levels of UGGT2 compared to UGGT1) [130]. In mammalian cells, both isoforms have glucosylating activity [131, 132]. The selenoprotein Sep15 binds UGGT, forming a tight 1:1 complex [133, 134] that remains intact during non-reducing native gel electrophoresis for long time periods (our unpublished observations). This interaction is mediated by an N-terminal cysteine-rich domain of Sep15 that is necessary and sufficient to form a complex with UGGT [134]. Sep15 has been shown to significantly enhance the glucosylation activity of UGGT towards a misfolded substrate in vitro [131]. Studies in murine fibroblasts indicate that Sep15 expression is upregulated in response to sub-acute UPR (induced by the anterograde transport inhibitor brefeldin A or N-glycosylation inhibitor tunicamycin), but Sep15 is rapidly degraded in reponse to acute stress induced by the reducing agent dithiothreitol (DTT) or thapsigargin, a non-competitive inhibitor of the ER calcium pump (SERCA) [135]. This suggests that Sep15 may have a role in modulating UGGT activity to increase quality control stringency under moderate stress, but upon acute stress, degradation of substrates may be favored by removal of its UGGT-enhancing activity. However, Sep15 deficiency does not itself lead to UPR, suggesting that Sep15 has a somewhat limited substrate specificity, or that its functions can be compensated for by other yet undiscovered modulators of UGGT [135].

In *trypanosoma cruzi*, deletion of UGGT impedes interaction of *T. cruzi* cathepsin L (TcrCATL), a lysosomal protease, with calreticulin as expected. The efficiency of TcrCATL folding is drastically reduced in UGGT-null cells where the protein forms intermolecular disulfide bonded aggregates. Perhaps the most probable explanation for this result is that calreticulin-mediated recruitment of a *T. cruzi* homolog of ERp57 to the protein is not achieved (glucosylation of TcrCATL is performed exclusively by UGGT in *T. cruzi*), and incorrectly formed disulfide bonds cannot be resolved. However, ERp57 is one of the more oxidizing PDI family members while Sep15 is much more reducing, with a biochemical standard reduction potential of −225 mV (−157 mV for the a' domain of ERp57). This reduction potential is similar to that of ERdj5 which appears to function primarily to reduce disulfides of terminally misfolded proteins prior to retrotranslocation and proteasomal degradation. It is interesting to note that the *T. cruzi* genome does not appear to contain a clearly recognizeable homolog of Sep15, although a related trypanosome does (hypothetical protein STCU_06695 of *Strigomonas culicis*). There is at least one *T. cruzi* protein that shares 30 % identity with the *S. culicis* protein, and may function like Sep15, but this remains to be determined.

A function for the Sep15-UGGT quality control complex in mammals may be not only to reglucosylate proteins with recognized flaws in folding but also to reduce or isomerise incorrect disulfide bonds [136]. Reduced disulfide bonds that require more oxidative "power" for fresh attempts to form correct bonds are then recruited back to more oxidizing ERp57 through Glc1-N-glycan interaction with calnexin/calreticulin. This may allow a superficial partitioning of the oxidative and reductive functions during protein folding, as appears to be the trend with the differing redox potentials of the PDIs described above, and their respectively distinct complexes.

The very tight 1:1 UGGT-Sep15 complex (K_d~40 nM) may serve to sequester Sep15 and ensure that Sep15 reductive activity is targeted to substrate proteins that interact with UGGT. Knockdown or overexpression of Sep15 in NIH-3T3 cells imbalances redox homeostasis, suggesting a pivotal role for this protein in ER redox [136]. Other Selenocysteine-containing proteins in the ER include SelM [137] and Fep15 [138], but their functions are not yet clear although they are unlikely to interact with UGGT because they lack the cysteine-rich N-terminus of Sep15 that mediates this interaction.

Other than membership in a multi-protein complex in the ER [6], there are few other known interactions of UGGT. How UGGT is recruited to this multi-protein complex is not clear. It appears that ERdj3 may play a role as it was recently found to interact with UGGT using an ER-tailored yeast two-hybrid system [13]. The functional significance of the ERdj3-UGGT interaction is not known.

8.2 Complexes that Mediate Exit from the Calnexin Cycle

Ultimately, proteins must either achieve a correct fold and bypass recognition by UGGT, bypass recognition while still aberrant due to UGGT saturation or other unknown mechanisms, or be deemed terminally misfolded and targeted for ERAD.

8.3 Complexes of the Mannose Trimming Timer

The most well-established model for N-glycoprotein folding quality control involves the successive slow trimming of mannose moeities from the glycan [139], resulting in eventual extraction of the protein from the calnexin cycle. In yeast, initial trimming is mediated by mannosidase 1 (Mns1p) followed by additional trimming carried out by Htm1p that forms a specific complex with Pdi1p [140]. Pdi1p interaction with Htm1p is mediated by a specific intermolecular disulfide bond with the C-terminal domain of Htm1p, and enhances the mannosidase activity of Htm1p [140]. The resultant trimmed glycan is recognized by the mannose-6-phosphate homology (MRH) domain of Yos9p that targets the protein for ERAD. Mammalian cells have a similar but more complex system involving the same initial trimming by ER mannosidase 1 followed by putative further trimming of mannoses by ER-degradation enhancing mannosidase-like (EDEM) proteins EDEM1, EDEM2, or EDEM3 (in vitro demonstration of EDEM mannosidase activities is still lacking). Similarly to yeast, the mammalian EDEMs appear to cooperate with PDI family proteins through interactions where EDEM3 interacts with ERp46 [90], and EDEM1 interacts with PDI [13]. EDEM1 also interacts with the reductive PDI family member ERdj5 [51] and EDEM2 and EDEM3 appear to also [13], suggesting they may cooperate in processing substrates for eventual retrotranslocation. The precise functional consequences of each PDI-EDEM interaction remain uncharacterized, but in the case of ERdj5-EDEM1, it appears to target PDI activity to terminally misfolded proteins to prepare them for retrotranslocation [51]. BiP is also found in the complex with EDEM1 and ERdj5, converging four major functions in a single unit (J-domain co-chaperone, PDI, chaperone, and lectin; Fig. 10.2). EDEM1, EDEM2, and EDEM3 have been found to interact with GRp94 [13] while GRp94 also interacts with the mannose-binding lectin XTP3-B [4] that is tightly linked to the ERAD dislocation machineries through SEL1 L (Fig. 10.2). Interestingly BiP also interacts with OS-9, another lectin that binds SEL1 L. Trimmed glycans are recognized by the MRH domains of OS-9 and XTP3-B that can recruit misfolded proteins to the retrotranslocation channel through interactions with SEL1 L [4, 141]. However, interactions of OS-9 and XTP3-B with glycan independent chaperones, and their functional involvement in retrotranslocation of non-glycosylated substrates along with the EDEMs suggest that glycan dependent and independent functions converge at the retrotranslocon and share components.

8.4 Anterograde Trafficking Complexes for Folded Proteins

It is clear that trafficking of correctly folded proteins from the ER to the Golgi for further processing occurs with the involvement of cargo receptors including vesicle integral membrane protein VIP36, VIPL, and ER Golgi intermediate compartment protein-53 (ERGIC-53) in association with coat protein complex II (COPII) and COPI [142–144]. However, there are many details that remain to be worked out, and a family of closely similar abundant membrane proteins termed p24 s has long been implicated in ER-Golgi trafficking. The sequence of the p24 s (EMP24/GP25 LErp) and the members of the family are highly conserved from yeast to humans and they are located in the ER, COPI, and COPII vesicles and the cis-Golgi. While this high degree of conservation implies that they are involved in an essential cellular function, a multiple knock out of the 8 family members in *S.cerevisiae* yielded viable cells that showed a decreased rate of secretion of GPI-anchored proteins and increased leakage of BiP [144]. In contrast, a knock out of a single member in mammalian cells is lethal [146]. Recent results have shown that they are important in secretion of some of the Wnt ligands [147]. Increasing, but not yet definitive evidence is that they function as originally speculated, as cargo receptors for secreted proteins [145, 148]. A greater knowledge of their functional and physical interactions with other proteins will aid in defining their functions.

8.5 ER Mannosidase-I/COPI Interactions and Protein Retrieval

An apparent paradox in ER protein folding quality control has been the localization of all quality control components in the same intracellular organelle as components involved in ERAD. However, there have been plenty of indications that ER quality control and productive folding is somehow partitioned from ERAD [149, 150].

Recently, careful study of the localization of ER mannosidase I (ERManI) revealed that it primarily localizes to the Golgi where it is O-glycosylated [151]. Furthermore, ERManI has been shown to physically interact with the γ-COP subunit of COPI, responsible for Golgi-to-ER retrograde protein transport [152]. Mutations that disrupted this interaction rendered ERManI incapable of mediating efficient disposal of the ERAD substrate Null Hong Kong (NHK) AAT. These functionally significant interactions underscore the importance of understanding partitioning of ER quality control and the PPIs involved.

9 Interactions and Functions in ERAD

The half-lives of the component proteins of the ER varies considerably, indicating that there are underlying mechanisms that maintain the homeostasis of the ER. While the mechanisms that signal the increased need for proteins have been recognized by their participation in the unfolded protein response, those involved in the removal of ER proteins have not been as well studied. The role of autophagy in ER homeostasis has yet to be fully explored but the process known as ERAD is known in detail. The proteins that participate in the ERAD of both membrane and soluble secretory proteins have been identified principally through functional studies mainly in yeast [34].

The behavior of mutant proteins that cause protein trafficking diseases has generated interest in the mechanism of ERAD. More precisely, how does this quality control system function and how might it be subverted? For example, in the archetypical protein trafficking disease cystic fibrosis the mutant protein F508del-CFTR is recognized as misfolded, retained in the ER and retrotranslocated from the ER to be degraded by the proteasome. However, if cells are incubated at low temperature the mutant protein traffics to the plasma membrane and is functional. The precise mechanism of recognition of the misfolded F508del-CFTR protein is unknown, and there are no mutations that decrease the fidelity of the cellular protein quality control mechanism in the same way that the trafficking of F508del-CFTR is corrected by compounds [153, 154].

The ER Calnexin cycle that recognizes secreted N-glycosylated proteins and assists in their folding is well characterized [70, 155] as is the system for non-glycosylated secreted proteins [156]. The main components of this quality control system for recognition of misfolded proteins UGGT and BiP are well characterized. Misfolded glycoproteins show prolonged association with calnexin but compounds, such as deoxynojirimycin, that divert glycoproteins from the Calnexin Cycle do not correct the ER retention of F508del-CFTR. In addition, the F508del mutation is in the nucleotide-binding domain 1 of the CFTR molecule and located in the cytosol. Although there has been a comprehensive catalogue of the proteins that bind to CFTR [157] only, a knockdown of Aha1, which is a cochaperone of Hsp90, had any functional effect on the trafficking of F508del CFTR. Thus, although the role of the Calnexin Cycle in the quality control of soluble secretory proteins is clear, the recognition mechanism of misfolded membrane proteins is not well understood.

The ER luminal components and the cytosolic ATPase of the retrotranslocation machinery are known. The connection of the most reductive PDI ERdj5, BiP, and EDEMs is mentioned above. This complex functionally links with the misfolded glycoprotein recognition receptors Hrd3 and Yos9 by an as yet unknown mechanism.

There are several candidates for the protein channel that facilitates exit of proteins from the ER. One is the translocon Sec61, or perhaps a version thereof, and more recently Derlin1 (Der1) has been recognized in yeast to form a complex together with the luminal Hrd3 that recognizes substrates, and Hrd1 that ubiquitylates them [158].

The cast of characters has probably been all identified, and some of the complexes that they form identified, but establishing how they function in ERAD needs a global approach.

A comprehensive approach to defining the components and organization of ERAD networks used several experimental approaches to define high confidence interactions [4]. The authors identified potential interacting ER proteins involved in ERAD with tagged proteins expressed in cells, and used purification protocols with different detergents. The complexes were analyzed by mass spectrometry and validated by combining the data from several approaches. They also integrated the data by detecting interactions of the ERAD components with a set of misfolded secreted and membrane proteins while correlating with gene expression data. These studies identified and confirmed complexes that perform the steps in the recognition, retrotranslocation, dislocation, and ubiquitination of misfolded proteins and also showed links to the 26 S proteasome. The details of these complexes has been extensively reviewed elsewhere [141]. This type of integrated approach promises to unravel the mechanism of how misfolded membrane proteins are recognized and how the high fidelity of the cellular protein quality control system is achieved.

10 Regulation of Interactions in the ER

10.1 Chaperone–Substrate Interactions

Regulation of interactions between substrates and chaperone proteins in the ER remains poorly understood although there are several well-characterized mechanisms such as glycan trimming, and emerging novel ones such as ADP-ribosylation of BiP [49]. Glycan trimming functions as both a mechanism for targeting proteins to their respective fates, and as a means to bring together ERAD machineries in the ER have been recently reviewed [139]. ADP-ribosylation was discussed above as a rapid regulator of BiP-substrate interactions to prevent non-productive binding during normal or minimal substrate load. A recent study uncovered a mechanism in yeast whereby substrate proteins are rapidly and irreversibly tagged for degradation by O-mannosylation [159]. O-mannosylation reduces interaction of substrates with Kar2p, the yeast homolog of BiP. This tagging mechanism provides an alternative to glycan trimming to regulate chaperone-substrate interactions, freeing up chaperones for useful engagement of substrates not past the point of no return. Whether this same mechanism operates in mammalian cells remains to be explored. There is also evidence to suggest that calcium can regulate interactions of smooth ER with mitochondria [160] probably having a broader impact on chaperone-substrate interactions although the mechanisms that dictate these apparent generalized effects remain poorly understood.

10.2 Chaperone–Chaperone Interactions

The regulation of interactions between ER chaperones is far less understood than the regulation of chaperone-substrate interactions. Glycans play a role that is becoming increasingly appreciated (XTP-3B, SEL1 L, and OS-9 are brought together through glycan specific interactions). Due to the electrostatic nature of many ER chaperone-chaperone interactions (that don't involve unfolded protein-binding sites), it is likely that these interactions can be regulated by cation or anion concentrations in the ER, but how exactly this regulation could occur in a specified manner remains unclear. While novel interactions between ER chaperones and foldases continue to be uncovered, defining their respective regulatory mechanisms will become increasingly important to understand. One fascinating area of study is the use of small molecules to correct trafficking defects of mutant proteins linked to human diseases, and some correctors may work by disrupting or modulating ER chaperone complexes. These studies will undoubtedly reveal novel regulations in the ER, as observed for F508del-CFTR correctors such as latonduine, and PARP inhibitors [161].

11 Conclusions and Future Outlook

Our understanding of ER protein folding machineries and their interactions in the ER has advanced greatly in recent years. While earlier studies appeared to indicate a certain degree of separation between functional classes of ER proteins, particularly between the glycan-dependent lectins and glycan-independent BiP chaperones, it is now apparent that chaperones and foldases of very different functional annotations interact promiscuously. Far from being nonspecific and random, many of these novel interactions have specific functional implications, illustrating how tightly interconnected the ER folding machineries are. Beyond the specific interactions, there are many fascinating questions regarding ER complexes to be addressed in future work. Recent high-resolution microscopy and image reconstruction has revealed a "parking garage"- like structure of the ER that can be modulated by beta sheet-rich proteins such as p180, Kinectin, and Climp63 [1]. There have also been a significant number of studies on the reticulons and their impacts on membrane curvature in the ER [2] while mitochondria associated ER membranes (MAMs) are becoming hotbeds of function related to apoptosis [162]. One obvious question will be—how do ER complexes localize within the ER? Are there particular complexes that contribute to, or are most associated with, particular morphologies/membrane arrangements? It seems clear that many ER chaperones can freely diffuse throughout the ER, but this free diffusion is slowed with increases in substrate load. Are there subregions in the ER that constitute specific targeting areas for chaperones and foldases, arranged spatially to best accommodate the folding needs of the cell? While versatile, these basic complexes may provide a backbone for additional regulatory interactions and organization so that, as in the case of the PDIs, oxidative protein folding is orchestrated such that thiol oxidation, reduction, and isomerisation are most efficient to fold proteins without compromising speed, fidelity, or undue production of reactive oxygen species. Perhaps the ER will emerge, as has the mitochondria, with complexes dedicated to specific functions in electron transport that can be arranged into supercomplexes that determine electron flux under different metabolic circumstances [163]. Certainly, there are many questions to be asked, and with the advent of enhanced resolution, sensitivity, and interaction validation methods, many novel ER chaperone interactions and functions will be discovered in the coming years.

References

1. Terasaki M, Shemesh T, Kasthuri N, Klemm RW, Schalek R, Hayworth KJ, Hand AR, Yankova M, Huber G, Lichtman JW, Rapoport TA, Kozlov MM (2013) Stacked endoplasmic reticulum sheets are connected by helicoidal membrane motifs. Cell 154(2):285–296. doi:10.1016/j.cell.2013.06.031, S0092-8674(13)00770-8 [pii]
2. Chen S, Novick P, Ferro-Novick S (2013) ER structure and function. Curr Opin Cell Biol 25(4):428–433. doi:10.1016/j.ceb.2013.02.006, S0955-0674(13)00029-X [pii]
3. Liu TY, Bian X, Sun S, Hu X, Klemm RW, Prinz WA, Rapoport TA, Hu J (2012) Lipid interaction of the C terminus and association of the transmembrane segments facilitate atlastin-mediated homotypic endoplasmic reticulum fusion. Proc Natl Acad Sci U S A 109(32):E2146–E2154. doi:10.1073/pnas.1208385109, 1208385109 [pii]
4. Christianson JC, Olzmann JA, Shaler TA, Sowa ME, Bennett EJ, Richter CM, Tyler RE, Greenblatt EJ, Harper JW, Kopito RR (2012) Defining human ERAD networks through an integrative mapping strategy. Nat Cell Biol 14(1):93–105. doi:10.1038/ncb2383
5. Landry CR, Levy ED, Abd Rabbo D, Tarassov K, Michnick SW (2013) Extracting insight from noisy cellular networks. Cell 155(5):983–989. doi:10.1016/j.cell.2013.11.003
6. Meunier L, Usherwood YK, Chung KT, Hendershot LM (2002) A subset of chaperones and folding enzymes form multiprotein complexes in endoplasmic reticulum to bind nascent proteins. Mol Biol Cell 13(12):4456–4469
7. Vandenbroeck K, Martens E, Alloza I (2006) Multi-chaperone complexes regulate the folding of interferon-gamma in the endoplasmic reticulum. Cytokine 33(5):264–273
8. Zhang J, Herscovitz H (2003) Nascent lipidated apolipoprotein B is transported to the Golgi as an incompletely folded intermediate as probed by its association with network of endoplasmic reticulum molecular chaperones, GRP94, ERp72, BiP, calreticulin, and cyclophilin B. J Biol Chem 278(9):7459–7468

9. Gavin AC, Aloy P, Grandi P, Krause R, Boesche M, Marzioch M, Rau C, Jensen LJ, Bastuck S, Dumpelfeld B, Edelmann A, Heurtier MA, Hoffman V, Hoefert C, Klein K, Hudak M, Michon AM, Schelder M, Schirle M, Remor M, Rudi T, Hooper S, Bauer A, Bouwmeester T, Casari G, Drewes G, Neubauer G, Rick JM, Kuster B, Bork P, Russell RB, Superti-Furga G (2006) Proteome survey reveals modularity of the yeast cell machinery. Nature 440(7084):631–636
10. Babu M, Vlasblom J, Pu S, Guo X, Graham C, Bean BD, Burston HE, Vizeacoumar FJ, Snider J, Phanse S, Fong V, Tam YY, Davey M, Hnatshak O, Bajaj N, Chandran S, Punna T, Christopolous C, Wong V, Yu A, Zhong G, Li J, Stagljar I, Conibear E, Wodak SJ, Emili A, Greenblatt JF (2012) Interaction landscape of membrane-protein complexes in *Saccharomyces cerevisiae*. Nature 489(7417):585–589. doi:10.1038/nature11354
11. Dejgaard K, Theberge JF, Heath-Engel H, Chevet E, Tremblay ML, Thomas DY (2010) Organization of the Sec61 translocon, studied by high resolution native electrophoresis. J Proteome Res 9(4):1763–1771. doi:10.1021/pr900900x
12. Jansen G, Maattanen P, Denisov AY, Scarffe L, Schade B, Balghi H, Dejgaard K, Chen LY, Muller WJ, Gehring K, Thomas DY (2012) An interaction map of endoplasmic reticulum chaperones and foldases. MCP 11(9):710–723. doi:10.1074/mcp.M111.016550
13. Tarassov K, Messier V, Landry CR, Radinovic S, Serna Molina MM, Shames I, Malitskaya Y, Vogel J, Bussey H, Michnick SW (2008) An in vivo map of the yeast protein interactome. Science 320(5882):1465–1470. doi:10.1126/science.1153878, 1153878 [pii]
14. Miller JP, Lo RS, Ben-Hur A, Desmarais C, Stagljar I, Noble WS, Fields S (2005) Large-scale identification of yeast integral membrane protein interactions. Proc Natl Acad Sci U S A 102(34):12123–12128. doi:10.1073/pnas.0505482102
15. Gilchrist A, Au CE, Hiding J, Bell AW, Fernandez-Rodriguez J, Lesimple S, Nagaya H, Roy L, Gosline SJ, Hallett M, Paiement J, Kearney RE, Nilsson T, Bergeron JJ (2006) Quantitative proteomics analysis of the secretory pathway. Cell 127(6):1265–1281. doi:10.1016/j.cell.2006.10.036
16. Ewing RM, Chu P, Elisma F, Li H, Taylor P, Climie S, McBroom-Cerajewski L, Robinson MD, O'Connor L, Li M, Taylor R, Dharsee M, Ho Y, Heilbut A, Moore L, Zhang S, Ornatsky O, Bukhman YV, Ethier M, Sheng Y, Vasilescu J, Abu-Farha M, Lambert JP, Duewel HS, Stewart II, Kuehl B, Hogue K, Colwill K, Gladwish K, Muskat B, Kinach R, Adams SL, Moran MF, Morin GB, Topaloglou T, Figeys D (2007) Large-scale mapping of human protein-protein interactions by mass spectrometry. Mol Syst Biol 3:89. doi:10.1038/msb4100134
17. Rutkevich LA, Cohen-Doyle MF, Brockmeier U, Williams DB (2010) Functional relationship between protein disulfide isomerase family members during the oxidative folding of human secretory proteins. Mol Biol Cell 21(18):3093–3105. doi:10.1091/mbc.E10-04-0356
18. Rhee HW, Zou P, Udeshi ND, Martell JD, Mootha VK, Carr SA, Ting AY (2013) Proteomic mapping of mitochondria in living cells via spatially restricted enzymatic tagging. Science 339(6125):1328–1331. doi:10.1126/science.1230593
19. Guo F, Snapp EL (2013) ERdj3 regulates BiP occupancy in living cells. J Cell Sci 126(Pt 6):1429–1439. doi:10.1242/jcs.118182, jcs.118182 [pii]
20. Verissimo F, Pepperkok R (2013) Imaging ER-to-Golgi transport: towards a systems view. J Cell Sci 126(Pt 22):5091–5100. doi:10.1242/jcs.121061, 126/22/5091 [pii]
21. Brunstein M, Wicker K, Herault K, Heintzmann R, Oheim M (2013) Full-field dual-color 100-nm super-resolution imaging reveals organization and dynamics of mitochondrial and ER networks. Opt Express 21(22):26162–26173. doi:10.1364/OE.21.026162, 270010 [pii]
22. Gao Y, Berciu C, Kuang Y, Shi J, Nicastro D, Xu B (2013) Probing nanoscale self-assembly of nonfluorescent small molecules inside live mammalian cells. ACS Nano 7(10):9055–9063. doi:10.1021/nn403664n
23. White KA, Zegelbone PM (2013) Directed evolution of a probe ligase with activity in the secretory pathway and application to imaging intercellular protein-protein interactions. Biochemistry 52(21):3728–3739. doi:10.1021/bi400268m
24. Johnson AE, van Waes MA (1999) The translocon: a dynamic gateway at the ER membrane. Annu Rev Cell Dev Biol 15:799–842. doi:10.1146/annurev.cellbio.15.1.799

25. Meacock SL, Greenfield JJ, High S (2000) Protein targeting and translocation at the endoplasmic reticulum membrane-through the eye of a needle? Essays Biochem 36:1–13
26. Mitra K, Frank J, Driessen A (2006) Co- and post-translational translocation through the protein-conducting channel: analogous mechanisms at work? Nat Struct Mol Biol 13(11):957–964. doi:10.1038/nsmb1166, nsmb1166 [pii]
27. Park E, Rapoport TA (2012) Mechanisms of Sec61/SecY-mediated protein translocation across membranes. Annu Rev Biophys 41:21–40. doi:10.1146/annurev-biophys-050511-102312
28. Nyathi Y, Wilkinson BM, Pool MR (2013) Co-translational targeting and translocation of proteins to the endoplasmic reticulum. Biochim Biophys Acta 1833(11):2392–2402. doi:10.1016/j.bbamcr.2013.02.021, S0167-4889(13)00079-7 [pii]
29. Shao S, Hegde RS (2011) Membrane protein insertion at the endoplasmic reticulum. Annu Rev Cell Dev Biol 27:25–56. doi:10.1146/annurev-cellbio-092910-154125
30. Kraut-Cohen J, Afanasieva E, Haim-Vilmovsky L, Slobodin B, Yosef I, Bibi E, Gerst JE (2013) Translation- and SRP-independent mRNA targeting to the endoplasmic reticulum in the yeast *Saccharomyces cerevisiae*. Mol Biol Cell 24(19):3069–3084. doi:10.1091/mbc.E13-01-0038, mbc.E13-01-0038 [pii]
31. Ast T, Cohen G, Schuldiner M (2013) A network of cytosolic factors targets SRP-independent proteins to the endoplasmic reticulum. Cell 152(5):1134–1145. doi:10.1016/j.cell.2013.02.003, S0092-8674(13)00151-7 [pii]
32. Johnson N, Powis K, High S (2013) Post-translational translocation into the endoplasmic reticulum. Biochim Biophys Acta 1833(11):2403–2409. doi:10.1016/j.bbamcr.2012.12.008, S0167-4889(12)00386-2 [pii]
33. Ng W, Sergeyenko T, Zeng N, Brown JD, Romisch K (2007) Characterization of the proteasome interaction with the Sec61 channel in the endoplasmic reticulum. J Cell Sci 120(Pt 4):682–691. doi:10.1242/jcs.03351, jcs.03351 [pii]
34. Needham PG, Brodsky JL (2013) How early studies on secreted and membrane protein quality control gave rise to the ER associated degradation (ERAD) pathway: the early history of ERAD. Biochim Biophys Acta 1833(11):2447–2457. doi:10.1016/j.bbamcr.2013.03.018, S0167-4889(13)00119-5 [pii]
35. Hampton RY, Sommer T (2012) Finding the will and the way of ERAD substrate retrotranslocation. Curr Opin Cell Biol 24(4):460–466. doi:10.1016/j.ceb.2012.05.010, S0955-0674(12)00081-6 [pii]
36. Park E, Rapoport TA (2011) Preserving the membrane barrier for small molecules during bacterial protein translocation. Nature 473(7346):239–242. doi:10.1038/nature10014, nature10014 [pii]
37. Tyedmers J, Lerner M, Wiedmann M, Volkmer J, Zimmermann R (2003) Polypeptide-binding proteins mediate completion of co-translational protein translocation into the mammalian endoplasmic reticulum. EMBO Rep 4(5):505–510. doi:10.1038/sj.embor.embor826, embor826 [pii]
38. Dudek J, Greiner M, Muller A, Hendershot LM, Kopsch K, Nastainczyk W, Zimmermann R (2005) ERj1p has a basic role in protein biogenesis at the endoplasmic reticulum. Nat Struct Mol Biol 12(11):1008–1014. doi:10.1038/nsmb1007, nsmb1007 [pii]
39. Blau M, Mullapudi S, Becker T, Dudek J, Zimmermann R, Penczek PA, Beckmann R (2005) ERj1p uses a universal ribosomal adaptor site to coordinate the 80 S ribosome at the membrane. Nat Struct Mol Biol 12(11):1015–1016. doi:10.1038/nsmb998, nsmb998 [pii]
40. Tyedmers J, Lerner M, Bies C, Dudek J, Skowronek MH, Haas IG, Heim N, Nastainczyk W, Volkmer J, Zimmermann R (2000) Homologs of the yeast Sec complex subunits Sec62p and Sec63p are abundant proteins in dog pancreas microsomes. Proc Natl Acad Sci U S A 97(13):7214–7219. doi:10.1073/pnas.97.13.7214
41. Davila S, Furu L, Gharavi AG, Tian X, Onoe T, Qian Q, Li A, Cai Y, Kamath PS, King BF, Azurmendi PJ, Tahvanainen P, Kaariainen H, Hockerstedt K, Devuyst O, Pirson Y, Martin RS, Lifton RP, Tahvanainen E, Torres VE, Somlo S (2004) Mutations in SEC63 cause autosomal dominant polycystic liver disease. Nat Genet 36(6):575–577. doi:10.1038/ng1357, ng1357 [pii]

42. Reithinger JH, Kim JE, Kim H (2013) Sec62 protein mediates membrane insertion and orientation of moderately hydrophobic signal anchor proteins in the endoplasmic reticulum (ER). J Biol Chem 288(25):18058–18067. doi:10.1074/jbc.M113.473009, M113.473009 [pii]
43. Lakkaraju AK, Thankappan R, Mary C, Garrison JL, Taunton J, Strub K (2012) Efficient secretion of small proteins in mammalian cells relies on Sec62-dependent posttranslational translocation. Mol Biol Cell 23(14):2712–2722. doi:10.1091/mbc.E12-03–0228, mbc.E12-03-0228 [pii]
44. Vembar SS, Jonikas MC, Hendershot LM, Weissman JS, Brodsky JL (2010) J domain co-chaperone specificity defines the role of BiP during protein translocation. J Biol Chem 285(29):22484–22494. doi:10.1074/jbc.M110.102186, M110.102186 [pii]
45. Lang S, Benedix J, Fedeles SV, Schorr S, Schirra C, Schauble N, Jalal C, Greiner M, Hassdenteufel S, Tatzelt J, Kreutzer B, Edelmann L, Krause E, Rettig J, Somlo S, Zimmermann R, Dudek J (2012) Different effects of Sec61alpha, Sec62 and Sec63 depletion on transport of polypeptides into the endoplasmic reticulum of mammalian cells. J Cell Sci 125(Pt 8):1958–1969. doi:10.1242/jcs.096727, jcs.096727 [pii]
46. Mades A, Gotthardt K, Awe K, Stieler J, Doring T, Fuser S, Prange R (2012) Role of human sec63 in modulating the steady-state levels of multi-spanning membrane proteins. PLoS ONE 7(11):e49243. doi:10.1371/journal.pone.0049243, PONE-D-12-09156 [pii]
47. Nicchitta CV, Blobel G (1993) Lumenal proteins of the mammalian endoplasmic reticulum are required to complete protein translocation. Cell 73(5):989–998. 0092-8674(93)90276-V [pii]
48. Ast T, Schuldiner M (2013) All roads lead to Rome (but some may be harder to travel): SRP-independent translocation into the endoplasmic reticulum. Crit Rev Biochem Mol Biol 48(3):273–288. doi:10.3109/10409238.2013.782999
49. Chambers JE, Petrova K, Tomba G, Vendruscolo M, Ron D (2012) ADP ribosylation adapts an ER chaperone response to short-term fluctuations in unfolded protein load. J Cell Biol 198(3):371–385. doi:10.1083/jcb.201202005, jcb.201202005 [pii]
50. Jin Y, Zhuang M, Hendershot LM (2009) ERdj3, a luminal ER DnaJ homologue, binds directly to unfolded proteins in the mammalian ER: identification of critical residues. Biochemistry 48(1):41–49. doi:10.1021/bi8015923,10.1021/bi8015923 [pii]
51. Ushioda R, Hoseki J, Araki K, Jansen G, Thomas DY, Nagata K (2008) ERdj5 is required as a disulfide reductase for degradation of misfolded proteins in the ER. Science 321(5888):569–572. doi:10.1126/science.1159293, 321/5888/569 [pii]
52. Oka OB, Pringle MA, Schopp IM, Braakman I, Bulleid NJ (2013) ERdj5 is the ER reductase that catalyzes the removal of non-native disulfides and correct folding of the LDL receptor. Mol Cell 50(6):793–804. doi:10.1016/j.molcel.2013.05.014, S1097-2765(13)00377-8 [pii]
53. Hagiwara M, Maegawa K, Suzuki M, Ushioda R, Araki K, Matsumoto Y, Hoseki J, Nagata K, Inaba K (2011) Structural basis of an ERAD pathway mediated by the ER-resident protein disulfide reductase ERdj5. Mol Cell 41(4):432–444. doi:10.1016/j.molcel.2011.01.021, S1097-2765(11)00048-7 [pii]
54. Svard M, Biterova EI, Bourhis JM, Guy JE (2011) The crystal structure of the human co-chaperone P58(IPK). PLoS ONE 6(7):e22337. doi:10.1371/journal.pone.0022337, PONE-D-11-04244 [pii]
55. Jin Y, Awad W, Petrova K, Hendershot LM (2008) Regulated release of ERdj3 from unfolded proteins by BiP. EMBO J 27(21):2873–2882. doi:10.1038/emboj.2008.207, emboj2008207 [pii]
56. Marcinowski M, Holler M, Feige MJ, Baerend D, Lamb DC, Buchner J (2011) Substrate discrimination of the chaperone BiP by autonomous and cochaperone-regulated conformational transitions. Nat Struct Mol Biol 18(2):150–158. doi:10.1038/nsmb.1970, nsmb.1970 [pii]
57. Kozlov G, Maattanen P, Thomas DY, Gehring K (2010) A structural overview of the PDI family of proteins. FEBS J 277(19):3924–3936. doi:10.1111/j.1742-4658.2010.07793.x, EJB7793 [pii]

58. Denisov AY, Maattanen P, Dabrowski C, Kozlov G, Thomas DY, Gehring K (2009) Solution structure of the bb' domains of human protein disulfide isomerase. FEBS J 276(5):1440–1449. doi:10.1111/j.1742-4658.2009.06884.x, EJB6884 [pii]
59. Shadidy M, Caubit X, Olsen R, Seternes OM, Moens U, Krauss S (1999) Biochemical analysis of mouse FKBP60, a novel member of the FKPB family. Biochim Biophys Acta 1446(3):295–307. S0167-4781(99)00080-9 [pii]
60. Buck TM, Kolb AR, Boyd CR, Kleyman TR, Brodsky JL (2010) The endoplasmic reticulum-associated degradation of the epithelial sodium channel requires a unique complement of molecular chaperones. Mol Biol Cell 21(6):1047–1058. doi:10.1091/mbc.E09-11-0944, E09-11-0944 [pii]
61. Behnke J, Hendershot LM (2013) The large Hsp70 Grp170 binds to unfolded protein substrates in vivo with a regulation distinct from conventional Hsp70s. J Biol Chem 289(5):2899–2907. doi:10.1074/jbc.M113.507491, M113.507491 [pii]
62. Buck TM, Plavchak L, Roy A, Donnelly BF, Kashlan OB, Kleyman TR, Subramanya AR, Brodsky JL (2013) The Lhs1/GRP170 chaperones facilitate the endoplasmic reticulum-associated degradation of the epithelial sodium channel. J Biol Chem 288(25):18366–18380. doi:10.1074/jbc.M113.469882, M113.469882 [pii]
63. Kozlov G, Pocanschi CL, Rosenauer A, Bastos-Aristizabal S, Gorelik A, Williams DB, Gehring K (2010) Structural basis of carbohydrate recognition by calreticulin. J Biol Chem 285(49):38612–38620. doi:10.1074/jbc.M110.168294, M110.168294 [pii]
64. Lakkaraju AK, Abrami L, Lemmin T, Blaskovic S, Kunz B, Kihara A, Dal Peraro M, van der Goot FG (2012) Palmitoylated calnexin is a key component of the ribosome-translocon complex. EMBO J 31(7):1823–1835. doi:10.1038/emboj.2012.15, emboj201215 [pii]
65. Zapun A, Darby NJ, Tessier DC, Michalak M, Bergeron JJ, Thomas DY (1998) Enhanced catalysis of ribonuclease B folding by the interaction of calnexin or calreticulin with ERp57. J Biol Chem 273(11):6009–6012
66. Kozlov G, Maattanen P, Schrag JD, Pollock S, Cygler M, Nagar B, Thomas DY, Gehring K (2006) Crystal structure of the bb' domains of the protein disulfide isomerase ERp57. Structure 14(8):1331–1339. doi:10.1016/j.str.2006.06.019, S0969-2126(06)00303-0 [pii]
67. Kozlov G, Bastos-Aristizabal S, Maattanen P, Rosenauer A, Zheng F, Killikelly A, Trempe JF, Thomas DY, Gehring K (2010) Structural basis of cyclophilin B binding by the calnexin/calreticulin P-domain. J Biol Chem 285(46):35551–35557. doi:10.1074/jbc.M110.160101, M110.160101 [pii]
68. Kimura T, Hosoda Y, Sato Y, Kitamura Y, Ikeda T, Horibe T, Kikuchi M (2005) Interactions among yeast protein-disulfide isomerase proteins and endoplasmic reticulum chaperone proteins influence their activities. J Biol Chem 280(36):31438–31441. doi:10.1074/jbc.M503377200, M503377200 [pii]
69. Maattanen P, Gehring K, Bergeron JJ, Thomas DY (2010) Protein quality control in the ER: the recognition of misfolded proteins. Semin Cell Dev Biol 21(5):500–511. doi:10.1016/j.semcdb.2010.03.006, S1084-9521(10)00068-6 [pii]
70. Rutkevich LA, Williams DB (2011) Participation of lectin chaperones and thiol oxidoreductases in protein folding within the endoplasmic reticulum. Curr Opin Cell Biol 23(2):157–166. doi:10.1016/j.ceb.2010.10.011, S0955-0674(10)00182-1 [pii]
71. Chapman DC, Williams DB (2010) ER quality control in the biogenesis of MHC class I molecules. Semin Cell Dev Biol 21(5):512–519. doi:10.1016/j.semcdb.2009.12.013, S1084-9521(09)00258-4 [pii]
72. Saunders PM, van Endert P (2011) Running the gauntlet: from peptide generation to antigen presentation by MHC class I. Tissue Antigens 78(3):161–170. doi:10.1111/j.1399-0039.2011.01735.x
73. Jessop CE, Tavender TJ, Watkins RH, Chambers JE, Bulleid NJ (2009) Substrate specificity of the oxidoreductase ERp57 is determined primarily by its interaction with calnexin and calreticulin. J Biol Chem 284(4):2194–2202. doi:10.1074/jbc.M808054200, M808054200 [pii]

74. Ishikawa Y, Vranka JA, Boudko SP, Pokidysheva E, Mizuno K, Zientek K, Keene DR, Rashmir-Raven AM, Nagata K, Winand NJ, Bachinger HP (2012) Mutation in cyclophilin B that causes hyperelastosis cutis in American Quarter Horse does not affect peptidylprolyl cis-trans isomerase activity but shows altered cyclophilin B-protein interactions and affects collagen folding. J Biol Chem 287(26):22253–22265. doi:10.1074/jbc.M111.333336, M111.333336 [pii]
75. Park S, You KH, Shong M, Goo TW, Yun EY, Kang SW, Kwon OY (2005) Overexpression of ERp29 in the thyrocytes of FRTL-5 cells. Mol Biol Rep 32(1):7–13
76. Horibe T, Gomi M, Iguchi D, Ito H, Kitamura Y, Masuoka T, Tsujimoto I, Kimura T, Kikuchi M (2004) Different contributions of the three CXXC motifs of human protein-disulfide isomerase-related protein to isomerase activity and oxidative refolding. J Biol Chem 279(6):4604–4611. doi:10.1074/jbc.M310922200, M310922200 [pii]
77. Vinaik R, Kozlov G, Gehring K (2013) Structure of the non-catalytic domain of the protein disulfide isomerase-related protein (PDIR) reveals function in protein binding. PLoS ONE 8(4):e62021. doi:10.1371/journal.pone.0062021, PONE-D-12-40270 [pii]
78. Barak NN, Neumann P, Sevvana M, Schutkowski M, Naumann K, Malesevic M, Reichardt H, Fischer G, Stubbs MT, Ferrari DM (2009) Crystal structure and functional analysis of the protein disulfide isomerase-related protein ERp29. J Mol Biol 385(5):1630–1642. doi:10.1016/j.jmb.2008.11.052, S0022-2836(08)01485-X [pii]
79. Mkrtchian S, Sandalova T (2006) ERp29, an unusual redox-inactive member of the thioredoxin family. Antioxid Redox Signal 8(3–4):325–337. doi:10.1089/ars.2006.8.325
80. Suaud L, Miller K, Alvey L, Yan W, Robay A, Kebler C, Kreindler JL, Guttentag S, Hubbard MJ, Rubenstein RC (2011) ERp29 regulates DeltaF508 and wild-type cystic fibrosis transmembrane conductance regulator (CFTR) trafficking to the plasma membrane in cystic fibrosis (CF) and non-CF epithelial cells. J Biol Chem 286(24):21239–21253. doi:10.1074/jbc.M111.240267, M111.240267 [pii]
81. Zhang D, Richardson DR (2011) Endoplasmic reticulum protein 29 (ERp29): an emerging role in cancer. Int J Biochem Cell Biol 43(1):33–36. doi:10.1016/j.biocel.2010.09.019, S1357-2725(10)00339-0 [pii]
82. Alanen HI, Salo KE, Pirneskoski A, Ruddock LW (2006) pH dependence of the peptide thiol-disulfide oxidase activity of six members of the human protein disulfide isomerase family. Antioxid Redox Signal 8(3–4):283–291. doi:10.1089/ars.2006.8.283
83. Lakkaraju AK, van der Goot FG (2013) Calnexin controls the STAT3-mediated transcriptional response to EGF. Mol Cell 51(3):386–396. doi:10.1016/j.molcel.2013.07.009, S1097-2765(13)00508-X [pii]
84. Delom F, Emadali A, Cocolakis E, Lebrun JJ, Nantel A, Chevet E (2007) Calnexin-dependent regulation of tunicamycin-induced apoptosis in breast carcinoma MCF-7 cells. Cell Death Differ 14(3):586–596. doi:10.1038/sj.cdd.4402012, 4402012 [pii]
85. Takizawa T, Tatematsu C, Watanabe K, Kato K, Nakanishi Y (2004) Cleavage of calnexin caused by apoptotic stimuli: implication for the regulation of apoptosis. J Biochem 136(3):399–405. doi:10.1093/jb/mvh133, 136/3/399 [pii]
86. Turano C, Gaucci E, Grillo C, Chichiarelli S (2011) ERp57/GRP58: a protein with multiple functions. Cell Mol Biol Lett 16(4):539–563. doi:10.2478/s11658-011-0022-z
87. Santana-Codina N, Carretero R, Sanz-Pamplona R, Cabrera T, Guney E, Oliva B, Clezardin P, Olarte OE, Loza-Alvarez P, Mendez-Lucas A, Perales JC, Sierra A (2013) A transcriptome-proteome integrated network identifies endoplasmic reticulum thiol oxidoreductase (ERp57) as a hub that mediates bone metastasis. Mol Cell Proteomics 12(8):2111–2125. doi:10.1074/mcp.M112.022772, M112.022772 [pii]
88. Gaucci E, Altieri F, Turano C, Chichiarelli S (2013) The protein ERp57 contributes to EGF receptor signaling and internalization in MDA-MB-468 breast cancer cells. J Cell Biochem 114(11):2461–2470. doi:10.1002/jcb.24590
89. Frand AR, Kaiser CA (1999) Ero1p oxidizes protein disulfide isomerase in a pathway for disulfide bond formation in the endoplasmic reticulum. Mol Cell 4(4):469–477. doi:S1097-2765(00)80198-7 [pii]

90. Jessop CE, Watkins RH, Simmons JJ, Tasab M, Bulleid NJ (2009) Protein disulphide isomerase family members show distinct substrate specificity: P5 is targeted to BiP client proteins. J Cell Sci 122(Pt 23):4287–4295. doi:10.1242/jcs.059154, jcs.059154 [pii]
91. Birk J, Meyer M, Aller I, Hansen HG, Odermatt A, Dick TP, Meyer AJ, Appenzeller-Herzog C (2013) Endoplasmic reticulum: reduced and oxidized glutathione revisited. J Cell Sci 126(Pt 7):1604–1617. doi:10.1242/jcs.117218, jcs.117218 [pii]
92. Lundstrom J, Holmgren A (1993) Determination of the reduction-oxidation potential of the thioredoxin-like domains of protein disulfide-isomerase from the equilibrium with glutathione and thioredoxin. Biochemistry 32(26):6649–6655
93. Aslund F, Berndt KD, Holmgren A (1997) Redox potentials of glutaredoxins and other thiol-disulfide oxidoreductases of the thioredoxin superfamily determined by direct protein-protein redox equilibria. J Biol Chem 272(49):30780–30786
94. Frickel EM, Frei P, Bouvier M, Stafford WF, Helenius A, Glockshuber R, Ellgaard L (2004) ERp57 is a multifunctional thiol-disulfide oxidoreductase. J Biol Chem 279(18):18277–18287. doi:10.1074/jbc.M314089200, M314089200 [pii]
95. Araki K, Inaba K (2012) Structure, mechanism, and evolution of Ero1 family enzymes. Antioxid Redox Signal 16(8):790–799. doi:10.1089/ars.2011.4418
96. Pollard MG, Travers KJ, Weissman JS (1998) Ero1p: a novel and ubiquitous protein with an essential role in oxidative protein folding in the endoplasmic reticulum. Mol Cell 1(2):171–182. doi:S1097-2765(00)80018-0 [pii]
97. Tu BP, Weissman JS (2002) The FAD- and O(2)-dependent reaction cycle of Ero1-mediated oxidative protein folding in the endoplasmic reticulum. Mol Cell 10(5):983–994. doi:S1097276502006962 [pii]
98. Sevier CS, Qu H, Heldman N, Gross E, Fass D, Kaiser CA (2007) Modulation of cellular disulfide-bond formation and the ER redox environment by feedback regulation of Ero1. Cell 129(2):333–344. doi:10.1016/j.cell.2007.02.039, S0092-8674(07)00325-X [pii]
99. Appenzeller-Herzog C, Riemer J, Zito E, Chin KT, Ron D, Spiess M, Ellgaard L (2010) Disulphide production by Ero1alpha-PDI relay is rapid and effectively regulated. EMBO J 29(19):3318–3329. doi:10.1038/emboj.2010.203, emboj2010203 [pii]
100. Benham AM, van Lith M, Sitia R, Braakman I (2013) Ero1-PDI interactions, the response to redox flux and the implications for disulfide bond formation in the mammalian endoplasmic reticulum. Philos Trans R Soc Lond B Biol Sci 368(1617):20110403. doi:10.1098/rstb.2011.0403, rstb.2011.0403 [pii]
101. Zito E, Chin KT, Blais J, Harding HP, Ron D (2010) ERO1-beta, a pancreas-specific disulfide oxidase, promotes insulin biogenesis and glucose homeostasis. J Cell Biol 188(6):821–832. doi:10.1083/jcb.200911086, jcb.200911086 [pii]
102. Ruddock LW (2012) Low-molecular-weight oxidants involved in disulfide bond formation. Antioxid Redox Signal 16(10):1129–1138. doi:10.1089/ars.2011.4481
103. Appenzeller-Herzog C, Riemer J, Christensen B, Sorensen ES, Ellgaard L (2008) A novel disulphide switch mechanism in Ero1alpha balances ER oxidation in human cells. EMBO J 27(22):2977–2987. doi:10.1038/emboj.2008.202, emboj2008202 [pii]
104. Inaba K, Masui S, Iida H, Vavassori S, Sitia R, Suzuki M (2010) Crystal structures of human Ero1alpha reveal the mechanisms of regulated and targeted oxidation of PDI. EMBO J 29(19):3330–3343. doi:10.1038/emboj.2010.222, emboj2010222 [pii]
105. Tavender TJ, Bulleid NJ (2010) Molecular mechanisms regulating oxidative activity of the Ero1 family in the endoplasmic reticulum. Antioxid Redox Signal 13(8):1177–1187. doi:10.1089/ars.2010.3230
106. Rutkevich LA, Williams DB (2012) Vitamin K epoxide reductase contributes to protein disulfide formation and redox homeostasis within the endoplasmic reticulum. Mol Biol Cell 23(11):2017–2027. doi:10.1091/mbc.E12-02-0102, mbc.E12-02-0102 [pii]
107. Araki K, Iemura S, Kamiya Y, Ron D, Kato K, Natsume T, Nagata K (2013) Ero1-alpha and PDIs constitute a hierarchical electron transfer network of endoplasmic reticulum oxidoreductases. J Cell Biol 202(6):861–874. doi:10.1083/jcb.201303027, jcb.201303027 [pii]
108. Tu BP, Weissman JS (2004) Oxidative protein folding in eukaryotes: mechanisms and consequences. J Cell Biol 164(3):341–346. doi:10.1083/jcb.200311055, jcb.200311055 [pii]

109. Gross E, Sevier CS, Heldman N, Vitu E, Bentzur M, Kaiser CA, Thorpe C, Fass D (2006) Generating disulfides enzymatically: reaction products and electron acceptors of the endoplasmic reticulum thiol oxidase Ero1p. Proc Natl Acad Sci U S A 103(2):299–304. doi:10.1073/pnas.0506448103, 0506448103 [pii]
110. Ramming T, Appenzeller-Herzog C (2013) Destroy and exploit: catalyzed removal of hydroperoxides from the endoplasmic reticulum. Int J Cell Biol 2013:180906. doi:10.1155/2013/180906
111. Tavender TJ, Sheppard AM, Bulleid NJ (2008) Peroxiredoxin IV is an endoplasmic reticulum-localized enzyme forming oligomeric complexes in human cells. Biochem J 411(1):191–199. doi:10.1042/BJ20071428, BJ20071428 [pii]
112. Tavender TJ, Bulleid NJ (2010) Peroxiredoxin IV protects cells from oxidative stress by removing H2O2 produced during disulphide formation. J Cell Sci 123(Pt 15):2672–2679. doi:10.1242/jcs.067843, jcs.067843 [pii]
113. Eisner V, Csordas G, Hajnoczky G (2013) Interactions between sarco-endoplasmic reticulum and mitochondria in cardiac and skeletal muscle—pivotal roles in Ca(2)(+) and reactive oxygen species signaling. J Cell Sci 126(Pt 14):2965–2978. doi:10.1242/jcs.093609, jcs.093609 [pii]
114. Laurindo FR, Araujo TL, Abrahao TB (2014) Nox NADPH oxidases and the endoplasmic reticulum. Antioxid Redox Signal 20(17):2755–75 doi:10.1089/ars.2013.5605
115. Tavender TJ, Springate JJ, Bulleid NJ (2010) Recycling of peroxiredoxin IV provides a novel pathway for disulphide formation in the endoplasmic reticulum. EMBO J 29(24):4185–4197. doi:10.1038/emboj.2010.273, emboj2010273 [pii]
116. Park SW, Zhen G, Verhaeghe C, Nakagami Y, Nguyenvu LT, Barczak AJ, Killeen N, Erle DJ (2009) The protein disulfide isomerase AGR2 is essential for production of intestinal mucus. Proc Natl Acad Sci U S A 106(17):6950–6955. doi:10.1073/pnas.0808722106, 0808722106 [pii]
117. Chevet E, Fessart D, Delom F, Mulot A, Vojtesek B, Hrstka R, Murray E, Gray T, Hupp T (2013) Emerging roles for the pro-oncogenic anterior gradient-2 in cancer development. Oncogene 32(20):2499–2509. doi:10.1038/onc.2012.346, onc2012346 [pii]
118. Maslon MM, Hrstka R, Vojtesek B, Hupp TR (2010) A divergent substrate-binding loop within the pro-oncogenic protein anterior gradient-2 forms a docking site for Reptin. J Mol Biol 404(3):418–438. doi:10.1016/j.jmb.2010.09.035, S0022-2836(10)01023-5 [pii]
119. Xu S, Sankar S, Neamati N (2013) Protein disulfide isomerase: a promising target for cancer therapy. Drug Discov Today 19(3):222–240 doi:10.1016/j.drudis.2013.10.017, S1359-6446(13)00384-X [pii]
120. Feige MJ, Groscurth S, Marcinowski M, Shimizu Y, Kessler H, Hendershot LM, Buchner J (2009) An unfolded CH1 domain controls the assembly and secretion of IgG antibodies. Mol Cell 34(5):569–579. doi:10.1016/j.molcel.2009.04.028, S1097-2765(09)00304-9 [pii]
121. Feige MJ, Hendershot LM, Buchner J (2010) How antibodies fold. Trends Biochem Sci 35(4):189–198. doi:10.1016/j.tibs.2009.11.005, S0968-0004(09)00229-1 [pii]
122. Alanen HI, Williamson RA, Howard MJ, Hatahet FS, Salo KE, Kauppila A, Kellokumpu S, Ruddock LW (2006) ERp27, a new non-catalytic endoplasmic reticulum-located human protein disulfide isomerase family member, interacts with ERp57. J Biol Chem 281(44):33727–33738. doi:10.1074/jbc.M604314200, M604314200 [pii]
123. Kober FX, Koelmel W, Kuper J, Drechsler J, Mais C, Hermanns HM, Schindelin H (2013) The crystal structure of the protein-disulfide isomerase family member ERp27 provides insights into its substrate binding capabilities. J Biol Chem 288(3):2029–2039. doi:10.1074/jbc.M112.410522, M112.410522 [pii]
124. Vavassori S, Cortini M, Masui S, Sannino S, Anelli T, Caserta IR, Fagioli C, Mossuto MF, Fornili A, van Anken E, Degano M, Inaba K, Sitia R (2013) A pH-regulated quality control cycle for surveillance of secretory protein assembly. Mol Cell 50(6):783–792. doi:10.1016/j.molcel.2013.04.016, S1097-2765(13)00297-9 [pii]
125. Grubb S, Guo L, Fisher EA, Brodsky JL (2012) Protein disulfide isomerases contribute differentially to the endoplasmic reticulum-associated degradation of apolipoprotein B and

other substrates. Mol Biol Cell 23(4):520–532. doi:10.1091/mbc.E11-08-0704, mbc.E11-08-0704 [pii]
126. Lee J, Choi TG, Ha J, Kim SS (2012) Cyclosporine A suppresses immunoglobulin G biosynthesis via inhibition of cyclophilin B in murine hybridomas and B cells. Int Immunopharmacol 12(1):42–49. doi:10.1016/j.intimp. 2011.10.007, S1567-5769(11)00401-2 [pii]
127. Bernasconi R, Solda T, Galli C, Pertel T, Luban J, Molinari M (2010) Cyclosporine A-sensitive, cyclophilin B-dependent endoplasmic reticulum-associated degradation. PLoS ONE 5(9) doi:10.1371/journal.pone.0013008
128. Lai CW, Aronson DE, Snapp EL (2010) BiP availability distinguishes states of homeostasis and stress in the endoplasmic reticulum of living cells. Mol Biol Cell 21(12):1909–1921. doi:10.1091/mbc.E09-12-1066, E09-12-1066 [pii]
129. Sakono M, Seko A, Takeda Y, Hachisu M, Ito Y (2012) Biophysical properties of UDP-glucose:glycoprotein glucosyltransferase, a folding sensor enzyme in the ER, delineated by synthetic probes. Biochem Biophys Res Commun 426(4):504–510. doi:10.1016/j.bbrc.2012.08.112, S0006-291×(12)01680-4 [pii]
130. Buzzi LI, Simonetta SH, Parodi AJ, Castro OA (2011) The two Caenorhabditis elegans UDP-glucose:glycoprotein glucosyltransferase homologues have distinct biological functions. PLoS ONE 6(11):e27025. doi:10.1371/journal.pone.0027025, PONE-D-11-08015 [pii]
131. Takeda Y, Seko A, Hachisu M, Daikoku S, Izumi M, Koizumi A, Fujikawa K, Kajihara Y, Ito Y (2014) Both isoforms of Human UDP-glucose:glycoprotein glucosyltransferase are enzymatically active. Glycobiology 24(4):344–350 doi:10.1093/glycob/cwt163, cwt163 [pii]
132. Prados MB, Caramelo JJ, Miranda SE (2013) Progesterone regulates the expression and activity of two mouse isoforms of the glycoprotein folding sensor UDP-Glc: Glycoprotein glucosyltransferase (UGGT). Biochim Biophys Acta 1833(12):3368–3374. doi:10.1016/j.bbamcr.2013.09.022, S0167-4889(13)00344-3 [pii]
133. Korotkov KV, Kumaraswamy E, Zhou Y, Hatfield DL, Gladyshev VN (2001) Association between the 15-kDa selenoprotein and UDP-glucose:glycoprotein glucosyltransferase in the endoplasmic reticulum of mammalian cells. J Biol Chem 276(18):15330–15336. doi:10.1074/jbc.M009861200, M009861200 [pii]
134. Labunskyy VM, Ferguson AD, Fomenko DE, Chelliah Y, Hatfield DL, Gladyshev VN (2005) A novel cysteine-rich domain of Sep15 mediates the interaction with UDP-glucose:glycoprotein glucosyltransferase. J Biol Chem 280(45):37839–37845. doi:10.1074/jbc.M508685200, M508685200 [pii]
135. Labunskyy VM, Yoo MH, Hatfield DL, Gladyshev VN (2009) Sep15, a thioredoxin-like selenoprotein, is involved in the unfolded protein response and differentially regulated by adaptive and acute ER stresses. Biochemistry 48(35):8458–8465. doi:10.1021/bi900717p
136. Ferguson AD, Labunskyy VM, Fomenko DE, Arac D, Chelliah Y, Amezcua CA, Rizo J, Gladyshev VN, Deisenhofer J (2006) NMR structures of the selenoproteins Sep15 and SelM reveal redox activity of a new thioredoxin-like family. J Biol Chem 281(6):3536–3543. doi:10.1074/jbc.M511386200, M511386200 [pii]
137. Korotkov KV, Novoselov SV, Hatfield DL, Gladyshev VN (2002) Mammalian selenoprotein in which selenocysteine (Sec) incorporation is supported by a new form of Sec insertion sequence element. Mol Cell Biol 22(5):1402–1411
138. Novoselov SV, Hua D, Lobanov AV, Gladyshev VN (2006) Identification and characterization of Fep15, a new selenocysteine-containing member of the Sep15 protein family. Biochem J 394(Pt 3):575–579. doi:10.1042/BJ20051569, BJ20051569 [pii]
139. Hebert DN, Molinari M (2012) Flagging and docking: dual roles for N-glycans in protein quality control and cellular proteostasis. Trends Biochem Sci 37(10):404–410. doi:10.1016/j.tibs.2012.07.005, S0968-0004(12)00109-0 [pii]
140. Gauss R, Kanehara K, Carvalho P, Ng DT, Aebi M (2011) A complex of Pdi1p and the mannosidase Htm1p initiates clearance of unfolded glycoproteins from the endoplasmic reticulum. Mol Cell 42(6):782–793. doi:10.1016/j.molcel.2011.04.027, S1097-2765(11)00373-X [pii]

141. Olzmann JA, Kopito RR, Christianson JC (2013) The mammalian endoplasmic reticulum-associated degradation system. Cold Spring Harb Perspect Biol 5(9). doi:10.1101/cshperspect.a013185, cshperspect.a013185 [pii]
142. Reiterer V, Nyfeler B, Hauri HP (2010) Role of the lectin VIP36 in post-ER quality control of human alpha1-antitrypsin. Traffic 11(8):1044–1055. doi:10.1111/j.1600-0854.2010.01078.x, TRA1078 [pii]
143. Brandizzi F, Barlowe C (2013) Organization of the ER-Golgi interface for membrane traffic control. Nat Rev Mol Cell Biol 14(6):382–392. doi:10.1038/nrm3588, nrm3588 [pii]
144. Venditti R, Wilson C, De Matteis MA (2014) Exiting the ER: what we know and what we don't. Trends Cell Biol 24(1):9–18. doi:10.1016/j.tcb.2013.08.005, S0962-8924(13)00141-4 [pii]
145. Marzioch M, Henthorn DC, Herrmann JM, Wilson R, Thomas DY, Bergeron JJ, Solari RC, Rowley A (1999) Erp1p and Erp2p, partners for Emp24p and Erv25p in a yeast p24 complex. Mol Biol Cell 10(6):1923–1938
146. Denzel A, Otto F, Girod A, Pepperkok R, Watson R, Rosewell I, Bergeron JJ, Solari RC, Owen MJ (2000) The p24 family member p23 is required for early embryonic development. Curr Biol 10(1):55–58. doi:S0960-9822(99)00266-3 [pii]
147. Buechling T, Chaudhary V, Spirohn K, Weiss M, Boutros M (2011) p24 proteins are required for secretion of Wnt ligands. EMBO Rep 12(12):1265–1272. doi:10.1038/embor.2011.212, embor2011212 [pii]
148. Strating JR, Martens GJ (2009) The p24 family and selective transport processes at the ER-Golgi interface. Biol Cell 101(9):495–509. doi:10.1042/BC20080233, BC20080233 [pii]
149. Lederkremer GZ (2009) Glycoprotein folding, quality control and ER-associated degradation. Curr Opin Struct Biol 19(5):515–523. doi:10.1016/j.sbi.2009.06.004, S0959-440×(09)00094-3 [pii]
150. Leitman J, Ron E, Ogen-Shtern N, Lederkremer GZ (2013) Compartmentalization of endoplasmic reticulum quality control and ER-associated degradation factors. DNA Cell Biol 32(1):2–7. doi:10.1089/dna.2012.1889
151. Pan S, Wang S, Utama B, Huang L, Blok N, Estes MK, Moremen KW, Sifers RN (2011) Golgi localization of ERManI defines spatial separation of the mammalian glycoprotein quality control system. Mol Biol Cell 22(16):2810–2822. doi:10.1091/mbc.E11-02-0118, mbc.E11-02-0118 [pii]
152. Pan S, Cheng X, Sifers RN (2013) Golgi-situated endoplasmic reticulum alpha-1, 2-mannosidase contributes to the retrieval of ERAD substrates through a direct interaction with gamma-COP. Mol Biol Cell 24(8):1111–1121. doi:10.1091/mbc.E12-12-0886, mbc.E12-12-0886 [pii]
153. Hanrahan JW, Sampson HM, Thomas DY (2013) Novel pharmacological strategies to treat cystic fibrosis. Trends Pharmacol Sci 34(2):119–125. doi:10.1016/j.tips.2012.11.006, S0165-6147(12)00198-8 [pii]
154. Carlile GW, Keyzers RA, Teske KA, Robert R, Williams DE, Linington RG, Gray CA, Centko RM, Yan L, Anjos SM, Sampson HM, Zhang D, Liao J, Hanrahan JW, Andersen RJ, Thomas DY (2012) Correction of F508del-CFTR trafficking by the sponge alkaloid latonduine is modulated by interaction with PARP. Chem Biol 19(10):1288–1299. doi:10.1016/j.chembiol.2012.08.014, S1074-5521(12)00323-7 [pii]
155. Kamiya Y, Satoh T, Kato K (2012) Molecular and structural basis for N-glycan-dependent determination of glycoprotein fates in cells. Biochim Biophys Acta 1820(9):1327–1337. doi:10.1016/j.bbagen.2011.12.017, S0304-4165(11)00317-5 [pii]
156. Gidalevitz T, Stevens F, Argon Y (2013) Orchestration of secretory protein folding by ER chaperones. Biochim Biophys Acta 1833(11):2410–2424. doi:10.1016/j.bbamcr.2013.03.007, S0167-4889(13)00098-0 [pii]
157. Wang X, Venable J, LaPointe P, Hutt DM, Koulov AV, Coppinger J, Gurkan C, Kellner W, Matteson J, Plutner H, Riordan JR, Kelly JW, Yates JR 3rd, Balch WE (2006) Hsp90 cochaperone Aha1 downregulation rescues misfolding of CFTR in cystic fibrosis. Cell 127(4):803–815. doi:10.1016/j.cell.2006.09.043, S0092-8674(06)01378-X [pii]

158. Mehnert M, Sommer T, Jarosch E (2014) Der1 promotes movement of misfolded proteins through the endoplasmic reticulum membrane. Nat Cell Biol 16(1):77–86. doi:10.1038/ncb2882, ncb2882 [pii]
159. Xu C, Wang S, Thibault G, Ng DT (2013) Futile protein folding cycles in the ER are terminated by the unfolded protein O-mannosylation pathway. Science 340(6135):978–981. doi:10.1126/science.1234055, 340/6135/978 [pii]
160. Wang HJ, Guay G, Pogan L, Sauve R, Nabi IR (2000) Calcium regulates the association between mitochondria and a smooth subdomain of the endoplasmic reticulum. J Cell Biol 150(6):1489–1498
161. Birault V, Solari R, Hanrahan J, Thomas DY (2013) Correctors of the basic trafficking defect of the mutant F508del-CFTR that causes cystic fibrosis. Curr Opin Chem Biol 17(3):353–360. doi:10.1016/j.cbpa.2013.04.020, S1367-5931(13)00075-6 [pii]
162. Grimm S (2012) The ER-mitochondria interface: the social network of cell death. Biochim Biophys Acta 1823(2):327–334. doi:10.1016/j.bbamcr.2011.11.018, S0167-4889(11)00320-X [pii]
163. Lapuente-Brun E, Moreno-Loshuertos R, Acin-Perez R, Latorre-Pellicer A, Colas C, Balsa E, Perales-Clemente E, Quiros PM, Calvo E, Rodriguez-Hernandez MA, Navas P, Cruz R, Carracedo A, Lopez-Otin C, Perez-Martos A, Fernandez-Silva P, Fernandez-Vizarra E, Enriquez JA (2013) Supercomplex assembly determines electron flux in the mitochondrial electron transport chain. Science 340(6140):1567–1570. doi:10.1126/science.1230381, 340/6140/1567 [pii]

Chapter 11
Chaperones of the Endoplasmic Reticulum Associated Degradation (ERAD) Pathway

Johan C. Sunryd, Abla Tannous, Lydia Lamriben and Daniel N. Hebert

Abstract The endoplasmic reticulum (ER) is a diverse organelle with multiple specialized functions including the folding and assembly of secretory proteins. The ER harbors a number of chaperones, foldases, and covalent modifiers that aid in these processes. Due to the high protein throughput of the ER, inevitably some of the secreted proteins fail to reach their native state and therefore must be degraded. In a process known as ER-associated degradation (ERAD), misfolded proteins are recognized, translocated out of the ER into the cytoplasm, ubiquitinated, extracted from the ER membrane, deubiquitinated, and finally degraded by the 26S proteasome to prevent toxic build-up of non-native proteins. To account for the different properties and topologies of misfolded secretory cargo, a number of ER protein complexes are employed for the distinct steps of ERAD. Some of these complexes are shared between yeast and metazoan, while other appear unique to metazoa, reflecting their increased secretory load.

1 Introduction

Polypeptides enter the eukaryotic secretory pathway in an unfolded state through Sec61 translocons in the endoplasmic reticulum (ER) membrane. The ER is a specialized organelle which houses factors that assist in the maturation of soluble and membrane proteins. Key protein maturation events include protein folding, signal sequence cleavage, disulfide bond formation, glycan addition and trim-

D. N. Hebert (✉) · J. C. Sunryd · A. Tannous · L. Lamriben
Department of Biochemistry and Molecular Biology, Program in Molecular and Cellular Biology, University of Massachusetts, Amherst, MA, USA
e-mail: dhebert@biochem.umass.edu

J. C. Sunryd
e-mail: jsunryd@mcb.umass.edu

A. Tannous
e-mail: atannous@mcb.umass.edu

L. Lamriben
e-mail: llamribe@mcb.umass.edu

W. A. Houry (ed.), *The Molecular Chaperones Interaction Networks in Protein Folding and Degradation,* Interactomics and Systems Biology, DOI 10.1007/978-1-4939-1130-1_11,

ming, and assembly [1, 2]. An important quality control process is also in place in the ER that evaluates the integrity of the protein product to ideally allow only functional, properly folded, and assembled proteins to progress through the secretory pathway. Defective or non-native products are targeted for degradation through a process known as ER-associated degradation or ERAD. Given the large load and diversity of substrates, the ER is equipped with a number of pathways that evaluate proteins based on exhibiting common aberrations associated with misfolded, non-native, or unassembled proteins. As proteins fold in the ER, hydrophobic patches are buried in the protein core and disulfide bonds are commonly formed given the oxidizing environment found in the ER lumen. Miscuing of these maturation events provides signals of protein aberration including exposed hydrophobic patches or free thiols that can be recognized by quality control factors. N-linked glycans are also used as quality control codes to signal the integrity of the maturing protein for assistance in the targeting process [3]. ERAD can also be engaged as a regulatory mechanism to control the level of proteins. A group of ER resident molecular chaperones play central roles in ERAD through selection, preparation, and targeting of secretory cargo for retrotranslocation from the ER lumen to the cytoplasm where they are degraded by the proteasome (Table 11.1).

2 Traditional Chaperones and Oxidoreductases

Traditional molecular chaperones are ubiquitously expressed proteins that bind exposed hydrophobic regions, thus protecting client proteins from aggregation. Their substrate binding is regulated by adenine nucleotides. The metazoan ER contains heat shock protein (Hsp)70 (binding immunoglobulin protein, BiP) and Hsp90 (glucose regulated protein of 90 kD, GRP94) family members but lacks an Hsp60 protein. Hsp70s contain a substrate-binding domain (SBD) connected by a linker to an N-terminal nucleotide-binding domain (NBD) [4]. When adenosine triphosphate (ATP) is bound to the NBD, the SBD adopts an open conformation with a low substrate-binding affinity. In the adenosine diphosphate (ADP)-bound form, the SBD is in a closed conformation with a high affinity for substrates. After ATP hydrolysis, nucleotide exchange factors allow exchange of ADP for ATP, regenerating the open conformation and resulting in substrate release. Hsp70s associate with J domain-containing proteins that stimulate ATP hydrolysis. BiP and its yeast ortholog Kar2p have been referred to as master regulators of the ER because of their diverse roles including aiding in translocation, folding and ERAD [5]. BiP is one of the most abundant proteins in the ER, and its levels are regulated by the unfolded protein response (UPR) signaling cascade, to cope with various ER stress conditions [6]. The centrality of BiP is illustrated by infant mortality, defects in brain development, and impaired ER quality control of BiP mutant knock-in mice [7].

Table 11.1 Chaperones, quality control factors, and other proteins involved in ERAD are listed with their human and *Saccharomyces cerevisiae* homologues denoted

Category	Human	Yeast	Function
Glycan processing and binding	Calnexin	Cne1p	Carbohydrate binding chaperone
	Calreticulin	Cne1p	Carbohydrate binding chaperone
	EDEM1	Htm1p	ER mannosidase or lectin
	EDEM2	–	ER mannosidase or lectin
	EDEM3	–	ER mannosidase or lectin
	ERMan1/MAN1B1	Mns1p	ER mannosidase
	Malectin	–	Lectin chaperone
	OS-9	Yos9p	MRH domain-containing ER quality control receptor
	UGT1	–	Glucosyl transferase and folding sensor
	XTP3-B	–	MRH domain-containing ER quality control receptor
Hsp70/90	BiP	Kar2p	Adenine nucleotide regulated chaperone
	GRP94	–	Adenine nucleotide regulated chaperone
Oxidoreductase	ERdj5	–	Oxidoreductase
	ERp57	–	Oxidoreductase
	PDI	Pdi1p	Oxidoreductase
Retrotranslocation	BAG6	–	Interacts with retrotranslocated proteins in cytoplasm
	Derlin-1	Der1p	Holds retrotranslocated substrates prior to ubiquitination
	Derlin-2	Der1p	Interacts with HRD1 and GP78
	Derlin-3	Der1p	Interacts with HRD1
	Erlin1–2	–	Interacts with GP78
	FAM8A1	–	Aids HRD1
	Herp	Usa1p	Interacts with HRD1
	NPL4	Npl4p	p97 co-factor
	p97	Cdc48p	AAA ATPase that provides the mechanical force for retrotranslocation
	SEL1L	Hrd3p	Adapter for the HRD1/Hrd1p complex
	TorsinA	–	AAA ATPase in ER lumen
	UBDX8	–	ER membrane protein that recruits p97
	UFD1	Ufd1p	p97 co-factor
	VIMP	–	ER membrane protein that recruits p97
Translocon	Sec61	Sec61p	ER translocon
Ubiquitination	–	Ufd2p	E4 ligase that extends ubiquitin chains after Cdc48p mediated extraction
	ATAXIN-3	–	A deubiquitinating enzyme found in complex with p97
	GP78	–	E3 ligase that ubiquitinates soluble and membrane proteins with luminal lesions
	HRD1	Hrd1p	E3 ligase that ubiquitinates soluble and membrane proteins with luminal lesions
	RMA1	–	E3 ligase that ubiquitinates membrane proteins with cytoplasmic lesions
	TEB4	Doa10p	E3 ligase that ubiquitinates membrane proteins with cytoplasmic lesions

Table 11.1 (continued)

Category	Human	Yeast	Function
	UBE2G2	Ubc7p	E2 conjugating enzyme
	UBE2J1	Ubc6p	E2 conjugating enzyme
	YOD1	–	A deubiquitinating enzyme found in complex with p97

ER endoplasmic reticulum, *ERAD* ER-associated degradation, *EDEM* ER-degradation enhancing α-mannosidase-like proteins

The involvement of BiP in ERAD was suggested from the observation that BiP associated with non-secreted immunoglobulin light chains that are targeted for degradation in the absence of immunoglobulin heavy chains [8]. Degradation of these substrates was concomitant with their dissociation from BiP, suggesting that BiP remains engaged with the misfolded substrates for delivery for degradation. In yeast, a temperature sensitive mutation in Kar2p inhibited ERAD [9, 10]. The function of Kar2p in enhancing ERAD resides in its ability to maintain substrates in a retrotranslocation competent form by preventing their aggregation [11]. Furthermore, BiP/Kar2p is a member of a downstream ERAD membrane complex containing several key players of the ERAD pathway including ER lectins, adapters, and an ubiquitin ligase [12, 13].

BiP/Kar2p associates with ER localized J domain-containing proteins that act as cofactors and help to define the function of BiP, and its conformations and substrate-binding specificity. BiP/Kar2p cofactors have also been shown to recognize ERAD substrates and promote their degradation. In yeast, Scj1p and Jem1p, two J-domain containing Kar2p co-chaperones assist in maintaining ERAD substrates in soluble retrotranslocation-compatible forms [11]. In mammals, among the seven known J-domain containing BiP cofactors, ERdj4 and ERdj5 have been implicated in ERAD [14–16]. The ERAD activity of ERdj4 is proposed to occur via its interaction with Derlin-1, an integral membrane protein that is part of the ERAD retrotranslocation complex [16]. ERdj5 is proposed to act as a reductase thereby helping to create a translocation competent substrate that can be threaded through a retrotranslocon [14, 15].

A recent study showed that unassembled subunits of a complex containing transmembrane segments with lower hydrophobicity failed to integrate into ER membranes and thus became luminal prior to being degraded [17]. BiP directly bound to the transmembrane segment of these subunits once they became luminal. These results support a mechanism whereby BiP searches for exposed hydrophobic domains, which includes hydrophobic transmembrane domains that have not stably partitioned into the membrane, and then helps target these bound proteins for destruction. The process by which BiP switches from its role as a folding factor to a degradative factor is not clearly defined although it has been proposed that these opposing functions could be regulated by BiP cofactors [18].

The ER Hsp90 family member, GRP94, has also been found to play a role in ERAD in metazoa even though it is absent from lower eukaryotes such as yeast. It associates with an ERAD quality control lectin, and the knockdown of GRP94 stabilized the ERAD substrate α-1 antitrypsin null Hong Kong (NHK) [19]. Its association with ERAD machinery and the stabilization of ERAD substrates in its absence are supportive of a role for GRP94 in substrate selection for ERAD.

The ER also harbors a large family of protein disulfide isomerases that catalyze the formation, reduction, or isomerization of disulfides bonds to assist proteins in reaching their tertiary structure. Yeast and mammals have five and twenty protein disulfide isomerase family members, respectively [20]. The redox activity of these oxidoreductases occurs via their catalytic Cys-X-X-Cys motifs that reside in a domain that adopts a thioredoxin-like fold. The number and position of these thioredoxin domains vary. In addition to possessing oxidoreductase activity, some of these family members also have chaperone-like activity [20, 21].

Protein disulfide isomerases help in the preparation of misfolded substrates for degradation. In yeast, this function was dependent on either the redox or the chaperone activity of Pdi1p [22, 23]. Supporting the requirement for the redox activity for degradation, mutating the enzymatic site of Pdi1p resulted in impaired degradation of a disulfide containing ERAD substrate. Conversely, the chaperone activity of Pdi1p has been described for the degradation of the non-glycosylated cysteine free substrate pro-α factor [22]. Substrate turnover was defective in a yeast strain harboring a mutation in the peptide-binding region of Pdi1p. Even the degradation of the cysteine-rich apolipoprotein B did not depend on the redox activity of Pdi1p but rather required its chaperone activity underscoring the importance of both the oxidoreductase and chaperone-like activities for Pdi1p in yeast ERAD [23].

The most abundant protein disulfide isomerase family member in the mammalian ER is PDI. It contains four thioredoxin domains (abb′a′), of which the two domains a and a′ have active Cys-X-X-Cys motifs [20]. In addition to its prominent role in oxidative folding of maturing substrates, PDI has been described as a factor implicated in substrate retrotranslocation and degradation similar to its yeast counterpart, Pdi1p. PDI was associated with reduced BACE457 prior to degradation suggesting that it could act to reduce and unfold the substrate in preparation for retrotranslocation [24]. PDI was also implicated in the degradation of the major histocompatibility complex (MHC) class I heavy chains by the human cytomegalovirus protein US2 [25]. However, this function was independent of the catalytic activity of PDI since catalytically inactive mutants of PDI retained their ability to facilitate degradation. Mutations that reduce the peptide-binding affinity of PDI in the b′ substrate-binding domain inhibited MHC class I heavy chain degradation. This hinted at a role for the chaperone activity of PDI in facilitating ERAD.

Additionally, some toxins rely upon PDI for their retrotranslocation across the ER membrane [26]. Once retrotranslocated, these toxins bypass degradation and cause intoxication in the cytoplasm. PDI can contribute to the retrotranslocation of these toxins by reducing intermolecular disulfides of the subunits or through its chaperone activity [27, 28]. The PDI family members of ERp57, ERp72, and ER flavoprotein-associated with degradation (ERFAD) have also been proposed to facilitate degradation of ERAD substrates in mammals [23, 29, 30]. ERFAD interacts with the PDIs, ERp90, and ERdj5, and it was postulated to be a potential reductase for ERdj5 [29, 30], thereby ensuring that ERdj5 returns to its functional reduced state in order to act on other ERAD substrates and render them translocation competent.

The mechanism through which PDIs are capable of differentiating between proteins on the way to maturation and proteins on the way to destruction is not known. The number of PDI family members that appear to play a role in ERAD continues to grow. Whether different PDIs play backup roles in facilitating ERAD or they have different specificity for different ERAD substrates remains an open question.

3 N-linked Glycans as Quality Control Signals

Nascent proteins co-translationally entering the rough ER are commonly modified by the oligosaccharyltransferase that transfers pre-assembled glycans from dolicholpyrophosphate onto asparagine found in the consensus sequence of Asn-X-Thr/Ser/Cys [2]. The glycans, which are transferred en bloc, initially comprised two N-acetyl glucosamines, nine mannoses, and three glucoses ($GlcNAc_2$-Man_9-Glc_3). The mannoses are distributed throughout the three branches of the glycan termed A, B, and C and the glucoses are linked to the A branch (Fig. 11.1). The composition of N-linked glycans is dynamic as it provides information about the integrity of the protein and helps to direct the fate of the protein by mediating interactions with various carbohydrate-binding chaperones and quality control receptors that recognize proteins possessing certain glycan compositions [3].

The first two glucose residues are rapidly trimmed co-translationally by glucosidases I and II shortly after the glycans are attached generating monoglucosylated species [31, 32] (Fig. 11.2). Glucosidase II also cleaves the final glucose from the glycans. The cleavage of the second glucose by glucosidase II occurs faster than the final or third glucose as they have contrasting roles [33]. The initial cleavage by glucosidase II results in the creation of monoglucosylated glycans that allow binding to the lectin chaperones calnexin, a type I membrane protein, and its soluble paralog, calreticulin while further glucose cleavage results in the release of the nonglucosylated glycoprotein from these chaperones [33–35].

Proteins that have not attained their native structure after complete de-glucosylation can be targeted for rebinding to calnexin and calreticulin through reglucosylation by uridine diphosphate (UDP)-glucose: glycoprotein glucosyltransferase 1 (UGT1) [36]. Similar to the classical chaperones, UGT1 appears to recognize exposed hydrophobic regions on incompletely folded proteins and deploys this binding selectivity to direct its transferase activity toward nearby unglucosylated glycans [37, 38]. In addition to recognizing folding defects, UGT1 recognizes folded but unassembled subunits of a complex and thus plays a role in quality control of unassembled oligomeric proteins [39]. The lectin chaperone-binding cycle promotes folding and assembly of glycoproteins by: (1) delaying the folding of chaperone bound domains; (2) supporting the recruitment of their associated oxidoreductase ERp57 and (3) proline isomerase CyB to assist in proline rearrangement, (4) inhibiting aggregation; and (5) mediating the retention of non-native proteins in the ER[40–45]

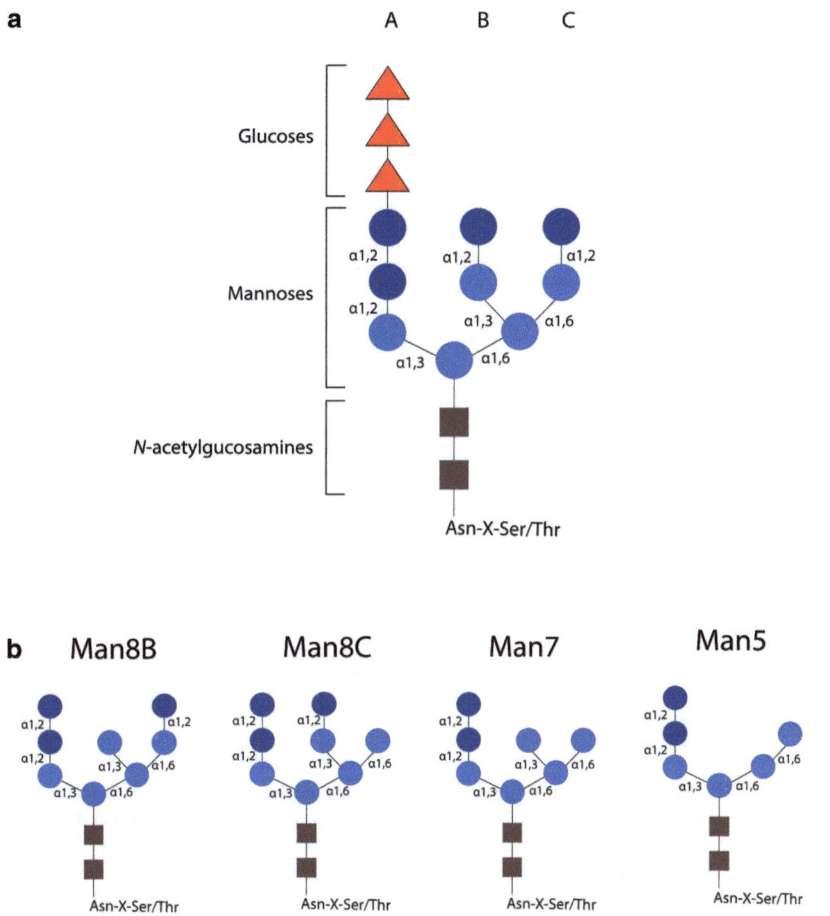

Fig. 11.1 Structure of an N-linked glycan. **a** Illustration of an asparagine-linked glycan composed of two N-acetylglucosamines (*brown squares*), nine mannoses (*dark and light blue circles*), and three glucoses (*red triangles*). The different branches are labeled *A*, *B*, and *C*. α(1,2)-linked mannoses are shown in *dark blue* while the α(1,6)-linked mannoses are in *light blue*. **b** Glycoforms generated as a result of glycan trimming in the endoplasmic reticulum (*ER*)

Calnexin and calreticulin can bind to proteins deemed misfolded to aid their folding reactions. Several lines of evidence suggest that the binding of misfolded proteins to lectin chaperones protects them from degradation by ERAD and retains them in the ER. Degradation of NHK required release from calnexin and was increased in presence of castanospermine, a glucosidase I and II inhibitor that supports the accumulation of tri-glucosylated glycans, thereby preventing binding to lectin chaperones [46]. A similar outcome was observed for BACE457 where glucosidase inhibition with deoxynojirimycin accelerated the degradation of BACE457 by ERAD [24]. Deletion of the calnexin homolog (Cne1p) in *Saccharomyces cerevisiae* resulted in accelerated degradation of a hypoglycosylated carboxypeptidase

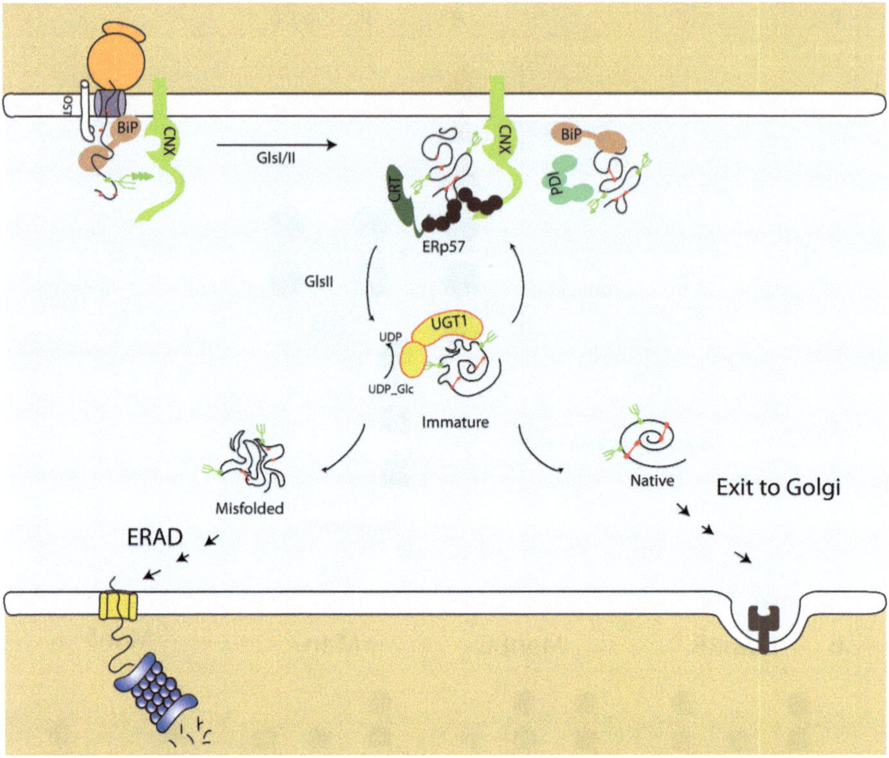

Fig. 11.2 Early ER protein maturation events. After proteins are translocated into the ER, glucosidases I and II trim the first and second glucoses resulting in monoglucosylated glycans that are recognized by the lectin chaperones calnexin and calreticulin. Binding to these lectin chaperones also recruits their associated factors ERp57, to assist in disulfide bond formation depicted by the *red lines*. Maturing substrates can also bind to BiP and the protein disulfide isomerase (*PDI*), as well as other PDI family members, to facilitate protein maturation. Trimming of the final glucose releases the substrate from calnexin/calreticulin. Substrates that reached their native fold are targeted for anterograde trafficking while immature substrates can be reglucosylated by (UDP)-glucose: glycoprotein glucosyltransferase 1 (*UGT1*) for rebinding to calnexin/calreticulin. Proteins deemed misfolded will be selected for degradation by ER-associated degradation (*ERAD*)

mutant, CPY* [47]. Conversely, lectin chaperones have been implicated in favoring or enhancing the degradation of ERAD substrates. For example, the degradation of the ERAD substrate pro-α factor was delayed in microsomes prepared from yeast strains with a defective *Cne1* gene [48]. Similarly, inhibiting the lectin chaperone-binding properties resulted in decreased degradation of the Z variant of α-1 antitrypsin [49].

The process by which substrates are eventually extracted from the calnexin/calreticulin cycle remains under investigation. One model proposes that trimming of mannose residues by ER resident mannosidases could contribute to the disassociation of ERAD substrates from calnexin/calreticulin since UGT1 has reduced affinity for $Man_{8-7}GlcNAc_2$ [50]. Alternatively, removal of the terminal mannose residue on the A-branch could stop the reglucosylation cycle. Finally, other machinery components could ultimately recognize substrates, irrespective of their glucosylation status, after spending sufficient time in the cycling process.

The ER contains an additional lectin that interacts with glucose-containing antennary of the glycan. Malectin is an ER stress-induced carbohydrate-binding chaperone that associates with di-glucosylated proteins [51, 52]. As di-glucosylated glycoforms are present shortly after glycan addition, malectin was expected to recognize proteins early during the maturation process. Interestingly, malectin overexpression reduced the secretion of the α-1-antitrypsin mutant NHK by enhancing its turnover through ERAD [53, 54]. This led to the hypothesis that upregulation of malectin by stress helped to inhibit the secretion of aberrant proteins.

It is uncertain how to reconcile the di-glucosylated-binding specificity of malectin with its apparent role in quality control that involves a decision on destruction. This decision is expected to occur at a later point in the maturation process after a chance to mature properly has been deemed futile. Perhaps the action of glucosidase II is inhibited for some nascent chains that are rapidly evaluated as misfolded. Alternatively, malectin may preferentially interact with glycans that are added posttranslationally by an oligosaccharyl transferase complex containing STT3B as malectin has also been shown to interact with ribophorin I, a component of the oligosaccharyl transferase complex [55]. Here, malectin may help to recruit consensus sites for modification as glycosylation sites are more likely to be recognized if there is an upstream site that has been previously modified. Further studies are needed to understand the function of malectin in the ER, as well as the complete picture of the role of N-linked glycans in protein folding and ERAD.

4 Sweet-Toothed Receptors

The mannose residues on the glycan also act as sorting signals as glycan demannosylation serves to target native proteins for trafficking through the secretory pathway and misfolded proteins for degradation by ERAD [2, 3]. It has been proposed that proteins are directed to either of these disparate pathways depending on the extent of glycan demannosylation. Evidence that illustrates the importance of glycan demannosylation for ERAD signaling emerged from studies in which mammalian cells treated with α(1,2)-mannosidase inhibitors exhibited delays in degradation [56–58]. Furthermore, native glycoproteins possess high mannose glycoforms ($Man_8GlcNAc_2$ or $Man_7GlcNAc_2$), which are later trimmed and modified with complex sugars once the proteins have reached the Golgi, while ERAD-targeted glycoproteins were shown to possess $Man_5GlcNAc_2$ or $Man_6GlcNAc_2$ glycoforms [57, 59, 60]. Together, these results implicated a potential role for ER resident mannosidases in the early stages of the ERAD pathway that serve to trim $Man_9GlcNAc_2$ to $Man_{8\text{-}5}GlcNAc_2$ in order to terminate futile folding cycles and target terminally misfolded proteins for destruction (Fig. 11.1).

Genetic studies performed in the budding yeast confirmed that an ER mannosidase was responsible for the demannosylation of the B-branch on ERAD-targeted glycoproteins, and it was later characterized and named Mns1p [61]. Likewise, in the mammalian system, $Man_8GlcNAc_2$ glycoforms accumulated upon overexpression of the mammalian Mns1p ortholog, ERManI, and conversely decreased upon treatment with α(1,2)-mannosidase inhibitors and knockdown of ERManI [62]. Furthermore, isolated ERManI preferentially removed the terminal B-branch mannose moiety (Fig. 11.1) [63].

Although data acquired over decades of study implicates ERManI in ERAD, its role in the pathway is not completely understood. It is unclear whether the sole contribution of ERManI to ERAD is demannosylation of the B branch terminal mannose or more extensive trimming is involved. Likewise, conflicting data has emerged regarding the subcellular localization of ERManI [64]. ERManI, as its name states, was originally classified as an ER membrane protein based on similar findings with Mns1p in yeast [61, 62, 65]. However, ERManI exhibits high sequence homology to the Golgi-resident α(1,2)-mannosidases, and the *arabidopsis thaliana* ERManI ortholog was shown to localize to the Golgi [63, 66]. Recently, localization studies of ERManI using an antibody able to visualize endogenous protein demonstrated that ERManI is an O-linked glycosylated protein resident to the Golgi [64]. Previous studies reliant upon the overexpression of ERManI apparently resulted in the mislocalization of the mannosidase to the ER. It was proposed that the localization of ERManI in the Golgi serves to separate the process of glycan trimming that is necessary for protein folding from that of ERAD targeting. This model implies that misfolded proteins reach the Golgi and are subsequently returned to the ER for ERAD. A requirement for retrieval from the Golgi has been previously proposed for the turnover of soluble ERAD substrates in yeast [67].

The proposed role of the $Man_8GlcNAc_2$ trimmed glycoform as an ERAD signal generated a search for Man8-binding lectins that aid in sorting the demannosylated substrates for ERAD and resulted in the discovery of a mannosidase-like protein (Htm1p) in *S. cerevisiae* [68]. Htm1p was originally thought to lack mannosidase activity, resulting in the conclusion that Htm1p was an ER lectin and Mns1p was sufficient for ERAD substrate demannosylation in yeast [68, 69]. However, Htm1p was later reported to be catalytically active and necessary for exposing the α(1,6)-linked mannose residues on the carbohydrate C-branch, which is a crucial step for ERAD substrate labeling and targeting in yeast (Fig. 11.1) [70, 71]. The mannosidase activity of Htm1p was weak, required co-expression and association with Pdi1p and the substrate to be denatured [70].

Htm1p orthologs were later identified in mammalian cells and termed ER-degradation enhancing α-mannosidase-like proteins (EDEMs). These putative mannosidases were discovered as a result of a search for Man8-binding lectins that were upregulated upon ER stress [72]. The EDEM family of proteins, EDEM1, EDEM2, and EDEM3, belong to the glycosylhydrolase 47 protein family, which includes ERManI as well as a number of Golgi α(1,2)-mannosidases that are involved in glycan demannosylation deeper in the secretory pathway [73]. Members of the glycosylhydrolase 47 family proteins share a conserved tertiary structure as well as several catalytic amino acid residues [74, 75].

EDEM1 was initially characterized as a Man8-binding lectin candidate due to its apparent lack of mannosidase activity [72]. It was found to extract glycosylated

ERAD substrates from a futile calnexin-binding cycle [76, 77]. These findings supported a carbohydrate-binding recognition step for the selection of ERAD substrates by EDEM1. However, further analysis revealed that $Man_{7-5}GlcNAc_2$ glycoforms were also necessary for targeting misfolded proteins to ERAD [57, 78]. Since ERManI and Mns1p appeared to generate Man8B glycoforms, the mannosidase activity of EDEM1/Htm1p was re-evaluated. Exogenously expressed EDEM1 accelerated ERAD substrate demannosylation, a process that was abolished upon mutation of the conserved, proposed catalytic residues [79, 80]. EDEM1 interacted with a number of ERAD components including ERMan1, SEL1L ERdj5, and Derlin2/3 [14, 81–83]. Together, these results supported a role for EDEM1 in ERAD.

Although EDEM1 preferentially bound misfolded proteins, glycans were not required for substrate binding, suggestive of a possible role for EDEM1 in recognition and targeting of non-glycosylated ERAD substrates [82, 84, 85]. Upon additional characterization of the substrate/machinery interaction between NHK and EDEM1, the interaction persisted regardless of the glycosylation state of NHK or under conditions in which the putative mannosidase-like activity of EDEM1 was inhibited [82]. A glycan-independent interaction between EDEM1 and the ERAD substrate H2a, the uncleaved precursor of the asialoglycoprotein receptor, was also reported, suggesting that the glycan-independent substrate-binding property of EDEM1 may apply to a larger set of ERAD substrates [84].

EDEM1 was found in a complex with BiP through its association with ERdj5 [14, 20, 86]. ERdj5 contains four catalytically active thioredoxin domains, and its reductase activity enhanced the degradation of misfolded substrates by reducing their disulfide bonds rendering the substrate retrotranslocation competent [14, 15, 59]. In this model, BiP bound the misfolded substrate and ERdj5 reduced the substrate's aberrant disulfide bonds suggesting that EDEM1 may recognize misfolded substrates through interacting machinery partners [14, 15]. A predicted N-terminal intrinsically disordered region of EDEM1 has been implicated in mediating protein-protein interactions between EDEM1 and a substrate. The intrinsically disordered domain of EDEM1 was required for binding to tyrosinase, but not for its association with the ER machinery components calnexin or SEL1L [87].

If glycans are not required for EDEM1 binding to substrates, then a question arises of what is the role of its mannosidase-like domain? The mannosidase-like domain of EDEM1 appeared to mediate the interactions with the glycoprotein SEL1L a member of the downstream ERAD retrotranslocation machinery [82]. Treatment with the class I mannosidase inhibitor, kifunensine, disrupted EDEM1 binding to SEL1L but not binding to substrates. A similar glycan-dependent machinery interaction was reported between SEL1L and the quality control receptors osteosarcoma 9 (OS-9) and XTP3-B [19]. These findings revealed a possible delivery role for glycans supporting interactions between EDEM1/OS-9/XTP3-B and the downstream glycosylated SEL1L containing ERAD complex [3].

EDEM2 and EDEM3 are less-characterized homologues, and their roles in the ERAD pathway are incompletely understood. Like EDEM1, EDEM2 and EDEM3 are both induced upon ER stress, and their overexpression accelerated the degradation of misfolded ERAD glycoproteins [88, 89]. EDEM2 and EDEM3 are soluble ER resident proteins, with EDEM3 possessing a C-terminal ER-retention sequence

(Fig. 11.3). The mechanism by which EDEM2 remains ER localized likely involves interacting with another ER-resident protein and may be mediated by the region C-terminal to its mannosidase-like domain [90]. Although sequence identity, including the catalytic triad, is high between EDEM2 and EDEM1, EDEM2 does not appear to be catalytically active (Fig. 11.3) [88]. In contrast, EDEM3 overexpression increased ERAD substrate demannosylation [89, 91]. Like EDEM1, EDEM3 interacted with SEL1L and this interaction was lost in the presence of kifunensine [92]. Demannosylation using purified EDEM1 and EDEM3 has not been shown yet. It is unclear why multiple mannosidase-like proteins exist in the ER and what specific functions each contribute to ERAD.

5 Substrate Recognition and Targeting

Trimming of the α(1,2)-linked mannose sugars by EDEM1/Htm1p results in exposure of an α(1,6)-linked terminal mannose on the C-branch (Fig. 11.1). Glycoforms exhibiting a terminal C-branch α(1,6)-linked mannose have been proposed to mark misfolded glycoproteins for ERAD as they are a product of extensive demannosylation [57, 93, 94]. It is possible that these glycoforms contribute to substrate recognition, since two key α(1,6)-linked mannose-binding lectins have been identified as ERAD machinery components that are required for ERAD of misfolded glycoproteins.

Two mannose 6-phosphate receptor homology domain (MRH)-containing lectins exist in mammalian cells, OS-9 and XTP3-B, and a yeast OS-9 ortholog (Yos9p) has also been identified [95–97] (Fig. 11.3). OS-9/Yos9p contains a single MRH domain while XTP3-B contains two domains; however, only one of these is believed to recognize glycans [98, 99]. Yos9p possesses a C-terminal ER-retention signal that is absent in OS-9 and XTP3-B, suggesting that interactions with ER-resident proteins are likely necessary for ER retention of OS-9 and XTP3-B.

The OS-9/Yos9p MRH domain oligosaccharide-binding affinity was determined in vitro using frontal affinity chromatography. The OS-9 MRH domain exhibited high affinity for glycoforms with exposed C-branch α(1,6)-linked mannose and weak affinity for Man9 glycoforms [95]. In addition to exhibiting a high affinity for α(1,6)-linked glycoforms, OS-9/Yos9p also exhibited a high affinity for $Man_{7-5}GlcNAc_2$ glycoforms [95, 100]. Additionally, recognition of misfolded glycoproteins by Yos9p required Htm1p function, supporting an upstream role for Htm1p for trimming substrates to the proper glycoforms containing exposed α(1,6)-linked mannose residues [71, 101]. The C-terminal MRH domain of XTP3-B exhibited lectin activity although this activity appeared weak compared to that of OS-9/Yos9p [98, 99]. As a result, the N-glycan-binding specificity of XTP3-B is not yet resolved; it is unclear whether XTP3-B binds glycoforms with exposed C-branch α(1,6)-linked mannose or untrimmed N-glycans [98, 99].

Deletion of the *yos9* gene demonstrated that Yos9p was required for the degradation of glycoproteins [101]. Furthermore, Yos9p overexpression enhanced the

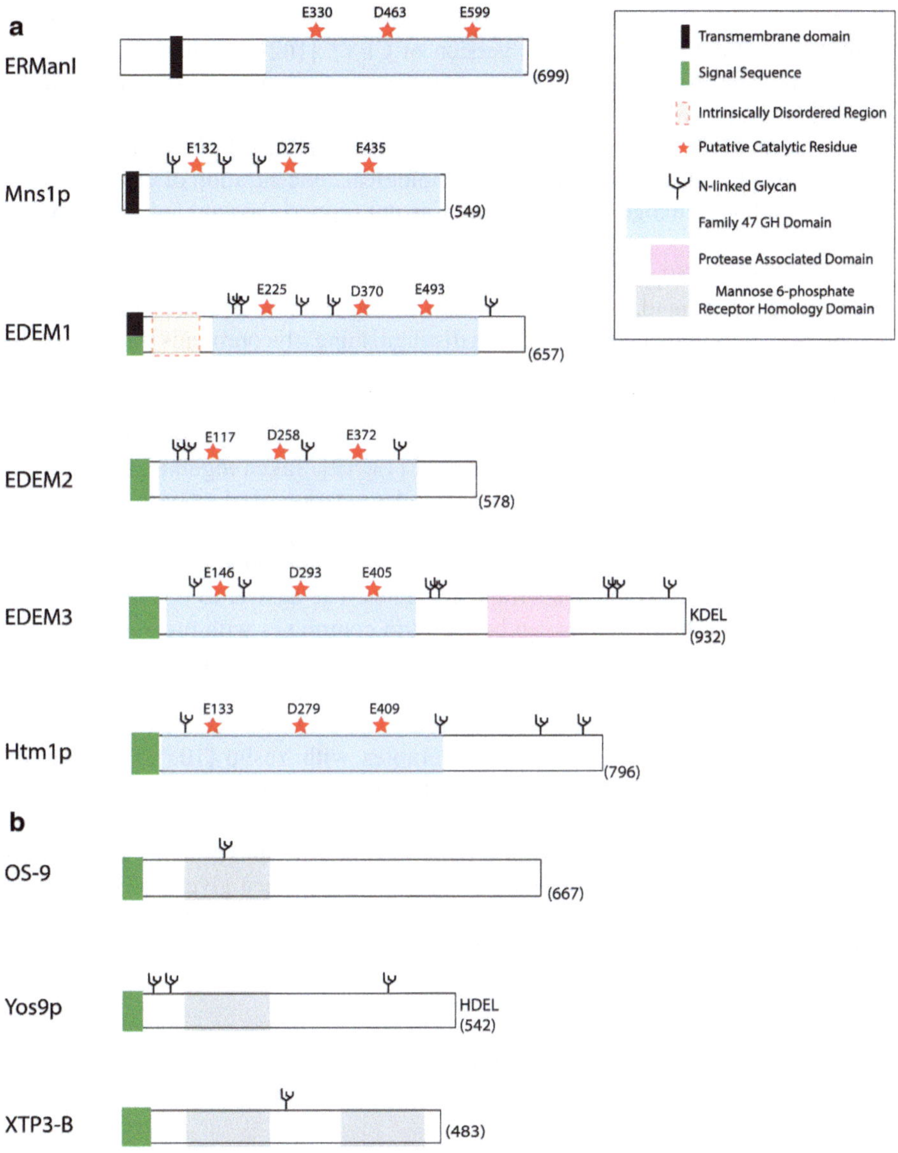

Fig. 11.3 Organization of ERAD machinery. **a** Six members of the glycosylhydrolase 47 protein family have been characterized in the mammalian and yeast endoplasmic reticulum (*ER*). ERManI, Mns1p, ER-degradation enhancing α-mannosidase-like proteins (*EDEM*)1–3 and Htm1p share a common mannosidase-like domain (*light blue*) with conserved putative catalytic residues (*red*). ERManI and Mns1p are type I membrane proteins with a single transmembrane domain (*black rectangle*). EDEM2–3 and Htm1p are soluble proteins with cleavable signal sequences (*green*), whereas EDEM1 possess a dual topology [166]. N-linked glycosylation sites are shown in *black*. EDEM3 possess a protease-associated domain (*pink*) and an ER retention sequence. **b** The mammalian and yeast ERAD lectins OS-9/Yos9p and XTP3-B are characterized by the presence of conserved mannose 6-phosphate receptor homology domain(s) (*grey*). OS-9/Yos9p and XTP3-B are soluble glycoproteins. Only Yos9p possesses an ER retention sequence

degradation of the glycosylated ERAD substrate CPY*; yet, it had no effect on the degradation of a non-glycosylated version of CPY* [102]. Similarly, silencing of OS-9 stabilized NHK [19, 103]. However, overexpression of OS-9 enhanced the stability of NHK and disrupted the retrotranslocation of RI_{332}, a truncated variant of ribophorin [103, 104]. The unexpected stabilizing effect of overexpressed OS-9 on NHK may be attributed to the non-physiological concentration of OS-9 resulting in NHK binding to exogenous OS-9 that was not properly assembled with its other factors.

Unlike OS-9/Yos9p, silencing of XTP3-B had no effect on NHK turnover [19, 99]. On the other hand, XTP3-B delayed the degradation of NHK bearing Man9 glycoforms, implying a possible role in distinguishing glycoproteins that were mistakenly targeted for ERAD; however, further validation is necessary as the binding of XTP3-B to Man9 glycoforms was weak [99]. XTP3-B also interacted with the non-glycosylated ERAD substrate, transthyretin, and a non-glycosylated variant of NHK in vivo and hindered their degradation [13, 19]. Taken together, these results suggest that XTP3-B might serve to prevent degradation of glycoproteins possessing Man9 glycoforms for a subset of substrates or non-glycosylated substrates.

In addition to their lectin-like properties, a question remains as to how OS-9/Yos9p and XTP3-B recognize aberrant substrates. The answer to this question may lie in the fact that these lectins can be found in complexes with molecular chaperones. Reports of GRP94 associating with OS-9 and enhancing degradation of NHK could provide insight into an alternative mechanism of ERAD substrate recognition [19]. Similarly, Yos9p interacts with misfolded ERAD substrates with Kar2p, the yeast BiP that also forms an E3 ligase complex with Yos9p [105]. OS-9/Yos9p, XTP3-B and Htm1p/EDEM1-3 are ERAD lectins or mannosidases with varying oligosaccharide and substrate affinities. However, there is still some question about the location of the glycans that support interactions with these quality control factors. Conflicting results are found as to whether the trimmed glycan recognized by the carbohydrate-binding quality control factors are located on ERAD substrates or on downstream ERAD machinery [3, 106, 107].

6 The SEL1L-HRD1 and Hrd3p-Hrd1p Membrane Complexes

The receptors OS-9/Yos9p and XTP3-B prepare and present misfolded proteins to an ER membrane complex that contains adapter proteins to nucleate interactions with the ER luminal quality control receptors, as well as a protein conduit to support retrotranslocation and an E3 ligase for ubiquitination followed by degradation by the 26S proteasome [1, 94]. A range of complexes has been identified to cope with the variety of secretory cargo.

The first ERAD membrane complex adapters identified were the yeast type I membrane protein Hrd3p and its human ortholog SEL1L [108, 109]. They provide a link between ERAD receptors and their respective E3 ligases, Hrd1p, and

HRD1 [12, 19, 110]. The centrality of Hrd3p/SEL1L in ERAD was supported by the observation that deletion or knockdown of Hrd3p and SEL1L, respectively, stabilized a range of ERAD substrates [104, 109, 111, 112]. Furthermore, *Sel1L* gene knockout in mice was embryonic lethal and UPR was constitutively activated in *Sel1l*$^{-/-}$ embryos [113]. While Hrd3p recognized misfolded proteins directly, SEL1L appeared to do so through auxiliary factors [19, 110].

The N-terminal domain of Hrd3p/SEL1L contains 5 N-linked glycans and 7 and 11 tetratricopeptide repeats (TPR), respectively. TPR motifs mediate protein-protein interactions and a functional unit exists as clusters of three or more repeats [114]. The TPR domains of Hrd3p/SEL1L are organized into three separate clusters, each containing various numbers of sequentially positioned TPR motifs. Deletion of the two most C-terminal TPR clusters in Hrd3p diminished binding to Yos9p and a similar deletion in SEL1L abolished binding to OS-9 and XTP3-B [19, 110]. SEL1L interacted with XTP3-B, OS-9, EDEM1 and HRD1 at the ER membrane through its luminal domain; however, its transmembrane domain was dispensable for the HRD1 interaction, implying that the TPR domains were required for this interaction [19]. Furthermore, deletion of any of the three TPR clusters in SEL1L inhibited degradation of NHK in chicken cells [115], demonstrating the importance of TPR domains for ERAD.

The N-linked glycans of SEL1L also appear to mediate protein-protein interactions by serving as docking sites for the ERAD quality control receptors EDEM1 and XTP3-B [19, 82]. This was supported by the observations that mutating the mannosidase-like domain of EDEM1 or the MRH domain of XTP3-B reduced their interaction with SEL1L but not with ERAD substrates, implying that N-linked glycans could be used along with TPR-mediated interactions to organize ERAD complexes. Whether OS-9 interacts with the SEL1L-HRD1 complex in a lectin-like manner remains unclear as results suggest this interaction required a functional OS-9 MRH domain, while other data implicated the MRH domain in the OS-9/ERAD substrate interactions [19, 95]. Taken together, the TPR domains and N-linked glycans support the nucleation of large membrane embedded protein complexes needed for ubiquitination of a number of soluble and membrane bound ERAD substrates in yeast and metazoa. Whether the components of the Hrd3p-Hrd1p/SEL1L-HRD1 retrotranslocation complex are constant or dynamic remains unclear (Fig. 11.4).

The stability of this retrotranslocation complex is tightly controlled. In yeast, deletion of Hrd3p lead to auto-ubiquitination and degradation of Hrd1p [111]. In mammals, knockdown of HRD1 stabilized SEL1L [116]. An additional method of regulating SEL1L levels is through ERAD tuning, a process in which the short cytoplasmic C-terminus of SEL1L recruits autophagy machinery to target excess ERAD factors for autophagy-like degradation [117]. ERAD tuning could explain the short half-life of SEL1L [109]. Taken together, the stability of the ERAD retrotranslocation complex is dependent on the adapters Hrd3p and SEL1L

Retrotranslocation is a key event in ERAD; however, the exact nature of the conduit required for this step is controversial. Multiple conduit candidates have been proposed, ranging from the Sec61 translocon, Hrd1p, lipid droplets to Derlin-1 [118–123]. Sec61p is essential for yeast viability, inhibiting the utility of deletion

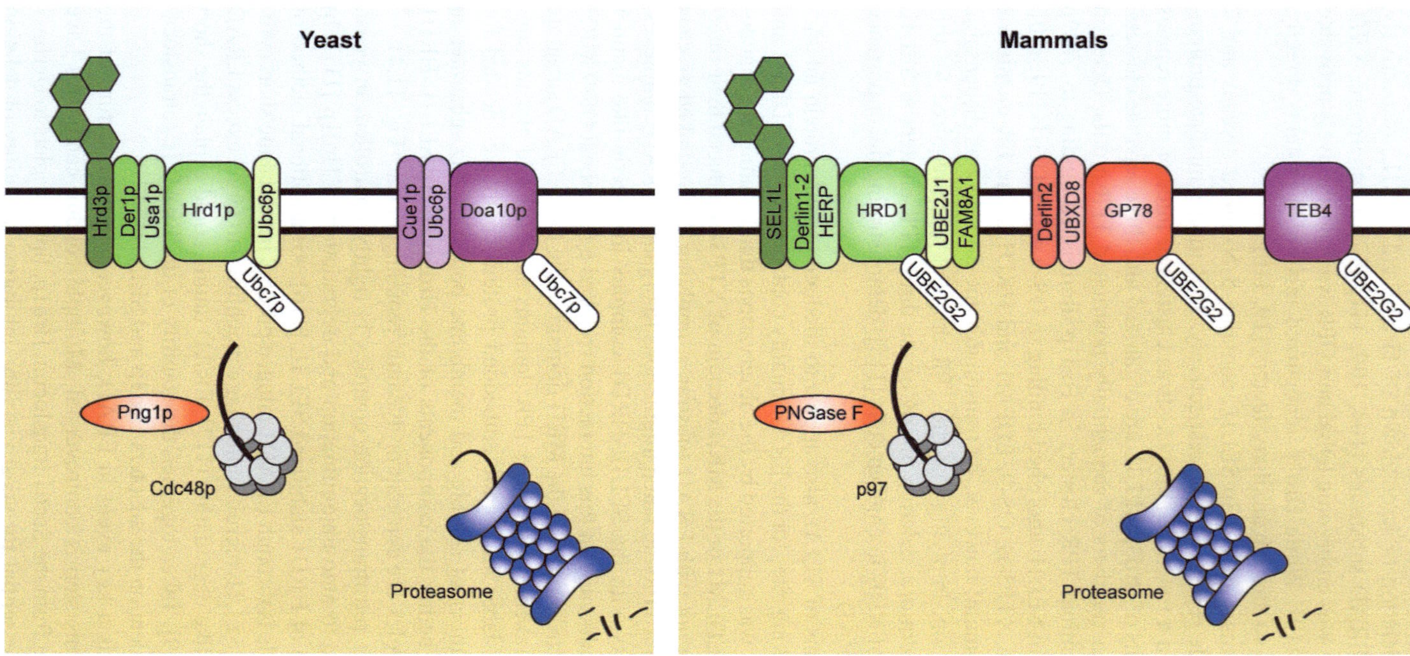

Fig. 11.4 Yeast and mammalian endoplasmic reticulum (*ER*) membrane E3 ligases complexes. Misfolded proteins that have been selected for ER-associated degradation (*ERAD*) are presented to ER membrane complexes that are centered around E3 ligases. Depicted are the yeast (*left*) Hrd1p and Doa10p and the mammalian (*right*) HRD1, GP78, and TEB4 E3 ligase complexes. These membrane complexes facilitate retrotranslocation of ERAD substrates into the cytoplasm where they are ubiquitinated. In yeast, Hrd1p and Doa10p are two ERAD E3 ligases that utilize the E2 conjugating enzymes Ubc6p and Ubc7p, and ubiquitinate different sets of substrates. A greater number of ERAD E3 ligases exist in mammals, of these HRD1, GP78, and TEB4 are the best characterized. Although these E3 ligases reside in unique complexes, they nevertheless share some auxiliary factors, such as the E2 conjugating UBE2G2. Once a substrate has been polyubiquitinated, the cytoplasmic adenosine triphosphatases (*ATPases*) associated with diverse cellular activities (AAA-ATPase) Cdc48p/p97 is recruited to extract it from the ER membrane. However, before it can interact with Cdc48p/p97, N-linked glycans are removed by Png1p/PNGase F and ubiquitin chains are trimmed. After the substrate is completely extracted by CDC48p/p97, the ubiquitin chains are re-extended to recruit the 26S proteasome for substrate destruction

studies. Studies that used viable Sec61p mutants resulted in conflicting conclusions, thus it remains unclear if Sec61p is needed for retrotranslocation of ERAD substrates [9, 124]. The oligomerization of Hrd1p with Usa1p, an ER membrane protein in the Hrd3p-Hrd1p complex, could explain how Hrd1p, which possesses six putative transmembrane domains, may form a pore of sufficient size to retrotranslocate a glycoprotein [123]. However, the retrotranslocation of 3-hydroxy-3-methylglutaryl (HMG)-CoA reductase occurred in the absence of Hrd1p, suggesting that if Hrd1p forms a conduit through the ER membrane it is unlikely to be the only route for retrotranslocation [125]. Lipid droplets have also been proposed to serve as an alternative route for ERAD substrates to reach the cytoplasm [122]. Some ERAD factors have been localized to lipid droplets and ubiquitination of HMG-CoA reductase has been proposed to occur in lipid droplets [126]. However, ERAD of soluble and membrane-bound substrates were unaffected in yeast strains deficient in lipid droplets raising the uncertainty of the role of lipid droplets in ERAD [127]. Derlin-1 was also proposed to be the retrotranslocon since it possesses multiple transmembrane domains and is associated with HRD1 [120, 121]. However, recent data placed the function of Derlin-1 downstream of the retrotranslocation step in ERAD, making it an unlikely retrotranslocon candidate [128]. Derlin-1 interacted with ERAD substrates after retrotranslocation but before ubiquitination, thus it was proposed that Derlin-1 held substrates in an unfolded state prior to HRD1 mediated ubiquitination. It is also possible that members of a proteinaceous retrotranslocation pore complex individually contribute to the formation of the pore through their respective transmembrane domains. One could envision a scenario where the size of the pore is not constant, but rather adapts to the size of a given substrate as is proposed from peroxisomal translocation [129].

In addition to its postulated role as a translocon, Derlin-1 and its yeast ortholog Der1p are implicated in ERAD through their interaction with HRD1/Hrd1p [121, 130]. Mammals possess two additional Derlin-1 orthologs, Derlin-2 and Derlin-3, which also interact with HRD1. Der1p and Derlin-1–3 belong to the 6 transmembrane domain rhomboid protein-family although they appear to be catalytically inactive rhomboids [128]. Deletion of *Derlin-1* in mice was embryonic lethal while the deletion of *Derlin-3* was not [131]. Conditional knockout mice of *Derlin-2* had a constitutively active UPR and a high infant mortality rate [132]. Derlin-1 was found to interact with a novel membrane complex termed mammalian ER membrane complex (mEMC) [133]. Though no clear function for this complex was identified, a role in ERAD has been proposed. Further studies are required to determine the role of Der1p and Derlin-1–3 in ERAD.

After the substrate has been retrotranslocated, ubiquitination labels it for destruction. Ubiquitination depends on a series of sequential enzymes classified as ubiquitin activating enzymes (E1), ubiquitin conjugationng enzymes (E2), and ubiquitin ligases (E3). In some cases, an E4 enzyme is required for the extension of the ubiquitin chain on the substrate. The complexity of the metazoan secretory pathway is illustrated by the increased number of ERAD E3 ligases compared to that of yeast, which has only two known ER membrane ERAD E3 ligases, Hrd1p and Doa10p. The human genome encodes two E1s, approximately 40 E2s, and more than 600

E3 ligases [134]. Of these, three E2 and approximately forty E3 ligases have been localized to the ER [135, 136]. Like HRD1, these E3 ligases reside in unique protein complexes that share common components and overlapping functions.

Studies found that Hrd1p/HRD1 was the dedicated E3 ligase for Hrd3p/SEL1L and has been shown to ubiquitinate a number of luminal and membrane bound substrates [12, 19, 109, 110, 111]. The specificity of HRD1 for ERAD substrates is regulated by its auxiliary factors such as SEL1L and FAM8A1, a recently identified ER membrane adapter [133]. FAM8A1 was necessary for the degradation of membrane-bound HRD1 substrates whereas SEL1L was required for luminal substrates. However, it is unclear how other auxiliary factors aid in the substrate selection for HRD1.

ER localized E3 ligases select substrates based on the location of their lesions. Some ER E3 ligases act on luminal and membrane substrates possessing ER luminal lesions, while others ubiquitinated cytoplasmic proteins as well as membrane proteins with cytoplasmic lesions [19, 137]. GP78 (also known as AMFR) is among the known E3 ligases required for ubiquitination of substrates with luminal lesions. GP78 has five putative transmembrane domains and functionally overlaps with HRD1 since in some cases they were found to ubiquitinate the same substrates [138, 139]. GP78 interacted with Derlin-1–2, UBE2G2, and UBDX8 at the ER membrane [133, 140]. UBE2G2 is an E2 conjugating enzyme and UBXD8 is needed for substrate extraction. These interactions are shared with HRD1; however, the complexes containing GP78 and HRD1 are separate (Fig. 11.4). Another common feature between GP78 and HRD1 is that both appear to interact with the 26S proteasome [133]. Unlike HRD1, GP78 currently lacks an adapter protein that could bridge luminal and membrane ERAD components. Additional studies will further our understanding of how a luminal substrate is presented to the GP78 complex. It is possible that the ERAD receptors ERLIN1–2 (also known as SPHF1–2) serve this function. In support of this, substrate presentation to the E3 ligase RNF170, which was responsible for the ubiquitination of inositol 1,4,5-trisphosphate receptor, depended on the ERLIN1–2 [141, 142].

ERAD substrates with cytoplasmic lesions are ubiquitinated by Doa10p in yeast and TEB4 (also called MARCH6) in metazoan (Fig. 11.4) [143, 144]. Doa10p is composed of 14 transmembrane domains and resides in a complex with two different E2 conjugating enzymes, Ubc6p and Ubc7p [110]. In addition to ubiquitinating ERAD substrates Doa10p/TEB4 may contribute to sterol regulation as it ubiquitinated enzymes needed for sterol synthesis [137]. Additionally, RMA1 (also known as RNF5), a mammalian tail-anchored E3 ligase, was found in a complex that recruited the Hsp40, DNAJB12, and Hsc70 for degradation of misfolded cystic fibrosis transmembrane conductance regulator (CFTR) with a cytoplasmic lesion [145]. Together these studies indicate that E3 ligases and their auxiliary factors are capable of distinguishing between ERAD substrates depending on the location of their lesions.

While a growing body of work illuminates the ubiquitination of ERAD substrates, a number of ER E3 ligases remain either poorly characterized or uncharacterized. In addition to the ER localized E3 ligases, cytoplasmic E3 ligases have also been implicated in ERAD, one of which is the SCF^{Fbx2} E3 ligase complex that in part selects substrates based on their N-linked glycans[136, 145–147]. These findings suggest that we have only scratched the surface of how ERAD substrates are ubiquitinated.

Lysine 48 and Lysine 11 linkages were traditionally thought to mediate degradation or act as sorting signals, respectively. Recent results have challenged this view [148, 149]. A number of studies have found that ERAD substrates can also be ubiquitinated on serine, threonine and cysteine residues, opening up the possibility that the nature of the modified residue could serve as a downstream signal [150–152]. Luminal ERAD substrates are selected for degradation prior to encountering their respective E3 ligases, thus ubiquitination of multiple amino acids could be used to ensure ERAD fidelity and prevent the substrates from escaping degradation. Once a substrate has been selected for degradation and polyubiquitinated in the cytoplasm, it must first be extracted from the membrane and deubiquitinated prior to its eventual degradation by the 26S proteasome.

7 Cytoplasmic Events

ERAD substrates are extracted from the membrane by the cytoplasmic adenosine triphosphatases (ATPases) associated with diverse cellular activities (AAA-ATPase) Cdc48p and p97 (also known as VCP) in yeast and metazoan, respectively [153]. Cdc48p/p97 resides in cytoplasmic complexes with the co-factors Npl4p/NPL4 and Ufd1p/UFD1 [154, 155]. The mechanical force for retrotranslocation is provided by ATP hydrolysis. Reducing the levels of p97 alone impaired the degradation of membrane and soluble substrates, as well as glycosylated and non-glycosylated ERAD substrates thus highlighting the central role of p97 in ERAD [133]. Furthermore, Derlin-1 interacted with deglycosylated and non-ubiquitinated NHK, which implies that the substrate was first retrotranslocated, possibly held in an unfolded state by Derlin-1, ubiquitinated by HRD1 and then extracted by p97 [128].

In order to extract misfolded proteins from the ER, p97 must be recruited to the ERAD membrane complex. There are multiple proteins that recruit p97; among these are UBDX8 and VIMP (VCP interacting membrane protein) [120]. The cytoplasmic side of VIMP possessed reductase activity in vitro, although its function is unclear [156]. Perhaps it serves to reduce disulfides that were missed by ER luminal oxidoreductases, helping to keep the substrates in an unfolded state and facilitate p97 extraction.

Based on current models that describe active extraction of ERAD substrates by p97, a substrate could be at most mono-ubiquitinated to fit through the p97 central pore. Another model proposed that the substrates associated with the outer ring of the p97 hexamer without passing through the central pore [153]. Although more evidence is required to confirm whether or not a substrate passes through the p97 central pore, it is clear that in both cases an issue of spatial restriction arises. It is unlikely that a substrate can possess more than a single ubiquitin during p97-mediated extraction, implying that substrate deubiquitination is a necessary prerequisite. Additionally, p97 recruits peptide: N-glycanase to remove N-linked glycans from ERAD substrates implying that further processing is required to facilitate extraction [157].

In support of these models, ATAXIN-3 and YOD1, two cytoplasmic deubiquitinating enzymes, were found in complex with p97 [158, 159]. Moreover, catalytically inactive ATAXIN-3 and YOD1 caused an accumulation of polyubiquitinated ERAD substrates implying a need for deubiquitinases prior to p97 extraction. While ERAD substrates need to be polyubiquitinated in order to initially interact with p97, they have to be deubiquitinated to facilitate their extraction by p97. Furthermore, p97 provided a "pulling" force from the cytoplasmic side; however, it is unclear whether there is a need for a "pushing" force from the ER lumen. TorsinA, an ER resident AAA-ATPase, is a candidate to serve this function because it interacted with VIMP and Derlin-1, implying a role in ERAD [160]. Overall, it is not clear whether p97 prefers to interact with ubiquitinated or unmodified regions of a substrate.

While deubiquitination was needed for p97 extraction, other deubiquitinases appear to counteract nonspecific ubiquitination of properly folded proteins. ER luminal proteins are protected from E3 ligases by the ER lipid bilayer; however, ER membrane proteins run the risk of overzealous ubiquitination through random movement in the membrane. To this end, are deubiquitinases capable of removing ubiquitins from erroneously ubiquitinated proteins?

The extent of polyubiquitination depended on the strength of the interaction between the E3 ligase complex and the substrates as shown by E3 ligase $SCF^{\beta TrCP}$, which is recruited to the ER membrane by the human immunodeficiency virus (HIV) viral protein, Vpu [161]. Although the rate of polyubiquitination for strongly interacting substrates was faster than weakly interacting substrates, the rate of deubiquitination was constant implying that non-intended substrates are unlikely to become polyubiquitinated. As a result, ERAD substrates with weak E3 ligase interactions are unlikely to accumulate the required ubiquitin chain length needed for degradation, whereas substrates with strong E3 ligase interactions would rapidly achieve a polyubiquitinated state. It would be of interest to determine whether additional ERAD E3 ligases regulate substrate ubiquitination in a similar fashion.

Considering the process of deubiquitination at the ER membrane prior to p97 engagement as well as deubiquitination prior to p97-mediated extraction, the question of whether the cell can discriminate between a properly folded mono-ubiquitinated protein versus a misfolded protein that was deubiquitinated to a mono-ubiquitinated state, remains unanswered. It is possible that a mono-ubiquitinated properly folded ER membrane protein failed to recruit p97 since it lacked the appropriate auxiliary factors. Alternatively, the mono-ubiquitin signal on a properly folded protein might be too short lived to allow p97 recruitment and subsequent extraction of properly folded proteins.

Substrate extraction can create some eminent problems since transmembrane domains and exposed hydrophobic residues can lead to protein aggregation, thus a mechanism must be in place to prevent aggregation by keeping the extracted substrates in soluble states. In support of this, ubiquitination and interactions with cytoplasmic chaperones can prevent ERAD substrates from aggregation [139, 145]. BAG6, a cytoplasmic protein that interacted with HRD1 and GP78, was originally implicated in the insertion of tail-anchored proteins [162]. Recently, BAG6 was also proposed to recognize hydrophobic stretches of ERAD substrates without promoting their refolding after retrotranslocation, thus preventing aggregation. The point at which BAG6 interacts with ERAD substrates relative to p97 recruitment remains unclear. However, ERAD of CFTR, but not ER luminal substrate, was found to be dependent on the cytoplasmic chaperone Hsp70, implying that cytoplasmic chaperones can target ERAD substrate for destruction [163]. Taken together, multiple cytoplasmic factors cooperate to ensure degradation of ERAD substrates by the 26S proteasome.

The 26S proteasome does not interact with mono-ubiquitinated substrates. It has thus been proposed that a second round of ubiquitin chain extension is needed to engage the 26S proteasome after p97 extraction, implying the need of yet unidentified E4 ligases in ERAD [164]. In yeast, the E4 ligase Ufd2p interacted with Cdc48p and promoted the degradation of ERAD substrates by extending their ubiquitin chains [165]. In metazoa, the situation is less clear. While GP78 is an E3 ligase, it was proposed to also act as an E4 ligase for CFTR [138]. Moreover, a few cytoplasmic E3 ligases have been linked to ERAD making them possible candidates for ubiquitin chain extension of ERAD substrates [136, 145, 147]. Altogether, the combined roles of ubiquitination, deubiquitination, extraction from the ER membrane and re-ubiquitination serve to prepare ERAD substrates for their eventual destruction by the 26S proteasome.

8 Concluding Remarks

The large amount of proteins that traverse the secretory pathway and the inherent difficulties of protein folding dictates that eukaryotic cells are highly dependent on ERAD for maintaining cellular homeostasis. The significance of this pathway is further underscored by the observation that mice in which individual ERAD components have been knocked out are frequently embryonic lethal. To cope with the wide range of ERAD substrates, metazoa have evolved highly specialized ERAD protein complexes. Due to the irreversible nature of protein degradation, cells employ a multitude of chaperones, foldases, and covalent modifiers for substrate evaluation before committing them to degradation. However, once substrates are selected for degradation, the decision must be readily carried to completion to prevent accumulation of misfolded proteins or aggregation. Although decades of work have expanded our understanding of ERAD, a number of questions remain. How can the folding machinery play dual roles in folding and degradation? Do different N-linked glycan compositions relay disparate information? What is the nature of the ERAD conduit? What other cytoplasmic factors aid in the final steps of ERAD? Future studies will aim to address these fundamental questions to expand our understanding of how eukaryotic cells dispose aberrant secretory proteins and how misfolded proteins can be corrected.

References

1. Tamura T, Sunryd JC, Hebert DN (2010) Sorting things out through endoplasmic reticulum quality control. Mol Membr Biol 27(8):412–427
2. Helenius A, Aebi M (2004) Roles of N-linked glycans in the endoplasmic reticulum. Annu Rev Biochem 73:1019–1049
3. Hebert DN, Molinari M (2012) Flagging and docking: dual roles for N-glycans in protein quality control and cellular proteostasis. Trends Biochem Sci 37(10):404–410
4. Bukau B, Horwich AL (1998) The hsp70 and hsp60 chaperone machines. Cell 92:351–366
5. Hendershot LM (2004) The ER function BiP is a master regulator of ER function. Mt Sinai J Med 71(5):289–297
6. Ron D, Walter P (2007) Signal integration in the endoplasmic reticulum unfolded protein response. Nat Rev Mol Cell Biol 8(7):519–529
7. Mimura N, Yuasa S, Soma M, Jin H, Kimura K, Goto S, Koseki H, Aoe T (2008) Altered quality control in the endoplasmic reticulum causes cortical dysplasia in knock-in mice expressing a mutant BiP. Mol Cell Biol 28(1):293–301
8. Knittler MR, Dirks S, Haas IG (1995) Molecular chaperones involved in protein degradation in the endoplasmic reticulum: quantitative interaction of the heat shock cognate protein BiP with partially folded immunoglobulin light chains that are degraded in the endoplasmic reticulum. Proc Natl Acad Sci U S A 92(5):1764–1768
9. Plemper RK, Bohmler S, Bordallo J, Sommer T, Wolf DH (1997) Mutant analysis links the translocon and BiP to retrograde protein transport for ER degradation. Nature 388(6645):891–895
10. Brodsky JL, Werner ED, Dubas ME, Goeckeler JL, Kruse KB, McCracken AA (1999) The requirement for molecular chaperones during endoplasmic reticulum-associated protein degradation demonstrates that protein export and import are mechanistically distinct. J Biol Chem 274(6):3453–3460
11. Nishikawa SI, Fewell SW, Kato Y, Brodsky JL, Endo T (2001) Molecular chaperones in the yeast endoplasmic reticulum maintain the solubility of proteins for retrotranslocation and degradation. J Cell Biol 153(5):1061–1070
12. Denic V, Quan EM, Weissman JS (2006) A luminal surveillance complex that selects misfolded glycoproteins for ER-associated degradation. Cell 126(2):349–359
13. Hosokawa N, Wada I, Nagasawa K, Moriyama T, Okawa K, Nagata K (2008) Human XTP3-B forms an endoplasmic reticulum quality control scaffold with the HRD1-SEL1 Lubiquitin ligase complex and BiP. J Biol Chem 283(30):20914–20924
14. Ushioda R, Hoseki J, Araki K, Jansen G, Thomas DY, Nagata K (2008) ERdj5 is required as a disulfide reductase for degradation of misfolded proteins in the ER. Science 321(5888):569–572
15. Hagiwara M, Maegawa K, Suzuki M, Ushioda R, Araki K, Matsumoto Y, Hoseki J, Nagata K, Inaba K (2011) Structural basis of an ERAD pathway mediated by the ER-resident protein disulfide reductase ERdj5. Mol Cell 41(4):432–444
16. Lai CW, Otero JH, Hendershot LM, Snapp E (2012) ERdj4 protein is a soluble endoplasmic reticulum (ER) DnaJ family protein that interacts with ER-associated degradation machinery. J Biol Chem 287(11):7969–7978
17. Feige MJ, Hendershot LM (2013) Quality control of integral membrane proteins by assembly-dependent membrane integration. Mol Cell 51(3):297–309
18. Otero JH, Lizak B, Hendershot LM (2010) Life and death of a BiP substrate. Semin Cell Dev Biol 21(5):472–478
19. Christianson JC, Shaler TA, Tyler RE, Kopito RR (2008) OS-9 and GRP94 deliver mutant alpha1-antitrypsin to the Hrd1-SEL1 Lubiquitin ligase complex for ERAD. Nat Cell Biol 10(3):272–282
20. Oka OB, Bulleid NJ (2013) Forming disulfides in the endoplasmic reticulum. Biochim Biophys Acta 1833(11):2425–2429

21. Cai H, Wang CC, Tsou CL (1994) Chaperone-like activity of protein disulfide isomerase in the refolding of a protein with no disulfide bonds. J Biol Chem 269(40):24550–24552
22. Gillece P, Luz JM, Lennarz WJ, de La CFJ, Romisch K (1999) Export of a cysteine-free misfolded secretory protein from the endoplasmic reticulum for degradation requires interaction with protein disulfide isomerase. J Cell Biol 147(7):1443–1456
23. Grubb S, Guo L, Fisher EA, Brodsky JL (2012) Protein disulfide isomerases contribute differentially to the endoplasmic reticulum-associated degradation of apolipoprotein B and other substrates. Mol Biol Cell 23(4):520–532
24. Molinari M, Galli C, Piccaluga V, Pieren M, Paganetti P (2002) Sequential assistance of molecular chaperones and transient formation of covalent complexes during protein degradation from the ER. J Cell Biol 158:247–257
25. Lee SO, Cho K, Cho S, Kim I, Oh C, Ahn K (2010) Protein disulphide isomerase is required for signal peptide peptidase-mediated protein degradation. EMBO J 29(2):363–375
26. Walczak CP, Bernardi KM, Tsai B (2012) Endoplasmic reticulum-dependent redox reactions control endoplasmic reticulum-associated degradation and pathogen entry. Antioxid Redox Signal 16(8):809–818
27. Moore P, Bernardi KM, Tsai B (2010) The Ero1alpha-PDI redox cycle regulates retro-translocation of cholera toxin. Mol Biol Cell 21(7):1305–1313
28. Spooner RA, Watson PD, Marsden CJ, Smith DC, Moore KA, Cook JP, Lord JM, Roberts LM (2004) Protein disulphide-isomerase reduces ricin to its A and B chains in the endoplasmic reticulum. Biochem J 383(Pt 2):285–293
29. Riemer J, Appenzeller-Herzog C, Johansson L, Bodenmiller B, Hartmann-Petersen R, Ellgaard L (2009) A luminal flavoprotein in endoplasmic reticulum-associated degradation. Proc Natl Acad Sci U S A 106(35):14831–14836
30. Riemer J, Hansen HG, Appenzeller-Herzog C, Johansson L, Ellgaard L (2011) Identification of the PDI-family member ERp90 as an interaction partner of ERFAD. PloS ONE 6(2):e17037
31. Deprez P, Gautschi M, Helenius A (2005) More than one glycan is needed for ER glucosidase II to allow entry of glycoproteins into the calnexin/calreticulin cycle. Mol Cell 19(2):183–195
32. Chen W, Helenius J, Braakman I, Helenius A (1995) Cotranslational folding and calnexin binding of influenza hemagglutinin in the endoplasmic reticulum. Proc Natl Acad Sci U S A 92:6229–6233
33. Helenius A (1994) How N-linked oligosaccharides affect glycoprotein folding in the endoplasmic reticulum. Mol Biol Cell 5:253–265
34. Hammond C, Braakman I, Helenius A (1994) Role of N-linked oligosaccharides, glucose trimming and calnexin during glycoprotein folding in the endoplasmic reticulum. Proc Natl Acad Sci USA 91:913–917
35. Hebert DN, Foellmer B, Helenius A (1995) Glucose trimming and reglucosylation determine glycoprotein association with calnexin in the endoplasmic reticulum. Cell 81(3):425–433
36. Caramelo JJ, Parodi AJ (2008) Getting in and out from calnexin/calreticulin cycles. J Biol Chem 283(16):10221–10225.
37. Caramelo JJ, Castro OA, Alonso LG, de Prat-Gay G, Parodi AJ (2003) UDP-Glc:glycoprotein glucosyltransferase recognizes structured and solvent accessible hydrophobic patches in molten globule-like folding intermediates. Proc Natl Acad Aca Sci U S A 100(1):86–91
38. Sousa M, Parodi AJ (1995) The molecular basis for the recognition of misfolded glycoproteins by the UDP-Glc: glycoprotein glucosyltransferase. EMBO J 14(17):4196–4203
39. Keith N, Parodi AJ, Caramelo JJ (2005) Glycoprotein tertiary and quaternary structures are monitored by the same quality control mechanism. J Biol Chem 280(18):18138–18141
40. Hebert DN, Foellmer B, Helenius A (1996) Calnexin and calreticulin promote folding, delay oligomerization and suppress degradation of influenza hemagglutinin in microsomes. EMBO J 15(12):2961–2968
41. Jackson MR, Cohendoyle MF, Peterson PA, Williams DP (1994) Regulation of MHC Class I transport by the molecular chaperone, calnexin (p88, IP90). Science 263:384–387

42. Frickel EM, Riek R, Jelesarov I, Helenius A, Wuthrich K, Ellgaard L (2002) TROSY-NMR reveals interaction between ERp57 and the tip of the calreticulin P-domain. Proc Natl Acad Sci U S A 99(4):1954–1959
43. Zapun A, Darby NJ, Tessier DC, Michalak M, Bergeron JJ, Thomas DY (1998) Enhanced catalysis of ribonuclease B folding by the interaction of calnexin or calreticulin with ERp57. J Biol Chem 273(11):6009–6012
44. Kozlov G, Bastos-Aristizabal S, Maattanen P, Rosenauer A, Zheng F, Killikelly A, Trempe JF, Thomas DY, Gehring K (2010) Structural basis of cyclophilin B binding by the calnexin/calreticulin P-domain. J Biol Chem 285(46):35551–35557
45. Pearse BR, Hebert DN (2010) Lectin chaperones help direct the maturation of glycoproteins in the endoplasmic reticulum. Biochim Biophys Acta 1803(6):684–693
46. Liu Y, Choudhury P, Cabral CM, Sifers RN (1997) Intracellular disposal of incompletely folded human alpha1-antitrypsin involves release from calnexin and post-translational trimming of asparagine-linked oligosaccharides. J Biol Chem 272(12):7946–7951
47. Kostova Z, Wolf DH (2005) Importance of carbohydrate positioning in the recognition of mutated CPY for ER-associated degradation. J Cell Science 118(Pt 7):1485–1492
48. McCracken AA, Brodsky JL (1996) Assembly of ER-associated protein degradation in vitro: dependence of cytosol, calnexin and ATP. J Cell Biol 132(3):291–298
49. Qu DF, Teckman JH, Omura S, Perlmutter DH (1996) Degradation of a mutant secretory protein, a_1-antitrypsin Z, in the endoplasmic reticulum requires proteasome activity. J Biol Chem 271:22791–22795
50. Sousa MC, Ferrero-Garcia MA, Parodi AJ (1992) Recognition of the oligosaccharide and protein moieties of glycoproteins by the UDP-Glc:glycoprotein glucosyltransferase. BioChemistry 31:97–105
51. Schallus T, Feher K, Sternberg U, Rybin V, Muhle-Goll C (2010) Analysis of the specific interactions between the lectin domain of malectin and diglucosides. Glycobiology 20(8):1010–1020
52. Schallus T, Jaeckh C, Feher K, Palma AS, Liu Y, Simpson JC, Mackeen M, Stier G, Gibson TJ, Feizi T, Pieler T, Muhle-Goll C (2008) Malectin: a novel carbohydrate-binding protein of the endoplasmic reticulum and a candidate player in the early steps of protein N-glycosylation. Mol Biol Cell 19(8):3404–3414
53. Chen Y, Hu D, Yabe R, Tateno H, Qin SY, Matsumoto N, Hirabayashi J, Yamamoto K (2011) Role of malectin in Glc(2)Man(9)GlcNAc(2)-dependent quality control of alpha1-antitrypsin. Mol Biol Cell 22(19):3559–3570
54. Galli C, Bernasconi R, Solda T, Calanca V, Molinari M (2011) Malectin participates in a backup glycoprotein quality control pathway in the mammalian ER. PloS ONE 6(1):e16304
55. Qin SY, Hu D, Matsumoto K, Takeda K, Matsumoto N, Yamaguchi Y, Yamamoto K (2012) Malectin forms a complex with ribophorin I for enhanced association with misfolded glycoproteins. J Biol Chem 287(45):38080–38089
56. Su K, Stoller T, Rocco J, Zemsky J, Green R (1993) Pre-Golgi degradation of yeast prepro-a-factor in a mammalian cell. J Biol Chem 268:14301–14309
57. Lederkremer GZ, Glickman MH (2005) A window of opportunity: timing protein degradation by trimming of sugars and ubiquitins. Trends Biochem Sci 30(6):297–303
58. Molinari M (2007) N-glycan structure dictates extension of protein folding or onset of disposal. Nat Chem Biol 3 (6):313–320
59. Ermonval M, Kitzmuller C, Mir AM, Cacan R, Ivessa NE (2001) N-glycan structure of a short-lived variant of ribophorin I expressed in the MadIA214 glycosylation-defective cell line reveals the role of a mannosidase that is not ER mannosidase I in the process of glycoprotein degradation. Glycobiology 11(7):565–576
60. Frenkel Z, Gregory W, Kornfeld S, Lederkremer GZ (2003) Endoplasmic reticulum-associated degradation of mammalian glycoproteins involves sugar chain trimming to Man6–5GlcNAc2. J Biol Chem 278:34119–33424.

61. Burke J, Lipari F, Igdoura S, Herscovics A (1996) The Saccharomyces cerevisiae processing alpha 1,2-mannosidase is localized in the endoplasmic reticulum, independently of known retrieval motifs. Eur J Cell Biol 70(4):298–305
62. Avezov E, Frenkel Z, Ehrlich M, Herscovics A, Lederkremer GZ (2008) Endoplasmic reticulum (ER) mannosidase I Is compartmentalized and required for N-glycan trimming to Man5 6GlcNAc2 in glycoprotein ER-associated degradation. Mol Biol Cell 19(1):216–225
63. Tremblay LO, Herscovics A (1999) Cloning and expression of a specific human alpha 1,2-mannosidase that trims Man9GlcNAc2 to Man8GlcNAc2 isomer B during N-glycan biosynthesis. Glycobiology 9(10):1073–1078
64. Pan S, Wang S, Utama B, Huang L, Blok N, Estes MK, Moremen KW, Sifers RN (2011) Golgi localization of ERManI defines spatial separation of the mammalian glycoprotein quality control system. Mol Biol Cell 22(16):2810–2822
65. Esmon B, Esmon P C, Scheckman R (1984) Early steps in processing of yeast glycoproteins. J Biol Chem 259:10322–10327
66. Liebminger E, Huttner S, Vavra U, Fischl R, Schoberer J, Grass J, Blaukopf C, Seifert GJ, Altmann F, Mach L, Strasser R (2009) Class I alpha-mannosidases are required for N-glycan processing and root development in Arabidopsis thaliana. Plant Cell 21(12):3850–3867
67. Vashist S, Kim W, Belden WJ, Spear ED, Barlowe C, Ng DT (2001) Distinct retrieval and retention mechanisms are required for the quality control of endoplasmic reticulum protein folding. J Cell Biol 155(3):355–368
68. Jakob CA, Bodmer D, Spirig U, Battig P, Marcil A, Dignard D, Bergeron JJ, Thomas DY, Aebi M (2001) Htm1p, a mannosidase-like protein, is involved in glycoprotein degradation in yeast. EMBO Rep 2(5):423–430
69. Gnann A, Riordan JR, Wolf DH (2004) Cystic fibrosis transmembrane conductance regulator degradation depends on the lectins Htm1p/EDEM and the Cdc48 protein complex in yeast. Mol Biol Cell 15(9):4125–4135
70. Clerc S, Hirsch C, Oggier DM, Deprez P, Jakob C, Sommer T, Aebi M (2009) Htm1 protein generates the N-glycan signal for glycoprotein degradation in the endoplasmic reticulum. J Cell Biol 184(1):159–172
71. Gauss R, Kanehara K, Carvalho P, Ng DT, Aebi M (2011) A complex of pdi1p and the mannosidase htm1p initiates clearance of unfolded glycoproteins from the endoplasmic reticulum. Mol Cell 42(6):782–793
72. Hosokawa N, Wada I, Hasegawa K, Yorihuzi T, Tremblay LO, Herscovics A, Nagata K (2001) A novel ER alpha-mannosidase-like protein accelerates ER-associated degradation. EMBO Rep 2(5):415–422
73. Moremen KW, Molinari M (2006) N-linked glycan recognition and processing: the molecular basis of endoplasmic reticulum quality control. Curr Opin Struct Biol 16(5):592–599
74. Karaveg K, Siriwardena A, Tempel W, Liu ZJ, Glushka J, Wang BC, Moremen KW (2005) Mechanism of class 1 (glycosylhydrolase family 47) {alpha}-mannosidases involved in N-glycan processing and endoplasmic reticulum quality control. J Biol Chem 280(16):16197–16207
75. Mast SW, Moremen KW (2006) Family 47 alpha-mannosidases in N-glycan processing. Methods Enzymol 415:31–46
76. Molinari M, Calanca V, Galli C, Lucca P, Paganetti P (2003) Role of EDEM in the release of misfolded glycoproteins from the calnexin cycle. Science 299(5611):1397–1400
77. Oda Y, Hosokawa N, Wada I, Nagata K (2003) EDEM as an acceptor of terminally misfolded glycoproteins released from calnexin. Science 299(5611):1394–1397
78. Hosokawa N, Tremblay LO, You Z, Herscovics A, Wada I, Nagata K (2003) Enhancement of endoplasmic reticulum (ER) degradation of misfolded null Hong Kong alpha1-antitrypsin by human ER mannosidase I. J Biol Chem 278:26287–26294
79. Olivari S, Cali T, Salo KE, Paganetti P, Ruddock LW, Molinari M (2006) EDEM1 regulates ER-associated degradation by accelerating de-mannosylation of folding-defective polypeptides and by inhibiting their covalent aggregation. Biochem Biophys Res Commun 349(4):1278–1284

80. Hosokawa N, Tremblay LO, Sleno B, Kamiya Y, Wada I, Nagata K, Kato K, Herscovics A (2010) EDEM1 accelerates the trimming of alpha1,2-linked mannose on the C branch of N-glycans. Glycobiology 20(5):567–575
81. Termine DJ, Moremen KW, Sifers RN (2009) The mammalian UPR boosts glycoprotein ERAD by suppressing the proteolytic downregulation of ER mannosidase I. J Cell Sci 122(Pt 7):976–984
82. Cormier JH, Tamura T, Sunryd JC, Hebert DN (2009) EDEM1 recognition and delivery of misfolded proteins to the SEL1 Lcontaining ERAD complex. Mol Cell 34(5):627–633
83. Oda Y, Okada T, Yoshida H, Kaufman RJ, Nagata K, Mori K (2006) Derlin-2 and Derlin-3 are regulated by the mammalian unfolded protein response and are required for ER-associated degradation. J Cell Biol 172(3):383–393
84. Ron E, Shenkman M, Groisman B, Izenshtein Y, Leitman J, Lederkremer GZ (2011) Bypass of glycan-dependent glycoprotein delivery to ERAD by up-regulated EDEM1. Mol Biol Cell 22(21):3945–3954
85. Shenkman M, Groisman B, Ron E, Avezov E, Hendershot LM, Lederkremer GZ (2013) A shared endoplasmic reticulum-associated degradation pathway involving the EDEM1 protein for glycosylated and nonglycosylated proteins. J Biol Chem 288(4):2167–2178
86. Cunnea PM, Miranda-Vizuete A, Bertoli G, Simmen T, Damdimopoulos AE, Hermann S, Leinonen S, Huikko MP, Gustafsson JA, Sitia R, Spyrou G (2003) ERdj5, an endoplasmic reticulum (ER)-resident protein containing DnaJ and thioredoxin domains, is expressed in secretory cells or following ER stress. J Biol Chem 278(2):1059–1066
87. Marin MB, Ghenea S, Spiridon LN, Chiritoiu GN, Petrescu AJ, Petrescu SM (2012) Tyrosinase degradation is prevented when EDEM1 lacks the intrinsically disordered region. PloS ONE 7(8):e42998
88. Olivari S, Galli C, Alanen H, Ruddock L, Molinari M (2005) A novel stress-induced EDEM variant regulating endoplasmic reticulum-associated glycoprotein degradation. J Biol Chem 280(4):2424–2428
89. Hirao K, Natsuka Y, Tamura T, Wada I, Morito D, Natsuka S, Romero P, Sleno B, Tremblay LO, Herscovics A, Nagata K, Hosokawa N (2006) EDEM3, a soluble EDEM homolog, enhances glycoprotein endoplasmic reticulum-associated degradation and mannose trimming. J Biol Chem 281(14):9650–9658
90. Mast SW, Diekman K, Karaveg K, Davis A, Sifers RN, Moremen KW (2005) Human EDEM2, a novel homolog of family 47 glycosidases, is involved in ER-associated degradation of glycoproteins. Glycobiology 15(4):421–436
91. Olivari S, Molinari M (2007) Glycoprotein folding and the role of EDEM1, EDEM2 and EDEM3 in degradation of folding-defective glycoproteins. FEBS Lett 581(19):3658–3664
92. Saeed M, Suzuki R, Watanabe N, Masaki T, Tomonaga M, Muhammad A, Kato T, Matsuura Y, Watanabe H, Wakita T, Suzuki T (2011) Role of the endoplasmic reticulum-associated degradation (ERAD) pathway in degradation of hepatitis C virus envelope proteins and production of virus particles. J Biol Chem 286(43):37264–37273
93. Lederkremer GZ (2009) Glycoprotein folding, quality control and ER-associated degradation. Curr Opin Struct Biol 19(5):515–523
94. Hebert DN, Bernasconi R, Molinari M (2010) ERAD substrates: which way out? Semin Cell Dev Biol 21(5):526–532
95. Hosokawa N, Kamiya Y, Kamiya D, Kato K, Nagata K (2009) Human OS-9, a lectin required for glycoprotein ERAD, recognizes mannose-trimmed N-glycans. J Biol Chem 284(25):17061–17068
96. Cruciat CM, Hassler C, Niehrs C (2006) The MRH protein Erlectin is a member of the endoplasmic reticulum synexpression group and functions in N-glycan recognition. J Biol Chem 281(18):12986–12993
97. Groisman B, Shenkman M, Ron E, Lederkremer GZ (2011) Mannose trimming is required for delivery of a glycoprotein from EDEM1 to XTP3-B and to late endoplasmic reticulum-associated degradation steps. J Biol Chem 286(2):1292–1300

98. Yamaguchi D, Hu D, Matsumoto N, Yamamoto K (2010) Human XTP3-B binds to alpha1-antitrypsin variant null(Hong Kong) via the C-terminal MRH domain in a glycan-dependent manner. Glycobiology 20(3):348–355
99. Fujimori T, Kamiya Y, Nagata K, Kato K, Hosokawa N (2013) Endoplasmic reticulum lectin XTP3-B inhibits endoplasmic reticulum-associated degradation of a misfolded alpha1-antitrypsin variant. FEBS J 280(6):1563–1575
100. Quan EM, Kamiya Y, Kamiya D, Denic V, Weibezahn J, Kato K, Weissman JS (2008) Defining the glycan destruction signal for endoplasmic reticulum-associated degradation. Mol Cell 32(6):870–877
101. Szathmary R, Bielmann R, Nita-Lazar M, Burda P, Jakob CA (2005) Yos9 protein is essential for degradation of misfolded glycoproteins and may function as lectin in ERAD. Mol Cell 19(6):765–775
102. Jaenicke LA, Brendebach H, Selbach M, Hirsch C (2011) Yos9p assists in the degradation of certain nonglycosylated proteins from the endoplasmic reticulum. Mol Biol Cell 22(16):2937–2945
103. Bernasconi R, Pertel T, Luban J, Molinari M (2008) A dual task for the Xbp1-responsive OS-9 variants in the mammalian endoplasmic reticulum: inhibiting secretion of misfolded protein conformers and enhancing their disposal. J Biol Chem 283(24):16446–16454
104. Mueller B, Klemm EJ, Spooner E, Claessen JH, Ploegh HL (2008) SEL1 Lnucleates a protein complex required for dislocation of misfolded glycoproteins. Proc Natl Acad Sci U S A 105(34):12325–12330
105. Kanehara K, Xie W, Ng DT (2010) Modularity of the Hrd1 ERAD complex underlies its diverse client range. J Cell Biol 188(5):707–716
106. Hosokawa N, Kamiya Y, Kato K (2010) The role of MRH domain-containing lectins in ERAD. Glycobiology 20(6):651–660
107. Yoshida Y, Tanaka K (2010) Lectin-like ERAD players in ER and cytosol. Biochim Biophys Acta 1800(2):172–180
108. Hampton RY, Gardner RG, Rine J (1996) Role of 26 S proteasome and HRD genes in the degradation of 3-hydroxy-3-methylglutaryl-CoA reductase, an integral endoplasmic reticulum membrane protein. Mol Biol Cell 7(12):2029–2044
109. Mueller B, Lilley BN, Ploegh HL (2006) SEL1 L the homologue of yeast Hrd3p, is involved in protein dislocation from the mammalian ER. J Cell Biol 175(2):261–270
110. Carvalho P, Goder V, Rapoport TA (2006) Distinct ubiquitin-ligase complexes define convergent pathways for the degradation of ER proteins. Cell 126(2):361–373
111. Gardner RG, Swarbrick GM, Bays NW, Cronin SR, Wilhovsky S, Seelig L, Kim C, Hampton RY (2000) Endoplasmic reticulum degradation requires lumen to cytosol signaling. Transmembrane control of Hrd1p by Hrd3p. J Cell Biol 151(1):69–82
112. Plemper RK, Bordallo J, Deak PM, Taxis C, Hitt R, Wolf DH (1999) Genetic interactions of Hrd3p and Der3p/Hrd1p with Sec61p suggest a retro-translocation complex mediating protein transport for ER degradation. J Cell Sci 112(Pt 22):4123–4134
113. Francisco AB, Singh R, Li S, Vani AK, Yang L, Munroe RJ, Diaferia G, Cardano M, Biunno I, Qi L, Schimenti JC, Long Q (2010) Deficiency of suppressor enhancer Lin12 1 like (SEL1 L in mice leads to systemic endoplasmic reticulum stress and embryonic lethality. J Biol Chem 285(18):13694–13703
114. D'Andrea LD, Regan L (2003) TPR proteins: the versatile helix. Trends Biochem Sci 28(12):655–662
115. Ninagawa S, Okada T, Takeda S, Mori K (2011) SEL1 Lis required for endoplasmic reticulum-associated degradation of misfolded luminal proteins but not transmembrane proteins in chicken DT40 cell line. Cell Struct Funct 36(2):187–195
116. Iida Y, Fujimori T, Okawa K, Nagata K, Wada I, Hosokawa N (2011) SEL1 Lprotein critically determines the stability of the HRD1-SEL1 Lendoplasmic reticulum-associated degradation (ERAD) complex to optimize the degradation kinetics of ERAD substrates. J Biol Chem 286(19):16929–16939

117. Bernasconi R, Galli C, Noack J, Bianchi S, de Haan CAM, Reggiori F, Molinari M (2012) Role of the SEL1 LLC3-I complex as an ERAD tuning receptor in the mammalian ER. Mol Cell 46(6):809–819
118. Plemper RK, Deak PM, Otto RT, Wolf DH (1999) Re-entering the translocon from the lumenal side of the endoplasmic reticulum. Studies on mutated carboxypeptidase yscY species. FEBS Lett 443(3):241–245
119. Stanley AM, Carvalho P, Rapoport T (2011) Recognition of an ERAD-L substrate analyzed by site-specific in vivo photocrosslinking. FEBS Lett 585(9):1281–1286
120. Ye Y, Shibata Y, Yun C, Ron D, Rapoport TA (2004) A membrane protein complex mediates retro-translocation from the ER lumen into the cytosol. Nature 429(6994):841–847
121. Lilley BN, Ploegh HL (2004) A membrane protein required for dislocation of misfolded proteins from the ER. Nature 429(6994):834–840
122. Ploegh HL (2007) A lipid-based model for the creation of an escape hatch from the endoplasmic reticulum. Nature 448(7152):435–438
123. Horn SC, Hanna J, Hirsch C, Volkwein C, Schutz A, Heinemann U, Sommer T, Jarosch E (2009) Usa1 functions as a scaffold of the HRD-ubiquitin ligase. Mol Cell 36(5):782–793
124. Wahlman J, DeMartino GN, Skach WR, Bulleid NJ, Brodsky JL, Johnson AE (2007) Real-time fluorescence detection of ERAD substrate retrotranslocation in a mammalian in vitro system. Cell 129(5):943–955
125. Garza RM, Sato BK, Hampton RY (2009) In vitro analysis of Hrd1p-mediated retrotranslocation of its multispanning membrane substrate 3-hydroxy-3-methylglutaryl (HMG)-CoA reductase. J Biol Chem 284(22):14710–14722
126. Hartman IZ, Liu P, Zehmer JK, Luby-Phelps K, Jo Y, Anderson RG, DeBose-Boyd RA (2010) Sterol-induced dislocation of 3-hydroxy-3-methylglutaryl coenzyme A reductase from endoplasmic reticulum membranes into the cytosol through a subcellular compartment resembling lipid droplets. J Biol Chem 285(25):19288–19298
127. Olzmann JA, Kopito RR (2011) Lipid droplet formation is dispensable for endoplasmic reticulum-associated degradation. J Biol Chem 286(32):27872–27874
128. Greenblatt EJ, Olzmann JA, Kopito RR (2011) Derlin-1 is a rhomboid pseudoprotease required for the dislocation of mutant alpha-1 antitrypsin from the endoplasmic reticulum. Nat Struct Mol Biol 18(10):1147–1152
129. Schliebs W, Girzalsky W, Erdmann R (2010) Peroxisomal protein import and ERAD: variations on a common theme. Nat Rev Mol Cell Biol 11(12):885–890
130. Knop M, Finger A, Braun T, Hellmuth K, Wolf DH (1996) Der1, a novel protein specifically required for endoplasmic reticulum degradation in yeast. EMBO J 15(4):753–763
131. Eura Y, Yanamoto H, Arai Y, Okuda T, Miyata T, Kokame K (2012) Derlin-1 deficiency is embryonic lethal, Derlin-3 deficiency appears normal, and Herp deficiency is intolerant to glucose load and ischemia in mice. PloS ONE 7(3):e34298
132. Dougan SK, Hu CC, Paquet ME, Greenblatt MB, Kim J, Lilley BN, Watson N, Ploegh HL (2011) Derlin-2-deficient mice reveal an essential role for protein dislocation in chondrocytes. Mol Cell Biol 31(6):1145–1159
133. Christianson JC, Olzmann JA, Shaler TA, Sowa ME, Bennett EJ, Richter CM, Tyler RE, Greenblatt EJ, Harper JW, Kopito RR (2012) Defining human ERAD networks through an integrative mapping strategy. Nat Cell Biol 14(1):93–105
134. Deshaies RJ, Joazeiro CA (2009) RING domain E3 ubiquitin ligases. Annu Rev Biochem 78:399–434
135. Neutzner A, Neutzner M, Benischke AS, Ryu SW, Frank S, Youle RJ, Karbowski M (2011) A systematic search for endoplasmic reticulum (ER) membrane-associated RING finger proteins identifies Nixin/ZNRF4 as a regulator of calnexin stability and ER homeostasis. J Biol Chem 286(10):8633–8643
136. Claessen JH, Kundrat L, Ploegh HL (2012) Protein quality control in the ER: balancing the ubiquitin checkbook. Trends Cell Biol 22(1):22–32

137. Foresti O, Ruggiano A, Hannibal-Bach HK, Ejsing CS, Carvalho P (2013) Sterol homeostasis requires regulated degradation of squalene monooxygenase by the ubiquitin ligase Doa10/Teb4. Elife 2:e00953
138. Morito D, Hirao K, Oda Y, Hosokawa N, Tokunaga F, Cyr DM, Tanaka K, Iwai K, Nagata K (2008) Gp78 cooperates with RMA1 in endoplasmic reticulum-associated degradation of CFTRDeltaF508. Mol Biol Cell 19(4):1328–1336
139. Shen Y, Ballar P, Fang S (2006) Ubiquitin ligase gp78 increases solubility and facilitates degradation of the Z variant of alpha-1-antitrypsin. Biochem Biophys Res Commun 349(4):1285–1293
140. Liang J, Yin C, Doong H, Fang S, Peterhoff C, Nixon RA, Monteiro MJ (2006) Characterization of erasin (UBXD2): a new ER protein that promotes ER-associated protein degradation. Journal Cell Sci 119(Pt 19):4011–4024
141. Lu JP, Wang Y, Sliter DA, Pearce MM, Wojcikiewicz RJ (2011) RNF170 protein, an endoplasmic reticulum membrane ubiquitin ligase, mediates inositol 1,4,5-trisphosphate receptor ubiquitination and degradation. J Biol Chem 286(27):24426–24433
142. Wang Y, Pearce MM, Sliter DA, Olzmann JA, Christianson JC, Kopito RR, Boeckmann S, Gagen C, Leichner GS, Roitelman J, Wojcikiewicz RJ (2009) SPFH1 and SPFH2 mediate the ubiquitination and degradation of inositol 1,4,5-trisphosphate receptors in muscarinic receptor-expressing HeLa cells. Biochim Biophys Acta 1793(11):1710–1718
143. Ravid T, Kreft SG, Hochstrasser M (2006) Membrane and soluble substrates of the Doa10 ubiquitin ligase are degraded by distinct pathways. EMBO J 25(3):533–543
144. Kreft SG, Wang L, Hochstrasser M (2006) Membrane topology of the yeast endoplasmic reticulum-localized ubiquitin ligase Doa10 and comparison with its human ortholog TEB4 (MARCH-VI). J Biol Chem 281(8):4646–4653
145. Younger JM, Chen L, Ren HY, Rosser MF, Turnbull EL, Fan CY, Patterson C, Cyr DM (2006) Sequential quality-control checkpoints triage misfolded cystic fibrosis transmembrane conductance regulator. Cell 126(3):571–582
146. Yoshida Y, Chiba T, Tokunaga F, Kawasaki H, Iwai K, Suzuki T, Ito Y, Matsuoka K, Yoshida M, Tanaka K, Tai T (2002) E3 ubiquitin ligase that recognizes sugar chains. Nature 418:438–442
147. Stolz A, Besser S, Hottmann H, Wolf DH (2013) Previously unknown role for the ubiquitin ligase Ubr1 in endoplasmic reticulum-associated protein degradation. Proc Natl Acad Sci U S A 110(38):15271–15276
148. Xu P, Duong DM, Seyfried NT, Cheng D, Xie Y, Robert J, Rush J, Hochstrasser M, Finley D, Peng J (2009) Quantitative proteomics reveals the function of unconventional ubiquitin chains in proteasomal degradation. Cell 137(1):133–145
149. Saeki Y, Kudo T, Sone T, Kikuchi Y, Yokosawa H, Toh-e A, Tanaka K (2009) Lysine 63-linked polyubiquitin chain may serve as a targeting signal for the 26S proteasome. EMBO J 28(4):359–371
150. Burr ML, van den Boomen DJ, Bye H, Antrobus R, Wiertz EJ, Lehner PJ (2013) MHC class I molecules are preferentially ubiquitinated on endoplasmic reticulum luminal residues during HRD1 ubiquitin E3 ligase-mediated dislocation. Proc Natl Acad Sci U S A 110(35):14290–14295
151. Ishikura S, Weissman AM, Bonifacino JS (2010) Serine residues in the cytosolic tail of the T-cell antigen receptor alpha-chain mediate ubiquitination and endoplasmic reticulum-associated degradation of the unassembled protein. J Biol Chem 285(31):23916–23924
152. Shimizu Y, Okuda-Shimizu Y, Hendershot LM (2010) Ubiquitylation of an ERAD substrate occurs on multiple types of amino acids. Mol Cell 40(6):917–926
153. Wolf DH, Stolz A (2012) The Cdc48 machine in endoplasmic reticulum associated protein degradation. Biochim Biophys Acta 1823(1):117–124
154. Ye Y, Meyer HH, Rapoport TA (2003) Function of the p97-Ufd1-Npl4 complex in retrotranslocation from the ER to the cytosol: dual recognition of nonubiquitinated polypeptide segments and polyubiquitin chains. J Cell Biol 162(1):71–84

155. Braun S, Matuschewski K, Rape M, Thomas S, Jentsch S (2002) Role of the ubiquitin-selective CDC48 (UFD1/NPL4) chaperone segregase in ERAD of OLE1 and other substrates. EMBO J 21:615–621
156. Christensen LC, Jensen NW, Vala A, Kamarauskaite J, Johansson L, Winther JR, Hofmann K, Teilum K, Ellgaard L (2012) The human selenoprotein VCP-interacting membrane protein (VIMP) is non-globular and harbors a reductase function in an intrinsically disordered region. J Biol Chem 287(31):26388–26399
157. Blom D, Hirsch C, Stern P, Tortorella D, Ploegh HL (2004) A glycosylated type I membrane protein becomes cytosolic when peptide: N-glycanase is compromised. EMBO J 23(3):650–658
158. Ernst R, Mueller B, Ploegh HL, Schlieker C (2009) The otubain YOD1 is a deubiquitinating enzyme that associates with p97 to facilitate protein dislocation from the ER. Mol Cell 36(1):28–38
159. Wang Q, Li L, Ye Y (2006) Regulation of retrotranslocation by p97-associated deubiquitinating enzyme ataxin-3. J Cell Biol 174(7):963–971
160. Nery FC, Armata IA, Farley JE, Cho JA, Yaqub U, Chen P, da Hora CC, Wang Q, Tagaya M, Klein C, Tannous B, Caldwell KA, Caldwell GA, Lencer WI, Ye Y, Breakefield XO (2011) TorsinA participates in endoplasmic reticulum-associated degradation. Nat Commun 2:393
161. Zhang ZR, Bonifacino JS, Hegde RS (2013) Deubiquitinases sharpen substrate discrimination during membrane protein degradation from the ER. Cell 154(3):609–622
162. Wang Q, Liu Y, Soetandyo N, Baek K, Hegde R, Ye Y (2011) A ubiquitin ligase-associated chaperone holdase maintains polypeptides in soluble states for proteasome degradation. Mol Cell 42(6):758–770
163. Zhang Y, Nijbroek G, Sullivan ML, McCracken AA, Watkins SC, Michaelis S, Brodsky JL (2001) Hsp70 molecular chaperone facilitates endoplasmic reticulum-associated protein degradation of cystic fibrosis transmembrane conductance regulator in yeast. Mol Biol Cell 12(5):1303–1314
164. Finley D (2009) Recognition and processing of ubiquitin-protein conjugates by the proteasome. Annu Rev Biochem 78:477–513
165. Nakatsukasa K, Huyer G, Michaelis S, Brodsky JL (2008) Dissecting the ER-associated degradation of a misfolded polytopic membrane protein. Cell 132(1):101–112
166. Tamura T, Cormier JH, Hebert DN (2011) Characterization of early EDEM1 protein maturation events and their functional implications. J Biol Chem 286(28):24906–24915

Chapter 12
Chaperones and Proteases of Mitochondria: From Protein Folding and Degradation to Mitophagy

Wolfgang Voos, Cornelia Rüb and Michael Bruderek

Abstract In eukaryotic cells, mitochondria fulfill a multitude of essential functions. Under both normal and stress conditions, a complex system of molecular chaperones and protease enzymes is at work to maintain mitochondrial biogenesis and protein quality control (PQC), summarized as "mitochondrial protein homeostasis." Mitochondrial chaperones of the heat shock protein Hsp60 and Hsp70 families play main roles in the import and folding of nuclear-encoded polypeptides, representing the majority of the mitochondrial proteome. These enzymes together with chaperones of the Clp family prevent the accumulation of aggregated or superfluous polypeptides in a close cooperation with specific PQC proteases of the AAA+ (adenosine triphosphatases associated with diverse cellular activities) family. Recent evidence demonstrated the removal of terminally damaged mitochondria as a whole by a variation of autophagy, termed mitophagy, as an additional level of mitochondrial quality control. The details of mitochondrial protein homeostasis, comprising the functional contribution of this broad set of chaperones and proteases will be discussed here.

1 General Introduction

Roughly 2 billion years ago, the uptake of a bacterial cell and the subsequent endosymbiosis profoundly changed the metabolic properties of the ancient eukaryotic ancestor cell. This incident lead to the evolution of modern eukaryotes where the descendents of these endosymbiotic bacteria developed into mitochondria. Mitochondrial functions are now indispensible for the survival of eukaryotic cells, providing energy and essential nutrients but also mediating important regulative

W. Voos (✉) · C. Rüb · M. Bruderek
Institut für Biochemie und Molekularbiologie (IBMB), Friedrich-Wilhelms-Universität Bonn, Bonn, Germany
e-mail: wolfgang.voos@uni-bonn.de

C. Rüb
e-mail: crueb@uni-bonn.de

M. Bruderek
e-mail: michael-bruderek@uni-bonn.de

W. A. Houry (ed.), *The Molecular Chaperones Interaction Networks in Protein Folding and Degradation,* Interactomics and Systems Biology, DOI 10.1007/978-1-4939-1130-1_12,

interactions, e.g., apoptosis. The sum of biochemical reactions leading to the proper maintenance of mitochondrial enzymatic activities under normal and stress conditions is summarized under the expression “mitochondrial protein homeostasis” [1, 2]. Successful protein homeostasis requires a complex and intricate system of proteinaceous mediators that deal with the biochemical reactions involved in protein biogenesis and quality control (Fig. 12.1). In the course of the last decade, an increasing amount of evidence demonstrated the important role of mitochondrial defects, broadly represented by the breakdown of mitochondrial homeostasis, during the etiology of many human pathologies [3], in particular neurodegenerative diseases and aging [4].

In general, protein homeostasis is achieved by an equilibrium between cellular polypeptide biogenesis and their removal by proteolytic reactions when they become superfluous or damaged. In case of mitochondria as separate organelles, both processes have their own complications. Concerning biogenesis, mitochondrial polypeptides have a dual origin, most are nuclear encoded but a small subset is provided by internal expression of mitochondrial genes. In addition, nuclear-encoded polypeptides, synthesized in the cytosol, have to pass, or insert into the mitochondrial membrane system. On the other hand, the proteolytic system in mitochondria (except for outer membrane proteins) is physically separated from the cytosolic ubiquitin-proteasome system [5] and requires its own set of degradation enzymes. Due to their endosymbiontic origin, the enzymatic mediators involved in mitochondrial protein homeostasis are closely related to their bacterial counterparts. Mitochondria contain the full set of typical molecular chaperones, or heat shock proteins (Hsp) of the classes Hsp40, Hsp60, Hsp70, Hsp90, and Hsp100, named after their average molecular mass [6]. However, no member of the small Hsp chaperone family has been identified in mitochondria so far. The mitochondrial chaperones participate in all types of protein biogenesis reactions, ranging from protein import through the membranes to the subsequent folding and/or assembly reactions. In addition, they also assist refolding reactions of denatured polypeptides under stress conditions or prevent their aggregation. Terminally damaged or surplus polypeptides are removed by a set of dedicated mitochondrial proteases that are, again similar to bacteria, members of the LON, Clp, and FtsH/adenosine triphosphatases associated with diverse cellular activities (AAA) protease families in most cases belonging to the adenosine triphosphate (ATP)-dependent AAA+ protein family. In forming an intrinsic mitochondrial protein quality control (PQC) system, these proteolytic enzymes usually closely collaborate with the respective chaperones, in particular during substrate polypeptide recognition and unfolding.

Recent studies demonstrate the importance of an additional level of mitochondrial quality control. In addition to maintaining the function of individual proteins, cellular mechanisms have emerged that are involved in keeping the organellar content of the cell as such intact. In this context, mitochondria, which have been severely damaged on the organellar level, are recognized and removed as a whole by a variation of the autophagy mechanism called mitophagy. In particular, a defective mitophagy process seems to contribute to the progress of neurodegenerative disorders like Parkinson disease [7].

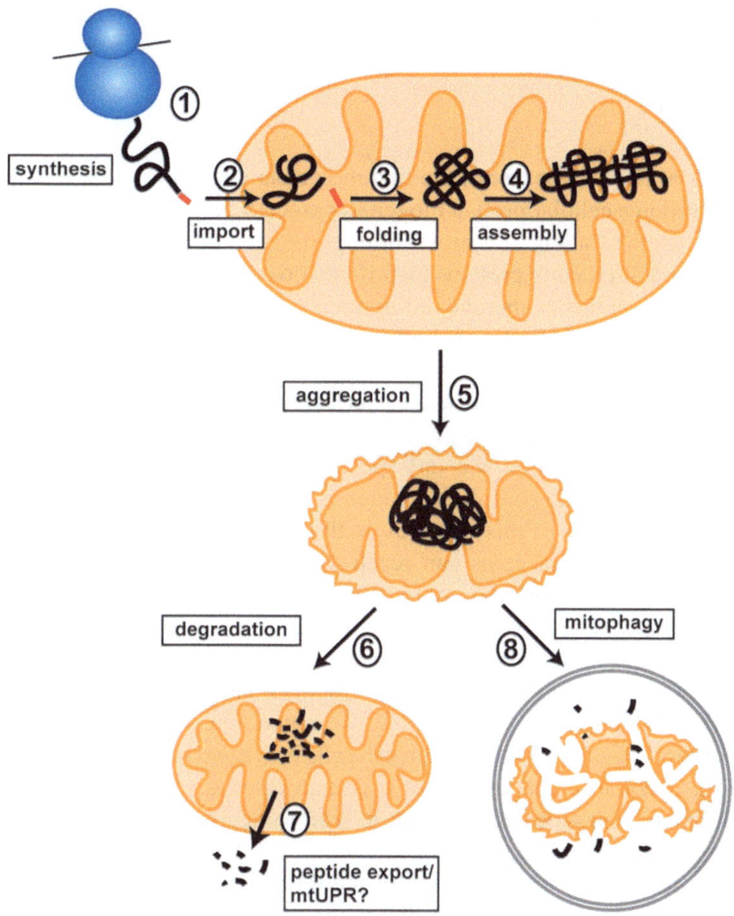

Fig. 12.1 Principles of mitochondrial protein homeostasis. The vast majority of mitochondrial polypeptides are synthesized at cytosolic ribosomes as precursor forms (*1*). The unfolded precursor protein, usually containing an N-terminal targeting signal is then imported into the mitochondrion via the outer membrane translocase complex/inner membrane translocase complex (TOM/TIM) translocase machinery (*2*). Import is assisted by a matrix-localized Hsp70 chaperone. Upon import, the targeting sequence is cleaved off, giving rise to the mature protein. Proteins fold into their native conformation, assisted by mitochondrial Hsp70 and Hsp60 chaperone systems (*3*). In case of multi-subunit protein complexes, e.g., the respiratory chain complexes, several monomers are assembled to form the functional enzyme complex (*4*). Under stress or disease conditions, proteins become damaged or denatured and potentially toxic protein aggregates may form, leading to functional defects of the affected mitochondria (*5*). These aggregates can be removed by one of the following mechanisms. Aggregates are degraded by mitochondrial AAA+ proteases (*6*) and the degradation products are then exported into the cytosol, restoring a healthy mitochondrion. In addition, mitochondrial protein stress signaling elicits a mitochondrial unfolded protein response (mtUPR) reaction, increasing the expression of mitochondrial chaperones (*7*). Terminally damaged mitochondria may be discarded by mitophagy, a mitochondria-specific type of selective autophagy (*8*). The mitochondrion is surrounded by a double membrane structure, which then fuses with a lysosome, delivering its complete contents to degradation

2 Chaperone Systems

2.1 Hsp60

Mitochondrial Hsp60 was one of the first molecular chaperones that was identified to be involved in a cellular protein biogenesis process. It is located in the matrix and responsible for the folding of a subset of imported polypeptides [8, 9]. A proteomic study of Hsp60 interacting proteins identified about 30 % of mitochondrial proteins to be dependent on this chaperone during their folding process [10]. Mitochondrial Hsp60 and its cofactor Hsp10 are both indispensable for the full functional activity of mitochondria and indirectly for the survival of the cell. The role of Hsp60 strongly resembles the activity of the bacterial GroEL chaperone that is involved in the folding of newly synthesized proteins after their release from the ribosome. All Hsp60-type chaperones employ a very specific biochemical mechanism. They form a large homo-oligomeric complex arranged in two stacked rings of seven subunits each. The inner cavity of this complex is thought to provide a separate "folding cage" where unfolded proteins can reach their native conformation without being exposed to the cellular environment. The formation of the folding cage is also dependent on a specific cofactor, Hsp10, which forms a lid structure that covers the opening of the cavity. Hsp60 chaperones undergo an enzymatic reaction cycle where the binding and hydrolysis of ATP regulate and coordinate conformational changes in the Hsp60 complex as well as the interaction with Hsp10 [11].

Due to their essential function for the cell, it is relatively rare to associate defects of mitochondrial chaperones with human diseases. However, mutant forms of Hsp60 have been repeatedly shown to lead to neurodegenerative diseases like hereditary spastic paraplegia in human patients [12]. Hsp60 has also been shown to be directly involved in the regulation of the mitochondrial permeability transition pore, which is an important event in triggering apoptosis [13].

2.2 Hsp70

Mitochondria contain usually one member of the Hsp70 family, which is closely related to the bacterial Hsp70 DnaK. The representative in human mitochondria is named alternatively mitochondrial Hsp70 (mtHsp70), Grp75, or mortalin, encoded by the gene *HSPA9*. Most biochemical studies on this chaperone have been performed in the model organism *Saccharomyces cerevisiae* (yeast). However, yeast mitochondria represent a special situation where due to ancient gene duplications, the Hsp70 activity is split up into three different proteins. The main mtHsp70 is named Ssc1, the second, with a more specialized function Ssq1, and a third with unclear functional relevance Ssc3. Ssc1 is an essential protein mediating the typical chaperone functions like protein import, folding, and stress protection. In contrast, Ssq1 seems to be specialized on a supportive function in the biogenesis of mitochondrial enzymes containing iron-sulfur (Fe/S) cluster as prosthetic groups. The cellular role of Ssc3 (encoded by *ECM10*) is not known. It is not expressed under normal conditions and the deletion mutant exhibits no phenotype.

In general, Hsp70 chaperones consist of two functionally distinct domains, an N-terminal ATPase domain and a C-terminal substrate-binding domain. The substrate-binding activity is regulated by a specific allosteric mechanism that is based on the nucleotide state of the ATPase domain [14]. In the ATP-bound state, substrate affinity is low while the adenosine diphosphate (ADP) state is characterized by high affinity to bound polypeptides. The intrinsic ATPase activity of Hsp70 chaperones is regulated by specific cofactors. Proteins of the J-domain family, derived from the bacterial DnaJ, assist the initial substrate interaction and induce a "locked-in" state of the substrate by stimulating ATP hydrolysis. On the other hand, substrate release is initiated by nucleotide-exchange factors (NEF), members of the GrpE protein family, which induce the release of bound phosphate and ADP, thereby facilitating binding of ATP.

In general, the interaction with specific J-proteins is a major determinant of the functional diversity of Hsp70 chaperones [15]. Due to their intrinsic substrate interaction ability, which is independent of the Hsp70 partner, these J-proteins are often listed as a separate chaperone family, termed Hsp40 according to their average molecular mass. Mitochondria in yeast contain at least 4 types of J-domain containing proteins. Mdj1 in yeast is the genuine DnaJ homolog and involved in polypeptide folding reactions in the mitochondrial matrix in cooperation with its respective Hsp70, Ssc1. Other family members are Mdj2 and Pam18, both inner membrane-integrated proteins with a role in protein import, and Jac1 that cooperates with Ssq1 (see below). Based on its assistance in general folding reactions, Mdj1 is involved in mitochondrial biogenesis reactions, both for proteins newly imported from the cytosol as well as proteins synthesized by mitochondrial ribosomes [16]. It also has some stress-protective function to prevent mitochondrial polypeptides from aggregation [17, 18]. Interestingly, the human homolog of Mdj1, named human tumourous imaginal disc (TID1; encoded by DNAJA3) has also been implicated in signaling pathways [19], indicating an extended role on cellular activities. The J-domain protein Jac1 supports the activity of Ssq1, the specialized mtHsp70 that is involved in the mitochondrial synthesis of iron-sulfur cluster [20]. The only known substrate of Ssq1 is the protein Isu that forms the scaffold for the cluster assembly [21]. Deletions of Ssq1 in yeast show a cold-sensitive growth phenotype but are not as deleterious as null mutations of Ssc1, which result in a lethal phenotype. This observation also indicates that a certain functional overlap between Ssq1 and Ssc1 exists [22, 23]. In fact, the activity of Ssq1 is taken over by the single mitochondrial Hsp70 in metazoan species.

The nucleotide exchange factor, in case of mitochondria the protein Mge1, is an essential cofactor of Hsp70 chaperones. Mge1 is indispensable for the completion of the enzymatic cycle by facilitating the exchange of ADP for ATP, resulting in large changes in substrate affinity. Hence, Mge1 is both crucially involved in protein import [24, 25] and in subsequent protein folding events [26]. Mge1 is also able to interact with the second yeast mtHsp70, Ssq1 [23]. Mge1 is present as a dimer in vivo that directly interacts with the ATPase domain of mtHsp70. The binding of nucleotide exchange factors like Mge1 is thought to slightly open the nucleotide-binding cleft in the ATPase domain, thereby reducing the affinity to the bound nucleotide. The dimer formation has been recently shown to be sensitive to reactive oxygen species (ROS), thereby compromising the essential mtHsp70 activities under oxidative stress conditions [27].

Although the structures of the individual Hsp70 domains have been solved since several years, the biochemical basis of the allosteric mechanism and the nucleotide-dependent regulation of Hsp70 substrate binding remained enigmatic. Only very recent structural studies on Hsp70 chaperones (although not including the mitochondrial variant) provided progress in this question [14]. Application of these novel structures, combined with modern biophysical methods [28], will allow to shed new light on the mechanistic details of mtHsp70 activity and also its special functions in mitochondria. Again, as already argued in the case of Hsp60, the important and essential role strongly reduces the chances to identify human patients with diseases caused by functional defects in mtHsp70. However, mtHsp70 mutants were recently described as susceptibility factors in Parkinson disease [29]. Here, an enhanced J-protein interaction and compromised protein stability of mtHsp70 variants lead to mitochondrial dysfunction that is connected with this disease. Arguably the most important role of mtHsp70 is its requirement for the import of matrix-targeted preproteins from the cytosol. In yeast mitochondria, and most likely its equivalents in all other species, Ssc1 is the central component of the import motor complex, which provides the driving force for the completion of the polypeptide translocation into the matrix compartment (Fig. 12.2). Together with the outer membrane translocase complex (TOM) the inner membrane translocase complex (TIM23) provides the pore structures for the translocation of unfolded precursor proteins through the mitochondrial membrane system [30]. Ssc1 is attached to the TIM23 complex via an interaction with the translocase component TIM44. At the exit of the translocase, Ssc1 interacts with the incoming (unfolded) polypeptide and drives its full translocation into the matrix by an ATP-dependent reaction. For its import activity, it also requires the inner membrane J-protein Pam18, a component of the TIM23 translocase, and the NEF Mge1. Pam18 forms a heterodimer with a degenerate J-protein, Pam16, which is required for its activity regulation and interaction with the TIM23 translocase [31, 32]. Different and controversial models for the biochemical basis of the mtHsp70 role in the translocation process have been discussed [33, 34]. While the basic mechanism relies on a Brownian ratchet model, where the translocation process is driven by diffusion, it has been experimentally proven that the complete import of preproteins containing conformational restriction, i.e., folded C-terminal domains, requires a more active mechanism [35]. The generation of a pulling force needs three prerequisites, (i) a conformational change in the chaperone, driven by nucleotide hydrolysis, combined with (ii) a stable and functional interaction with the translocase and (iii) the preprotein polypeptide [36, 37]. All these prerequisites are fulfilled in case of the import motor complex, consisting of Ssc1, Tim44, Pam18, and Mge1. Recently, a novel hypothesis has been proposed that attributes the generation of the pulling force by the import motor to an entropic mechanism

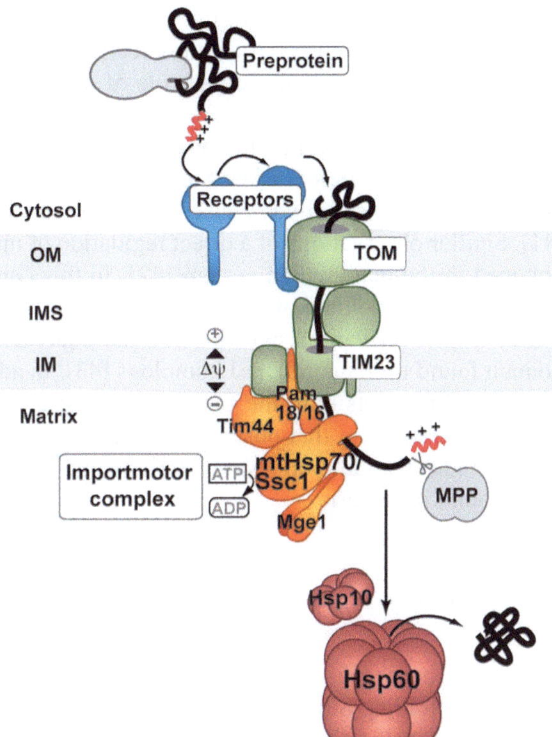

Fig. 12.2 Protein import and folding reactions. The targeting signal sequence of mitochondrial precursor proteins is recognized and bound by receptor proteins at the mitochondrial surface, which transfer the precursor to the translocase complex of the outer membrane (TOM). After insertion into the inner membrane translocase complex (TIM23) in a membrane potential ($\Delta\psi$)-dependent manner, preproteins start to interact with the mitochondrial Hsp70 (mtHsp70), in yeast encoded by *SSC1*. Ssc1 is anchored at the TIM23 complex by an interaction with Tim44 and activated by the membrane-integrated J-domain protein Pam18 that itself is regulated by an interaction with Pam16. In cooperation with the nucleotide-exchange factor Mge1, Ssc1 actively completes the full translocation of the unfolded polypeptide chain into the matrix compartment in an adenosine triphosphate (*ATP*)-dependent reaction. After removal of the N-terminal targeting sequence by the matrix processing peptidase (*MPP*), folding of many preproteins to the native conformation proceeds with the help of the Hsp60 complex in coordination with the specific co-chaperone Hsp10

[38]. However, a final clarification of the biochemical mechanism of the import motor and its core component mtHsp70 awaits the in vitro reconstitution of the mitochondrial preprotein import system.

A novel twist to the regulation of mtHsp70 activity by co-chaperones is provided by new studies on the mitochondrial protein Zim17 (Hep1). Zim17, present in mitochondria of all eukaryotic species but not in bacteria, was initially described as a Ssc1-interacting protein that supports the solubility of the chaperone [39]. As Ssc1 and Ssq1 fulfill important functions in mitochondria, a depletion of these mtHsp70s by aggregation would result in the accumulation of many secondary defects, like a loss of membrane potential and a defective protein import [40]. However, novel experiments indicate a more direct role of Zim17 on the activity of mtHsp70 that is independent of an aggregate prevention. Studies using new conditional mutants of Zim17 demonstrated that the observed loss of respiratory activity is most likely due to a functional defect of Fe/S cluster synthesis by Ssq1 while a defective protein translocation system is caused by a prolonged substrate interaction of Ssc1 [41]. Similar observations of a direct regulation of mtHsp70 activity have also been obtained in the mammalian system [42]. In this context, it was hypothesized that the zinc-binding domain of Zim17 complements the activity that is provided by the J-domain protein Pam18, which lacks the equivalent of a C-terminal zinc-finger domain found in standard DnaJ homologs [43]. In addition, in vitro studies showed that Zim17 interacts with the C-terminal substrate-binding domain of Ssc1, a behavior different from that observed with traditional J-proteins [44]. Hence, Zim17 may provide an additional level of activity and stability regulation of mtHsp70 that is not found in bacteria.

2.3 Hsp90

Molecular chaperones of the Hsp90 family are involved in profound cellular processes, like folding of proteins for signal transduction, protein trafficking, receptor maturation, innate and adaptive immunity, and morphological evolution [45]. While the bacterial family member is termed high temperature protein G (HtpG), eukaryotic cells contain several members of this chaperone class, two major cytosolic isoforms, the inducible Hsp90α and the constitutively expressed Hsp90β, as well as an endoplasmic reticulum (ER)-specific form named Grp94 (94 kDa Glucose-regulated protein), a chloroplast-specific isoform in plants, and a mitochondrial protein. The mitochondrial Hsp90 homolog was initially identified by two independent yeast-two-hybrid screens and named tumor-necrosis factor receptor associated protein 1 (TRAP1) according to the bait protein used in the screen [46] or Hsp75, which was found due to its interaction with the retinoblastoma (Rb) protein [47]. It should be noted, however, that the significance of these interactions has not been demonstrated to date. TRAP1 was later described as a single protein, which carries an N-terminal mitochondrial targeting signal and is localized predominantly in the mitochondrial matrix of mammalian cells [48, 49]. An extra-mitochondrial localization of a minor fraction of the protein has also been proposed. Hsp90 type chaperones are highly conserved homodimers, with each protomer consisting of structurally and functionally distinct N-terminal-, middle-, and C-terminal domain [50, 51]. TRAP1 binds ATP and this binding can be blocked by the classic Hsp90 inhibitors radicicol and geldanamycin [48]. In a comprehensive in vitro analysis, TRAP1 exhibited enzymatic activities similar to other Hsp90 type proteins [52]. Despite these similarities, cytosolically expressed TRAP1 is not able to complement the loss of other Hsp90 isoforms [48] and was thus proposed to fulfill specific, yet largely unknown functions within mitochondria.

The few available functional studies on TRAP1 point towards a stress-protective role against ROS production and/or ROS related apoptosis [53–56]. TRAP1 expression was shown to be upregulated in several tumor cell lines [57, 58], and the proposed cytoprotective effects of TRAP1 might contribute to cancer cell survival and drug-resistance [59]. However, information about specific mitochondrial substrates and possible co-chaperones remains scarce. First insights into a potential mode of action emerged as TRAP1 was identified as a member of a chaperone network, comprising also cytosolic Hsp90, and Hsp60. The combined action of the chaperones was proposed to antagonize the pro-apoptotic function of the immunophilin cyclophilin D, via regulation of the mitochondrial permeability transition pore [58]. More recent findings relate TRAP1 to mitochondrial respiration and tumorigenesis. It was proposed that TRAP1 functions as a metabolic switch between mitochondrial respiration and aerobic glycolysis [60]. Providing a possible mechanistic explanation, TRAP1 was found to bind and inhibit succinate dehydrogenase (SDH), the complex II of the mitochondrial respiratory chain [61, 62]. The authors propose an intruiging mechanism by which upregulation of TRAP1 may promote tumorigenesis. In this model, inhibition of SDH by TRAP1 would result in the accumulation of the SDH substrate succinate, which in turn stabilizes the pro-neoplastic transcription factor hypoxia-inducible factor 1 (HIF1α). However, the molecular details of the connection between SDH and TRAP1 remain to be established. In this context, TRAP1 seems to work in a quite distinct fashion from its cytosolic counterparts with only limited relations to classical chaperone functions.

2.4 Hsp100

In contrast to the ubiquitous chaperones of the Hsp70 family that occupy a central role in the PQC system, members of the Hsp100 chaperone family accomplish more distinct tasks in a cell. Due to their ability to remodel proteins in an ATP-dependent manner, they mostly mediate unfolding and disassembly reactions of proteins and protein-complexes. Since this remodelling activity of Hsp100 chaperones could lead either to refolding or degradation of substrate proteins, they are primarily involved in the protection against proteotoxic stress conditions. Members of the Hsp100 family are characterized by a common ATPase activity, which is carried out by one or two AAA+ modules including a basic core of 200–250 amino acids that comprises a α-helical part and a nucleotide-binding part. Hence, Hsp100 proteins belong to the class of AAA+ ATPases. The number of nucleotide-binding domains (NBD) divides the Hsp100 family into two classes characterized either by two conserved (class 1 Hsp100 s) or by just one NBD (class 2 Hsp100 s) [63]. Each NBD consists of two specific sequence motifs, referred to as Walker A and Walker B motifs. While Walker A is required for ATP hydrolysis, Walker B is mainly responsible for nucleotide binding. A specific structural feature of class 1 Hsp100 chaperones is the formation of a large hexameric ring-shaped protein complex containing an intramolecular pore. Formation and molecular function of this complex is largely governed by the interaction of the two NBD domains. The ATPase activity and the correlated allosteric communication between the NBD domains enable these chaperones to act as ATP-dependent protein unfoldases [64]. It is thought that a concerted ATP hydrolysis in the hexameric complex allows to exert a pulling force on proteins in order to translocate polypeptide chains through the central pore and unfold them at the same time. Typically, specific interaction partners determine the final fate of unfolded proteins interacting with Hsp100-type chaperones. In bacteria, proteins of this class constitute subunits of the caseinolytic peptidase (Clp). The proteins ClpA, ClpC, and ClpX are typical unfoldases and act in the degradation of terminally damaged proteins in cooperation with the AAA+ protease ClpP. An exception is the bacterial protein ClpB that acts together with the Hsp70/Hsp40 system in refolding and reactivation of misfolded and aggregated proteins and is hence to designate as disaggregase [65, 66].

2.4.1 Hsp78

To counteract protein aggregation processes upon cellular stress, fungi, and plants contain homologs of the bacterial ClpB, both in the cytosol, named Hsp104 and Hsp101, respectively, but also in their organelles [65, 67]. In fungi, the protein Hsp78 is the mitochondrial homolog of ClpB (Fig. 12.3). Hsp78 consists of the two common nucleotide-binding domains of Hsp100 proteins, which mark Hsp78 as a member of the class 1 Hsp100 protein family. Hsp78 contains a conserved carboxy-terminal domain and the middle domain typical for ClpB homologs but lacks the amino-terminal domain completely [68], which has been described to be involved in substrate interaction. Instead of the amino-terminal domain, the mitochondrial disaggregase contains a mitochondrial import targeting sequence, which is cleaved off by the matrix processing peptidase (MPP) upon import of Hsp78 into the mitochondrial matrix, accounting for the smaller size of Hsp78 compared to other typical Hsp100 proteins. Although Hsp78 protomers form ring-shaped hexameric protein complexes in the presence of ATP in vitro [69], in vivo studies revealed only a trimeric complex of Hsp78 at least under normal conditions [70], indicating a high degree of structural dynamics in ClpB-type chaperones.

Like its bacterial counterpart, ClpB, the activity of Hsp78 in yeast mitochondria is mainly required in remodelling of aggregated proteins to unfolded polypeptide chains [71, 72]. Due to the close relation to bacterial ClpB, it is assumed that the Hsp78 unfolding activity is based on the unravelling of the polypeptide substrate through its intramolecular pore. Hsp78 has no crucial role in protection of mitochondrial protein homeostasis during stress, but it is rather required for efficient refolding and reactivation of inactivated and aggregated proteins caused by stress conditions and therefore it is upregulated during such situations [68, 73]. Due to its disaggregation activity, Hsp78 is involved for example in maintenance of the integrity of the mitochondrial genome by reactivation of key players of the mitochondrial

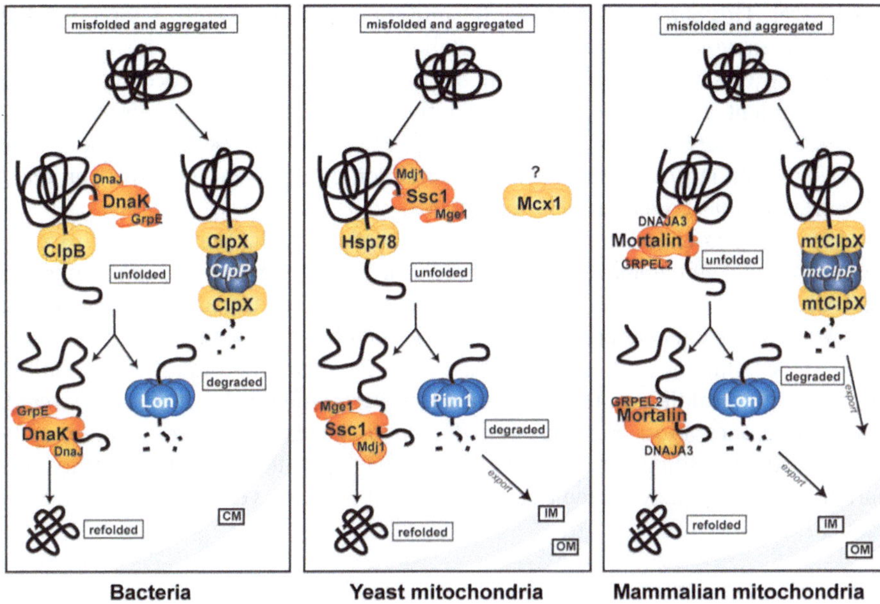

Fig. 12.3 Schematic illustration of clearance of aggregated and misfolded proteins in bacteria, yeast and mammalian mitochondria. In bacteria, misfolded and aggregated proteins are cleared by one of two mechanisms: Either aggregated proteins are solubilized by the disaggregase machinery ClpB/Dnak/DnaJ followed by a refolding to the native state or the disaggregated proteins become degraded by the protease complex ClpX/ClpP. In mitochondria of yeast, the ClpB homolog Hsp78 disaggregates aggregated proteins in cooperation with the Hsp70 chaperone system Ssc1/Mdj1 to unfolded polypeptide chains. Subsequently, such unfolded proteins are either refolded by Ssc1/Mdj or degraded by the matrix protease Pim1 depending on the degree of damage. Finally, degraded peptides are exported to the cytosol of yeast. Notably, a homologous protein of mtClpP is not present in mitochondria of yeast, while a ClpX homolog, Mcx1, still exists. In mammalian mitochondria, aggregated proteins are either degraded by mtClpX/ClpP or unfolded likely due to the activity of the Hsp70 system components Grp75 (Mortalin) and DnaJA1. Furthermore, unfolded proteins get either refolded to functional proteins by Mortalin/DnaJA1 or they are degraded by the protease activity of LON. Finally, degraded peptides are exported to the cytosol through the inner and outer membrane

deoxyribonucleic acid (DNA) replication system [74]. Hence, a deletion of Hsp78 will eventually result in a loss of respiratory components and an impaired oxidative phosphorylation. Furthermore, Hsp78 also acts in maintaining the membrane potential of the inner membrane, which is required for import of newly synthesized proteins [71, 73] that is in turn essential for the mitochondrial protein homeostasis. Similar to the behavior of ClpB, Hsp78 functions as a bichaperone system requiring the cooperation with the mtHsp70 chaperone Ssc1 [71, 73]. This interaction seems to be crucial for the disaggregation activity, since inactivation of Ssc1 by temperature-sensitive mutations lead to a complete loss of refolding activity [75].

In addition to its function in refolding, Hsp78 also plays a role in efficient degradation of terminally damaged proteins together with the mitochondrial protease

Pim1/LON. This activity was shown for imported proteins as well as for a mutated form of an enzyme of the amino acid metabolism [72, 76]. Although the functionality is unknown so far, the participation of Hsp78 in degradation processes seems to be independent from the disaggregation activity, since all identified substrates were already completely soluble. Interestingly, this activity of Hsp78 seems to be unique to yeast mitochondria, since no similar interaction of ClpB and the AAA+ protease LON has been described for other organisms.

However, mitochondria of metazoan organisms lack the presence of a Hsp78-like chaperone, which appears unusual regarding the important role of disaggregation reactions. How the lack of a Hsp78 activity is compensated is not clear so far, but two explanations are possible. On the one hand, higher organisms can actively avoid stress situations, like heat stress, due to their mobility and therefore might not require a refolding system specialized in thermotolerance. On the other hand, other chaperones, most likely the Hsp70-Hsp40 chaperone system, could directly take over the refolding activity of Hsp78. In principle, the second possibility was partially shown already for the Hsp70 systems of bacteria, yeast and higher eukaryotes [77, 78].

3 Protease Systems

Mitochondria contain their own endogenous proteolytic system that is largely independent of the cytosolic ubiquitin-proteasome system. Only some outer membrane proteins seem to be degraded by the proteasome, in a mechanism reminiscent of the Endoplasmic-reticulum-associated protein degradation (ERAD) process [79]. Of the different types of endogenous proteolytic activities in mitochondria, we have to distinguish processing enzymes that remove parts of mitochondrial polypeptides in the course of the biogenesis process and PQC-related proteases that are responsible for the complete degradation of surplus or damaged polypeptides. In this context, only the latter will be discussed. Similar to the situation in bacteria, mitochondrial PQC proteases belong to the AAA+ protein family (ATPases associated with a large variety of cellular activities). These enzymes form large ring-shaped protein complexes that degrade polypeptides in an ATP-dependent reaction in an internal cavity. Hence, they are also termed "chambered proteases." In the matrix compartment, three different proteolytic activities have been identified (Fig. 12.3). Of these, the LON/Pim1 and ClpP proteases are soluble enzymes of the mitochondrial matrix and responsible for the degradation of soluble substrates. In contrast, the m-AAA-protease is integrated into the inner membrane and mainly responsible for the degradation of membrane proteins, but its active site faces the matrix. An equivalent membrane-integrated protease, the i-AAA protease, faces the intermembrane space (IMS) compartment. It needs to be noted that apart from PQC reactions the listed proteases are also important for mitochondrial genome maintenance and adaptation of nucleoid components to different environmental conditions [80].

3.1 LON

The soluble LON protease (in yeast encoded by *PIM1*) was the first identified mitochondrial PQC protease [81]. It is the major protease system that degrades soluble damaged polypeptides in the matrix compartment. The protease works in a close cooperation with molecular chaperones of the Hsp70 system [82]. Under normal conditions, the turnover of mitochondrial proteins is relatively low. Here, the protease is mainly involved in the degradation of newly imported polypeptides that fail to reach their native conformation or to assemble into their respective enzyme complexes [83]. The role of the LON protease in PQC is particularly prominent in case of polypeptides, which are damaged by ROS. Due to their covalent nature, ROS modifications lead to the denaturation of affected polypeptides that require their proteolytic removal to avoid their accumulation as insoluble species [84, 85]. Hence, the yeast LON homolog, Pim1, supports cell growth under oxidative stress conditions and has been shown to be a major factor in the protection against heat-induced protein aggregates [17]. Pim1 has also been shown to affect the mitochondrial Fe/S cluster biosynthesis, one of the essential functions of the organelle [86]. Based on the general PQC activity of the LON protease, it is not surprising that its mutants display a decreased life span in fungal aging models [87, 88]. Although the precise biochemical context of the long-term effect of LON defects needs to be established, a contribution of LON to pathological reactions has also recently been shown in the human model [89].

In addition to its role during PQC in the matrix, Pim1 is also involved in maintaining the integrity of the mitochondrial genome. Deletion mutants of Pim1 in yeast are not lethal but exhibit a loss of respiratory competence due to an accumulation of lesions in the mitochondrial genome [90]. A direct connection between Pim1 activity and mitochondrial genome function was supported by the observation that LON is responsible for the degradation of the human mitochondrial transcription factor TFAM, a protein that assists both transcription and genome replication. Pim1 mutants also exhibit defects in the splicing of certain transcripts resulting in the inability to produce the respective mitochondrially encoded proteins [90].

3.2 AAA Proteases

The membrane-bound mitochondrial proteases of the AAA-type are homologs of the bacterial FtsH protease [91]. Typically, they consist of an ATP-binding AAA domain connected to additional conserved structural elements, in particular either single or double membrane spanning segments [92]. The mitochondrial AAA proteases degrade predominantly membrane integrated substrate proteins in a process that involves an active extraction of the polypeptide from the membrane [93]. Potential substrates are primarily non-assembled subunits of inner membrane respiratory chain components. Here, the membrane topology of the substrate, i.e., on which side the bulk of the polypeptide chain resides, determines the type of protease responsible for degradation [94]. Recently, defects in mitochondrial AAA proteases were shown to cause a variety of human neurodegenerative diseases [95].

The m-AAA protease, facing the matrix compartment, is anchored in the inner membrane by two transmembrane segments. Recent structural studies confirmed the formation of a hexameric ring structure and gave first information on the domain orientation in the complex [96, 97]. In yeast, the m-AAA protease is formed by two proteins, Yta10 and Yta12. In humans, these subunits are termed AFG3L2 and paraplegin (encoded by *SPG7*). However, the subunit composition can be more complex since AFG3L2 is also able to form homo-oligomeric complexes. Mutations in the m-AAA protease subunits are associated with certain neurodegenerative diseases like spastic paraplegia (mutation in *SPG7*) or a form of spinocerebellar ataxia (mutations in *AFG3L2*) [95]. In addition to the general PQC activity, the m-AAA protease has also been shown to participate in processing events that are required to remove import targeting signal sequences and are a prerequisite for the acquisition of a native structure after import has been completed. A typical example is the ribosomal protein MrpL32 [98]. It can be expected that the defective biogenesis of a single protein with a central function like MrpL32 leads to a major and general defect in mitochondrial activities.

The i-AAA protease consists of a homo-oligomeric ring of six Yme1 subunits YME1 in humans. Its proteolytic site faces the IMS and mainly degrades inner membrane polypeptides, but is also capable of degrading soluble proteins of the IMS [99]. A biochemical analysis of the i-AAA protease activity directly revealed the dual nature of the AAA-type proteases. The N-terminal domain preceding the AAA domain is responsible for the recognition and binding of substrate polypeptides. When expressed alone, this domain exhibited chaperone-like activity [100]. Mechanistically, AAA proteases therefore represent a protease together with a chaperone activity on the same polypeptide. A recent analysis of folding and aggregation reactions in the IMS supported the role of the i-AAA protease as an IMS chaperone [101]. Similar to most proteins involved in mitochondrial protein homeostasis, also mutants of the i-AAA protease exhibit reduced lifespan, in this case also influenced by the ambient temperature [102].

4 ClpP

Mitochondria of higher eukaryotes contain a second soluble protease of the AAA+ family, the mitochondrial ClpP (mtClpP). In contrast to Pim1/LON, this protease acts as a two-component system built up by the proteolytic core mtClpP and the Hsp100 chaperone mtClpX (Fig. 12.3). MtClpP forms a two-stack heptameric complex of, in total, 14 protomers, shielding the catalytic site in the inside of the complex from the mitochondrial environment [103]. For the bacterial counterpart ClpP, it has been shown that substrate proteins become degraded to small peptides of about 5–20 amino acids in length [104]. However, ClpP alone has only low degradation efficiency for full-length proteins due to its weak unfolding activity [105].

Therefore, it is assumed that also mtClpP needs the unfolding activity of the Hsp100 chaperone component, in this case mtClpX. Like other Hsp100 proteins, mtClpX forms the common hexameric ring-shaped complex, forming an intramolecular pore that most likely enables the chaperone to act as an unfoldase [106]. In addition to this unfolding activity, ClpX-type chaperones are also involved in the recognition of substrate proteins and their translocation into the proteolytic chamber of ClpP [65].

Up to this date, no comprehensive data are available on the function of mtClpP in mitochondria. Based on its close relation to the bacterial protease system, mtClpP is assumed to be involved in PQC functions in the matrix although no specific substrates are known so far. Very recently, a knockout model of mtClpP in mice was published, exhibiting pronounced growth retardation and failure of certain organ functions [107]. Another recent finding revealed that a knockdown of the corresponding chaperone component, mtClpX, resulted in alterations of mitochondrial nucleoid structure in human cells [108]. Notably, in mitochondria from the yeast model S*accharomyces cerevisiae*, the protease mtClpP does not exist while the respective Hsp100 chaperone, here called Mcx1, is still present [109]. However, since the deletion of Mcx1 has no observable phenotype, the function of Mcx1 remains unclear. It might fulfill a so far unidentified independent chaperone function in mitochondria.

5 The Mitochondrial Unfolded Protein Response

In the recent years, experimental evidence has emerged that the mitochondrial network of chaperones and proteases underlies a regulatory mechanism adapting its activity to proteotoxic stress conditions, at least in certain model systems. An accumulation of damaged and unfolded polypeptides results in an increased expression of some or all components of the PQC system, a process that is termed "unfolded protein response (UPR)." Well studied in the ER, the UPR reaction comprises a specific signal-transduction pathway from the respective subcellular compartment to the nucleus in order to induce the expression of Hsps upon stress conditions [110]. In case of mitochondria, only limited information about the signalling pathways is available to date. Studies in *C. elegans* revealed that matrix proteases act not only in degradation of misfolded proteins, but also as a crucial factor in the mitochondrial UPR (mtUPR). Here, the intrinsic proteolytic activity of mtClpP, but not the activity of LON, seems to be important for the induction of this signalling pathway. It is speculated that during stress conditions, an overrepresentation of small peptides, resulting from increased degradation processes by mtClpP somehow leads to the activation of cytosolic signaling factors, which in turn induce the expression of mitochondrial members of the PQC system [111]. This model was supported by the finding that the inner membrane protein HAF-1, responsible for the active export of small peptides from the matrix into the cytosol, is a crucial component of the mitochondria-nucleus stress signaling pathway [112]. Although mammalian cells also exhibit a mtUPR response, it is not clear if they employ the same signal process. In particular, human ClpP together with other chaperones seems to be rather a substrate than mediator of the mtUPR reaction [113]. Remarkably, at least in *C. elegans*, a mtUPR reaction elicited in mitochondria of one organ, in this case caused by the accumulation of non-assembled components of respiratory chain complexes, can be mediated by so far unknown signals to other organs that were initially not affected by the stress condition [114]. Although the details of this stress signaling process on the level of the organism still need to be established, it might provide an additional level of stress adaptation.

6 Mitophagy

What happens if the above-discussed biochemical mechanisms of mitochondrial PQC are exhausted? An accumulation of denatured polypeptides or even protein aggregates would constitute a grave danger to mitochondrial functions, for example causing the generation of toxic ROS. More directly, damaged mitochondria can release apoptotic factors that can lead to a premature cell death. Thus, cells have evolved a more radical mechanism to remove whole mitochondria (Fig. 12.1), a process that utilizes the reactions of macroautophagy. Autophagy generally describes the sequestration of a portion of the cytoplasm, protein aggregates or whole organelles in a double membrane structure termed autophagosome, which then fuses with a lysosome, thereby delivering its cargo to degradation by lysosomal enzymes. While nonspecific autophagy of intracellular components occurs in response to nutrient deprivation, a second main type of autophagy can be highly selective for specific organelles, namely peroxisomes (pexophagy), ribosomes (ribophagy), or mitochondria (mitophagy). By removing superfluous or damaged organelles, selective autophagy helps adjusting organelle number to the cell's metabolic demands and maintaining organelle quality. The mechanism of mitophagy has been intensively studied in yeast. Like other types of selective autophagy, mitophagy utilizes the core autophagy machinery, comprising numerous of the 35 known so-called autophagy-related (Atg) proteins [115]. Using yeast genetic screens, two independent groups recently identified the outer mitochondrial membrane (OMM) protein Atg32 as a mitophagy-specific receptor that interacts with both Atg8 and Atg11 to mediate recruitment of mitochondria to the autophagosome. No mammalian homolog of Atg32 has been identified so far but the Bcl-2 family protein Nix is a candidate functional counterpart. Like Atg32, Nix is located in the OMM and binds Atg8 [116, 117].

Mitochondrial dysfunction was observed to be a common feature of various neurodegenerative diseases like Parkinson disease (PD), Alzheimer disease, Huntington disease, or amyotrophic lateral sclerosis [118]. In this regard, mitophagy has recently come into focus, since two PD-related genes were shown to be involved in mitophagy signaling (Fig. 12.4). Mutations in the genes encoding the mitochondrial serine/threonine kinase PARK6/PINK1 and the cytoplasmic E3 ubiquitin ligase PARK2/Parkin, respectively, can cause autosomal recessive PD [7]. Although many mechanistic details of the Pink1-Parkin pathway are still unclear, it may represent a model for the roles of mitophagy during neurodegenerative diseases. Genetic studies in Drosophila first suggested that PINK1 functions upstream of Parkin in a molecular pathway to suppress mitochondrial damage [119, 120]. In human cell lines, cytoplasmic Parkin was selectively localized to depolarized mitochondria followed by mitochondrial disappearance. Co-localization of mitochondria with structures positive for lysosomal markers suggested that uncoupled mitochondria are indeed removed by mitophagy [121]. In the emerging model, PINK1 acts as a sensor of mitochondrial damage and selectively recruits Parkin to depolarized mitochondria. It was furthermore proposed that the ubiquitination of mitochondrial proteins by Parkin results in the activation of the downstream autophagic pathway. So how does PINK1 distinguish between healthy and damaged mitochondria? This question was partially answered by the observation that under normal conditions, the full-length PINK1 protein is imported into mitochondria in a membrane potential ($\Delta\psi$) dependent manner. The subsequent N-terminal cleavage of PINK1 gives rise to a shorter form of PINK1, which in turn is rapidly degraded, resulting in a low steady state level of PINK1. Upon loss of $\Delta\psi$, PINK1 turnover is arrested, resulting in its accumulation on mitochondria and activation of the not yet understood downstream mitophagy pathway. A critical question is if degradation actually is the sole purpose of the short form of PINK1. Intriguingly, a recently proposed model suggests that the cleaved PINK1 moves back to the cytoplasm to bind Parkin, thereby preventing Parkin translocation to mitochondria and hence mitophagy in healthy mitochondria [122]. New insight into the role of Parkin comes also from recent studies suggesting a broad proteasomal degradation of dozens of OMM proteins following Parkin recruitment to mitochondria [123, 124]. Thus, Parkin might have a dual function in mitophagy: (i) The labeling of specific proteins for downstream mitohagy signalling and (ii) The ubiquitylation of a different set of OMM proteins for proteasomal degradation, which in turn promotes mitophagy. Taken together, a steadily increasing amount of details is known about PINK1/Parkin-mediated mitophagy but for a comprehensive understanding of the pathway, the current model will sure need to be extended and refined.

7 Outlook

Research of the recent decade has yielded much information on the biochemical mechanism of the different chaperone and protease species involved in protein homeostasis. Detailed biochemical and structural studies, mainly on the bacterial relatives, have revealed mechanisms of substrate recognition, refolding reactions, aggregate prevention by disaggregation, and proteolytic removal. By applying this information on the situation in mitochondria, significant progress has been made in characterizing the unique reactivity of chaperones and proteases in this endosymbiotic organelle. These classes of enzymes have been demonstrated to provide the important functions throughout the full life cycle of organellar proteins. The

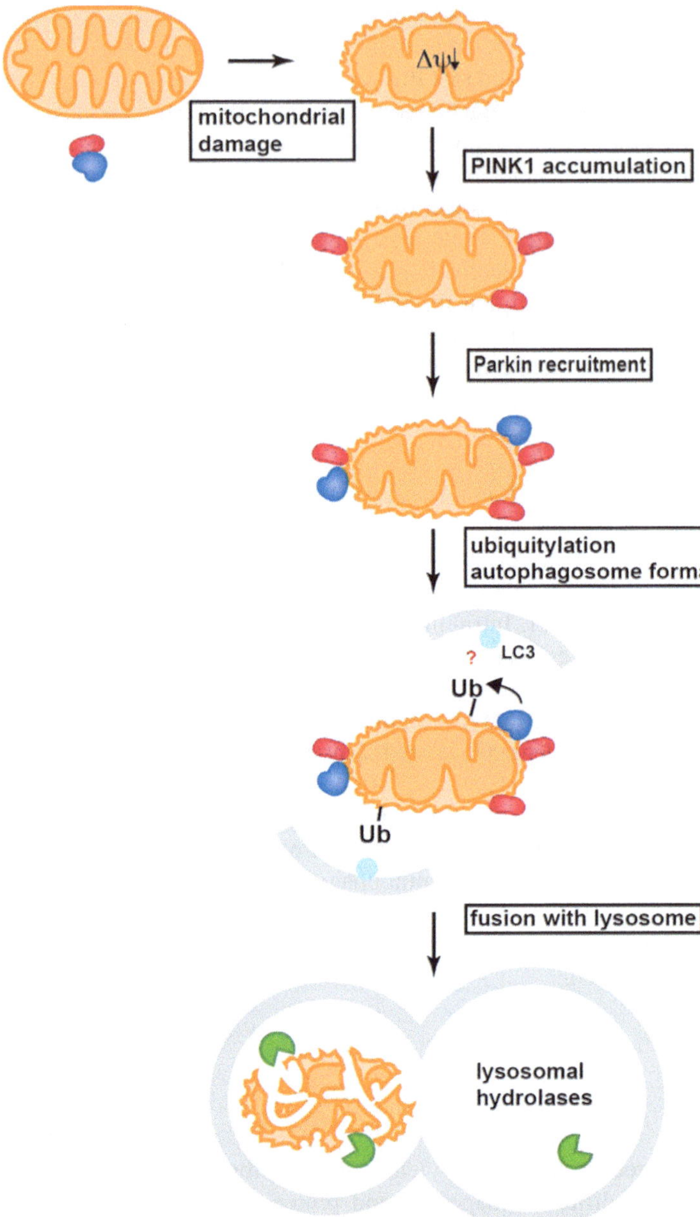

Fig. 12.4 Mechanism of PINK1/Parkin mediated mitophagy. PINK1 (*red*) associates with healthy mitochondria and is rapidly processed to a smaller form. Processed PINK1 in the cytosol possibly prevents Parkin (*blue*) translocation to mitochondria and hence mitophagy of healthy mitochondria. In damaged mitochondria, characterized by a low or absent inner membrane potential ($\Delta\psi$), full length, unprocessed PINK1 accumulates on the outer mitochondrial membrane (*OMM*) and recruits the E3 ubiquitin ligase Parkin (*blue*). In turn, Parkin conjugates ubiquitin (*Ub*) to various mitochondrial proteins. In a process probably involving ubiquitin-binding adaptor proteins and the autophagic protein LC3 (*light blue*), the mitochondrion is successively engulfed by an autophagosomal membrane (*grey lines*). The autophagosome fuses with a lysosome, delivering its complete contents to degradation by lysosomal hydrolases (*green*)

main future challenge will be to define the causal involvement of the components of the mitochondrial protein homeostasis system into processes leading to human diseases, in particular their contribution to the pathogenesis of neurodegenerative diseases.

References

1. Voos W (2009) Mitochondrial protein homeostasis: the cooperative roles of chaperones and proteases. Res Microbiol 160(9):718–725
2. Baker MJ, Tatsuta T, Langer T (2011) Quality control of mitochondrial proteostasis. Cold Spring Harb Perspect Biol 3(7). doi:10.1101/cshperspect.a007559
3. Chan DC (2006) Mitochondria: dynamic organelles in disease, aging, and development. Cell 125(7):1241–1252
4. Lin MT, Beal MF (2006) Mitochondrial dysfunction and oxidative stress in neurodegenerative diseases. Nature 443(7113):787–795
5. Amm I, Sommer T, Wolf DH (2013) Protein quality control and elimination of protein waste: The role of the ubiquitin-proteasome system. Biochim Biophys Acta doi:10.1016/j.bbamcr.2013.06.031
6. Voos W (2012) Chaperone-protease networks in mitochondrial protein homeostasis. Biochim Biophys Acta 1833(2):388–399
7. Vives-Bauza C, Przedborski S (2011) Mitophagy: the latest problem for Parkinson's disease. Trends Mol Med 17(3):158–165
8. Cheng MY, Hartl FU, Martin J, Pollock RA, Kalusek F, Neupert W, Hallberg EM, Hallberg RL, Horwich AL (1989) Mitochondrial heat-shock protein hsp60 is essential for assembly of proteins imported into yeast mitochondria. Nature 337:620–625
9. Lubben T, Gatenby A, Donaldson G, Lorimer G, Viitanen P (1990) Identification of a groES-like chaperonin in mitochondria that facilitates protein folding. Proc Natl Acad Sci U S A 87:7683–7687
10. Rospert S, Looser R, Dubaquié Y, Matouschek A, Glick BS, Schatz G (1996) Hsp60-independent protein folding in the matrix of yeast mitochondria. EMBO J 15(4):764–774
11. Hartl FU, Bracher A, Hayer-Hartl M (2011) Molecular chaperones in protein folding and proteostasis. Nature 475(7356):324–332
12. Bross P, Magnoni R, Bie AS (2012) Molecular chaperone disorders: defective Hsp60 in neurodegeneration. Curr Top Med Chem 12(22):2491–2503
13. Ghosh JC, Siegelin MD, Dohi T, Altieri DC (2010) Heat shock protein 60 regulation of the mitochondrial permeability transition pore in tumor cells. Cancer Res 70(22):8988–8993
14. Mayer MP (2013) Hsp70 chaperone dynamics and molecular mechanism. Trends Biochem Sci 38(10):507–514
15. Kampinga HH, Craig EA (2010) The HSP70 chaperone machinery: J proteins as drivers of functional specificity. Nat Rev Mol Cell Biol 11(8):579–592
16. Westermann B, Gaume B, Herrmann JM, Neupert W, Schwarz E (1996) Role of the mitochondrial DnaJ homolog Mdj1p as a chaperone for mitochondrially synthesized and imported proteins. Mol Cell Biol 16(12):7063–7071
17. Bender T, Lewrenz I, Franken S, Baitzel C, Voos W (2011) Mitochondrial enzymes are protected from stress-induced aggregation by mitochondrial chaperones and the Pim1/LON protease. Mol Biol Cell 22(5):541–554
18. Prip-Buus C, Westermann B, Schmitt M, Langer T, Neupert W, Schwarz E (1996) Role of the mitochondrial DnaJ homologue, Mdj1p, in the prevention of heat-induced protein aggregation. FEBS Lett 380(1-2):142–146

19. Lu B, Garrido N, Spelbrink JN, Suzuki CK (2006) Tid1 isoforms are mitochondrial DnaJ-like chaperones with unique carboxyl termini that determine cytosolic fate. J Biol Chem 281(19):13150–13158
20. Lill R, Mühlenhoff U (2008) Maturation of iron-sulfur proteins in eukaryotes: mechanisms, connected processes, and diseases. Annu Rev Biochem 77:669–700. doi:10.1146/annurev.biochem.76.052705.162653
21. Dutkiewicz R, Schilke B, Cheng S, Knieszner H, Craig EA, Marszalek J (2004) Sequence-specific interaction between mitochondrial Fe-S scaffold protein Isu and Hsp70 Ssq1 is essential for their in vivo function. J Biol Chem 279(28):29167–29174
22. Dutkiewicz R, Marszalek J, Schilke B, Craig EA, Lill R, Muhlenhoff U (2006) The Hsp70 chaperone Ssq1p is dispensable for iron-sulfur cluster formation on the scaffold protein Isu1p. J Biol Chem 281(12):7801–7808
23. Schmidt S, Strub A, Röttgers K, Zufall N, Voos W (2001) The two mitochondrial heat shock proteins 70, Ssc1 and Ssq1, compete for the cochaperone Mge1. J Mol Biol 313(1):13–26
24. Schneider HC, Westermann B, Neupert W, Brunner M (1996) The nucleotide exchange factor MGE exerts a key function in the ATP- dependent cycle of mt-Hsp70-Tim44 interaction driving mitochondrial protein import. EMBO J 15(21):5796–5803
25. Voos W, Gambill BD, Laloraya S, Ang D, Craig EA, Pfanner N (1994) Mitochondrial GrpE is present in a complex with hsp70 and preproteins in transit across membranes. Mol Cell Biol 14(10):6627–6634
26. Westermann B, Prip-Buus C, Neupert W, Schwarz E (1995) The role of the GrpE homologue, Mge1p, in mediating protein import and protein folding in mitochondria. EMBO J 14(14):3452–3460
27. Marada A, Allu PK, Murari A, PullaReddy B, Tammineni P, Thiriveedi VR, Danduprolu J, Sepuri NB (2013) Mge1, a nucleotide exchange factor of Hsp70, acts as an oxidative sensor to regulate mitochondrial Hsp70 function. Mol Biol Cell 24(6):692–703
28. Sikor M, Mapa K, von Voithenberg LV, Mokranjac D, Lamb DC (2013) Real-time observation of the conformational dynamics of mitochondrial Hsp70 by spFRET. EMBO J 32(11):1639–1649
29. Goswami AV, Samaddar M, Sinha D, Purushotham J, D'Silva P (2012) Enhanced J-protein interaction and compromised protein stability of mtHsp70 variants lead to mitochondrial dysfunction in Parkinson's disease. Hum Mol Genet 21(15):3317–3332
30. Chacinska A, Koehler CM, Milenkovic D, Lithgow T, Pfanner N (2009) Importing mitochondrial proteins: machineries and mechanisms. Cell 138(4):628–644
31. Mokranjac D, Berg A, Adam A, Neupert W, Hell K (2007) Association of the Tim14.Tim16 subcomplex with the TIM23 translocase is crucial for function of the mitochondrial protein import motor. J Biol Chem 282(25):18037–18045
32. Li Y, Dudek J, Guiard B, Pfanner N, Rehling P, Voos W (2004) The presequence translocase-associated protein import motor of mitochondria. Pam16 functions in an antagonistic manner to Pam18. J Biol Chem 279(36):38047–38054
33. Matouschek A, Pfanner N, Voos W (2000) Protein unfolding by mitochondria. The Hsp70 import motor. EMBO Rep 1:404–410
34. Voos W, Martin H, Krimmer T, Pfanner N (1999) Mechanisms of protein translocation into mitochondria. Biochim Biophys Acta 1422(3):235–254
35. Voisine C, Craig EA, Zufall N, von Ahsen O, Pfanner N, Voos W (1999) The protein import motor of mitochondria: unfolding and trapping of preproteins are distinct and separable functions of matrix Hsp70. Cell 97:565–574
36. Becker D, Krayl M, Strub A, Li Y, Mayer MP, Voos W (2009) Impaired interdomain communication in mitochondrial Hsp70 results in the loss of inward-directed translocation force. J Biol Chem 284(5):2934–2946
37. D'Silva PR, Schilke B, Hayashi M, Craig EA (2008) Interaction of the J-protein heterodimer Pam18/Pam16 of the mitochondrial import motor with the translocon of the inner membrane. Mol Biol Cell 19(1):424–432

38. De Los Rios P, Ben-Zvi A, Slutsky O, Azem A, Goloubinoff P (2006) Hsp70 chaperones accelerate protein translocation and the unfolding of stable protein aggregates by entropic pulling. Proc Natl Acad Sci U S A 103(16):6166–6171
39. Sichting M, Mokranjac D, Azem A, Neupert W, Hell K (2005) Maintenance of structure and function of mitochondrial Hsp70 chaperones requires the chaperone Hep1. EMBO J 24(5):1046–1056
40. Sanjuan Szklarz LK, Guiard B, Rissler M, Wiedemann N, Kozjak V, van derLM, Lohaus C, Marcus K, Meyer HE, Chacinska A, Pfanner N, Meisinger C (2005) Inactivation of the mitochondrial heat shock protein Zim17 leads to aggregation of matrix Hsp70 s followed by pleiotropic effects on morphology and protein biogenesis. J Mol Biol 351(1):206–218
41. Lewrenz I, Rietzschel N, Guiard B, Lill R, van der Laan M, Voos W (2013) The functional interaction of mitochondrial Hsp70 s with the escort protein Zim17 is critical for Fe/S biogenesis and substrate interaction at the inner membrane preprotein translocase. J Biol Chem. doi:10.1074/jbc.M113.465997
42. Vu MT, Zhai P, Lee J, Guerra C, Liu S, Gustin MC, Silberg JJ (2012) The DNLZ/HEP zinc-binding subdomain is critical for regulation of the mitochondrial chaperone HSPA9. Protein Sci 21 (2):258–267. doi:10.1002/pro.2012 [doi]
43. Burri L, Vascotto K, Fredersdorf S, Tiedt R, Hall MN, Lithgow T (2004) Zim17, a novel zinc finger protein essential for protein import into mitochondria. J Biol Chem 279(48):50243–50249
44. Goswami AV, Chittoor B, D'Silva P (2010) Understanding the functional interplay between mammalian mitochondrial Hsp70 chaperone machine components. J Biol Chem 285(25):19472–19482
45. Taipale M, Jarosz DF, Lindquist S (2010) HSP90 at the hub of protein homeostasis: emerging mechanistic insights. Nat Rev Mol Cell Biol 11(7):515–528
46. Song HY, Dunbar JD, Zhang YX, Guo D, Donner DB (1995) Identification of a protein with homology to hsp90 that binds the type 1 tumor necrosis factor receptor. J Biol Chem 270(8):3574–3581
47. Chen CF, Chen Y, Dai K, Chen PL, Riley DJ, Lee WH (1996) A new member of the hsp90 family of molecular chaperones interacts with the retinoblastoma protein during mitosis and after heat shock. Mol Cell Biol 16(9):4691–4699
48. Felts SJ, Owen BA, Nguyen P, Trepel J, Donner DB, Toft DO (2000) The hsp90-related protein TRAP1 is a mitochondrial protein with distinct functional properties. J Biol Chem 275(5):3305–3312
49. Cechetto JD, Gupta RS (2000) Immunoelectron microscopy provides evidence that tumor necrosis factor receptor-associated protein 1 (TRAP-1) is a mitochondrial protein which also localizes at specific extramitochondrial sites. Exp Cell Res 260(1):30–39
50. Meyer P, Prodromou C, Hu B, Vaughan C, Roe SM, Panaretou B, Piper PW, Pearl LH (2003) Structural and functional analysis of the middle segment of hsp90: implications for ATP hydrolysis and client protein and cochaperone interactions. Mol Cell 11(3):647–658
51. Pearl LH, Prodromou C (2006) Structure and mechanism of the Hsp90 molecular chaperone machinery. Annu Rev Biochem 75:271–294
52. Leskovar A, Wegele H, Werbeck ND, Buchner J, Reinstein J (2008) The ATPase cycle of the mitochondrial Hsp90 analog Trap1. J Biol Chem 283(17):11677–11688
53. Masuda Y, Shima G, Aiuchi T, Horie M, Hori K, Nakajo S, Kajimoto S, Shibayama-Imazu T, Nakaya K (2004) Involvement of tumor necrosis factor receptor-associated protein 1 (TRAP1) in apoptosis induced by beta-hydroxyisovalerylshikonin. J Biol Chem 279(41):42503–42515
54. Hua G, Zhang Q, Fan Z (2007) Heat shock protein 75 (TRAP1) antagonizes reactive oxygen species generation and protects cells from granzyme M-mediated apoptosis. J Biol Chem 282(28):20553–20560
55. Im CN, Lee JS, Zheng Y, Seo JS (2007) Iron chelation study in a normal human hepatocyte cell line suggests that tumor necrosis factor receptor-associated protein 1 (TRAP1) regulates production of reactive oxygen species. J Cell Biochem 100(2):474–486

56. Montesano Gesualdi N, Chirico G, Pirozzi G, Costantino E, Landriscina M, Esposito F (2007) Tumor necrosis factor-associated protein 1 (TRAP-1) protects cells from oxidative stress and apoptosis. Stress 10(4):342–350
57. Costantino E, Maddalena F, Calise S, Piscazzi A, Tirino V, Fersini A, Ambrosi A, Neri V, Esposito F, Landriscina M (2009) TRAP1, a novel mitochondrial chaperone responsible for multi-drug resistance and protection from apoptotis in human colorectal carcinoma cells. Cancer Lett 279(1):39–46
58. Kang BH, Plescia J, Dohi T, Rosa J, Doxsey SJ, Altieri DC (2007) Regulation of tumor cell mitochondrial homeostasis by an organelle-specific Hsp90 chaperone network. Cell 131(2):257–270
59. Altieri DC, Stein GS, Lian JB, Languino LR (2012) TRAP-1, the mitochondrial Hsp90. Biochim Biophys Acta 1823(3):767–773
60. Yoshida S, Tsutsumi S, Muhlebach G, Sourbier C, Lee MJ, Lee S, Vartholomaiou E, Tatokoro M, Beebe K, Miyajima N, Mohney RP, Chen Y, Hasumi H, Xu W, Fukushima H, Nakamura K, Koga F, Kihara K, Trepel J, Picard D, Neckers L (2013) Molecular chaperone TRAP1 regulates a metabolic switch between mitochondrial respiration and aerobic glycolysis. Proc Natl Acad Sci U S A 110(17):E1604–E1612
61. Richter J, Rüb C, Schöler S, Kunz W, Voos W (2013) The mitochondrial chaperone TRAP1 interacts with subunit A of the succinate dehydrogenase (SDH) and influences SDH complex formation and activity, submitted
62. Sciacovelli M, Guzzo G, Morello V, Frezza C, Zheng L, Nannini N, Calabrese F, Laudiero G, Esposito F, Landriscina M, Defilippi P, Bernardi P, Rasola A (2013) The mitochondrial chaperone TRAP1 promotes neoplastic growth by inhibiting succinate dehydrogenase. Cell Metab 17(6):988–999
63. Schirmer EC, Glover JR, Singer MA, Lindquist S (1996) HSP100/Clp proteins: a common mechanism explains diverse functions. Trends Biochem Sci 21(8):289–296
64. Weibezahn J, Tessarz P, Schlieker C, Zahn R, Maglica Z, Lee S, Zentgraf H, Weber-Ban EU, Dougan DA, Tsai FT, Mogk A, Bukau B (2004) Thermotolerance requires refolding of aggregated proteins by substrate translocation through the central pore of ClpB. Cell 119(5):653–665
65. Sauer RT, Baker TA (2011) AAA+ proteases: ATP-fueled machines of protein destruction. Annu Rev Biochem 80:587–612
66. Winkler J, Tyedmers J, Bukau B, Mogk A (2012) Chaperone networks in protein disaggregation and prion propagation. J Struct Biol 179(2):152–160
67. Doyle SM, Wickner S (2009) Hsp104 and ClpB: protein disaggregating machines. Trends Biochem Sci 34(1):40–48
68. Leonhardt SA, Fearon K, Danese PN, Mason TL (1993) *HSP78* encodes a yeast mitochondrial heat shock protein in the Clp family of ATP-dependent proteases. Mol Cell Biol 13(10):6304–6313
69. Krzewska J, Konopa G, Liberek K (2001) Importance of Two ATP-binding Sites for Oligomerization, ATPase Activity and Chaperone Function of Mitochondrial Hsp78 Protein. J Mol Biol 314(4):901–910
70. Leidhold C, Janowsky BV, Becker D, Bender T, Voos W (2006) Structure and function of Hsp78, the mitochondrial ClpB homolog. J Struct Biol 156:149–164
71. Germaniuk A, Liberek K, Marszalek J (2002) A bi-chaperone (Hsp70-Hsp78) system restores mitochondrial DNA synthesis following thermal inactivation of Mip1p polymerase. J Biol Chem 277(31):27801–27808
72. Röttgers K, Zufall N, Guiard B, Voos W (2002) The ClpB homolog Hsp78 is required for the efficient degradation of proteins in the mitochondrial matrix. J Biol Chem 277:45829–45837
73. Moczko M, Schönfisch B, Voos W, Pfanner N, Rassow J (1995) The mitochondrial ClpB homolog Hsp78 cooperates with matrix Hsp70 in maintenance of mitochondrial function. J Mol Biol 254(4):538–543
74. Schmitt M, Neupert W, Langer T (1996) The molecular chaperone Hsp78 confers compartment-specific thermotolerance to mitochondria. J Cell Biol 134(6):1375–1386

75. von Janowsky B, Major T, Knapp K, Voos W (2006) The disaggregation activity of the mitochondrial ClpB homolog Hsp78 maintains Hsp70 function during heat stress. J Mol Biol 357(3):793–807
76. Bateman JM, Iacovino M, Perlman PS, Butow RA (2002) Mitochondrial DNA instability mutants of the bifunctional protein Ilv5p have altered organization in mitochondria and are targeted for degradation by Hsp78 and the Pim1p protease. J Biol Chem 277(49):47946–47953
77. Ben-Zvi A, De Los Rios P, Dietler G, Goloubinoff P (2004) Active solubilization and refolding of stable protein aggregates by cooperative unfolding action of individual hsp70 chaperones. J Biol Chem 279(36):37298–37303
78. Rampelt H, Kirstein-Miles J, Nillegoda NB, Chi K, Scholz SR, Morimoto RI, Bukau B (2012) Metazoan Hsp70 machines use Hsp110 to power protein disaggregation. EMBO J 31(21):4221–4235
79. Heo JM, Livnat-Levanon N, Taylor EB, Jones KT, Dephoure N, Ring J, Xie J, Brodsky JL, Madeo F, Gygi SP, Ashrafi K, Glickman MH, Rutter J (2010) A stress-responsive system for mitochondrial protein degradation. Mol Cell 40(3):465–480
80. Ambro L, Pevala V, Bauer J, Kutejova E (2012) The influence of ATP-dependent proteases on a variety of nucleoid-associated processes. J Struct Biol 179(2):181–192
81. Suzuki CK, Suda K, Wang N, Schatz G (1994) Requirement for the yeast gene *LON* in intramitochondrial proteolysis and maintenance of respiration. Science 264:273–276
82. Wagner I, Arlt H, van Dyck L, Langer T, Neupert W (1994) Molecular chaperones cooperate with PIM1 protease in the degradation of misfolded proteins in mitochondria. EMBO J 13(21):5135–5145
83. Major T, von Janowsky B, Ruppert T, Mogk A, Voos W (2006) Proteomic analysis of mitochondrial protein turnover: identification of novel substrate proteins of the matrix protease Pim1. Mol Cell Biol 26(3):762–776
84. Bota DA, Davies KJ (2002) Lon protease preferentially degrades oxidized mitochondrial aconitase by an ATP-stimulated mechanism. Nat Cell Biol 4(9):674–680
85. Bender T, Leidhold C, Ruppert T, Franken S, Voos W (2010) The role of protein quality control in mitochondrial protein homeostasis under oxidative stress. Proteomics 10(7):1426–1443
86. Song JY, Marszalek J, Craig EA (2012) Cysteine desulfurase Nfs1 and Pim1 protease control levels of Isu, the Fe-S cluster biogenesis scaffold. Proc Natl Acad Sci U S A 109(26):10370–10375
87. Adam C, Picard M, Dequard-Chablat M, Sellem CH, Hermann-Le Denmat S, Contamine V (2012) Biological roles of the Podospora anserina mitochondrial Lon protease and the importance of its N-domain. PLoS ONE 7(5):e38138
88. Erjavec N, Bayot A, Gareil M, Camougrand N, Nystrom T, Friguet B, Bulteau AL (2012) Deletion of the mitochondrial Pim1/Lon protease in yeast results in accelerated aging and impairment of the proteasome. Free Radic Biol Med, in press. doi:10.1016/j.freeradbiomed.2012.11.019
89. Cheng CW, Kuo CY, Fan CC, Fang WC, Jiang SS, Lo YK, Wang TY, Kao MC, Lee AY (2013) Overexpression of Lon contributes to survival and aggressive phenotype of cancer cells through mitochondrial complex I-mediated generation of reactive oxygen species. Cell Death Dis 4:e681
90. van Dyck L, Neupert W, Langer T (1998) The ATP-dependent PIM1 protease is required for the expression of intron-containing genes in mitochondria. Genes Dev 12(10):1515–1524
91. Janska H, Kwasniak M, Szczepanowska J (2013) Protein quality control in organelles–AAA/FtsH story. Biochim Biophys Acta 1833(2):381–387
92. Gerdes F, Tatsuta T, Langer T (2012) Mitochondrial AAA proteases-towards a molecular understanding of membrane-bound proteolytic machines. Biochim Biophys Acta 1823(1):49–55
93. Leonhard K, Guiard B, Pellecchia G, Tzagoloff A, Neupert W, Langer T (2000) Membrane protein degradation by AAA proteases in mitochondria: extraction of substrates from either membrane surface. Mol Cell 5(4):629–638

94. Leonhard K, Herrmann JM, Stuart RA, Mannhaupt G, Neupert W, Langer T (1996) AAA proteases with catalytic sites on opposite membrane surfaces comprise a proteolytic system for the ATP-dependent degradation of inner membrane proteins in mitochondria. EMBO J 15(16):4218–4229
95. Martinelli P, Rugarli EI (2010) Emerging roles of mitochondrial proteases in neurodegeneration. Biochim Biophys Acta 1797(1):1–10
96. Lee S, Augustin S, Tatsuta T, Gerdes F, Langer T, Tsai FT (2011) Electron cryomicroscopy structure of a membrane-anchored mitochondrial AAA protease. J Biol Chem 286(6):4404–4411
97. Karlberg T, van den Berg S, Hammarstrom M, Sagemark J, Johansson I, Holmberg-Schiavone L, Schuler H (2009) Crystal structure of the ATPase domain of the human AAA+ protein paraplegin/SPG7. PLoS ONE 4(10):e6975
98. Bonn F, Tatsuta T, Petrungaro C, Riemer J, Langer T (2011) Presequence-dependent folding ensures MrpL32 processing by the m-AAA protease in mitochondria. EMBO J 30(13):2545–2556
99. Baker MJ, Mooga VP, Guiard B, Langer T, Ryan MT, Stojanovski D (2012) Impaired folding of the mitochondrial small TIM chaperones induces clearance by the i-AAA protease. J Mol Biol 424(5):227–239
100. Leonhard K, Stiegler A, Neupert W, Langer T (1999) Chaperone-like activity of the AAA domain of the yeast Yme1 AAA protease. Nature 398(6725):348–351
101. Schreiner B, Westerburg H, Forne I, Imhof A, Neupert W, Mokranjac D (2012) Role of the AAA protease Yme1 in folding of proteins in the intermembrane space of mitochondria. Mol Biol Cell 23(22):4335–4346
102. Weil A, Luce K, Drose S, Wittig I, Brandt U, Osiewacz HD (2011) Unmasking a temperature-dependent effect of the P. anserina i-AAA protease on aging and development. Cell Cycle 10(24):4280–4290
103. Corydon TJ, Bross P, Holst HU, Neve S, Kristiansen K, Gregersen N, Bolund L (1998) A human homologue of *Escherichia coli* ClpP caseinolytic protease: recombinant expression, intracellular processing and subcellular localization. Biochem J 331(Pt 1):309–316
104. Thompson MW, Singh SK, Maurizi MR (1994) Processive degradation of proteins by the ATP-dependent Clp protease from *Escherichia coli*. Requirement for the multiple array of active sites in ClpP but not ATP hydrolysis. J Biol Chem 269(27):18209–18215
105. Jennings LD, Lun DS, Medard M, Licht S (2008) ClpP hydrolyzes a protein substrate processively in the absence of the ClpA ATPase: mechanistic studies of ATP-independent proteolysis. BioChemistry 47(44):11536–11546
106. Corydon TJ, Wilsbech M, Jespersgaard C, Andresen BS, Borglum AD, Pedersen S, Bolund L, Gregersen N, Bross P (2000) Human and mouse mitochondrial orthologs of bacterial ClpX. Mamm Genome 11(10):899–905
107. Gispert S, Parganlija D, Klinkenberg M, Drose S, Wittig I, Mittelbronn M, Grzmil P, Koob S, Hamann A, Walter M, Buchel F, Adler T, Hrabe deAM, Busch DH, Zell A, Reichert AS, Brandt U, Osiewacz HD, Jendrach M, Auburger G (2013) Loss of mitochondrial peptidase Clpp leads to infertility, hearing loss plus growth retardation via accumulation of CLPX, mtDNA and inflammatory factors. Hum Mol Genet: doi:10.1093/hmg/ddt338
108. Kasashima K, Sumitani M, Endo H (2012) Maintenance of mitochondrial genome distribution by mitochondrial AAA+ protein ClpX. Exp Cell Res 318(18):2335–2343
109. van Dyck L, Dembowski M, Neupert W, Langer T (1998) Mcx1p, a ClpX homologue in mitochondria of *Saccharomyces cerevisiae*. FEBS Lett 438(3):250–254
110. Ron D, Walter P (2007) Signal integration in the endoplasmic reticulum unfolded protein response. Nat Rev Mol Cell Biol 8 (7):519–529. doi:nrm2199 [pii], 10.1038/nrm2199
111. Haynes CM, Petrova K, Benedetti C, Yang Y, Ron D (2007) ClpP mediates activation of a mitochondrial unfolded protein response in C. elegans. Dev Cell 13(4):467–480
112. Haynes CM, Yang Y, Blais SP, Neubert TA, Ron D (2010) The matrix peptide exporter HAF-1 signals a mitochondrial UPR by activating the transcription factor ZC376.7 in C. elegans. Mol Cell 37(4):529–540

113. Zhao Q, Wang J, Levichkin IV, Stasinopoulos S, Ryan MT, Hoogenraad NJ (2002) A mitochondrial specific stress response in mammalian cells. EMBO J 21(17):4411–4419
114. Durieux J, Wolff S, Dillin A (2011) The cell-non-autonomous nature of electron transport chain-mediated longevity. Cell 144(1):79–91
115. Wang K, Klionsky DJ (2011) Mitochondria removal by autophagy. Autophagy 7 (3):297–300. doi:14502 [pii]
116. Ding WX, Ni HM, Li M, Liao Y, Chen X, Stolz DB, Dorn GW 2nd, Yin XM (2010) Nix is critical to two distinct phases of mitophagy, reactive oxygen species-mediated autophagy induction and Parkin-ubiquitin-p62-mediated mitochondrial priming. J Biol Chem 285(36):27879–27890
117. Novak I, Kirkin V, McEwan DG, Zhang J, Wild P, Rozenknop A, Rogov V, Lohr F, Popovic D, Occhipinti A, Reichert AS, Terzic J, Dotsch V, Ney PA, Dikic I (2010) Nix is a selective autophagy receptor for mitochondrial clearance. EMBO Rep 11(1):45–51
118. Dupuis L (2013) Mitochondrial quality control in neurodegenerative diseases. Biochimie. doi:10.1016/j.biochi.2013.1007.1033
119. Clark IE, Dodson MW, Jiang C, Cao JH, Huh JR, Seol JH, Yoo SJ, Hay BA, Guo M (2006) Drosophila pink1 is required for mitochondrial function and interacts genetically with parkin. Nature 441(7097):1162–1166
120. Park J, Lee SB, Lee S, Kim Y, Song S, Kim S, Bae E, Kim J, Shong M, Kim JM, Chung J (2006) Mitochondrial dysfunction in Drosophila PINK1 mutants is complemented by parkin. Nature 441(7097):1157–1161
121. Narendra D, Tanaka A, Suen DF, Youle RJ (2008) Parkin is recruited selectively to impaired mitochondria and promotes their autophagy. J Cell Biol 183(5):795–803
122. Fedorowicz MA, de Vries-Schneider R, D. B, Huang Y, Zhou C, Alessi D, Voos W, Liu Y, Przedborski S (2013) The dual localization of PINK1 modulates Parkin translocation to the mitochondria. EMBO Rep 15(1):86–93
123. Chan NC, Salazar AM, Pham AH, Sweredoski MJ, Kolawa NJ, Graham RL, Hess S, Chan DC (2011) Broad activation of the ubiquitin-proteasome system by Parkin is critical for mitophagy. Hum Mol Genet 20(9):1726–1737
124. Sarraf SA, Raman M, Guarani-Pereira V, Sowa ME, Huttlin EL, Gygi SP, Harper JW (2013) Landscape of the PARKIN-dependent ubiquitylome in response to mitochondrial depolarization. Nature 496(7445):372–376

Part VII
The Ubiquitin-Proteasome System Network

Chapter 13
The Biogenesis of the Eukaryotic Proteasome

Andrew R. Kusmierczyk

Abstract Proteasomes are at the heart of the ubiquitin-proteasome system and are responsible for the degradation of the majority of intracellular proteins in eukaryotes. Consequently, proteasome activity impacts all cellular processes. Recent structural advances have provided the most informative view yet of the architecture of the proteasome, enriching our already detailed knowledge of its function. An interesting question, which continues to draw much experimental attention, is how these large molecular machines are put together in the first place. Over the last decade, the picture that has emerged is that of an ordered, efficient, and highly regulated process. Assembly of the proteasome is dependent upon intrinsic structural features distributed among the ~33 subunits that comprise it, and upon extrinsic protein factors, some of which function as dedicated proteasome chaperones. This chapter summarizes our understanding of the mechanism by which cells bring proteasomes into existence.

Abbreviations

UPS	Ubiquitin proteasome system
CP	Core particle
RP	Regulatory particle
Rpt	Regulatory particle ATPase
Rpn	Regulatory particle non-ATPase
HbYX	Hydrophobic-tyrosine-any amino acid
PHP	Preholoproteasome
ER	Endoplasmic reticulum
PAC	Proteasome assembly chaperone
RAC	RP assembly chaperone

A. R. Kusmierczyk (✉)
Department of Biology, Indiana University-Purdue University Indianapolis,
723 West Michigan Street, Indianapolis, IN 46202, USA
e-mail: akusmier@iupui.edu

W. A. Houry (ed.), *The Molecular Chaperones Interaction Networks in Protein Folding and Degradation,* Interactomics and Systems Biology, DOI 10.1007/978-1-4939-1130-1_13,

1 Introduction

The existence of the majority of intracellular proteins in eukaryotes is bookended by two large macromolecular assemblies: the ribosome and the proteasome. The former is a 4+ MDa behemoth of ribonucleic acid (RNA) and protein that orchestrates an intricate series of events bringing new proteins into existence [1]. The latter, though somewhat smaller at 2.5 MDa, is at the heart of a complex system, the ubiquitin-proteasome system (UPS). Its role is to identify, target, and ultimately remove from existence these very same proteins once their purpose has been served or their function compromised [2–4]. The reason for the juxtaposition of these two complexes here has less to do with their function, and more to do with how such large complexes themselves are put together. Ribosome assembly is a carefully choreographed dance of hundreds of accessory proteins, assembly factors, posttranslational modifications, and cleavage events spanning multiple cellular compartments [5, 6] (and references therein). Proteasome assembly is a considerably younger field of study. But the progress that has been made, especially in the last decade, certainly makes it plausible that assembly of the proteasome could reach the same order of magnitude in complexity as that of the ribosome. This chapter will review our current understanding of the various mechanisms that contribute to proteasome biogenesis in eukaryotes.

2 Overview of Proteasome Structure

The eukaryotic proteasome is comprised of two major subassemblies, a 20S proteasome, or core particle (CP), which is capped on one or both ends by the 19S regulatory particle (RP) (Fig. 13.1). The association of CP and RP brings together at least 33 different proteins, giving rise to the 2.5 MDa 26S proteasome (Table 13.1). The 20S proteasome is the actual protease component of the 26S proteasome, consisting of four stacked heptameric rings assembled from two types of structurally related subunits, α and β [7–10]. The two outer rings are identical and each consist of 7 unique α subunits; similarly, the two identical inner rings consist of seven unique β subunits, three of which possess proteolytic activity. The proteasome belongs to the N-terminal nucleophile (Ntn) hydrolase class of enzymes with a catalytic N-terminal threonine [11]. Ntn hydrolases are synthesized as proproteins, requiring a self-cleavage event in order for their active site nucleophile to become exposed [12], and the 20S proteasome is no exception. The three catalytic subunits, β1, β2, and β5, possess propeptides of varying lengths that are removed during assembly [13, 14] giving rise to active subunits that cleave following acidic, basic, and hydrophobic amino acids, respectively [15]. The 14 α and β subunits referred to above comprise the "constitutive" proteasome. However, four additional β subunits have been described in mammals: β1i, β2i, β5i, and β5t (where i = immuno, t = thymo; reviewed in [16, 17]). These subunits can substitute their constitutive counterparts

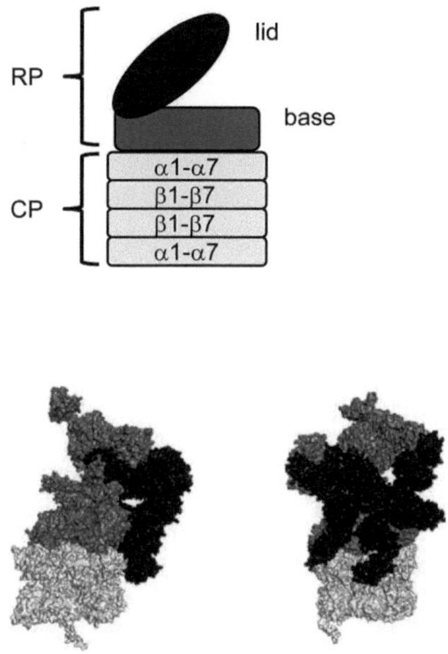

Fig. 13.1 Structure of the 26S proteasome (*Top panel*) The eukaryotic 26S proteasome is comprised of a centrally positioned 20S core particle (*CP*) consisting of heteroheptameric α and β rings. The CP can be capped on either one or both ends by the 19S regulatory particle (*RP*). The RP can be subdivided into two autonomously stable subassemblies, the base and lid. This "traditional" view of the 26S proteasome is still useful from a conceptual point of view. However, recent structural studies (see text) have revealed a much more complex, and interconnected architecture (*Bottom panel*). The color scheme matches that of the top panel, with shading of subunits based on their assignment into the CP (*light gray*), RP base (*medium gray*), and RP lid (*dark gray*). Only one pair of α and β rings in the CP is shown, with the C-terminal tail of the β7 subunit clearly protruding downwards (see text). The two structures are related to each other by a 90° rotation about their vertical axis. The figures were generated using POLYVIEW-3D with PyMol rendering [193] using protein data bank (PDB) coordinates 4B4T from [48]

and endow the resulting "immunoproteasomes" and "thymoproteasomes" with altered proteolytic activity that is necessary for proper function in antigen presentation by major histocompatibility complex (MHC) I class molecules and in T cell development, respectively [18–20]. A paralog of α4 is also present in mammalian testes and gives rise to "spermatoproteasomes" that function in spermatogenesis [21]. Compositionally, simpler 20S proteasomes are found in archaea and the actinomycete lineage of bacteria generally made up of only one or two types of α and β subunit. These proteasomes will not be discussed directly in this chapter, though experimental work relevant to eukaryotic proteasome assembly will be referenced. For a recent review on non-eukaryotic proteasomes, the reader is directed to [22].

The quaternary structure of the 20S proteasome results in active sites that are located in a central cavity formed by the stacked β rings meaning substrates must enter through entry pores located on either end of the cylinder. However, the isolated 20S proteasome is a closed structure [10], with N-terminal extensions of certain

Table 13.1 Subunits of the 26S proteasome

20S core particle			
Subunit	Notes	Subunit	Notes
α1		β1	Catalytic; cleaves after acidic residues. Has propeptide
α2		β2	Catalytic; cleaves after basic residues. Has propeptide. One of the "early" β subunits during assembly
α3	Not essential in yeast. Major gating subunit.	β3	One of the "early" β subunits during assembly
α4	Can replace α3 in the α ring.	β4	One of the "early" β subunits during assembly
α5		β5	Catalytic; cleaves after hydrophobic residues. Has propeptide that can function in trans
α6		β6	Has propeptide
α7		β7	Has propeptide. The last β subunit to incorporate
19S regulatory particle—base		*19S regulatory particle—lid*	
Rpt1	AAA+ ATPase.	Rpn3	
Rpt2	AAA+ ATPase. HbYX motif; inserts into α3/α4 pocket in mature 26S.	Rpn5	
Rpt3	AAA+ ATPase. HbYX motif; inserts into α1/α2 pocket in mature 26S.	Rpn6	
Rpt4	AAA+ ATPase.	Rpn7	
Rpt5	AAA+ ATPase. HbYX motif; inserts into α5/α6 pocket in mature 26S.	Rpn8	
Rpt6	AAA+ ATPase.	Rpn9	Not essential in yeast.
Rpn1	Probable scaffold.	Rpn11	Deubiquitinating enzyme; zinc metalloprotease
Rpn2	Probable scaffold.	Rpn12	Promotes lid-base joining
Rpn10	Ubiquitin receptor. Not essential in yeast. Promotes lid-base joining.	Rpn15	Also called Sem1. Not essential in yeast
Rpn13	Ubiquitin receptor. Not essential in yeast.		

ATPase Adenosine triphosphatase

α subunits serving as gates that occlude the pores [23]; thus access to the central proteolytic cavity is strictly controlled. However, even when fully open, the entry pores are too narrow to permit a substrate to enter in a folded state. On the one hand, this structural arrangement is advantageous as it prevents unintended degradation of folded proteins. On the other, it necessitates the existence of a functionality that not only opens the 20S gate but also unfolds proteins to enable their entry into the central cavity. In eukaryotes, this functionality is provided by the 19S RP.

The 19S RP is a 700 kDa complex that receives substrate proteins and processes them for degradation. In eukaryotes, targeting proteins for degradation by the proteasome generally entails the covalent attachment of multiple copies of the highly conserved, small protein modifier ubiquitin [3]. The RP must be capable of recognizing ubiquitinated cargo, removing the ubiquitins, unfolding the substrate protein, opening the entry pores of the CP, and translocating the unfolded substrate into the CP for degradation. These different functions of the RP are spread out among the 19 subunits that comprise it. Based on biochemical separation methods, these 19 subunits have been assigned into two subassemblies, the lid and the base [24–26]. The base consists of nine proteins. Six are members of the AAA+ adenosine triphosphatase (ATPase) superfamily, are referred to as regulatory particle ATPase (Rpt)1–6, and form a hexameric ring [27]. This ring supplies the mechanical force necessary to unfold proteins and translocate them into the CP [28, 29], a role that requires it to be in direct contact with the CP. These RP-CP contacts are themselves mediated by C-terminal tails of the Rpt subunits, some of which contain conserved tripeptide motifs. These so-called HbYX motifs (for hydrophobic-tyrosine (or phenylalanine)-any amino acid) insert into binding pockets located at the interface of two neighboring α subunits and form a stabilizing salt-bridge between the HbYX carboxylate and a conserved lysine located at the base of the pocket [30–32]. In addition to serving as points of attachment, insertion of the HbYX-containing C-terminal tails leads to opening of the CP entry pores which allows substrates to enter [33]. In addition to the six Rpt proteins, the base also contains regulatory particle, non-ATPase (Rpn)1, Rpn2, and Rpn13. The first two are large, structurally related proteins [34] that can be thought to function as scaffolds for the proper positioning of functional components [35–37], such as Rpn13, which together with Rpn10 constitute the two intrinsic ubiquitin-binding receptors of the 26S proteasome [38–41].

The lid of the RP comprises nine proteins, Rpn3, Rpn5–9, Rpn11–12, and Rpn15 (Sem1). The major functionality contributed by the lid is the deubiquitinating (DUB) activity of Rpn11 [42, 43]. The traditional view of the lid is that of an independently stable RP subcomplex, interacting with the base but located distal to the CP (like an actual "lid" or "cap" of the 26S proteasome); this view was supported by early lower resolution structural studies of the 26S proteasome [44] and held to be true even through later, more refined attempts [45–47]. However, the most recent structural determinations have turned the traditional view sideways, literally and figuratively, and revealed an RP structure that is as beautiful as it is interconnected [35–37, 48] (Fig. 13.1). The 26S proteasome is no longer a "three-layer cake" of lid stacked on base, stacked on CP. Rather, the base contains subunits that are both most intimately associated with (the Rpt hexamer), and the most distal from (Rpn2), the CP. The lid is no longer a cap, but a "hand" located on the side of the 26S, grasping the base in its palm and fingers and, most unexpected of all, making direct contacts with the CP. These structural studies reveal a remarkable logic in the organization of the subunits responsible for the various RP functionalities (ubiquitin binding, deubiquitination, substrate unfolding, translocation, and opening of the CP entry pore). This organization allows all the events, from substrate delivery through eventual proteolysis, to occur seamlessly and efficiently [49].

In addition to structural determinations of the 26S proteasome, numerous other structural studies of isolated components and co-complexes have been carried out. Together, they have contributed significantly to our understanding of the architecture of the proteasome and its associated proteins. For detailed structural descriptions, the reader is directed to one of the following excellent reviews of proteasome structure as a starting point [50, 51]. Discussion in this chapter will focus instead on the biogenesis of the "canonical" 26S proteasome (i.e., containing the constitutively expressed β subunits), beginning with separate sections for the CP and RP, and ending with a discussion of 26S assembly along with additional considerations. When referring to proteasome assembly in general, yeast terminology will be used for simplicity; however, mammalian terminology will be employed when appropriate (e.g., when comparing the results of yeast and mammalian studies).

3 Assembly of the 20S Core Particle

The 14 different α and β subunits that make up the 20S CP assemble in a highly ordered process [52], driven by a combination of cis-acting elements and trans-acting assembly factors (Fig. 13.2). The process begins with the coalescing of α subunits into an α ring. This is assisted by two heterodimeric assembly factors, Pba1-Pba2 and Pba3-Pba4, which occupy opposite faces of the ring, outer and inner, respectively (Table 13.2 and see below). The α ring serves as a platform for the orderly entry of β subunits [53], with their propeptides firmly in place. The β subunits can be subdivided into two sub-populations, early and late, based on their arrival order [54–58]. In mammalian cells, the early β subunits can assemble in the order β2, β3, β4 [58] though it is also possible that they arrive as a unit; in yeast the smallest β-subunit containing intermediate that can stably be isolated contains the full complement of α subunits and β2, β3, β4 [57]. This species is referred to as the 15S complex (Fig. 13.2). Formation of the 15S species is accompanied by the departure of Pba3-Pba4 and the arrival of another assembly factor, Ump1. Next, three of the four late β subunits are added presumably in the order β5, β6, β1; however, β1 can enter earlier if the incorporation of β4, β5, or β6 is blocked by small interfering RNA (siRNA) knockdown [58]. What is certain is that β7 is the last to incorporate in both yeast and mammals [57, 58]; the precursor lacking β7, called the "−β7-half-mer," is stable enough to be isolated [57]. Entry of β7 to form a complete half-proteasome is followed quickly by dimerization of two half-proteasomes into a transient

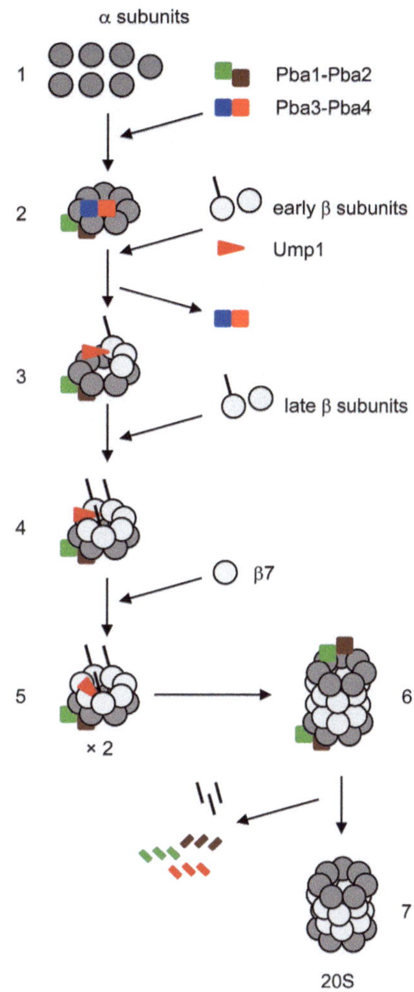

Fig. 13.2 Assembly of the 20S core particle. Individual α subunits (*1*), assisted by the assembly chaperones Pba1-Pba2 and Pba3-Pba4, coalesce into α rings (*2*). These serve as a scaffold for the entry of β subunits. The arrival of the early β subunits (β2, β3, β4) and Ump1 results in the displacement of Pba3-Pba4 and the formation of the 15S complex (*3*). Subsequent incorporation of the late β subunits, except β7, leads to the–β7-half-mer (*4*). The propeptides of the catalytically active β subunits are denoted by the angled black lines. The slow entry of β7 completes the half-proteasome (*5*), and dimerization of two half-proteasomes quickly follows, giving rise to the preholo-proteasome, or PHP (*6*). The PHP undergoes maturation to the fully functional core particle (*7*), a process that involves the autocatalytic processing of the β subunit propeptides (*black lines*) and degradation of the remaining associated chaperones (*small color-coded strips*). *PHP* preholoproteasome

species termed the preholoproteasome (PHP), that is, essentially a 20S particle but with β subunit propeptides still intact [57, 59, 60]. The propeptides of β subunits are removed after (or during) the formation of the PHP in a maturation process that involves both autocatalytic activation and trimming of neighboring subunits by autoactivated subunits [13, 14]. Ump1 is encapsulated inside the proteasome during PHP formation and is degraded following maturation of the β subunits [61]. Pba1-Pba2 is also degraded following completion of assembly [62], though probably first released from the 20S that it helped to assemble [63]. The fully mature 20S is now ready to bind the 19S RP. With this framework of 20S assembly in mind, attention is focused on the role that cis- and trans-acting components contribute to the process.

Table 13.2 Assembly chaperones and select associated proteins

Yeast name(s)	Mammalian name(s)	Notes
20S core particle assembly chaperones		
Pba1, Poc1	PAC1	HbYX motif; inserts into α5/α6 pocket. Not essential in yeast
Pba2, Poc2, Add66	PAC2	HbYX motif in fungi, HbY/F in mammals; inserts into α6/α7 pocket. Not essential in yeast
Pba3, Poc3, Dmp2	PAC3	Binds α5. Promotes correct placement of α3. Not essential in yeast
Pba4, Poc4, Dmp1	PAC4	Binds α5. Promotes correct placement of α3. Not essential in yeast
Ump1	hUmp1; POMP; proteassemblin	Binds β5 propeptide. Checkpoint function; ensures all β subunits enter half-proteasome. Not essential in yeast
19S regulatory particle assembly chaperones (RACs)		
Hsm3	S5b	Binds C domain of Rpt1. Not essential in yeast. Part of Hsm3 module: Hsm3-Rpt1-Rpt2-Rpn1
Nas2	p27	Binds HbYX motif of Rpt5. Not essential in yeast. Part of Nas2 module: Rpt4-Rpt5-Nas2
Nas6	Gankyrin; p28	Binds C domain of Rpt3. Not essential in yeast. Part of Nas6 module: Rpn14-Rpt6-Rpt3-Nas6.
Rpn14	PAAF1	Binds C domain of Rpt6. Not essential in yeast. Part of Nas6 module: Rpn14-Rpt6-Rpt3-Nas6.
Blm10	PA200	HbYX motif; inserts into α5/α6 pocket. Can function as activator of 20S. Not essential in yeast.
Ecm29	Ecm29	Potential role in quality control of 26S proteasome assembly. Not essential in yeast.

PAC proteasome assembly chaperone

3.1 Cis-acting Elements in 20S CP Assembly

Highly conserved structural features of α and β subunits are critical determinants of proteasome assembly. Proteasome α and β subunits are nearly super imposable [9, 10, 64] owing in part to a common evolutionary origin [65]. However, a major difference between α and β subunits is the presence of an N-terminal extension in the former that is approximately 35 amino acids (depending on the subunit) [9, 10].

Part of the extension forms an α helix (H0) and experiments with archaeal α subunits expressed heterologously in bacteria demonstrated that the integrity of this helix allows α subunits to spontaneously assemble into rings [53]. Expressed on their own, recombinant archaeal β subunits were incapable of self-assembly, but when coexpressed with α subunits gave rise to fully functional proteasomes [53]. This was the basis of the α-ring-as-template model described above and later shown to be true in eukaryotes. Certain eukaryotic α subunits, such as human α7, can assemble into heptameric rings when expressed in bacteria as well [66]. Interestingly human α7 recruits its neighboring subunits, which do not form rings on their own, into rings though these rings vary in their subunit stoichiometry [67]. This implies that in eukaryotes, presence of an N-terminal extension is not sufficient (though it is likely necessary) for the assembly of all α subunits into rings, and that additional elements contribute to ensure that only correctly ordered α rings form (see Sect. 13.3.2 below).

Whereas N-terminal extensions with structurally conserved H0 helices differentiate α subunits from β subunits, five of the seven β subunits (β1, β2, β5, β6, β7) contain N-terminal propeptides of varying length. The propeptides serve several purposes. Those found on catalytic subunits protect the N-terminal threonine nucleophile from N-terminal acetylation until assembly is complete, a modification that would inactivate the subunit [68]. The β5 propeptide is required for viability in both yeast [13] and mammalian cells [58] and its deletion impairs proteasome assembly at nearly identical points in both species. In yeast, β5 lacking its propeptide (β5Δpro) is not incorporated efficiently into proteasomes [13], whereas in mammalian cells β5Δpro can incorporate but entry of the next subunit, β6, is prevented [58]. Intriguingly, expression of the β5 propeptide in trans restores viability and normal proteasome assembly in yeast [13], arguing for an intramolecular chaperone function during assembly. Given the considerable size of this propeptide in yeast (75 amino acids) and mammals (59 amino acids), one can readily envision them as autonomously folding and functioning units. Viability of a β5Δpro yeast strain is also restored by deletion of the assembly factor Ump1 leading to the suggestion that the propeptide modulates the function of this protein [61]. An alternate, though not mutually exclusive, interpretation is that Ump1 antagonizes the essential function of the β5 propeptide, postulated to be mediating the dimerization of two half-proteasomes [57].

The propeptide of β2 is next in the hierarchy of importance. Although the 29 amino acid yeast version is dispensable for viability under normal growth conditions, where its deletion causes defects in the processing of the β5 propeptide, its absence severely limits growth at elevated temperatures [68]. By contrast, the 43 amino acid counterpart appears essential in a mammalian cell model where it is required for β3 incorporation [58]. The propeptide of β1 is not essential in either yeast or mammalian cells though its deletion in yeast also impairs β5 propeptide processing and results in strong growth defects when combined with a deletion of the β2 propeptide [58, 68]. The propeptides of non-catalytic β6 and β7 undergo processing by the catalytically active β subunits during the final stages of 20S maturation, and their role in assembly is not readily discerned [69]. However, deletion analysis of the yeast β6 propeptide reveals specific genetic interactions with the β5 propeptide and with Ump1, arguing that it participates directly in late assembly events that lead up to half-proteasome dimerization [57].

In addition to N-terminal propeptides, certain β subunits also contain lengthy C-terminal tails used in assembly. The β2 subunit possesses a ~30 amino acid long tail that wraps around the neighboring β3 subunit within the same β ring [10, 64]. This tail is required for viability in yeast and mammalian cells and its absence prevents β3 incorporation in the latter [58, 70]. It is possible that this intimate interaction between β2 and β3 results in these two "early" subunits entering as a unit to help form the 15S intermediate during assembly. Compared to the within-ring interactions of the β2 tail, the ~19 amino acid tail of β7 from one β ring extends into a groove between the β1 and β2 subunits of the opposite ring [10, 64]. Predictably, this tail is required to stabilize the 20S CP in yeast and mammalian cells, and its absence results in accumulation of the "−β7-half-mer," a half-proteasome species that contains all subunits except β7 [58, 70, 71]. The β7 tail functions in concert with the β5 propeptide in facilitating half-proteasome dimerization, since high copy β7 can suppress the lethality of β5Δpro yeast strains in a tail-dependent fashion [57]. The tail-dependent insertion of β7 is the rate limiting step of half-proteasome dimerization [57, 71] and is also necessary for β1 maturation and catalytic activity.

3.2 *Trans-acting Factors in 20S CP Assembly*

Ump1/hUmp1 A genetic screen in yeast looking to identify mutants that could not properly degrade a number of test substrates in a ubiquitin-dependent manner identified a 16 kDa protein called Ump1 [61]. Its deletion increased accumulation of immature CP subparticles and decreased β subunit propeptide processing, suggesting its function was to promote assembly and maturation. The human ortholog of Ump1, called hUmp1 (or proteassemblin or proteasome maturation protein (POMP)), was discovered based on weak sequence similarity to the yeast protein [72–74]. The weak sequence conservation probably explains why hUmp1 cannot replace Ump1 in yeast [72] though they likely carry out functionally equivalent roles. Ump1 and hUmp1 associate with proteasome species smaller than the CP that also contain unprocessed β subunits [54, 57, 72–74]. In yeast, the smallest such species that can be stably isolated contains a full α ring as well as β2, β3 and β4 [57]; in mammalian cells, a complete α ring with β2 only is sufficient enough to recruit hUmp1 [58]. As additional β subunits are added, both Ump1 and hUmp1 remain associated with these growing intermediates, but are not found in fully assembled proteasomes. In yeast, Ump1 is degraded upon completion of assembly as it becomes trapped inside the CP following half-proteasome dimerization and subsequent processing of β subunit propeptides [61]; hUmp1 likely meets a similar fate. There are some minor biochemical differences that become apparent when Ump1 is deleted in yeast versus when hUmp1 is knocked down in mammalian cells. In yeast, ump1Δ cells accumulate proteasome precursors that contain unprocessed β subunits [61]. By contrast, β-containing intermediates do not accumulate in mammalian cells when hUmp1 is knocked down [58, 75]; what accumulates instead is a species comprised of an α ring, 20S assembly factors, and potentially Hsp90 and Hsc70 [58]. Based on this difference, a role for hUmp1 in facilitating the incorporation of "early" β subunits (starting with β2) and in maintaining the stability of β-subunit-containing intermediates has been proposed and suggested to be an example of how hUmp1 function may differ from Ump1 function [58]. However, an alternate explanation is that Ump1 can carry out the same roles described above as hUmp1 but that yeast β-subunit-containing intermediates are more stable and less sensitive to Ump1 loss, thereby accounting for their ability to be detected/isolated.

The requirement of the β5 subunit propeptide for viability [13] can be bypassed by the deletion of Ump1 [61]. This led to the proposal that the β5 propeptide drives half-proteasome dimerization and that Ump1 is a negative regulator of this process [57]. Specifically, Ump1 carries out a "checkpoint" function to inhibit such dimerization until all β subunits have been incorporated, with β7 incorporation being the rate limiting step [57, 71]. By contrast, knockdown of hUmp1 did not rescue the lethality of mammalian cells in which a propeptide-lacking β5 subunit was expressed in the context of a knocked down endogenous β5 subunit [58]. However, this does not imply that hUmp1 cannot fulfill a similar "checkpoint" function in mammalian cells. It may simply reflect subtle differences in the role of the β5 propeptides between species, e.g., in mammalian cells, the β5 propeptide is required for the incorporation of β6 [58].

Given its encapsulation during the final stages of assembly, it is reasonable to surmise that Ump1 is located on interior, or cavity-facing, surfaces of proteasome precursors though which subunits it contacts is not known. Various experimental approaches have yielded mixed results. Yeast-two-hybrid analyses suggested Ump1 interacts with yeast α1 and β5 subunits [76] while hUmp1 interacted with mammalian β1, β6, β7 subunits in one study [77] and all β subunits plus α3, α4, and α7 in another [78]. Coexpression experiments in reticulocyte lysate revealed binding was possible between hUmp1 and α2, α3, α5, α7, β2, and β3 [58]. It is similarly not known if Ump1 functions as a monomer, or higher order species in vivo. Dimerization and tetramerization of bacterially expressed recombinant hUmp1 has been observed in vitro [79] and a dimeric (or larger) species could allow it to contact multiple 20S subunits, accounting for the disparate binding results above though interior cavity size constraints will likely favor smaller (if any) quaternary structure.

An intriguing report suggests hUmp1 is able to bind membranes and specifically recruits proteasome precursors to the endoplasmic reticulum (ER) membrane [78]. These precursors may be as early as isolated α rings. The authors propose that recruitment of β subunits would progress in a stepwise fashion [78], consistent with what is described by others [57, 58], but with all events occurring at the ER membrane. A recent report [80] describing the involvement in proteasome assembly of components of the pathway by which tail-anchored proteins are inserted into the ER is consistent with such a role for hUmp1 (see below). Though the details remain to be elucidated, it appears that these two reports have uncovered an interesting new angle to how (and where) proteasomes assemble.

Pba1-Pba2/PAC1-PAC2 Seven years following the discovery of Ump1, the description of a new heterodimeric protein complex in mammalian cells [62] would usher in a period of rapid-fire discovery that quickly expanded the known repertoire of these dedicated chaperones. The new protein complex was called proteasome assembling chaperone 1 and 2 (PAC1-PAC2) and identified on the basis of its association with proteasomes in immunoprecipitations from cells expressing epitope-tagged β1i subunits. Glycerol gradient centrifugation of cell lysates revealed that PAC1-PAC2 was present in fractions lighter than those containing hUmp1, associating specifically with species presumed to be α rings. Knockdowns of PAC1-PAC2 resulted in the shift of these α subunit fractions containing putative rings to much higher molecular weight. Given the tendency of α subunits to form ring dimers, at least when expressed recombinantly in bacteria [53, 66, 67], the authors suggested one function of PAC1-PAC2 was to prevent the formation of ring dimers which would presumably be non-productive (or dead-end) species. The ability of PAC1-PAC2 to associate with subsets of α subunits also led to the suggestion that the complex may promote α ring assembly.

PAC2 exhibited very weak homology to a yeast protein encoded by the YKL206C locus, also called ADD66, the deletion of which resulted in mild ER associated degradation (ERAD) defects [81] and a reduction of the chymotryptic activity of the proteasome in yeast [82]. No clear counterpart to PAC1 was detected, leading to speculation that perhaps yeast lacked functional orthologs of PAC1-PAC2 [62]. However, two subsequent analyses demonstrated that the numerous parallels in proteasome biology between yeast and mammalian systems continued to hold, including the conservation of this assembly factor. In the first study, several new 20S assembly intermediates were identified in yeast, all of which contained the protein products encoded by YKL206C and another unknown gene, YLR199C [57]. Genetic interactions were observed when the deletion of either unknown gene was combined with deletions known to perturb proteasome assembly (ump1Δ and α3Δ) and the proteins encoded by these two genes were shown to form a heterodimer when expressed in E. coli. The second study confirmed the genetic association with proteasome function, the physical association with proteasome precursor species, and the ability of the two unknown proteins to form an in vivo complex [83] that had initially been identified in a genome-wide proteomic analysis [84]. The ability of this complex to interact with subsets of α subunits, just like PAC1-PAC2, was also demonstrated [83] and iterative sequence analysis suggested a weak similarity between YLR199C and PAC1 [57, 83]. The unknown proteins encoded by YLR199C and YKL206C were rechristened *p*roteasome *b*iogenesis *a*ssociated (Pba)1 and Pba2, respectively ([57] also referred to as Poc1-Poc2 [83]), and suggested to be the yeast equivalents of PAC1-PAC2.

Given the similarities between Pba1-Pba2 and PAC1-PAC2, does the former function similarly in the prevention of α ring dimers? Owing to the structural similarities between α and β subunits, α ring dimers and α/β ring pairs (i.e., half proteasomes) share a remarkably similar architecture, including identical displacements of one ring relative to the other [9, 53]. This suggests that inhibition of α ring dimer formation by PAC1-PAC2 could be envisioned most readily if the heterodimer bound to the α ring surface that faces β rings, because this should be the same surface with which two α rings interact. However, recent biochemical [63] and structural [85] evidence demonstrates that Pba1-Pba2 binds α rings on the opposite surface, the surface that proteasome activators bind to. This mode of binding is mediated by a C-terminal HbYX motif present in both Pba1 and Pba2 [63]. The HbYX motif was originally described in the archaeal proteasome-activating nucleotidase (PAN) [31], but is also present at the C-termini of Rpt subunits in the 19S base [86], the Blm10/PA200 activator [87, 88], as well as Cdc48 and its related archaeal orthologs [89–91]. The Pba1 HbYX motif inserts into the pocket formed by α5 and α6, and salt-bridges with the a6 "pocket lysine," whereas the Pba2 HbYX motif inserts into the α6/α7 pocket and contacts the α7 "pocket lysine" [85]. The HbYX motif in Pba1 is conserved in PAC1 across all species while the Pba2 HbYX motif is conserved in fungi [63]. However, PAC2 proteins end in a conserved HbY/F motif [63]. This may be sufficient to bind their cognate proteasomes based on the observation that C-terminal HbY/F dipeptides are found in 11S regulator subunits [92] which use the same inter-α-subunit pockets to bind proteasomes [93]. Intriguingly, archaeal orthologs of Pba1-Pba2 have also been described (PbaA and PbaB) and shown to use the same HbYX-mediated binding to interact with α rings [63]. This remarkable conservation of binding mode suggests that whatever the purpose of this interaction may be, it holds the key to the function of this assembly factor.

In general, HbYX containing proteins activate 20S proteasomes, meaning that they stimulate its peptidase/protease activity. Do Pba1-Pba2/PAC1-PAC2 possess this function as well? Multiple lines of evidence suggest that the answer is, no. Neither Pba1-Pba2 nor PAC1-PAC2 bind appreciably to fully mature 20S proteasomes in vivo; they were always isolated in fractions containing proteasome precursors [62, 63, 75, 83] or bound directly to proteasome precursors isolated from native polyacrylamide gels [57, 94]. Moreover, Pba1-Pba2 cannot stimulate peptide hydrolysis in vitro [63] and the 20S substrate pore in the Pba1-Pba2–20S cocrystal structure is closed [85]. Returning to the hypothesis of α ring dimer suppression, the binding of Pba1-Pba2 to the outer α ring surface makes this functionality less certain given that α ring dimers likely interact via their inner (i.e., β subunit-facing) surfaces [53]. One can surmise that the suppression occurs allosterically, which is possible but would need to be demonstrated experimentally. Alternatively, one can consider the possibility that the shift of α subunits to a higher molecular weight species observed when PAC1-PAC2 is knocked down [62] is not the result of ring dimerization per se but the result of some other aberrant association of α subunits with each other, or even the association of α rings with other proteins. This last possibility leads to an alternative model of Pba1-Pba2 function: Given that the outer α ring surface is bound by a number of activators using the inter α subunit pockets, and that these activators function with mature 20S proteasomes, it may be beneficial to prevent the premature association of these activators with α rings found in proteasome precursors.

Such a "safety" function, whereby occupation of the outer α ring surface by Pba1-Pba2 during assembly sterically prevents association with proteasome activators, was recently proposed [63]. This mechanism requires the ability of Pba1-Pba2 to detect the completion of 20S assembly. The final steps of proteasome assembly, following dimerization of two half-proteasomes, involve the autocatalytic removal of β subunit propeptides (Fig. 13.2). This event could trigger Pba1-Pba2 release, but would necessitate allosteric signaling between the active site and the outer α ring surface, specifically by modulating the inter α subunit/HbYX interaction. Long-range communication has been demonstrated between the active site and a bound 19S [95]. And a key finding in support of this mechanism was the observation of an inverse relationship between the affinity of archaeal PbaA proteins for immature proteasome precursors and the extent of β subunit propeptide processing in these precursors [63]. Whether this can be extended to Pba1-Pba2/PAC1-PAC2 proteins remains to be seen. An alternate view is that, like Ump1, this assembly factor is degraded upon completion of assembly by the very same 20S it helps to chaperone [62]. This possibility still permits the "safety" function and differs only in the mechanism by which Pba1-Pba2 vacates the outer α ring surface. Consistent with this view is the instability of PAC1-PAC2 and Pba2 proteins [62, 82]. And, the similarity in structure between unliganded 20S and 20S bound by Pba1-Pba2 also argues against allostery being the signal [85]. However, neither Pba1-Pba2 nor their archaeal orthologs are degraded when incubated with their respective, fully mature, 20S in vitro [63, 85]. This suggests that the degradation might not occur in cis (i.e., by the same 20S molecule it helps to assemble, as is the case for Ump1) but in trans, perhaps in a ubiquitin-dependant manner. Moreover, dense crystal packing can alter proteasome structure [96] and solution methods, such as methyl TROSY NMR, support the existence of allosteric communication between proteasome active sites and activator-binding sites [97]. The final verdict in this, and in other aspects of Pba1-Pba2 function, remains to be delivered. Having structural information in hand for both Pba1-Pba2 [85] and its archaeal orthologs, PbaA (3GAA) and PbaB [98], will be of great value in designing and interpreting future experiments.

Pba3-Pba4/PAC3-PAC4 A third assembly factor, PAC3, was described shortly following the discovery of PAC1-PAC2 [75]. Unlike PAC1-PAC2 and hUmp1, PAC3 did not associate with later assembly intermediates. Instead, this homodimer was found in glycerol gradient fractions containing only α subunits [75], and later shown to be α rings [58]. Combined knockdowns of PAC3 and PAC1-PAC2 were additive in their negative effects on assembly [75]. Together, this pointed to an earlier and distinct role in assembly for PAC3 compared to the other assembly factors. Not long after the discovery of PAC3, the yeast ortholog of PAC3 (called Pba3, Poc3, or Dmp2) was identified and shown to function as a heterodimer with the protein product of the YPL144W [83, 99–101].

Pba3-Pba4, like PAC3, associated with much earlier assembly intermediates than Pba1-Pba2 or Ump1. Immunoprecipitation experiments showed this heterodimer to be present in precursor species containing the full complement of α subunits (α ring) and unprocessed β2 [83, 100], findings that were mirrored in mammalian cells [58]. In vitro binding assays indicate this binding is mediated via its interaction with α5 [63, 100]. Knockdown of PAC3, or deletion of Pba3 or Pba4, results in inefficient assembly of α subunits into rings, leading to decreased levels of 20S [75, 83, 99–101]. Pba3-Pba4 and PAC3 are stable suggesting they dissociate prior to completion of assembly [75, 83]. The inability of Pba3-Pba4 to effectively precipitate Ump1 and the largely nonoverlapping profiles of these two assembly factors in glycerol gradient fractions argues that this dissociation occurs right around the time of Ump1 incorporation [83, 100]. The crystal structure of Pba3-Pba4 complexed with α5 provided important additional perspectives on these findings [100]. Pba3-Pba4 was bound to helices 1 and 2 of α5 which would place it on the inner α ring surface. Modeling of the Pba3-Pba4-α5 complex onto a fully formed α ring as it exists in the CP confirmed this and suggested that Pba3-Pba4 binding would be incompatible with β subunit binding to the same α ring surface, especially β4 and β5 whose spots it occupied. This would explain the early exit of Pba3-Pba4 during assembly. As β4, one of the early β subunits, enters a nascent proteasome (either alone, or as a complex with other early β subunits) it would displace Pba3-Pba4 from its binding site. Such displacement of PAC3 by early β subunits has been observed in mammalian cells [58].

The similarities between Pba3-Pba4 and PAC3 function were strengthened with identification of PAC4 as the mammalian ortholog of Pba4 [83]. PAC3-PAC4 formed a heterodimer in vivo, could associate with α subunits and, more importantly, were stable only if present together. Targeting of either by siRNA resulted in the knockdown of both, a situation also true of Pba3-Pba4 in yeast [101], arguing that the heterodimer is the functional unit in all species. These findings continued to reinforce the precise parallels between yeast and human proteasome assembly. However, they only established that Pba3-Pba4 is important for assembly, though not in what way. Genetic experiments in yeast suggested a link between Pba3-Pba4 function and α3 [99]. PRE9, which encodes α3, is the only non-essential 20S gene in yeast and its deletion gives rise to an alternative 20S particle in which a second copy of α4 occupies the spot left vacant by the absence of α3 [102]. Surprisingly, deletion of Pba3-Pba4 also leads to the accumulation of these α4-α4 proteasomes even though α3 is still present [99]. The levels of these alternative proteasomes in Pba3-Pba4-deleted cells are estimated to be between 20 and 50 % of total proteasomes. Thus, the function of Pba3-Pba4 appears to be to ensure the generation of "canonical" 20S proteasomes, in which all α subunits are properly incorporated, by promoting correct placement of α3. It is not clear how Pba3-Pba4 accomplishes this, given that it preferentially binds to α5 but not α3 [99, 100]. However, one possibility is that it functions as a scaffold for α ring assembly and influences assembly to proceed in a direction that favors α3 incorporation ahead of α4. Consistent with this is the observation that Pba3-Pba4 can exist in a complex with α5, α6, and α7 in vitro [99], which could act as a nucleus for additional α subunit entry. Nevertheless, complete understanding of Pba3-Pba4 function awaits further study.

Blm10/PA200. Blm10 and its mammalian ortholog PA200 are large (*H*untingtin, elongation factor 3 (*E*F3), protein phosphatase 2A (PP2*A*), and the yeast kinase *T*OR1) HEAT-repeat proteins and function as proteasome activators [103, 104]. They are capable of binding CP alone, or in a hybrid with RP. Blm10-CP complexes have been reported to have a role in deoxyribonucleic acid (DNA) repair, genomic stability, spermatogenesis, ribosome biogenesis, and histone degradation among others [21, 103, 105–107]. The large dome-shaped structure of Blm10 does have an opening wide enough to accommodate unfolded substrates [87], consistent with an ability to target unstructured substrates or peptides [88]. However, Blm10 also appears to play a role in CP assembly [108]. Blm10 can be isolated with a number of CP assembly intermediates [57] and deletion of Blm10 in yeast results in a profound assembly defect when combined with deletion of the β7 tail [71]. Blm10 binds to the outer α ring surface and inserts its HbYX motif into the α5/α6 pocket [87]. Presumably it interacts with assembly precursors in the same manner. A recent report describes a role for Blm10 in CP import into the nucleus [109]; it would be interesting to determine if Blm10 interaction with CP precursors could also mediate similar transport events (see 6.2 below).

4 Assembly of the 19S Regulatory Particle

Despite the interwoven nature of lid and base subunit interactions in the RP (Fig. 13.1), it is beneficial to discuss these two subcomplexes separately given evidence that the two can form independently of each other [110, 111]. Much like assembly of the CP, assembly of the base subcomplex depends on cis-acting elements and trans-acting factors (Table 13.2). Since considerably more is known about this aspect of RP biogenesis, it will be examined first.

4.1 Cis-acting Elements in 19S RP Base Assembly

The distinguishing feature of the base is the hexameric ring formed by the six AAA+ ATPase Rpt subunits. Heterologous expression of subsets of Rpt subunits in bacteria leads to the formation of high molecular weight complexes, some of which hydrolyze adenosine triphosphate (ATP) and stimulate protein degradation when paired with archaeal proteasomes [112]. Thus, in a situation that mimics α subunits of the CP, Rpt subunits have the ability to self-associate. And, just as α subunits assemble into a single defined ring structure, the unique positioning of the Rpt subunits within their hexameric ring (Rpt1-Rpt2-Rpt6-Rpt3-Rpt4-Rpt5) [27] necessitates the existence of an ordered assembly pathway. This ordered pathway is initiated by sequences at the N-termini of the Rpt subunits that mediate interactions between pairs of Rpt subunits and are predicted to form coiled coils [113]. Structural studies of proteasomal AAA+ ATPases from prokaryotes confirmed that all proteasomal AAA+ ATPases are held together by two conserved N-terminal elements, a coiled coil domain (CC) followed by a domain that adopts an oligonucleotide/oligosaccharide binding fold (OB); this pair of elements can hexamerize independently [114, 115]. The compositional simplicity of prokaryotic AAA+ ATPases (i.e., homohexamers versus eukaryotic heterohexamers) nevertheless belies another important structural caveat. A highly conserved proline residue located between the CC and OB domains is peptide bonded in cis, or in trans, with its preceding amino acid in alternating monomers of the archaeal AAA+ ATPase, PAN [114, 115]. The cis configuration of one subunit, and the conformational change it produces, is what enables its associated CC domain to reach over and interact with that of its trans-configured neighbor. With three of the six eukaryotic Rpt subunits (Rpt2, 3, and 5) containing highly conserved Pro residues at this position, this cis-trans coupling is conserved in eukaryotes and lays the foundation for the assembly of the base (see below).

Beyond the CC and OB domains, the primary sequence of Rpt subunits continues on to the ATPase domain followed by a C-domain, and ending in the C-terminal tail. Although C terminal deletions do not impair the formation of specific Rpt subunit pairs [113], the C-terminal regions of Rpt subunits nevertheless feature prominently in assembly. As will be explained in more detail below, the C-terminal tails of Rpt subunits mediate CP-RP docking to enable the formation of the 26S proteasome [31, 86, 116, 117]. The C-domain, a defining feature of AAA+ ATPases [118], provides a platform for the binding of several structurally distinct yet functionally related proteins that chaperone assembly of the base.

4.2 *Trans-acting Factors in 19S RP Base Assembly*

Long before the discovery that Rpt subunits of the base organize into the cis-trans pairs described above, certain combinations of Rpt subunits were shown to be capable of forming very stable complexes, either by themselves [113] or with additional binding partners. Mammalian Rpt4 and Rpt5 were found in a trimer with p27, a complex termed "modulator" that could stimulate 19S activation of 20S, whereas Rpt1 and Rpt2 formed a tetramer with Rpn1 and S5b, and Rpt3 could be bound by p28 (also known as gankyrin) [119–122]; subsequent work uncovered additional interaction between Rpt6 and *p*roteasomal *A*TPase-*a*ssociated *f*actor 1 (PAAF1) [123]. These mammalian Rpt-interacting proteins were known to have yeast counterparts, and one orthologous pair (p28 in mammals, Nas6 in yeast) bound to the C-domain of its cognate Rpt3 in nearly identical fashion arguing for strong conservation of function [124, 125]. But the function of these Rpt-interacting proteins, and the reason behind the stability of the associated Rpt complexes, remained obscure. In 2009, a flurry of papers from five groups in the space of a couple of months provided the needed breakthrough [126–131]. In a series of elegant biochemical and genetic experiments, these groups demonstrated that the Rpt-interacting proteins are dedicated chaperones for the RP base (henceforth referred to as RP assembly chaperones, RACs) and the Rpt complexes they associate with reflect the highly modular nature of base assembly (Fig. 13.3). In brief, the key findings can be summarized as follows:

i RACs are conserved from yeast (Hsm3, Nas2, Nas6, Rpn14) to man (S5b, p27, p28 or gankyrin, PAAF1, respectively).
ii The absence of RACs results in decreased levels of 26S proteasome, increased levels of free CP, increased levels of free lid, and accumulation of various base subcomplexes.

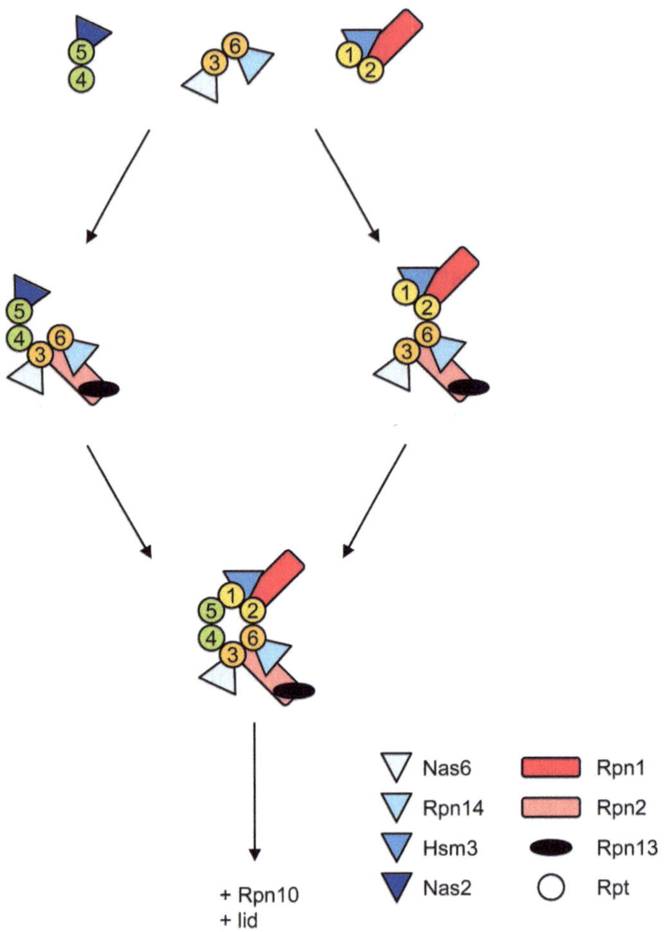

Fig. 13.3 Assembly of the 19S base. Base subunits associate with respective regulatory particle assembly chaperone (RAC) proteins to form the three modules (Hsm3, Nas2, and Nas6; see text). In yeast, the Nas2 and Nas6 modules assemble first (*left pathway*) while in mammalian cells the pathway might involve the Hsm3 module binding to the Nas6 module (*right pathway*). In either case, Rpn2 and Rpn13 are recruited in the process. Addition of the remaining module (Hsm3 in yeast, Nas2 in mammals) results in the formation of base accompanied by release of Nas2

iii RACs interact with specific Rpt subunits (Hsm3-Rpt1, Nas2-Rpt5, Nas6-Rpt3, Rpn14-Rpt6). For Hsm3, Nas6, and Rpn14, the interaction is with the C-domain of the cognate Rpt.

iv RACs and base subunits are organized into stable modules (Hsm3 module: Hsm3-Rpt1-Rpt2-Rpn1; Nas2 module: Rpt4-Rpt5-Nas2; Nas6 module: Nas6-Rpt3-Rpt6-Rpn14).

Based on these, and subsequent experiments, the mechanism of base assembly that emerges starts with the formation of the three modules though the role of RACs in this process is likely to be module dependent. The Rpt pairs in each of the modules are the aforementioned cis-trans dimers, but they are not equally stable. The Rpt4-Rpt5 pair can form in absence of Nas2 [127], whereas the Rpt1-Rpt2 interaction requires Hsm3 [132]. In yeast, the Nas2 and Nas6 modules are thought to come together, along with Rpn2 and Rpn13, prior to the addition of the Hsm3 module [27]. Around this time, Nas2 is released and the resulting complex is ready to accept the lid. In mammals, the Hsm3 module might interact with the Nas6 module first [129]. Ring formation appears to require ATP binding [133].

An interesting question is whether or not the CP can function as a base chaperone. The idea arose from observations that yeast mutants defective in CP assembly accumulated base subcomplexes and free lid [99]. Support for this "template model" of RP base assembly comes from observations that perturbing the CP-RP interaction, via C-terminal tail mutations, is detrimental to RP formation in yeast and mammalian cells [131, 133]. Another report in mammalian cells also suggests a template function for the CP though the nature of this mechanism, in which a complex consisting of several lid and base subunits is bound to CP, is less clear [134]. Conversely, the ability to reconstitute mammalian RP from three purified RP precursors, in the absence of CP, argues against templating [135]. More likely, it is possible that both CP-dependent and CP-independent assembly pathways operate simultaneously. Regardless of the role of CP in base assembly, what is clear is that the interaction of CP with base is incompatible with the interaction of base with the RACs [131]. The former requires the C-terminal tails of Rpt subunits to dock into the inter-α-subunit pockets [33], and the latter is mediated via RAC binding to the C-domain of the cognate Rpt [125, 136, 137]. The proximity of the C-domain to the C-terminal tail would result in a steric clash between CP and RAC if both tried to engage the base simultaneously. This mutual antagonism has been demonstrated both by the ability of free CP to release RACs from base and by RACs being able to disrupt a base-CP complex [117, 131]. Despite the advances of the past 4 years with respect to base assembly, a precise definition of RAC function remains elusive.

4.3 Assembly of the Lid

Understanding the assembly of this subcomplex has lagged behind that of the CP or base. But a number of recent reports have begun to close the gap. The ability to reconstitute functional lid in recombinant form argues that dedicated chaperones, like those for the CP and base, may not be required by the lid [35]. Biochemical analysis in yeast revealed a number of stable lid subcomplexes that accumulate in various lid subunit mutants [111, 138–140]. This led to the prediction that an early event in lid assembly is the formation of an Rpn5-Rpn8-Rpn9-Rpn11 complex which combines with Rpn6 to give rise to a stable species called module 1 [50]. Module 1 combines with a trimer of Rpn3-Rpn7-Rpn15, called LP3 (lid particle 3), to give rise to a large complex called lid particle 2 (LP2) which lacks only Rpn12. Addition of Rpn12, in an association dependent on its C-terminus, completes the lid [111]. A very recent report identifies a unique helical bundle comprised of 12 helices from 8 of the 9 lid subunits and responsible for the integrity and assembly of the lid [141]. Mutational analysis of the bundle revealed a hierarchal assembly pathway entirely consistent with the aforementioned scheme and including the additional detail that formation of the Rpn5-Rpn8-Rpn9-Rpn11 complex begins with a stable Rpn8-Rpn11 dimer. Once the lid is formed, joining of base and lid occurs with the assistance of Rpn10 in a mechanism that is also dependent on the Rpn12 C-terminus [111].

5 Assembly of the 26S Proteasome

The CP can function in protein degradation, especially the degradation of oxidatively damaged proteins, without the RP [142]. Similarly, the RP has roles that are independent of proteolysis (though not necessarily requiring physical separation from the CP) in DNA repair and transcription [143, 144]. Nevertheless, biogenesis of the eukaryotic 26S proteasome necessitates the joining of CP and RP. This interaction of RP and CP requires ATP binding [28, 145], which is likely the physiological condition in vivo, even though binding without nucleotide can be retained under certain conditions in vitro [95]. As mentioned earlier, the RP-CP interaction is mediated by the insertion of Rpt subunit C-termini into inter-α-subunit pockets [30–32]. However, there is an inherent register mismatch between the six tails of the hexameric Rpt ring and the seven pockets of the α ring. Which tails go into which pockets? Cross-linking results suggested an asymmetric interaction between RP and CP, with three neighboring Rpt subunits on one side of the ATPase ring being cross-linked to one pocket each: Rpt2 to α3/α4, Rpt6 to α2/α3, and Rpt3 to α1/α2 [116]. Conversely, the three neighboring subunits on the other side of the ring could be cross-linked to multiple pockets. Heterogeneity in Rpt-pocket interactions could also be seen for all Rpt tails, except Rpt6, in an in vitro formed base-CP complex; the Rpt6 tail occupied the α2/α3 pocket [117]. However, in a recent structure of the complete 26S proteasome, only the tails of three Rpt subunits were visualized bound to defined pockets: Rpt2 to α3/α4, Rpt3 to α1/α2, and Rpt5 to α5/α6 [35]. This implies that as a base-CP complex is joined to the lid, the RP-CP interface undergoes extensive remodeling to enable defined occupancy of the pockets as seen in the structure. Interestingly, Rpt2, Rpt3, and Rpt5 alternate in the ring and also happen to be the three HbYX-containing subunits with roles in gating the 20S [31, 86]. The inability to visualize the other three tails in the structure could simply reflect that they are mobile.

The importance of maintaining correct RP-CP contacts is illustrated by the existence of a highly conserved protein, Ecm29, which is recruited to proteasomes whenever the RP-CP interaction is perturbed [94, 133, 146]. Ecm29 inhibits the ATPase activity and substrate degradation of 26S proteasomes, suggesting an ability to shut off defective proteasomes [147]. The ability of Ecm29 to effectively mark defective 26S proteasomes begs the question, mark them for what? One possibility is repair. Ecm29 was found associated with aberrant 26S proteasomes that lacked β3 and contained unprocessed β5 [146]. Addition of exogenous β3 resulted in completion of assembly and Ecm29 release. Other studies have failed to detect an absence of β3 in aberrant 26S proteasomes, but Ecm29 nevertheless remained bound to defective 26S proteasomes consistent with a proteasomal quality control function [94, 133].

6 Additional Considerations in Proteasome Biogenesis

Proteasome biogenesis is more than just the sum of the assembly reactions that bring together the 65+ proteins into a doubly capped 26S particle. Additional layers of complexity impinge upon the process. And though some of these could be discussed in the context of the earlier sections, they are given their due attention here to recognize the growing body of data that attests to their increasing importance.

6.1 Regulation of Biogenesis

The central role of the UPS under normal growth, and especially in times of stress, necessitates an ability to respond efficiently to changing conditions. This includes altering the levels of the proteasome inside the cell. Proteasome synthesis is controlled transcriptionally in a manner that ensures that all subunits are expressed at near-stoichiometric levels, which in yeast have been estimated to be between 10,000 and 30,000 copies per cell [24, 148, 149]. This mechanism is best understood in yeast cells where Rpn4, a transcription factor with a C2H2 zinc finger motif, binds to a unique 9 base pair proteasome-associated control element (PACE) sequence found in the promoters of almost all proteasome subunits in yeast [150, 151]. Rpn4 is a short-lived protein, degraded by the proteasome in both ubiquitin dependent and ubiquitin independent pathways [151, 152]; its levels rise dramatically if proteasome activity is compromised [153, 154]. The result is an elegant negative feedback loop that ties Rpn4 levels to cellular demand for proteolytic activity. In brief, conditions that impair activity of the proteasome, or that overwhelm it with competing substrates, slow down the normally rapid rate of degradation of Rpn4. The resulting increased levels of Rpn4 lead to increased transcription of proteasome genes and to more proteasome synthesis. The additional proteolytic capacity restores the rapid degradation of Rpn4, shutting off the signal for additional proteasome synthesis. Rpn4 levels are also regulated by transcription factors that transduce a variety of upstream stress signals, including heat shock and oxidative stress, thereby allowing proteasome biogenesis to be responsive to a wide variety of extrinsic and intrinsic stimuli [155, 156]. The importance of coordinated synthesis of proteasome subunits is illustrated in a yeast strain where just a single subunit was rendered unresponsive to Rpn4 via mutation of its PACE sequence [157]. Proteasome levels in these cells were greatly reduced, resulting in disruption of the cell cycle and sensitization to various stresses.

Higher eukaryotes possess neither any gene with sequence similarity to Rpn4 nor PACE sequences, yet their proteasome genes are also regulated by coordinated feedback mechanisms [158]. This argues for the existence of functional, if not structural, homologs of Rpn4 in higher eukaryotes. Evidence has been presented for the role of Nrf1 [159] and Nrf2 transcription factors [160, 161] in stress-responsive regulation of proteasome synthesis, though the relative importance of each may be cell-type dependent [162]. A major difference between yeast and mammalian cells also seems to be the observation that Rpn4 regulates induced and basal levels of proteasomes while the Nrf proteins do not [151, 153, 159]. Basal proteasome levels in mammalian cells might be set by a third transcription factor, NF-Y [163].

6.2 Localization of Assembly

Proteasomes are found in both cytoplasmic and nuclear compartments, though the proportion in each may differ somewhat among the eukarya. For instance, whereas mammalian proteasomes are more or less evenly distributed between these two compartments [164], yeast proteasomes are enriched in the nucleus and nuclear periphery [148, 165, 166]. The relationship between assembly and localization is an important, and thus far underexplored, issue in the proteasome field. Subsets of CP and RP subunits contain nuclear localization sequences (NLSs) [167–169] which argues that non-NLS-containing subunits would rely on assembly events taking place prior to transport, in either direction, across the nuclear envelope (NE). Preformed base and lid subcomplexes can be imported into the nucleus independent of the CP [110, 169]; the latter can be imported in a complex with some associated proteins (Rpn1 and 2, Hsp90, importin β) in a reconstituted in vitro import assay in Xenopus extract [170], though other experimental systems suggest import of CP alone may also be possible [171, 172]. A very recent report identifies Blm10 as a CP-specific nuclear import factor in yeast [109]. Nevertheless, the ability of intact subcomplexes (CP, base, lid) to traverse the NE does not address whether or not nucleocytoplasmic transport is relevant/required for (earlier) assembly events. For instance, it has been argued that during CP assembly in yeast, 20S precursors are imported into the nucleus where assembly is presumably completed [173]. And, the idea that assembly precursors traffic between compartments is consistent with the observed cellular distribution of assembly chaperones as well. In yeast, Pba1-Pba2 and Pba3-Pba4 are localized throughout the cell, as are the four 19S base assembly chaperones while Ump1 appears to be mainly nuclear [83, 128, 173, 174]. Interestingly, in ump1Δ yeast cells, Pba3-Pba4 redistribute to the nucleus [83].

Embedded deeper in the discussion of proteasome localization, or assembly precursor trafficking, is the question of whether or not assembly takes place within discrete regions/subcompartments within a cell. Upon glucose depletion, yeast enter a quiescent state during which a massive reorganization of proteasomes takes place [175]. Subunits of both CP and RP coalesce into one or two cytoplasmic foci called proteasome storage granules (PSGs). The assembly status of proteasomes within PSGs, whether as intact 26S particles [176] or dissociated into CP and RP [109], remains to be resolved. Nevertheless, PSGs illustrate that it is possible to rapidly and efficiently confine proteasomes into discrete subcompartments. Similar spatial restriction could be at work during assembly as well, with the ER emerging as an intriguing candidate for an assembly compartment. As mentioned earlier, hUmp1 has been observed to interact directly with membranes and recruit proteasome precursors to the ER [78]. In addition, knockdowns of components of the mammalian *T*ransmembrane *R*ecognition *C*omplex (TRC) pathway, by which tail-anchored proteins are inserted into ER membranes [177], give rise to very specific proteasome assembly defects [80]. Specifically, knockdowns of TRC40 or Bag6 result in accumulation of immature CP precursors in which β subunit assembly is perturbed. Neither TRC40, nor Bag6, bind directly to CP subunits suggesting the knockdown effect is indirect and may be the result of improper insertion of a yet-to-be-identified tail-anchored protein that does bind to proteasomes directly. Additionally, this study demonstrated direct binding of Bag6 to RP precursors; in particular, Bag6 was required for the stability of the Nas2 (p27) module in base assembly. Hence, ER localization could be important for both CP and RP assembly. These findings may extend to all eukaryotes as strong genetic interactions between proteasome assembly factors and components of the guided entry of tail-anchored proteins (GET) pathway (the yeast equivalent of the TRC pathway [178]) were also observed [80].

6.3 Alternative Subunits, Alternative Proteasomes, Alternative Pathways

Duplication of proteasome genes giving rise to paralogous subunits is not unique to mammalian cells. Drosophila melanogaster and Arabidopsis thaliana provide wonderful examples of proteasome subunit heterogeneity [179–181]. As is the case with the mammalian β1i, β2i, β5i and β5t subunits, paralogous subunits in both of these species give rise to alternative proteasomes with unique function. The Drosophila paralogs are exclusively expressed in the testes where they are required for proper sperm development [182]. The Arabidopsis paralogs can share some redundant functionality, but otherwise play unique roles in a wide variety of processes among them flowering, development, nucleosome function, and leaf size [183–185]. The existence of paralogous subunits presents an interesting question with regards to assembly: what determines the incorporation of one subunit versus another? In the simplest case, the paralogs are not expressed at the same time, as occurs for some of the subunits in Drosophila testes [179]. But this is not usually so. For example, in cells of the mammalian immune system, synthesis of the inducible β subunits sets up a situation where the paralogs compete for assembly with constitutive β subunits, and the inducible subunits are preferentially incorporated [186]. Several factors account for this preference. First, cooperativity among the inducible subunits ensures that βi subunits recruit other βi subunits. Unlike the assembly of constitutive proteasomes where β1 is a late subunit [58], β1i is an early subunit in immunoproteasome assembly and facilitates entry of β2i [187]. The presence of β1i and β2i on proteasome precursors results in the preferential recruitment of β5i over β5 [188]. Second, the propeptides of β2i and β5i direct these subunits to the nascent immunoproteasome and, remarkably, their function is transplantable [188, 189]. For example, replacement of the β2i propeptide with that of β2 will send the modified β2i into constitutive proteasomes, and grafting the β2i propeptide onto β5 will cause that constitutive subunit to incorporate into immunoproteasomes. Finally, the assembly factor hUmp1 contributes to immunoproteasome formation because it preferentially binds to β5i versus β5 [190]. The net result is an altered assembly pathway that is more efficient and thus faster, relative to that of the constitutive proteasome.

The ability of alternative subunits to give rise to alternative proteasomes appears to rest within particular structural features of the alternative subunit; these features endow it with properties that enable it to compete effectively with its "normal" (or constitutive) counterpart for incorporation during assembly. The corollary to this observation is that in absence of alternative subunits, all "normal" subunits arrive at their single, predetermined final destination. Yet this is not always the case. As described earlier, deletion of the only non-essential CP subunit in yeast, α3, gives rise to alternative proteasomes in which a second copy of α4 occupies the position normally held by α3 [102]. Such altered proteasomes are also generated in the absence of Pba3-Pba4, though at lower levels [99]. Thus, the same set of subunits can give rise to an entirely different assembly pathway. Whether the formation of α4-α4 proteasomes occurs under physiological conditions is not known, although the non-essential nature of α3 may hold in other organisms [191]. Moreover, the absolute conservation of Pba3-Pba4 across eukarya suggests an interest in maintaining one assembly pathway (i.e., the one that leads to normal α3-containing proteasomes), but leaves open the possibility that downregulation of Pba3-Pba4 could allow the alternate pathway to operate.

It is perhaps fitting that a chapter devoted to understanding the pathway that gives rise to functional proteasomes should end with a discussion of the possibility that there may be multiple pathways involved. In addition to the examples in this section, additional support for multiple pathways comes from evidence for both CP-dependent and CP-independent assembly of the base (see 4.2) and from the non-essential nature of all of the known dedicated assembly chaperones in yeast. If shepherding of Rpt subunits to their final positions in the ATPase ring occurs only along a route prescribed by the RACs, it would not be possible to generate a viable yeast strain completely devoid of them [127]. However, the possibility of alternative pathways does not imply that these are equally likely to occur. The fourfold faster assembly of immunoproteasomes relative to constitutive proteasomes, despite the simultaneous presence of all 6 β subunit paralogs, is an excellent example of this [190]. Yeast cells lacking dedicated chaperones provide another illustration. For instance, pba4Δ cells are viable because Rpn4 activity results in increased synthesis of proteasome subunits. But the assembly pathway in the absence of Pba4, which gives rise to α4-α4 proteasomes, must be less efficient because coincident deletion of Rpn4, which gets rid of the compensatory effect of increased subunit levels, results in severe growth impairment [99]. Conversely, PBA4 cells do not synthesize α4-α4 proteasomes (to any detectable extent) because assembly in the presence of the dedicated chaperone takes a different, more efficient route. The βi paralogs offer one final informative example. Although generally described in all-or-none terms, the incorporation of βi subunits exhibits surprising heterogeneity across tissues and gives rise to proteasome populations with different ratios of the six paralogous β subunits [192]. Thus, although proteasome biogenesis conjures up an image of a single, highly ordered, sequential pathway, we should be mindful of parallel pathways that can become significantly populated.

7 Concluding Remarks

Proteasome biogenesis is elaborate. The assembly mechanism itself is an intricate interplay of subunit association guided by intrinsic structural features of the ~33 unique subunits and a growing cadre of dedicated chaperones. Additional layers of complexity in terms of cellular localization and trafficking, the existence of alternative subunits, and a proteolytically regulated transcriptional feedback loop add to the picture of a highly ordered yet highly efficient process. Ensuring the efficiency and accuracy of the process is of paramount importance owing to the key roles of the proteasome in numerous cellular functions. Each new set of observations that broadens our understanding brings with it new questions and a virtual guarantee of more exciting discoveries to come.

References

1. Ben-Shem A, Garreau de Loubresse N, Melnikov S, Jenner L, Yusupova G, Yusupov M (2011) The structure of the eukaryotic ribosome at 3.0 Å resolution. Science 334(6062):1524–1529. doi:10.1126/science.1212642science, 1212642 [pii]
2. Finley D, Ulrich HD, Sommer T, Kaiser P (2012) The ubiquitin-proteasome system of *Saccharomyces cerevisiae*. Genetics 192(2):319–360. doi:10.1534/genetics,112.140467192/2/319 [pii]
3. Komander D, Rape M (2012) The ubiquitin code. å81:203–229. doi:10.1146/annurev-biochem-060310-170328
4. Hochstrasser M, Deng M, Kusmierczyk AR, Li X, Kreft SG, Ravid T, Funakoshi M, Kunjappu M, Xie Y (2008) Molecular genetics of the ubiquitin-proteasome system: lessons from yeast. Ernst Schering Found Symp Proc (1):41–66
5. Staley JP, Woolford JL Jr. (2009) Assembly of ribosomes and spliceosomes: complex ribonucleoprotein machines. Curr Opin Cell Biol 21(1):109–118. doi:10.1016/j.ceb.2009.01.003, S0955-0674(09)00005-2 [pii]
6. Kressler D, Hurt E, Bassler J (2010) Driving ribosome assembly. Biochim Biophys Acta 1803(6):673–683. doi:10.1016/j.bbamcr.2009.10.009, S0167-4889(09)00265−1 [pii]
7. Heinemeyer W, Trondle N, Albrecht G, Wolf DH (1994) PRE5 and PRE6, the last missing genes encoding 20S proteasome subunits from yeast? Indication for a set of 14 different subunits in the eukaryotic proteasome core. BioChemistry 33(40):12229–12237
8. Chen P, Hochstrasser M (1995) Biogenesis, structure and function of the yeast 20S proteasome. EMBO J 14(11):2620–2630
9. Lowe J, Stock D, Jap B, Zwickl P, Baumeister W, Huber R (1995) Crystal structure of the 20S proteasome from the archaeon *T. acidophilum* at 3.4 A resolution. Science 268(5210):533–539

10. Groll M, Ditzel L, Lowe J, Stock D, Bochtler M, Bartunik HD, Huber R (1997) Structure of 20S proteasome from yeast at 2.4 A resolution. Nature 386(6624):463–471. doi:10.1038/386463a0
11. Seemuller E, Lupas A, Stock D, Lowe J, Huber R, Baumeister W (1995) Proteasome from *Thermoplasma acidophilum*: a threonine protease. Science 268(5210):579–582
12. Brannigan JA, Dodson G, Duggleby HJ, Moody PC, Smith JL, Tomchick DR, Murzin AG (1995) A protein catalytic framework with an N-terminal nucleophile is capable of self-activation. Nature 378(6555):416–419. doi:10.1038/378416a0
13. Chen P, Hochstrasser M (1996) Autocatalytic subunit processing couples active site formation in the 20S proteasome to completion of assembly. Cell 86(6):961–972. doi:10.1016/S0092-8674(00)80171-3 [pii]
14. Witt S, Kwon YD, Sharon M, Felderer K, Beuttler M, Robinson CV, Baumeister W, Jap BK (2006) Proteasome assembly triggers a switch required for active-site maturation. Structure 14(7):1179–1188. doi:10.1016/j.str.2006.05.019, S0969-2126(06)00258-9 [pii]
15. Arendt CS, Hochstrasser M (1997) Identification of the yeast 20S proteasome catalytic centers and subunit interactions required for active-site formation. Proc Natl Acad Sci U S A 94(14):7156–7161
16. Basler M, Kirk CJ, Groettrup M (2013) The immunoproteasome in antigen processing and other immunological functions. Curr Opin Immunol 25(1):74–80. doi:10.1016/j.coi.2012.11.004, S0952-7915(12)00185-9 [pii]
17. Takahama Y, Takada K, Murata S, Tanaka K (2012) beta5t-containing thymoproteasome: specific expression in thymic cortical epithelial cells and role in positive selection of CD8+ T cells. Curr Opin Immunol 24(1):92–98. doi:10.1016/j.coi.2012.01.006, S0952-7915(12)00008-8 [pii]
18. Driscoll J, Brown MG, Finley D, Monaco JJ (1993) MHC-linked LMP gene products specifically alter peptidase activities of the proteasome. Nature 365(6443):262–264. doi:10.1038/365262a0
19. Gaczynska M, Rock KL, Goldberg AL (1993) Gamma-interferon and expression of MHC genes regulate peptide hydrolysis by proteasomes. Nature 365(6443):264–267. doi:10.1038/365264a0
20. Murata S, Sasaki K, Kishimoto T, Niwa S, Hayashi H, Takahama Y, Tanaka K (2007) Regulation of CD8+ T cell development by thymus-specific proteasomes. Science 316(5829):1349–1353. doi:10.1126/science.1141915, 316/5829/1349[pii]
21. Qian MX, Pang Y, Liu CH, Haratake K, Du BY, Ji DY, Wang GF, Zhu QQ, Song W, Yu Y, Zhang XX, Huang HT, Miao S, Chen LB, Zhang ZH, Liang YN, Liu S, Cha H, Yang D, Zhai Y, Komatsu T, Tsuruta F, Li H, Cao C, Li W, Li GH, Cheng Y, Chiba T, Wang L, Goldberg AL, Shen Y, Qiu XB (2013) Acetylation-mediated proteasomal degradation of core histones during DNA repair and spermatogenesis. Cell 153(5):1012–1024. doi:10.1016/j.cell.2013.04.032, S0092-8674(13)00508-4 [pii]
22. Maupin-Furlow J (2012) Proteasomes and protein conjugation across domains of life. Nat Rev Microbiol 10(2):100–111. doi:10.1038/nrmicro2696nrmicro2696 [pii]
23. Groll M, Bajorek M, Kohler A, Moroder L, Rubin DM, Huber R, Glickman MH, Finley D (2000) A gated channel into the proteasome core particle. Nat Struct Biol 7(11):1062–1067. doi:10.1038/80992
24. Glickman MH, Rubin DM, Fried VA, Finley D (1998) The regulatory particle of the Saccharomyces cerevisiae proteasome. Mol Cell Biol 18(6):3149–3162
25. Glickman MH, Rubin DM, Coux O, Wefes I, Pfeifer G, Cjeka Z, Baumeister W, Fried VA, Finley D (1998) A subcomplex of the proteasome regulatory particle required for ubiquitin-conjugate degradation and rêlated to the COP9-signalosome and eIF3. Cell 94(5):615-623. doi: 10.1016/S0092-8674(00)81603-7 [pii]
26. Glickman MH, Rubin DM, Fu H, Larsen CN, Coux O, Wefes I, Pfeifer G, Cjeka Z, Vierstra R, Baumeister W, Fried V, Finley D (1999) Functional analysis of the proteasome regulatory particle. Mol Biol Rep 26(1–2):21–28
27. Tomko RJ Jr, Funakoshi M, Schneider K, Wang J, Hochstrasser M (2010) Heterohexameric ring arrangement of the eukaryotic proteasomal ATPases: implications for proteasome structure and assembly. Mol Cell 38(3):393–403. doi:10.1016/j.molcel.2010.02.035, S1097-2765(10)00314-X [pii]

28. Smith DM, Kafri G, Cheng Y, Ng D, Walz T, Goldberg AL (2005) ATP binding to PAN or the 26S ATPases causes association with the 20S proteasome, gate opening, and translocation of unfolded proteins. Mol Cell 20(5):687–698. doi:10.1016/j.molcel.2005.10.019, S1097-2765(05)01714-4 [pii]
29. Zhang F, Wu Z, Zhang P, Tian G, Finley D, Shi Y (2009) Mechanism of substrate unfolding and translocation by the regulatory particle of the proteasome from *Methanocaldococcus jannaschii*. Mol Cell 34(4):485–496. doi:10.1016/j.molcel.2009.04.022, S1097-2765(09)00275-5 [pii]
30. Forster A, Masters EI, Whitby FG, Robinson H, Hill CP (2005) The 1.9 A structure of a proteasome-11S activator complex and implications for proteasome-PAN/PA700 interactions. Mol Cell 18(5):589–599. doi:10.1016/j.molcel.2005.04.016, S1097-2765(05)01279-7 [pii]
31. Smith DM, Chang SC, Park S, Finley D, Cheng Y, Goldberg AL (2007) Docking of the proteasomal ATPases' carboxyl termini in the 20S proteasome's alpha ring opens the gate for substrate entry. Mol Cell 27(5):731–744. doi:10.1016/j.molcel.2007.06.033, S1097-2765(07)00445-5 [pii]
32. Yu Y, Smith DM, Kim HM, Rodriguez V, Goldberg AL, Cheng Y (2010) Interactions of PAN's C-termini with archaeal 20S proteasome and implications for the eukaryotic proteasome-ATPase interactions. EMBO J 29(3):692–702. doi:10.1038/emboj.2009.382, emboj2009382 [pii]
33. Stadtmueller BM, Hill CP (2011) Proteasome activators. Mol Cell 41(1):8–19. doi:10.1016/j.molcel.2010.12.020, S1097-2765(10)01005-1 [pii]
34. Kajava AV (2002) What curves alpha-solenoids? Evidence for an alpha-helical toroid structure of Rpn1 and Rpn2 proteins of the 26S proteasome. J Biol Chem 277(51):49791–49798. doi:10.1074/jbc, M204982200M204982200 [pii]
35. Lander GC, Estrin E, Matyskiela ME, Bashore C, Nogales E, Martin A (2012) Complete subunit architecture of the proteasome regulatory particle. Nature 482(7384):186–191. doi:10.1038/nature10774, nature10774 [pii]
36. Lasker K, Forster F, Bohn S, Walzthoeni T, Villa E, Unverdorben P, Beck F, Aebersold R, Sali A, Baumeister W (2012) Molecular architecture of the 26S proteasome holocomplex determined by an integrative approach. Proc Natl Acad Sci U S A 109(5):1380–1387. doi:10.1073/pnas.1120559109, 1120559109 [pii]
37. da Fonseca PC, He J, Morris EP (2012) Molecular Model of the human 26S proteasome. Mol Cell 46(1):54–66. doi:10.1016/j.molcel.2012.03.026, S1097-2765(12)00263-8 [pii]
38. Deveraux Q, Ustrell V, Pickart C, Rechsteiner M (1994) A 26S protease subunit that binds ubiquitin conjugates. J Biol Chem 269(10):7059–7061
39. van Nocker S, Deveraux Q, Rechsteiner M, Vierstra RD (1996) Arabidopsis MBP1 gene encodes a conserved ubiquitin recognition component of the 26S proteasome. Proc Natl Acad Sci U S A 93(2):856–860
40. Husnjak K, Elsasser S, Zhang N, Chen X, Randles L, Shi Y, Hofmann K, Walters KJ, Finley D, Dikic I (2008) Proteasome subunit Rpn13 is a novel ubiquitin receptor. Nature 453(7194):481–488. doi:10.1038/nature06926, nature06926 [pii]
41. Schreiner P, Chen X, Husnjak K, Randles L, Zhang N, Elsasser S, Finley D, Dikic I, Walters KJ, Groll M (2008) Ubiquitin docking at the proteasome through a novel pleckstrin-homology domain interaction. Nature 453(7194):548–552. doi:10.1038/nature06924, nature06924 [pii]
42. Verma R, Aravind L, Oania R, McDonald WH, Yates JR 3rd, Koonin EV, Deshaies RJ (2002) Role of Rpn11 metalloprotease in deubiquitination and degradation by the 26S proteasome. Science 298(5593):611–615. doi:10.1126/science, 10758981075898 [pii]
43. Yao T, Cohen RE (2002) A cryptic protease couples deubiquitination and degradation by the proteasome. Nature 419(6905):403–407. doi:10.1038/nature01071, nature01071 [pii]
44. Walz J, Erdmann A, Kania M, Typke D, Koster AJ, Baumeister W (1998) 26S proteasome structure revealed by three-dimensional electron microscopy. J Struct Biol 121(1):19–29. doi:10.1006/jsbi.1998.3958, S1047-8477(98)93958-2 [pii]
45. da Fonseca PC, Morris EP (2008) Structure of the human 26S proteasome: subunit radial displacements open the gate into the proteolytic core. J Biol Chem 283(34):23305–23314. doi:10.1074/jbc.M802716200, M802716200 [pii]
46. Nickell S, Beck F, Scheres SH, Korinek A, Forster F, Lasker K, Mihalache O, Sun N, Nagy I, Sali A, Plitzko JM, Carazo JM, Mann M, Baumeister W (2009) Insights into the molecu-

lar architecture of the 26S proteasome. Proc Natl Acad Sci U S A 106(29):11943–11947. doi:10.1073/pnas.0905081106, 0905081106 [pii]
47. Bohn S, Beck F, Sakata E, Walzthoeni T, Beck M, Aebersold R, Forster F, Baumeister W, Nickell S (2010) Structure of the 26S proteasome from Schizosaccharomyces pombe at subnanometer resolution. Proc Natl Acad Sci U S A 107(49):20992–20997. doi:10.1073/pnas.1015530107, 1015530107 [pii]
48. Beck F, Unverdorben P, Bohn S, Schweitzer A, Pfeifer G, Sakata E, Nickell S, Plitzko JM, Villa E, Baumeister W, Forster F (2012) Near-atomic resolution structural model of the yeast 26S proteasome. Proc Natl Acad Sci U S A 109(37):14870–14875. doi:10.1073/pnas.1213333109, 1213333109 [pii]
49. Matyskiela ME, Lander GC, Martin A (2013) Conformational switching of the 26S proteasome enables substrate degradation. Nat Struct Mol Biol 20(7):781–788. doi:10.1038/nsmb.2616, nsmb.2616 [pii]
50. Tomko RJ Jr, Hochstrasser M (2013) Molecular architecture and assembly of the eukaryotic proteasome. Annu Rev Biochem 82:415–445. doi:10.1146/annurev-biochem-060410-150257
51. Kish-Trier E, Hill CP (2013) Structural biology of the proteasome. Annu Rev Biophys 42:29–49. doi:10.1146/annurev-biophys-083012-130417
52. Kunjappu MJ, Hochstrasser M (2013) Assembly of the 20S proteasome. Biochim Biophys Acta. doi:10.1016/j.bbamcr.2013.03.008, S0167-4889(13)00099-2 [pii]
53. Zwickl P, Kleinz J, Baumeister W (1994) Critical elements in proteasome assembly. Nat Struct Biol 1(11):765–770
54. Frentzel S, Pesold-Hurt B, Seelig A, Kloetzel PM (1994) 20S proteasomes are assembled via distinct precursor complexes. Processing of LMP2 and LMP7 proproteins takes place in 13-16S preproteasome complexes. J Mol Biol 236(4):975–981. doi:10.1016/0022-2836(94)90003-5 [pii]
55. Nandi D, Woodward E, Ginsburg DB, Monaco JJ (1997) Intermediates in the formation of mouse 20S proteasomes: implications for the assembly of precursor beta subunits. EMBO J 16(17):5363–5375. doi:10.1093/emboj/16.17.5363
56. Schmidtke G, Schmidt M, Kloetzel PM (1997) Maturation of mammalian 20S proteasome: purification and characterization of 13S and 16S proteasome precursor complexes. J Mol Biol 268(1):95–106. doi:10.1006/jmbi.1997.0947, S0022-2836(97)90947-5 [pii]
57. Li X, Kusmierczyk AR, Wong P, Emili A, Hochstrasser M (2007) beta-Subunit appendages promote 20S proteasome assembly by overcoming an Ump1-dependent checkpoint. EMBO J 26(9):2339–2349. doi:10.1038/sj.emboj.7601681, 7601681 [pii]
58. Hirano Y, Kaneko T, Okamoto K, Bai M, Yashiroda H, Furuyama K, Kato K, Tanaka K, Murata S (2008) Dissecting beta-ring assembly pathway of the mammalian 20S proteasome. EMBO J 27(16):2204–2213. doi:10.1038/emboj.2008.148, emboj2008148 [pii]
59. Baumeister W, Walz J, Zuhl F, Seemuller E (1998) The proteasome: paradigm of a self-compartmentalizing protease. Cell 92(3):367–380. doi:10.1016/S0092-8674(00)80929-0 [pii]
60. Groll M, Brandstetter H, Bartunik H, Bourenkow G, Huber R (2003) Investigations on the maturation and regulation of archaebacterial proteasomes. J Mol Biol 327(1):75–83. doi:10.1016/S0022-2836(03)00080-9 [pii]
61. Ramos PC, Hockendorff J, Johnson ES, Varshavsky A, Dohmen RJ (1998) Ump1p is required for proper maturation of the 20S proteasome and becomes its substrate upon completion of the assembly. Cell 92(4):489–499. doi:S0092-8674(00)80942-3 [pii]
62. Hirano Y, Hendil KB, Yashiroda H, Iemura S, Nagane R, Hioki Y, Natsume T, Tanaka K, Murata S (2005) A heterodimeric complex that promotes the assembly of mammalian 20S proteasomes. Nature 437(7063):1381–1385. doi:10.1038/nature04106, nature04106 [pii]
63. Kusmierczyk AR, Kunjappu MJ, Kim RY, Hochstrasser M (2011) A conserved 20S proteasome assembly factor requires a C-terminal HbYX motif for proteasomal precursor binding. Nat Struct Mol Biol 18(5):622–629. doi:10.1038/nsmb.2027, nsmb.2027 [pii]
64. Unno M, Mizushima T, Morimoto Y, Tomisugi Y, Tanaka K, Yasuoka N, Tsukihara T (2002) The structure of the mammalian 20S proteasome at 2.75 A resolution. Structure 10(5):609–618. doi: 10.1016/S0969-2126(02)00748-7 [pii]
65. Volker C, Lupas AN (2002) Molecular evolution of proteasomes. Curr Top Microbiol Immunol 268:1–22

66. Gerards WL, Enzlin J, Haner M, Hendriks IL, Aebi U, Bloemendal H, Boelens W (1997) The human alpha-type proteasomal subunit HsC8 forms a double ringlike structure, but does not assemble into proteasome-like particles with the beta-type subunits HsDelta or HsBPROS26. J Biol Chem 272(15):10080–10086
67. Gerards WL, de Jong WW, Bloemendal H, Boelens W (1998) The human proteasomal subunit HsC8 induces ring formation of other alpha-type subunits. J Mol Biol 275(1):113–121. doi:10.1006/jmbi.1997.1429, S0022-2836(97)91429-7 [pii]
68. Arendt CS, Hochstrasser M (1999) Eukaryotic 20S proteasome catalytic subunit propeptides prevent active site inactivation by N-terminal acetylation and promote particle assembly. EMBO J 18(13):3575–3585. doi:10.1093/emboj/18.13.3575
69. Groll M, Heinemeyer W, Jager S, Ullrich T, Bochtler M, Wolf DH, Huber R (1999) The catalytic sites of 20S proteasomes and their role in subunit maturation: a mutational and crystallographic study. Proc Natl Acad Sci U S A 96(20):10976–10983
70. Ramos PC, Marques AJ, London MK, Dohmen RJ (2004) Role of C-terminal extensions of subunits beta2 and beta7 in assembly and activity of eukaryotic proteasomes. J Biol Chem 279(14):14323–14330. doi:10.1074/jbc.M308757200, M308757200 [pii]
71. Marques AJ, Glanemann C, Ramos PC, Dohmen RJ (2007) The C-terminal extension of the beta7 subunit and activator complexes stabilize nascent 20S proteasomes and promote their maturation. J Biol Chem 282(48):34869–34876. doi:10.1074/jbc.M705836200, M705836200 [pii]
72. Burri L, Hockendorff J, Boehm U, Klamp T, Dohmen RJ, Levy F (2000) Identification and characterization of a mammalian protein interacting with 20S proteasome precursors. Proc Natl Acad Sci U S A 97(19):10348–10353. doi:10.1073/pnas.190268597, 190268597 [pii]
73. Griffin TA, Slack JP, McCluskey TS, Monaco JJ, Colbert RA (2000) Identification of proteassemblin, a mammalian homologue of the yeast protein, Ump1p, that is required for normal proteasome assembly. Mol Cell Biol Res Commun 3(4):212–217. doi:10.1006/mcbr.2000.0213, S1522472400902137 [pii]
74. Witt E, Zantopf D, Schmidt M, Kraft R, Kloetzel PM, Kruger E (2000) Characterisation of the newly identified human Ump1 homologue POMP and analysis of LMP7(beta 5i) incorporation into 20S proteasomes. J Mol Biol 301(1):1–9. doi:10.1006/jmbi.2000.3959, S0022-2836(00)93959-7 [pii]
75. Hirano Y, Hayashi H, Iemura S, Hendil KB, Niwa S, Kishimoto T, Kasahara M, Natsume T, Tanaka K, Murata S (2006) Cooperation of multiple chaperones required for the assembly of mammalian 20S proteasomes. Mol Cell 24(6):977–984. doi:10.1016/j.molcel.2006.11.015, S1097-2765(06)00786-6 [pii]
76. Cagney G, Uetz P, Fields S (2001) Two-hybrid analysis of the Saccharomyces cerevisiae 26S proteasome. Physiol Genomics 7(1):27–34.
77. Jayarapu K, Griffin TA (2004) Protein-protein interactions among human 20S proteasome subunits and proteassemblin. Biochem Biophys Res Commun 314(2):523–528. doi:10.1016/j.bbrc.2003.12.119 [pii]
78. Fricke B, Heink S, Steffen J, Kloetzel PM, Kruger E (2007) The proteasome maturation protein POMP facilitates major steps of 20S proteasome formation at the endoplasmic reticulum. EMBO Rep 8(12):1170–1175. doi:10.1038/sj.embor.7401091, 7401091 [pii]
79. Hoefer MM, Boneberg EM, Grotegut S, Kusch J, Illges H (2006) Possible tetramerisation of the proteasome maturation factor POMP/proteassemblin/hUmp1 and its subcellular localisation. Int J Biol Macromol 38(3–5):259–267. doi:10.1016/j.ijbiomac.2006.03.015, S0141-8130(06)00096-1 [pii]
80. Akahane T, Sahara K, Yashiroda H, Tanaka K, Murata S (2013) Involvement of Bag6 and the TRC pathway in proteasome assembly. Nat Commun 4:2234. doi:10.1038/ncomms3234, ncomms3234 [pii]
81. Palmer EA, Kruse KB, Fewell SW, Buchanan SM, Brodsky JL, McCracken AA (2003) Differential requirements of novel A1PiZ degradation deficient (ADD) genes in ER-associated protein degradation. J Cell Sci 116(Pt 11):2361–2373. doi:10.1242/jcs.00439, jcs.00439 [pii]
82. Scott CM, Kruse KB, Schmidt BZ, Perlmutter DH, McCracken AA, Brodsky JL (2007) ADD66, a gene involved in the endoplasmic reticulum-associated degradation of alpha-1-an-

titrypsin-Z in yeast, facilitates proteasome activity and assembly. Mol Biol Cell 18(10):3776–3787. doi:10.1091/mbc.E07-01-0034, E07-01-0034 [pii]
83. Le Tallec B, Barrault MB, Courbeyrette R, Guerois R, Marsolier-Kergoat MC, Peyroche A (2007) 20S proteasome assembly is orchestrated by two distinct pairs of chaperones in yeast and in mammals. Mol Cell 27(4):660–674. doi:10.1016/j.molcel.2007.06.025, S1097-2765(07)00415-7 [pii]
84. Krogan NJ, Cagney G, Yu H, Zhong G, Guo X, Ignatchenko A, Li J, Pu S, Datta N, Tikuisis AP, Punna T, Peregrin-Alvarez JM, Shales M, Zhang X, Davey M, Robinson MD, Paccanaro A, Bray JE, Sheung A, Beattie B, Richards DP, Canadien V, Lalev A, Mena F, Wong P, Starostine A, Canete MM, Vlasblom J, Wu S, Orsi C, Collins SR, Chandran S, Haw R, Rilstone JJ, Gandi K, Thompson NJ, Musso G, St Onge P, Ghanny S, Lam MH, Butland G, Altaf-Ul AM, Kanaya S, Shilatifard A, O'Shea E, Weissman JS, Ingles CJ, Hughes TR, Parkinson J, Gerstein M, Wodak SJ, Emili A, Greenblatt JF (2006) Global landscape of protein complexes in the yeast Saccharomyces cerevisiae. Nature 440(7084):637–643. doi:10.1038/nature04670, nature04670 [pii]
85. Stadtmueller BM, Kish-Trier E, Ferrell K, Petersen CN, Robinson H, Myszka DG, Eckert DM, Formosa T, Hill CP (2012) Structure of a proteasome Pba1-Pba2 complex: implications for proteasome assembly, activation, and biological function. J Biol Chem 287(44):37371–37382. doi:10.1074/jbc.M112.367003, M112.367003 [pii]
86. Gillette TG, Kumar B, Thompson D, Slaughter CA, DeMartino GN (2008) Differential roles of the COOH termini of AAA subunits of PA700 (19S regulator) in asymmetric assembly and activation of the 26S proteasome. J Biol Chem 283(46):31813–31822. doi:10.1074/jbc.M805935200, M805935200 [pii]
87. Sadre-Bazzaz K, Whitby FG, Robinson H, Formosa T, Hill CP (2010) Structure of a Blm10 complex reveals common mechanisms for proteasome binding and gate opening. Mol Cell 37(5):728–735. doi:10.1016/j.molcel.2010.02.002, S1097-2765(10)00116-4 [pii]
88. Dange T, Smith D, Noy T, Rommel PC, Jurzitza L, Cordero RJ, Legendre A, Finley D, Goldberg AL, Schmidt M (2011) Blm10 protein promotes proteasomal substrate turnover by an active gating mechanism. J Biol Chem 286(50):42830–42839. doi:10.1074/jbc.M111.300178, M111.300178 [pii]
89. Barthelme D, Sauer RT (2012) Identification of the Cdc48*20S proteasome as an ancient AAA+ proteolytic machine. Science 337(6096):843–846. doi:10.1126/science.1224352, science.1224352 [pii]
90. Forouzan D, Ammelburg M, Hobel CF, Stroh LJ, Sessler N, Martin J, Lupas AN (2012) The archaeal proteasome is regulated by a network of AAA ATPases. J Biol Chem 287(46):39254–39262. doi:10.1074/jbc.M112.386458, M112.386458 [pii]
91. Barthelme D, Sauer RT (2013) Bipartite determinants mediate an evolutionarily conserved interaction between Cdc48 and the 20S peptidase. Proc Natl Acad Sci U S A 110(9):3327–3332. doi:10.1073/pnas.1300408110, 1300408110 [pii]
92. Realini C, Jensen CC, Zhang Z, Johnston SC, Knowlton JR, Hill CP, Rechsteiner M (1997) Characterization of recombinant REGalpha, REGbeta, and REGgamma proteasome activators. J Biol Chem 272(41):25483–25492
93. Whitby FG, Masters EI, Kramer L, Knowlton JR, Yao Y, Wang CC, Hill CP (2000) Structural basis for the activation of 20S proteasomes by 11S regulators. Nature 408(6808):115–120. doi:10.1038/35040607
94. Park S, Kim W, Tian G, Gygi SP, Finley D (2011) Structural defects in the regulatory particle-core particle interface of the proteasome induce a novel proteasome stress response. J Biol Chem 286(42):36652–36666. doi:10.1074/jbc.M111.285924, M111.285924 [pii]
95. Kleijnen MF, Roelofs J, Park S, Hathaway NA, Glickman M, King RW, Finley D (2007) Stability of the proteasome can be regulated allosterically through engagement of its proteolytic active sites. Nat Struct Mol Biol 14(12):1180–1188. doi:10.1038/nsmb1335, nsmb1335 [pii]
96. Osmulski PA, Hochstrasser M, Gaczynska M (2009) A tetrahedral transition state at the active sites of the 20S proteasome is coupled to opening of the alpha-ring channel. Structure 17(8):1137–1147. doi:10.1016/j.str.2009.06.011, S0969-2126(09)00253-6 [pii]

97. Ruschak AM, Kay LE (2012) Proteasome allostery as a population shift between interchanging conformers. Proc Natl Acad Sci U S A 109(50):E3454–E3462. doi:10.1073/pnas.1213640109, 1213640109 [pii]
98. Kumoi K, Satoh T, Murata K, Hiromoto T, Mizushima T, Kamiya Y, Noda M, Uchiyama S, Yagi H, Kato K (2013) An archaeal homolog of proteasome assembly factor functions as a proteasome activator. PLoS ONE 8(3):e60294. doi:10.1371/journal.pone.0060294, PONE-D-12-23454 [pii]
99. Kusmierczyk AR, Kunjappu MJ, Funakoshi M, Hochstrasser M (2008) A multimeric assembly factor controls the formation of alternative 20S proteasomes. Nat Struct Mol Biol 15(3):237–244. doi:10.1038/nsmb.1389, nsmb.1389 [pii]
100. Yashiroda H, Mizushima T, Okamoto K, Kameyama T, Hayashi H, Kishimoto T, Niwa S, Kasahara M, Kurimoto E, Sakata E, Takagi K, Suzuki A, Hirano Y, Murata S, Kato K, Yamane T, Tanaka K (2008) Crystal structure of a chaperone complex that contributes to the assembly of yeast 20S proteasomes. Nat Struct Mol Biol 15(3):228–236. doi:10.1038/nsmb.1386, nsmb.1386 [pii]
101. Hoyt MA, McDonough S, Pimpl SA, Scheel H, Hofmann K, Coffino P (2008) A genetic screen for Saccharomyces cerevisiae mutants affecting proteasome function, using a ubiquitin-independent substrate. Yeast 25(3):199–217. doi:10.1002/yea.1579
102. Velichutina I, Connerly PL, Arendt CS, Li X, Hochstrasser M (2004) Plasticity in eucaryotic 20S proteasome ring assembly revealed by a subunit deletion in yeast. EMBO J 23(3):500–510. doi:10.1038/sj.emboj.7600059, 7600059 [pii]
103. Ustrell V, Hoffman L, Pratt G, Rechsteiner M (2002) PA200, a nuclear proteasome activator involved in DNA repair. EMBO J 21(13):3516–3525. doi:10.1093/emboj/cdf333
104. Schmidt M, Haas W, Crosas B, Santamaria PG, Gygi SP, Walz T, Finley D (2005) The HEAT repeat protein Blm10 regulates the yeast proteasome by capping the core particle. Nat Struct Mol Biol 12(4):294–303. doi:10.1038/nsmb914, nsmb914 [pii]
105. Khor B, Bredemeyer AL, Huang CY, Turnbull IR, Evans R, Maggi LB Jr, White JM, Walker LM, Carnes K, Hess RA, Sleckman BP (2006) Proteasome activator PA200 is required for normal spermatogenesis. Mol Cell Biol 26(8):2999–3007. doi:10.1128/MCB.26.8.2999-3007.2006, 26/8/2999 [pii]
106. Blickwedehl J, Agarwal M, Seong C, Pandita RK, Melendy T, Sung P, Pandita TK, Bangia N (2008) Role for proteasome activator PA200 and postglutamyl proteasome activity in genomic stability. Proc Natl Acad Sci U S A 105(42):16165–16170. doi:10.1073/pnas.0803145105, 0803145105 [pii]
107. Lopez AD, Tar K, Krugel U, Dange T, Ros IG, Schmidt M (2011) Proteasomal degradation of Sfp1 contributes to the repression of ribosome biogenesis during starvation and is mediated by the proteasome activator Blm10. Mol Biol Cell 22(5):528–540. doi:10.1091/mbc.E10-04–0352, mbc.E10-04-0352 [pii]
108. Fehlker M, Wendler P, Lehmann A, Enenkel C (2003) Blm3 is part of nascent proteasomes and is involved in a late stage of nuclear proteasome assembly. EMBO Rep 4(10):959–963. doi:10.1038/sj.embor.embor938, embor938 [pii]
109. Weberruss MH, Savulescu AF, Jando J, Bissinger T, Harel A, Glickman MH, Enenkel C (2013) Blm10 facilitates nuclear import of proteasome core particles. EMBO J. doi:10.1038/emboj.2013.192, emboj2013192 [pii]
110. Isono E, Nishihara K, Saeki Y, Yashiroda H, Kamata N, Ge L, Ueda T, Kikuchi Y, Tanaka K, Nakano A, Toh-e A (2007) The assembly pathway of the 19S regulatory particle of the yeast 26S proteasome. Mol Biol Cell 18(2):569–580. doi:10.1091/mbc.E06-07-0635, E06-07-0635 [pii]
111. Tomko RJ Jr, Hochstrasser M (2011) Incorporation of the Rpn12 subunit couples completion of proteasome regulatory particle lid assembly to lid-base joining. Mol Cell 44(6):907–917. doi:10.1016/j.molcel.2011.11.020, S1097-2765(11)00939-7 [pii]
112. Takeuchi J, Tamura T (2004) Recombinant ATPases of the yeast 26S proteasome activate protein degradation by the 20S proteasome. FEBS Lett 565(1–3):39–42. doi:10.1016/j.febslet.2004.03.073, S0014579304003783 [pii]

113. Richmond C, Gorbea C, Rechsteiner M (1997) Specific interactions between ATPase subunits of the 26S protease. J Biol Chem 272(20):13403–13411
114. Zhang F, Hu M, Tian G, Zhang P, Finley D, Jeffrey PD, Shi Y (2009) Structural insights into the regulatory particle of the proteasome from *Methanocaldococcus jannaschii*. Mol Cell 34(4):473–484. doi:10.1016/j.molcel.2009.04.021, S1097-2765(09)00274-3 [pii]
115. Djuranovic S, Hartmann MD, Habeck M, Ursinus A, Zwickl P, Martin J, Lupas AN, Zeth K (2009) Structure and activity of the N-terminal substrate recognition domains in proteasomal ATPases. Mol Cell 34(5):580–590. doi:10.1016/j.molcel.2009.04.030, S1097-2765(09)00306-2 [pii]
116. Tian G, Park S, Lee MJ, Huck B, McAllister F, Hill CP, Gygi SP, Finley D (2011) An asymmetric interface between the regulatory and core particles of the proteasome. Nat Struct Mol Biol 18(11):1259–1267. doi:10.1038/nsmb.2147, nsmb.2147 [pii]
117. Park S, Li X, Kim HM, Singh CR, Tian G, Hoyt MA, Lovell S, Battaile KP, Zolkiewski M, Coffino P, Roelofs J, Cheng Y, Finley D (2013) Reconfiguration of the proteasome during chaperone-mediated assembly. Nature 497(7450):512–516. doi:10.1038/nature12123, nature12123 [pii]
118. Ammelburg M, Frickey T, Lupas AN (2006) Classification of AAA+ proteins. J Struct Biol 156(1):2–11. doi:10.1016/j.jsb.2006.05.002, S1047-8477(06)00165-1 [pii]
119. DeMartino GN, Proske RJ, Moomaw CR, Strong AA, Song X, Hisamatsu H, Tanaka K, Slaughter CA (1996) Identification, purification, and characterization of a PA700-dependent activator of the proteasome. J Biol Chem 271(6):3112–3118
120. Watanabe TK, Saito A, Suzuki M, Fujiwara T, Takahashi E, Slaughter CA, DeMartino GN, Hendil KB, Chung CH, Tanahashi N, Tanaka K (1998) cDNA cloning and characterization of a human proteasomal modulator subunit, p27(PSMD9). Genomics 50(2):241–250. doi:10.1006/geno.1998.5301 [pii]
121. Gorbea C, Taillandier D, Rechsteiner M (2000) Mapping subunit contacts in the regulatory complex of the 26S proteasome. S2 and S5b form a tetramer with ATPase subunits S4 and S7. J Biol Chem 275(2):875–882
122. Dawson S, Apcher S, Mee M, Higashitsuji H, Baker R, Uhle S, Dubiel W, Fujita J, Mayer RJ (2002) Gankyrin is an ankyrin-repeat oncoprotein that interacts with CDK4 kinase and the S6 ATPase of the 26 S proteasome. J Biol Chem 277(13):10893–10902. doi:10.1074/jbc.M107313200, M107313200 [pii]
123. Park Y, Hwang YP, Lee JS, Seo SH, Yoon SK, Yoon JB (2005) Proteasomal ATPase-associated factor 1 negatively regulates proteasome activity by interacting with proteasomal ATPases. Mol Cell Biol 25(9):3842–3853. doi:10.1128/MCB.25.9.3842-3853.2005, 25/9/3842 [pii]
124. Nakamura Y, Nakano K, Umehara T, Kimura M, Hayashizaki Y, Tanaka A, Horikoshi M, Padmanabhan B, Yokoyama S (2007) Structure of the oncoprotein gankyrin in complex with S6 ATPase of the 26S proteasome. Structure 15(2):179–189. doi:10.1016/j.str.2006.11.015, S0969-2126(07)00028-7 [pii]
125. Nakamura Y, Umehara T, Tanaka A, Horikoshi M, Padmanabhan B, Yokoyama S (2007) Structural basis for the recognition between the regulatory particles Nas6 and Rpt3 of the yeast 26S proteasome. Biochem Biophys Res Commun 359(3):503–509. doi:10.1016/j.bbrc.2007.05.138, S0006-291×(07)01091-1 [pii]
126. Le Tallec B, Barrault MB, Guerois R, Carre T, Peyroche A (2009) Hsm3/S5b participates in the assembly pathway of the 19S regulatory particle of the proteasome. Mol Cell 33(3):389–399. doi:10.1016/j.molcel.2009.01.010, S1097-2765(09)00037-9 [pii]
127. Funakoshi M, Tomko RJ Jr, Kobayashi H, Hochstrasser M (2009) Multiple assembly chaperones govern biogenesis of the proteasome regulatory particle base. Cell 137(5):887–899. doi:10.1016/j.cell.2009.04.061, S0092-8674(09)00526-1 [pii]
128. Saeki Y, Toh EA, Kudo T, Kawamura H, Tanaka K (2009) Multiple proteasome-interacting proteins assist the assembly of the yeast 19S regulatory particle. Cell 137(5):900–913. doi:10.1016/j.cell.2009.05.005, S0092-8674(09)00528-5 [pii]

129. Kaneko T, Hamazaki J, Iemura S, Sasaki K, Furuyama K, Natsume T, Tanaka K, Murata S (2009) Assembly pathway of the Mammalian proteasome base subcomplex is mediated by multiple specific chaperones. Cell 137(5):914–925. doi:10.1016/j.cell.2009.05.008, S0092-8674(09)00565-0 [pii]
130. Roelofs J, Park S, Haas W, Tian G, McAllister FE, Huo Y, Lee BH, Zhang F, Shi Y, Gygi SP, Finley D (2009) Chaperone-mediated pathway of proteasome regulatory particle assembly. Nature 459(7248):861–865. doi:10.1038/nature08063, nature08063 [pii]
131. Park S, Roelofs J, Kim W, Robert J, Schmidt M, Gygi SP, Finley D (2009) Hexameric assembly of the proteasomal ATPases is templated through their C termini. Nature 459(7248):866–870. doi:10.1038/nature08065, nature08065 [pii]
132. Barrault MB, Richet N, Godard C, Murciano B, Le Tallec B, Rousseau E, Legrand P, Charbonnier JB, Le Du MH, Guerois R, Ochsenbein F, Peyroche A (2012) Dual functions of the Hsm3 protein in chaperoning and scaffolding regulatory particle subunits during the proteasome assembly. Proc Natl Acad Sci U S A 109(17):E1001–E1010. doi:10.1073/pnas.1116538109, 1116538109 [pii]
133. Lee SH, Moon JH, Yoon SK, Yoon JB (2012) Stable incorporation of ATPase subunits into 19S regulatory particle of human proteasome requires nucleotide binding and C-terminal tails. J Biol Chem 287(12):9269–9279. doi:10.1074/jbc.M111.316208, M111.316208 [pii]
134. Hendil KB, Kriegenburg F, Tanaka K, Murata S, Lauridsen AM, Johnsen AH, Hartmann-Petersen R (2009) The 20S proteasome as an assembly platform for the 19S regulatory complex. J Mol Biol 394(2):320–328. doi:10.1016/j.jmb.2009.09.038, S0022-2836(09)01165-6 [pii]
135. Thompson D, Hakala K, DeMartino GN (2009) Subcomplexes of PA700, the 19S regulator of the 26S proteasome, reveal relative roles of AAA subunits in 26S proteasome assembly and activation and ATPase activity. J Biol Chem 284(37):24891–24903. doi:10.1074/jbc.M109.023218, M109.023218 [pii]
136. Takagi K, Kim S, Yukii H, Ueno M, Morishita R, Endo Y, Kato K, Tanaka K, Saeki Y, Mizushima T (2012) Structural Basis for Specific Recognition of Rpt1p, an ATPase Subunit of 26S Proteasome, by Proteasome-dedicated Chaperone Hsm3p. J Biol Chem 287(15):12172–12182. doi:10.1074/jbc.M112.345876, M112.345876 [pii]
137. Ehlinger A, Park S, Fahmy A, Lary JW, Cole JL, Finley D, Walters KJ (2013) Conformational dynamics of the Rpt6 ATPase in proteasome assembly and Rpn14 binding. Structure 21(5):753–765. doi:10.1016/j.str.2013.02.021, S0969-2126(13)00079-8 [pii]
138. Isono E, Saeki Y, Yokosawa H, Toh-e A (2004) Rpn7 Is required for the structural integrity of the 26S proteasome of Saccharomyces cerevisiae. J Biol Chem 279(26):27168–27176. doi:10.1074/jbc.M314231200, M314231200 [pii]
139. Isono E, Saito N, Kamata N, Saeki Y, Toh EA (2005) Functional analysis of Rpn6p, a lid component of the 26S proteasome, using temperature-sensitive rpn6 mutants of the yeast Saccharomyces cerevisiae. J Biol Chem 280(8):6537–6547. doi:10.1074/jbc.M409364200, M409364200 [pii]
140. Fukunaga K, Kudo T, Toh-e A, Tanaka K, Saeki Y (2010) Dissection of the assembly pathway of the proteasome lid in *Saccharomyces cerevisiae*. Biochem Biophys Res Commun 396(4):1048–1053. doi:10.1016/j.bbrc.2010.05.061, S0006-291×(10)00952−6 [pii]
141. Estrin E, Lopez-Blanco JR, Chacon P, Martin A (2013) Formation of an intricate helical bundle dictates the assembly of the 26S proteasome lid. Structure 21(9):1624–1635. doi:10.1016/j.str.2013.06.023, S0969-2126(13)00247−5 [pii]
142. Pickering AM, Davies KJ (2012) Degradation of damaged proteins: the main function of the 20S proteasome. Prog Mol Biol Transl Sci 109:227–248. doi:10.1016/B978-0-12-397863-9.00006-7 [pii]
143. Russell SJ, Reed SH, Huang W, Friedberg EC, Johnston SA (1999) The 19S regulatory complex of the proteasome functions independently of proteolysis in nucleotide excision repair. Mol Cell 3(6):687–695. doi:10.1016/S1097-2765(01)80001-0 [pii]
144. Lee D, Ezhkova E, Li B, Pattenden SG, Tansey WP, Workman JL (2005) The proteasome regulatory particle alters the SAGA coactivator to enhance its interactions with transcriptional activators. Cell 123(3):423–436. doi:10.1016/j.cell.2005.08.015, S0092-8674(05)00819-6 [pii]

145. Eytan E, Ganoth D, Armon T, Hershko A (1989) ATP-dependent incorporation of 20S protease into the 26S complex that degrades proteins conjugated to ubiquitin. Proc Natl Acad Sci U S A 86(20):7751–7755
146. Lehmann A, Niewienda A, Jechow K, Janek K, Enenkel C (2010) Ecm29 fulfils quality control functions in proteasome assembly. Mol Cell 38(6):879–888. doi:10.1016/j.molcel.2010.06.016, S1097-2765(10)00457-0 [pii]
147. De LMota-PeynadoA, Lee SY, Pierce BM, Wani P, Singh CR, Roelofs J (2013) The proteasome-associated protein Ecm29 inhibits proteasomal ATPase activity and in vivo protein degradation by the proteasome. J Biol Chem. doi:10.1074/jbc.M113.491662, M113.491662 [pii]
148. Russell SJ, Steger KA, Johnston SA (1999) Subcellular localization, stoichiometry, and protein levels of 26S proteasome subunits in yeast. J Biol Chem 274(31):21943–21952
149. Ghaemmaghami S, Huh WK, Bower K, Howson RW, Belle A, Dephoure N, O'Shea EK, Weissman JS (2003) Global analysis of protein expression in yeast. Nature 425(6959):737–741. doi:10.1038/nature02046, nature02046 [pii]
150. Mannhaupt G, Schnall R, Karpov V, Vetter I, Feldmann H (1999) Rpn4p acts as a transcription factor by binding to PACE, a nonamer box found upstream of 26S proteasomal and other genes in yeast. FEBS Lett 450(1–2):27–34. doi:10.1016/S0014-5793(99)00467-6 [pii]
151. Xie Y, Varshavsky A (2001) RPN4 is a ligand, substrate, and transcriptional regulator of the 26S proteasome: a negative feedback circuit. Proc Natl Acad Sci U S A 98(6):3056–3061. doi:10.1073/pnas, 07102229898/6/3056 [pii]
152. Ju D, Xie Y (2004) Proteasomal degradation of RPN4 via two distinct mechanisms, ubiquitin-dependent and -independent. J Biol Chem 279(23):23851–23854. doi:10.1074/jbc.C400111200, C400111200 [pii]
153. Ju D, Wang L, Mao X, Xie Y (2004) Homeostatic regulation of the proteasome via an Rpn4-dependent feedback circuit. Biochem Biophys Res Commun 321(1):51–57. doi:10.1016/j.bbrc.2004.06.105, S0006-291 × (04)01385-3 [pii]
154. London MK, Keck BI, Ramos PC, Dohmen RJ (2004) Regulatory mechanisms controlling biogenesis of ubiquitin and the proteasome. FEBS Lett 567(2–3):259–264. doi:10.1016/j.febslet.2004.04.078, S001457930400554X [pii]
155. Owsianik G, Balzi lL, Ghislain M (2002) Control of 26S proteasome expression by transcription factors regulating multidrug resistance in Saccharomyces cerevisiae. Mol Microbiol 43(5):1295–1308. doi:10.1046/j.1365-2958.2002.02823.x [pii]
156. Hahn JS, Neef DW, Thiele DJ (2006) A stress regulatory network for co-ordinated activation of proteasome expression mediated by yeast heat shock transcription factor. Mol Microbiol 60(1):240–251. doi:10.1111/j.1365-2958.2006.05097.x, MMI5097 [pii]
157. Wang X, Xu H, Ju D, Xie Y (2008) Disruption of Rpn4-induced proteasome expression in *Saccharomyces cerevisiae* reduces cell viability under stressed conditions. Genetics 180(4):1945–1953. doi:10.1534/genetics.108.094524, genetics.108.094524 [pii]
158. Meiners S, Heyken D, Weller A, Ludwig A, Stangl K, Kloetzel PM, Kruger E (2003) Inhibition of proteasome activity induces concerted expression of proteasome genes and de novo formation of mammalian proteasomes. J Biol Chem 278(24):21517–21525. doi:10.1074/jbc.M301032200, M301032200 [pii]
159. Radhakrishnan SK, Lee CS, Young P, Beskow A, Chan JY, Deshaies RJ (2010) Transcription factor Nrfl mediates the proteasome recovery pathway after proteasome inhibition in mammalian cells. Mol Cell 38(1):17–28. doi:10.1016/j.molcel.2010.02.029, S1097-2765(10)00240-6 [pii]
160. Kraft DC, Deocaris CC, Wadhwa R, Rattan SI (2006) Preincubation with the proteasome inhibitor MG-132 enhances proteasome activity via the Nrf2 transcription factor in aging human skin fibroblasts. Ann N Y Acad Sci 1067:420–424. doi:10.1196/annals.1354.060, 1067/1/420 [pii]
161. Kapeta S, Chondrogianni N, Gonos ES (2010) Nuclear erythroid factor 2-mediated proteasome activation delays senescence in human fibroblasts. J Biol Chem 285(11):8171–8184. doi:10.1074/jbc.M109.031575, M109.031575 [pii]

162. Xie Y (2010) Structure, assembly and homeostatic regulation of the 26S proteasome. J Mol Cell Biol 2(6):308–317. doi:10.1093/jmcb/mjq030, mjq030 [pii]
163. Xu H, Fu J, Ha SW, Ju D, Zheng J, Li L, Xie Y (2012) The CCAAT box-binding transcription factor NF-Y regulates basal expression of human proteasome genes. Biochim Biophys Acta 1823(4):818–825. doi:10.1016/j.bbamcr.2012.01.002, S0167-4889(12)00004-3 [pii]
164. Reits EA, Benham AM, Plougastel B, Neefjes J, Trowsdale J (1997) Dynamics of proteasome distribution in living cells. EMBO J 16(20):6087–6094. doi:10.1093/emboj/16.20.6087
165. Enenkel C, Lehmann A, Kloetzel PM (1998) Subcellular distribution of proteasomes implicates a major location of protein degradation in the nuclear envelope-ER network in yeast. EMBO J 17(21):6144–6154. doi:10.1093/emboj/17.21.6144
166. Wilkinson CR, Wallace M, Morphew M, Perry P, Allshire R, Javerzat JP, McIntosh JR, Gordon C (1998) Localization of the 26S proteasome during mitosis and meiosis in fission yeast. EMBO J 17(22):6465–6476. doi:10.1093/emboj/17.22.6465
167. Tanaka K, Yoshimura T, Tamura T, Fujiwara T, Kumatori A, Ichihara A (1990) Possible mechanism of nuclear translocation of proteasomes. FEBS Lett 271 (1–2):41–46. doi:10.1016/0014-5793(90)80367-R [pii]
168. Nederlof PM, Wang HR, Baumeister W (1995) Nuclear localization signals of human and Thermoplasma proteasomal alpha subunits are functional in vitro. Proc Natl Acad Sci U S A 92(26):12060–12064
169. Wendler P, Lehmann A, Janek K, Baumgart S, Enenkel C (2004) The bipartite nuclear localization sequence of Rpn2 is required for nuclear import of proteasomal base complexes via karyopherin alphabeta and proteasome functions. J Biol Chem 279(36):37751–37762. doi:10.1074/jbc.M403551200, M403551200 [pii]
170. Savulescu AF, Shorer H, Kleifeld O, Cohen I, Gruber R, Glickman MH, Harel A (2011) Nuclear import of an intact preassembled proteasome particle. Mol Biol Cell 22(6):880–891. doi:10.1091/mbc.E10-07-0595, mbc.E10-07-0595 [pii]
171. Wang HR, Kania M, Baumeister W, Nederlof PM (1997) Import of human and Thermoplasma 20S proteasomes into nuclei of HeLa cells requires functional NLS sequences. Eur J Cell Biol 73(2):105–113
172. Mayr J, Wang HR, Nederlof P, Baumeister W (1999) The import pathway of human and Thermoplasma 20S proteasomes into HeLa cell nuclei is different from that of classical NLS-bearing proteins. Biol Chem 380(10):1183–1192. doi:10.1515/BC.1999.150
173. Lehmann A, Janek K, Braun B, Kloetzel PM, Enenkel C (2002) 20S proteasomes are imported as precursor complexes into the nucleus of yeast. J Mol Biol 317(3):401–413. doi:10.1006/jmbi.2002.5443, S0022283602954434 [pii]
174. Huh WK, Falvo JV, Gerke LC, Carroll AS, Howson RW, Weissman JS, O'Shea EK (2003) Global analysis of protein localization in budding yeast. Nature 425(6959):686–691. doi:10.1038/nature02026, nature02026 [pii]
175. Laporte D, Salin B, Daignan-Fornier B, Sagot I (2008) Reversible cytoplasmic localization of the proteasome in quiescent yeast cells. J Cell Biol 181(5):737–745. doi:10.1083/jcb.200711154, jcb.200711154 [pii]
176. Saunier R, Esposito M, Dassa EP, Delahodde A (2013) Integrity of the *Saccharomyces cerevisiae* Rpn11 protein is critical for formation of proteasome storage granules (PSG) and survival in stationary phase. PLoS ONE 8(8):e70357. doi:10.1371/journal.pone.0070357, PONE-D-13-16012 [pii]
177. Stefanovic S, Hegde RS (2007) Identification of a targeting factor for posttranslational membrane protein insertion into the ER. Cell 128(6):1147–1159. doi:10.1016/j.cell.2007.01.036, S0092-8674(07)00195-X [pii]
178. Schuldiner M, Metz J, Schmid V, Denic V, Rakwalska M, Schmitt HD, Schwappach B, Weissman JS (2008) The GET complex mediates insertion of tail-anchored proteins into the ER membrane. Cell 134(4):634–645. doi:10.1016/j.cell.2008.06.025, S0092-8674(08)00777-0 [pii]
179. Belote JM, Zhong L (2009) Duplicated proteasome subunit genes in Drosophila and their roles in spermatogenesis. Heredity 103(1):23–31. doi:10.1038/hdy.2009.23, hdy200923 [pii]

180. Fu H, Doelling JH, Arendt CS, Hochstrasser M, Vierstra RD (1998) Molecular organization of the 20S proteasome gene family from *Arabidopsis thaliana*. Genetics 149(2):677–692
181. Yang P, Fu H, Walker J, Papa CM, Smalle J, Ju YM, Vierstra RD (2004) Purification of the Arabidopsis 26S proteasome: biochemical and molecular analyses revealed the presence of multiple isoforms. J Biol Chem 279(8):6401–6413. doi:10.1074/jbc.M311977200, M311977200 [pii]
182. Zhong L, Belote JM (2007) The testis-specific proteasome subunit Prosalpha6T of *D. melanogaster* is required for individualization and nuclear maturation during spermatogenesis. Development 134(19):3517–3525. doi:10.1242/dev.004770, dev.004770 [pii]
183. Book AJ, Smalle J, Lee KH, Yang P, Walker JM, Casper S, Holmes JH, Russo LA, Buzzinotti ZW, Jenik PD, Vierstra RD (2009) The RPN5 subunit of the 26S proteasome is essential for gametogenesis, sporophyte development, and complex assembly in Arabidopsis. Plant Cell 21(2):460–478. doi:10.1105/tpc.108.064444, tpc.108.064444 [pii]
184. Lee KH, Minami A, Marshall RS, Book AJ, Farmer LM, Walker JM, Vierstra RD (2011) The RPT2 subunit of the 26S proteasome directs complex assembly, histone dynamics, and gametophyte and sporophyte development in Arabidopsis. Plant Cell 23(12):4298–4317. doi:10.1105/tpc.111.089482, tpc.111.089482 [pii]
185. Sonoda Y, Sako K, Maki Y, Yamazaki N, Yamamoto H, Ikeda A, Yamaguchi J (2009) Regulation of leaf organ size by the Arabidopsis RPT2a 19S proteasome subunit. Plant J 60(1):68–78. doi:10.1111/j.1365-313X.2009.03932.x, TPJ3932 [pii]
186. Griffin TA, Nandi D, Cruz M, Fehling HJ, Kaer LV, Monaco JJ, Colbert RA (1998) Immunoproteasome assembly: cooperative incorporation of interferon gamma (IFN-gamma)-inducible subunits. J Exp Med 187(1):97–104
187. Groettrup M, Standera S, Stohwasser R, Kloetzel PM (1997) The subunits MECL-1 and LMP2 are mutually required for incorporation into the 20S proteasome. Proc Natl Acad Sci U S A 94(17):8970–8975
188. Kingsbury DJ, Griffin TA, Colbert RA (2000) Novel propeptide function in 20S proteasome assembly influences beta subunit composition. J Biol Chem 275 (31):24156–24162. doi:10.1074/jbc.M001742200, M001742200 [pii]
189. De M, Jayarapu K, Elenich L, Monaco JJ, Colbert RA, Griffin TA (2003) Beta 2 subunit propeptides influence cooperative proteasome assembly. J Biol Chem 278(8):6153–6159. doi:10.1074/jbc.M209292200, M209292200 [pii]
190. Heink S, Ludwig D, Kloetzel PM, Kruger E (2005) IFN-gamma-induced immune adaptation of the proteasome system is an accelerated and transient response. Proc Natl Acad Sci U S A 102(26):9241–9246. doi:10.1073/pnas.0501711102, 0501711102 [pii]
191. Lee SC, Shaw BD (2007) A novel interaction between N-myristoylation and the 26S proteasome during cell morphogenesis. Mol Microbiol 63(4):1039–1053. doi:10.1111/j.1365-2958.2006.05575.x, MMI5575 [pii]
192. Gomes AV, Young GW, Wang Y, Zong C, Eghbali M, Drews O, Lu H, Stefani E, Ping P (2009) Contrasting proteome biology and functional heterogeneity of the 20S proteasome complexes in mammalian tissues. Mol Cell Proteomics 8(2):302–315. doi:10.1074/mcp.M800058-MCP200, M800058-MCP200 [pii]
193. Porollo A, Meller J (2007) Versatile annotation and publication quality visualization of protein complexes using POLYVIEW-3D. BMC Bioinformatics 8:316. doi:10.1186/1471-2105-8-316, 1471-2105-8-316 [pii]

Chapter 14
Systems-Wide Analysis of Protein Ubiquitylation: We Finally Have the Tiger by the Tail

Nancy N. Fang, Razvan F. Albu and Thibault Mayor

Abstract Ubiquitylation is a posttranslational modification in which ubiquitin, a conserved polypeptide, is conjugated to targeted proteins. The human genome encodes for over a thousand proteins that can either mediate ubiquitylation, remove the modification, or bind to modified proteins. Hence, the ubiquitin system is expected to affect a large portion of the proteome. Since its discovery, our understanding of the ubiquitin system has come a long way. Great efforts made by several prominent members of the scientific community have paid off, with leaps and bounds being made in areas such as the discovery of ubiquitylated proteins and ubiquitylation sites. Boosting the success of classical proteomic approaches, the recent introduction of a new method (by which ubiquitylated peptides are captured by antibodies) enabled the identification of a large number of ubiquitylation sites. In this chapter, we will review the different proteomic strategies that have been established in order to uncover which proteins are ubiquitylated in the cell, and then further discuss the novel biological insights revealed by these systems-wide studies.

Abbreviations

AQUA	Absolute quantification (of protein)
diGly tail	Two glycine from the C terminus of ubiquitin covalently attached to target's lysine
DUB	De-ubiquitylating enzyme
E1	Ubiquitin-activating enzyme
E2	Ubiquitin-conjugating enzyme
E3	Ubiquitin ligase
ERAD	Endoplasmic reticulum-associated degradation
ISG15	Interferon-stimulated gene of 15 kDa

T. Mayor (✉) · N. N. Fang · R. F. Albu
Department of Biochemistry and Molecular Biology, Centre for High-throughput Biology, University of British Columbia, Vancouver, BC, Canada
e-mail: mayor@mail.ubc.ca

R. F. Albu
e-mail: ralbu@mail.ubc.ca

N. N. Fang
e-mail: nancynfang@gmail.com

W. A. Houry (ed.), *The Molecular Chaperones Interaction Networks in Protein Folding and Degradation,* Interactomics and Systems Biology, DOI 10.1007/978-1-4939-1130-1_14,

MHC I	Histocompatibility complex class I
MS/MS	Tandem mass spectrometry
NEDD8	Neural-precursor-cell-expressed and developmentally down-regulated gene 8
PSAQ	Protein standard absolute quantification
PTM	Posttranslational modification
SEPTM	Serial enrichments of different posttranslational modifications
SILAC	Stable isotope labeling with amino acids in cell culture
SUMO	Small ubiquitin-like modifier
TUBEs	Tandem ubiquitin-binding entities
UBA	Ubiquitin-associated domain
UBD	Ubiquitin-binding domain
Ubl	Ubiquitin-like protein

1 Introduction: The Ubiquitin System and Ubiquitylation

The proteome is far more complex than the genome from which it is derived due to alternative splicing and extensive protein posttranslational modifications (PTMs). Importantly, PTMs fulfill key roles by controlling protein-protein interactions, protein localization, enzymatic activities, and protein turnover. To date, ubiquitylation (i.e., modification by ubiquitin conjugation) is one of the most abundantly identified PTM after phosphorylation, of which there are more than 50,000 reported protein modification sites (www.phosphosite.org) [1]. The remarkable progress in the enrichment and proteomic techniques that we will review in this chapter have enabled great advancements in the identification of novel ubiquitylation sites, which now offers a unique opportunity to better understand the ubiquitin system and its impact on the proteome.

Ubiquitylation stands apart from other PTMs in that a small protein (ubiquitin), instead of small functional groups (like phosphate and acetyl), is used as the modifier. Ubiquitin was first discovered in 1975 [2] and was named after its ubiquitous expression in eukaryotes. Ubiquitin is a 76 amino acid, highly structured protein with a molecular weight of about 8.5 kDa. It is covalently attached to proteins through the formation of an isopeptide bond between the carboxyl group of its last glycine residue and, typically, the epsilon amino group of a target lysine residue. There are also other similar protein modifiers, called ubiquitin-like proteins (Ubls), such as small ubiquitin-like modifier (SUMO), interferon-stimulated gene of 15 kDa (ISG15), and neural-precursor-cell-expressed and developmentally down-regulated gene (NEDD8). These Ubls have both a sequence and a structural homology to ubiquitin and are attached by similar conjugation mechanisms. However, proteomic analyses of these modifications will not be reviewed in this chapter.

Three classes of enzymes are required for the ubiquitylation cascade that leads to substrate modification, namely an ubiquitin-activating enzyme (E1), an ubiquitin-conjugating enzyme (E2), and an ubiquitin ligase (E3). Prior to substrate conjugation, ubiquitin first needs to be activated by the E1 enzyme through an adenosine triphosphate (ATP)-dependent reaction, in which a thioester linkage is formed between the C-terminus of ubiquitin and a cysteine residue of the E1 [3, 4]. The E1 then mediates the transfer of ubiquitin to a cysteine residue of an E2 enzyme via a trans-thioesterification reaction [5, 6]. Finally, an E3 ligase will recruit the target substrate for the last ligation step, either by facilitating the direct transfer of the ubiquitin molecule from the E2 to the associated substrate, or by transiently accepting the ubiquitin from the E2 prior to transferring it to the substrate [7]. Ubiquitylation mostly occurs on lysine residues of substrate proteins. However, in some cases the N-terminus of a protein is conjugated instead of a lysine [8–10]. Other residues, like cysteine, threonine, and serine, have also been reported as possible ubiquitylation sites [11–13].

Ubiquitylation is a highly dynamic process that relies on a complex and modular network of proteins which comprise the ubiquitin system. There are more than 600 E3 ligases encoded in the human genome that are assisted by about 40 E2s [14]. To maintain balance of the system, ubiquitylation can also be reversed by a class of enzymes called de-ubiquitylating enzymes (DUBs), of which around 80 have so far been identified in humans. DUBs are involved in maintaining ubiquitin homeostasis by processing newly synthesized ubiquitin and recycling used ubiquitin as well as controlling or modulating the fate of ubiquitylated proteins [15].

Ubiquitylation is a versatile PTM due to its built-in ability to generate linkages of diverse architecture, which can each be used in different signalling pathways [16]. Substrate proteins can be modified by a single ubiquitin on a single lysine (monoubiquitylation) or multiple lysine residues (multi-monoubiquitylation). Alternatively, lysine residues within ubiquitin itself can be used to covalently attach subsequent ubiquitin molecules, forming multimeric chains conjugated to a single lysine residue on the targeted protein (polyubiquitylation). The seven lysine residues (K6, 11, 27, 29, 33, 48, and 63) and the amino-terminus (M1) of ubiquitin can be used to generate ubiquitin chains that are either homogenous (one linkage type throughout), mixed (with different linkage types), or branched (arising when two ubiquitins are conjugated to separate lysines on the same ubiquitin moiety, thus creating a branching point). To make matters even more complicated, ubiquitin chains can also be built upon other Ubl modifications. For example, the RNF4 E3 ligase specifically extends the SUMO modification with a poly-ubiquitin chain [17]. To recognize and distinguish between different ubiquitin modifications, over 20 different families of ubiquitin-binding domains (UBDs) have evolved, which can be specific either to mono-ubiquitin or to one type of poly-ubiquitin chain [18–20].

Due to high proteomic penetrance, ubiquitylation affects most cellular pathways in some way. A major function of ubiquitylation is to target substrates for proteasome degradation. In addition (or in tandem), a plethora of different processes are regulated by ubiquitylation, including endocytosis, selective macro-autophagy, cell cycle control, inflammation and nuclear factor kappa-light-chain-enhancer of activated B cells (NF-κB) activation, deoxyribonucleic acid (DNA) repair, transcription, antigen processing, viral infection, and ribosome and peroxisome biogenesis

[16, 21, 22]. While our general understanding of the enzymatic mechanisms for ubiquitylation has advanced impressively thanks to detailed biochemical analyses, a more comprehensive view of the relationships among the different components of the ubiquitin system is still lacking. Therefore, there is an increasing need to integrate more systems-wide approaches, such as mass spectrometry-based proteomics, which would offer a broader understanding of this intricate system.

2 Proteomic Approaches for Studying Ubiquitylation

To date, systems-wide analyses of ubiquitylated proteins (hereafter referred to as the *ubiquitome)* are mainly based on three major types of proteomic approaches: mass spectrometry analysis, in vitro protein/peptide arrays, and the newly arising bioinformatics methods. More and more studies combine two or, indeed, all three of these methods. In this chapter, we will mainly focus on the study of the ubiquitome using mass spectrometry and give a brief overview of the other two approaches.

2.1 The Systems-Wide Study of the Ubiquitome by Mass Spectrometry

Most mass spectrometry-based protein analyses use a bottom up approach in which proteins are digested into peptides, which are then separated by liquid chromatography prior to their analysis by the mass spectrometer. Using tandem mass spectrometry (MS/MS), peptides are further fragmented into second order mass spectra, which act as "fingerprints" that are subsequently searched against a database to obtain a potential sequence match. The peptides can, in turn, be used to identify the proteins from which they are derived. Mass spectrometry can thus be used to identify the protein composition of the analyzed biological sample, and, in some cases, also to quantify differences between two or more samples [23].

Analysis of PTMs is well suited for mass spectrometry, as in most cases modified peptides can be characterized by a difference in mass (compared to their unmodified counterparts) that can be detected by the instrument. In the case of ubiquitylation, a characteristic amino acid "tail" is left on the lysine residue of the substrate peptide, which was first used to detect in vivo ubiquitylation sites by mass spectrometry in 1993 [24]. Different proteases have been used to generate peptides with remnant ubiquitin tails for mass spectrometry analysis [24]. However, due to its robust catalytic activity and cleavage site specificity, trypsin is now the most commonly used enzyme (among many other applications) for proteomic studies of ubiquitylation. Trypsin is a serine protease that cuts peptide chains on the carboxyl side of lysine and arginine. Because the carboxy-terminal end of ubiquitin has the amino acid sequence Arg-Gly-Gly, tryptic digestion of ubiquitylated peptides will generate a "tail" consisting of the last two glycine residues of ubiquitin. This

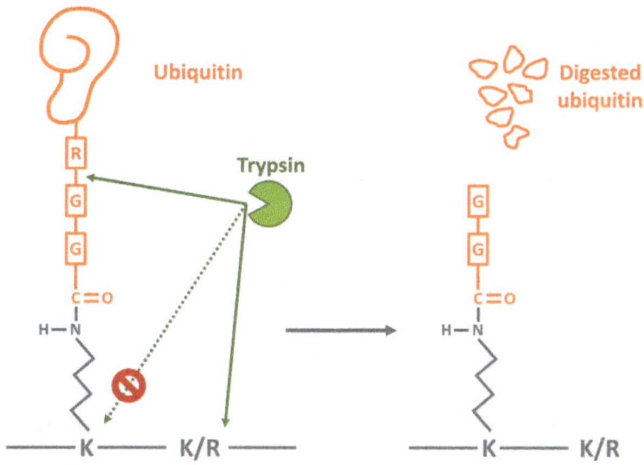

Fig. 14.1 The diGly signature. Trypsin digestion of ubiquitylated proteins leaves a distinctive di-glycine (*diGly*) signature attached to the modified lysine residues owing to the presence within ubiquitin of an arginine residue directly preceding the two carboxyl-terminal glycines. The diGly-containing peptide will also include a missed-cleavage site, since ubiquitylation blocks trypsin digestion next to the modified lysine

di-glycine "tail" (hereafter referred to as the remnant, or diGly tail) adds a signature mass shift of + 114 Da on the modified lysine residues. At the same time, trypsin is unable to cleave ubiquitylated lysines, and thus modified peptides also contain a missed-cleavage site (Fig. 14.1). One potential problem is that the same mass shift of 114 Da can also occur due to the alkylation of lysine residues during sample preparation when using iodoacetamide (commonly used to protect reduced cysteine residues prior to mass spectrometry) [25]. This is favored even more when using high concentrations of alkylating reagents and heat [26]. Therefore, sample alkylation is commonly performed at no more than room temperature, and iodoacetamide is substituted by chloroacetamide, which generally does not cross-react as readily.

By the early 2000s, most of the effector-enzymes (E1s, E2s, E3s, DUBs) in the ubiquitin system had been uncovered [27]. However, only a small fraction of their substrates were characterized. To fill in the gap—and taking advantage of major progresses in mass spectrometry instrumentation—systems-wide proteomic approaches were developed. Given the transient, and mostly low-abundant nature of ubiquitylation in the cell, conjugated proteins and their ubiquitylation sites are difficult to detect by mass spectrometry analysis without any pre-enrichment. Therefore, the depth and coverage of the ubiquitome is tightly dependent on the enrichment method. In this section, we will present the hallmark systems-wide approaches developed to study the ubiquitome, including the most current methods, with an emphasis on the enrichment techniques.

2.1.1 Enrichment of Ubiquitylated Proteins Using Tagged Ubiquitin

Taking advantage of yeast genetics, Steven Gygi and colleagues were among the first to intracellularly express tagged ubiquitin, enabling them to enrich for ubiquitin conjugates for large-scale analysis of the ubiquitome [28]. The authors used a yeast strain in which an ectopic 6xHis-myc-ubiquitin was expressed instead of wild-type ubiquitin to purify ubiquitin conjugates under denaturing conditions using nickel-affinity chromatography (Fig. 14.2a). In this landmark study, 1075 proteins were identified by mass spectrometry, including 110 ubiquitylation sites (i.e., diGly-containing peptides) in 72 proteins. Because a certain level of contaminating proteins still co-purified along with the ubiquitylated proteins, even under denaturing conditions, the identification of ubiquitylation sites using the characteristic remnant ubiquitin tail was key in this study. One concern which could be raised regarding this method—and several other approaches described below—is that it is based on the ectopic expression of a tagged ubiquitin, which could result in the disruption of ubiquitin levels and/or hindrance of its amino terminus, thereby potentially differentially influencing the ubiquitylation rates in distinct pathways or substrates (depending on which ubiquitylating enzymes are affected).

Analysis of ubiquitin conjugates is not constrained to single cell model organisms. In order to pull down ubiquitylated proteins from a specific tissue of a whole animal, Ugo Mayor and colleagues used the GAL4/UAS tissue-targeted expression system in *Drosophila melanogaster* [29]. In this study, ubiquitin tagged with the BirA recognition sequence was over-expressed solely in the nervous system, together with the *Escherichia colibirA* gene that encodes a biotin ligase, to biotinylate tagged ubiquitin. Consequently, 48 novel neuronal-specific ubiquitylation substrates were identified in this pioneering proteomic study conducted in a multicellular organism.

To further reduce the levels of non-specifically bound proteins (binding to the tag or to the affinity column) during the enrichment of tagged ubiquitin, which is notoriously problematic with the histidine-tag system, different approaches have subsequently been used. For instance, Peter Kaiser's group developed a tandem affinity tag approach by employing a two-step purification, using 6xhistine and biotin tags [30]. In the second step, the high affinity between biotin and streptavidin allows the enrichment of targets under very stringent conditions, such as 2% sodium dodecyl sulfate (SDS) and 8 M urea. Over 150 ubiquitylated proteins were identified under these conditions. This approach was then applied to mammalian tissue cultures to identify over 600 ubiquitylated conjugates in HeLa cells [31]. In our lab, we instead utilized an 8xhistidine tag that enabled us to use up to 0.5% SDS for washing in a single-step purification scheme (Fig. 14.2a) [32]. We performed several control experiments in which we mixed differentially labeled cells (using either ^{14}N or ^{15}N stable isotopes) that expressed (or did not express) the tagged ubiquitin, and we showed that, under these conditions, most purified proteins were, indeed, specifically conjugated to the tagged ubiquitin. We identified on average 200 ubiquitylated proteins within a 4-h mass spectrometry analysis. However, only a small number of ubiquitylated peptides containing the diGly remnant tail were identified in both of the above-mentioned studies (mostly derived from ubiquitin and only a few other proteins). It is not entirely clear why this is the case. One possibility is that the sample fractionation conditions used in these two studies were not optimal for peptides containing the ubiquitin remnant tail. By contrast, strong cation exchange was used by Stanley Fields and colleagues in order to further fractionate ubiquitylated peptides after nickel chromatography (also using a 8xhistidine tagged ubiquitin), as the amino

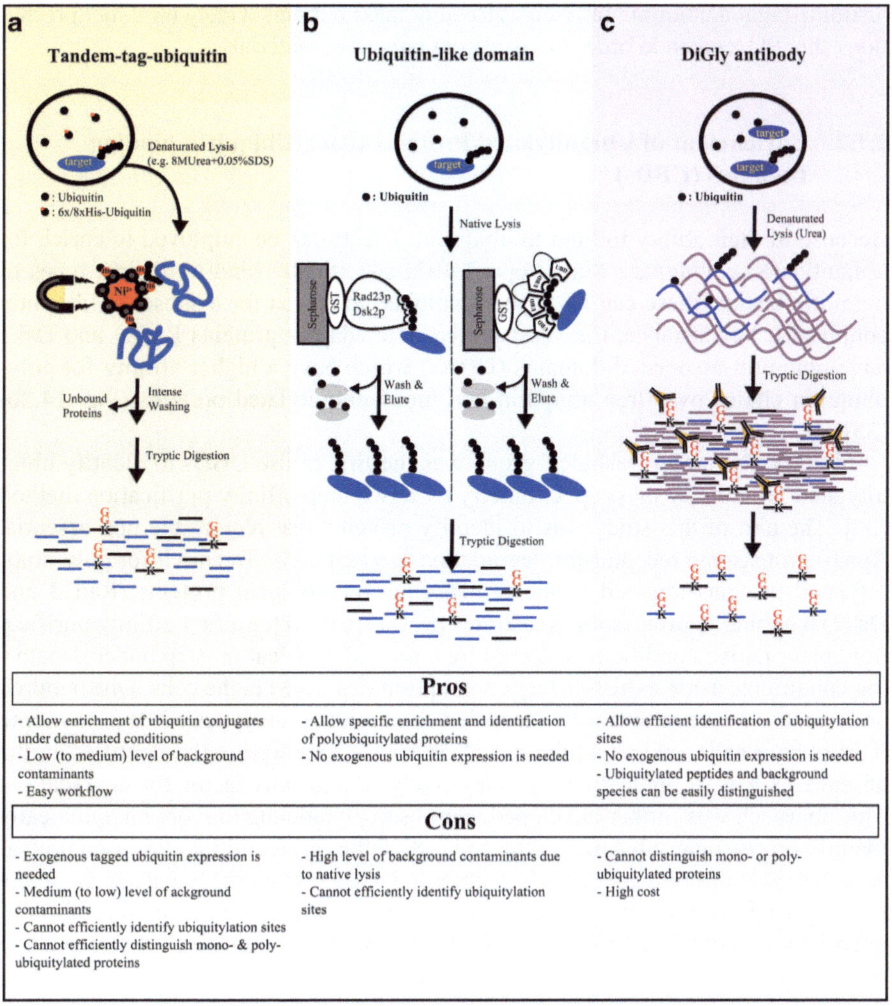

Fig. 14.2 Workflow diagram for three different basic enrichment methods. The main advantages and disadvantages of each method are shown below each diagram. **a** Tandem histidine tagged-ubiquitin-based enrichment of ubiquitylated proteins. **b** UBD-based (Ubiquitin-binding domain-based) enrichment of ubiquitylated proteins. **c** Anti-diGly antibody-based enrichment of ubiquitylated peptides

group of the remnant ubiquitin tail adds an additional positive charge at low pH in comparison to other peptides [33]. In their study, 870 ubiquitylation sites were identified among 438 proteins. The number of analyzed fractions may also influence the output. After tagging ubiquitin with a tandem tag (streptavidin and hemagglutinin, HA) and separating the purified proteins through gel electrophoresis, Danielsen and colleagues identified over 700 ubiquitylation sites after analyzing 20 gel fractions [34]. Overall, due to its relative simplicity, the purification of proteins conjugated to histidine-tagged ubiquitin (as well as to other tags) remains widely used, but precautions should be taken in order to avoid contaminating materials.

2.1.2 Enrichment of Ubiquitylated Proteins Using Ubiquitin-binding Domains (UBDs)

Because of their ability to bind to ubiquitin, UBDs can be employed to enrich for ubiquitylated conjugates. Since some UBDs specifically bind to distinct types of ubiquitin chains, these can be further exploited to enrich for a subset of ubiquitin conjugates. For instance, the yeast proteasome adaptor proteins Rad23 and Dsk2 have ubiquitin-associated domains (UBAs) which have a higher affinity for poly-ubiquitin chains over free ubiquitin and monoubiquitylated proteins (Fig. 14.2b) [35].

In 2005, Raymond Deshaies' group was the first to use UBDs to identify ubiquitylated proteins by mass spectrometry by a two-step affinity purification method [36]. The aim of this study was to identify proteins that require the non-essential Rpn10 proteasome receptor for degradation in yeast cells. To enrich for poly-ubiquitylated proteins targeted to the proteasome, recombinant proteins (Rad23 and Dsk2) that bind to proteasome substrates [37] were used for a first affinity purification under native conditions followed by a second purification step under denaturing conditions, using 6xhistine tagged ubiquitin expressed in the cells. One hundred and twenty-seven proteins were identified as candidate substrates of the proteasome in wild type cells, and an additional 50 or so proteins were only identified in the absence of Rpn10, indicating that they likely require this factor for degradation. This approach was further developed using isotope labeling to more unequivocally identify proteasome substrates in the cell [38]. Other proteomic studies used proteasome adaptor proteins to enrich for ubiquitylated conjugates. For instance, using this approach, Ron Kopito and colleagues found that the overexpression of a mutated fragment of huntingtin (which has been linked to Huntington's disease) led to the accumulation of several poly-ubiquitin chain types in tissue culture cells [39].

The group of Manuel Rodriguez developed tandem ubiquitin-binding entities (TUBEs) in order to enrich for ubiquitylated proteins [40]. Notably, this approach relies solely on endogenous ubiquitin to identify conjugated proteins. TUBEs were developed by fusing several UBAs together to increase affinity for proteins conjugated to poly-ubiquitin chains (Fig. 14.2b). TUBEs can also inhibit the activity of DUBs and of the proteasome, preserving ubiquitin chains during the native purification process. Using this approach, Rodriguez and colleagues identified a total of 643 proteins in two biological replicates from human breast adenocarcinoma cells treated with the DNA damage-inducing agent Adriamycin [41]. The ability to enrich for ubiquitin conjugates without ectopic expression of a tagged ubiquitin has great potential, especially for animal models.

One major issue is that many other proteins may interact with ubiquitylated proteins or the recombinant UBD-containing proteins under native conditions. While a high salt concentration was used (2 M NaCl) in the first study mentioned above, a second step purification was required to further enrich for ubiquitylated proteins [36]. It is actually challenging to directly circumvent this issue, since the ubiquitin-UBD interaction is mainly mediated by hydrophobic interactions [19], thus the usage of anionic detergents or chaotropic reagents would be detrimental. Therefore, UBDs may be better suited in the future to analyze subsets of the ubiquitome (taking advantage of the specificity of the recombinant UBD for a particular chain type) in combination with a second purification step.

2.1.3 Enrichment of Ubiquitylated Peptides Using α-diGly Antibodies

A major breakthrough was the introduction of antibodies that directly bind to ubiquitylated peptides. After the first 110 ubiquitylation sites identified by Gygi and colleagues [28], the uncovering of new ubiquitylation sites had largely stalled. One issue was that in that particular analysis in-depth study required 5 days of mass spectrometer instrument time, which is difficult to secure with equipment that is typically oversubscribed. By contrast, identification of phosphorylation sites benefited from tremendous activity starting in 2006, after TiO_2 beads began to be widely used to enrich for phosphorylated peptides [42–44]. Fortunately, the ubiquitin field was not idle, and monoclonal antibodies were soon developed to specifically enrich for ubiquitylated peptides.

In the antibody-based approach, ubiquitylated peptides of low abundance are greatly enriched prior to identification by mass spectrometry using antibodies that recognize the ubiquitin remnant tail left on trypsin-cleaved peptides. The laboratory of Samie Jaffrey was the first to publish an antibody-based approach to enrich for diGly peptides in 2010 (Fig. 14.2c) [45]. In this study, 374 ubiquitylation sites on 236 proteins were identified from HEK293 cells. To generate the antibodies, the authors synthesized a lysine-rich protein antigen (histone) containing multiple K-ε-GG that was then injected into mice. In the following year, both Steven Gygi's and Chuna Ram Choudhary's groups successfully and independently conducted notable large scale studies: around 19,000 ubiquitylation sites in ~5000 human proteins and ~11,000 sites in ~4200 proteins were mapped, respectively [46, 47]. Several additional studies followed up using the same approach, including analyses of the ubiquitome in rat brain and murine tissues [48, 49]. In addition, several modifications have been made in order to increase the yield of this method [50]. One reason the diGly-antibody approach is so potent is that trypsin digestion essentially abolishes most other protein-protein interactions and, combined with the high-affinity of the antibody, modified peptides are effectively enriched, despite their low abundance in the cell. Other major advantages are that no extra experimental controls are required to distinguish ubiquitylated sites from the rest of the identified peptides in the sample (the detection of the +114 Da mark is sufficient), it does not rely on ectopic expression of a tagged ubiquitin, it is applicable to all eukaryotic organisms or tissues and it is commercially available in the form of a kit.

While the antibody-based approach is very potent and already widely used, it also has a few shortcomings. Since the NEDD8 and ISG15 Ubls share the same carboxyl-terminal sequence (RGG) with ubiquitin, proteins conjugated to these Ubls will also generate indistinguishable remnant tails after tryptic digestion. Therefore, some ubiquitylated peptides may be mis-assigned. Fortunately, levels of ISG15 are usually undetectable in cells, unless stimulated by interferon (IFN)-α/β [51]. Furthermore, Gygi's group determined that more than 95 % of the sites which they identified were conjugated to ubiquitin and not NEDD8 (which is mainly conjugated to cullins). The monoclonal antibodies against the ubiquitin remnant may also introduce a bias for some sites, since Choudhary's group found that these antibodies have a slight sequence preference [49]. Our lab also noted that, using this approach, proteins ubiquitylated at multiple sites may be less prevalent compared to proteins conjugated at a single lysine, as peptide ion intensities would be lower (and thereby possibly not detected by the mass spectrometer) in the former case [52]. Another consideration is that the information regarding the chain linkage on a particular conjugation site (for poly-ubiquitylation) is lost when using this approach. In addition, atypical sites (N-terminal, for instance) are also not selected. Nevertheless, the antibody-based approach has been the key to the recent advancements and has now been adopted by many groups around the world.

2.1.4 An Alternative to the Antibody-Based Approach

The price of commercially available antibodies can be an obstacle, prompting researchers to look for more cost-accessible alternatives. One approach which has shown promise in preliminary trials makes use of the fact that diGly-containing peptides have two N-terminal primary amines (one from the peptide N-terminus and the second from the diGly remnant) that can be modified and exploited further for enrichment. Primary amines can be fluorinated using perfluorinated compounds, while the ε-amino groups of lysines are protected by guanidination [53, 54]. The doubly labeled (and thus containing a diGly signature) peptides have a significantly higher affinity than the singly labeled ones for a matrix which retains fluorinated compounds, thus allowing the separation of the two species by eluting with different concentrations of organic solvent [55]. Fluorous affinity tag enrichment was successfully used to isolate diGly containing peptides from a tryptic digest of pure poly-ubiquitin chains [56]. Later studies have employed reversed-phase chromatography to enrich peptides with large fluorinated moieties attached to cysteine residues [57, 58], but this approach has not yet been used for diGly enrichment. The challenge is now to apply this method to the isolation of diGly-containing peptides from complex samples of biological origin.

2.1.5 Systems-Wide Analysis of Ubiquitin Linkages Using Selective Reaction Monitoring (SRM)

In addition to identifying substrate ubiquitylation sites, it is also important to determine the modification type (i.e., mono- vs. poly-ubiquitylation). One particularly effective approach utilizes selective reaction monitoring (SRM) to determine, for instance, which chain linkages are present in a cell extract, or synthesized by a given E3 ligase in an in vitro experiment. In SRM, the mass spectrometry instrument is set up to specifically monitor one or several preselected peptide(s), greatly improving both the sensitivity and identification rate from complex biological samples [59–63]. In combination with SRM, artificial peptides (of known concentration) can be added in for absolute quantification (AQUA) [64]. The synthetic peptides are identical to the endogenous peptides of interest, but carry a specific isotopic mark (typically "heavy" lysine or arginine residues, giving a +8 or +10 Da mark, respectively). The AQUA peptides co-elute with their counterparts within the sample, and their intensities can be compared to calculate the concentrations (or the absolute quantification) of the peptides of interest. To quantify chain type, typically three peptides are required for each of the seven lysine residues of ubiquitin: one longer peptide, which includes the ubiquitylated lysine with a remnant diGly tail, and two peptides corresponding to the N- and C-terminal parts of the former, on either side of the (in this case non-ubiquitylated) lysine. In this way, a quantification of the relative amounts of modified and unmodified peptide is possible (Fig. 14.3).

The AQUA approach was pioneered in the workgroup of Steven Gygi [65] and has since then been successfully used to quantify ubiquitin chain linkages under a variety of conditions and model systems. For instance, it was used to determine which ubiquitin chains were conjugated in vitro onto the cell cycle-regulated protein cyclin B1 [66]. The method was used in numerous studies to quantify in vivo ubiquitin chain linkages in yeast cells [67], in cultured mammalian cells [68–70] and in mice [71]. Notably, AQUA and related methods have helped to explore the role of ubiquitin chain linkages whose biological role were not well understood (see below in Sect. 3.2). These methods have been adopted and further developed by numerous groups. However, implementation of this approach on each proteomic platform and instrument still required dedicated researchers to perform those types of experiments.

Kopito's group developed an interesting, closely related method that employed spiked-in proteins for Protein Standard Absolute Quantification (PSAQ) [70]. Compared to AQUA, this method comes with a twist, namely that the spiked "standards" are differentially labeled ubiquitins instead of peptides. This allows the monitoring of a different pool of proteins within a sample, namely free ubiquitin, monoubiquitylated and poly-ubiquitylated proteins. Using this approach, the authors were able to show that the majority of conjugated proteins in the cell are, surprisingly, monoubiquitylated and not poly-ubiquitylated.

2.1.6 Systems-Wide Analysis of the Ubiquitome Using Approaches that are Not Based on Mass Spectrometry

Protein array technology can also be used to perform proteomic analysis of the ubiquitin system. One major task is to delineate the substrate specificity of the different enzymes involved in ubiquitylation. This is often difficult to tackle in vivo, because

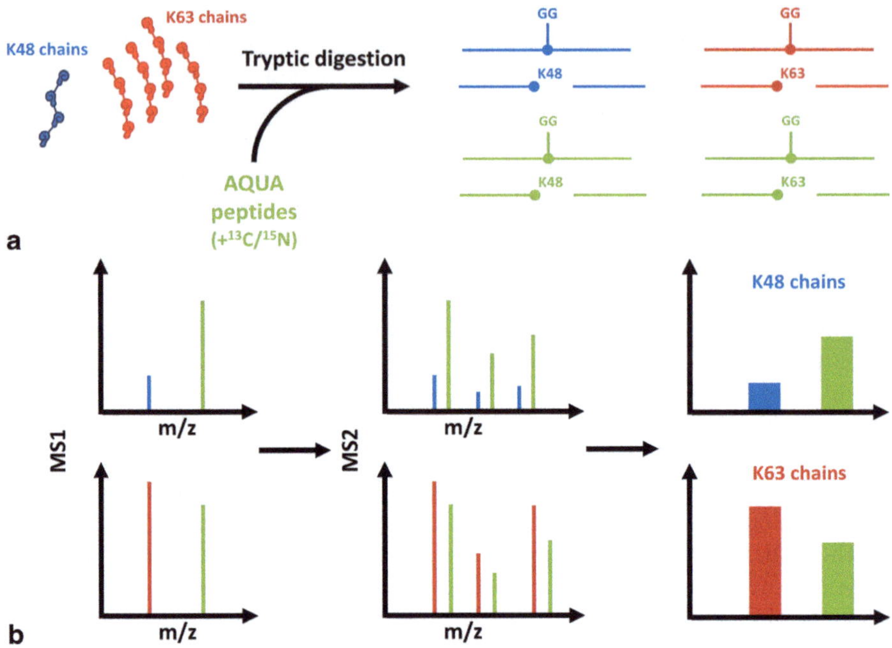

Fig. 14.3 The absolute quantification (AQUA) peptide-SRM (selective reaction monitoring) approach to the quantification of ubiquitin chain linkages. **a** A mixture of different ubiquitin chains obtained from a biological sample is digested along with known amounts of absolute quantification (*AQUA*) peptides. The AQUA peptides are isotopically labeled ($^{13}C/^{15}N$) to distinguish them from their endogenous counterparts. Three peptides are monitored for each linkage type: the diGly-containing peptide and the N- and C-terminal parts of the former that occur by tryptic digestion of an non-ubiquitylated parent peptide. **b** Tandem mass spectrometry through which only specific charge-to-mass (m/z) ratios are monitored—termed selected reaction monitoring (SRM)—is used in order to obtain a ratio between the endogenous peptides and the artificial standards. By using AQUA peptides that cover both the modified (diGly containing) and the unmodified states, an absolute quantification of the amount of a particular linkage, as well as the comparison between different linkage types, is possible

of the diversity of the ubiquitin system and the transient nature of ubiquitylation (due to de-ubiquitylation). The recent development of protein arrays containing purified recombinant proteins from a variety of organisms, ranging from yeast to human, allows researchers to take an in vitro approach in a systems-wide manner. So far, this strategy has been applied to the study of substrates of ligases such as Rsp5 in yeast [72, 73], Nedd4 − 1 and Nedd4 − 2 and the anaphase-promoting complex (APC) in humans [74, 75], and a panel of human DUBs [76]. Its cell-free nature endows the in vitro protein microarray analysis with a number of advantages, but it also has the weakness of decoupling the ubiquitylation reactions from a cellular context (e.g., localization, physiological concentration). In a recent study, Marc Kirschner and colleagues used protein arrays incubated with a cell extract to identify which proteins are conjugated to ubiquitin and other probed Ubls (e.g., SUMO, NEDD8, and ISG15) [77]. Surprisingly, almost entirely distinct subsets of proteins were targeted by each conjugation system. Apart from proteins, peptide arrays have also been used to identify substrate candidates, as many E3s rely on distinct domains to recruit their targets, like the ligand of Numb protein-X (LNX) E3 ligase that contains a PDZ domain [78]. Additional genome-wide approaches have also been developed that do not rely on protein arrays. For instance, Stephen Elledge and colleagues have used a green fluorescent protein (GFP)-fusion library in tissue culture cells to identify substrates of cullin-based E3 enzymes, which form a major ligase family [79].

Another approach to deciphering the ubiquitome is based on bioinformatic site prediction. The first algorithm, called UbiPred, was developed by Ho's group in 2008. It used 531 physicochemical properties and a limited training set of 157 ubiquitylation sites and 3676 putative non-ubiquitylation sites from 105 proteins for building the prediction algorithm [80]. Recent advances in experimental investigation have helped to develop several additional ubiquitylation site prediction programs, such as UbPred, weighted passive nearest neighbor algorithm (WPNNA), and composition of k-spaced amino acid pairs (CKSAAP)_UbSite, Ubiprober [81–84]. These prediction algorithms have an overall reasonable accuracy. However, due to the fact that there is no universal target sequence for ubiquitylation sites, their usage may be limited. It may be more appropriate to use these approaches to identify candidate substrates for a given ubiquitin ligase. For instance, several algorithms can be used to predict substrates of the APC E3 ligase based on the presence of specific motifs [85, 86]. In one case, a combination of both mass spectrometry and bioinformatics analyses was used to narrow down a list of candidate substrates for the Fbw7 cullin-based E3 ligase [87]. More general computational approaches have increasingly been used to probe the ubiquitome in the past few years, as researchers began to obtain more and more data. We will highlight several of these studies in the next section.

3 Contributions of Systems-Wide Proteomic Approaches to Gaining Insights into the Ubiquitin System

The diverse systems-wide approaches which we have described so far have led to a significantly better understanding of the ubiquitin system, often by providing unique information that could not have been obtained by other means. We will review several key contributions in the last section of this chapter.

3.1 Ubiquitylation Sites

As mentioned earlier, there is no specific motif that directs ubiquitylation. However, several residues are more frequently present (or absent) at certain positions near to the ubiquitylation sites. The absence of a universal ubiquitylation motif can be explained by the fact that there is a large variety of different E3 ligases, each of which are likely to recognize their target in a distinctive way. In addition, E3 ligases do not directly bind to the ubiquitylation sites, but typically to motifs that are adjacent to the conjugated lysine residues. Because the lysine residue directs the nucleophilic attack that leads to its conjugation, nearby residues may chemically influence the reaction. Indeed, Kim and colleagues found that several acidic residues were enriched while basic residues were instead depleted in the vicinity of ubiquitylation sites [46]. Cysteine residues were also less prevalent, since they would likely compete as ubiquitin acceptor residues. Ubiquitylation sites are also slightly more conserved than unmodified sites and, interestingly, often appeared earlier in evolution (i.e., prior to the adoption of ubiquitylation) [88]. The latter observation indicates that the ubiquitin system likely accommodates preexisting sites rather than triggering the appearance of new sites. Intriguingly, ubiquitylation sites were also often found in ordered regions in contrast to phosphorylation sites that are more prevalent in regions predicted to be disordered (i.e., without a specific secondary structure) [89]. It will be important to reexamine some of these observations in the near future as more data is generated.

3.2 Ubiquitin Chains and the Ubiquitin Code

Ubiquitylation supports numerous cellular responses or functions, and the formation of different chain types has long been thought to play a major role in directing conjugates to their final intracellular fates. The targeting of short-lived proteins for degradation is a major function of the ubiquitin system [90], and the K48-linked poly-ubiquitin chain was the first chain type associated with the turnover of these proteins [91]. In contrast, K63-linked ubiquitin chains have distinct cellular functions, like DNA repair and endocytosis [92, 93]. These first landmark studies were key to laying down the now well-accepted hypothesis that chain linkage types determine, at least partially, the fate of the substrates. However, major questions remain unanswered: *What is the function of the other chain types (apart from K48 and K63)?* and *How do chain linkage amounts fluctuate in the cell, especially in response to stress?* As these questions are difficult to investigate solely by using conventional approaches (e.g., ubiquitin mutants that cannot form certain chain types), the development of new proteomic approaches will potentially open up new horizons.

Several chain types participate in proteasome degradation. Gygi's group demonstrated for the very first time that all seven lysines in ubiquitin contribute to the assembly of poly-ubiquitin chains in vivo [28]. Subsequently, it was also found that linear ubiquitin chains (linked via the ubiquitin amino terminus; M1) were present in mammalian cells [94]. Using a SRM-based method, Peng and co-workers first determined the presence of each chain linkage type in yeast cells and found that K11 chains were almost as prevalent as K48 chains [67]. To assess the role of each chain type in proteasomal degradation, the authors then inhibited the proteasome and found out that all chain types rapidly accumulated in the cell, except for K63 chains, and K27 and K33 were affected to a lesser extent. To confirm the role of K11 in proteolysis, they also showed that K11-linked chains were important for the endoplasmic reticulum-associated degradation (ERAD) pathway, which targets misfolded endoplasmic reticulum (ER) proteins [67]. K11 linkages were also found to be important for the proteasomal degradation of cell cycle-regulated proteins in an independent study [95]. However, this linkage is less prevalent in mammalian tissue culture cells [69]. Indeed, in higher eukaryotes, K48 linkages make up about 80 % of the poly-ubiquitin chains targeted to the proteasome, while K29, K11 and possibly K6 were found (in this order) in decreasing concentrations [69]. Nonetheless, K11 chains further accumulate in brain tissues from patients suffering from Alzheimer's disease (as do K48 and K63 chains, albeit to a lesser extent), thus confirming that these chain types play a prevalent role in the cell. Furthermore, the function of K11 linkages may not be limited to proteasomal degradation. For instance, the major histocompatibility complex class I (MHC I) was found conjugated to both K11- and K63-linked ubiquitin chains that are involved in the endocytosis of MHC I [96, 97]. In another study, the presence of K33-linked chains was confirmed by mass spectrometry on T cell receptor-zeta, another membrane protein that is regulated by endocytosis [98]. This is intriguing, because endocytosis was originally thought to be mediated by mono- and K63-linked poly-ubiquitylation. While these studies greatly expanded our knowledge, one question that remains is whether more complicated chain architectures (i.e., branched chains) are also present in the cell. New techniques will have to be developed to answer this question, as this information is lost with current sample preparation methods.

3.3 Protein Quality Control and the Ubiquitin Proteasome System

A major function of the ubiquitin proteasome system is to eliminate misfolded and damaged proteins from the cell with the assistance of several protein quality control pathways [99]. One striking observation is that the majority of proteins targeted for degradation are newly synthesized [46]. It has already been shown that a large fraction (up to 30 %) of newly translated proteins in tissue culture cells is rapidly degraded by the proteasome, presumably because these proteins were not properly folded [100]. A recent study from Jon Huibregtse's group confirmed that about 12 % of nascent polypeptides are ubiquitylated [101]. The ubiquitylation of these nascent polypeptides occurs on both stalled and actively translating ribosome complexes. To further characterize these conjugates, an elegant approach was used, in which tandem affinity purification of ubiquitylated nascent polypepides was achieved from purified polysomes by using both flag-tagged ubiquitin and biotinylated puromycin, which forms a covalent bond with the carboxyl terminus of nascent polypeptides prior to releasing them from ribosomes. Using mass spectrometry, the authors found that the majority of these ubiquitylated nascent polypeptides corresponded to cytosolic proteins [101].

The impairment of the ubiquitin proteasome system has been associated to many aggregation-related disorders, especially neurodegenerative diseases [102, 103], which typically feature ubiquitin-containing aggregates [104]. We found that inhibition of the proteasome can cause the formation of large ubiquitin-containing aggresomes (which are amorphous structures mainly composed of misfolded proteins) even in the absence of disease specific proteins (e.g., mutated cystic fibrosis transmembrane conductance regulator, CFTR, involved in cystic fibrosis) [105]. To further characterize these ubiquitin-containing protein aggregates induced by proteasome inhibition, we combined a biochemical isolation method with quantitative proteomic analysis using stable isotope labeling with amino acids in cell culture (SILAC) [106]. We identified more than 500 proteins in these aggregates, including the p62/sequestosome, several E3s and DUBs, and chaperones. One possibility is that, in addition to those involved in the aggregation process, many of these proteins are aggregation-prone and would normally be efficiently eliminated by the proteasome.

The different protein quality control pathways form an intricate network and keep the proteome in check. While many studies at the time relied on the assessment of a few model substrates, we instead focused on the effect of the heat-stress response on the ubiquitinome in a systems-wide manner. Heat shock that causes protein misfolding has long been known to cause an increase in poly-ubiquitylation in the cell [107]. However, we only recently found that proteins ubiquitylated after heat shock were cytosolic and in part targeted by the Hul5 ubiquitin ligase [32]. By using bioinformatics tools to further analyze our proteomic datasets, we found that several features were shared by the heat-induced ubiquitylated proteins [52]. Notably, these proteins are longer on average and contain fewer hydrophobic residues. Interestingly, intrinsically disordered proteins (that contain large regions predicted not to fold in a specific secondary structure) are also prominently identified among proteins ubiquitylated upon heat shock [52]. We also found that these proteins while generally not essential and less abundant in the cell, contain more predicted interaction motifs in their disordered regions. Presumably, these proteins are ubiquitylated upon heat shock due to the loss of interactions with their binding partners. It is not clear whether this is triggered by a specific protein quality control pathway, or whether these disordered proteins may themselves also be important for the heat-shock response. In addition to heat-shock response, the study of the ubiquitome may be particularly interesting when assessing client proteins of a given chaperone protein. In absence of the chaperone's activity, its client proteins should misfold and at least a portion of them will be ubiquitylated for proteasome targeting.

3.4 The Quest to Find Ubiquitin Ligase Substrates

One of the major challenges to the study of the ubiquitin proteasome system is to assign substrates to ubiquitin ligases. Several proteomic approaches have been developed in which individual E3s—or components of E3s—are pulled down to identify their substrates, a method that was notably championed by the group of Michele Pagano [108–114]. In each case, follow-up studies are required to verify that the interacting proteins are indeed E3 targets. To further enrich for substrates, inactive E3 ligases were also employed [115]. In another case, the E3 ligase was re-engineered to modify its specificity and conjugate NEDD8 instead of ubiquitin [116]. Alternatively, the substrates of a specific E3s could be identified by analyzing which proteins are no longer ubiquitylated upon the inhibition or deletion of the investigated ligase. For instance, Kim and colleagues monitored changes in the ubiquitome to identify substrates of cullin-RING ubiquitin ligases by comparing cells with and without treatment with the MLN4924 inhibitor that blocks NEDD8 conjugation (neddylation is required to maintain cullin-RING ligases in an active state) [46]. They consequently identified over 4000 ubiquitylation sites that were dependent on cullin-RING ligase activity.

However, cullin-RING ligases form the largest family of E3s. Therefore, the identified proteins may be targeted by any of the 350 or so E3s. Our lab instead focused on identifying candidate substrates of a single ubiquitin ligase, the yeast HECT ligase Hul5. By comparing wild-type and deletion mutant yeast cells, we discovered over 90 putatively misfolded substrates of Hul5, and thereby revealed that this ubiquitin ligase plays an important role in cytosolic protein quality control [32].

More recently, the Parkin-dependent ubiquitome was analyzed under different stress conditions [117]. The Parkin ubiquitin ligase has been linked to Parkinson's disease (PD) and several Parkin mutations are associated to a familial and early-onset form of PD. In this study, ubiquitylated peptides were enriched using the anti-diGly antibodies prior to quantitative mass spectrometric analysis to compare cells that either did or did not express Parkin under mitochondrial depolarization conditions. The study revealed that Parkin expression dramatically altered the ubiquitylation status of the mitochondrial proteome [117], consistent with its role in mitophagy [118]. These studies illustrate the great potential of analyses of the ubiquitome in linking components of the ubiquitin machinery to their functional roles and biological substrates.

3.5 Chasing Lost Ubiquitylation Events and DUB Functions

DUBs that cleave off the C-terminal carboxyl group of ubiquitin from the ε-amino group of lysine side chains or the α-amino group of conjugated proteins are essential to the ubiquitin system [15]. There are over 80 genes in the human genome predicted to have de-ubiquitylating activity. Because of their role in processing ubiquitin precursors and recycling ubiquitin, DUBs play a key role in maintaining ubiquitin homeostasis. In addition, many DUBs display specificity both for particular ubiquitin chain types, as well as to certain substrates. Altered DUB activity has been linked to several diseases, such as the Machado-Josephdisease (MJD), microcephaly-capillary malformation (MIC-CAP) syndrome, and coronary artery disease(CAD), as well as numerous forms of cancer [119–121]. Therefore, DUBs are considered to have great therapeutic potential as drug targets [122] and the identification of the substrate repertoires and functions of DUBs is urgently needed.

Mass spectrometry has been applied to the study of DUBs in a systems-wide manner. For instance, Wade Harper and colleagues expressed and pulled down 75 human DUBs to identify their interacting partners by mass spectrometry [123]. A similar study was performed with the 20 DUBs of *Saccharomyces pombe* [124]. More recently, Gardner's group used in vivo cross-linking prior to co-immunoprecipitation of the yeast DUB Ubp10 in order identify transiently interacting proteins, including ribonucleic acid (RNA) polymerase I, which was later confirmed to be a substrate of Ubp10 [125]. Alternatively, the change in proteome composition was analyzed by quantitative mass spectrometry to identify potential DUB substrates accumulating upon deletion of each of the 20 yeast DUBs [126].

Another approach consists in trapping DUBs using so-called suicide substrates, in which a thiol-reactive group is placed at the C-terminus of ubiquitin to ligate it irreversibly to the DUB's active site cysteine. Kessler and colleagues used this method to identify DUBs and their interacting proteins by mass spectrometry from mouse tissue [127]. This method, together with quantitative mass spectrometry, was also used to characterize chemical DUB inhibitors to gauge their selectivity, as inhibition of a specific DUB results in less binding to the activity-based probe [128]. One of the tested compounds, P22077, was selective to Ubiquitin-specific-processing protease 7 (USP7), thereby causing a lower rate of association of USP7 to the suicide substrate. Using TUBEs and mass spectrometry, Kessler and colleagues then identified potential substrates of USP7 upon its inhibition [128]. However, researchers have not yet taken full advantage of newly developed proteomic approaches to study DUBs and deubiquitylation in vivo. In the near future, combining SRM and diGly approaches to examine ubiquitin chain linkages and ubiquitylated proteins upon the inhibition or down-regulation of a specific DUB (or, conversely, its over-expression) will help greatly with building a comprehensive knowledge base of DUBs.

3.6 Cross-Talk Between Ubiquitylation and Other PTMs

Multiple PTMs may occur on the same protein and the cross-talk between different functional modulators may be important for protein activity, function, and localization. Histone tails are a well-known example: Being particularly lysine-rich amino acid sequences, they are subjected to a variety of modifications, such as acetylation, methylation and ubiquitylation (on lysine), methylation of arginine, and phosphorylation of serine and tyrosine residues. Different combinations of these PTMs generate the "epigenetic code" and have distinct functions in regulating chromatin organization and accessibility. Thus, the modifications are set in a highly controlled manner [129]. Similarly, the selectivity of the ubiquitin ligases for their substrates is also tightly regulated. In many cases, substrates first need to be modified to trigger their recognition by an E3 ligase and phosphorylation plays a prominent role in this process.

The group of Steven Carr recently established a method (called serial enrichments of different posttranslational modifications, SEPTM) for the serial enrichment of PTMs, such as phosphorylation, ubiquitylation, and acetylation, from the same sample [130]. Phosphorylated peptides were first isolated using the IMAC phosphor-enrichment method followed by the diGly antibody-based method for ubiquitylation enrichment. The same sample was then subjected to the enrichment of acetylated peptides, also using an antibody-based approach. SEPTM enabled the identification of more than 20,000 phosphorylation, 15,000 ubiquitylation, and 3000 acetylation sites from around 8000 proteins and provides a unique opportunity to study the cross-talk between these three different PTMs. Using quantitative mass spectrometry, the authors analyzed changes incurred by proteasome inhibition and showed that the number of phosphorylation sites, as well as the number of ubiquitylation sites, were highly increased. These results confirmed that phosphorylation plays a major role in regulating proteolysis. Similar data were obtained in an independent study from Villen's group [131]. Here the authors first enriched ubiquitylated proteins using His-tagged ubiquitin followed by the isolation of diGly-containing peptides and phosphorylated peptides. Using this approach, they identified a total of 466 proteins that were ubiquitylated and phosphorylated. While the non-ubiquitylated forms of these proteins were also found to be phosphorylated, proteins that are both ubiquitylated and phosphorylated, and accumulate upon proteasome degradation, have, on average, shorter half-lives, indicating that phosphorylation is likely involved in the regulation of their turnover. By contrast, the acetylome analyzed in the former study was not significantly affected by proteasome inhibition. However, around 400 of the acetylation sites were also found to be ubiquitylated, indicating that cells contain distinctly modified populations of the same proteins. Therefore, cross-talk can be cooperative (multiple PTMs on the same molecule) or competitive (modifying the same site). It will be interesting to further integrate additional PTMs to identify which other modifications may impact ubiquitylation positively or negatively.

4 Concluding Remarks

In summary, systems-wide analyses of the ubiquitome have uniquely contributed to and shaped our understanding of the ubiquitin system. Within a few years, we gained tremendous insights into the ubiquitome. A broad panel of different methods is now available to the scientific community. The further integration of these methods will undoubtedly play an important role in deciphering disease mechanisms linked to ubiquitylation, and potentially make great contributions towards future therapeutic development.

Acknowledgments We thank Sophie Comyn and Gerard Chan, and all other members of the Mayor lab for discussion and suggestions. This work is supported by grants from the Natural Sciences and Engineering Research Council of Canada (NSERC) and the Canadian Institutes of Health Research (CIHR). TM is supported by New Investigator Career Awards from CIHR and the Michael Smith Foundation for Health Research.

References

1. Hornbeck PV, Kornhauser JM, Tkachev S et al (2012) PhosphoSitePlus: a comprehensive resource for investigating the structure and function of experimentally determined post-translational modifications in man and mouse. Nucleic Acids Res 40(Database issue):D261–D270
2. Goldstein G, Scheid M, Hammerling U et al (1975) Isolation of a polypeptide that has lymphocyte-differentiating properties and is probably represented universally in living cells. Proc Natl Acad Sci U S A 72(1):11–15
3. Pickart CM (2001) Mechanisms underlying ubiquitination. Annu Rev Biochem 70:503–533
4. Schulman BA, Harper JW (2009) Ubiquitin-like protein activation by E1 enzymes: the apex for downstream signalling pathways. Nat Rev Mol Cell Biol 10(5):319–331
5. van Wijk SJ, Timmers HT (2010) The family of ubiquitin-conjugating enzymes (E2s): deciding between life and death of proteins. FASEB J 24(4):981–993
6. Ye Y, Rape M (2009) Building ubiquitin chains: E2 enzymes at work. Nat Rev Mol Cell Biol 10(11):755–764
7. Metzger MB, Hristova VA, Weissman AM (2012) HECT and RING finger families of E3 ubiquitin ligases at a glance. J Cell Sci 125(Pt 3):531–537
8. Breitschopf K, Bengal E, Ziv T et al (1998) A novel site for ubiquitination: the N-terminal residue, and not internal lysines of MyoD, is essential for conjugation and degradation of the protein. EMBO J 17(20):5964–5973
9. Bloom J, Amador V, Bartolini F et al (2003) Proteasome-mediated degradation of p21 via N-terminal ubiquitinylation. Cell 115(1):71–82
10. Scaglione KM, Basrur V, Ashraf NS et al (2013) The ubiquitin-conjugating enzyme (E2) Ube2w ubiquitinates the N terminus of substrates. J Biol Chem 288(26):18784–18788
11. Cadwell K, Coscoy L (2005) Ubiquitination on nonlysine residues by a viral E3 ubiquitin ligase. Science 309(5731):127–130
12. Wang X, Herr RA, Chua WJ et al (2007) Ubiquitination of serine, threonine, or lysine residues on the cytoplasmic tail can induce ERAD of MHC-I by viral E3 ligase mK3. J Cell Biol 177(4):613–624
13. Shimizu Y, Okuda-Shimizu Y, Hendershot LM (2010) Ubiquitylation of an ERAD substrate occurs on multiple types of amino acids. Mol Cell 40(6):917–926
14. Deshaies RJ, Joazeiro CA (2009) RING domain E3 ubiquitin ligases. Annu Rev Biochem 78:399–434
15. Komander D, Clague MJ, Urbe S (2009) Breaking the chains: structure and function of the deubiquitinases. Nat Rev Mol Cell Biol 10(8):550–563
16. Komander D, Rape M (2012) The ubiquitin code. Annu Rev Biochem 81:203–229
17. Tatham MH, Geoffroy MC, Shen L et al. (2008) RNF4 is a poly-SUMO-specific E3 ubiquitin ligase required for arsenic-induced PML degradation. Nat Cell Biol 10(5):538–546
18. Dikic I, Wakatsuki S, Walters KJ (2009) Ubiquitin-binding domains—from structures to functions. Nat Rev Mol Cell Biol 10(10):659–671
19. Winget JM, Mayor T (2010) The diversity of ubiquitin recognition: hot spots and varied specificity. Mol Cell 38(5):627–635
20. Husnjak K, Dikic I (2012) Ubiquitin-binding proteins: decoders of ubiquitin-mediated cellular functions. Annu Rev Biochem 81:291–322
21. Mukhopadhyay D, Riezman H (2007) Proteasome-independent functions of ubiquitin in endocytosis and signaling. Science 315(5809):201–205
22. Grabbe C, Husnjak K, Dikic I (2011) The spatial and temporal organization of ubiquitin networks. Nat Rev Mol Cell Biol 12(5):295–307
23. Armirotti A, Damonte G (2010) Achievements and perspectives of top-down proteomics. Proteomics 10(20):3566–3576
24. Morishima-Kawashima M, Hasegawa M, Takio K et al (1993) Ubiquitin is conjugated with amino-terminally processed tau in paired helical filaments. Neuron 10(6):1151–1160
25. Nielsen ML, Vermeulen M, Bonaldi T et al (2008) Iodoacetamide-induced artifact mimics ubiquitination in mass spectrometry. Nat Methods 5(6):459–460

26. Kaiser P, Mayor T (2011) Gold for ubiquitin in Vancouver: first conference on proteomics of protein degradation and ubiquitin pathways held June 6–8, 2010 in Vancouver, University of British Columbia, organized by Lan Huang, Thibault Mayor, and Peipei Ping. Mol Cell Proteomics 10(5):R110.003863
27. Semple CA, RIKEN GER Group, GSL Members (2003) The comparative proteomics of ubiquitination in mouse. Genome Res 13(6B):1389–1394
28. Peng J, Schwartz D, Elias JE et al (2003) A proteomics approach to understanding protein ubiquitination. Nat Biotechnol 21(8):921–926
29. Franco M, Seyfried NT, Brand AH et al (2011) A novel strategy to isolate ubiquitin conjugates reveals wide role for ubiquitination during neural development. Mol Cell Proteomics 10(5):M110.002188
30. Tagwerker C, Flick K, Cui M et al (2006) A tandem affinity tag for two-step purification under fully denaturing conditions: application in ubiquitin profiling and protein complex identification combined with in vivocross-linking. Mol Cell Proteomics 5(4):737–748
31. Meierhofer D, Wang X, Huang L et al (2008) Quantitative analysis of global ubiquitination in HeLa cells by mass spectrometry. J Proteome Res 7(10):4566–4576
32. Fang NN, Ng AH, Measday V et al (2011) Hul5 HECT ubiquitin ligase plays a major role in the ubiquitylation and turnover of cytosolic misfolded proteins. Nat Cell Biol 13(11):1344–1352
33. Starita LM, Lo RS, Eng JK et al (2012) Sites of ubiquitin attachment in *Saccharomyces cerevisiae*. Proteomics 12(2):236–240
34. Danielsen JM, Sylvestersen KB, Bekker-Jensen S et al (2011) Mass spectrometric analysis of lysine ubiquitylation reveals promiscuity at site level. Mol Cell Proteomics 10(3):M110.003590
35. Lowe ED, Hasan N, Trempe JF et al (2006) Structures of the Dsk2 UBL and UBA domains and their complex. Acta Crystallogr D Biol Crystallogr 62(Pt 2):177–188
36. Mayor T, Lipford JR, Graumann J et al (2005) Analysis of polyubiquitin conjugates reveals that the Rpn10 substrate receptor contributes to the turnover of multiple proteasome targets. Mol Cell Proteomics 4(6):741–751
37. Verma R, Oania R, Graumann J et al (2004) Multiubiquitin chain receptors define a layer of substrate selectivity in the ubiquitin-proteasome system. Cell 118(1):99–110
38. Mayor T, Graumann J, Bryan J et al (2007) Quantitative profiling of ubiquitylated proteins reveals proteasome substrates and the substrate repertoire influenced by the Rpn10 receptor pathway. Mol Cell Proteomics 6(11):1885–1895
39. Bennett EJ, Shaler TA, Woodman B et al (2007) Global changes to the ubiquitin system in Huntington's disease. Nature 448(7154):704–708
40. Hjerpe R, Aillet F, Lopitz-Otsoa F et al (2009) Efficient protection and isolation of ubiquitylated proteins using tandem ubiquitin-binding entities. EMBO Rep 10(11):1250–1258
41. Lopitz-Otsoa F, Rodriguez-Suarez E, Aillet F et al (2012) Integrative analysis of the ubiquitin proteome isolated using tandem ubiquitin binding entities (TUBEs). J Proteomics 75(10):2998–3014
42. Olsen JV, Blagoev B, Gnad F et al (2006) Global, in vivo, and site-specific phosphorylation dynamics in signaling networks. Cell 127(3):635–648
43. Villen J, Beausoleil SA, Gerber SA et al (2007) Large-scale phosphorylation analysis of mouse liver. Proc Natl Acad Sci U S A 104(5):1488–1493
44. Bodenmiller B, Mueller LN, Mueller M et al (2007) Reproducible isolation of distinct, overlapping segments of the phosphoproteome. Nat Methods 4(3):231–237
45. Xu G, Paige JS, Jaffrey SR (2010) Global analysis of lysine ubiquitination by ubiquitin remnant immunoaffinity profiling. Nat Biotechnol 28(8):868–873
46. Kim W, Bennett EJ, Huttlin EL et al (2011) Systematic and quantitative assessment of the ubiquitin-modified proteome. Mol Cell 44(2):325–340
47. Wagner SA, Beli P, Weinert BT et al (2011) A proteome-wide, quantitative survey of in vivo ubiquitylation sites reveals widespread regulatory roles. Mol Cell Proteomics 10(10):M111.013284

48. Na CH, Jones DR, Yang Y et al (2012) Synaptic protein ubiquitination in rat brain revealed by antibody-based ubiquitome analysis. J Proteome Res 11(9):4722–4732
49. Wagner SA, Beli P, Weinert BT et al (2012) Proteomic analyses reveal divergent ubiquitylation site patterns in murine tissues. Mol Cell Proteomics 11(12):1578–1585
50. Udeshi ND, Svinkina T, Mertins P et al (2013) Refined preparation and use of anti-diglycine remnant (K-epsilon-GG) antibody enables routine quantification of 10,000s of ubiquitination sites in single proteomics experiments. Mol Cell Proteomics 12(3):825–831
51. Skaug B, Chen ZJ (2010) Emerging role of ISG15 in antiviral immunity. Cell 143(2):187–190
52. Ng AH, Fang NN, Comyn SA et al (2013) System-wide analysis reveals intrinsically disordered proteins are prone to ubiquitylation after misfolding stress. Mol Cell Proteomics 12(9):2456–2467
53. Hale JE, Butler JP, Knierman MD et al (2000) Increased sensitivity of tryptic peptide detection by MALDI-TOF mass spectrometry is achieved by conversion of lysine to homoarginine. Anal Biochem 287(1):110–117
54. Peters EC, Horn DM, Tully DC et al (2001) A novel multifunctional labeling reagent for enhanced protein characterization with mass spectrometry. Rapid Commun Mass Spectrom 15(24):2387–2392
55. Nakamura H, Linclau B, Curran DP (2001) Fluorous triphasic reactions: transportative deprotection of fluorous silyl ethers with concomitant purification. J Am Chem Soc 123(41):10119–10120
56. Brittain SM, Ficarro SB, Brock A et al (2005) Enrichment and analysis of peptide subsets using fluorous affinity tags and mass spectrometry. Nat Biotechnol 23(4):463–468
57. Go EP, Uritboonthai W, Apon JV et al (2007) Selective metabolite and peptide capture/mass detection using fluorous affinity tags. J Proteome Res 6(4):1492–1499
58. Ying W, Perlman DH, Li L et al (2009) Highly efficient and selective enrichment of peptide subsets combining fluorous chemistry with reversed-phase chromatography. Rapid Commun Mass Spectrom 23(24):4019–4030
59. Baty JD, Robinson PR (1977) Single and multiple ion recording techniques for the analysis of diphenylhydantoin and its major metabolite in plasma. Biomed.Mass Spectrom 4(1):36–41
60. Aebersold R, Mann M (2003) Mass spectrometry-based proteomics. Nature 422(6928):198–207
61. Domon B, Aebersold R (2006) Mass spectrometry and protein analysis. Science 312(5771):212–217
62. Gallien S, Duriez E, Domon B (2011) Selected reaction monitoring applied to proteomics. J Mass Spectrom 46(3):298–312
63. Picotti P, Aebersold R (2012) Selected reaction monitoring-based proteomics: workflows, potential, pitfalls and future directions. Nat Methods 9(6):555–566
64. Gerber SA, Rush J, Stemman O et al (2003) Absolute quantification of proteins and phosphoproteins from cell lysates by tandem MS. Proc Natl Acad Sci U S A 100(12):6940–6945
65. Kirkpatrick DS, Gerber SA, Gygi SP (2005) The absolute quantification strategy: a general procedure for the quantification of proteins and post-translational modifications. Methods 35(3):265–273
66. Kirkpatrick DS, Hathaway NA, Hanna J et al (2006) Quantitative analysis of in vitro ubiquitinated cyclin B1 reveals complex chain topology. Nat Cell Biol 8(7):700–710
67. Xu P, Duong DM, Seyfried NT et al (2009) Quantitative proteomics reveals the function of unconventional ubiquitin chains in proteasomal degradation. Cell 137(1):133–145
68. Mirzaei H, Rogers RS, Grimes B et al (2010) Characterizing the connectivity of poly-ubiquitin chains by selected reaction monitoring mass spectrometry. Mol Biosyst 6(10):2004–2014
69. Dammer EB, Na CH, Xu P et al (2011) Polyubiquitin linkage profiles in three models of proteolytic stress suggest the etiology of Alzheimer disease. J Biol Chem 286(12):10457–10465
70. Kaiser SE, Riley BE, Shaler TA et al (2011) Protein standard absolute quantification (PSAQ) method for the measurement of cellular ubiquitin pools. Nat Methods 8(8):691–696

71. Bedford L, Layfield R, Mayer RJ et al (2011) Diverse polyubiquitin chains accumulate following 26 S proteasomal dysfunction in mammalian neurones. Neurosci Lett 491(1):44–47
72. Gupta R, Kus B, Fladd C et al (2007) Ubiquitination screen using protein microarrays for comprehensive identification of Rsp5 substrates in yeast. Mol Syst Biol 3:116
73. Lu JY, Lin YY, Qian J et al (2008) Functional dissection of a HECT ubiquitin E3 ligase. Mol Cell Proteomics 7(1):35–45
74. Persaud A, Alberts P, Amsen EM et al (2009) Comparison of substrate specificity of the ubiquitin ligases Nedd4 and Nedd4-2 using proteome arrays. Mol Syst Biol 5:333
75. Merbl Y, Kirschner MW (2009) Large-scale detection of ubiquitination substrates using cell extracts and protein microarrays. Proc Natl Acad Sci U S A 106(8):2543–2548
76. Loch CM, Strickler JE (2012) A microarray of ubiquitylated proteins for profiling deubiquitylase activity reveals the critical roles of both chain and substrate. Biochim Biophys Acta 1823(11):2069–2078
77. Merbl Y, Refour P, Patel H et al (2013) Profiling of ubiquitin-like modifications reveals features of mitotic control. Cell 152(5):1160–1172
78. Guo Z, Song E, Ma S et al (2012) Proteomics strategy to identify substrates of LNX, a PDZ domain-containing E3 ubiquitin ligase. J Proteome Res 11(10):4847–4862
79. Emanuele MJ, Elia AE, Xu Q et al (2011) Global identification of modular cullin-RING ligase substrates. Cell 147(2):459–474
80. Tung CW, Ho SY (2008) Computational identification of ubiquitylation sites from protein sequences. BMC Bioinformatics 9:310. doi:10.1186/1471-2105-9-310
81. Radivojac P, Vacic V, Haynes C et al (2010) Identification, analysis, and prediction of protein ubiquitination sites. Proteins 78(2):365–380
82. Chen Z, Chen YZ, Wang XF et al (2011) Prediction of ubiquitination sites by using the composition of k-spaced amino acid pairs. PLoS ONE 6(7):e22930
83. Feng KY, Huang T, Feng KR et al (2013) Using WPNNA classifier in ubiquitination site prediction based on hybrid features. Protein Pept Lett 20(3):318–323
84. Chen X, Qiu JD, Shi SP et al (2013) Incorporating key position and amino acid residue features to identify general and species-specific ubiquitin conjugation sites. Bioinformatics 29(13):1614–1622
85. Michael S, Trave G, Ramu C et al (2008) Discovery of candidate KEN-box motifs using cell cycle keyword enrichment combined with native disorder prediction and motif conservation. Bioinformatics 24(4):453–457
86. Liu Z, Yuan F, Ren J et al (2012) GPS-ARM: computational analysis of the APC/C recognition motif by predicting D-boxes and KEN-boxes. PLoS ONE 7(3):e34370
87. Arabi A, Ullah K, Branca RM et al (2012) Proteomic screen reveals Fbw7 as a modulator of the NF-kappaB pathway. Nat Commun 3:976
88. Hagai T, Toth-Petroczy A, Azia A et al (2012) The origins and evolution of ubiquitination sites. Mol Biosyst 8(7):1865–1877
89. Hagai T, Azia A, Toth-Petroczy A et al (2011) Intrinsic disorder in ubiquitination substrates. J Mol Biol 412(3):319–324
90. Hershko A, Ciechanover A, Heller H et al (1980) Proposed role of ATP in protein breakdown: conjugation of protein with multiple chains of the polypeptide of ATP-dependent proteolysis. Proc Natl Acad Sci U S A 77(4):1783–1786
91. Chau V, Tobias JW, Bachmair A et al (1989) A multiubiquitin chain is confined to specific lysine in a targeted short-lived protein. Science 243(4898):1576–1583
92. Spence J, Sadis S, Haas AL et al (1995) A ubiquitin mutant with specific defects in DNA repair and multiubiquitination. Mol Cell Biol 15(3):1265–1273
93. Galan JM, Haguenauer-Tsapis R (1997) Ubiquitin lys63 is involved in ubiquitination of a yeast plasma membrane protein. EMBO J 16(19):5847–5854
94. Tokunaga F, Sakata S, Saeki Y et al (2009) Involvement of linear polyubiquitylation of NEMO in NF-kappaB activation. Nat Cell Biol 11(2):123–132
95. Jin L, Williamson A, Banerjee S et al (2008) Mechanism of ubiquitin-chain formation by the human anaphase-promoting complex. Cell 133(4):653–665

96. Boname JM, Thomas M, Stagg HR et al (2010) Efficient internalization of MHC I requires lysine-11 and lysine-63 mixed linkage polyubiquitin chains. Traffic 11(2):210–220
97. Goto E, Yamanaka Y, Ishikawa A et al (2010) Contribution of lysine 11-linked ubiquitination to MIR2-mediated major histocompatibility complex class I internalization. J Biol Chem 285(46):35311–35319
98. Huang H, Jeon MS, Liao L et al (2010) K33-linked polyubiquitination of T cell receptor-zeta regulates proteolysis-independent T cell signaling. Immunity 33(1):60–70
99. Comyn SA, Chan GT, Mayor T (2013) False start: Cotranslational protein ubiquitination and cytosolic protein quality control. J Proteomics
100. Schubert U, Anton LC, Gibbs J et al (2000) Rapid degradation of a large fraction of newly synthesized proteins by proteasomes. Nature 404(6779):770–774
101. Wang F, Durfee LA, Huibregtse JM (2013) A cotranslational ubiquitination pathway for quality control of misfolded proteins. Mol Cell 50(3):368–378
102. Lecker SH, Goldberg AL, Mitch WE (2006) Protein degradation by the ubiquitin-proteasome pathway in normal and disease states. J Am Soc Nephrol 17(7):1807–1819
103. Ross CA, Poirier MA (2004) Protein aggregation and neurodegenerative disease. Nat Med (10 Suppl):S10–S17
104. Alves-Rodrigues A, Gregori L, Figueiredo-Pereira ME (1998) Ubiquitin, cellular inclusions and their role in neurodegeneration. Trends Neurosci 21(12):516–520
105. Gies E, Wilde I, Winget JM et al (2010) Niclosamide prevents the formation of large ubiquitin-containing aggregates caused by proteasome inhibition. PLoS ONE 5(12):e14410
106. Wilde IB, Brack M, Winget JM et al (2011) Proteomic characterization of aggregating proteins after the inhibition of the ubiquitin proteasome system. J Proteome Res 10(3):1062–1072
107. Carlson N, Rogers S, Rechsteiner M (1987) Microinjection of ubiquitin: changes in protein degradation in HeLa cells subjected to heat-shock. J Cell Biol 104(3):547–555
108. Bennett EJ, Rush J, Gygi SP et al (2010) Dynamics of cullin-RING ubiquitin ligase network revealed by systematic quantitative proteomics. Cell 143(6):951–965
109. Busino L, Millman SE, Scotto L et al. (2012) Fbxw7alpha- and GSK3-mediated degradation of p100 is a pro-survival mechanism in multiple myeloma. Nat Cell Biol 14(4):375–385
110. Guardavaccaro D, Frescas D, Dorrello NV et al (2008) Control of chromosome stability by the beta-TrCP-REST-Mad2 axis. Nature 452(7185):365–369
111. Peschiaroli A, Dorrello NV, Guardavaccaro D et al (2006) SCFbetaTrCP-mediated degradation of claspin regulates recovery from the DNA replication checkpoint response. Mol Cell 23(3):319–329
112. Dorrello NV, Peschiaroli A, Guardavaccaro D et al (2006) S6K1- and betaTRCP-mediated degradation of PDCD4 promotes protein translation and cell growth. Science 314(5798):467–471
113. Liu B, Zheng Y, Wang TD et al (2012) Proteomic identification of common SCF ubiquitin ligase FBXO6-interacting glycoproteins in three kinds of cells. J Proteome Res 11(3):1773–1781
114. Teixeira FR, Yokoo S, Gartner CA et al (2010) Identification of FBXO25-interacting proteins using an integrated proteomics approach. Proteomics 10(15):2746–2757
115. Yumimoto K, Matsumoto M, Oyamada K et al. (2012) Comprehensive identification of substrates for F-box proteins by differential proteomics analysis. J Proteome Res 11(6):3175–3185
116. Zhuang M, Guan S, Wang H et al (2013) Substrates of IAP ubiquitin ligases identified with a designed orthogonal E3 ligase, the NEDDylator. Mol Cell 49(2):273–282
117. Sarraf SA, Raman M, Guarani-Pereira V et al (2013) Landscape of the PARKIN-dependent ubiquitylome in response to mitochondrial depolarization. Nature 496(7445):372–376
118. Narendra D, Tanaka A, Suen DF et al. (2008) Parkin is recruited selectively to impaired mitochondria and promotes their autophagy. J Cell Biol 183(5):795–803

119. Luise C, Capra M, Donzelli M et al (2011) An atlas of altered expression of deubiquitinating enzymes in human cancer. PLoS ONE 6(1):e15891
120. McDonell LM, Mirzaa GM, Alcantara D et al. (2013) Mutations in STAMBP, encoding a deubiquitinating enzyme, cause microcephaly-capillary malformation syndrome. Nat Genet 45(5):556–562
121. Cilenti L, Balakrishnan MP, Wang XL et al. (2011) Regulation of Abro1/KIAA0157 during myocardial infarction and cell death reveals a novel cardioprotective mechanism for Lys63-specific deubiquitination. J Mol Cell Cardiol 50(4):652–661
122. Fraile JM, Quesada V, Rodriguez D et al (2012) Deubiquitinases in cancer: new functions and therapeutic options. Oncogene 31(19):2373–2388
123. Sowa ME, Bennett EJ, Gygi SP et al (2009) Defining the human deubiquitinating enzyme interaction landscape. Cell 138(2):389–403
124. Kouranti I, McLean JR, Feoktistova A et al (2010) A global census of fission yeast deubiquitinating enzyme localization and interaction networks reveals distinct compartmentalization profiles and overlapping functions in endocytosis and polarity. PLoS Biol 8(9). doi:10.1371/journal.pbio.1000471
125. Richardson LA, Reed BJ, Charette JM et al (2012) A conserved deubiquitinating enzyme controls cell growth by regulating RNA polymerase I stability. Cell Rep 2(2):372–385
126. Poulsen JW, Madsen CT, Young C et al (2012) Comprehensive profiling of proteome changes upon sequential deletion of deubiquitylating enzymes. J Proteomics 75(13):3886–3897
127. Borodovsky A, Ovaa H, Kolli N et al (2002) Chemistry-based functional proteomics reveals novel members of the deubiquitinating enzyme family. Chem Biol 9(10):1149–1159
128. Altun M, Kramer HB, Willems LI et al (2011) Activity-based chemical proteomics accelerates inhibitor development for deubiquitylating enzymes. Chem Biol 18(11):1401–1412
129. Lennartsson A, Ekwall K (2009) Histone modification patterns and epigenetic codes. Biochim Biophys Acta 1790(9):863–868
130. Mertins P, Qiao JW, Patel J et al (2013) Integrated proteomic analysis of post-translational modifications by serial enrichment. Nat Methods 10(7):634–637
131. Swaney DL, Beltrao P, Starita L et al (2013) Global analysis of phosphorylation and ubiquitylation cross-talk in protein degradation. Nat Methods 10(7):676–682

Part VIII
The Chaperone and Protease Networks in Model Bacteria and Parasites

Chapter 15
The Interaction Networks of *E. coli* Chaperones

Hideki Taguchi

Abstract Protein folding is often hampered by protein aggregation, which can be prevented by a variety of chaperones in the cell. In this review, I summarize recent topics on *in vitro* and *in vivo* approaches to understand the role of *Escherichia coli* chaperones to prevent protein aggregations.

1 Introduction

A protein must fold into the correct tertiary structure after emerging from the ribosome. The unique native structure of a protein is encoded in its amino acid sequence [1]. However, protein folding is often hampered by protein aggregation, which is generally prevented by a variety of chaperone proteins in the cell [2]. Chaperones are also involved in other multiple cellular processes associated with the conformational changes of proteins, such as stress responses [2].

Many efforts over the past two decades have been dedicated to elucidate the mechanism of molecular chaperones. The best-characterized chaperones are those in *Escherichia coli*. In *E. coli*, three major chaperone systems are involved in the folding of newly synthesized proteins in the cytoplasm [2]. The first is trigger factor (TF), which directly associates with ribosome and interacts with nascent chains co-translationally [3]. TF associates with L23 ribosomal protein in a monomer with the dragon-shaped structure [3]. The second is DnaK, which is one of the Hsp70 family conserved widely in all kingdoms of life. DnaK, as other Hsp70 proteins, consists of two domains, an adenosine triphosphatase (ATPase) domain and a polypeptide-binding domain [2, 4]. The interaction of DnaK with the substrate polypeptide is ATP dependent. DnaK binds substrate proteins with high affinity in the adenosine diphosphate (ADP) state, and with low affinity in the ATP state. The function of DnaK is regulated by cofactors, DnaJ and GrpE. DnaJ binds denatured proteins in an ATP-independent manner to be targeted to DnaK. GrpE is a nucleotide exchange factor to regulate the ATPase cycle of DnaK. The third is GroEL, which belongs to

H. Taguchi (✉)
Department of Biomolecular Engineering, Graduate School of Biosciences and Biotechnology, Tokyo Institute of Technology, Yokohama, Kanagawa, Japan
e-mail: taguchi@bio.titech.ac.jp

W. A. Houry (ed.), *The Molecular Chaperones Interaction Networks in Protein Folding and Degradation,* Interactomics and Systems Biology, DOI 10.1007/978-1-4939-1130-1_15,

conserved chaperonin family. GroEL forms a large cylindrical complex with co-chaperonin GroES in the presence of ATP, which encapsulates substrate proteins into its cavity to assist the folding [2, 4, 5]. The bacterial GroEL and GroES are the only indispensable chaperones for the viability of *E. coli* [6, 7]. GroEL consists of two heptameric rings of 57 kDa subunits, and provides binding sites for non-native substrate proteins. GroES, a dome-shaped heptameric ring of ~10 kDa subunits, caps GroEL in the presence of adenine nucleotides, forming a central cavity that can accommodate substrate proteins up to ~60 kDa in size [8, 9].

It has been known that these three chaperone systems act cooperatively; TF and DnaK are known to exhibit overlapping co-translational roles *in vivo* [10–12], whereas GroEL is believed to be implicated in folding after the polypeptides are released from the ribosome, albeit that possible cotranslational involvement of GroEL has also been reported [13–16].

In addition to the "classical" functions in the protein folding, chaperones are proposed to promote protein evolution by buffering the destabilization of proteins caused by harmful genetic mutations [17–21]. The absolute requirement of chaperones for cellular functions and for protein evolution might account for the fact that an organism lacking chaperones has not been identified.

To understand protein folding in the cell, we at first need to know which proteins are aggregation-prone. Then, we have to tackle the question of which aggregation-prone proteins are rescued by chaperones. In this review, I summarize recent topics on *in vitro* and *in vivo* approaches to understand the role of *E. coli* chaperones to assist folding by preventing aggregates formation. Main topics include global *in vitro* analysis of protein aggregation and chaperone effects on aggregation-prone proteins, and *in vivo* substrates of chaperonin GroEL and chaperone DnaK.

2 *In Vitro* Analyses Using a Reconstituted Cell-Free Translation System

Chaperones are required to prevent protein aggregations [2]. Although we empirically know some proteins are aggregation-prone, there is no systematic global analysis on which proteins are actually aggregation-prone or not. In general, it is very difficult to evaluate the aggregation propensity of a protein of interest after the expression of the protein in cells, since cells contain a variety of chaperone proteins. Similar limitation would be applicable for conventional cell-free analysis since cell lysates contain endogenous chaperones. In this context, chaperone-free translation system would be ideal to evaluate the aggregation-prone propensity of proteins of interest. A highly controllable cell-free translation system called the Protein synthesizing Using Recombinant Elements (PURE) system was successfully reconstituted by using only the essential purified factors and enzymes that are responsible for gene expression in *E. coli* [22, 23]. The protein synthesis in this system has been proven to be highly manipulatable. Importantly, PURE system does not contain any chaperone [22, 24]. Therefore, we can evaluate the inherent aggregation propensities of proteins of interest in a translation-coupled manner under the chaperone-free conditions. In addition, this reconstituted translation system represents a straightforward approach for studying the chaperone-assisted folding of newly synthesized polypeptides, as the roles these chaperones play can be determined by simply adding them to the translation system.

2.1 *Global Aggregation Analysis of the Entire Ensemble of* E. coli *Proteins*

Understanding the mechanism underlying aggregate formation is required for the development of a wide variety of protein sciences. However, the relationship between the protein aggregation propensities and the primary sequences remains poorly understood. Since it is empirically known that some proteins tend to aggregate, several groups systematically studied the effects of mutations in proteins of interest that caused the formation of insoluble aggregates [25–28]. Aggregation-prone properties of thousands of *E. coli* proteins were evaluated using PURE system [29].

2.1.1 Experiment

The scheme of the global analysis is shown in Fig. 15.1: the one-by-one synthesis of individual *E. coli* proteins, the quantification of solubility by a centrifugation-based assay, and the statistical analyses of the collected data. This is an "*in vitro* (reconstituted) proteome" approach, in which the properties of thousands of proteins, including proteins with extremely low abundance in cells, are investigated individually after cell-free translation.

A comprehensive analysis, in which the complete *E. coli* open reading frame (ORF) library (ASKA library) [30] was translated in the PURE system under the same conditions, was conducted. The ASKA library consists of all predicted ORFs of the *E. coli* genome, including membrane proteins [30]. A total of 4132 ORFs were individually amplified by polymerase chain reaction (PCR) using a common primer set, and then were used for protein synthesis in the PURE system at 37 °C for 60 min.

The ^{35}S-methionine-labeled proteins were quantified after electrophoresis of the translation products. Approximately 70 % of the *E. coli* ORFs (3173 proteins out of 4132) were successfully quantified. The remainder was not quantified, due to insufficient translation and trouble during the electrophoresis. The unquantifiable group contained ~60 % of the inner membrane proteins (435/754), whereas more than 80 % of the cytoplasmic proteins (2277/2688) were quantified.

The propensity for protein aggregation was examined by a centrifugation assay [15, 24]. An aliquot of the translation mixture was centrifuged. The proportion of the supernatant fraction, which was obtained after the centrifugation of the translation mixture, to the uncentrifuged total protein was defined as the solubility, the index of the aggregation propensity.

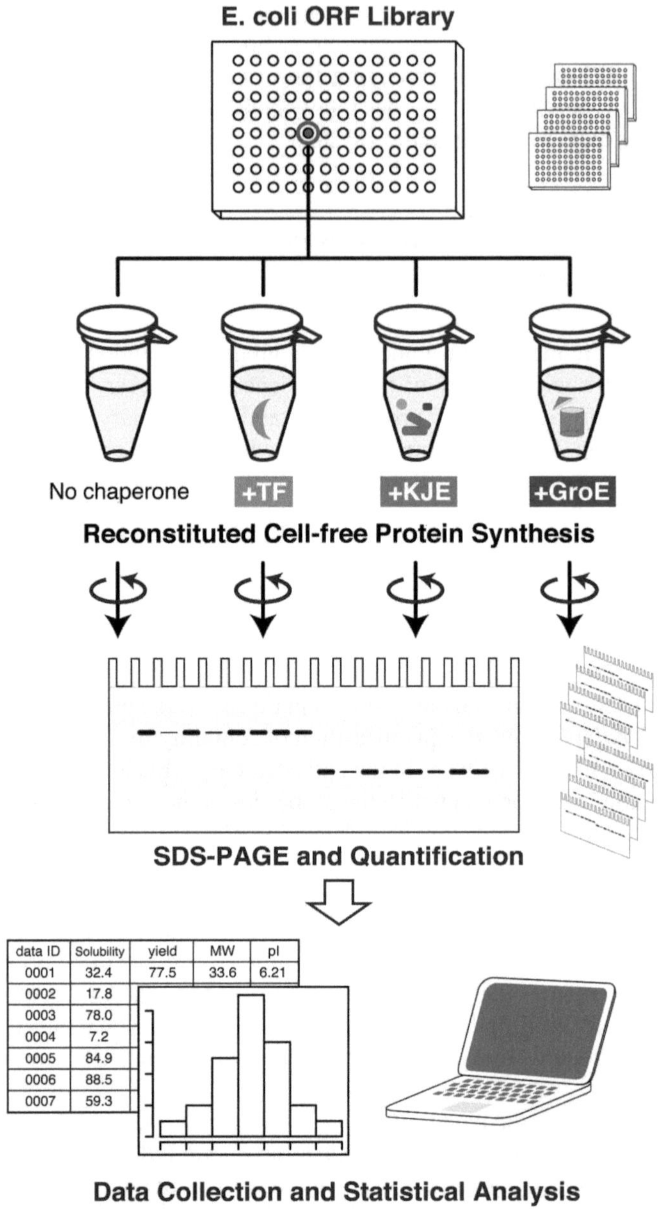

Fig. 15.1 An *in vitro* expressed proteome approach for global aggregation analysis. Schematic illustration of the experiment. *Escherichia coli* proteins were separately expressed with a reconstituted cell-free translation system, the PURE system, in the absence and the presence of the major *E. coli* chaperones (trigger factor, TF; GroEL/GroES, GroE; DnaK/DnaJ/GrpK, KJE). Each translation product was labeled with [^{35}S]methionine. After translation, the uncentrifuged total fraction (*Total*) and the supernatant fraction after centrifugation (*Sup*) were electrophoresed and quantified by autoradiography. The ratio of the translation products in the Total and Sup fractions was defined as the solubility, which represented the aggregation propensity of the protein. Reprinted from ref [44]

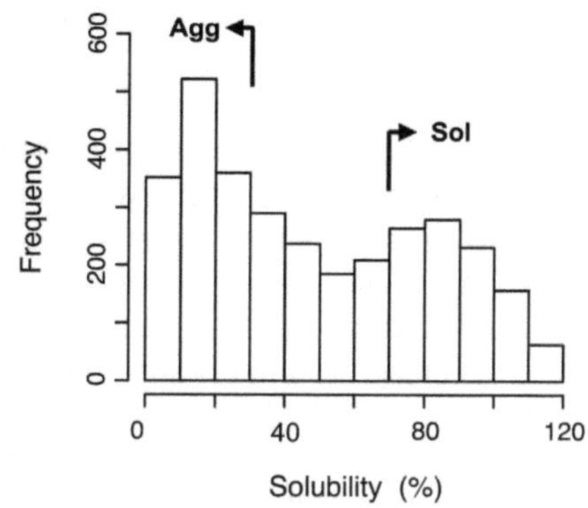

Fig. 15.2 Histogram of solubility for the 3173 quantified proteins. The proteins with solubilities below 30 % and above 70 % were defined as the aggregation-prone (*Agg*) and soluble (*Sol*) groups, respectively. Adapted from ref [29]

2.1.2 Bimodal Solubility Distribution

A histogram of the individual solubilities, based on data from 3173 translated proteins, showed a clear bimodal, rather than normal Gaussian, distribution (Fig. 15.2), indicating that the aggregation propensities are not evenly distributed across a continuum. Subtraction of the predicted integral membrane proteins from the data did not change the bimodal distribution, suggesting that the cytoplasmic proteins can be categorized into an aggregation-prone group and a highly soluble one. To elucidate which characteristics of the protein influence this bimodality, we compared a variety of protein properties in the aggregation-prone (Agg: defined as less than 30 %) and highly soluble (Sol: defined as more than 70 %) groups. One might expect that the bimodal distribution in the histogram is simply due to the difference in the synthesized yield of proteins, since it has been generally believed that higher protein concentrations generate more protein aggregates. However, this is not the case, since there is no apparent correlation between the solubilities and the yields.

The essential proteins tended to be enriched in the high solubility group, suggesting that the essential proteins might have evolved to be soluble for their irreplaceable properties. In addition to the essentiality, the solubilities are strongly dependent on the functions. For example, structural component group, which is mainly composed of ribosomal proteins, and factor group, which includes transcription or translation factors, chaperones, and proteases, showed a strong bias to the high solubility group. In contrast, the proteins in the transporter group tended to be aggregation-prone. Regarding the oligomeric states of the proteins, preliminary analysis shows that hetero-oligomers seem to be aggregation-prone although we cannot say the tendency is statistically significant due to the incomplete database on the oligomeric states.

2.1.3 Physicochemical Properties

The physicochemical properties of the proteins, such as the molecular weights, the deduced isoelectric points (pI) and the amino acid residue content were compared to address the relationship between solubility and amino acid sequence. The distribution of molecular weights in the soluble group was shifted to smaller sizes as compared with the total histogram. Regarding the pI, an enrichment of low pI (5–7) proteins in the high solubility distribution was observed, whereas the aggregation-prone proteins showed a somewhat broader pI distribution (ranging from 5 to 10). Analysis to test whether the amino acid residue content affected the solubility revealed that higher contents of negatively charged residues (Asp and Glu) tended to be soluble. Higher contents of aromatic residues (Phe, Tyr, and Trp) were slightly biased to be aggregation-prone. In contrast, no significant difference was observed in the contents of hydrophobic residues (Val, Leu, and Ile) and positively charged residues (Lys, Arg, and His). Since it has been believed that the hydrophobic interaction is a critical driving force in aggregate formation, the lack of an apparent correlation in the hydrophobic residue content was unexpected. Other attempts to detect a bias between the solubility and the hydrophobicity, including a well-known hydropathy plot analysis [31, 32], which shows clusters of hydrophobic residues in the primary amino acid sequences, or several hydrophobic-polar alternates analyses also failed. We note that Gln/Asn-rich sequences including polyglutamine repeats, which tend to form amyloid fibrils, are very rare in the *E. coli* ORFs [33].

Several analyses related to the secondary structures were subsequently conducted. The secondary structure contents were predicted by using popular prediction methods, such as Chou-Fasman [34] and Psi-blast based secondary structure prediction (PSIPRED) [35, 36]. However, a notable correlation between the predicted secondary structure content and the solubility was not detected.

To address the correlation between the solubilities and the tertiary structures, the solubilities with the Structural Classification of Proteins (SCOP) database, which is a comprehensive ordering of all proteins with known structures, according to their evolutionary and structural relationships [37] were compared. The classification is based on hierarchical levels: class, fold, superfamily and family. Superfamilies and families are defined as having a common fold if their proteins have the same major secondary structures in the same arrangement and with the same topological connections. Roughly, the secondary structures did not correlate with the aggregation propensities. Regarding the SCOP folds some of the SCOP folds were extremely biased toward their solubilities. For example, in the periplasmic-binding protein-like II fold (SCOP fold: c94) group, which is largely dominated by deoxyribonucleic acid (DNA)-binding transcriptional regulator proteins, 83 % of the members were low solubility proteins (35 out of 42 assigned proteins), whereas only one protein was in a soluble group. Other low soluble folds included Pyridoxal-phosphate (PLP)-dependent transferases fold (c67), DNA/ribonucleic acid (RNA)-binding 3-helical bundle fold (a4), triose phosphate isomerase (TIM) β/α-barrel fold (c1) and P-loop containing nucleoside triphosphate hydrolases (c37). For the highly soluble folds, we assigned Flavodoxin-like fold (c23), oligonucleotide/oligosaccharide-binding (OB)-fold (b40), and Thioredoxin fold (c47).

2.1.4 Aggregation Propensity Prediction

These data can be applicable to several recently developed web tools to predict protein aggregation. However, none of the tools including TANGO [38], AGGRESCAN [39], and PASTA programs [40] extracted a notable positive correlation between our datasets and the predicted results, probably due to the fact that the algorithms used in those programs basically relied on data from amyloid aggregates in eukaryotes. Other attempts to predict the solubilities based on this dataset have been developed [41–43].

2.2 *Global Analyses of Chaperone Effects on Aggregation-Prone Proteins*

Global aggregation analysis under chaperone-free conditions has been extended to that in the presence of the major *E. coli* chaperones, TF, DnaK/DnaJ/GrpE (DnaKJE), and GroEL/GroES (GroE) [44].

In previous global aggregation analysis, aggregation-prone group was defined as the proteins with less than 30 % solubility [29]. Effects of chaperones were examined for the aggregation-prone *E. coli* proteins that are predicted to reside in the cytosol (792 proteins).

2.2.1 Experiment

All of the cytosolic aggregation-prone proteins were synthesized by the PURE system at 37 °C for 60 min in the absence or presence of each chaperone. Each chaperone was added at the approximate physiological concentration based on previous assessments of chaperone activities under cell-free conditions [15, 45, 46]. The ^{35}S-methionine-labeled proteins were electrophoresed on Sodium dodecyl sulfate (SDS) gels and quantified. The aggregation propensity was examined by the centrifugation assay [29]. Typical results are shown in Fig. 15.3a. Almost all of the proteins (788/792) were quantified for their solubilities under each condition.

In total, more than 3000 assays (788 proteins × 4 conditions = 3152) were conducted. Overall, the chaperones tested effectively increased the solubility of the aggregation-prone proteins (Fig. 15.3b). Overall, the solubilities of two-thirds of the proteins (526/788) were drastically increased, defined as more than a 50 % increase, in the presence of any one of the chaperones. The proteins that were not rescued by any one of the chaperones, defined as less than a 20 % increase in the solubility, represented only 3 % of the total (24 out of 788). Taken together, this comprehensive analysis has explicitly confirmed the global role of chaperones in preventing the aggregation of hundreds of proteins.

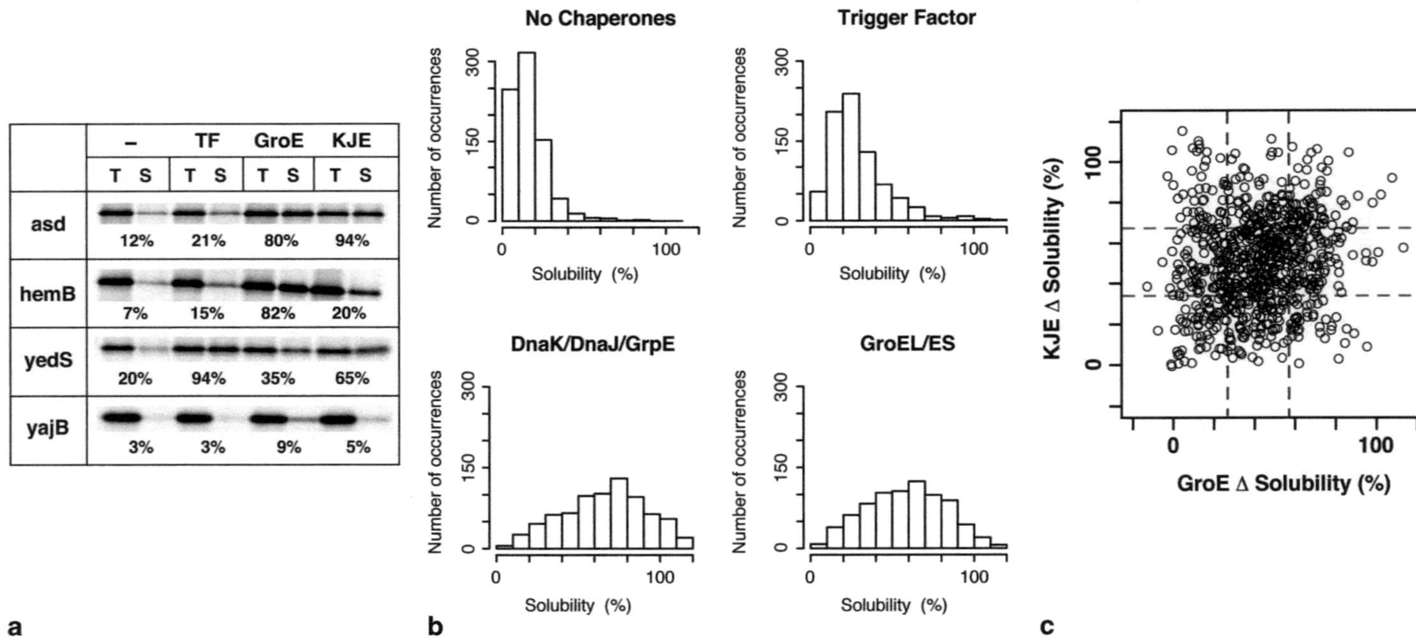

Fig. 15.3 Global analysis of chaperone effects on the prevention of aggregate formation. **a** Typical examples. SDS-gels of four aggregation-prone *E. coli* cytosolic proteins (asd, hemB, yedS, and yajB) in the absence and the presence of the chaperones are shown. The numbers below the electrophoretic pattern indicate the solubility values, calculated by the ratio of the amount of translation products in the Sup (*S*) and Total (*T*) fractions. **b** Histograms of solubilities in the absence or the presence of three *E. coli* chaperones. **c** Two-dimensional distribution plot of Δsolubilities for DnaKJE and GroE. Dashed lines represent the boundaries of the lower and upper quartiles (34, 67 % and 26, 58 % solubility values for DnaKJE and GroE, respectively). Reprinted from ref [44]

2.2.2 Each Chaperone Effects

The effects of each chaperone were compared. It is noteworthy that TF had only a marginal effect. On the other hand, DnaKJE and GroE increased the solubilities of many proteins. The solubilities of 409 and 287 proteins with DnaKJE and GroE, respectively, were drastically (>50 %) increased. Approximately 30 % of the proteins with >50 % increase in solubility (175 proteins) were common between DnaKJE and GroE, indicating that these overlapping proteins were rescued well by either DnaKJE or GroE. Taken together, the data clearly show the global effects of DnaKJE and GroE in preventing aggregation.

2.2.3 Physicochemical Properties

To further investigate the effects of DnaKJE and GroE, the data on DnaKJE and GroE were plotted in two dimensions. A substantial fraction of the proteins was biased toward DnaKJE or GroE. These biases suggest that DnaKJE and GroE could have different recognition modes for substrates. To extract the possible preferences of DnaKJE and GroE, physicochemical properties of the proteins were analyzed. Some biases were observed in molecular weights when the analysis was limited in the proteins that were well-solubilized by either DnaK or GroE, defined as the upper quartile (≥75th percentile) in the distribution. GroE is biased toward lower molecular weight proteins (20~50 kDa), whereas DnaKJE is effective for larger ones (>60 kDa).

To address the correlation between the chaperone effects and the tertiary or quaternary structures, the SCOP database (class and fold) [37] and the oligomeric states of proteins were compared, although only a small number of proteins was analyzed, due to the limited database size. When classified by the SCOP classes (all-α, all-β, α/β, and α+β), DnaKJE was effective for the α+β class, whereas GroE was not effective for the all-α class. Furthermore, there are some biases for DnaKJE and GroE in several SCOP folds. GroE was biased toward the c1 (TIM barrel) fold, which is plausible since the most abundant fold in the *in vivo* obligate GroE substrates is the TIM barrel fold [47, 48]. Neither DnaKJE nor GroE was effective for the a4 (DNA/RNA-binding 3-helical bundle fold) and c94 (periplasmic-binding protein-like II) folds.

2.2.4 Cooperative Effects of Chaperones

Neither DnaKJE nor GroE could rescue a subset of proteins mapped around the lower left area in the plot (Fig. 15.3c), termed recalcitrant proteins. Various combinations of chaperones were examined to investigate whether these recalcitrant proteins could be solubilized. All of the recalcitrant proteins, which were defined as the proteins categorized in the lower quartiles in DnaKJE and GroE (53 proteins), were translated in the presence of chaperone combinations: TF+DnaKJE, TF+GroE,

DnaKJE+GroE, and TF+DnaKJE+GroE. The solubilities under the TF+DnaKJE and TF+GroE conditions were slightly increased, whereas the combination of DnaKJE and GroE was more effective for some of these recalcitrant proteins, reflecting the consensus that GroE and DnaKJE synergistically assist with the folding of nascent polypeptides [45, 49]. Strikingly, the addition of all three chaperones to the recalcitrant proteins drastically changed the solubility distribution: More than 70 % of the recalcitrant proteins (38/53) showed significantly improved solubilities in the presence of all chaperones. These results suggest that TF also has the potential to act cooperatively with DnaKJE and GroE, although TF itself was not very effective in preventing aggregation.

2.3 Solubility Database of all E. coli *Proteins (eSOL)*

The dataset of aggregation analyses including chaperone effects are freely accessible at online database (*e*SOL database: http://www.tanpaku.org/tp-esol).

3 *In Vivo* Substrates of Chaperonin GroEL

An important goal in chaperonin biology is to identify a subset of obligate GroEL/GroES (GroE) substrates that absolutely require GroE for folding in cells. Precise identification of the obligate GroE substrates should contribute to the identification of a distinctive role for GroE among chaperones, reveal the structural features of the obligate substrates, and shed light on the role of GroE in protein evolution.

3.1 Phenotype Analysis Using GroE-Knockdown Strain

One approach to identify obligate GroE substrates is a detailed analysis of the phenotypes of GroE-depleted cells. Since chaperonin GroE is the only indispensable chaperone for the viability of *E. coli* [6, 7], GroE-deletion strain is not available. However, a conditional GroE expression strain, MGM100, in which the native *groE* chromosomal promoter region has been replaced with the *araC* gene and the *araBAD* promoter [50]. When the sugar in the growth medium is changed from arabinose to glucose, the GroE levels decrease by 90 % within 2 h in this strain [50]. Investigations of *E. coli* phenotype after the GroE-depletion have identified DapA and FtsE as obligate GroE substrates in the cell lysis and filamentous morphology phenotypes, respectively [50, 51]. Although a detailed phenotypic analysis can precisely identify obligate GroE substrates, this approach is limited, in that the substrates can only be identified one by one, and only in the cells with experimentally tractable phenotypes.

3.2 *Mass Spectrometry-based Proteomics of GroEL Interacting Proteins*

Another approach to substrate identification is a proteome-wide analysis. Hundreds of GroEL substrates have been identified using mass spectrometry (MS) [47, 52]. In particular, Kerner et al. have identified ~250 substrates that interact with GroE in *E. coli* and categorized them into three classes depending on their enrichment in the GroE complex: Class I substrates as spontaneous folders, Class II as partial GroEL-dependent substrates, and Class III as the potential obligate GroE substrates [47]. Notably, ~84 Class III substrates are estimated to occupy ~80% of the available GroEL capacity in the cell. This classification was primarily based on the proteomics of the GroE interactors. However, except for DapA, GatY, MetK, ADD, and YajO, which were verified as requiring GroE for folding, the *in vivo* GroE dependency of other Class III substrates has not been tested [47].

Impairment of GroE function in *E. coli* results in the accumulation or degradation of newly translated polypeptides due to misfolding. Wholesale accumulation of aggregates was observed in the severe temperature-sensitive GroE strain, which harbors GroEL (E461K) mutant instead of wild-type GroEL [53]. The proteomic analysis using MS of the aggregated proteins identified ~300 proteins [53]. Most of the identified proteins were cytoplasmic proteins, many known to be highly abundant [53].

Note that similar approach to identify *in vivo* GroEL interactors has been applied to other bacteria besides *E. coli*. 24~28 GroEL interacting proteins were identified by MS-based proteomics in *Thermus thermophilus* and *Bacillus subtils* [54, 55].

3.3 *Combined Approach to Identify Obligate GroE-Dependent Substrates*

Previous proteome-wide analysis of *E. coli* chaperonin GroEL interactors predicted obligate chaperonin substrates, which were termed Class III substrates. However, the requirement of chaperonins for *in vivo* folding has not been fully examined. In fact, one of the Class III proteins, ParC, was functional even under GroE-depleted conditions [51], raising the possibility that the predicted Class III proteins are not necessarily obligate substrates of GroE [48].

3.3.1 Proteomics of the Soluble Fraction in GroE-Depleted *E. coli*

In GroE-depleted cells, the known obligate GroE substrates either aggregate (e.g., MetK) or are degraded (e.g., DapA, FtsE, and GatY) [47, 50, 51]. Thus, the abundance of other potential GroE obligate substrates would also be reduced in the soluble fraction of GroE-depleted cells. A proteome-wide analysis of the soluble fraction of GroE-depleted cells, using a conditional GroE expression strain

MGM100 [50], was therefore conducted to find candidate *in vivo* obligate GroE substrates. For proteomics, cells were subjected to a 2 h depletion of GroE in LB medium, during which the level of GroEL was reduced to less than 10 % of that in undepleted cells [50], and, as a control, cells with a normal level of GroE were also prepared. Note that diaminopimelic acid (DAP) was added to the medium to prevent cell lysis, a known consequence of DapA (a Class III substrate) deficiency in GroE-depleted cells [50]. The abundance of each protein was quantified by the exponentially modified Protein Abundance Index (emPAI), which provides an estimate of protein abundance by quantitating non-redundant peptides identified by MS [56, 57]. As a control for the sugar-associated changes in the proteome, MG1655, the wild-type parent strain of MGM100, was also examined by the same procedures.

The proteomics quantified a total of 986 proteins in MGM100 cells under glucose and arabinose conditions. The relative abundances of proteins defined as $emPAI_{glucose}/emPAI_{arabinose}$ shows that ~33 % of proteins were reduced by more than 50 % in MGM100 cells. The drastic reduction of many proteins during glucose growth in MGM100 cells was caused by the GroE-depletion. We note that a significant number of proteins were increased in the GroE-depleted cells, including methionine biosynthetic enzymes, such as MetE [7, 53, 58, 59], and certain chaperones, such as DnaK and SecB.

The proteome data were used to roughly choose candidate GroE substrates by the following criteria. First, the proteins with soluble abundance that was reduced during depletion to less than 50 % of that found during arabinose growth in MGM100 (as a genetic control) were chosen. The cutoff value of 50 % was set to minimize false negatives, as the highest solubility of known *in vivo* obligate GroE substrates was 46 %, as found with MetK. The 347 proteins chosen by the first criterion contained many false positives due to a sugar-associated reduction in their levels, and thus were filtered by a second criterion, in which the proteins with expression in MGM100 during glucose growth reduced to less than 50 % of that found during glucose growth in MG1655 were chosen. Using the genetic and sugar controls, 252 proteins among the detected 986 proteins met both criteria for rough candidate GroE substrates. The candidates included all of the *in vivo* tested obligate GroE substrates (MetK, GatY, and DapA), except for FtsE, which was not quantified in the proteomics, confirming the reasonableness of the selected threshold for the growth conditions used here. Then, the percentages of the candidates showing protein reductions in each of the GroEL substrate classes defined by Kerner et al. [47] were calculated. Eight percent of Class I, 32 % of Class II, and 56 % of Class III substrates were reduced in the GroE-depleted cells. The fraction of class members showing reduced protein amounts increased with the degree of GroEL dependence. It is also noteworthy that about 44 % of the Class III substrates (24 out of the 43 quantified proteins) did not meet the criteria for GroE obligate substrates. This again suggests that a significant fraction of Class III members, in addition to ParC, are not obligate substrates *in vivo*.

3.3.2 About 40 % of Class III Substrates Do Not Require GroE for Solubility

To assess whether ~40 % of the Class III substrates are not actually obligate GroE substrates, we developed methods, independent of proteomics, to verify their GroE requirement for solubility. The methods also aimed to comprehensively cover all of the Class III proteins suggested by Kerner et al. [47], since the proteomics of MGM100 detected only about half of the Class III substrates (43 of 84). The expression of individual target proteins was induced from a *tac* promoter in MGM100 cells after a 2 h depletion of GroE, and their total amounts and the proportion in the soluble fraction were measured. The obligate GroE substrates would be expected to become insoluble or be degraded. To validate the strategy, proteins for which the status of GroE-dependence had already been verified were examined: Enolase (spontaneously folding *in vitro*, Class I), GatD (partial GroE dependent-folding *in vitro*, Class II), MetK, FtsE, DapA (the *in vivo* obligate GroE substrates, Class III), and ParC (assigned as Class III, but functional in the GroE-depleted cells) [47, 50, 51]. Enolase, GatD, and ParC were soluble irrespective of the GroE level, whereas MetF, MetK, FtsE, and DapA aggregated in the GroE-depleted cells. The disappearance or persistence of the bands under GroE-depleted conditions was almost complete, enabling easy and clear discrimination.

Next, the method was extended to all of the essential genes in the three GroEL substrate classes, to test the GroE requirement for solubility (G.R.) in cells. The solubility of the essential Class I and Class II proteins (proteins with low expression levels were not measured) was independent of the GroE levels, confirming that Classes I and II were not dependent on GroE for folding. The GroE-independence of Ppa (Class I), GatD, LpxA, HemL, and FabG (Class II), which were candidates of GroE substrates identified by the proteomics, indicated that not all of the candidates predicted by the proteomics are *in vivo* obligate GroE substrates. The results showed that the *in vivo* obligate GroE substrates were enriched in Class III, but not in Class I and II proteins. More importantly, the results also indicated that approximately half of the Class III proteins did not require GroE for solubility, as already suggested by the above data. Depending on the *in vivo* GroE requirement, the Kerner's Class III substrates were divided into Class III$^+$ (plus; GroE dependent for solubility *in vivo*) and Class III$^-$ (minus; not GroE dependent for solubility *in vivo*). Finally, all of the remaining Class III substrates were tested for an *in vivo* GroE requirement. When all of the solubility assays for the Class III substrates were combined, Class III was divided into 49 Class III$^+$ and 34 Class III$^-$ substrates.

GroE-independent folding of Class III$^-$ proteins under GroE-depleted conditions was demonstrated for several representative Class III$^-$ proteins. First, the intracellular thymidine concentration was not decreased in the GroE-depleted cells, implying that FolE, one of the essential Class III$^-$ proteins, is functional in GroE-depleted cells, since FolE-defective cells only grow in thymidine-supplemented rich medium [60]. Second, the enzymatic activities of the Class III$^-$ proteins were directly assayed in the *E. coli* lysates. The activities of two essential Class III$^-$ proteins, an inositol monophosphatase, SuhB, and a tRNA methylase, TrmD, were measured in the lysates after the overexpression of Class III$^-$ proteins. The enzymes were active in both the GroE-depleted and -normal cells, indicating that the enzymes are both soluble and functional in the GroE-depleted cells. Among the Class III$^-$ substrates, only 4 proteins (ParC, FolE, SuhB, and TrmD) are essential. Although the functionality of the remaining Class III$^-$ proteins was not tested, at least all of the essential Class III$^-$ proteins were physiologically functional even in the GroE-depleted cells, further supporting the validity of the Class III$^+$ and III$^-$ grouping.

3.3.3 Identification of Other *In Vivo* Obligate GroE Substrates that Were Not Previously Assigned as Class III Substrates

After the complete survey of the Class III substrates, other novel GroE obligate substrates besides the identified Class III proteins were searched. The metabolomics data showed that the level of O-phosphoserine, the product of a Class II substrate, SerC, was reduced in the GroE-depleted cells, suggesting that SerC was not active in the cells. In addition, the proteomics data also suggested that SerC was reduced in the supernatant of the GroE-depleted cells. SerC was aggregated in the GroE-depleted cells, strongly suggesting that the *in vivo* obligate GroE substrates are not confined to the identified Class III substrates.

Other putative GroE substrates were also identified. First, candidate proteins were selected based on the proteomics data under GroE-depleted conditions. The GroE requirement for a dozen drastically reduced proteins in the GroE-depleted cells was verified. These included 3 Class II proteins (KdsA, PyrC, and NuoC) and 6 proteins that had not appeared among the GroEL interactors (GuaC, ThiL, SdaB, PyrD, NemA, GdhA). The solubility assays of these candidate proteins revealed that all of the Class II candidates and 2 of the other 6 candidates (PyrD and GdhA) behaved as *in vivo* GroE obligate substrates.

In addition, the homologs of Class III$^+$ substrates were searched on a database and were evaluated for the GroE requirement in GroE-depleted cells. From this strategy, TatD and YjhH were identified as *in vivo* GroE obligate substrates.

3.3.4 Class IV Substrates as *In Vivo* Obligate GroE Substrates

Eight new proteins were added to the list of *in vivo* GroE obligate substrates. We have combined these 8 *in vivo* substrates with the 49 Class III$^+$ members and now suggest their classification together to form a new group, the Class IV substrates (Table 15.1).

Taken together, a comprehensive assessment of the GroE requirement using a conditional GroE expression strain (MGM100) revealed that only ~60 % of Class III substrates are bona fide obligate GroE substrates *in vivo,* and renamed as Class III$^+$ (49 proteins). The *in vivo* obligate substrates, combined with newly identified 8 GroE obligate substrates, were termed Class IV substrates, for which folding is obligatorily dependent on GroE *in vivo* (Table 15.1) [48].

Table 15.1 Obligate GroE substrates (Class IV substrates). (Source: Data from [48])

Metabolic reactions						
Gene	b num	Sol[a]	MW[b]	pI	Folds[c]	Function
yqaB	b2690	13 %	20757	5.5	c.108	Fructose-1-phosphatase
rfbC	b2038	n.d.	21246	5.5	b.82	dTDP-4-deoxyrhamnose-3,5-epimerase
acpH	b0404	14 %	22938	5.9		Acyl carrier protein phosphodiesterase
serC	b0907	17 %	28177	5.4	c.67	3-phosphoserine/phosphohydroxythreonine aminotransferase
gatY	b2096	11 %	30782	5.9	c.1	D-tagatose 1,6-bisphosphate aldolase 2, catalytic subunit
dapA	b2478	n.d.	31238	6.0	c.1	Dihydrodipicolinate synthase
nanA	b3225	16 %	32556	5.6	c.1	N-acetylneuraminate lyase
metF	b3941	26 %	33068	6.0	c.1	5,10-methylenetetrahydrofolate reductase
dusC	b2140	10 %	35162	6.1	c.1	tRNA-dihydrouridine synthase C
hemB	b0369	7 %	35580	5.3	c.1	Porphobilinogen synthase
dusB	b3260	16 %	35830	6.3	c.1	tRNA-dihydrouridine synthase B
lipA	b0628	9 %	36043	8.1		Lipoate synthase
add	b1623	15 %	36355	5.4	c.1	Adenosine deaminase
yajO	b0419	5 %	36374	5.2	c.1	2-carboxybenzaldehyde reductase
ltaE	b0870	10 %	36455	5.8	c.67	L-allo-threonine aldolase, PLP-dependent
pyrD	b0945	13 %	36775	7.7	c.1	Dihydro-orotate oxidase, FMN-linked
nagZ	b1107	12 %	37556	5.9	c.1	Beta N-acetyl-glucosaminidase
fbaB	b2097	5 %	38071	6.2	c.1	Fructose-bisphosphate aldolase class I
kdsA	b1215	30 %	38808	6.3	c.1	3-deoxy-D-manno-octulosonate 8-phosphate synthase
pyrC	b1062	4 %	38817	5.8	c.1	Dihydro-orotase
dadX	b1190	3 %	38842	6.6	c.1; b.49	Alanine racemase 2, PLP-binding
asd	b3433	19 %	39970	5.4	c.2; d.81	Aspartate-semialdehyde dehydrogenase, (nicotinamide adenine dinucleotide (phosphate)) NAD(P)-binding
fadA	b3845	10 %	40872	6.3	c.95	3-ketoacyl-CoA thiolase (thiolase I)
bioF	b0776	3 %	41557	6.6	c.67	8-amino-7-oxononanoate synthase
metK	b2942	48 %	41898	5.1	d.130	Methionine adenosyltransferase 1
argE	b3957	42 %	42301	5.5	d.58; c.56	Acetylornithine deacetylase
lldD	b3605	4 %	42683	6.3	c.1	L-lactate dehydrogenase, FMN-linked
fabF	b1095	8 %	42999	5.7	c.95	3-oxoacyl-[acyl-carrier-protein] synthase II
thiH	b3990	n.d.	43279	6.6		Thiamin biosynthesis ThiGH complex subunit
csdB	b1680	11 %	44390	5.9	c.67	Selenocysteine lyase, PLP-dependent
rspA	b1581	4 %	45919	5.7	d.54; c.1	Mannonate/altronate dehydratase

Table 15.1 (continued)

Metabolic reactions						
Gene	b num	Sol[a]	MW[b]	pI	Folds[c]	Function
deoA	b4382	13%	47148	5.2	d.41; a.46; c.27	Thymidine phosphorylase
dadA	b1189	12%	47558	6.2	c.5; d.16; c.3; c.4; c.2	D-amino acid dehydrogenase
gdhA	b1761	22%	48530	6.0	c.2; c.58	Glutamate dehydrogenase
eutB	b2441	13%	49334	4.8	a.105; c.1	Ethanolamine ammonia-lyase, large subunit, heavy chain
xylA	b3565	n.d.	49691	5.8	c.1	D-xylose isomerase
uxaC	b3092	8%	53925	5.4	c.1	Uronate isomerase
araA	b0062	8%	56021	6.1	c.118; b.71	L-arabinose isomerase
aldB	b3588	n.d.	56306	5.4	c.82	Aldehyde dehydrogenase
sdhA	b0723	10%	64355	5.9	a.7; c.3; d.168	Succinate dehydrogenase, flavoprotein subunit
frdA	b4154	51%	65904	5.9	a.7; c.3; d.168	Fumarate reductase (anaerobic) catalytic and NAD/flavoprotein subunit
nuoC	b2286	18%	68683	6.0	e.18	NADH:ubiquinone oxidoreductase, chain C,D
Other processes						
ftsE	b3463	12%	24425	9.4	c.37	Predicted transporter subunit: ATP-binding component of ABC superfamily
fucR	b2805	36%	27342	7.8	a.4; c.35	DNA-binding transcriptional activator
tatD[‡]	b4483	n.d.	28961	5.2	c.1	DNase, magnesium-dependent
nfo	b2159	14%	31444	5.4	c.1	Endonuclease IV with intrinsic 3′-5′ exonuclease activity
argP	b2916	17%	33438	6.4	c.94; a.4	DNA-binding transcriptional activator, replication initiation inhibitor
tldE	b4235	16%	48313	5.4		Protease involved in Microcin B17 maturation and in sensitivity to the DNA gyrase inhibitor LetD
pepQ	b3847	15%	50122	5.6	d.127	Proline dipeptidase
tldD	b3244	91%	51295	4.9		Protease involved in Microcin B17 maturation and in sensitivity to the DNA gyrase inhibitor LetD
Unknown						
ycfH	b1100	n.d.	29772	5.2	c.1	Predicted metallodependent hydrolase
yafD	b0209	10%	29972	9.6	d.151	Conserved protein
ybjS	b0868	7%	38089	8.8	c.2	Predicted NAD(P)H-binding oxidoreductase
yneB	b1517	34%	31859	6.1	c.1	Predicted aldolase

Table 15.1 (continued)

Metabolic reactions						
Gene	b num	Sol[a]	MW[b]	pI	Folds[c]	Function
yjhH[‡]	b4298	8%	32714	5.3	c.1	Predicted lyase/synthase
yjjU	b4377	7%	39794	8.7		Predicted esterase
yfbQ	b2290	21%	45468	5.9	c.67	Predicted aminotransferase

dTDP thymidine diphosphate glucose, *tRNA* transfer ribonucleic acid, *PLP* pyridoxal-phosphate, *FMN* flavin mononucleotide, *ATP* adenosine triphosphate, *DNA* deoxyribose nucleic acid
n.d. not determined in [29]
[a] Sol: solubility in a reconstituted cell-free translation without any chaperone [29]
[b] MW: molecular weight (Da)
[c] SCOP fold ID of the proteins

Class IV substrates are restricted to proteins with molecular weights that could be encapsulated in the chaperonin cavity, are enriched in alanine/glycine residues, and have a strong structural preference for aggregation-prone folds. Notably, ~70% of the Class IV substrates appear to be metabolic enzymes, supporting a hypothetical role of GroE in enzyme evolution.

3.3.5 GroE-Dependency of *In Vitro* Translated Class III⁻ and Newly Identified Class IV Proteins.

To elucidate the *in vitro* GroE-dependency of the newly identified substrates, a Class III$^-$ protein (FolE) and several Class IV proteins (DapA as a known obligate substrate, and SerC and KdsA as newly identified Class IV substrates) were translated by PURE system [15, 22, 29]. The requirements of the DnaK (DnaK, DnaJ, and GrpE) and GroE (GroEL and GroES) chaperone systems on the folding were monitored by the solubility and the appearance of folded structures, defined as a sharp band in native polyacrylamide gel electrophoresis (PAGE). FolE (Class III$^-$) was soluble and formed a folded structure even in the absence of chaperones. In contrast, all of the Class IV proteins tested (DapA, SerC, and KdsA) were aggregation-prone without chaperones. The addition of the DnaK system increased the solubilities of the Class IV proteins to a greater or lesser extent, but the folded structures were not detected in native PAGE, implying that the soluble but unfolded structures in the presence of DnaK might be easily degraded *in vivo*. The Class IV proteins were soluble and formed folded structures only in the presence of GroE. The *in vitro* folding assay further confirmed our conclusion, in which the Class IV substrates, including the substrates that were not originally assigned as Class III (SerC and KdsA), stringently require GroE for correct folding.

3.3.6 Amino Acid Sequence Features of Class IV and III$^-$ Proteins

To define the features that are correlated with *in vivo* GroE dependency, physicochemical properties of the Class IV and III$^-$ proteins were compared. First, the molecular weights of the Class IV substrates were distributed normally, with a peak around 40 kDa, and ranging from 21 to 68 kDa, whereas the molecular weights of the 34 Class III$^-$ proteins ranged broadly, including 5 proteins smaller than 20 kDa and 4 proteins larger than 70 kDa, including ParC (84 kDa).

Second, the pI distribution and the hydrophobicity of Classes IV and III$^-$ were compared. The pI values of the Class IV substrates were distributed with a single peak around pI 5.8, whereas the pI distribution of the Class III$^-$ members was bimodal and similar to that of all cytosolic proteins of *E. coli.* The hydrophobicity distribution of the Class IV substrates was similar to that of all cytosolic proteins, whereas the Class III$^-$ proteins had lower hydrophobicity than either Class IV or all cytosolic proteins.

Third, the amino acid compositions of the Class IV and III$^-$ proteins were analyzed. As expected from the differences in the pI distributions and hydrophobicity, the amino acid compositions also differed between Classes IV and III$^-$. Specifically, positively charged amino acids (arginine, lysine, and histidine) were enriched among the Class III$^-$ members. Neither hydrophobic amino acids (phenylalanine, tyrosine, tryptophan, isoleucine, leucine, and valine) nor other amino acids (negative, polar, neutral, and sulfur-containing amino acids) were enriched in either Class IV or III$^-$.

3.3.7 Class IV Substrates Are Inherently Aggregation-Prone

Inherent solubility of Class IV substrates under chaperone-free conditions were already evaluated by global aggregation analysis using PURE system [29]. The histograms of the inherent solubilities of Classes IV and III$^-$ indicated a striking difference. The Class IV substrates were inherently highly aggregation-prone, whereas the Class III$^-$ proteins were broadly distributed from soluble to aggregation-prone.

3.3.8 Structural Features of Class IV Substrates

Previous studies on GroE substrates revealed that TIM barrel folds were substantially enriched in Class III proteins [47]. Surprisingly, all of the TIM barrel folds identified in the Class III proteins, except one (GatZ), were within the Class IV substrates (Table 15.1). As a result, the TIM barrel folds were further enriched in the Class IV substrates with 25 out of 57 Class IV substrates identified in this study possessing one (Table 15.1). This enrichment further supports the notion that the TIM barrel fold is correlated with GroE dependency [47, 58]. Not only TIM barrel folds (c.1 in SCOP database terminology [37]), but also FAD/NAD(P)-binding domains (c.3), PLP-dependent transferase like folds (c.67), and thiolase folds (c.95) were highly enriched in the class IV substrates, as compared with the frequency of their appearance in all cytosolic proteins. As Masters et al. pointed out in the review [58], the aldolase superfamily (c.1.10) subclass of TIM barrel folds is preferred among Class III proteins. In addition, metallo-dependent hydrolase (c.1.9) superfamily folds is also enriched within the Class IV members.

3.3.9 Class IV Homologs in a GroE-Lacking Organism

The genome of *Ureaplasma urealyticum* lacks the groELS gene [61]. Therefore, it is worth testing whether the Class IV homologs in *Ureaplasma* are GroE dependent. Five homologs of Class IV members (*Uu*DeoA, *Uu*CsdB, *Uu*GatY, *Uu*MetK, and *Uu*YcfH) were found in *Ureaplasma*, by BLAST search. The genes encoding *Uu*DeoA, *Uu*MetK, and *Uu*YcfH were cloned and overexpressed in MGM100, and their solubilities were assessed under GroE-depleted conditions. Strikingly, all of the Class IV homologs tested were soluble, even in the GroE-depleted cells, indicating that the GroE dependency was not conserved among the homologs. In addition, the S-adenosylmethionine synthase activity of *Uu*MetK in the lysate of the GroE-depleted cells was comparable to that of the GroE-normal cells. Moreover, the leaky *Uu*MetK expression suppressed the overexpression of MetE, which is one of the hallmarks of GroE-depleted cells [53, 58]. Collectively, *Uu*MetK is active in the GroE-depleted cells. Intriguingly, the amino acid compositions of the *Ureaplasma* Class IV homologs revealed that the Ala/Gly fractions in all of the homologs were lower than those in their *E. coli* counterparts, whereas the contents of other amino acid groups, including aromatic, hydrophobic and positive amino acids, were indistinguishable from those in the *E. coli* counterparts. The amino acid content analysis again raises the possibility that a high Ala/Gly content might be involved in GroE dependency.

3.3.10 Possibility to Create GroE-Lacking *E. coli*

Verified obligate GroE substrates included only 6 genes (DapA, ASD, MetK, FtsE, HemB, and KdsA) essential for the viability of *E. coli* in rich medium. Although unidentified essential Class IV substrates may exist among the proteins that were not tested, we can predict the possible phenotypic defects caused by their inactivation in the GroE-depleted cells. If there are no further essential GroE-dependent proteins, then the complementation of these 6 essential genes by some means should generate an *E. coli* strain that can grow without *groEL/ES*. Such viable groEL/ES-knockout *E. coli* would provide the answer to the long-standing question of why GroE is essential for cell viability. Alternatively, the complementation of the 6 essential genes in *E. coli* lacking groEL/ES could be still lethal, due to the presence of unidentified essential Class IV substrates. In such a case, we can extend the phenotypic analysis to find the unidentified Class IV members, using the engineered *E. coli*.

3.3.11 Implication to Protein Evolution

Chaperones are known to provide a buffering system for genetic mutations, and thus promote genetic diversity [19, 62]. A recent quantitative assessment clearly showed that GroE promotes enzyme evolution by buffering the destabilizing mutations that confer improved enzymatic activities [21]. Since the destabilization of proteins generally results in their intracellular aggregation, aggregation-prone proteins, such as the Class IV substrates, could survive mutations if their aggregation is prevented by GroE, leading to the acquisition of diversity and/or the potential improvement of the enzymes in the Class IV substrates.

As already mentioned, many chaperonin GroE substrates are metabolic enzymes. However, the relationship between chaperonins and metabolism is still unclear. The distribution of GroE substrate enzymes in the metabolic network was investigated using network analysis techniques. Bioinformatics analysis revealed that as GroE requirement increases, substrate enzymes are more laterally distributed in the metabolic pathways [63]. In addition, comparative genome analysis showed that the GroE-dependent substrates were less conserved, suggesting that these substrates were acquired later on in evolutionary history [63]. This result implies the expansion of metabolic networks due to this chaperonin, and it supports the existing hypothesis of acceleration of evolution by chaperonins [64].

4 *In Vivo* Substrates of *E. coli* DnaK and Trigger Factor

DnaK, the major bacterial Hsp70 family is one of most abundant chaperones in the cytosol of *E. coli.* After an initial attempt to identify DnaK substrates in *E. coli* lacking trigger factor upon DnaK depletion [12], the direct isolation of DnaK-substrate complexes was conducted [65]. Quantitative MS-based proteomics revealed that DnaK interacts with at least ~700 proteins, including ~180 relatively aggregation-prone proteins. The study also focused on the role of DnaK in the *E. coli* chaperone network. Individual deletion of TF or depletion of GroE leads to specific changes in the DnaK interactome, suggesting that DnaK and other chaperones cooperates each other in *E. coli* cells. Taken together, Calloni et al. concluded that DnaK is a central organized of the chaperone network [65].

Regarding TF, modified ribosome profiling, which sequences messenger ribonucleic acid (mRNA) fragments covered by translating ribosomes, combined with a procedure to affinity purify ribosomes whose nascent polypeptides are associated with TF, called selective ribosome profiling, was used to reveal *in vivo* target proteins of TF [66]. Although TF can interact with many polypeptides, β-barrel outer membrane proteins are the most prominent substrates of TF [66].

5 Simulation of *E. coli* Chaperone Network

Proteins must fold into the unique native structures in a crowded environment in the cell. To understand the role of chaperones such as GroE or DnaK in the cell, we have to consider a balance between folding assisted by chaperones and degradation by proteases to maintain protein homeostasis, called proteostasis [2]. To gain insight into the interplay of processes and chaperones that maintain a functional proteome, a computational model called FoldEco was developed by Powers et al. [67].

FoldEco (http://foldeco.scripps.edu.) is the result of a joint effort to model how the proteostasis network affects protein folding in *E. coli*. *E. coli* was chosen as a model organism because its proteostasis networks are better characterized biochemically than those of any other organism. FoldEco simulates what happens to soluble proteins of interest as they are produced in the *E. coli* cytosol with five systems: protein synthesis and folding, DnaK system, GroE system, ClpB/DnaK disaggregation system, and protein degradation system. We can easily change parameter such as concentrations of chaperones in FoldEco to predict the fate of proteins of interest.

6 Concluding Remarks

After the emergence of chaperone concept in the late 1980s [68], extensive efforts have been dedicated to elucidate the molecular mechanism of chaperone proteins such as GroE or DnaK [2]. Compared to the understanding of chaperones as molecular machine, the role of chaperones in the cell still remains to be elucidated. Recent proteome-wide approaches from cell-free to *in vivo* would open the door to understand cellular functions of chaperones.

References

1. Anfinsen CB (1973) Principles that govern the folding of protein chains. Science 181:223–230
2. Hartl FU, Bracher A, Hayer-Hartl M (2011) Molecular chaperones in protein folding and proteostasis. Nature 475:324–332
3. Hoffmann A, Bukau B, Kramer G (2010) Structure and function of the molecular chaperone trigger factor. Biochim Biophys Acta 1803:650–661
4. Richter K, Haslbeck M, Buchner J (2010) The heat shock response: life on the verge of death. Mol Cell 40:253–266
5. Taguchi H (2005) Chaperonin GroEL meets the substrate protein as a "load" of the rings. J Biochem 137:543–549
6. Fayet O, Ziegelhoffer T, Georgopoulos C (1989) The groES and groEL heat shock gene products of *Escherichia coli* are essential for bacterial growth at all temperatures. J Bacteriol 171:1379–1385
7. Horwich AL, Low KB, Fenton WA, Hirshfield IN, Furtak K (1993) Folding in vivo of bacterial cytoplasmic proteins: role of GroEL. Cell 74:909–917

8. Xu Z, Horwich AL, Sigler PB (1997) The crystal structure of the asymmetric GroEL-GroES-(ADP)7 chaperonin complex. Nature 388:741–750
9. Sakikawa C, Taguchi H, Makino Y, Yoshida M (1999) On the maximum size of proteins to stay and fold in the cavity of GroEL underneath GroES. J Biol Chem 274:21251–21256
10. Deuerling E, Schulze-Specking A, Tomoyasu T, Mogk A, Bukau B (1999) Trigger factor and DnaK cooperate in folding of newly synthesized proteins. Nature 400:693–696
11. Teter SA, Houry WA, Ang D, Tradler T, Rockabrand D, Fischer G, Blum P, Georgopoulos C, Hartl FU (1999) Polypeptide flux through bacterial Hsp70: DnaK cooperates with trigger factor in chaperoning nascent chains. Cell 97:755–765
12. Deuerling E, Patzelt H, Vorderwulbecke S, Rauch T, Kramer G, Schaffitzel E, Mogk A, Schulze-Specking A, Langen H, Bukau B (2003) Trigger factor and DnaK possess overlapping substrate pools and binding specificities. Mol Microbiol 47:1317–1328
13. Vorderwulbecke S, Kramer G, Merz F, Kurz TA, Rauch T, Zachmann-Brand B, Bukau B, Deuerling E (2004) Low temperature or GroEL/ES overproduction permits growth of *Escherichia coli* cells lacking trigger factor and DnaK. FEBS Lett 559:181–187
14. Genevaux P, Keppel F, Schwager F, Langendijk-Genevaux PS, Hartl FU, Georgopoulos C (2004) In vivo analysis of the overlapping functions of DnaK and trigger factor. EMBO Rep 5:195–200
15. Ying BW, Taguchi H, Kondo M, Ueda T (2005) Co-translational involvement of the chaperonin GroEL in the folding of newly translated polypeptides. J Biol Chem 280:12035–12040
16. Ying BW, Taguchi H, Ueda T (2006) Co-translational binding of GroEL to nascent polypeptides is followed by post-translational encapsulation by GroES to mediate protein folding. J Biol Chem 281:21813–21819
17. Jenkins AJ, March JB, Oliver IR, Masters M (1986) A DNA fragment containing the *groE* genes can suppress mutations in the *Escherichia coli* dnaA gene. Mol Gen Genet 202(3):446–454
18. Van Dyk TK, Gatenby AA, LaRossa RA (1989) Demonstration by genetic suppression of interaction of GroE products with many proteins. Nature 342:451–453
19. Rutherford SL, Lindquist S (1998) Hsp90 as a capacitor for morphological evolution. Nature 396:336–342
20. Fares MA, Ruiz-Gonzalez MX, Moya A, Elena SF, Barrio E (2002) Endosymbiotic bacteria: groEL buffers against deleterious mutations. Nature 417:398
21. Tokuriki N, Oldfield CJ, Uversky VN, Berezovsky IN, Tawfik DS (2009) Do viral proteins possess unique biophysical features? Trends Biochem Sci 34:53–59
22. Shimizu Y, Inoue A, Tomari Y, Suzuki T, Yokogawa T, Nishikawa K, Ueda T (2001) Cell-free translation reconstituted with purified components. Nat Biotechnol 19:751–755
23. Shimizu Y, Kanamori T, Ueda T (2005) Protein synthesis by pure translation systems. Methods 36:299–304
24. Ying BW, Taguchi H, Ueda H, Ueda T (2004) Chaperone-assisted folding of a single-chain antibody in a reconstituted translation system. Biochem Biophys Res Commun 320:1359–1364
25. Chiti F, Taddei N, Baroni F, Capanni C, Stefani M, Ramponi G, Dobson CM (2002) Kinetic partitioning of protein folding and aggregation. Nat Struct Biol 9:137–143
26. Chiti F, Stefani M, Taddei N, Ramponi G, Dobson CM (2003) Rationalization of the effects of mutations on peptide and protein aggregation rates. Nature 424:805–808
27. Williams AD, Portelius E, Kheterpal I, Guo JT, Cook KD, Xu Y, Wetzel R (2004) Mapping abeta amyloid fibril secondary structure using scanning proline mutagenesis. J Mol Biol 335:833–842
28. de Groot NS, Aviles FX, Vendrell J, Ventura S (2006) Mutagenesis of the central hydrophobic cluster in Abeta42 Alzheimer's peptide. Side-chain properties correlate with aggregation propensities. FEBS J 273:658–668
29. Niwa T, Ying BW, Saito K, Jin W, Takada S, Ueda T, Taguchi H (2009) Bimodal protein solubility distribution revealed by an aggregation analysis of the entire ensemble of *Escherichia coli* proteins. Proc Natl Acad Sci U S A 106:4201–4206

30. Kitagawa M, Ara T, Arifuzzaman M, Ioka-Nakamichi T, Inamoto E, Toyonaga H, Mori H (2005) Complete set of ORF clones of *Escherichia coli* ASKA library (a complete set of *E. coli* K-12 ORF archive): unique resources for biological research. DNA Res 12:291–299
31. Hopp TP, Woods KR (1981) Prediction of protein antigenic determinants from amino acid sequences. Proc Natl Acad Sci U S A 78:3824–3828
32. Kyte J, Doolittle RF (1982) A simple method for displaying the hydropathic character of a protein. J Mol Biol 157:105–132
33. Michelitsch MD, Weissman JS (2000) A census of glutamine/asparagine-rich regions: implications for their conserved function and the prediction of novel prions. Proc Natl Acad Sci U S A 97:11910–11915
34. Chou PY, Fasman GD (1974) Conformational parameters for amino acids in helical, beta-sheet, and random coil regions calculated from proteins. BioChemistry 13:211–222
35. Jones DT (1999) Protein secondary structure prediction based on position-specific scoring matrices. J Mol Biol 292:195–202
36. Bryson K, McGuffin LJ, Marsden RL, Ward JJ, Sodhi JS, Jones DT (2005) Protein structure prediction servers at University College London. Nucleic Acids Res 33:W36–W38
37. Murzin AG, Brenner SE, Hubbard T, Chothia C (1995) SCOP: a structural classification of proteins database for the investigation of sequences and structures. J Mol Biol 247:536–540
38. Fernandez-Escamilla AM, Rousseau F, Schymkowitz J, Serrano L (2004) Prediction of sequence-dependent and mutational effects on the aggregation of peptides and proteins. Nat Biotechnol 22:1302–1306
39. Conchillo-Sole O, de Groot NS, Aviles FX, Vendrell J, Daura X, Ventura S (2007) AGGRESCAN: a server for the prediction and evaluation of "hot spots" of aggregation in polypeptides. BMC Bioinformatics 8:65
40. Trovato A, Seno F, Tosatto SC (2007) The PASTA server for protein aggregation prediction. Protein Eng Des Sel 20:521–523
41. Agostini F, Vendruscolo M, Tartaglia GG (2012) Sequence-based prediction of protein solubility. J Mol Biol 421:237–241
42. Stiglic G, Kocbek S, Pernek I, Kokol P (2012) Comprehensive decision tree models in bioinformatics. PLoS ONE 7:e33812
43. Fang Y, Fang J (2013) Discrimination of soluble and aggregation-prone proteins based on sequence information. Mol Biosyst 9:806–811
44. Niwa T, Kanamori T, Ueda T, Taguchi H (2012) Global analysis of chaperone effects using a reconstituted cell-free translation system. Proc Natl Acad Sci U S A 109:8937–8942
45. Agashe VR, Guha S, Chang HC, Genevaux P, Hayer-Hartl M, Stemp M, Georgopoulos C, Hartl FU, Barral JM (2004) Function of trigger factor and DnaK in multidomain protein folding: increase in yield at the expense of folding speed. Cell 117:199–209
46. Hoffmann A, Merz F, Rutkowska A, Zachmann-Brand B, Deuerling E, Bukau B (2006) Trigger factor forms a protective shield for nascent polypeptides at the ribosome. J Biol Chem 281:6539–6545
47. Kerner MJ, Naylor DJ, Ishihama Y, Maier T, Chang HC, Stines AP, Georgopoulos C, Frishman D, Hayer-Hartl M, Mann M, Hartl FU (2005) Proteome-wide analysis of chaperonin-dependent protein folding in *Escherichia coli*. Cell 122:209–220
48. Fujiwara K, Ishihama Y, Nakahigashi K, Soga T, Taguchi H (2010) A systematic survey of in vivo obligate chaperonin-dependent substrates. EMBO J 29:1552–1564
49. Langer T, Lu C, Echols H, Flanagan J, Hayer MK, Hartl FU (1992) Successive action of DnaK, DnaJ and GroEL along the pathway of chaperone-mediated protein folding. Nature 356:683–689
50. McLennan N, Masters M (1998) GroE is vital for cell-wall synthesis. Nature 392:139
51. Fujiwara K, Taguchi H (2007) Filamentous morphology in GroE-depleted *Escherichia coli* induced by impaired folding of FtsE. J Bacteriol 189:5860–5866
52. Houry WA, Frishman D, Eckerskorn C, Lottspeich F, Hartl FU (1999) Identification of in vivo substrates of the chaperonin GroEL. Nature 402:147–154

53. Chapman E, Farr GW, Usaite R, Furtak K, Fenton WA, Chaudhuri TK, Hondorp ER, Matthews RG, Wolf SG, Yates JR, Pypaert M, Horwich AL (2006) Global aggregation of newly translated proteins in an *Escherichia coli* strain deficient of the chaperonin GroEL. Proc Natl Acad Sci U S A 103:15800–15805
54. Shimamura T, Koike-Takeshita A, Yokoyama K, Masui R, Murai N, Yoshida M, Taguchi H, Iwata S (2004) Crystal structure of the native chaperonin complex from *Thermus thermophilus* revealed unexpected asymmetry at the cis-cavity. Structure 12:1471–1480
55. Endo A, Kurusu Y (2007) Identification of in vivo substrates of the chaperonin GroEL from *Bacillus subtilis*. Biosci Biotechnol Biochem 71:1073–1077
56. Ishihama Y, Oda Y, Tabata T, Sato T, Nagasu T, Rappsilber J, Mann M (2005) Exponentially modified protein abundance index (emPAI) for estimation of absolute protein amount in proteomics by the number of sequenced peptides per protein. Mol Cell Proteomics 4:1265–1272
57. Ishihama Y, Schmidt T, Rappsilber J, Mann M, Hartl FU, Kerner MJ, Frishman D (2008) Protein abundance profiling of the *Escherichia coli* cytosol. BMC Genomics 9:102
58. Masters M, Blakely G, Coulson A, McLennan N, Yerko V, Acord J (2009) Protein folding in *Escherichia coli*: the chaperonin GroE and its substrates. Res Microbiol 160:267–277
59. Fujiwara K, Taguchi H (2012) Mechanism of methionine synthase overexpression in chaperonin-depleted *Escherichia coli*. Microbiology 158:917–924
60. El Yacoubi B, Bonnett S, Anderson JN, Swairjo MA, Iwata-Reuyl D, de Crecy-LagardV (2006) Discovery of a new prokaryotic type I GTP cyclohydrolase family. J Biol Chem 281:37586–37593
61. Glass JI, Lefkowitz EJ, Glass JS, Heiner CR, Chen EY, Cassell GH (2000) The complete sequence of the mucosal pathogen *Ureaplasma urealyticum*. Nature 407:757–762
62. Queitsch C, Sangster TA, Lindquist S (2002) Hsp90 as a capacitor of phenotypic variation. Nature 417:618–624
63. Takemoto K, Niwa T, Taguchi H (2011) Difference in the distribution pattern of substrate enzymes in the metabolic network of *Escherichia coli*, according to chaperonin requirement. BMC Syst Biol 5:98
64. Bogumil D, Dagan T (2012) Cumulative impact of chaperone-mediated folding on genome evolution. BioChemistry 51:9941–9953
65. Calloni G, Chen T, Schermann SM, Chang HC, Genevaux P, Agostini F, Tartaglia GG, Hayer-Hartl M, Hartl FU (2012) DnaK functions as a central hub in the *E. coli* chaperone network. Cell Rep 1:251–264
66. Oh E, Becker AH, Sandikci A, Huber D, Chaba R, Gloge F, Nichols RJ, Typas A, Gross CA, Kramer G, Weissman JS, Bukau B (2011) Selective ribosome profiling reveals the cotranslational chaperone action of trigger factor in vivo. Cell 147:1295–1308
67. Powers ET, Powers DL, Gierasch LM (2012) FoldEco: a model for proteostasis in *E. coli*. Cell Rep 1:265–276
68. Ellis J (1987) Proteins as molecular chaperones. Nature 328:378–379

Chapter 16
Chaperone-Proteases of Mycobacteria

Juerg Laederach, Julia Leodolter, Jannis Warweg and Eilika Weber-Ban

Abstract Energy-dependent protein degradation ensures protein homeostasis in all organisms. It requires the concerted action of unfoldase and protease components to achieve the processive cleavage of substrate proteins into small peptides. Rapid changes in nutritional sources and oxygen levels are constant challenges of the mycobacterial habitat, and survival under these conditions requires a multitude of adaptive response mechanisms including—as one important cornerstone—protein degradation. Three energy-dependent protease systems invariantly exist in all mycobacteria: the membrane-associated FtsH, the Clp protease complexes, and a bacterial proteasome assembly that is unique to the actinobacterial phylum. In this chapter, structural features, substrate recruitment principles, and cellular roles of all three systems are discussed.

1 Introduction

The secret to success for bacteria is quick adaptation to their surroundings. Mycobacteria, both pathogenic and non-pathogenic, encounter many challenges in their environment, living in the rapidly changing and often highly competitive milieu of soil and marshes or under the constant onslaught of a host's defensive warfare inside macrophages [1–3]. Their survival thus depends on a fast and variable response to a range of conditions that allows them to adjust their metabolism to long periods of starvation and to enter dormant, low-replicative states [4, 5]. While protein degradation is only one facet of this response and adaptability, it is an important one, since it affords the organism rapid changes in protein composition without ad-

E. Weber-Ban (✉) · J. Laederach · J. Leodolter · J. Warweg
Institute of Molecular Biology & Biophysics, ETH Zurich, Zurich, Switzerland
e-mail: eilika@mol.biol.ethz.ch

J. Laederach
e-mail: ljuerg@mol.biol.ethz.ch

J. Leodolter
e-mail: lejulia@mol.biol.ethz.ch

J. Warweg
e-mail: jannis.warweg@mol.biol.ethz.ch

W. A. Houry (ed.), *The Molecular Chaperones Interaction Networks in Protein Folding and Degradation,* Interactomics and Systems Biology, DOI 10.1007/978-1-4939-1130-1_16,

ditional synthetic effort and can remain active under nutrient stress or even relieve it. The molecular machines that carry out this fundamental cellular activity, the chaperone-proteases, are capable of great selective substrate discrimination while at the same time processing a large range of substrates, a feat accomplished with the help of their modular architecture as well as their cooperation with adaptors and regulators. Like all bacteria, mycobacteria possess more than one of these degradation complexes. The minimal set of energy-dependent protein degradation systems encountered in this genus consists of the membrane-bound FtsH protease, the cytosolic ClpCP and ClpXP complexes, and a bacterial proteasome that is a special occurrence in the actinobacterial phylum and is not encountered in other bacteria [6] (Fig. 16.1). Under standard culture conditions, only the Clp system and FtsH are essential [7–9], whereas the proteasome and its associated AAA unfoldase are not required and thus appear to present a degradation pathway that is important or acts mostly under specific environmental pressures [10, 11]. The highly pathogenic *Mycobacterium tuberculosis* employs this degradation pathway as part of its survival strategy inside host macrophages [10]. The recruitment of substrate proteins for degradation by the bacterial proteasome occurs by an ubiquitin-like modification pathway termed pupylation [12–14]. Despite the functional analogy, this posttranslational modification has evolved independently in bacteria and the modification enzymes are not related to the eukaryotic E3 ligases [14–17]. The ClpP protease in association with its AAA regulators likely plays a broader role in mycobacteria and relatives than in bacteria outside the actinobacterial phylum, since it appears to be essential, and the broad-specificity protease Lon is missing in many members of mycobacteria. However, some roles are clearly conserved and amongst the substrates are a number of "old friends" already described for the *E. coli* Clp system, as for example ssrA-tagged proteins [18, 19], the cell division protein FtsZ [20], or the anti-σ factor RseA [21].

Given the importance of these proteolytic systems for the pathogenic members of mycobacteria, either for general viability or for infection, it is obvious that the development of drugs based on inhibition of chaperone-proteases offers powerful new avenues in the treatment of one of mankind's oldest scourges.

2 Unfold and Degrade—A Powerful Principle for Protein Destruction

In all organisms, energy-dependent protein degradation represents an essential activity that ensures protein homeostasis and participates in regulatory processes [22–24]. As proteins have evolved to withstand the random access of small, energy-independent proteases present in the cytosol, large molecular machines referred to as chaperone-proteases are required for protein breakdown. They combine adenosine triphosphate (ATP)-dependent unfolding of the substrate protein with processive degradation into small peptides inside a sequestered molecular space [22, 25] (Fig. 16.1). The general cylindrical architecture of these multi-enzyme assemblies

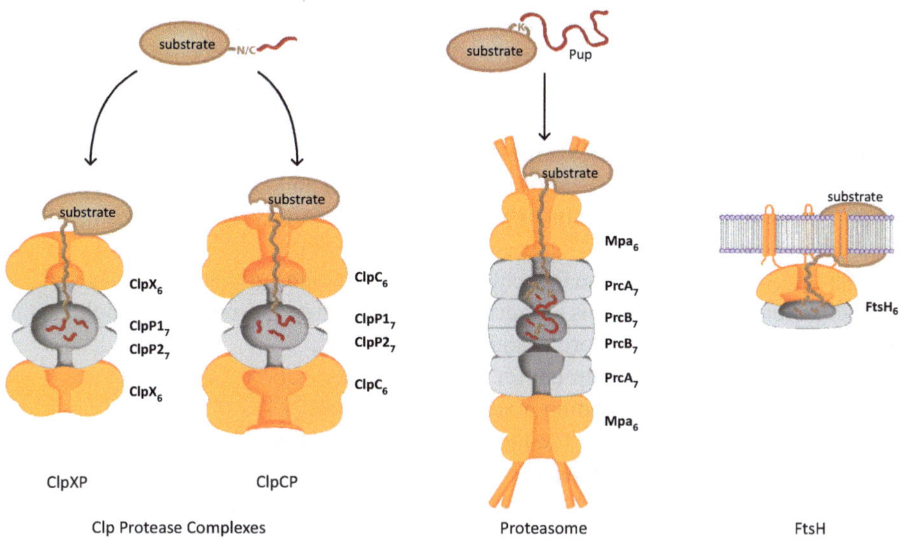

Fig. 16.1 The minimal set of chaperone-proteases present in mycobacteria. Different chaperone-proteases with common architectural and mechanistic features coexist in mycobacteria. A barrel-shaped proteolytic particle enclosing a sequestered space for degradation (*grey*) stacks together with an ATPase ring (*orange*) to form the active protease complex. The ATPase ring recruits substrates (*beige*) through interaction with a recognition element (*red*), unfolds them in an ATPase-dependent fashion and translocates them into the protease core for degradation into small peptides. Recruitment to the proteasome complex occurs by a posttranslational modification pathway with functional analogies to ubiquitination

is conserved between all three domains of life. The ATPase and protease subunits are arranged around a central axis into six- or seven-membered rings that stack to form an enclosed space. The proteolytic core cylinder displays the protease-active sites on the inside walls of its cavity and is closed off by narrow entrance pores at both ends. The ATPase rings stack to both faces of the core cylinder, thereby controlling access to the protease active sites. They all belong to the family of AAA+ proteins (*A*TPases *a*ssociated with various cellular *a*ctivities), defined by the presence of the so-called AAA+ module that consists of a P-loop ATPase domain followed by a small helical domain [26].

Despite a similar architecture, different chaperone-proteases exist in the different domains of life. In eukaryotes, the 26S proteasome represents the main extralysosomal degradation pathway [27, 28]. It is built of the 20S core cylinder, a four-layer stacked-ring structure of two inner β-subunit-rings sandwiched by two outer α-subunit-rings, and associates at both ends with the ATPase-ring of the regulatory particle [29]. Likewise, in archaea, the proteasome in complex with a AAA ring is responsible for energy-dependent protein degradation [30]. In bacteria, however, instead of a single main chaperone-protease, multiple and partially redundant degradation systems exist [22, 31] (Fig. 16.1). These complexes either carry the protease and unfoldase activity on different subunits, like ClpP that associates with the AAA unfoldases ClpA, ClpC, or ClpX, and HslV that associates with the AAA ring HslU, or on separate domains of a single polypeptide, like the cytosolic Lon protease and the membrane-anchored FtsH. As a general rule, bacteria do not harbor proteasomes; however, one large phylum makes an exception. Actinobacteria, in addition to the usual bacterial chaperone-proteases, contain also proteasomes [32] (Table 16.1). This large and varied phylum occupies a special place in microbial evolution as one of the earliest prokaryotic lineages [33, 34]. Interestingly, aside from encoding a proteasome, its members share other traits with eukaryotes, as for example exospore formation or the eukaryotic-like fatty acid synthase I (FASI). This has led to the speculation that eukaryotes and archaea might have evolved from actinobacteria [35]. Under this scenario, the actinobacterial proteasome could be the precursor from which other proteasomes derived [36].

Modes of substrate recruitment for degradation by chaperone-proteases are as diverse as the molecular shredding machines themselves [24, 37]. The common principle that can be discerned is the targeting of the substrate first to the AAA unfoldase ring followed by the engagement of a loosely structured part of the substrate, a disordered N- or C-terminus for example, into the AAA-ring pore where conserved loops protruding into the pore associate with the threaded polypeptide stretch and directionally move it along the pore toward the proteolytic core by ATPase-driven up and down movements [22, 25, 31]. This energy-dependent threading through the AAA-ring pore results in unfolding of the protein substrate and translocation toward the protease active sites [38, 39].

3 Mycobacteria: Their Degradation Needs and Arsenal

Mycobacteria are some of the most prominent members of the actinobacterial phylum owing largely to the notoriety of the pathogenic members, in particular the obligate pathogen *Mycobacterium tuberculosis* (Mtb), a microbe that plagues mankind to this day causing more than a million deaths each year [1, 40–42]. The genus Mycobacterium encompasses more than 100 species and its members are widely distributed in the environment, mostly water and soil, with some acting as opportunistic pathogens [1, 2]. According to their growth phenotype, mycobacteria fall into two groups, fast-growing members with doubling times of 2–3 h, like for example *Mycobacterium smegmatis* (*M. smegmatis*), and slow-growing members with doubling times of 24 h or more, like Mtb. All mycobacteria can enter dormant states where growth is minimal [5]. Posttranslational regulation mechanisms including protein degradation are likely particularly important under such conditions. It was noted early on that the repertoire of energy-dependent proteases in mycobacteria differs among individual members and that the occurrence is unrelated to growth phenotype or genome size [6]. Like all actinobacteria, mycobacteria lack the HslUV chaperone-protease that is generally present in other bacterial genomes (Table 16.1). The minimal set of mycobacterial chaperone-proteases consists of FtsH, the Clp proteases, and the proteasome [43], all three also present in *Mycobacterium leprae* (*M. leprae*), which has undergone extensive reductive evolution and differs from

Table 16.1 Occurrence of chaperone-proteases in actinobacteria. The core set of energy-dependent proteases shared by all actinobacteria consists of the Clp-chaperone proteases and FtsH. HslUV is absent in all actinobacteria and the Lon protease occurs only in some members. Proteasomes (PrcBA) are an unusual feature found in most, but not all actinobacteria. Interestingly, the actinobacteria that carry the pupylation genes without harboring proteasomal core subunits nevertheless maintain the proteasomal adenosine triphosphatase (ATPase; ARC)

	FtsH	ClpP1/P2	ClpC	ClpX	ARC	PrcBA	Lon	HslUV
Mycobacterium tuberculosis	+	+	+	+	+	+	–	–
Mycobacterium leprae	+	+	+	+	+	+	–	–
Mycobacterium smegmatis	+	+	+	+	+	+	+	–
Mycobacterium avium	+	+	+	+	+	+	+	–
Mycobacterium marinum	+	+	+	+	+	+	+	–
Corynebacterium glutamicum	+	+	+	+	+	–	–	–
Corynebacterium diphtheriae	+	+	+	+	+	–	–	–
Corynebacterium efficiens	+	+	+	+	+	–	–	–
Bifidobacterium adolescentis	+	+	+	+	+	–	–	–
Bifidobacterium animalis	+	+	+	+	+	–	–	–
Bifidobacterium longum	+	+	+	+	+	–	–	–
Rhodococcus equi	+	+	+	+	+	+	–	–
Rhodococcus opacus	+	+	+	+	+	+	–	–
Rhodococcus sp.	+	+	+	+	+	+	–	–
Streptomyces coelicolor	+	+	+	+	+	+	+	–
Streptomyces avermitili	+	+	+	+	+	+	+	–
Nocardia brasiliensis	+	+	+	+	+	+	+	–
Nocardia farcinica	+	+	+	+	+	+	+	–
Frankia alni	+	+	+	+	+	+	+	–
Frankia sp.	+	+	+	+	+	+	+	–
Arthrobacter aurescens	+	+	+	+	+	+	–	–
Arthrobacter arilaitensis	+	+	+	+	+	–	–	–
Propionibacterium acnes	+	+	+	+	–	+	–	–
Propionibacterium propionicum	+	+	+	+	–	–	–	–
Kineococcus radiotolerans	+	+	+	+	+	+	–	–
Janibacter sp.	+	+	+	+	+	+	–	–
Acidimicrobium ferrooxidans	+	+	+	+	+	+	–	–
Kocuria rhizophila	+	+	+	+	+	–	–	–
Micrococcus luteus	+	+	+	+	+	–	–	–

the maximal set present for example in *M. smegmatis* only by the absence of the Lon protease. This suggests that the three "minimal" proteolytic systems have at least some specific functions and that the proteasome, although unlike FtsH and ClpP nonessential under normal growth conditions, might be rendered essential under specific environmental conditions for mycobacteria. This is not the case for all actinobacteria, however, since a number of them lack the proteasomal core genes thus featuring an even further reduced set of chaperone-proteases. Corynebacteria and Bifidobacteria lack both Lon and the proteasome core and rely entirely on FtsH and the Clp protease complexes for their degradation needs. Interestingly, despite a lack of proteasomal core genes, the gene for the chaperone component is nevertheless maintained in most proteasome-deficient members (Table 16.1).

4 Clp Proteases of Mycobacterium tuberculosis

ClpP protease complexes are universal in the bacterial kingdom, the only exception being *Mycoplasma* and relatives, which feature an exceptionally reduced genome [44]. The ClpP cylinder is built of two back-to-back stacked heptameric rings enclosing the serine-protease active sites [45]. In order to degrade proteins, the ClpP double-ring complex must associate with an ATPase chaperone ring that unfolds the protein substrate and translocates it into the ClpP proteolysis chamber [46]. Usually, at least two alternative ATPase partners co-occur in all bacterial organisms, allowing formation of different Clp-proteases according to a modular design principle. While ClpX is the most ubiquitous binding partner of ClpP and occurs in almost all bacteria, ClpA and ClpC are orthologs of which bacteria feature either ClpA (gram-negatives) or ClpC (gram-positives and cyanobacteria). ClpC differs from ClpA by the presence of an inserted "linker" domain homologous to (albeit shorter than) the middle domain found in ClpB, a homologue of ClpA/C that does not associate with a protease partner [47]. Actinobacteria encode two ClpP-binding ATPase chaperones, ClpX and ClpC. Both are members of the AAA+ family, but they differ in the number of AAA+ modules. The ClpX protein has only one AAA+-module, whereas ClpC (like the orthologous ClpA) contains two such modules. Since the AAA+ modules arrange in a separate ring each, the ClpC ring has a two-tiered appearance [48] in contrast to the single-tiered topology of ClpX [49] (Fig. 16.1). Although the AAA+-modules are highly homologous, members of the AAA+ protein family show pronounced differences in their N-terminal domains preceding the AAA+-modules. These domains provide an additional level of selectivity and specificity, as they are often involved directly in substrate recruitment or modulate activity and/or binding of substrates via enlistment of adaptor proteins [50]. It has been shown that ClpP particles recruit their ATPase partners via two binding elements on their ring faces: hydrophobic surface patches formed by aromatic residues from two neighboring subunits, and the seven N-terminal β-hairpins extending from the ring surface [51].

Most bacteria encode only a single ClpP subunit and as a result, the ClpP particle in those organisms is a homo-tetradecamer with twofold symmetry across the ring-ring interface [52, 53]. However, mycobacteria (and actinobacteria in general) encode two ClpP protease subunits, ClpP1 and ClpP2, from a bicistronic operon (Fig. 16.2), leading to more complex ClpP assemblies [54–56] (Fig. 16.1). The *clpP*

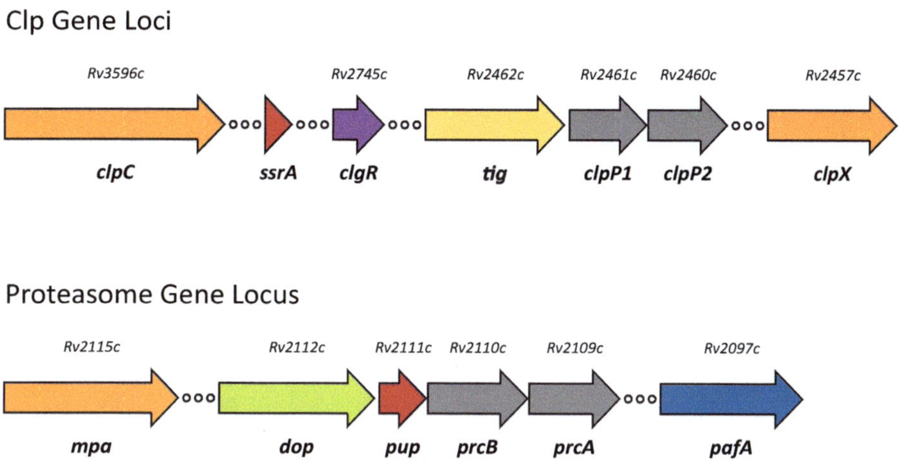

Fig. 16.2 The Clp-protease and proteasomal gene loci in *Mycobacterium tuberculosis H37Rv*. Gene names are shown below the arrows denoting the gene and the according numbers of the genes in the genome are displayed above. It should be noted that ssrA refers not to a regular gene, but to the sequence stretch coding for the transfer-messenger ribonucleic acid (*tmRNA*) molecule that is later translated into the 11-amino acid ssrA-tag during the process of trans-translation

operon is preceded by the trigger factor gene *tig* and is followed by the gene encoding the unfoldase partner of ClpP, *clpX*, either directly or separated by a few unrelated genes. ClpC, the alternative unfoldase partner of ClpP, is encoded in a separate locus distant from the *clpP* operon (Fig. 16.2).

Unlike in other bacterial genomes, adaptor proteins of ClpC, as for example MecA, do not appear to be present in the mycobacterial genomes. Adaptors like MecA usually influence ClpC assembly and activity [57, 58], but for the Mtb ClpC chaperone it was shown that it can self-oligomerize in presence of ATP in vitro [59]. Expression of the *clpP* operon is controlled by the transcriptional regulator *clp* gene *r*egulator (ClgR) [60, 61], which is also a substrate of ClpP [61, 62] (Table 16.2 and Fig. 16.3). ClgR has a five-helix-bundle fold with a helix-turn-helix deoxyribonucleic acid (DNA)-binding motif [63] and likely acts as a dimer binding to a pseudo-palindromic DNA sequence [64]. It was demonstrated in *Corynebacterium glutamicum* (*C. glutamicum*) to upregulate proteolysis-related genes (among them *clpC*, *clpP1*, and *clpP2*) as well as genes connected to DNA repair [64]. In the same organism, it was shown that environmental factors also stimulate *clp* gene expression, as *clpC*, *clpP1*, *clpP2*, and *clgR* expression is under the control of the sigma factor σ^H (Fig. 16.3) [55]. In Mtb, disruption of the *clgR* gene deprives the bacteria of their ability to prevent the pH-drop in the phagosome and impairs survival inside macrophages while under standard culture conditions, no clear phenotype is observed [65].

Some actinobacteria encode two bicistronic *clpP* operons, like for example the *clpP1/P2* and *clpP3/P4* operons in *Streptomyces lividans (S. lividans)* [66]. However, it appears that in those cases one is constitutively expressed, whereas the other

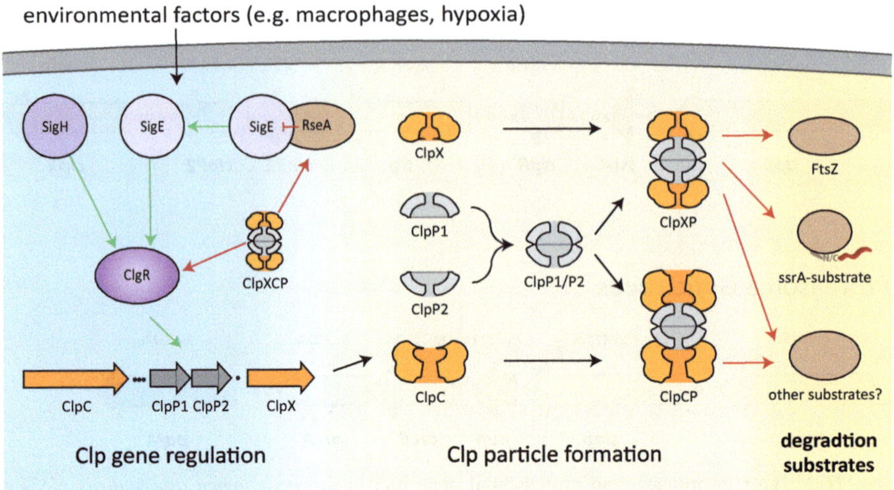

Fig. 16.3 Gene regulation and assembly of the Clp chaperone-protease. (*left side*) *Clp* gene regulation. Environmental factors influence *clp* gene expression by activation of the sigma factors SigE and SigH, which in turn activate the *clp* gene regulator (*ClgR*), a transcription factor that was shown to upregulate the *clp* gene expression. Feedback mechanisms are in place: ClpP negatively regulates its own transcription by degrading ClgR [61] while, on the other hand, ClpCP activates the sigmaE pathway by degrading the anti-sigma factor RseA [71]. (*middle*) Clp particle formation. ClpP1 and ClpP2 form a 14-meric particle consisting of a heptameric ring of N-terminally processed ClpP1 and ClpP2 each. The ClpP1/P2 proteolytic particle then associates with the AAA ATPases ClpX or ClpC to form protein degradation-competent particles. (*right side*) The ClpXP particle was implicated in the degradation of the cell division protein FtsZ [69] as well as in the degradation of ssrA-tagged substrates [88]. The ClpCP particle is involved in the degradation of the anti-sigma factor RseA. Many other as of yet unidentified substrates are expected to be degraded through the Clp-protease pathway

acts as a backup and is induced when the constitutive system is failing. This has for example been demonstrated for *S. lividans* ClpP3/P4, which is induced by the transcriptional regulator PopR (a paralog of ClgR) [67]. Itself a substrate of the ClpP1/P2 protease, PopR accumulates and induces the backup system in absence of ClpP1/P2.

ClpP subunits appear to be highly expressed in Mtb, as ClpP2 was detected as one of the ten most abundant proteins in a recent mass spectrometric analysis of the Mtb proteome [68]. Assuming an average cell size of about 0.4 fl and estimating the total protein to about 50 fg per cell, ClpP2 would be present at roughly 3000 and ClpP1 at around 1000 copies per cell, resulting in 100−200 ClpP double-ring particles per cell. The exact amount present, however, will likely depend also on transcriptional control via ClgR in response to environmental factors. For example, ClpX messenger ribonucleic acid (mRNA) levels were reported to increase 20-fold for Mtb inside macrophages and were also shown to be increased after reaeration of hypoxic cultures [62, 69]. For *M. leprae* and Mtb it was suggested that ClpC is expressed at high levels when the organisms reside in host macrophages, since patients with active leprosy or tuberculosis were found to produce antibodies against

Table 16.2 Mtb Clp-protease substrates. The substrates of Mtb Clp protease complexes, which have been identified to date are listed along with the according literature reference

Suggested Clp protease substrate	Cellular function	Involved chaperone	Reference
clp gene regulator (ClgR)	Transcriptional activator of *clp* genes	Not investigated	Sherrid et al. [62], Estorninho et al. [65]
RseA	σ^E-specific anti-σ factor	ClpC	Barik et al. [71]
FtsZ	Tubulin-like cell division protein; assembles into Z-ring at midcell division site	ClpX	Dziedzic et al. [69]
ssrA-tagged proteins	ssrA-tagging is part of tmRNA-mediated rescue of stalled ribosomes and marks translation truncations for degradation	Unknown	Raju et al. [77]

tmRNA transfer-messenger ribonucleic acid

ClpC [70]. The existence of a positive feedback loop has been suggested through a complex network involving the sigma factor σ^E that is induced following infection of macrophages. ClgR is induced in presence of σ^E, leading to increased levels of ClpCP that in turn recognizes the σ^E-anti-sigma factor RseA as a substrate and degrades it, thereby amplifying the σ^E-mediated regulation (Fig. 16.3) [71].

While in *E. coli* and many other gram-negative bacteria the *clpP* gene can be disrupted without affecting viability, it is essential in mycobacteria [7] and possibly most actinobacteria. It is therefore likely that it adopted a broader role in these organisms. This notion would also be supported by the lack of the broad-specificity protease Lon in many members of actinobacteria (Table 16.1).

In many bacteria, ClpP chaperone-proteases were found to play regulatory roles that affect the cell cycle, developmental processes as well as morphology, like for example regulation of sporulation and competence development in *Bacillus subtilis* [72, 73], inhibition of cell division in *E. coli* or control of cell cycle progression in *Bacillus crescentus* [74]. This appears to also be the case for the actinobacterial ClpP protease complexes. *ClpP* genes are involved in the formation of aerial mycelium in *Streptomyces*, as inactivation of *clpP* genes leads to an arrest of differentiation at the basal mycelium step and overexpression accelerates differentiation into aerial mycelium [54]. It has been suggested by principle of exclusion that ClpC might be the chaperone component involved in this regulation, since a *clpX*-disrupted strain does not exhibit the same phenotype [75]. Mtb ClpX, like its *E. coli* homolog, also plays a role in regulating cell division as ClpX overexpression inhibits Z-ring formation at the mid-cell division site, which suggests that Mtb ClpX might also be involved in degradation of the bacterial tubulin-homolog FtsZ [69] (Table 16.2).

Mycobacterial genomes encode the rescue tmRNA, an RNA molecule involved in rescuing ribosomes from translationally stalled polypeptide chains. The mRNA part of the tmRNA molecule codes for a short peptide tag that is added C-terminally to the stalled nascent chains, thereby not only releasing the chains from the ribosome but also marking them for degradation by several chaperone-proteases including Clp proteases. It was shown that tmRNA-dependent degradation takes place in Mtb, likely also involving the Clp proteases [76, 77] (Table 16.2).

Due to the presence of two ClpP subunits, ClpP1 and ClpP2, formation of the ClpP proteolytic particle in mycobacteria is more complex. Bacterial ClpP particles characterized crystallographically so far are all homo-oligomeric, either from bacteria encoding a single *clpP* subunit gene (*E. coli*, *Streptococcus pneumonia, Staphylococcus aureus, Helicobacter pylori, Bacillus subtilis*) [53, 78–85] or particles built from only one type of subunit of bacteria encoding more than one subunit gene (Mtb) [86]. All structures share the overall topology: a double-ring structure of two back-to-back stacked heptameric rings enclosing an almost spherical cavity with two narrow axial pores [87]. The individual subunits have been likened to a hatchet, with a head-region and handle-region easily distinguishable and the serine protease catalytic triade located at the junction of the two domains. Contacts within the heptameric rings are formed through the head-domains while the inter-ring contacts are formed mainly through interdigitating handle-domains [53]. However, the extent of contact between the rings and the amount of order in the handle region vary between the structures. Less ordered, shorter-handle regions, as present in the Mtb $ClpP1_{14}$-structure [86], lead to a more compact particle with weaker inter-ring contacts and mostly disordered catalytic triades. It has been suggested that these different states occur during assembly or even during ATPase-associated substrate processing to allow product release [82]. The only available structure of a ClpP particle from Mtb is a homo-tetradecamer of ClpP1 subunits that displays the typical hallmarks of an inactive particle: a condensed overall structure, disordered handle regions, and a disordered catalytic triade [86]. It is not altogether surprising that the homo-oligomeric particle would adopt an inactive conformation, since it has been shown both in vitro [88] and in vivo [77] that ClpP1 and ClpP2 assemble into a hetero-oligomeric ClpP1/P2 particle that is likely the physiologically active form (Fig. 16.3). ClpP1 co-purifies with affinity-tagged ClpP2 and vice versa when expressed in *M. smegmatis*. Growth of a tetracycline-dependent *clpP* knockout strain could only be restored by a plasmid expressing both ClpP1 and ClpP2 and not by expression of either subunit alone [77]. The in vitro behavior of the mycobacterial ClpP proteolytic particle remains somewhat puzzling, however, since even a mixture of recombinantly produced ClpP1 and ClpP2 displays only very low peptidase activity against fluorogenic peptides both in absence and presence of the ClpC chaperone ring [88]. When Goldberg and colleagues tested potential inhibitory peptides for their ability to suppress whatever small peptidase activity was displayed by the ClpP1/P2 complex, they surprisingly found that rather than acting as inhibitors, some of these peptides (in particular the dipeptide aldehyde protease inhibitor benzyloxycarbonyl(Z)-Leu-leucinal) dramatically stimulated ClpP1/P2 peptidase activity [88]. This stimulation only took place with both subunits present and only as long as the activator peptide remained in the mixture, apparently not mimicking interaction with a regulatory ATPase partner, since addition of the ClpC ring did not change the observation. Activator peptide was still required to stimulate the peptidase activity and, furthermore, fluorescently labelled casein as a polypeptide model substrate could only be degraded by ClpC in conjunction with ClpP1 and ClpP2 when activator was present. A mechanism of activation was suggested by the authors postulating an effect of the activator peptide on the association constants between the single ClpP rings in the double-ring particle [88]. They suggest that the activator peptide weakens double-ring formation in the ClpP1 or ClpP2 homo-oligomer, but strengthens formation of a heterotetradecamer from a single ClpP1 heptameric ring and a single ClpP2 heptameric ring. It is unclear at this point, whether Mtb indeed harbors endogenous activators (either small molecules or even polypeptides), that would serve this function in an analogous manner in vivo. Until such bona fide activators are identified, the activator peptides act as a makeshift, allowing enzymatic measurements where none were possible before. It needs to be noted in this context that the role of ClpP inside the cell is not as a peptidase but as a protease [46, 89]. Hence the activity toward fluorogenic peptide substrates is an artificial system and the question of in vivo relevant ClpP activity still needs to be addressed in the context of a AAA+ partner and with protein substrates. Here at least, in vivo evidence certainly points to proteolytically active Mtb ClpP protease, since heterologously expressed GFPssrA accumulates in *M. smegmatis* when ClpP2 is depleted, indicating the involvement of ClpP protease in tmRNA-directed degradation in mycobacteria just as is the case in *E. coli* [77]. It remains to be seen if this proteolytic activity is dependent on an activator of the ClpP particle or not.

From the combined work on mycobacterial and other actinobacterial Clp proteases, it has become clear that many parallels to the well-investigated *E. coli* Clp network exist. However, there are some crucial differences in composition, assembly, and probably also substrate specificity that distinguish the mycobacterial Clp complexes from their *E. coli* homologs. Many questions remain unanswered, starting from assembly of the active Clp protease complexes to their regulation by potential unidentified adaptors and the elucidation of substrate profiles.

5 Mycobacterial Proteasomes and the Pupylation Pathway

Proteasomes are an unusual occurrence in the bacterial world and, unlike Clp proteases and FtsH, are restricted to actinobacteria. Their existence in this phylum has been known since the mid-1990s [32] and representatives from several species groups were investigated, including *Rhodococcus* [90, 91], *Streptomyces* [92], *Mycobacterium* [11], and *Frankia* [93, 94]. Structural studies confirmed that the bacterial 20S cylinder shows the canonical proteasome architecture with two inner β-rings, carrying the active sites, flanked by the outer α-rings, together forming a four-ring stack with sevenfold rotational symmetry around the rings [92, 94–96]. Early investigations demonstrated that proteasomes are nonessential under normal laboratory culture conditions [11]. However, the field gained fresh impetus from the discovery that one of the world's most notorious killers, Mtb—the causative organism of tuberculosis—makes use of this degradation pathway to persist in host macrophages [97, 98], a finding that sparked an interest in the mycobacterial proteasome as a drug target and led to screening for inhibitors with selective affinity for bacterial versus human proteasomes [99]. In 2008, two independent studies, one in Mtb and the other in *M. smegmatis*, revealed that recruitment of substrate proteins to the mycobacterial proteasomes is mediated through a posttranslational modification of lysine residues with a small protein modifier, termed prokaryotic ubiquitin-like protein (Pup) for its functional analogy to ubiquitin as a degradation tag [12, 13].

According to its molecular features, however, Pup is not very "ubiquitin-like," as it shares no sequence or structural homology with ubiquitin [100–102]. Pup was shown by circular dichroism and nuclear magnetic resonance (NMR) analysis to be intrinsically disordered in its free state. Commonalities between the two modifiers are their small size, 78 residues for ubiquitin and 64 residues for Pup, and a flexible di-glycine motif at or near the C-terminus. The analogy between ubiquitination and pupylation also does not extend to the existence of a related set of modification enzymes, as E3-like ligases are absent from the actinobacterial genomes.

Two enzymes responsible for modification of target proteins with Pup were biochemically identified from mycobacterial lysate in a pull-down experiment with immobilized Pup as bait [14]. The two homologous proteins were predicted carboxylate-amine ligases and are likely evolutionary descendants of glutamine synthetase [15, 16]. The disruption of the genes coding for either of the enzymes abolishes pupylation, as was shown in Mtb and *M. smegmatis,* respectively [13, 103]. Biochemical in vitro analysis however revealed that only one of them, proteasome accessory factor A (PafA), is able to ligate Pup to target lysines [14]. It could later be shown that the reaction involves induced folding of Pup upon binding into a conserved groove on the enzyme [104] and that it passes through a phosphorylated Pup-intermediate [105], a step that activates the carboxylate for the nucleophilic attack by the lysine side chain (Fig. 16.4). The isopeptide linkage is not made through the C-terminal carboxylate as in ubiquitin, but via the side chain carboxylate of the C-terminal residue, a glutamate [106]. This is a further difference to ubiquitin or ubiquitin-like modifiers that all feature a C-terminal Gly-Gly motif, whereas in Pup, the Gly-Gly motif is followed by a glutamate.

Interestingly, however, all mycobacteria and many other actinobacteria encode Pup with a C-terminal glutamine, thus producing it in a ligation-incompetent form during biogenesis, requiring deamidation of the glutamine side chain to produce the carboxylate for coupling (Fig. 16.4). The enzyme responsible for this activity and hence termed *d*eamidase *o*f *P*up (Dop) turned out to be the homolog of the Pup ligase also present in the gene locus (Fig. 16.2) [14].

Although rendering Pup ligation competent is strictly required in mycobacteria for pupylation and by consequence also for proteasomal degradation, this is not the only role played by the enzyme Dop. This fact is further underscored by the existence of actinobacterial members encoding Pup with a C-terminal glutamate but which nevertheless maintain the *dop* gene. Resolving this apparent puzzle, it

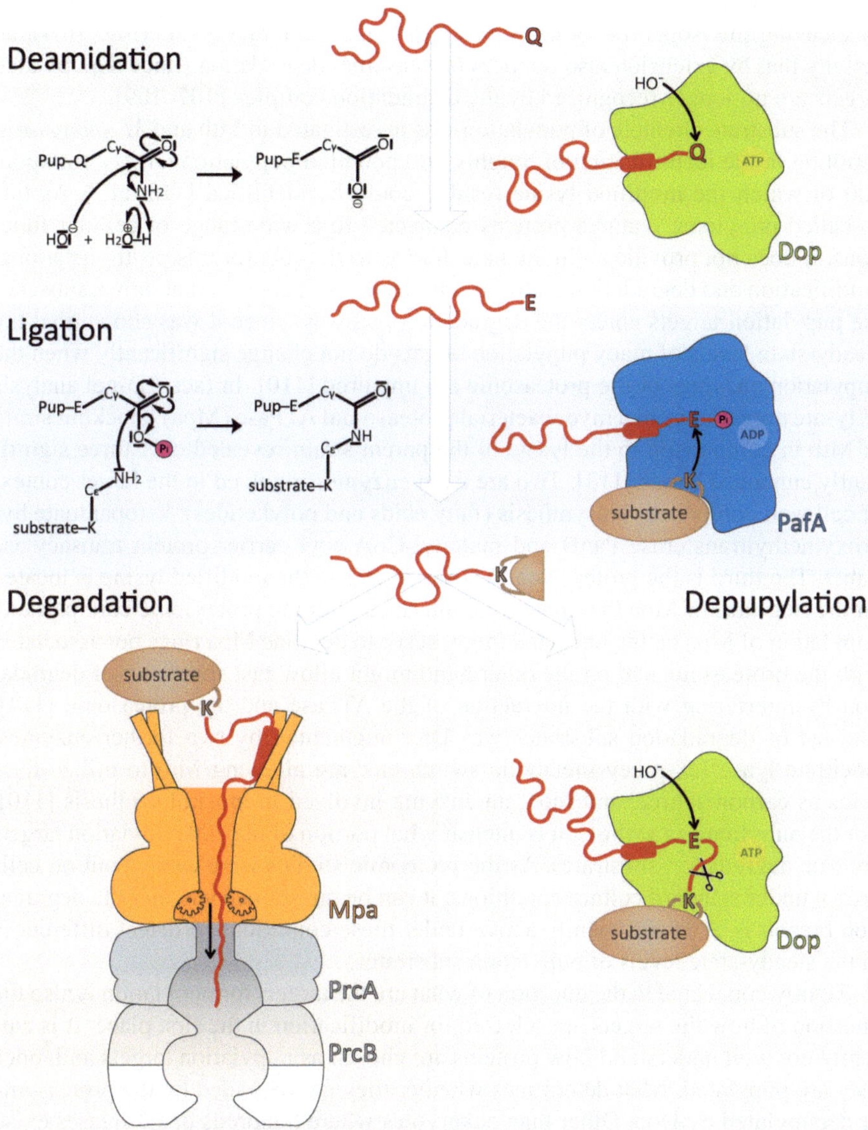

Fig. 16.4 The mycobacterial Pup-proteasome system. The C-terminal glutamine of prokaryotic ubiquitin-like protein (*Pup*) is deamidated by deamidase of Pup (*Dop*) resulting in Pup carrying a C-terminal glutamate, the ligation competent form of Pup. Proteasome accessory factor A (*PafA*) ligates the side chain carboxylate of the C-terminal glutamate to a lysine side-chain on the target protein, forming an isopeptide-bond. The substrate-Pup fusion can subsequently be recognized by the mycobacterial proteasomal ATPase (*Mpa*) that pulls the substrate through its central pore unfolding it in the process, and translocating it into the proteolytic core to be degraded. Alternatively, pupylated substrate can be bound by Dop and, in a reaction chemically very similar to deamidation, Pup can be cleaved off the substrate, thereby rescuing both from degradation.

was demonstrated in vitro and in vivo that Dop opposes the ligase activity of PafA by cleaving the isopeptide bond between Pup and the target lysine (Fig. 16.4), an activity that by extension also opposes proteasomal degradation, since depupylated targets are no longer recognized by the degradation complex [107–109].

The substrate clientele of pupylation was investigated in Mtb and *M. smegmatis*, resulting in the identification of roughly 700 potential pupylation targets, for about 100 of which the modified lysine residue could be identified [110–112]. As this so-called pupylome features proteins connected to a wide range of cellular functions, it does not provide an immediate clue as to the role for this posttranslational modification and degradation pathway. Interestingly, it appears that only a subset of the pupylation targets enters the degradation pathway, since it was shown that the steady-state levels of many pupylation targets do not change significantly when the pupylation enzymes or the proteasome are impaired [110]. In fact, 2D-gel analysis of lysate prepared from a mycobacterial proteasomal ATPase (Mpa) knockout strain of Mtb in comparison to the lysate of the parent strain revealed only three significantly enhanced bands [113]. Two are from enzymes involved in the larger context of cell wall component biosynthesis (fatty acids and polyketides): ketopantoate hydroxymethyltransferase PanB and malonyl-CoA acyl carrier protein transacylase FabD. The third is the proteasomal ATPase itself. As the modified lysine is located on the ring face of Mpa that makes the interface with the proteasome core particle, pupylation of Mpa on the one hand might serve to degrade Mpa rings not associated with the proteasome and on the other hand might allow fast shutdown of degradation by interfering with the interaction of the ATPase and the proteasome [114]. The list of degradation substrates was later augmented by two further enzymes, isocitrate lyase Icl, a key metabolic switch enzyme allowing Mtb to utilize fatty acids as carbon source, and Ino1, an enzyme involved in inositol synthesis [110]. For the pupylome as a whole, it is unclear what portion of the 700 pupylation targets are true degradation substrates. As the proteomic studies were carried out on cells grown under standard culture conditions, it can be questioned whether the degradation branch is even sufficiently active under these conditions to detect differences in the steady-state levels of pupylation substrates.

Tightly connected to the question of what are the targets for pupylation is also the question of how the targets are selected for modification in the first place. It is currently not well understood how proteins are chosen as pupylation targets and, once they are pupylated, what determines whether they are degraded by the proteasome or depupylated by Dop. Other than eukaryotes where hundreds of E3 ligases exist, PafA appears to be the only ligase in the pupylation pathway [12, 13]. No sequence or structural recognition features have been identified yet that would allow reliable prediction of pupylation sites in proteins.

Once pupylated, the modified proteins can be recognized on the one hand by the proteasomal ATPase Mpa and on the other hand by the depupylase enzyme (Fig. 16.4). Both binding partners feature interaction sites for Pup and both binding events likely induce different folds in Pup [104, 115]. Interaction with the proteasomal ATPase involves the binding of the C-terminal half of Pup to the N-terminal coiled-coil domains protruding from the Mpa ring surface [100]. Upon binding to

Mpa, Pup residues 21–51 form an extended helix that docks to one of the coiled coils forming a shared, three-stranded coil. This mode of binding points the N-terminal, disordered portion of Pup into the ATPase pore, providing a threading element along with the recognition tag [116, 117]. The fold adopted by Pup when it interacts with Mpa is different to the fold it adopts in the binding groove of the ligase where it forms two orthogonal helices connected by a linker [104]. The depupylase features a Pup-binding groove similar to the ligase and therefore likely invokes a similar fold in Pup upon binding of pupylated substrates [17]. A very intriguing observation is the fact that depupylation of folded substrates is enhanced about tenfold in rate when the ATPase Mpa is added to the reaction [107]. The reason might be that pupylated substrates, in absence of the protease, are still recruited by the ATPase and are unfolded. The unfolded substrate might provide better access to the depupylase enzyme active site for cleavage of the isopeptide bond than the native conformation.

It is clear that the different players in the pathway are highly interconnected functionally, both competitively and synergistically. However, how the different components of the Pup-proteasome pathway play together in the cell to determine the fate of a substrate is still not fully understood.

6 FtsH—Degradation at the Membrane

The membrane-associated FtsH chaperone-protease is ubiquitously present throughout all bacteria. The name FtsH derives from the observation that a certain mutation of the *ftsH1* gene in *E. coli* caused loss of viability at high temperature leading to a temperature-sensitive, filament-forming phenotype (*ftsH1*: *f*ilamentation *t*emperature-*s*ensitive) [118]. Even though FtsH is present in all bacteria, it is not always essential: in gram-positive bacteria, it is generally dispensable, whereas in most gram-negative bacteria, it is not. But even within actinobacteria exigence for FtsH appears to differ between members. In Mtb, FtsH was found to be essential [8, 9], whereas in *C. glutamicum,* no significant growth defect in standard minimal medium was observed upon gene deletion [119]. Despite this difference, the key architectural and mechanistic features of the FtsH chaperone-protease are highly conserved in all bacteria. It encodes both the chaperone and the protease domain on a single chain of usually about 600 − 800 amino acids. This results in a symmetrical 6:6 ratio between chaperone and protease, in contrast to the Clp and proteasome complexes where hexameric chaperone rings associate with sevenfold symmetric protease cylinders (Fig. 16.1). The main distinguishing feature, however, is that FtsH is attached to the membrane. Two N-terminal trans-membrane segments are followed by the cytosolic AAA-type ATPase module and C-terminal protease domain. Mechanistic studies of the *E. coli* FtsH ATPase found its unfolding activity to be noticeably weaker than that of other chaperone-proteases. For example, in contrast to the Clp system, it rarely degrades folded proteins. Rather, the weak ATPase activity appears to make FtsH specific for the degradation of already partly unfolded targets [120–122]. The FtsH protease domain belongs to the family of zinc metalloproteases (more specifically the Aspzincin family), with a characteristic HEXXH motif where the glutamate acts as the catalytic base and the zinc ion is coordinated by the two histidines as well as by a downstream aspartate residue [123]. The structure of the soluble part of FtsH has been solved for *Thermotoga maritima* [123] and *Thermus thermophilus* [124]. Like the two-component chaperone-proteases, FtsH features an AAA+ ring with a central pore, and protein substrates have to be translocated through the pore in an ATP-dependent manner to gain access to the protease active sites. While one structure study proposes the canonical proteolytic chamber architecture [123], the other advocates a set of tunnels leading into the individual metalloprotease sites [124]. In either case, the overall functional principle of the chaperone-proteases, the ATP-driven unfolding/threading through the chaperone-pore followed by translocation of the polypeptide to otherwise inaccessible protease-active sites is conserved. Exactly how and where protein substrates, either soluble or membrane associated, are recognized and interact with the FtsH chaperone-protease is still a matter of ongoing research.

While the key features of FtsH are highly conserved throughout different organisms, the C-terminal part of the protein varies in length and composition, and there is indication that this variable part confers substrate recognition specificity to different species. A feature unique to mycobacterial FtsH proteases is a domain of unknown function C-terminal of the protease domain, which might be involved in such kind of specific interaction [125]. Furthermore, *M. smegmatis*, *M. leprae*, and *M. avium* FtsH proteins have yet another unique sequence stretch at the C-terminus, a proline- and glutamine-rich region, which is not present in the Mtb protein [125]. The fact that FtsH is essential in many organisms but not in all and that mutations/deletions cause a multitude of phenotypes, also indicates that it has different substrate specificities in various organisms. For *E. coli* FtsH, a broad substrate specificity has already been discovered, encompassing cytosolic as well as membrane proteins. It recognizes ssrA and C-terminal non-polar tags on cytoplasmic proteins, but also N- and C-terminal cytosolic regions of membrane proteins (reviewed in [126]). Some of the substrates tested in in vitro experiments with Mtb, *M. smegmatis*, and *E. coli* FtsH demonstrate recognition overlap between the three species. Mtb FtsH for example was shown to successfully complement for *E. coli* FtsH [127] and, additionally, in vivo protease assays have shown that Mtb and *M. smegmatis* FtsH are active against heterologously expressed substrates from *E. coli*. Mtb FtsH is able to degrade the *E. coli* phage λ CII protein and both Mtb and *M. smegmatis* FtsH were shown to degrade the *E. coli* membrane protein SecY and the cytoplasmic σ^{32} (which has no homologues in mycobacteria) [125, 127]. Conversely, in vitro experiments have shown that *E. coli* FtsH degrades the cell division protein FtsZ from *E. coli* as well as from Mtb [128] although the in vivo relevance of this finding is not entirely clear, since FtsZ degradation in Mtb was attributed to the ClpXP system. However, Mtb FtsH was also implicated in the regulation of FtsZ, as an FtsH-overexpressing strain showed decreased levels of FtsZ [129].

In *C. glutamicum*, the role of FtsH substrates was investigated in detail. First, an influence of FtsH (together with the Clp proteases) on the signal transduction protein GlnK was shown. GlnK is a protein involved in the regulation of nitrogen uptake in *C. glutamicum* under starvation conditions and seems to be degraded under high ammonium conditions by FtsH (and the ClpXCP system) [130]. The substrate spectrum of *C. glutamicum* was further investigated on the proteome level by Lüdke et al. [119] in an *ftsH*-deletion strain. Of the protein substrates identified, at least one-third is involved in energy and carbon metabolism. Combined with the findings that deletion of *ftsH* does not result in a significant growth defect, neither under standard nor under heat-shock or osmotic stress conditions, these results implicate *C. glutamicum* FtsH rather in regulation of metabolic processes than in stress response. In Mtb on the other hand, FtsH was found to be upregulated during several stress conditions: during growth in macrophages as well as upon exposure to agents producing reactive oxygen and nitrogen species. In contrast, during starvation conditions and in stationary phase, FtsH is downregulated. However, even though FtsH is upregulated during growth in macrophages, overexpression of FtsH decreased viability in macrophages, indicating that balanced levels of FtsH are needed for optimal Mtb survival [129].

In Mtb, the FtsH promoter sequence was analyzed in detail. It contains repeat sequences that are polymorphic in different Mtb clinical isolates due to variations in the repeat copy number. Such repeats are referred to as *v*ariable-*n*umber *t*andem *r*epeats (VNTRs), and they can be used to distinguish between isolates. Several such VNTRs have been described in Mtb. One of them, VNTR4052, is located upstream of the *ftsH* gene within the *ftsH* promoter region (in the coding region of the putative protein Rv3611) [131]. Functional roles for VNTRs in gene regulation have been suggested for other bacteria (reviewed in [132, 133]). VNTR4052 might complement FtsH promoter activity, as multiple copies of the VNTR enhance promoter activity [134].

It is clear that the FtsH protease is an essential player in mycobacterial degradation. However, which substrates it degrades and under the control of which factors is still an unanswered question.

7 Chaperone-Proteases and Pathogenicity of Mtb: The Achilles Heel of a Powerful Enemy?

When bacteria enter the host, they are usually faced with unfavorable conditions, not only due to potentially triggered host-defensive pathways but also because of limitations in nutrient availability [41]. Survival under such conditions can become critically dependent upon the ability to processively break down proteins as a source of nitrogen and/or carbon. Another important aspect is the removal of impaired bacterial proteins, since the "load" of damaged proteins is increased under the onslaught of reactive oxygen species, lowered pH, or other host-defensive measures. Although chaperone-proteases cannot be considered true virulence factors that are directly involved in colonizing the host and are specifically expressed for that purpose, their activity can become an important component of survival and can thus be considered a virulence-supporting or boosting determinant.

Tuberculosis infection of the host starts with inhalation of the tubercle bacilli into the lungs where they are taken up via phagocytosis into macrophages that form the first line of host defence [135–137]. This marks the start of an intracellular existence in the host that can last for decades. Mtb takes up residence in the phagosome and effectively prevents fusion with lysosomes, eventually leading to aggregation of macrophages with lymphocytes and fibroblasts to form granuloma. There, Mtb persists in a low-replicating, dormant state until years or decades later it might progress into active tuberculosis.

It has been demonstrated that the ability of Mtb to persist and survive over long periods of time inside macrophages is dependent, among other factors, on the Pup-proteasome-system (PPS) [10]. Mutant strains of Mtb lacking either the proteasomal ATPase Mpa or the Pup ligase PafA are unable to proliferate inside macrophages and are strongly attenuated in a mouse infection model [10, 97]. Silencing of the 20S proteasome likewise leads to severely reduced lung bacterial counts [10]. Considering the wide functional range of identified pupylation substrates through proteomic studies in Mtb [110] and *M. smegmatis* [111, 112], a role played through general proteolysis for alleviating starvation conditions is a possibility. However, it is also possible that more specific proteolysis or regulatory effects through pupylation could influence the metabolic state of the bacterium, acting like a switch that adjusts the metabolic flux to different conditions, like for example in hypoxia. In either case, the involvement of the PPS in Mtb persistence provides a cluster of targets for drug development aimed at interfering with proteasomal degradation or the pupylation enzymes. Efforts are underway to develop drugs that inhibit mycobacterial proteasomes, and through large-scale screens compounds have already been identified that preferentially inhibit mycobacterial over human proteasomes [99]. The experience with the use and administering of proteasomal inhibitors available on the market already to treat other human diseases will no doubt accelerate the development of related drugs for use against certain cases of tuberculosis. With extremely drug resistant Mtb strains emerging, that no longer respond to any existing antibiotic therapy, it is imperative that we augment our arsenal of therapeutic weapons to fight these new editions of an old and powerful enemy.

As ClpP is essential already under normal laboratory culture conditions, it is difficult to judge the role played by this system during pathogenesis. However, from the fact that in certain pathogenic bacteria, where ClpP is nonessential, it plays an essential role during pathogenesis and stress [138], it can be concluded that it likely plays an important role in Mtb for pathogenesis as well. In fact, the absolute requirement for ClpP and ClpC for viability of Mtb under all conditions makes this proteolytic system a particularly attractive drug target [7]. Furthermore, to forestall the emergence of resistance against interference with protein breakdown, it might even be an especially powerful approach to inhibit more than one chaperone-protease system simultaneously, since bacterial chaperone-proteases can often act synergistically or evolve redundancy. Be it through the PPS pathway or by way of the Clp proteases, preventing the organism from degrading its cellular proteins might prove to be a potent means of interference with its survival in the host.

Interestingly, the ClpP proteolytic particle was shown to be the target of a class of antibiotics called acyldepsipeptides (ADEPs) [139]. ADEPs exert their bacteriotoxic effect by opening the ClpP entrance pores which in turn leads to unchecked degradation of polypeptides [140, 141]. ADEPs were found to be active against Mtb in vivo, further validating the ClpP proteolytic system as a drug target [7]. However, efflux mechanisms of Mtb appear to interfere with the efficacy of ADEPs, and efflux-pump inhibitors might be required simultaneously to achieve sufficient intracellular concentrations. The use of ClpP inhibitors that covalently modify the ClpP active sites might therefore be more effective. One class of inhibitors that specifically inhibits ClpP by covalent modification of the active site serine are di-substituted β-lactones, in particular those carrying an aromatic ring at the α-carbon [142–144]. It was recently shown that Mtb and *M. smegmatis* strains are also effectively inhibited by these compounds and that the mechanism of inhibition can be attributed to covalent modification of the ClpP active site [145], suggesting that they might be successfully included into the repertoire of chemicals used to combat mycobacterial infections.

In addition to the ClpP particle its AAA+ ATPase partners can also serve as a target. Certain strains of marine *Streptomyces* were found to produce cyclic heptapeptide antibiotics, referred to as cyclomarins, which are bactericidal against Mtb both in culture and in macrophages [146]. Surprisingly, the natural target of this class of antibiotics turned out to be ClpC. Structural analysis showed that Cyclomarin A binds tightly to the N-terminal domain of ClpC [147], potentially interfering with the recruitment of substrates. The potential of these compounds for drug development looks very promising, since they are active even against hypoxic, non-replicating Mtb. Furthermore, with ClpP and ClpC being essential in Mtb, their inhibition has the added advantage that resistance might not be so easily acquired, since deletion mutants are not viable.

The finding that energy-dependent protein degradation systems serve as natural targets of certain classes of antibiotics underlines the central importance of the activities carried out by these molecular machines. It is evident that energy-dependent proteases and the associated substrate recruitment pathways represent an attractive new target group for the development of novel antibiotics against Mtb and might offer a second line of defense against multi-drug-resistant (MDR-TB) or extremely-drug-resistant (XDR-TB) forms of tuberculosis infections.

Acknowledgments We thank members of the Weber-Ban group for critically reading the manuscript. This work was supported by the Swiss National Science Foundation (SNSF), the National Centre of Excellence in Research (NCCR) Structural Biology program of the SNSF and an ETH research grant.

References

1. Cook GM, Berney M, Gebhard S et al (2009) Physiology of *Mycobacteria*. Adv Microb Physiol 55:81–182, 318–189
2. Grange JM (1996) The biology of the genus *Mycobacterium*. Soc Appl Bacteriol Symp Ser 25:1S–9S
3. Mckinney JD, Gomez JE (2003) Life on the inside for *Mycobacterium tuberculosis*. Nat Med 9:1356–1357
4. Schnappinger D, Ehrt S, Voskuil MI et al (2003) Transcriptional adaptation of *Mycobacterium tuberculosis* within macrophages: insights into the phagosomal environment. J Exp Med 198:693–704
5. Rittershaus ES, Baek SH, Sassetti CM (2013) The normalcy of dormancy: common themes in microbial quiescence. Cell Host Microbe 13:643–651
6. Knipfer N, Seth A, Roudiak SG et al (1999) Species variation in ATP-dependent protein degradation: protease profiles differ between *Mycobacteria* and protease functions differ between *Mycobacterium smegmatis* and *Escherichia coli*. Gene 231:95–104
7. Ollinger J, O'malley T, Kesicki EA et al (2012) Validation of the essential ClpP protease in *Mycobacterium tuberculosis* as a novel drug target. J Bacteriol 194:663–668
8. Griffin JE, Gawronski JD, Dejesus MA et al (2011) High-resolution phenotypic profiling defines genes essential for mycobacterial growth and cholesterol catabolism. PLoS Pathog 7:e1002251
9. Sassetti CM, Boyd DH, Rubin EJ (2003) Genes required for mycobacterial growth defined by high density mutagenesis. Mol Microbiol 48:77–84
10. Gandotra S, Schnappinger D, Monteleone M et al (2007) In vivo gene silencing identifies the *Mycobacterium tuberculosis* proteasome as essential for the bacteria to persist in mice. Nat Med 13:1515–1520
11. Knipfer N, Shrader TE (1997) Inactivation of the 20S proteasome in *Mycobacterium smegmatis*. Mol Microbiol 25:375–383
12. Burns KE, Liu WT, Boshoff HI et al (2009) Proteasomal protein degradation in *Mycobacteria* is dependent upon a prokaryotic ubiquitin-like protein. J Biol Chem 284:3069–3075
13. Pearce MJ, Mintseris J, Ferreyra J et al (2008) Ubiquitin-like protein involved in the proteasome pathway of *Mycobacterium tuberculosis*. Science 322:1104–1107
14. Striebel F, Imkamp F, Sutter M et al (2009) Bacterial ubiquitin-like modifier Pup is deamidated and conjugated to substrates by distinct but homologous enzymes. Nat Struct Mol Biol 16:647–651
15. Iyer LM, Abhiman S, Maxwell Burroughs A et al (2009) Amidoligases with ATP-grasp, glutamine synthetase-like and acetyltransferase-like domains: synthesis of novel metabolites and peptide modifications of proteins. Mol Biosyst 5:1636–1660
16. Iyer LM, Burroughs AM, Aravind L (2008) Unraveling the biochemistry and provenance of pupylation: a prokaryotic analog of ubiquitination. Biol Direct 3:45
17. Ozcelik D, Barandun J, Schmitz N et al (2012) Structures of Pup ligase PafA and depupylase Dop from the prokaryotic ubiquitin-like modification pathway. Nat Commun 3:1014
18. Gottesman S, Roche E, Zhou Y et al (1998) The ClpXP and ClpAP proteases degrade proteins with carboxy-terminal peptide tails added by the SsrA-tagging system. Genes Dev 12:1338–1347
19. Roche ED, Sauer RT (1999) SsrA-mediated peptide tagging caused by rare codons and tRNA scarcity. EMBO J 18:4579–4589
20. Weart RB, Nakano S, Lane BE et al (2005) The ClpX chaperone modulates assembly of the tubulin-like protein FtsZ. Mol Microbiol 57:238–249
21. Flynn JM, Levchenko I, Sauer RT et al (2004) Modulating substrate choice: the SspB adaptor delivers a regulator of the extracytoplasmic-stress response to the AAA+ protease ClpXP for degradation. Genes Dev 18:2292–2301

22. Striebel F, Kress W, Weber-Ban E (2009) Controlled destruction: AAA+ ATPases in protein degradation from bacteria to eukaryotes. Curr Opin Struct Biol 19:209–217
23. Mogk A, Huber D, Bukau B (2011) Integrating protein homeostasis strategies in prokaryotes. Cold Spring Harb Perspect Biol 3:a004366
24. Schmidt R, Bukau B, Mogk A (2009) Principles of general and regulatory proteolysis by AAA+ proteases in *Escherichia coli*. Res Microbiol 160:629–636
25. Sauer RT, Baker TA (2011) AAA+ proteases: ATP-fueled machines of protein destruction. Annu Rev Biochem 80:587–612
26. Snider J, Thibault G, Houry WA (2008) The AAA+ superfamily of functionally diverse proteins. Genome Biol 9:216
27. Ciechanover A, Stanhill A (2013) The complexity of recognition of ubiquitinated substrates by the 26S proteasome. Biochim Biophys Acta
28. Schrader EK, Harstad KG, Matouschek A (2009) Targeting proteins for degradation. Nat Chem Biol 5:815–822
29. Lander GC, Martin A, Nogales E (2013) The proteasome under the microscope: the regulatory particle in focus. Curr Opin Struct Biol 23:243–251
30. Maupin-Furlow JA, Humbard MA, Kirkland PA et al (2006) Proteasomes from structure to function: perspectives from Archaea. Curr Top Dev Biol 75:125–169
31. Baker TA, Sauer RT (2006) ATP-dependent proteases of bacteria: recognition logic and operating principles. Trends Biochem Sci 31:647–653
32. Tamura T, Nagy I, Lupas A et al (1995) The first characterization of a *Eubacterial proteasome*: the 20S complex of *Rhodococcus*. Curr Biol (CB) 5:766–774
33. Gao B, Gupta RS (2012) Phylogenetic framework and molecular signatures for the main clades of the phylum Actinobacteria. Microbiol Mol Biol Rev 76:66–112
34. Ventura M, Canchaya C, Tauch A et al (2007) Genomics of Actinobacteria: tracing the evolutionary history of an ancient phylum. Microbiol Mol Biol Rev 71:495–548
35. Valas RE, Bourne PE (2011) The origin of a derived superkingdom: how a gram-positive bacterium crossed the desert to become an archaeon. Biol Direct 6:16
36. Valas RE, Bourne PE (2008) Rethinking proteasome evolution: two novel bacterial proteasomes. J Mol Evol 66:494–504
37. Dougan DA, Mogk A, Zeth K et al (2002) AAA+ proteins and substrate recognition, it all depends on their partner in crime. FEBS Lett 529:6–10
38. Reid BG, Fenton WA, Horwich AL et al (2001) ClpA mediates directional translocation of substrate proteins into the ClpP protease. Proc Natl Acad Sci U S A 98:3768–3772
39. Weber-Ban EU, Reid BG, Miranker AD et al (1999) Global unfolding of a substrate protein by the Hsp100 chaperone ClpA. Nature 401:90–93
40. Eisenstadt J, Hall GS (1995) Microbiology and classification of *Mycobacteria*. Clin Dermatol 13:197–206
41. Pieters J (2001) Entry and survival of pathogenic *Mycobacteria* in macrophages. Microbes Infect 3:249–255 (Institut Pasteur)
42. Rastogi N, Legrand E, Sola C (2001) The *Mycobacteria*: an introduction to nomenclature and pathogenesis. Rev Sci Tech 20:21–54
43. Ribeiro-Guimaraes ML, Pessolani MC (2007) Comparative genomics of mycobacterial proteases. Microb Pathog 43:173–178
44. Yu AY, Houry WA (2007) ClpP: a distinctive family of cylindrical energy-dependent serine proteases. FEBS Lett 581:3749–3757
45. Kress W, Maglica Z, Weber-Ban E (2009) Clp chaperone-proteases: structure and function. Res Microbiol 160:618–628
46. Katayama-Fujimura Y, Gottesman S, Maurizi MR (1987) A multiple-component, ATP-dependent protease from *Escherichia coli*. J Biol Chem 262:4477–4485
47. Desantis ME, Shorter J (2012) The elusive middle domain of Hsp104 and ClpB: location and function. Biochim Biophys Acta 1823:29–39
48. Liu J, Mei Z, Li N et al (2013) Structural dynamics of the MecA-ClpC complex: a type II AAA+ protein unfolding machine. J Biol Chem 288:17597–17608

49. Ortega J, Lee HS, Maurizi MR et al (2002) Alternating translocation of protein substrates from both ends of ClpXP protease. EMBO J 21:4938–4949
50. Kirstein J, Moliere N, Dougan DA et al (2009) Adapting the machine: adaptor proteins for Hsp100/Clp and AAA+ proteases. Nat Rev Microbiol 7:589–599
51. Kim YI, Levchenko I, Fraczkowska K et al (2001) Molecular determinants of complex formation between Clp/Hsp100 ATPases and the ClpP peptidase. Nat Struct Biol 8:230–233
52. Kessel M, Maurizi MR, Kim B et al (1995) Homology in structural organization between *E. coli* ClpAP protease and the eukaryotic 26S proteasome. J Mol Biol 250:587–594
53. Wang J, Hartling JA, Flanagan JM (1997) The structure of ClpP at 2.3 A resolution suggests a model for ATP-dependent proteolysis. Cell 91:447–456
54. De Crecy-Lagard VServant-Moisson, Viala J et al (1999) Alteration of the synthesis of the Clp ATP-dependent protease affects morphological and physiological differentiation in *Streptomyces*. Mol Microbiol 32:505–517
55. Engels S, Schweitzer JE, Ludwig C et al (2004) *clpC* and *clpP1P2* gene expression in *Corynebacterium glutamicum* is controlled by a regulatory network involving the transcriptional regulators ClgR and HspR as well as the ECF sigma factor sigmaH. Mol Microbiol 52:285–302
56. Benaroudj N, Raynal B, Miot M et al (2011) Assembly and proteolytic processing of mycobacterial ClpP1 and ClpP2. BMC Biochem 12:61
57. Kirstein J, Schlothauer T, Dougan DA et al (2006) Adaptor protein controlled oligomerization activates the AAA+ protein ClpC. EMBO J 25:1481–1491
58. Schlothauer T, Mogk A, Dougan DA et al (2003) MecA, an adaptor protein necessary for ClpC chaperone activity. Proc Natl Acad Sci U S A 100:2306–2311
59. Kar NP, Sikriwal D, Rath P et al (2008) *Mycobacterium tuberculosis* ClpC1: characterization and role of the N-terminal domain in its function. FEBS J 275:6149–6158
60. Bellier A, Mazodier P (2004) ClgR, a novel regulator of clp and lon expression in *Streptomyces*. J Bacteriol 186:3238–3248
61. Bellier A, Gominet M, Mazodier P (2006) Post-translational control of the *Streptomyces lividans* ClgR regulon by ClpP. Microbiology 152:1021–1027
62. Sherrid AM, Rustad TR, Cangelosi GA et al (2010) Characterization of a *Clp* protease gene regulator and the reaeration response in *Mycobacterium tuberculosis*. PloS ONE 5:e11622
63. Russo S, Schweitzer JE, Polen T et al (2009) Crystal structure of the caseinolytic protease gene regulator, a transcriptional activator in actinomycetes. J Biol Chem 284:5208–5216
64. Engels S, Ludwig C, Schweitzer JE et al (2005) The transcriptional activator ClgR controls transcription of genes involved in proteolysis and DNA repair in *Corynebacterium glutamicum*. Mol Microbiol 57:576–591
65. Estorninho M, Smith H, Thole J et al (2010) ClgR regulation of chaperone and protease systems is essential for *Mycobacterium tuberculosis* parasitism of the macrophage. Microbiology 156:3445–3455
66. Viala J, Rapoport G, Mazodier P (2000) The clpP multigenic family in *Streptomyces lividans*: conditional expression of the clpP3 clpP4 operon is controlled by PopR, a novel transcriptional activator. Mol Microbiol 38:602–612
67. Viala J, Mazodier P (2002) ClpP-dependent degradation of PopR allows tightly regulated expression of the clpP3 clpP4 operon in *Streptomyces lividans*. Mol Microbiol 44:633–643
68. Schubert OT, Mouritsen J, Ludwig C et al. (2013) The Mtb proteome library: a resource of assays to quantify the complete proteome of *Mycobacterium tuberculosis*. Cell Host Microbe 13:602–612
69. Dziedzic R, Kiran M, Plocinski P et al (2010) *Mycobacterium tuberculosis* ClpX interacts with FtsZ and interferes with FtsZ assembly. PloS ONE 5:e11058
70. Misra N, Habib S, Ranjan A et al (1996) Expression and functional characterisation of the *clpC* gene of *Mycobacterium leprae*: ClpC protein elicits human antibody response. Gene 172:99–104
71. Barik S, Sureka K, Mukherjee P et al (2010) RseA, the SigE specific anti-sigma factor of *Mycobacterium tuberculosis*, is inactivated by phosphorylation-dependent ClpC1P2 proteolysis. Mol Microbiol 75:592–606

72. Turgay K, Hahn J, Burghoorn J et al (1998) Competence in *Bacillus subtilis* is controlled by regulated proteolysis of a transcription factor. EMBO J 17:6730–6738
73. Moliere N, Turgay K (2013) General and regulatory proteolysis in *Bacillus subtilis*. Subcell Biochem 66:73–103
74. Jenal U, Fuchs T (1998) An essential protease involved in bacterial cell-cycle control. EMBO J 17:5658–5669
75. Viala J, Mazodier P (2003) The ATPase ClpX is conditionally involved in the morphological differentiation of *Streptomyces lividans*. Mol Genet Genomics (MGG) 268:563–569
76. Personne Y, Brown AC, Schuessler DL et al (2013) *Mycobacterium tuberculosis* ClpP proteases are co-transcribed but exhibit different substrate specificities. PloS ONE 8:e60228
77. Raju RM, Unnikrishnan M, Rubin DH et al (2012) *Mycobacterium tuberculosis* ClpP1 and ClpP2 function together in protein degradation and are required for viability in vitro and during infection. PLoS Pathog 8:e1002511
78. Bewley MC, Graziano V, Griffin K et al (2006) The asymmetry in the mature amino-terminus of ClpP facilitates a local symmetry match in ClpAP and ClpXP complexes. J Struct Biol 153:113–128
79. Geiger SR, Bottcher T, Sieber SA et al (2011) A conformational switch underlies ClpP protease function. Angew Chem 50:5749–5752
80. Gersch M, List A, Groll M et al (2012) Insights into structural network responsible for oligomerization and activity of bacterial virulence regulator caseinolytic protease P (ClpP) protein. J Biol Chem 287:9484–9494
81. Gribun A, Kimber MS, Ching R et al (2005) The ClpP double ring tetradecameric protease exhibits plastic ring-ring interactions, and the N termini of its subunits form flexible loops that are essential for ClpXP and ClpAP complex formation. J Biol Chem 280:16185–16196
82. Kimber MS, Yu AY, Borg M et al (2010) Structural and theoretical studies indicate that the cylindrical protease ClpP samples extended and compact conformations. Structure 18:798–808
83. Lee BG, Kim MK, Song HK (2011) Structural insights into the conformational diversity of ClpP from *Bacillus subtilis*. Mol Cells 32:589–595
84. Szyk A, Maurizi MR (2006) Crystal structure at 1.9 A of *E. coli* ClpP with a peptide covalently bound at the active site. J Struct Biol 156:165–174
85. Zhang J, Ye F, Lan L et al (2011) Structural switching of *Staphylococcus aureus* Clp protease: a key to understanding protease dynamics. J Biol Chem 286:37590–37601
86. Ingvarsson H, Mate MJ, Hogbom M et al (2007) Insights into the inter-ring plasticity of caseinolytic proteases from the X-ray structure of *Mycobacterium tuberculosis* ClpP1. Acta Crystallogr D Biol Crystallogr 63:249–259
87. Alexopoulos JA, Guarne A, Ortega J (2012) ClpP: a structurally dynamic protease regulated by AAA+ proteins. J Struct Biol 179:202–210
88. Akopian T, Kandror O, Raju RM et al (2012) The active ClpP protease from *M. tuberculosis* is a complex composed of a heptameric ClpP1 and a ClpP2 ring. EMBO J 31:1529–1541
89. Thompson MW, Singh SK, Maurizi MR (1994) Processive degradation of proteins by the ATP-dependent Clp protease from *Escherichia coli*. Requirement for the multiple array of active sites in ClpP but not ATP hydrolysis. J Biol Chem 269:18209–18215
90. De Mot R, Nagy I, Baumeister W (1998) A self-compartmentalizing protease in *Rhodococcus*: the 20S proteasome. Antonie Van Leeuwenhoek 74:83–87
91. Wolf S, Nagy I, Lupas A et al (1998) Characterization of ARC, a divergent member of the AAA ATPase family from *Rhodococcus erythropolis*. J Mol Biol 277:13–25
92. Nagy I, Tamura T, Vanderleyden J et al (1998) The 20S proteasome of *Streptomyces coelicolor*. J Bacteriol 180:5448–5453
93. Benoist P, Muller A, Diem HG et al (1992) High-molecular-mass multicatalytic proteinase complexes produced by the nitrogen-fixing actinomycete *Frankia* strain BR. J Bacteriol 174:1495–1504
94. Pouch MN, Cournoyer B, Baumeister W (2000) Characterization of the 20S proteasome from the actinomycete *Frankia*. Mol Microbiol 35:368–377

95. Kwon YD, Nagy I, Adams PD et al (2004) Crystal structures of the *Rhodococcus* proteasome with and without its pro-peptides: implications for the role of the pro-peptide in proteasome assembly. J Mol Biol 335:233–245
96. Hu G, Lin G, Wang M et al (2006) Structure of the *Mycobacterium tuberculosis* proteasome and mechanism of inhibition by a peptidyl boronate. Mol Microbiol 59:1417–1428
97. Darwin KH, Ehrt S, Gutierrez-Ramos JC et al (2003) The proteasome of *Mycobacterium tuberculosis* is required for resistance to nitric oxide. Science 302:1963–1966
98. Lamichhane G, Raghunand TR, Morrison NE et al (2006) Deletion of a *Mycobacterium tuberculosis* proteasomal ATPase homologue gene produces a slow-growing strain that persists in host tissues. J Infect Dis 194:1233–1240
99. Lin G, Li D, De Carvalho LP et al (2009) Inhibitors selective for mycobacterial versus human proteasomes. Nature 461:621–626
100. Sutter M, Striebel F, Damberger FF et al (2009) A distinct structural region of the prokaryotic ubiquitin-like protein (Pup) is recognized by the N-terminal domain of the proteasomal ATPase Mpa. FEBS lett 583:3151–3157
101. Chen X, Solomon WC, Kang Y et al (2009) Prokaryotic ubiquitin-like protein pup is intrinsically disordered. J Mol Biol 392:208–217
102. Liao S, Shang Q, Zhang X et al (2009) Pup, a prokaryotic ubiquitin-like protein, is an intrinsically disordered protein. Biochem J 422:207–215
103. Imkamp F, Rosenberger T, Striebel F et al (2010) Deletion of dop in *Mycobacterium* smegmatis abolishes pupylation of protein substrates in vivo. Mol Microbiol 75:744–754
104. Barandun J, Delley CL, Ban N et al (2013) Crystal structure of the complex between prokaryotic ubiquitin-like protein and its ligase PafA. J Am Chem Soc 135:6794–6797
105. Guth E, Thommen M, Weber-Ban E (2011) Mycobacterial ubiquitin-like protein ligase PafA follows a two-step reaction pathway with a phosphorylated pup intermediate. J Biol Chem 286:4412–4419
106. Sutter M, Damberger FF, Imkamp F et al (2010) Prokaryotic ubiquitin-like protein (Pup) is coupled to substrates via the side chain of its C-terminal glutamate. J Am Chem Soc 132:5610–5612
107. Imkamp F, Striebel F, Sutter M et al (2010) Dop functions as a depupylase in the prokaryotic ubiquitin-like modification pathway. EMBO Rep 11:791–797
108. Burns KE, Cerda-Maira FA, Wang T et al (2010) "Depupylation" of prokaryotic ubiquitin-like protein from mycobacterial proteasome substrates. Mol Cell 39:821–827
109. Cerda-Maira FA, Pearce MJ, Fuortes M et al (2010) Molecular analysis of the prokaryotic ubiquitin-like protein (Pup) conjugation pathway in *Mycobacterium tuberculosis*. Mol Microbiol 77:1123–1135
110. Festa RA, Mcallister F, Pearce MJ et al (2010) Prokaryotic ubiquitin-like protein (Pup) proteome of *Mycobacterium tuberculosis* [corrected]. PloS ONE 5:e8589
111. Poulsen C, Akhter Y, Jeon AH et al (2010) Proteome-wide identification of mycobacterial pupylation targets. Mol Syst Biol 6:386
112. Watrous J, Burns K, Liu WT et al (2010) Expansion of the mycobacterial "PUPylome". Mol Biosyst 6:376–385
113. Pearce MJ, Arora P, Festa RA et al (2006) Identification of substrates of the *Mycobacterium tuberculosis* proteasome. EMBO J 25:5423–5432
114. Delley CL, Striebel F, Heydenreich FM et al. (2012) Activity of the mycobacterial proteasomal ATPase Mpa is reversibly regulated by pupylation. J Biol Chem 287:7907–7914
115. Wang T, Darwin KH, Li H (2010) Binding-induced folding of prokaryotic ubiquitin-like protein on the *Mycobacterium* proteasomal ATPase targets substrates for degradation. Nat Struct Mol Biol 17:1352–1357
116. Striebel F, Hunkeler M, Summer H et al (2010) The mycobacterial Mpa-proteasome unfolds and degrades pupylated substrates by engaging Pup's N-terminus. EMBO J 29:1262–1271
117. Burns KE, Pearce MJ, Darwin KH (2010) Prokaryotic ubiquitin-like protein provides a two-part degron to *Mycobacterium* proteasome substrates. J Bacteriol 192:2933–2935

118. Begg KJ, Tomoyasu T, Donachie WD et al (1992) *Escherichia coli* mutant Y16 is a double mutant carrying thermosensitive ftsH and ftsI mutations. J Bacteriol 174:2416–2417
119. Lüdke A, Krämer R, Burkovski A et al (2007) A proteomic study of *Corynebacterium glutamicum* AAA+ protease FtsH. BMC Microbiol 7:6
120. Ayuso-Tejedor S, Nishikori S, Okuno T et al (2010) FtsH cleavage of non-native conformations of proteins. J Struct Biol 171:117–124
121. Herman C, Prakash S, Lu CZ et al (2003) Lack of a robust unfoldase activity confers a unique level of substrate specificity to the universal AAA protease FtsH. Mol cell 11:659–669
122. Koodathingal P, Jaffe NE, Kraut DA et al (2009) ATP-dependent proteases differ substantially in their ability to unfold globular proteins. J Biol Chem 284:18674–18684
123. Bieniossek C, Schalch T, Bumann M et al (2006) The molecular architecture of the metalloprotease FtsH. Proc Natl Acad Sci U S A 103:3066–3071
124. Suno R, Niwa H, Tsuchiya D et al (2006) Structure of the whole cytosolic region of ATP-dependent protease FtsH. Mol cell 22:575–585
125. Anilkumar G, Srinivasan R, Ajitkumar P (2004) Genomic organization and in vivo characterization of proteolytic activity of FtsH of *Mycobacterium smegmatis* SN2. Microbiology 150:2629–2639 (Reading England)
126. Ogura T, Okuno T, Suno R et al (2013) FtsH protease. Handbook of proteolytic enzymes, pp 685–692
127. Srinivasan R, Anilkumar G, Rajeswari H et al. (2006) Functional characterization of AAA family FtsH protease of *Mycobacterium tuberculosis*. FEMS Microbiol Lett 259:97–105
128. Anilkumar G, Srinivasan R, Anand S et al (2001) Bacterial cell division protein FtsZ is a specific substrate for the AAA family protease FtsH. Microbiology 147:515–517
129. Kiran M, Chauhan A, Dziedzic R et al (2009) *Mycobacterium tuberculosis* ftsH expression in response to stress and viability. Tuberculosis (Edinb, Scotland) 89(Suppl 1):S70–S73
130. Strösser J, Lüdke A, Schaffer S et al (2004) Regulation of GlnK activity: modification, membrane sequestration and proteolysis as regulatory principles in the network of nitrogen control in *Corynebacterium glutamicum*. Mol Microbiol 54:132–147
131. Skuce RA, McCorry TP, McCarroll JF et al (2002) Discrimination of *Mycobacterium tuberculosis* complex bacteria using novel VNTR-PCR targets. Microbiology 148:519–528
132. Belkum AV, Scherer S, Alphen LV et al (1998) Short-sequence DNA repeats in prokaryotic genomes short-sequence DNA repeats in prokaryotic genomes. Microbiol Mol Biol Rev 62:275–293
133. Moxon R, Bayliss C, Hood D (2006) Bacterial contingency loci: the role of simple sequence DNA repeats in bacterial adaptation. Annu Rev Genet 40:307–333
134. Tantivitayakul P, Panapruksachat S, Billamas P et al (2010) Variable number of tandem repeat sequences act as regulatory elements in *Mycobacterium tuberculosis*. Tuberculosis (Edinb) 90:311–318
135. Peloquin CA, Berning SE (1994) Infection caused by *Mycobacterium tuberculosis*. Ann Pharmacother 28:72–84
136. Tufariello JM, Chan J, Flynn JL (2003) Latent tuberculosis: mechanisms of host and bacillus that contribute to persistent infection. Lancet Infect Dis 3:578–590
137. Zahrt TC (2003) Molecular mechanisms regulating persistent *Mycobacterium tuberculosis* infection. Microbes Infect 5:159–167 (Institut Pasteur)
138. Michel A, Agerer F, Hauck CR et al (2006) Global regulatory impact of ClpP protease of *Staphylococcus aureus* on regulons involved in virulence, oxidative stress response, autolysis, and DNA repair. J Bacteriol 188:5783–5796
139. Kirstein J, Hoffmann A, Lilie H et al (2009) The antibiotic ADEP reprogrammes ClpP, switching it from a regulated to an uncontrolled protease. EMBO Mol Med 1:37–49
140. Lee BG, Park EY, Lee KE et al (2010) Structures of ClpP in complex with acyldepsipeptide antibiotics reveal its activation mechanism. Nat Struct Mol Biol 17:471–478
141. Li DH, Chung YS, Gloyd M et al (2010) Acyldepsipeptide antibiotics induce the formation of a structured axial channel in ClpP: a model for the ClpX/ClpA-bound state of ClpP. Chem Biol 17:959–969

142. Bottcher T, Sieber SA (2008) Beta-lactones as specific inhibitors of ClpP attenuate the production of extracellular virulence factors of *Staphylococcus aureus*. J Am Chem Soc 130:14400–14401
143. Bottcher T, Sieber SA (2009) Structurally refined beta-lactones as potent inhibitors of devastating bacterial virulence factors. Chembiochem 10:663–666 (A European Journal of Chemical Biology)
144. Gersch M, Gut F, Korotkov VS et al (2013) The mechanism of caseinolytic protease (ClpP) inhibition. Angew Chem 52:3009–3014
145. Compton CL, Schmitz KR, Sauer RT et al (2013) Antibacterial activity of and resistance to small molecule inhibitors of the ClpP peptidase. ACS Chem Biol 20:2669–2677
146. Schmitt EK, Riwanto M, Sambandamurthy V et al (2011) The natural product cyclomarin kills *Mycobacterium tuberculosis* by targeting the ClpC1 subunit of the caseinolytic protease. Angew Chem 50:5889–5891
147. Vasudevan D, Rao SP, Noble CG (2013) Structural basis of mycobacterial inhibition by Cyclomarin A. J Biol Chem 288:30883–30891

Chapter 17
The Interaction Networks of Hsp70 and Hsp90 in the *Plasmodium* and *Leishmania* Parasites

Thiago Vargas Seraphim, Carlos H. I. Ramos and Júlio César Borges

Abstract Tropical diseases affect the lives of at least one-tenth of the global population and are among the main global health priorities of the World Health Organization, the United Nations branch concerned with international public health. Most tropical diseases are caused by blood and tissue protozoal parasites, such as those belonging to the *Plasmodium* and *Leishmania* genera. Specifically, *Leishmania spp.* cause human leishmaniasis, and *Plasmodium spp.* cause malaria. Due to their overlapping geographical regions, these parasites are of great public health concern. They are microscopic, unicellular eukaryotes that are transmitted by blood-sucking insects, and this transmission can be stressful for both the parasites and the host. To cope with this stress, the parasites cycle between different stages in which many proteins are involved. Most of these proteins require assistance to reach their correct conformations; therefore, they are aided by molecular chaperones and Heat Shock Proteins (Hsps). Hsp70 and Hsp90 are the most important Hsps that assist in folding, and they are part of a Protein Quality Control system that helps maintain protein homeostasis. Furthermore, their functions in protozoa have expanded, as both Hsp70 and Hsp90 chaperones play important roles in cell growth and adaptation, allowing life cycle progression. Interaction networks of Hsp70 and Hsp90 in the *Plasmodium* and *Leishmania* genera are presented in this chapter.

J. C. Borges (✉) · T. V. Seraphim
Department of Chemistry and Molecular Physics,
Institute of Chemistry of São Carlos, University of São Paulo, São Carlos, SP, Brazil
e-mail: borgesjc@iqsc.usp.br

C. H. I. Ramos
Department of Organic Chemistry, Institute of Chemistry,
University of Campinas—UNICAMP, Campinas, SP, Brazil
e-mail: cramos@iqm.unicamp.br

T. V. Seraphim
e-mail: tvs0604@gmail.com

W. A. Houry (ed.), *The Molecular Chaperones Interaction Networks in Protein Folding and Degradation,* Interactomics and Systems Biology, DOI 10.1007/978-1-4939-1130-1_17,

1 Heat Shock Proteins and Chaperones in the Life Cycles of Protozoan Parasites

All cells, and consequently all living organisms, depend on proteins. These biomolecules are involved in almost all physiological functions, including catalysis, transport, signaling, and movement. The lifespan of each protein is generally controlled, as the expression and degradation of these biomolecules are well regulated. Transcription factors and the proteasome are established examples of regulators of these processes. Additionally, a protein's conformation is closely related to its function; thus, its folding state is also a target of regulation. Molecular chaperones and heat shock proteins assist in proper protein folding and prevent misfolding. A significant portion of the proteome is assisted by this class of proteins, which may be divided into foldases (assistants of folding and refolding), holders (preventers of unfolding and misfolding), and disaggregases (involved in solubilizing aggregates). This last function has become of great interest recently, as aggregates have been linked to cytotoxicity and consequently to many diseases. The chaperones from the heat shock protein (Hsp)70 and Hsp90 families are the best-known examples of foldases, whereas the chaperones from the Hsp40 and small Hsp (smHsp) families are the best known examples of holders. Hsp100 chaperones appear to be the unique examples of disaggregases. To learn more about molecular chaperones, the reader is referred to other chapters in this book and also to a large number of excellent reviews in the literature [1–5].

For intracellular protozoa, the roles of molecular chaperones and/or Hsps go beyond maintaining protein homeostasis; they are also involved in various developmental and pathogenesis-related processes [6]. Infection by intracellular pathogens is stressful for the host as well as for the parasite. The host is stressed from the symptoms caused by the infection, whereas the parasite becomes stressed by facing sudden changes in pH, temperature, reactive oxygen species, and enzymatic degradation. To respond to these challenges, the parasites produce molecular chaperones and Hsps to allow cell differentiation and consequently, adaptation [6, 7].

Leishmania spp. and *Plasmodium spp.* are intracellular kinetoplastids protozoa that cause leishmaniasis and malaria, respectively, which are both neglected tropical diseases [8]. Kinetoplastids are flagellated protozoa characterized by the presence of a structure named kinetoplast, a deoxyribonucleic acid (DNA)-containing region in the mitochondrion. Kinetoplastids cause different diseases, despite the similarity of their cellular and genomic organization, and have a complex life cycle [9]. A neglected tropical disease can result in physical disfiguration and incapacity, causing economical, social, and political problems that impact the health, education, agriculture, and local economies of poor countries [10].

1.1 Leishmania spp: Importance and Life Cycle

Ten *Leishmania spp.* distributed in 98 countries impact public health by causing three types of leishmaniasis (visceral, cutaneous, and mucocutaneous). Annually, this parasite is estimated to cause 0.9–1.6 million new cases of leishmaniasis, of which 0.2–0.4 million are visceral and 0.7–1.2 million are cutaneous or mucocutaneous [11]. Although the cutaneous and mucocutaneous leishmaniasis cases are not lethal, they cause a social stigma against the infected patients, whereas the visceral form leads to 20,000–50,000 deaths per year [8, 11].

Leishmania protozoa have a digenetic life cycle (Fig. 17.1a), and transmission occurs when female sand flies (*Phlebotomus* and *Lutzomyia* genera) bite a mammalian host, infecting it with flagellated metacyclic promastigotes that are phagocytosed by the immune system cells [12, 13]. The metacyclic promastigotes are engulfed by phagolysosomes and differentiate into amastigotes upon a pH decrease and temperature increase. This transformation is characterized by the loss of the flagellum and changes in the cell surface [12]. The amastigotes are metabolically more active and proliferate within the acidic phagolysosomal environment [12, 14]. When a female sand fly ingests blood containing amastigote-infected macrophages, the temperature and pH changes induce the transformation of the amastigotes into procyclic promastigotes. Additional transformations occur until the cells reach the infective metacyclic promastigote stage, which completes the *Leishmania* life cycle [13].

1.2 *Plasmodium falciparum*: Importance and Life Cycle

The second organism discussed in this chapter is the protozoan parasite *Plasmodium falciparum (P. falciparum)*, one of the etiologic agents of malaria, a disease that occurs in almost 100 countries. In 2010, an estimated 154–289 million new cases of malaria occurred, of which 490,000–836,000 led to death. In 2011, 3.3 billion of people were at risk of malaria in the world [15].

The protozoan *P. falciparum* has a complex life cycle (Fig. 17.1b) that includes exoerythrocytic, intraerythrocytic, and mosquito stages [16]. It is transmitted by the bite of the vector (female *Anopheles* mosquitoes), which initiates the exoerythrocytic stage in which the sporozoites pass from the vector to the blood stream of the mammalian host and migrate to the liver. This infective form enters the hepatocytes, replicates asexually and initiates schizogony. Each sporozoite gives rise to approximately 1,000 merozoites that are packed in vesicles called merosomes and then released into the blood stream to invade red blood cells (RBCs) [16–18]. In the RBCs, the intraerythrocytic stage takes place, and the merozoites differentiate into the early trophozoite stage (the ring-like form). The transformation continues to a later trophozoite form, a highly metabolically active phase in which the parasite replicates and forms schizonts. The mature schizonts are then disrupted, and the merozoites are released into the blood stream to infect additional RBCs [16, 17]. The release of the merozoites from the RBCs coincides with the peak of fever in patients with malaria [16]. A small fraction of the merozoites differentiate into gametocytes and can be ingested by the *Anopheles* females during blood feeding. At this point, the mosquito stage begins, male and female gametocytes fuse, and a series of events leads to the formation of the sporozoites, beginning a new life cycle [16, 17]. It is noteworthy that the intracellular phase of the *Leishmania* life cycle occurs in the phagolysosome, an acidic but nutrient-rich environment, whereas the *P. falciparum*

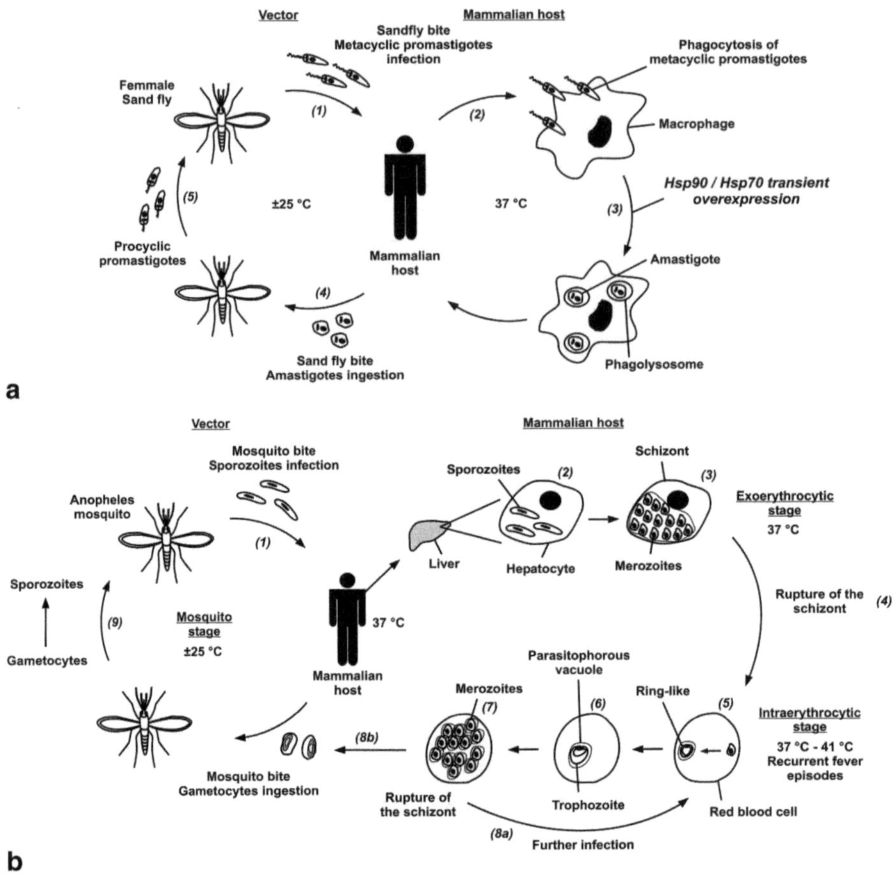

Fig. 17.1 Protozoan parasite life cycle. **a** *Leishmania* life cycle. (*1*) The mammalian host is bitten by a female sand fly and infected with metacyclic promastigotes, (*2*) which are phagocytosed by immune system cells and (*3*) differentiate into amastigotes within the phagolysosomes. (*4*) A female sand fly consumes blood containing infected macrophages and (*5*) amastigotes differentiate into procyclic promastigotes before the transformation into infective metacyclic promastigotes. **b** *P. falciparum* life cycle. (*1*) Sporozoites are transferred from the *Anopheles* female to the mammalian host blood stream, (*2*) migrate to the hepatocytes, (*3*) replicate and form schizonts. (*4*) The merozoites are released in the blood stream, invade the Red blood cells (*RBCs*) and (*5*) transform into early trophozoites (ring-like) and (*6*) trophozoites. (*7*) The latter replicate, form new schizonts and release merozoites into the blood stream to (8a) infect additional RBCs; (8b) a small portion of merozoites differentiate into gametocytes, which can be ingested by an *Anopheles* female and (*9*) give rise to new sporozoites. Hsp70 and Hsp90 are upregulated during phases 2 and 3 in *Leishmania*, as well as in the intraerythrocytic stage (phases 2 to 7) of the *P. falciparum* life cycle (see the text for details)

intracellular phase occurs in the parasitophorous vacuole, which has a physiological pH and a low nutrient supply [19].

Considering the widespread distribution of parasitic protozoa diseases, the limited treatment strategies, drug toxicity, and the resistance to therapy, it is important to understand the heat shock experienced by *Leishmania spp.* and *P. falciparum* upon infection and the roles of molecular chaperones in the prozotoa life cycles [20, 21]. These proteins are highly attractive targets for the development of new therapies, especially considering the reported malaria and leishmaniasis drug resistance cases [8, 16, 22]. However, one must identify the specificities of the protozoan molecular chaperones to develop selective drugs against the parasitic proteins and not the human orthologues [23].

1.3 The Importance of the Molecular Chaperones in the Protozoan Life Cycle

As discussed in the previous section, both *Leishmania spp.* and *P. falciparum* have a biphasic life cycle; the first stage occurs in a poikilothermic invertebrate (approximately 25 °C), and the second stage occurs in a homeothermic mammalian host (37 °C). Furthermore, these parasites experience even greater temperature fluctuations due to the fever response of the host (up to 41 °C). At first glance, the main difference between the vector and the host seems to be the temperature. However, the cellular environments are also quite different. Such dissimilarities, which cause environmental stress, drive morphological and metabolic changes in the protozoa. Thus, since the early 1980s, several studies have attempted to identify the pathways involved in parasite differentiation and the proteins involved in the heat shock and stress responses, such as the molecular chaperones, including Hsps.

To elucidate the molecular mechanisms involved in the *Leishmania spp.* transformation during the mammalian host invasion, as well as the role of Hsps in cell adaptation, several studies were conducted on axenic cultures. In this experimental approach, the promastigote cultures are subjected to temperature shifts similar to those occurring during the transfer of the parasites from an infected sand fly to a mammalian host that stimulates the differentiation into the amastigote-like forms. The temperature change is responsible for generating a heat shock response in these parasites, causing the overexpression of Hsps to facilitate the parasite adaptation to the new environment [24]. For some *Leishmania spp.*, the promastigote to amastigote transformation is achieved only upon temperature change although this is not true for all species. In spite of the upregulation of Hsps in all species upon heat shock, some additional factors may be required for the promastigote to amastigote transformation [24, 25].

Due to their essential roles in maintaining cellular protein homeostasis, Hsp70 and Hsp90 (also called Hsp83 in protozoa) constitute the main molecular chaperones in *Leishmania spp.*, making up 2.1 % and 2.8 %, respectively, of the total protein level in unstressed promastigotes [26]. In addition to their constitutive expression, these proteins are upregulated upon heat shock in all of the promastigote growth phases except for the stationary phase when the metabolic rate decreases [24, 25]. Nonetheless, Hsp70 and Hsp90 remain at high levels in the amastigote forms [27] and become phosphorylated after temperature adaptation [28].

Molecular chaperones function as environmental thermosensors and trigger developmental changes in *Leishmania*. For example, disturbances in Hsp90 homeostasis induce the synthesis of amastigote-specific proteins and promote the transformation from promastigotes to amastigotes [6, 27, 29]. Interestingly, the *in vitro Leishmania* transformation can be achieved by the direct inhibition of Hsp90, and 94 % of the proteins induced in this way (under geldanamycin (GA) treatment) are identical to those induced upon classical differentiation treatment (mammalian host temperature of 37 °C and mildly acidic medium) [27, 29]. Such observations point toward the requirement of molecular chaperones for parasite transformation.

In addition to the effects of temperature change on *Leishmania spp.* transformation and Hsp expression, other stressful conditions such as sand fly nutritional deprivation and oxidative products in the mammalian host also induce Hsp expression [25]. This increase prior to *Leishmania* transmission may pre-adapt the parasites to additional stresses [25]. Additionally, molecular chaperones directly influence the parasite's virulence and its survival inside the macrophage in the mammalian host [6]. Promastigote cells become more infective when they are exposed to heat shock due to the heat shock response [30]; therefore, *Leishmania* cells defective in producing Hsps have attenuated virulence [31]. Furthermore, molecular chaperones seem to have important roles in the *Leishmania* defense against oxidative stress in the macrophages and in drug resistance [6, 32].

P. falciparum also experiences several temperatures (in the insect vector, in the mammalian host, and during fever episodes at the intraerythrocyte stage) throughout its life cycle (Fig. 17.1b). Thus, a strong response against the heat shock is required [33]. Both Hsp90 and Hsp70 are directly involved in *P. falciparum* development and are upregulated during heat shock, e.g., after mammalian infection at 37 °C and during fever episodes at up to 41 °C [6, 33, 34], which suggests that they have thermosensor activity. Additionally, Hsp90 inhibition blocks the parasite development at specific stages, which suggests that it participates in the transformations. Interestingly, molecular chaperones are also involved in processes related to nutrient acquisition and drug resistance in *P. falciparum* [6].

Lastly, as *P. falciparum* goes through the intraerythrocytic stage, the protozoa molecular chaperones play essential roles in the host cell remodeling, which is essential for parasite survival. It has been reported that the secretion of protozoa Hsp70 and J-proteins into the erythrocyte cytosol directly affects host cell remodeling [6, 33, 35, 36]. Such cell remodeling events include the emergence of a knob apparatus on the infected RBC surface, which is important for cytoadherence and contributes to malaria pathology [6, 33]. Knobs are erythrocyte surface protrusions that mediate cytoadhesion to endothelial cells [37].

For a more detailed view of the stresses suffered by *Leishmania spp.* and *P. falciparum* as well as the importance of molecular chaperones in the protozoa thermo-driven transformation, we suggest consulting reviews in the literature [6, 33, 38–40].

2 The Foldosome Machinery

Hsp70, Hsp90 and several co-chaperones, among other proteins, associate transiently but coordinately in a heterocomplex called the foldosome, in which the Hsp70 and Hsp90 functional cycles work in a Synergistic manner [41–43]. The foldosome machinery will be summarized in this section, and it is depicted in Fig. 17.2. The sections that follow provide details on Hsp70, Hsp90 and the main co-chaperones that participate in this machinery, focusing on protozoal parasite organisms.

To initiate the Hsp70 molecular cycle, a J-protein binds to the hydrophobic surfaces of an unfolded, misfolded, or partially folded client protein and delivers it to Hsp70, which is in an adenosine triphosphate (ATP)-bound state (Fig. 17.2). As explained below, Hsp70 bound to ATP has a low affinity for client proteins. Both the client protein and the J-protein stimulate ATP hydrolysis by Hsp70, which switches Hsp70 to an ADP-bound state that has a high affinity for client proteins [3, 5]. At this stage, Hsp70-interacting protein (Hip) or small glutamine-rich tetratricopeptide repeat (TPR) protein (SGT) may associate with the Hsp70-ADP-client protein complex, stabilizing it to keep Hsp70 bound to the client protein and avoiding off-pathway protein folding [5, 44–46]. Nucleotide exchange factors (NEFs), such as GrpE, Hsp110, HspBP1, and BAG (Bcl-2 associated athanogene), interact with the adenosine diphosphate (ADP)-bound Hsp70 and stimulate the ADP exchange for ATP. This exchange reduces Hsp70's affinity for the client protein, which is released [3, 5].

Alternatively, the complex (called an early complex) formed by the Hsp70 in the ATP-bound state, the J-protein, and the client protein may follow another pathway involving the Hsp90 chaperone machinery [42]. Hip and/or SGT proteins may also participate in the early complex because they have been found in the foldosome assembly (see below) and we hypothesize that SGT functions as an adaptor because it is a dimer that binds both Hsp70 and Hsp90 [47–49]. Hsp70/Hsp90 organizing protein (Hop) is another adaptor that connects the Hsp70 and Hsp90 proteins (Fig. 17.2) [2, 41, 50].

The Hsp90-ATP-Hop/SGT complex associates with the early complex, forming the intermediate complex, and the client protein is transferred from Hsp70 to Hsp90 [41, 42]. The binding of co-chaperones, such as activator of Hsp90-ATPase activity 1 (Aha1) or p23, to Hsp90 disrupts the intermediate complex and gives rise to the Hsp90 late complex, releasing Hsp70, the J-protein, and Hop/SGT [42, 50]. Hsp90 has several conformational states that are driven by ATP hydrolysis and posttranslational modifications and are regulated by co-chaperones [2, 4]. Aha1 interacts with Hsp90, stabilizing it in the closed state and stimulating its ATPase activity [50, 51]. After ATP hydrolysis, Hsp90 switches to an ADP-bound, open conformational state from which the mature client protein is released [4, 50]. Hsp90 inhibition leads to the disruption of the intermediate complex, and the client protein may be quickly eliminated by the ubiquitin-proteasome system [4, 52, 53]. The co-chaperone carboxyl-terminus of Hsp70/Hsp90 interacting protein (CHIP), an E3-ubiquitin ligase, participates in this process by targeting the client protein for degradation [54].

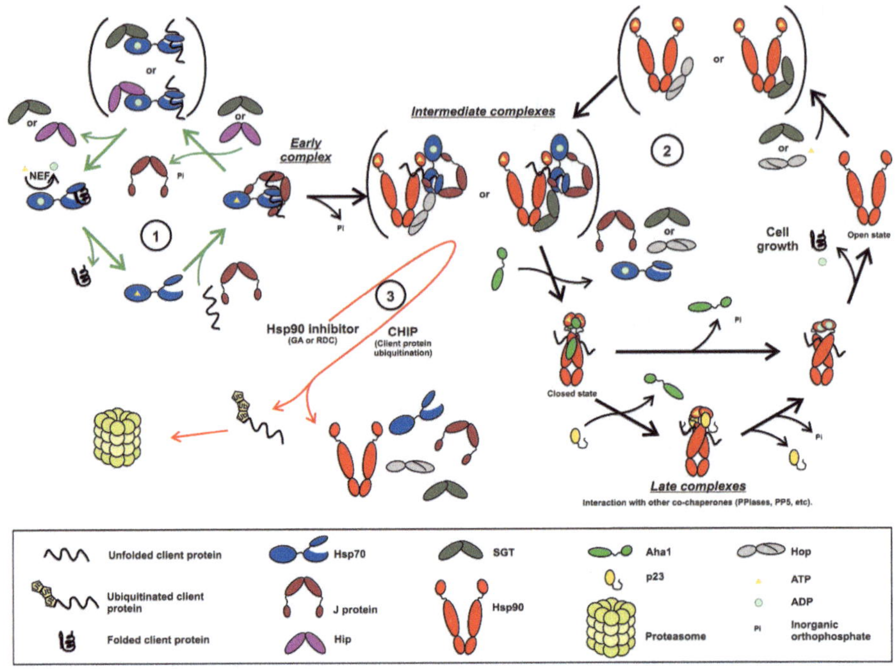

Fig. 17.2 Molecular chaperones interaction network: the foldosome. **1** The Hsp70 chaperone machinery. A J-protein binds to the unfolded client protein, interacts with it and delivers it to the ATP-bound Hsp70. After ATP hydrolysis, the J-protein leaves the complex and Hsp70-interacting protein (Hip) or Small glutamine-rich tetratricopeptide repeat protein (SGT) interacts with the Nucleotide-binding domain (NBD) of the ADP-bound Hsp70, maintaining it in this nucleotide-bound state. After Hip or SGT is released from the complex, a Nucleotide exchange factor (NEF) stimulates the exchange of ADP to ATP in the Hsp70 NBD, releasing the folded client protein for its own fate. **2** The Hsp90 chaperone machinery. The open conformation of ATP-bound Hsp90 interacts via its C-terminal with Hsp70/Hsp90 organizing protein (Hop) or SGT. The Hsp70-J-protein-unfolded client protein complex joins the Hsp90 complex with Hop or SGT, forming an intermediate complex. At this point, Hsp70 transfers the client protein to Hsp90 and Hsp70 together with the J-protein, Hop or SGT, leaves the complex. The late complex is formed when ATP-bound Hsp90 interacts with Aha1, stabilizing it in its closed conformation. After this step, Hsp90 can hydrolyze ATP to ADP, creating a more compact, ADP-bound state. Alternatively, the Hsp90-Aha1 interaction can be disrupted by p23, maintaining Hsp90 in the ATP-bound state. Additional co-chaperones, such as Peptidyl-prolyl *cis-trans* isomerases (PPIases) and PP5, can join the late complex. After the release of p23, ATP hydrolysis occurs, and Hsp90 forms the compact, ADP-bound state. After the release of ADP, Hsp90 returns to its open state, which starts a new cycle, and the folded/mature client protein is released to play its role to cell growth. **3** Targeting client proteins for degradation. The inhibition of the ATPase activity of Hsp90 allows Carboxyl-terminus of Hsp70/Hsp90 interacting protein (CHIP) to target the client protein for degradation and disrupts the intermediate complex. The polyubiquitinated client protein is recycled via the proteasome system. The individual figures represent the approximate structures of the molecular chaperones. See the text for more details on the foldosome machinery.

3 Hsp70

The Hsp70 molecular chaperone family acts in almost all events involving protein homeostasis, such as protein transportation through membranes, targeting client proteins for degradation, exocytosis, and signal transduction, among others [54, 55]. This wide spectrum of action is due to the ability of Hsp70 to handle unfolded, misfolded, or partially folded proteins by receiving them from and transferring them to other molecular chaperones, including Hsp90 (reviewed in [5, 43]). Along with this pivotal function, Hsp70 also has the following functional characteristics: (1) constitutive and stress-inducible isoforms, (2) isoforms with varying degrees of specialization, (3) activity, cycling and cellular functions modulated by co-chaperones, (4) organelle-specific isoforms, and (5) low-specificity interactions with hydrophobic segments normally exposed on the surfaces of client proteins [55]. Additionally, the Hsp70 cycle is governed by reciprocal heterotropic allostery (see below), which imposes several check points in the cycle and helps Hsp70 to determine the fate taken by the client protein along protein folding, degradation, membrane translocation, etc (reviewed in [5]).

Hsp70 has a molecular mass of approximately 70 kDa shared in two domains. At the N-terminal exists the nucleotide-binding domain (NBD) that binds and hydrolyzes ATP [5, 56]. The substrate-binding domain (SBD) is located in the C-terminus and interacts with client proteins. The SBD is formed by a β-sheet sandwich subdomain (SBDβ) that contains the peptide-binding site and a long α-helix at the C-terminus (SBDα) that closes on the unfolded client proteins bound to the SBDβ [5, 56–58]. Based on the domain architecture described above, one can conclude that Hsp70 is a modular protein. Nevertheless, unlike other molecular chaperones, Hsp70 functions as a monomer [43, 59]. It is well documented that Hsp70 alternates between two main states: an ATP-bound state in which the affinity for client proteins is low and an ADP-bound state in which the affinity for client proteins is high and the SBDα is closed on the SBDβ [56]. Therefore, the Hsp70 affinity for client proteins is regulated by the presence of either ADP or ATP in the NBD. When ATP binds the NBD, it leads to a reduction in the affinity of SBDβ for client proteins by increasing both the association and the dissociation rates. On the other hand, in the ADP-bound state, the affinity of SBDβ for client proteins increases as the association and dissociation rates decrease [5, 56, 57]. ATP hydrolysis, which can be stimulated by J-proteins and/or client proteins bound to the SBD, serves as a molecular switch between the ATP- and ADP-bound states of Hsp70 [5, 56, 57]. These two Hsp70 conformations might also be described based on the SBD open and closed states [5]. However, these two conformational states do not explain how partially folded or even folded client proteins interact with the SBD because SBDα should not be able to close on the client protein due to its own steric hindrance [58]. This observation suggests the existence of a third conformational state and underlines the Hsp70 plasticity required to interact with different client proteins, including unfolded, folded, partially folded, and aggregated proteins. The ability of Hsp70 to bind and hold client proteins is under the control of the reciprocal or bidirectional heterotropic allostery between the NBD and the SBD, which depends on the coupling between these domains [60]. In this mechanism, different domains of a protein act heterotropically as a signal source and regulate each other reciprocally. In the case of Hsp70, when ATP binds to NBD, conformational changes, which are transmitted to SBD, causes an overall reorganization of the Hsp70 structure resulting in client protein affinity decrement and releasing. On the other hand, client protein binding to SBD also induces conformational changes in the NBD, stimulating ATP hydrolysis and increasing substrate affinity [5]. This allostery also depends on the presence of an adenosine nucleotide in the NBD, the presence of client proteins in the SBD and on a cohort of co-chaperones (see below) [5]. The cycle of ATP binding, ATP hydrolysis, and exchange of ADP for ATP in the NBD is assisted by Hsp70 co-chaperones, J-proteins, and NEFs and is described in part 1 of Fig. 17.2 [5, 56, 61].

P. falciparum has a classical, cytosolic Hsp70 protein called PfHsp70-1; it is mostly monomeric, although dimer and trimer forms also exist [62, 63]. PfHsp70-1 is an ATP-dependent foldase that suppresses protein aggregation and refolds denatured client proteins, both in the absence and in the presence of J-proteins [63–65]. The ATPase activity of the PfHsp70 was found to be higher than that of human, bovine, and bacterial Hsp70s but lower than that of *T. cruzi* [62]. The NBD has lower ATPase activity than the full-length protein, implying that additional regions of the protein also modulate its ATPase activity. As showed for orthologous Hsp70s, the SBDα of PfHsp70-1 plays an important role in regulating the interdomain coupling of the NBD and the SBD as well as client protein binding [64, 65]. Interestingly, the C-terminal of the SBDα seems to interact with the NBD, contributing to the overall protein stability [63].

In the context of new malaria therapies, four compounds (malonganenones A and C, lapachol, and bromo-β-lapachone) were found to interact with PfHsp70-1 and inhibit *P. falciparum* growth with IC_{50} values lower than 20 μM [66]. In addition, these compounds interact with Hsp70s from other organisms with different specificities [66], pointing toward the possibility of developing specific inhibitors for protozoa Hsp70s.

Another *P. falciparum* Hsp70, PfHsp70-x, is found both in the erythrocyte cytosol and in the parasitophorous vacuole [35, 67]. PfHsp70-x is directly associated with at least two exported J-proteins (see below), and these proteins are part of large molecular complexes within erythrocytes [35]. There is strong evidence of the importance of PfHsp70-x in the transport of protozoa proteins through the erythrocyte cytosol [35].

Despite the lack of structural studies on Hsp70 from *Leishmania*, some studies were conducted to understand its functional role. Abnormal growth and morphology occur in *Leishmania* cells without an intact *hsp70-II* gene [68]. It has been reported that Hsp70 and Hsp90 work together in complexes that interact with proteins related to cell cycle control [69]. Thus, Hsp70 depletion likely disrupts those complexes [68]. The *hsp70-II* null cells exhibited significantly decreased infectivity (1,000-fold reduction in mice models) although their entry into macrophages was not affected [68]. This result suggested that Hsp70 may be involved in the amastigote cell cycle and replication, underlining the importance of Hsp70 in protozoa development [68].

Treatment-resistant *Leishmania* cells responded to antimonial therapy by overexpressing cytoplasmic inducible and cognate Hsp70s. However, it has been proposed that Hsp70s do not directly confer antimonial tolerance but may instead promote an initial, non-specific stress response, allowing the cells time to develop more efficient and specific mechanisms of resistance [70]. In contrast, inducible and cognate Hsp70s are directly involved in thermo tolerance after heat shock [70]. Additional information on Hsp70s from protozoa can be found in recent reviews: [6, 33, 65, 71–73].

4 Hsp90

The Hsp90 family is ubiquitous and is composed of 90 kDa heat shock proteins that represent up to 2.8 % of the soluble proteins in *Leishmania spp*. [20, 26]. Hsp90 (also called Hsp83) is essential to protozoa growth and differentiation, as it triggers changes during the life cycles of these organisms when inside the vertebrate host [21, 74, 75].

Hsp90 has three distinct domains: an N-terminal (N-domain), a Middle (M-domain), and a C-terminal domain (C-domain) [2, 4, 76]. The N-domain has an ATP-binding site that is essential for Hsp90 function because this protein's ATPase activity depends on N-domain dimerization [4, 77–80]. In addition, the N-domain interacts with client proteins and co-chaperones [4, 76, 80, 81] and is attached to the M-domain via a charged linker of variable length that has roles in both N-domain dimerization and ATPase activity regulation [4, 76, 82]. The M-domain has several sites for client proteins and co-chaperone binding. It also has two conserved amino acid residues (Arg380 and Gln384 in the yeast Hsp90 sequence) inside a catalytic loop that interact with the γ-phosphate of the ATP-bound N-domain and are essential for ATPase activity [4, 83, 84]. The C-domain is responsible for dimerization and possesses sites for client protein interactions. A MEEVD motif at the end of the C-terminus is responsible for binding to TPR domain proteins [2, 4, 23].

Hsp90 function strictly depends on its weak ATPase activity, which in turns is regulated by (1) N-domain dimerization, (2) its amino acids sequence, (3) a cohort of co-chaperones, and (4) conformational changes that interfere with the ATP hydrolysis rate [4, 78, 80, 81, 83, 85, 86]. The binding of ATP at the N-domain triggers the Hsp90 conformational change from an open to a closed state; the N-domain of each protomer dimerizes, couples to the M-domain and hydrolyzes ATP [76, 80, 83, 85–87]. Furthermore, the conformational cycle of some Hsp90s is thermally driven and not exclusively dependent on ATP, allowing the recruitment of several interacting proteins depending on the conformational state [87, 88]. During the Hsp90 conformational cycle, client proteins and co-chaperones associate with Hsp90 to assemble dynamic heterocomplexes that eventually result in protein folding and/or activation [2, 4, 52, 84]. From a mechanistic point of view, the Hsp90 cycle seems conserved; however, subtle differences exist between various organisms due to different (1) co-chaperones, (2) posttranslational modifications, (3) small ligands, and (4) client proteins [2, 4, 76, 82, 89]. In protozoa, Hsp90 posttranslational modifications, such as acetylation and phosphorylation, seem to play an important role in regulating Hsp90 complex assembly [7, 28]. Thus, protozoa parasites might exhibit differences in the Hsp90 cycle compared to human orthologous.

Mutations or inhibitors affecting the Hsp90 cycle might lead to cell death. To address this hypothesis, Hsp90 inhibitors, such as GA and radicicol (RDC), have been described that bind the Hsp90 nucleotide site, i.e., they mimic ATP [77, 78, 80]. *In vivo*, GA-bound Hsp90 targets client proteins for degradation by the ubiquitin-proteasome system after client protein maturation fails (Fig. 17.2) [4, 52, 53]. Because Hsp90 is highly important for the maturation of oncoproteins, this chaperone has become a promising target for cancer therapeutics [4, 90]. Similarly, because Hsp90 inhibition causes defects in cell growth and differentiation in *P. falciparum*, *T. cruzi* and *L. donovani*, it is a highly attractive target for therapeutics against protozoa parasites [7, 29, 74, 91, 92].

Studies on *P. falciparum* Hsp90 (PfHsp90) have shown that it binds ATP and ADP in a Mg^{2+}-dependent manner [91] and that it has both a greater affinity for ATP and higher ATPase activity than human Hsp90 [7]. The high rate of ATP hydrolysis implies a high degree of client protein cycling during the PfHsp90 cycle. Since *P. falciparum* experiences multiple stressful conditions during development, PfHsp90 might have evolved to contend with high-stress conditions that require high client protein turnover rates [7]. Furthermore, PfHsp90 is expressed in all stages of the protozoan life cycle although the level is low during the later schizont stage [91].

As a consequence of its greater affinity for ATP, as discussed above, PfHsp90 also possesses a greater affinity for GA and thus is more sensitive to inhibition by this compound than its human orthologous [7, 93]. Additionally, 17-AAG (a GA analogue) inhibits PfHsp90, but to a lesser extent; both GA and 17-AAG disrupt parasite growth [7]. GA is effective against all forms of the intraerythrocytic stage of *P. falciparum*; however, the effect is most pronounced during the transformation from the ring stage to the trophozoite stage [91, 92]. GA treatment induces PfHsp90 overexpression (as well as PfHsp70 overexpression), which counteracts the GA-mediated inhibition and maintains PfHsp90 in a dephosphorylated state [92]. Interestingly, GA can act synergistically with chloroquine in the treatment against malarial diseases, suggesting that the combined use of these compounds has great potential in treating malaria [91]. As expected, the *P. falciparum* Hsp90 N-domain, including the crucial amino acid residues for ATP and GA binding is similar to that of human and yeast orthologues [92, 94], however, the ATP-binding site shows slight stereochemical and hydrophobic differences that could affect the PfHsp90-ligand interactions. In fact, drug-screening approaches have led to the identification of inhibitor compounds that exhibit, to various extents, selectivity and specificity against PfHsp90 [95]. This observation indicates that the combined subtle differences in the structure of the PfHsp90 N-domain may favor the development of drugs that are selective against particular Hsp90s [94].

In *Leishmania* protozoa, the inhibition of Hsp90 by GA (or its analogues) affects cellular growth, but cells are able to evade the GA treatment by overexpressing Hsp90 [29]. In fact, several studies suggest that *Leishmania spp.* possess multiple Hsp90 genes. According to genomic sequencing results, *L. major* has 19 genes organized in clusters that encode essentially the same Hsp90 protein [72]. Having multiple Hsp90 copies is common among *Leishmania spp.* and allows high levels of protein synthesis, which is then subject to posttranscriptional regulation [96]. In these protozoa, GA treatment of axenic promastigote cultures induces differentiation to the amastigote-like forms. Thus, perturbations of Hsp90 homeostasis may trigger the promastigote to amastigote differentiation [29]. Analogously, the stress caused by temperature increases in *Leishmania* cells when they pass from the insect vector to the mammalian host could lead to the redistribution of Hsp90 molecules to activities such as the stabilization of labile client proteins, which would reduce the number of Hsp90 molecules available and trigger the protozoan transformation/adaptation [6, 29].

The *L. braziliensis* Hsp90 (LbHsp90) is highly conserved in comparison with Hsp90s from other trypanosomatids and humans. Recently, the structure-function relationship of this chaperone based on studies on truncated proteins was reported by our group [23]. LbHsp90 self-associates through its C-domain and is an elongated dimer in solution with secondary and tertiary structural elements similar to those of both yeast and human orthologues [23]. LbHsp90 also possesses potent chaperone activity and has binding sites for the client protein distributed along its structure. Furthermore, LbHsp90 binds ATP in a Mg^{2+}-independent manner, in contrast to PfHsp90 [23, 91], and it possesses weak ATPase activity that depends on its dimeric state [23]. The weak LbHsp90 ATPase activity is inhibited by GA. LbHsp90 is less sensitive to GA inhibition than PfHsp90 although it is more sensitive than yeast Hsp90 and comparable to the human orthologue [7, 23, 78]. In conclusion, LbHsp90 shares several functional and structural features with orthologous Hsp90s. However, LbHsp90 has some peculiar features in the N-domain that might be explored in the development of new therapeutic strategies against protozoan infections [23]. We observed that the LbHsp90 affinity for GA depends on the length of the protein, whereas the LbHsp90 nucleotide-binding properties are almost unchanged among different LbHsp90 deletion constructs. The truncated N-domain has a greater affinity for GA than the constructs consisting of the N- and M-domains or the full-length LbHsp90, likely because the full-length LbHsp90 dynamics disrupts the N-domain interaction with GA [23]. This observation may have important implications for drug development; some trials used only the N-domain construct for drug interaction tests and may have not taken into account the influence of the M- and C-domains on the N-domain interaction with inhibitor candidates.

For more information on the interactomes of molecular chaperones from protozoa, we suggest consulting a book chapter [97] and several review articles that cover more detailed views of many aspects of Hsp90 biology in these organisms [6, 20, 33, 72, 98].

5 The Hsp70 and Hsp90 Interaction Networks

In previous sections, the characteristics of Hsp70 and Hsp90 were introduced as if these chaperones functioned in isolation to facilitate comprehension. However, although these two chaperones have essential functions by themselves, they accomplish much more when acting together. The responsibility of coordinating their action lies heavily on the co-chaperones. Both Hsp90 and Hsp70 have co-chaperones that are involved in the ATP-driven cycle [2, 5, 20] and have either stimulating or inhibiting activities. Other co-chaperones, such as p23 and the J-proteins, also act as client protein scanning factors for one or both chaperones. These co-chaperones are discussed below.

Several Hsp70 and/or Hsp90 co-chaperones have a TPR motif that interacts with the C-terminus EEVD motifs of both Hsp70 and Hsp90, and they use these motifs to act as linkers between these chaperones. The helix-turn-helix TPR motif is a highly degenerated 34 amino acid repeat that folds into an antiparallel α-helical hairpin with a groove on the surface [99, 100]. The groove can be used for intermolecular [100] or intramolecular interactions [101]. TPR domains are conserved scaffolds that mediate protein-protein interactions in multiprotein complexes and are involved in a wide variety of cellular functions such as cell cycle control and transcription, among others (reviewed in [101]). Many known TPR-containing co-chaperones are present in protozoa and work on Hsp70/Hsp90 multichaperone foldosome assembly. Among them, three TPR proteins have already been characterized in protozoa and are discussed below. In fact, several Hsp90 and Hsp70 co-chaperones have been identified in protozoa. Table 17.1 shows an overview of Hsp70 and Hsp90 proteins as well as their co-chaperones present in the *L. braziliensis* and *P. falciparum* genomes; note the abundance of these proteins in protozoa. Much work remains to be done to ascertain their structural conformations as well as their roles in the Hsp70/Hsp90 network and in protozoa parasitism.

5.1 J-proteins

The main Hsp70 co-chaperones, Hsp40, or J-proteins are involved in protein-folding process by preventing client proteins from aggregating [43, 56, 102]. Both ATP hydrolysis stimulation and client protein presentation to Hsp70 are performed by the J-protein family [103, 104]; thus, the J-proteins function directly as chaperones and as client protein scanning factors for Hsp70 [105, 106]. J-proteins are characterized by an α-helical Hsp70-binding domain, called the J-domain, which stimulates Hsp70 ATPase activity [107, 108]. J-proteins form a ubiquitous and diverse family of co-chaperones. They differ in their domain organization and cellular localization [56].

The J-protein family is divided into four types based on their domain organization. Types I and II possess a flexible Gly/Phe-rich region that functions as a spacer between the J-domain and the C-terminal region (reviewed in [106, 109, 110]). Type I members have a Cys-rich region containing four repeats of the CxxGxGxG motif at the central region while type II members possess a Gly/Met-rich region at

Table 17.1 Summary of the Hsp70 and Hsp90 chaperones and their main co-chaperones identified in the *L. braziliensis* and *P. falciparum* genome databases

Molecular chaperone	Functions/observations	Accession number	
		L. braziliensis	*P. falciparum*
Hsp70	See the text for details	LbrM.28.2990 (71)	PF3D7_0831700 (75)
	‡mitochondrial protein—mtHsp70	LbrM.26.1260 (71)	PF3D7_0818900 (74)
	Φendoplasmic reticulum protein—Grp78	LbrM.28.1300 (72) Φ	PF3D7_0917900 (72)
		LbrM.30.2420 (70) ‡	PF3D7_1134000 (73)
		LbrM.30.2440 (71) ‡	PF3D7_0708800 (99)
J-proteins	Stimulate Hsp70 ATPase activity	LbrM.27.2610 (43) @	PF3D7_1437900 (48) @
	Act as client protein scanning factors		PF3D7_0409400 (76) @
		LbrM.25.0990 (44) @	PF3D7_0501100.2 (45) §
	May possess intrinsic chaperone activity		PF3D7_0201800 (48) §
		LbrM.34.2890 (51) @	PF3D7_0113700 (47) §
	May participate in the Hsp70/Hsp90 multichaperone machinery/foldosome in the early complex		PF3D7_0501100.1 (46) §
		LbrM.21.0550 (50) @	PF3D7_0213100 (37) §
			PF3D7_0629200 (44) §
	@Type I	LbrM.32.3590 (45) @	PF3D7_1211400 (28) †
	§Type II		PF3D7_1108700 (62) †
	†Type III	LbrM.04.0730 (50) @	PF3D7_1330300 (39) †
			PF3D7_0220100 (117) †
		LbrM.15.1170 (61) @	PF3D7_1413900 (45) †
			PF3D7_1136800 (44) †
		LbrM.35.6590 (38) §	PF3D7_1149600 (83) †
		LbrM.32.2130 (35) §	PF3D7_0806500 (76) †
		LbrM.27.0570 (19) †	PF3D7_1038800 (107) †
		LbrM.27.0500 (42) †	PF3D7_1356700 (24) †
		LbrM.35.4710 (31) †	PF3D7_0919100 (43) †
		LbrM.33.1090 (33) †	PF3D7_0523400 (62) †
		LbrM.31.3530 (24) †	PF3D7_0823800 (76) †
		LbrM.35.0590 (56) †	PF3D7_1318800 (76) †
		LbrM.19.0390 (50) †	PF3D7_1005600 (94) †
		LbrM.35.0730 (28) †	PF3D7_1149200 (126) †
		LbrM.32.3320 (36) †	PF3D7_1201100 (107) †
		LbrM.25.1770 (44) †	PF3D7_0831200 (114) †
		LbrM.20.5290 (74) †	PF3D7_1002800 (170) †
		LbrM.35.2340 (32) †	PF3D7_1039100 (67) †
			PF3D7_1102200 (67) †
			PF3D7_1473200 (52) †
			PF3D7_0114000 (170) †
			PF3D7_0102200 (126) †
			PF3D7_1253000 (56) †
			PF3D7_0201700 (106) †
			PF3D7_1149500 (99) †
			PF3D7_0220400 (72) †
			PF3D7_1401100 (57) †
			PF3D7_0502800 (16) †
			PF3D7_1307200 (39) †
			PF3D7_1143200 (39) †
			PF3D7_1216900 (111) †
			PF3D7_1422300 (43) †

Table 17.1 (continued)

Molecular chaperone	Functions/observations	Accession number	
		L. braziliensis	*P. falciparum*
GrpE	NEFs Participate in the Hsp70 ATP cycle	LbrM.30.0820 (24) LbrM.30.0830 (29)	PF3D7_1124700 (35)
Hsp110	GrpE is the mitochondrial protein	LbrM.18.1400 (92)	PF3D7_1344200 (108)
BAG	Hsp110 and HspBP1 are	Not identified	Not identified
HspBP1	cytoplasmic See details in [5]	LbrM.31.2540 (49)	PF3D7_0410600 (37) PF3D7_0441900 (31)
Hep1	Prevents mitochondrial Hsp70 aggregation	LbrM.25.0680 (23)	PF3D7_1420300 (35)
Hsp90	See the text for details.	LbrM.33.0340 (80) brM.33.0350 (44) LbrM.33.0330 (41) LbrM.29.0780 (89) LbrM.33.2670 (72)	PF3D7_0708400 (86) PF3D7_1443900 (107) PF3D7_1118200 (108) PF3D7_1222300 (kDa)
Aha1	Stimulates Hsp90 ATPase activity See the text for details	LbrM.18.0230 (38)	PF3D7_0308500 (16) PF3D7_0306200 (42)
p23	p23-like protein; inhibits Hsp90 ATPase activity Has intrinsic chaperone activity; see the text for details *Dissimilar domain organization	LbrM.34.4450 (22) LbrM.20.0220 (23)	PF3D7_1453700 (30) PF3D7_0927000 (18) PF3D7_0314000 (32)*
Hop/STI1	TPR-containing proteins;	LbrM.08.0880 (62)	PF3D7_1434300 (66)
Hip	May participate in the Hsp70/Hsp90	LbrM.29.0260 (41)	PF3D7_0527500 (51)
SGT	multichaperone machinery/foldosome in the early complex stage; see the text for details	LbrM.30.2700 (46)	Not identified
CHIP	Hsp90 and Hsp70 co-chaperone Targets client proteins to the proteasome	Not identified	PF3D7_0707700 (110)
Cyp	Cyclophilin-like domain-containing protein; primarily discovered as bound to cyclosporine A; Peptidyl-prolyl *cis-trans* isomerase (PPIase) activity, catalyzes the *cis-trans* isomerization of peptide bonds preceding proline residues See details in [187] #TPR-containing proteins	LbrM.06.0100 (20) LbrM.23.0140 (20) LbrM.33.1900 (21) LbrM.31.0050 (24) LbrM.22.1330 (25) LbrM.34.3570 (34) LbrM.24.1570 (26) LbrM.30.0030 (27) LbrM.35.3350 (22) LbrM.23.0060 (32) LbrM.01.0250 (37) LbrM.16.1240 (37) LbrM.34.1620 (49) LbrM.18.0940 (113) LbrM.34.4730 (39)# LbrM.25.0790 (19)	PF3D7_0322000 (19) PF3D7_1115600 (22) PF3D7_1116300 (19) PF3D7_0528700 (23) PF3D7_0804800 (25) PF3D7_1202400 (26) PF3D7_1215200 (32) PF3D7_1423200 (53) PF3D7_0803000 (81) PF3D7_0510200 (87)
FKBP	FKB-506-binding protein; Protein belonging to the Immunophilin family; Possesses PPIase activity; Some proteins may also present chaperone activity See details in [186]	LbrM.22.1310 (21) LbrM.35.0310 (12) LbrM.19.1790 (48) LbrM.10.0990 (16) LbrM.19.1230 (23)	PF3D7_1247400 (FKBP35–35) PF3D7_1313300 (26)

Table 17.1 (continued)

Molecular chaperone	Functions/observations	Accession number	
		L. braziliensis	*P. falciparum*
PDI	Protein disulfide isomerase; Disulfide-bound oxireductase/ isomerase protein that catalyzes rearrangements, breaks and the formation of disulfide bonds [&]Thioredoxin domain-containing protein See details in [188]	LbrM.35.7320 (52) LbrM.06.1030 (15) LbrM.26.0670 (42) LbrM.20.1700 (47)[&] LbrM.26.0680 (19)[&] LbrM.34.1250 (63)[&] LbrM.06.1170 (17)[&]	PF3D7_0827900 (55) PF3D7_1472600 (66) PF3D7_0919400 (61) PF3D7_1134100 (49) PF3D7_1127500 (51) PF3D7_1352500 (24)[&] PF3D7_1457200 (12)[&] PF3D7_1345100 (19)[&] PF3D7_0925500 (15)[&]
PP5	Ser/Thr phosphatase; member of the protein phosphatase family, subfamily phosphoprotein phosphatase (PPP); associates with Hsp90 in complexes during the maturation of client proteins See details in [189]	LbrM.18.0170 (105)	PF3D7_1355500 (77)

Proteins identified by a BLASTp search against the TritrypDB and PlasmoDB databases using human orthologues as templates

The numbers inside the brackets are the predicted molecular masses in kDa

ATP: Adenosine triphosphate, NEFs: Nucleotide exchange factors, Hsp: Heat shock protein, Aha1: Activator of Hsp90-ATPase activity 1, Hop: Hsp70/Hsp90 organizing protein, Hip: Hsp70-interacting protein, SGT: Small glutamine-rich tetratricopeptide repeat protein, CHIP: Carboxyl-terminus of Hsp70/Hsp90 interacting protein, Cyp: Cyclophilin-type, FKBP: FKB-506 drug-binding protein type, PDI: Protein disulfide isomerase, PP5: Ser/Thr phosphatases, STI1: Stress inducible 1, TPR: Tetratricopeptide repeatt

the same location. The C-terminal domain is formed by β-sheet structures that act together with the central regions of the J-proteins to provide chaperone activity and present client proteins to Hsp70 or Hsp70/Hsp90 foldosome [105, 106, 109, 110]. The type III J-proteins have a conserved J-domain located anywhere along the protein, whereas the J-domain in types I and II is located at the N-terminal region [56]. The type IV J-proteins have variations in the conserved HPD motif in the J-domain and are thus not expected to interact with Hsp70 chaperones and may function via a different mechanism [111]. Despite the similarities among type I and II J-proteins, they possess different chaperone functions and have specific Hsp70 regulatory functions [109, 110, 112, 113]. Interestingly, J-proteins are modular proteins with a highly elongated shape [109, 110, 113, 114], and their different structures, functions, localizations, and specificities allow them to directly influence Hsp70 function in numerous ways [106]. J-proteins can amplify Hsp70 efficiency by reducing the time spent searching for client proteins.

In contrast to the well-studied humans and yeast orthologues, J-proteins from protozoa are structurally uncharacterized, and only a few studies regarding their functions exist. Although studies on *P. falciparum* J-proteins are found in the literature, we found no studies on *Leishmania spp*. Table 17.1 indicates the presence of 21 predicted J-proteins in the *L. braziliensis* genome and twice as many as previously reported for the *P. falciparum* genome [98, 111].

The *P. falciparum* genome contains several genes that encode the J-proteins (Pfjs or PfHsp40s). Data mining suggests that these proteins are located in multiple organelles and in the cytosol [98, 111]. Interestingly, some *P. falciparum* J-proteins possess a *Plasmodium* export element (PEXEL) motif that directs protein secretion into the erythrocyte cytosol [33, 111]. The secreted PfHsp40s may participate in host cell remodeling together with the host Hsp70 [33, 65, 111]. This hypothesis is interesting because the parasite needs to modify the host environment to survive [33], and in this case, PfHsp40s and PfHsp70 would assemble in high molecular mass complexes within the infected erythrocytes [35]. In support of this hypothesis, a J-protein that possess a PEXEL motif in its N-terminus, called PfKAHsp40, is exported from the parasite at the trophozoite stage and associates with erythrocyte knobs and with PEXEL translocon components in the parasitophorous vacuole membrane [115]. Furthermore, two type II PfHsp40s are exported and form dot-like structures called J-dots; however, the function of these structures is unknown [36].

PfHsp40s might function in transcription, DNA damage repair, nucleosome assembly, and translation, among other processes [33]. In addition to the canonical type I J-protein domain organization, the type I J-protein Pfj1 contains an additional Pro/Lys-rich region of unknown function [116]. Pfj2 was the most abundant J-protein transcript, followed by Pfj4, Pfj3, and Pfj1. Pfj1, Pfj3, and Pfj4 transcription increases after heat shock, whereas Pfj2 transcription decreases [116]. Pfj4, which is a type II J-protein, forms large complex with PfHsp70 and is localized in the cytoplasm and the nucleus of *P. falciparum* [117]. Interestingly, the J-domains of both Pfj1 and Pfj4 are able to functionally replace the J-domain of the prokaryotic J-protein DnaJ, suggesting a conserved mechanism of J-domain function [118]. Another type I J-protein, PfHsp40I, is expressed similarly to PfHsp70 under normal and heat shock conditions. PfHsp40I localizes in the cytosol and stimulates the ATPase activity of both PfHsp70 and human Hsp70 [119]. In addition, PfHsp40I can suppress protein aggregation by itself and in cooperation with PfHsp70 [119]. Interestingly, the PfHsp40I-PfHsp70 pair interacts with different molecules than the human Hsp40-Hsp70 pair, making these proteins possible anti-malarial therapeutic targets [119].

As mentioned, a few Kinetoplastid J-proteins were characterized, mainly from *Trypanosoma spp*. In *T. cruzi*, the mitochondrial J-protein TcDj1 and the cytoplasmic J-protein family members (Tcj1, Tcj2, Tcj3, Tcj4, and Tcj6) were studied. The type I J-proteins include Tcj2, Tcj3, and Tcj4 while Tcj6 is a type II J-protein. Tcj1 and TcDj1 are type III J-proteins [71]. The Tcj1 protein is unable to stimulate Hsp70 ATPase activity; however, its orthologue in *T. brucei* (Tbj1) is able to cooperate with Hsp70 to avoid protein aggregation [71, 120]. Tcj2 is an essential protein in stressful conditions, and Tcj6 may associate with ribosomes during protein synthesis [71]. Furthermore, Tcj2 stimulates Hsp70 ATPase activity, and *in vivo* experiments suggested a cytoprotective effect as observed for the yeast type I J-protein orthologue [121]. As observed for *P. falciparum*, the Tcj3 J-domain can functionally replace the prokaryotic J-domain [118]. The mitochondrial TcDj1 expression profile fluctuates throughout the *T. cruzi* life cycle [71].

Little is known regarding the functional and structural properties of the J-proteins in protozoa parasites. These proteins are closely related to parasite survival and the remodeling of the host cell. Therefore, more studies are needed to elucidate the structure-function relationships of the protozoa J-proteins, especially because this family exhibits such large functional and structural diversity (Table 17.1). For a more detailed description of protozoa J-proteins, several recent reviews on functional and structural features are available (see [33, 72, 111, 122, 123]).

5.2 Activator of Hsp90-ATPase Activity 1—Aha1

The Hsp90 ATPase cycle is modulated by a cohort of co-chaperones [2, 4]. One of them, Aha1, stimulates weak Hsp90 ATPase activity [78, 79], as also reported for the protozoa orthologues [7, 23]. Typical Aha1 is a modular protein consisting of two domains at the N- and C-termini [83, 124, 125] that bind cooperatively to the Hsp90 dimer interface [51, 124] and are necessary for the maximum stimulation of Hsp90 ATPase activity [126]. However, a significant increase in Hsp90 ATPase activity can be reached by expressing only the Aha1 N-terminal domain, although at a high concentration, suggesting that this region is responsible for ATPase stimulation [125–127]. Additionally, the N-terminal domain of Aha1 is in close contact with the M-domain of Hsp90, where the R380 residue (as numbered in the yeast sequence) lies. This residue is stabilized and participates in the ATP γ-phosphate coordination during hydrolysis [126]. The C-terminal domain makes additional contacts with the Hsp90 N-domain in its dimerized state [51, 124]; thus, it also stimulates Hsp90 ATP hydrolysis. However, the extent to which the C-terminal domain of Aha1 stimulates Hsp90 ATPase activity is under debate [124].

The first report of a kinetoplastid Aha1 came from our effort [128] to understand the structural and functional properties of the *L. braziliensis* Hsp90 machinery. This protozoan possesses only one Aha1 gene (LbAha1; Table 17.1), and in spite of its low sequence conservation with the yeast (23 % identity) and human (30 % identity) Aha1 proteins, it likely possesses similar structural features. LbAha1 is a quite elongated monomeric protein formed by N- and C-terminal domains connected to each other by a flexible linker [128]. The N-terminal domain forms a slightly more elongated shape than the C-terminal domain [128]. Indeed, LbAha1 is large enough to interact with both the M- and N-domains of LbHsp90. Two LbAha1 molecules interact with each LbHsp90 dimer, suggesting that the LbAha1 molecules are positioned at the opposite interfaces of LbHsp90, as occurs in yeast [51]. Furthermore, the LbAha1-LbHsp90 interaction is thermodynamically driven by enthalpy and opposed by entropy, likely due to LbHsp90 conformational change restrictions [128]. In addition, the LbAha1-LbHsp90 interaction involves hydration events, hydrogen bonds, and long-range electrostatic interactions, similar to what was observed for yeast Aha1 [126]. The functional consequence of the LbAha1-LbHsp90 interaction is the stimulation of LbHsp90 ATPase activity by approximately 10-fold via a positive cooperative mechanism [128]. Aha1 was discovered as a stress-inducible protein in yeast and humans [127]; however, an *in vivo* analysis of the Aha1 expression profiles in *L. braziliensis*, *L. chagasi,* and *L. guyanensis* promastigote axenic cultures showed that the co-chaperone is a cognate protein and is expressed under both normal growth and heat-shock conditions [128]. LbAha1 has remarkable interdomain flexibility, which allows for a wide range of conformations, including the compact and extended conformations observed during its interaction with LbHsp90 [128]. Therefore, based on its flexibility and dimensions, the N-terminal domain of LbAha1 should first interact with the LbHsp90 M-domain followed by the LbAha1 C-terminal domain coupling to LbHsp90 N-domain.

Like *Leishmania* Aha1, little is known regarding the Aha1 proteins of the Apicomplexa phylum parasites. Our data mining results led to the identification of two Aha1 genes in the *P. falciparum* genome, one comprised of two Aha1 domains and one comprised of one Aha1 domain that resembled the yeast Hch1 protein (Table 17.1). Recently, Chua and co-workers [129] investigated the function of *P. falciparum* Aha1 (PfAha1), which shares 20 % identity with both the yeast and human orthologues, and possesses two domains, N- and C-termini. The structure of the PfAha1 N-terminal domain (PDB 3N72; unpublished data) was solved and found to be elongated like both yeast Aha1 and LbAha1 [126, 128]. The PfAha1 N- and C-terminal domains bind to the PfHsp90 M-domain in a reaction dependent on ATP and Mg^{2+}, suggesting that this interaction depends on the conformational state of PfHsp90 [129]. In addition, the contacts between these two proteins occur by interactions between polar and/or charged amino acid residues and result in PfHsp90 ATPase activity stimulation [129]. As reported for the yeast orthologues [130], PfAha1 and Pfp23 interact mutually exclusively with PfHsp90, suggesting that they have overlapping binding sites [129].

Aha1 co-chaperones share structural properties and act by stimulating weak Hsp90 ATPase activity; however, the molecular mechanisms used to achieve this stimulation are not fully understood. Because Aha1 stimulates Hsp90 ATPase activity, it can compensate for the effects of Hsp90 inhibitors that compete for the ATP-binding site, such as GA-based compounds, leading to drug desensitization [131, 132]. Therefore, the knowledge of how Aha1 regulates Hsp90 in protozoa is of great importance and may reveal an opportunity for drug design as it could lead to the development of selective inhibitors against the Aha1-Hsp90 complex.

5.3 Protozoa p23

One of the best studied Hsp90 interactors is the p23 co-chaperone, which is an acidic protein formed by a β-sheet domain connected to a C-terminal tail [133, 134]. The p23-Hsp90 interaction occurs at the p23 β-sheet domain, which appears to be sufficient for this interaction, and involves an extensive surface area and conformational changes [134–136]. The β-sheet domain is also responsible for binding partially unfolding proteins [133, 135], and several reports suggest that the p23 C-terminal tail is essential for chaperone activity [133–135]. p23 recognizes ATP-bound Hsp90 and stimulates Hsp90-client protein dissociation [137–139]. Human p23 binds Hsp90 cooperatively causing dimerization of the Hsp90 N-terminal domain [127, 140]. At physiological conditions, complexes with both 2:2 and 2:1 p23:Hsp90 stoichiometries occur [141].

The importance of p23 in Hsp90 cycling has been shown using Hsp90 inhibitors such as GA and RDC. These compounds prevent the proper interaction between Hsp90 and p23, blocking the Hsp90 late complex stage and targeting client proteins to the degradation machinery (Fig. 17.2) [142, 143]. On the other hand, p23 protects Hsp90 against inhibitors, preventing their entry to the ATP-binding site and acting as a desensitizing agent [135]. Some reports suggest that p23 stabilizes the interactions of Hsp90 with client proteins *in vitro* and reduces Hsp90 ATPase activity [139, 140, 144, 145]. Interestingly, the yeast p23/Sba1 has an interactome independent of yeast Hsp90, suggesting that it is involved in more processes than initially expected [146].

The *P. falciparum* genome has three predicted p23 proteins (Table 17.1): two of them with classical p23 domain organization and one in which the p23 β-domain is located at the C-terminal region. A *P. falciparum* phosphoprotein of 34 kDa was identified as a p23 orthologue (called Pfp23) because it consists of a highly conserved N-terminal domain with high identity to other p23 proteins, a Gly-rich tandem repeat composed of seven GNMGGL motifs, an acidic region, and a C-terminal region with low identity among *Plasmodium spp* [147]. Pfp23 is longer than the human and yeast orthologues due to the presence of the GNMGGL repeats and an extensive C-terminus; however, the meaning of such differences remains unclear [147, 148]. The binding of Pfp23 to PfHsp90 strictly depends on ATP and Mg^{2+} ions, suggesting the significance of the PfHsp90 conformational state for this interaction, in contrast to the interaction between Hsp90 and p23 in humans [147, 149]. The amino acid residues from positions 75 to 100 are crucial for the binding of Pfp23 to PfHsp90, and mutational studies revealed that substitutions of residues K91, H93, W94, and K96 reduced the interaction between these two proteins [147]. Conversely, deletion of the GNMGGL repeats did not interfere with the Pfp23-PfHsp90 interaction but impaired the ability to avoid protein aggregation. This result indicates that the C-terminal tail is critical for the Pfp23 intrinsic chaperone activity [147], as observed for the human and yeast orthologues [133–135]. The functional consequence of the Pfp23-PfHsp90 interaction is an approximately 80% reduction in PfHsp90 ATPase activity [147]. *In vivo*, the expression of Pfp23 messenger ribonucleic acid (mRNA) is detected after merozoite infection, increasing during the ring-like stage and reaching the maximum expression level at the end of the ring-like stage and the beginning of the trophozoite stage. The Pfp23 mRNA expression level remains high during the trophozoite stage [148]. Conversely, a large reduction in Pfp23 mRNA levels is observed in the schizogony form [148]. Interestingly, the level of Pfp23 phosphorylation is higher in the later trophozoite forms and schizonts than in the early trophozoite forms. Additionally, Pfp23 seems to be distributed throughout the *P. falciparum* cytoplasm, and its localization does not change during the course of parasite's development [148]. The changing expression profile of Pfp23 during the parasite life cycle and its interactions with Hsp90 and client proteins allow us to speculate that this small acidic protein might be involved in regulatory and differentiation processes in *P. falciparum*.

The p23 orthologue of *Toxoplasma gondii* (Tgp23), an Apicomplexa protozoan, has also been studied and has provided some insight into the function of p23 in the

Hsp90 conformational cycle. Tgp23 associates with Hsp90 at the late complex stage (Fig. 17.2), whereas Hsp70 participates in the early and intermediate complexes stages [150]. Furthermore, the Hsp90-Hsp70 and Hsp90-p23 complexes are mutually exclusive, indicating that Tgp23-specific interactions depend on the Hsp90 conformation [150]. As reported for yeast p23, the Tgp23 interactome consists of Hsp90-dependent and independent proteins [150]. This observation indicates that, as in other organisms, p23 has multiple functions in protozoa.

A bioinformatic analysis revealed that several Kinetoplastids possess two proteins with identity to human p23 (unpublished data and Table 17.1). In *L. braziliensis*, these orthologous proteins were named Lbp23A and Lbp23B; they are 22% identical to each other and share 31% and 20% identity, respectively, with human p23. Although Lbp23 proteins are not well conserved at the sequence level, their secondary and tertiary structures seem quite similar (unpublished data). *In vitro* studies suggested that both proteins are monomeric and have elongated shapes with similar dimensions. However, Lbp23A is significantly more chemically and thermally stable than Lbp23B (unpublished data). Both proteins possess chaperone activity, but they have different client protein specificities; they also inhibit LbHsp90 ATPase activity to different extents (unpublished data). Such functional differences might be important in both Hsp90 regulation and in their interactions with client proteins during the *Leishmania* life stage transformations. However, to support these assertions, more functional and *in vivo* studies are needed.

Based on the aforementioned studies, we conclude that p23 is an important Hsp90 regulator that also desensitizes Hsp90 to inhibitors [135]. Additionally, p23 has functions that are independent of Hsp90 [146, 150]. Furthermore, two Kinetoplastid p23 proteins share similar structures, despite low sequence identity and differences in protein stabilities and chaperone activities. Taken together, these data suggest that p23 proteins function in multiple cellular activities and not just as Hsp90 co-chaperones [151].

5.4 Hop/Sti1

The Hop, also called stress inducible protein 1 (Sti1) in yeast and p60, is a co-chaperone that directly associates with both Hsp70 and Hsp90 and facilitates their interaction [41, 100, 101, 152, 153]. Hop/Sti1 also modulates the Hsp70/Hsp90 multichaperone system [154] and stimulates refolding when bound to Hsp70, independently of the presence of Hsp90 [153]. Therefore, during the foldosome cycle, Hop/Sti1 may act as a hub in transferring client proteins between the Hsp70 and Hsp90 chaperones [153, 155, 156].

Hop/Sti1 is a modular protein containing three TPR domains: TPR1, TPR2a, and TPR2b [100, 101, 156, 157]. Hop/Sti1 also has two DP/Sti1 motifs, one between the TPR1 and TPR2a domains and another on the C-terminal side of the TPR2b domain [156, 158]. TPR1 binds to the C-terminal heptapeptide of Hsc70, which contains the PTIEEVD motif, by a dicarboxylate clamp while the TPR2a domain binds to the cytoplasmic Hsp90 C-terminal pentapeptide containing the MEEVD motif [100, 156, 159]. Nonetheless, the Hsp70 C-terminal consisting of 50 amino acids may be involved with Hop/Sti1 binding, and other regions of Hsp70 may interact with regions of Hop/Sti1, including TPR2b [101, 156, 157, 160]. Conversely, TPR2a and TPR2b are involved in the interaction with Hsp90 [100, 156, 157, 160, 161]. Hop/Sti1 interacts with Hsp70 both in the presence of ADP and in the absence of nucleotides, while ATP inhibits the Hop/Sti1-Hsp70 interaction. Nevertheless, Hop/Sti1 does not possess NEF activity or induce Hsp70 ATP hydrolysis [153, 155]. Unlike human Hop, yeast Sti1 stimulates ATPase activity of the yeast Hsp70 Ssa1, suggesting a different mechanism of action for mammalian Hop and yeast Sti1 [154].

Hop/Sti1 inhibits Hsp90 ATPase activity through the binding of the TPR2a-TPR2b module to the Hsp90 M-domain [156, 162]. This process blocks the dimerization of the Hsp90 N-domain as well as its coupling to the Hsp90 M-domain [161]. Interestingly, the isolated TPR2a and TPR2b domains did not inhibit Hsp90 ATPase activity, highlighting the importance of the rigid linker between these domains that also orientates them in the space [156].

A remaining question about Hop/Sti1 focuses on its quaternary structure. Both Sti1 and human Hop have been described as dimers in solution based on size exclusion chromatography [157, 160, 162, 163]; however, this technique is sensitive to protein shape and may lead to inaccurate molecular mass estimates [162]. When using the sedimentation equilibrium analytical ultracentrifugation technique, which is insensitive to particle shape effects [164], the results indicated that monomers and dimers are in equilibrium [162, 165].

Our studies led to the identification of Hop/Sti1 orthologues in both the *L. braziliensis* and the *P. falciparum* genomes (Table 17.1). We produced and characterized these recombinant proteins and showed that they share structural properties with the human and yeast orthologues. Recombinant Hop/Sti1 of *L. braziliensis* protozoa behaves only as an elongated monomer, as suggested by analytical ultracentrifugation and small angle X-ray scattering studies (unpublished data). Despite the aforementioned controversies, all of these data indicate that the Hop/Sti1 proteins are highly elongated. This shape likely facilitates the ability of Hop/Sti1 to connect Hsp70 with Hsp90 in its function as an adaptor of the foldosome assembly in the intermediate complex stage of the multichaperone cycle (Fig. 17.2). The Hop/Sti1 modular structure, as well as its elongated shape, may also be critical for this protein's function. Sti1 TPR2b directly interacts with the Hsp90 N-domain while TPR2a binds to the C-terminal pentapeptide [161], which is approximately 110 Å away from the Hsp90 N-domain. This structure allows for the long-range interactions mediated by Hop/Sti1 with both the Hsp90 C-terminus and the N-domain.

Hop/Sti1 proteins have also been characterized *in vivo* and *in vitro* in Kinetoplastids and *P. falciparum* [28, 166–169]. In *L. major* and *T. cruzi*, the Hop/Sti1 orthologues are constitutively expressed in both amastigotes and promastigotes, but they are transiently upregulated (along with the Hsp90 and Hsp70) in promastigotes after the temperature shift to 37 °C [166, 168]. Indeed, Hop/Sti1 orthologues from *P. falciparum*, *L. major*, *L. donovani*, and *T. cruzi* can be immunoprecipitated with both Hsp90 and Hsp70 [28, 166–168], and they co-localize with both of these chaperones in the cytoplasm and/or around nucleus, thus likely participating in the foldosome apparatus [167–169]. Interestingly, the depletion of *L. donovani* Hop/Sti1, Hsp90 or their interaction motifs led to growth impairment using a chemical knockout complementation assay [169]. Additionally, *L. donovani* Hop/Sti1 is a phosphoprotein found in complexes with other molecular chaperones and ribosomal client proteins in infective amastigotes, and its phosphorylation is critical for amastigote viability [28]. Interestingly, several members of the Hsp90 and Hsp70 families are also amastigote-specific phosphoproteins, suggesting that the phosphorylation of these chaperones is an important regulatory step [28]. All of these data suggest that Hop/Sti1 functions as an adaptor protein for Hsp90 and Hsp70 chaperones in multiple organisms, including protozoa, where it likely participates in the intermediate complex foldosome assembly.

5.5 Hip

Hip (also known as p48 in yeast) is a eukaryotic cytoplasmic 48 kDa protein that acts as a Hsp70 co-chaperone [44] and is present in Apicomplexa protozoa such as *T. gondii* and *Plasmodium berguei* [150, 170]. In *P. berguei*, it is produced as a phosphoprotein in the late ring and early trophozoite stages [171]. *Hip* genes were also identified in the *P. falciparum* (Table 17.1), *L. donovani* [172], and *L. braziliensis* genomes (Table 17.1). The Hip protein of *L. braziliensis* was investigated and found to have structural properties similar to other Hip orthologues [173].

Structurally, Hip can be divided into three domains. The N-terminal domain is 100 amino acids long and is responsible for dimerization. The central region contains a TPR domain, is 100 amino acids long and is involved in Hsp70 binding. The C-terminal domain has no clear function but appears to interact with hydrophobic amino acid sequences [174, 175]. Hip is a highly elongated dimer [46, 173, 176, 177] that interacts with two Hsp70-ADP molecules [44, 46, 175, 176]. It interacts with the NBD of Hsp70 through an α-helix at the C-terminal of the TPR domain [46, 174–176, 178]. Therefore, Hip competes with Hsp70 exchange factors for the same binding site in the NBD to stabilize the Hsp70-ADP state [46, 179]. It has been shown that Hip participates in glucocorticoid receptor maturation in cooperation with Hsp70 and Hsp90 [180] and can be co-immunoprecipitated with Hsp70-Hsp90 in *T. gondii* [150]. These data suggest that Hip participates in the early complex stage of the Hsp70-Hsp90 foldosome cycle (Fig. 17.2). However, in *L. donovani*, Hip does not seem to be essential for cell survival [172], possibly because other co-chaperones compensate for its absence (see below).

As previously mentioned, *L. braziliensis* Hip (LbHip) shares structural properties with orthologous Hips despite low sequence conservation. LbHip is a dimer formed by three domains with different degrees of stability [181]. However, the most intriguing structural property of LbHip and orthologous proteins is their highly elongated shape [46, 173, 177]. With such an elongated shape and a dimeric structure, Hip may act as a client protein scanning factor and/or a scaffolding protein that allows the interaction of two ADP-bound Hsp70 molecules with the same client protein, even in the presence of J-proteins. Therefore, Hip helps Hsp70 keep client proteins in a foldable state, preparing them to assume the correct folding and/or delivering them to downstream molecular chaperones. It also prevents the premature release of the client protein by Hsp70, thus preventing the incorrect protein folding. Finally, it may participate in targeting unsuccessfully folded client proteins to the proteasome [44, 46].

5.6 SGT

Although present in a variety of organisms, including *Leishmania spp.*, SGT appears to be absent from Apicomplexa protozoa, such as *P. falciparum* [47] (Table 17.1). SGT is a modular protein that forms elongated homodimers through a coiled coil α-helix located at the N-terminal domain [48, 49, 182] and contains three distinct domains, the N-terminal domain, the central TPR domain and the C-terminal domain [45, 48, 49]. The SGT TPR domain contains all essential residues for binding both the Hsp70 and the Hsp90 C-terminal EEVD motifs, likely via a dicarboxylate clamp, as reported for Hop [48, 182]. The comparison among *C. elegans*, yeast, and human SGT proteins suggests that SGT interacts with both Hsp70 and Hsp90 through a TPR domain on each protomer [48, 49, 182]. Hop uses two independent TPR domains to discriminate between Hsp70 and Hsp90, forming a bridge between these two chaperones [45, 100]. Conversely, SGT interacts indiscriminately with the C-terminal EEVD motif of both Hsp70 and Hsp90 [48, 49]. In human cells, SGT is located mainly in the cytoplasm in a complex with Hsp90; nevertheless, in the presence of GA or in the event of apoptosis, the interaction is disrupted, and SGT partially migrates to the nucleus [183]. Therefore, human SGT has pro-apoptotic properties [183]. The significance of the SGT-Hsp90 and SGT-Hsp70 interactions has been under discussion, and it has been shown that SGT negatively affects Hsp70 ATPase activity and client protein refolding efficiency [45].

There is evidence that SGT also interacts with the Hsp100 molecular chaperone through its TPR domain [184]. The C-terminal region of SGT possesses a Gln-rich (Q-rich) motif that interacts with peptides containing non-polar amino acid residues [49, 182]. This region is absent in the SGTs from Kinetoplastids, e.g., *Leishmania spp.*, and is substituted by a region containing charged amino acid residues, mainly Glu [47]. In addition to the reports that the TPR and C-terminal regions mediate SGT interactions, evidence suggests that the N-terminal region associates with partner proteins, which extends the role of this region beyond oligomerization [184].

SGT is likely an essential protein for *L. donovani* promastigote growth and viability, as it is expressed in amastigotes and is located in the cytoplasm where it co-localizes with Hsp70, Hsp90, and Hip [47]. Interestingly, *Leishmania* SGT interacts directly with Hip, forming a large complex with Hsp70, Hsp90, and Hop proteins [47]. In addition, SGT interacts genetically and physically, via the adaptor protein Mdy2, with the yeast J-protein Ydj1 [184]. The interactions between SGT and

Hsp70 or Hsp90 using the same SGT TPR domain suggest that the SGT dimer is likely a multivalent co-chaperone [182]. Therefore, the capacity of SGT to interact with chaperones, as observed in humans, yeast, and *Leishmania*, could implicate these proteins in complex assembly in order to execute some appropriately cellular function [184]. SGT could function as an adaptor protein that mediates the associations between molecular chaperones, such as Hsp90, Hsp70, Hip, Hop, J-proteins, and Hsp100 [184], and directly participates in the Hsp70/Hsp90 foldosome interactome (Fig. 17.2) in Kinetoplastid protozoa [47].

Based on these information, one can hypothesize that the SGT dimer interacts simultaneously with both Hsp90 and Hsp70 via the TPR domain of each protomer, resembling the structural mechanism that Hop uses as a monomer [49, 100]. Interestingly, SGT and Hip also share similar domain organization, both are dimers and have elongated shapes although Hip does not interact with the Hsp70 or Hsp90 C-terminal EEVD motifs (see above). As previously mentioned, it has been proposed that dimeric Hip may interact, through its TPR domains, with two ADP-bound Hsp70 molecules in a complex with the same client protein [44, 46]. One can hypothesize a convergent mechanism of action of SGT and Hip in the Hsp70 chaperone system and of SGT and Hop in connecting the Hsp70 and Hsp90 chaperone machineries (Fig. 17.2). Based on these observations, SGT could complement the roles of other co-chaperones, which would explain the reported non-essentiality of Hip in *L. donovani* [172].

5.7 Additional Protozoa Co-chaperones

Several other Hsp70/Hsp90 co-chaperones were identified in the *L. braziliensis* and *P. falciparum* genomes and are summarized in Table 17.1. In the Hsp70 ATP-driven cycle, the nucleotide exchange at the Hsp70 NBD is an important step [5]. This task is executed by the Hsp70 NEFs, and there are at least 4 types of NEFs in eukaryotes, indicating their importance in Hsp70 recycling [5]. Our data mining led to the identification of several NEF candidates in the *L. braziliensis* and *P. falciparum* genomes (Table 17.1). Like humans [185], *L. braziliensis* has two GrpE-like proteins while *P. falciparum* has only one. Additionally, both protozoa genomes have predicted *hsp110* and *hspBP1* orthologues, and curiously, both act as cytoplasmic Hsp70 NEFs [5]. However, neither *L. braziliensis* nor *P. falciparum* appear to have BAG orthologues (Table 17.1). BAG proteins can function as molecular levers to detect if the environmental conditions are stressful or favorable and activate protein folding or protein degradation accordingly [5, 54]. The reason that BAG protein orthologues are absent from these protozoa is unknown.

Like human mitochondrial Hsp70, the *L. braziliensis* orthologue is located in the mitochondria, tends to aggregate (unpublished data) and folds with the help of a mitochondrial co-chaperone called Hsp70-escort protein 1 (Hep1) that is essential for mitochondria biogenesis [181]. Interestingly, the *L. braziliensis* Hep1 orthologue is able to assist the human mitochondrial Hsp70 as well as the *L. braziliensis* mitochondrial Hsp70, despite the low sequence identity of Hep1 from these organisms (unpublished data). A putative *Hep1* orthologue was also identified in the *P. falciparum* genome (Table 17.1).

CHIP is other co-chaperone containing TPR domains that interacts with the EEVD motifs of the cytoplasmic Hsp70 and Hsp90 chaperones through a dicarboxylate clamp [54]. CHIP contributes to the balance of protein folding and degradation of client proteins (Fig. 17.2). It has a ubiquitin-box (U-box) domain that possesses ubiquitin ligase activity and targets Hsp70 and Hsp90 client proteins for proteasome degradation (reviewed in [54]). Our data mining results identified a CHIP-like gene only in the *P. falciparum* genome (Table 17.1), and to our knowledge no studies have reported on this protein.

Peptidyl-prolyl *cis-trans* isomerases (PPIases) are enzymes that accelerate protein folding by catalyzing the *cis-trans* isomerization of proline imidic/peptide bonds [186, 187]. These enzymes, together with the protein disulfide isomerase (PDI) family members and some phosphatases, form a large group of proteins that associate with Hsp90 during client protein maturation and actively act on these proteins. The best characterized PPIases belong to two families, the cyclophilin-type (Cyp) and the FKB-506 drug-binding protein type (FKBP). Our data mining led to the identification of several predicted proteins of both types in the *L. braziliensis* and *P. falciparum* genomes (Table 17.1). Proteins belonging to the PDI family were also identified in the *L. braziliensis* and *P. falciparum* genomes (Table 17.1). PDIs are enzymes that catalyze the formation and rearrangement of disulfide bonds in various proteins in an oxidizing environment. Thus, PDIs stabilize the tertiary and quaternary structures of proteins and facilitate protein folding [188].

Finally, Ser/Thr phosphatases (called PP5s) associate with Hsp90 in complexes during client protein maturation [189]. PP5s act as phosphatases that modify the phosphorylation status of Hsp90 client proteins [189]. These important proteins were also identified in both the *L. braziliensis* and *P. falciparum* genomes (Table 17.1). No studies on the *L. braziliensis* protein were conducted. However, *P. falciparum* PP5 was studied and found to have a similar structural organization to the human orthologue and to interact with Hsp90 in a GA-sensitive manner (reviewed in [190]).

6 Concluding Remarks

In the last 2 decades, experimental evidence has highlighted the participation of Hsp70 and Hsp90 (as well as their co-chaperones) in the thermally driven transformation of intracellular protozoa parasites. In mammalian cells, these molecular chaperones work together in the foldosome machinery, which has been targeted for cancer treatment and become a candidate protozoa target. In this chapter, the importance of molecular chaperones in *Plasmodium* and *Leishmania* was revisited with a focus on their interactome networks. Co-chaperones involved in foldosome machinery assembly and modulation were discussed, and interestingly, most of them are modular proteins with domains that operate independently. This type of structural organization is observed in proteins from several organisms, including those that diverged early during evolution. In several points of views, Hsp90, Hsp70, and their co-chaperones share several structural and functional features with the human orthologues; nevertheless, differences exist and offer opportunities for drug design.

In considering the use of Hsp70 and Hsp90 as targets for drug development against protozoa parasites, two main questions emerge: (1) Will the functional and structural similarities between the Hsp70 and Hsp90 molecules from humans and protozoa undermine drug specificity? (2) What are the effects of the protozoa co-chaperones on the foldosome machinery and its inhibition? Nevertheless, recent work indicates the feasibility of using protozoa chaperones as inhibition targets. In *P. falciparum*, the potential use of selective inhibitors against Hsp90 has been shown, and they could be used in a synergistic manner with other antimalarial compounds [97, 191]. For example, it has been shown that GA inhibits PfHsp90 with higher potency than human Hsp90 [7], and Shahinas et al. [191] identified three inhibitor compounds that also have greater affinity for PfHsp90 than human Hsp90. Interestingly, yeast strains complemented with PfHsp90 are more sensitive to both GA and RDC than when complemented with human Hsp90 [93]. All of this experimental evidence points towards the potential of Hsp90 inhibitors in treating malaria. Similar efforts have focused on *P. falciparum* Hsp70 [66].

However, it has been shown that co-chaperones desensitize Hsp90 to drug inhibition [131, 132, 135]. This finding could be a problem if one addresses only Hsp90 in a drug trial without taking into account the Hsp90 complex machinery as a whole. Nonetheless, it also represents an opportunity for drug design because it can lead to the identification and development of selective inhibitors against particular Hsp90 co-chaperones due to their lower conservation between protozoa and humans. Hsp90 possesses overlapping binding sites for co-chaperones, opening the possibility of using competing molecules to fix Hsp90 in a specific conformational state. This issue deserves close attention in the development of parasite-specific drugs.

Acknowledgments The authors thank the Fundação de Amparo à pesquisa do Estado de São Paulo (FAPESP) and the Conselho Nacional de Pesquisa e Desenvolvimento (CNPq) for financial support.

References

1. Tiroli-Cepeda AO, Ramos CHI (2011) An overview of the role of molecular chaperones in protein homeostasis. Protein Pept Lett 18(2):101–109
2. Li J, Soroka J, Buchner J (2012) The Hsp90 chaperone machinery: conformational dynamics and regulation by co-chaperones. BBA-Mol Cell Res 1823(3):624–635
3. Kampinga HH, Craig EA (2010) The HSP70 chaperone machinery: J proteins as drivers of functional specificity. Nat Rev Mol Cell Biol 11(8):579–592
4. Krukenberg KA, Street TO, Lavery LA, Agard DA (2011) Conformational dynamics of the molecular chaperone Hsp90. Q Rev Biophys 44(2):229–255

5. da Silva KP, Borges JC (2011) The molecular chaperone Hsp70 family members function by a bidirectional heterotrophic allosteric mechanism. Protein Pept Lett 18(2):132–142
6. Shonhai A, Maier A, Przyborski J, Blatch G (2011) Intracellular protozoan parasites of humans: the role of molecular chaperones in development and pathogenesis. Protein Pept Lett 18(2):143–157
7. Pallavi R, Roy N, Nageshan RK, Talukdar P, Pavithra SR, Reddy R, Venketesh S, Kumar R, Gupta AK, Singh RK, Yadav SC, Tatu U (2010) Heat shock protein 90 as a drug target against protozoan infections. J Biol Chem 285(49):37964–37975
8. WHO (2010) First WHO report on neglected tropical diseases: working to overcome the global impact of neglected tropical diseases, pp 91
9. Stuart K, Brun R, Croft S, Fairlamb A, rtler RE, McKerrow J, Reed S, Tarleton R (2008) Kinetoplastids: related protozoan pathogens, different diseases. J Clin Invest 118(4):1301–1310
10. Lindoso JA, Lindoso AA (2009) Neglected tropical diseases in Brazil. Rev I Med Trop 51:247–253
11. Alvar J, Vélez ID, Bern C, Herrero M, Desjeux P, Cano J, Jannin J, Boer MD, the WHO Leishmaniasis Control Team (2012) Leishmaniasis worldwide and global estimates of its incidence. PLoS ONE 7(5):e35671
12. Beattie L, Kaye PM (2011) Leishmania-host interactions: what has imaging taught us? Cell Microbiol 13(11):1659–1667
13. Dostalova A, Volf P (2012) *Leishmania* development in sand flies: parasite-vector interactions overview. Parasit Vectors 5(1):276
14. Leiriao P, Rodrigues CD, Albuquerque SS, Mota MM (2004) Survival of protozoan intracellular parasites in host cells. EMBO Rep 5(12):1142–1147
15. WHO (2012) World malaria report, 2012, pp 195
16. Tuteja R (2007) Malaria—an overview. FEBS J 274(18):4670–4679
17. Fujioka H, Aikawa M (2002) Structure and life cycle. In: Perlmann P, Aikawa M (eds) Malaria immunology, Karger, Basel, pp 1–26
18. Sturm A, Amino R, van de Sand C, Regen T, Retzlaff S, Rennenberg A, Krueger A, Pollok JM, Menard R, Heussler VT (2006) Manipulation of host hepatocytes by the malaria parasite for delivery into liver sinusoids. Science 313(5791):1287–1290
19. Baumeister S, Winterberg M, Przyborski J, Lingelbach K (2010) The malaria parasite *Plasmodium falciparum*: cell biological peculiarities and nutritional consequences. Protoplasma 240(1–4):3–12
20. Roy N, Nageshan RK, Ranade S, Tatu U (2012) Heat shock protein 90 from neglected protozoan parasites. BBA—Mol Cell Res 1823(3):707–711
21. Shonhai A (2010) Plasmodial heat shock proteins: targets for chemotherapy. FEMS Immunol Med Mic 58(1):61–74
22. WHO (2010) Guidelines for the treatment of malaria. 2nd edn. World Health Organization, pp 73. ISBN: 9789241547925
23. Silva KP, Seraphim TV, Borges JC (2013) Structural and functional studies of *Leishmania braziliensis* Hsp90. BBA—Proteins Proteom 1834(1):351–361
24. Van der Ploeg LH, Giannini SH, Cantor CR (1985) Heat shock genes: regulatory role for differentiation in parasitic protozoa. Science 228(4706):1443–1446
25. Shapira M, McEwen JG, Jaffe CL (1988) Temperature effects on molecular processes which lead to stage differentiation in *Leishmania*. EMBO J 7(9):2895–2901
26. Brandau S, Dresel A, Clos J (1995) High constitutive levels of heat-shock proteins in human-pathogenic parasites of the genus *Leishmania*. Biochem J 310(1):225–232
27. Bente M, Harder S, Wiesgigl M, Heukeshoven J, Gelhaus C, Krause E, Clos J, Bruchhaus I (2003) Developmentally induced changes of the proteome in the protozoan parasite *Leishmania donovani*. PROTEOMICS 3(9):1811–1829
28. Morales MA, Watanabe R, Dacher M, Chafey P, Fortéa J, Scott DA, Beverley SM, Ommen G, Clos J, Hem S, Lenormand P, Rousselle JC, Namane A, Spath GF (2010) Phosphoproteome dynamics reveal heat-shock protein complexes specific to the *Leishmania donovani* infectious stage. Proc Natl Acad Sci U S A 107(18):8381–8386

29. Wiesgigl M, Clos J (2001) Heat shock protein 90 homeostasis controls stage differentiation in *Leishmania donovani*. Mol Biol Cell 12(11):3307–3316
30. Smejkal RM, Wolff R, Olenick JG (1988) *Leishmania braziliensis panamensis*: increased infectivity resulting from heat shock. Exp Parasitol 65(1):1–9
31. Salotra P, Ralhan R, Bhatnagar R (1994) Differential expression of stress proteins in virulent and attenuated promastigotes of *Leishmania donovani*. Biochem Mol Biol Int 33(4):691–697
32. Matrangolo FSV, Liarte DB, Andrade LC, de Melo MF, Andrade JM, Ferreira RF, Santiago A, Pirovani CP, Silva-Pereira RA, Murta SMF (2013) Comparative proteomic analysis of antimony-resistant and -susceptible *Leishmania braziliensis* and *Leishmania infantum* chagasi lines. Mol Biochem Parasitol 190(2):63–75
33. Acharya P, Kumar R, Tatu U (2007) Chaperoning a cellular upheaval in malaria: heat shock proteins in *Plasmodium falciparum*. Mol Biochem Parasitol 153(2):85–94
34. Kumar N, Koski G, Harada M, Aikawa M, Zheng H (1991) Induction and localization of *Plasmodium falciparum* stress proteins related to the heat shock protein 70 family. Mol Biochem Parasitol 48(1):47–58
35. Külzer S, Charnaud S, Dagan T, Riedel J, Mandal P, Pesce ER, Blatch GL, Crabb BS, Gilson PR, Przyborski JM (2012) *Plasmodium falciparum*-encoded exported hsp70/hsp40 chaperone/co-chaperone complexes within the host erythrocyte. Cell Microbiol 14(11):1784–1795
36. Külzer S, Rug M, Brinkmann K, Cannon P, Cowman A, Lingelbach K, Blatch GL, Maier AG, Przyborski JM (2010) Parasite-encoded Hsp40 proteins define novel mobile structures in the cytosol of the *P. falciparum*-infected erythrocyte. Cell Microbiol 12(10):1398–1420
37. Leech JH, Barnwell JW, Aikawa M, Miller LH, Howard RJ (1984) *Plasmodium falciparum* malaria: association of knobs on the surface of infected erythrocytes with a histidine-rich protein and the erythrocyte skeleton. J Cell Biol 98(4):1256–1264
38. Vonlaufen N, Kanzok SM, Wek RC, Sullivan Jr WJ (2008) Stress response pathways in protozoan parasites. Cell Microbiol 10(12):2387–2399
39. Edkins AL, Boshoff A (2014) General structural and functional features of molecular chaperones. In: Shonhai A, Blatch GL (eds) Heat shock proteins of malaria. Springer, Netherlands, pp 5–45
40. Przyborski J (2014) The importance of molecular chaperones in survival and pathogenesis of the malaria parasite plasmodium falciparum. In: Shonhai A, Blatch GL (eds) Heat shock proteins of malaria. Springer, Netherlands, pp 1–3
41. Wegele H, Muller L, Buchner J (2004) Hsp70 and Hsp90—a relay team for protein folding. Rev Physiol Biochem Pharmacol 151:1–44
42. Cano LQ, Lavery DN, Bevan CL (2013) Mini-review: foldosome regulation of androgen receptor action in prostate cancer. Mol Cell Endocrinol 369(1–2):52–62
43. Borges JC, Ramos CHI (2005) Protein folding assisted by chaperones. Protein Pept Lett 12(3):257–261
44. Hohfeld J, Minami Y, Hartl FU (1995) Hip, a novel cochaperone involved in the eukaryotic Hsc70/Hsp40 reaction cycle. Cell 83(4):589–598
45. Angeletti PC, Walker D, anganiban AT (2002) Small glutamine-rich protein/viral protein U-binding protein is a novel cochaperone that affects heat shock protein 70 activity. Cell Stress Chaperon 7(3):258–268
46. Li Z, Hartl FU, Bracher A (2013) Structure and function of Hip, an attenuator of the Hsp70 chaperone cycle. Nat Struct Mol Biol 20(8):929–935
47. Ommen G, Chrobak M, Clos J (2010) The co-chaperone SGT of *Leishmania donovani* is essential for the parasite's viability. Cell Stress Chaperon 15(4):443–455
48. Worrall LJ, Wear MA, Page AP, Walkinshaw MD (2008) Cloning, purification and characterization of the *Caenorhabditis elegans* small glutamine-rich tetratricopeptide repeat-containing protein. BBA—Proteins Proteom 1784(3):496–503
49. Liou ST, Wang C (2005) Small glutamine-rich tetratricopeptide repeat-containing protein is composed of three structural units with distinct functions. Arch Biochem Biophys 435(2):253–263
50. Li J, Richter K, Reinstein J, Buchner J (2013) Integration of the accelerator Aha1 in the Hsp90 co-chaperone cycle. Nat Struct Mol Biol 20(3):326–331

51. Retzlaff M, Hagn F, Mitschke L, Hessling M, Gugel F, Kessler H, Richter K, Buchner J (2010) Asymmetric activation of the Hsp90 dimer by its cochaperone Aha1. Mol Cell 37(3):344–354
52. Pratt WB, Morishima Y, Osawa Y (2008) The Hsp90 chaperone machinery regulates signaling by modulating ligand binding clefts. J Biol Chem 283(34):22885–22889
53. Biamonte MA, Van de Water R, Arndt JW, Scannevin RH, Perret D, Lee WC (2009) Heat shock protein 90: inhibitors in clinical trials. J Med Chem 53(1):3–17
54. Arndt V, Rogon C, Hohfeld J (2007) To be, or not to be—molecular chaperones in protein degradation. Cell Mol Life Sci 64(19–20):2525–2541
55. Meimaridou E, Gooljar SB, Chapple JP (2009) From hatching to dispatching: the multiple cellular roles of the Hsp70 molecular chaperone machinery. J Mol Endocrinol 42(1–2):1–9
56. Mayer MP, Bukau B (2005) Hsp70 chaperones: cellular functions and molecular mechanism. Cell Mol Life Sci 62(6):670–684
57. Bertelsen EB, Chang L, Gestwicki JE, Zuiderweg ERP (2009) Solution conformation of wild-type *E. coli* Hsp70 (DnaK) chaperone complexed with ADP and substrate. Proc Natl Acad Sci U S A 106(21):8471–8476
58. Schlecht R, Erbse AH, Bukau B, Mayer MP (2011) Mechanics of Hsp70 chaperones enables differential interaction with client proteins. Nat Struct Mol Biol 18(3):345–135
59. Borges JC, Ramos CHI (2009) Characterization of nucleotide-induced changes on the quaternary structure of human 70 kDa heat shock protein Hsp70.1 by analytical ultracentrifugation. BMB Rep 42(3):166–171
60. Bhattacharya A, Kurochkin AV, Yip GNB, Zhang YB, Bertelsen EB, Zuiderweg ERP (2009) Allostery in Hsp70 chaperones is transduced by subdomain rotations. J Mol Biol 388(3):475–490
61. Bukau B, Weissman J, Horwich A (2006) Molecular chaperones and protein quality control. Cell 125(3):443–451
62. Matambo TS, Odunuga OO, Boshoff A, Blatch GL (2004) Overproduction, purification, and characterization of the *Plasmodium falciparum* heat shock protein 70. Protein Expres Purif 33(2):214–222
63. Misra G, Ramachandran R (2009) Hsp70−1 from *Plasmodium falciparum*: protein stability, domain analysis and chaperone activity. Biophys Chem 142(1–3):55−64
64. Shonhai A, Botha M, de Beer TAP, Boshoff A, Blatch GL (2008) Structure-function study of a *Plasmodium falciparum* Hsp70 using three dimensional modelling and in vitro analyses. Protein Pept Lett 15:1117–1125
65. Shonhai A, Boshoff A, Blatch GL (2007) The structural and functional diversity of Hsp70 proteins from *Plasmodium falciparum*. Protein Sci 16(9):1803–1818
66. Cockburn IL, Pesce E, Pryzborski JM, Davies-Coleman MT, Clark Peter GK, Keyzers RA, Stephens LL, Blatch GL (2011) Screening for small molecule modulators of Hsp70 chaperone activity using protein aggregation suppression assays: inhibition of the plasmodial chaperone PfHsp70-1. Biol Chem 392(5):431–438
67. Grover M, Chaubey S, Ranade S, Tatu U (2013) Identification of an exported heat shock protein 70 in *Plasmodium falciparum*. Parasite 20:2
68. Folgueira C, Carrión J, Moreno J, Saugar J, Cañavate C, Requena J (2008) Effects of the disruption of the HSP70-II gene on the growth, morphology, and virulence of *Leishmania infantum* promastigotes. Int Microbiol 11(2):81–89
69. Helmbrecht K, Zeise E, Rensing L (2000) Chaperones in cell cycle regulation and mitogenic signal transduction: a review. Cell Prolif 33(6):341–365
70. Brochu C, Haimeur A, Ouellette M (2004) The heat shock protein HSP70 and heat shock cognate protein HSC70 contribute to antimony tolerance in the protozoan parasite Leishmania. Cell Stress Chaperon 9(3):294–303
71. Louw CA, Ludewig MH, Mayer J, Blatch GL (2010) The Hsp70 chaperones of the tritryps are characterized by unusual features and novel members. Parasitol Inter 59(4):497–505
72. Folgueira C, Requena JM (2007) A postgenomic view of the heat shock proteins in kinetoplastids. FEMS Microbiol Rev 31(4):359–377
73. Shonhai A (2014). The role of Hsp70 s in the development and pathogenicity of plasmodium species. In: Shonhai A, Blatch GL (eds) Heat shock proteins of malaria. Springer, Netherlands, pp 47–69

74. Wiesgigl M, Clos J (2001) The heat shock protein 90 of *Leishmania donovani*. Med Microbiol Immun 190(1–2):27–31
75. Pavithra SR, Banumathy G, Joy O, Singh V, Tatu U (2004) Recurrent fever promotes *Plasmodium falciparum* development in human erythrocytes. J Biol Chem 279(45):46692–46699
76. Wandinger SK, Richter K, Buchner J (2008) The Hsp90 chaperone machinery. J Biol Chem 283(27):18473–18477
77. Stebbins CE, Russo AA, Schneider C, Rosen N, Hartl FU, Pavletich NP (1997) Crystal structure of an Hsp90-geldanamycin complex: targeting of a protein chaperone by an antitumor agent. Cell 89(2):239–250
78. Panaretou B, Prodromou C, Roe SM, O'Brien R, Ladbury JE, Piper PW, Pearl LH (1998) ATP binding and hydrolysis are essential to the function of the Hsp90 molecular chaperone in vivo. EMBO J 17(16):4829–4836
79. Obermann WMJ, Sondermann H, Russo AA, Pavletich NP, Hartl FU (1998) In Vivo Function of Hsp90 Is Dependent on ATP Binding and ATP Hydrolysis. J Cell Biol 143(4):901–910
80. Prodromou C, Panaretou B, Chohan S, Siligardi G, O'Brien R, Ladbury JE, Roe SM, Piper PW, Pearl LH (2000) The ATPase cycle of Hsp90 drives a molecular 'clamp' via transient dimerization of the N-terminal domains. EMBO J 19(16):4383–4392
81. Ali MMU, Roe SM, Vaughan CK, Meyer P, Panaretou B, Piper PW, Prodromou C, Pearl LH (2006) Crystal structure of an Hsp90-nucleotide-p23/Sba1 closed chaperone complex. Nature 440(7087):1013–1017
82. Owen BAL, Sullivan WP, Felts SJ, Toft DO (2002) Regulation of Heat Shock protein 90 ATPase activity by sequences in the carboxyl terminus. J Biol Chem 277(9):7086–7091
83. Meyer P, Prodromou C, Hu B, Vaughan C, Roe SM, Panaretou B, Piper PW, Pearl LH (2003) Structural and functional analysis of the middle segment of Hsp90: implications for ATP hydrolysis and client protein and cochaperone interactions. Mol Cell 11(3):647–658
84. Hawle P, Siepmann M, Harst A, Siderius M, Reusch HP, Obermann WMJ (2006) The middle domain of Hsp90 acts as a discriminator between different types of client proteins. Mol Cell Biol 26(22):8385–8395
85. Cunningham CN, Krukenberg KA, Agard DA (2008) Intra- and intermonomer interactions are required to synergistically facilitate ATP hydrolysis in Hsp90. J Biol Chem 283(30):21170–21178
86. Hessling M, Richter K, Buchner J (2009) Dissection of the ATP-induced conformational cycle of the molecular chaperone Hsp90. Nat Struct Mol Biol 16(3):287–293
87. Mickler M, Hessling M, Ratzke C, Buchner J, Hugel T (2009) The large conformational changes of Hsp90 are only weakly coupled to ATP hydrolysis. Nat Struct Mol Biol 16(3):281–286
88. Ratzke C, Berkemeier F, Hugel T (2012) Heat shock protein 90s mechanochemical cycle is dominated by thermal fluctuations. Proc Natl Acad Sci U S A 109(1):161–166
89. Hainzl O, Lapina MC, Buchner J, Richter K (2009) The charged linker region is an important regulator of Hsp90 function. J Biol Chem 284(34):22559–22567
90. Janin YL (2010) ATPase inhibitors of heat-shock protein 90 s season. Drug Discov Today 15(9–10):342–353
91. Kumar R, Musiyenko A, Barik S (2003) The heat shock protein 90 of *Plasmodium falciparum* and antimalarial activity of its inhibitor, geldanamycin. Mal J 2(1):30
92. Banumathy G, Singh V, Pavithra SR, Tatu U (2003) Heat shock protein 90 function is essential for *Plasmodium falciparum* growth in human erythrocytes. J Biol Chem 278(20):18336–18345
93. Wider D, Peli-Gulli MP, Briand PA, Tatu U, Picard D (2009) The complementation of yeast with human or *Plasmodium falciparum* Hsp90 confers differential inhibitor sensitivities. Mol Biochem Parasitol 164(2):147–152
94. Corbett KD, Berger JM (2010) Structure of the ATP-binding domain of *Plasmodium falciparum* Hsp90. Proteins 78(13):2738–2744
95. Shahinas D, Taldone T, Chiosis G, Pillai DR (2010) A structure based drug design approach to repurpose drugs against *Plasmodium Falciparum* Hsp90 (Pfhsp90). Am J Trop Med Hyg 83(5):77–78

96. Ivens AC, Peacock CS, Worthey EA, Murphy L, Aggarwal G, Berriman M, Sisk E, Rajandream MA, Adlem E, Aert R, Anupama A, Apostolou Z, Attipoe P, Bason N, Bauser C, Beck A, Beverley SM, Bianchettin G, Borzym K, Bothe G, Bruschi CV, Collins M, Cadag E, Ciarloni L, Clayton C, Coulson RMR, Cronin A, Cruz AK, Davies RM, De Gaudenzi J, Dobson DE, Duesterhoeft A, Fazelina G, Fosker N, Frasch AC, Fraser A, Fuchs M, Gabel C, Goble A, Goffeau A, Harris D, Hertz-Fowler C, Hilbert H, Horn D, Huang Y, Klages S, Knights A, Kube M, Larke N, Litvin L, Lord A, Louie T, Marra M, Masuy D, Matthews K, Michaeli S, Mottram JC, Müller-Auer S, Munden H, Nelson S, Norbertczak H, Oliver K, O'Neil S, Pentony M, Pohl TM, Price C, Purnelle B, Quail MA, Rabbinowitsch E, Reinhardt R, Rieger M, Rinta J, Robben J, Robertson L, Ruiz JC, Rutter S, Saunders D, Schäfer M, Schein J, Schwartz DC, Seeger K, Seyler A, Sharp S, Shin H, Sivam D, Squares R, Squares S, Tosato V, Vogt C, Volckaert G, Wambutt R, Warren T, Wedler H, Woodward J, Zhou S, Zimmermann W, Smith DF, Blackwell JM, Stuart KD, Barrell B, Myler PJ (2005) The genome of the kinetoplastid parasite, *Leishmania* major. Science 309(5733):436–442
97. Shahinas D, Pillai DR (2014). Role of Hsp90 in *Plasmodium falciparum* malaria. In: Shonhai A, Blatch GL (eds) Heat shock proteins of malaria. Springer, Netherlands, pp 87–97
98. Pavithra SR, Kumar R, Tatu U (2007) Systems analysis of chaperone networks in the malarial parasite *Plasmodium falciparum*. PLoS Comput Biol 3(9):e168
99. Lamb JR, Tugendreich S, Hieter P (1995) Tetratrico peptide repeat interactions—to Tpr or not to Tpr. Trends Biochem Sci 20(7):257–259
100. Scheufler C, Brinker A, Bourenkov G, Pegoraro S, Moroder L, Bartunik H, Hartl FU, Moarefi I (2000) Structure of TPR domain-peptide complexes: critical elements in the assembly of the Hsp70-Hsp90 multichaperone machine. Cell 101(2):199–210
101. Odunuga OO, Longshaw VM, Blatch GL (2004) Hop: more than an Hsp70/Hsp90 adaptor protein. Bioessays 26(10):1058–1068
102. Fan CY, Lee S, Cyr DM (2003) Mechanisms for regulation of Hsp70 function by Hsp40. Cell Stress Chaperon 8(4):309–316
103. Karzai AW, McMacken R (1996) A bipartite signaling mechanism involved in DnaJ-mediated activation of the *Escherichia coli* DnaK protein. J Biol Chem 271(19):11236–11246
104. Laufen T, Mayer MP, Beisel C, Klostermeier D, Mogk A, Reinstein J, Bukau B (1999) Mechanism of regulation of Hsp70 chaperones by DnaJ cochaperones. Proc Natl Acad Sci U S A 96(10):5452–5457
105. Rudiger S, Schneider-Mergener J, Bukau B (2001) Its substrate specificity characterizes the DnaJ co-chaperone as a scanning factor for the DnaK chaperone. EMBO J 20(5):1042–1050
106. Summers DW, Douglas PM, Ramos CHI, Cyr DM (2009) Polypeptide transfer from Hsp40 to Hsp70 molecular chaperones. Trends Biochem Sci 34(5):230–233
107. Szabo A, Langer T, Schroder H, Flanagan J, Bukau B, Hartl FU (1994) The Atp hydrolysis-dependent reaction cycle of the *Escherichia-Coli* Hsp70 system—Dnak, Dnaj, and Grpe. Proc Natl Acad Sci U S A 91(22):10345–10349
108. Greene MK, Maskos K, Landry SJ (1998) Role of the J-domain in the cooperation of Hsp40 with Hsp70. Proc Natl Acad Sci U S A 95(11):6108–6113
109. Borges JC, Seraphim TV, Mokry DZ, Almeida FCL, Cyr DM, Ramos CHI (2012) Identification of regions involved in substrate binding and dimer stabilization within the central domains of yeast Hsp40 Sis1. PLoS ONE 7(12):e50927
110. Silva JC, Borges JC, Cyr DM, Ramos CHI, Torriani IL (2011) Central domain deletions affect the SAXS solution structure and function of Yeast Hsp40 proteins Sis1 and Ydj1. BMC Struct Biol 11(1):40
111. Botha M, Pesce ER, Blatch GL (2007) The Hsp40 proteins of *Plasmodium falciparum* and other apicomplexa: regulating chaperone power in the parasite and the host. Inter J Biochem Cell B 39(10):1781–1803
112. Fan CY, Lee S, Ren HY, Cyr DM (2004) Exchangeable chaperone modules contribute to specification of type I and type II Hsp40 cellular function. Mol Biol Cell 15(2):761–773
113. Borges JC, Fischer H, Craievich AF, Ramos CHI (2005) Low resolution structural study of two human HSP40 chaperones in solution—DjA1 from subfamily A and DjB4 from subfamily B have different quaternary structures. J Biol Chem 280(14):13671–13681

114. 114. Ramos CHI, Oliveira CLP, Fan CY, Torriani IL, Cyr DM (2008) Conserved central domains control the quaternary structure of type I and type II Hsp40 molecular chaperones. J Mol Biol 383(1):155–166
115. Acharya P, Chaubey S, Grover M, Tatu U (2012) An exported heat shock protein 40 associates with pathogenesis-related knobs in *Plasmodium falciparum* infected erythrocytes. PLoS ONE 7(9):e44605
116. Watanabe J (1997) Cloning and characterization of heat shock protein DnaJ homologues from *Plasmodium falciparum* and comparison with ring infected erythrocyte surface antigen. Mol Biochem Parasitol 88(1–2):253–258
117. Pesce ER, Acharya P, Tatu U, Nicoll WS, Shonhai A, Hoppe HC, Blatch GL (2008) The *Plasmodium falciparum* heat shock protein 40, Pfj4, associates with heat shock protein 70 and shows similar heat induction and localisation patterns. Int J Biochem Cell B 40(12):2914–2926
118. Nicoll WS, Botha M, McNamara C, Schlange M, Pesce ER, Boshoff A, Ludewig MH, Zimmermann R, Cheetham ME, Chapple JP, Blatch GL (2007) Cytosolic and ER J-domains of mammalian and parasitic origin can functionally interact with DnaK. Int J Biochem Cell B 39(4):736–751
119. Botha M, Chiang A, Needham P, Stephens L, Hoppe H, Külzer S, Przyborski J, Lingelbach K, Wipf P, Brodsky J, Shonhai A, Blatch G (2011) *Plasmodium falciparum* encodes a single cytosolic type I Hsp40 that functionally interacts with Hsp70 and is upregulated by heat shock. Cell Stress Chaperon 16(4):389–401
120. Louw CA, Ludewig MH, Blatch GL (2010) Overproduction, purification and characterisation of Tbj1, a novel Type III Hsp40 from *Trypanosoma brucei*, the African sleeping sickness parasite. Protein Expres Purif 69(2):168–177
121. Edkins AL, Ludewig MH, Blatch GL (2004) A *Trypanosoma cruzi* heat shock protein 40 is able to stimulate the adenosine triphosphate hydrolysis activity of heat shock protein 70 and can substitute for a yeast heat shock protein 40. Int J Biochem Cell B 36(8):1585–1598
122. Rug M, Maier AG (2011) The heat shock protein 40 family of the malaria parasite *Plasmodium falciparum*. IUBMB Life 63(12):1081–1086
123. Pesce ER, Maier AG, Blatch GL (2014) Role of the Hsp40 family of proteins in the survival and pathogenesis of the malaria parasite. In: Shonhai A, Blatch GL (eds) Heat shock proteins of malaria. Springer, Netherlands, pp 71–85
124. Koulov AV, LaPointe P, Lu BW, Razvi A, Coppinger J, Dong MQ, Matteson J, Laister R, Arrowsmith C, Yates JR, Balch WE (2010) Biological and structural basis for Aha1 regulation of Hsp90 ATPase activity in maintaining proteostasis in the human disease cystic fibrosis. Mol Biol Cell 21(6):871–884
125. Lotz GP, Lin H, Harst A, Obermann WMJ (2003) Aha1 binds to the middle domain of Hsp90, contributes to client protein activation, and stimulates the ATPase activity of the molecular chaperone. J Biol Chem 278(19):17228–17235
126. Meyer P, Prodromou C, Liao C, Hu B, Mark Roe S, Vaughan CK, Vlasic I, Panaretou B, Piper PW, Pearl LH (2004) Structural basis for recruitment of the ATPase activator Aha1 to the Hsp90 chaperone machinery. EMBO J 23(3):511–519
127. Panaretou B, Siligardi G, Meyer P, Maloney A, Sullivan JK, Singh S, Millson SH, Clarke PA, Naaby-Hansen S, Stein R, Cramer R, Mollapour M, Workman P, Piper PW, Pearl LH, Prodromou C (2002) Activation of the ATPase activity of Hsp90 by the stress-regulated cochaperone Aha1. Mol Cell 10(6):1307–1318
128. Seraphim TV, Alves MM, Silva IM, Gomes FER, Silva KP, Murta SMF, Barbosa LRS, Borges JC (2013) Low resolution structural studies indicate that the activator of Hsp90 ATPase 1 (Aha1) of *Leishmania braziliensis* has an elongated shape which allows its interaction with both N- and M-domains of Hsp90. PLoS ONE 8(6):e66822
129. Chua CS, Low H, Lehming N, Sim TS (2012) Molecular analysis of *Plasmodium falciparum* co-chaperone Aha1 supports its interaction with and regulation of Hsp90 in the malaria parasite. Int J Biochem Cell B 44(1):233–245
130. Harst A, Lin HY, Obermann WMJ (2005) Aha1 competes with Hop, p50 and p23 for binding to the molecular chaperone Hsp90 and contributes to kinase and hormone receptor activation. Biochem J 387:789–796

131. Zurawska A, Urbanski J, Matulienė J, Baraniak J, Klejman MP, Filipek S, Matulis D, Bieganowski P (2010) Mutations that increase both Hsp90 ATPase activity in vitro and Hsp90 drug resistance in vivo. BBA-Mol Cell Res 1803(5):575–583
132. Holmes JL, Sharp SY, Hobbs S, Workman P (2008) Silencing of HSP90 cochaperone AHA1 expression decreases client protein activation and increases cellular sensitivity to the HSP90 inhibitor 17-allylamino-17-demethoxygeldanamycin. Cancer Res 68(4):1187–1196
133. Weikl T, Abelmann K, Buchner J (1999) An unstructured C-terminal region of the Hsp90 co-chaperone p23 is important for its chaperone function. J Mol Biol 293(3):685–691
134. Weaver AJ, Sullivan WP, Felts SJ, Owen BAL, Toft DO (2000) Crystal structure and activity of human p23, a heat shock protein 90 co-chaperone. J Biol Chem 275(30):23045–23052
135. Forafonov F, Toogun OA, Grad I, Suslova E, Freeman BC, Picard D (2008) p23/Sba1p protects against Hsp90 inhibitors independently of its intrinsic chaperone activity. Mol Cell Biol 28(10):3446–3456
136. Martinez-Yamout MA, Venkitakrishnan RP, Preece NE, Kroon G, Wright PE, Dyson HJ (2006) Localization of sites of interaction between p23 and Hsp90 in solution. J Biol Chem 281(20):14457–14464
137. Johnson JL, Toft DO (1994) A novel chaperone complex for steroid-receptors involving heat-shock proteins, immunophilins, and P23. J Biol Chem 269(40):24989–24993
138. Johnson JL, Beito TG, Krco CJ, Toft DO (1994) Characterization of a novel 23-kilodalton protein of unactive progesterone-receptor complexes. Mol Cell Biol 14(3):1956–1963
139. Kosano H, Stensgard B, Charlesworth MC, McMahon N, Toft D (1998) The assembly of progesterone receptor-hsp90 complexes using purified proteins. J Biol Chem 273(49):32973–32979
140. Richter K, Walter S, Buchner J (2004) The co-chaperone Sba1 connects the ATPase reaction of Hsp90 to the progression of the chaperone cycle. J Mol Biol 342(5):1403–1413
141. Karagoz GE, Duarte AMS, Ippel H, Uetrecht C, Sinnige T, van Rosmalen M, Hausmann J, Heck AJR, Boelens R, Rudiger SGD (2011) N-terminal domain of human Hsp90 triggers binding to the cochaperone p23. Proc Natl Acad Sci U S A 108(2):580–585
142. Hohfeld J, Cyr DM, Patterson C (2001) From the cradle to the grave: molecular chaperones that may choose between folding and degradation. EMBO Rep 2(10):885–890
143. Pearl LH, Prodromou C (2002) Structure, function, and mechanism of the Hsp90 molecular chaperone. Adv Protein Chem 59:157–186
144. McLaughlin SH, Smith HW, Jackson SE (2002) Stimulation of the weak ATPase activity of human Hsp90 by a client protein. J Mol Biol 315(4):787–798
145. Dittmar KD, Demady DR, Stancato LF, Krishna P, Pratt WB (1997) Folding of the glucocorticoid receptor by the heat shock protein (hsp) 90-based chaperone machinery - The role of p23 is to stabilize receptor-hsp90 heterocomplexes formed by hsp90-p60-hsp70. J Biol Chem 272(34):21213–21220
146. Echtenkamp FJ, Zelin E, Oxelmark E, Woo JI, Andrews BJ, Garabedian M, Freeman BC (2011) Global functional map of the p23 molecular chaperone reveals an extensive cellular network. Mol Cell 43(2):229–241
147. Chua CS, Low H, Goo KS, Sim TS (2010) Characterization of *Plasmodium falciparum* co-chaperone p23: its intrinsic chaperone activity and interaction with Hsp90. Cell Mol Life Sci 67(10):1675–1686
148. Wiser M (2003) A Plasmodium homologue of cochaperone p23 and its differential expression during the replicative cycle of the malaria parasite. Parasitol Res 90(2):166–170
149. McLaughlin SH, Sobott F, Yaol ZP, Zhang W, Nielsen PR, Grossmann JG, Laue ED, Robinson CV, Jackson SE (2006) The co-chaperone p23 arrests the Hsp90 ATPase cycle to trap client proteins. J Mol Biol 356(3):746–758
150. Echeverria PC, Figueras MJ, Vogler M, Kriehuber T, de Miguel N, Deng B, Dalmasso MC, Matthews DE, Matrajt M, Haslbeck M, Buchner J, Angel SO (2010) The Hsp90 co-chaperone p23 of *Toxoplasma gondii*: identification, functional analysis and dynamic interactome determination. Mol Biochem Parasitol 172(2):129–140

151. Felts SJ, Toft DO (2003) P23, a simple protein with complex activities. Cell Stress Chaperon 8(2):108–113
152. Pratt WB, Toft DO (2003) Regulation of signaling protein function and trafficking by the hsp90/hsp70-based chaperone machinery. Exp Biol Med 228(2):111–133
153. Johnson BD, Schumacher RJ, Ross ED, Toft DO (1998) Hop modulates hsp70/hsp90 interactions in protein folding. J Biol Chem 273(6):3679–3686
154. Wegele H, Haslbeck M, Reinstein J, Buchner J (2003) Sti1 is a novel activator of the Ssa proteins. J Biol Chem 278(28):25970–25976
155. Hernandez MP, Sullivan WP, Toft DO (2002) The assembly and intermolecular properties of the hsp70-Hop-hsp90 molecular chaperone complex. J Biol Chem 277(41):38294–38304
156. Schmid AB, Lagleder S, Grawert MA, Rohl A, Hagn F, Wandinger SK, Cox MB, Demmer O, Richter K, Groll M, Kessler H, Buchner J (2012) The architecture of functional modules in the Hsp90 co-chaperone Sti1/Hop. EMBO J 31(6):1506–1517
157. Carrigan PE, Nelson GM, Roberts PJ, Stoffer JN, Riggs DL, Smith DF (2004) Multiple domains of the co-chaperone Hop are important for Hsp70 binding. J Biol Chem 279(16):16185–16193
158. Nelson GM, Huffman H, Smith DF (2003) Comparison of the carboxy-terminal DP-repeat region in the co-chaperones Hop and Hip. Cell Stress Chaperon 8(2):125–133
159. Demand J, Luders J, Hohfeld J (1998) The carboxy-terminal domain of Hsc70 provides binding sites for a distinct set of chaperone cofactors. Mol Cell Biol 18(4):2023–2028
160. Flom G, Behal RH, Rosen L, Cole DG, Johnson JL (2007) Definition of the minimal fragments of Sti1 required for dimerization, interaction with Hsp70 and Hsp90 and in vivo functions. Biochem J 404:159–167
161. Lee CT, Graf C, Mayer FJ, Richter SM, Mayer MP (2012) Dynamics of the regulation of Hsp90 by the co-chaperone Sti1. EMBO J 31(6):1518–1528
162. Prodromou C, Siligardi G, O'Brien R, Woolfson DN, Regan L, Panaretou B, Ladbury JE, Piper PW, Pearl LH (1999) Regulation of Hsp90 ATPase activity by tetratricopeptide repeat (TPR)-domain co-chaperones. EMBO J 18(3):754–762
163. Onuoha SC, Couistock ET, Grossmann JG, Jackson SE (2008) Structural studies on the co-chaperone hop and its complexes with Hsp90. J Mol Biol 379(4):732–744
164. Borges JC, Ramos CHI (2011) Analysis of molecular targets of mycobacterium tuberculosis by analytical ultracentrifugation. Curr Med Chem 18(9):1276–1285
165. Yi F, Doudevski I, Regan L (2010) HOP is a monomer: investigation of the oligomeric state of the co-chaperone HOP. Protein Sci 19(1):19–25
166. Webb JR, Campos-Neto A, Skeiky YAW, Reed SG (1997) Molecular characterization of the heat-inducible LmSTI1 protein of *Leishmania* major. Mol Biochem Parasitol 89(2):179–193
167. Gitau G, Mandal P, Blatch G, Przyborski J, Shonhai A (2012) Characterisation of the *Plasmodium falciparum* Hsp70-Hsp90 organising protein (PfHop). Cell Stress Chaperon 17(2):191–202
168. Schmidt JC, Soares MJ, Goldenberg S, Pavoni DP, Krieger MA (2011) Characterization of TcSTI-1, a homologue of stress-induced protein-1, in *Trypanosoma cruzi*. Mem I Oswaldo Cruz 106:70–77
169. Hombach A, Ommen G, Chrobak M, Clos J (2013) The Hsp90-Sti1 interaction is critical for *Leishmania donovani* proliferation in both life cycle stages. Cell Microbiol 15(4):585–600
170. Wiser MF, Jennings GJ, Uparanukraw P, Van Belkum A, Van Doorn LJ, Kumar N (1996) Further characterization of a 58 kDa *Plasmodium berghei* phosphoprotein as a cochaperone. Mol Biochem Parasitol 83(1):25–33
171. Wiser MF, Plitt B (1987) *Plasmodium berghei*, *P. chabaudi*, and *P. falciparum*: similarities in phosphoproteins and protein kinase activities and their stage specific expression. Exp Parasitol 64(3):328–335
172. Ommen G, Lorenz S, Clos J (2009) One-step generation of double-allele gene replacement mutants in *Leishmania donovani*. Inter J Parasitol 39(5):541–546
173. Dores-Silva PR, Silva ER, Gomes FER, Silva KP, Barbosa LRS, Borges JC (2012) Low resolution structural characterization of the Hsp70-interacting protein—Hip—from *Leishmania braziliensis* emphasizes its high asymmetry. Arch Biochem Biophys 520(2):88–98

174. Prapapanich V, Chen SY, Toran EJ, Rimerman RA, Smith DF (1996) Mutational analysis of the hsp70-interacting protein hip. Mol Cell Biol 16(11):6200–6207
175. Irmer H, Hohfeld J (1997) Characterization of functional domains of the eukaryotic co-chaperone hip. J Biol Chem 272(4):2230–2235
176. Velten M, Villoutreix BO, Ladjimi MM (2000) Quaternary structure of the HSC70 cochaperone HIP. BioChemistry 39(2):307–315
177. Velten M, Gomez-Vrielynck N, Chaffotte A, Ladjimi MM (2002) Domain Structure of the HSC70 Cochaperone, HIP. J Biol Chem 277(1):259–266
178. Gebauer M, Zeiner M, Gehring U (1997) Proteins interacting with the molecular chaperone hsp70/hsc70: physical associations and effects on refolding activity. FEBS Lett 417(1):109–113
179. Hohfeld J, Jentsch S (1997) GrpE-like regulation of the Hsc70 chaperone by the anti-apoptotic protein BAG-1. EMBO J 16(20):6209–6216
180. Prapapanich V, Chen SY, Nair SC, Rimerman RA, Smith DF (1996) Molecular cloning of human p48, a transient component of progesterone receptor complexes and an hsp70-binding protein. Mol Endocrinol 10(4):420–431
181. Dores-Silva PR, Minari K, Ramos CHI, Barbosa LRS, Borges JC (2013) Structural and stability studies of the human mtHsp70-escort protein 1: an essential mortalin co-chaperone. Int J Biol Macromol 56(0):140–148
182. Tobaben S, Varoqueaux F, Brose N, Stahl B, Meyer G (2003) A brain-specific isoform of small glutamine-rich tetratricopeptide repeat-containing protein binds to Hsc70 and the cysteine string protein. J Biol Chem 278(40):38376–38383
183. Yin H, Wang H, Zong H, Chen X, Wang Y, Yun X, Wu Y, Wang J, Gu J (2006) SGT, a Hsp90 binding partner, is accumulated in the nucleus during cell apoptosis. Biochem Biophys Res Comm 343(4):1153–1158
184. Liou S, Cheng M, Wang C (2007) GT2 and MDY2 interact with molecular chaperone YDJ1 in *Saccharomyces cerevisiae*. Cell Stress Chaperon 12(1):59–70
185. Oliveira CLP, Borges JC, Torriani IL, Ramos CHI (2006) Low resolution structure and stability studies of human GrpE#2, a mitochondrial nucleotide exchange factor. Arch Biochem Biophys 449(1–2):77–86
186. Kang CB, Hong Y, Dhe-Paganon S, Yoon HS (2008) FKBP family proteins: immunophilins with versatile biological functions. Neurosignals 16(4):318–325
187. Lee J, Kim SS (2010) An overview of cyclophilins in human cancers. J Inter Med Res 38(5):1561–1574
188. Benham AM (2012) The protein disulfide isomerase family: key players in health and disease. Antioxid Redox Sign 16(8):781–789
189. Golden T, Swingle M, Honkanen R (2008) The role of serine/threonine protein phosphatase type 5 (PP5) in the regulation of stress-induced signaling networks and cancer. Cancer Metast Rev 27(2):169–178
190. Kutuzov MA, Andreeva AV (2008) Protein Ser/Thr phosphatases of parasitic protozoa. Mol Biochem Parasitol 161(2):81–90
191. Shahinas D, Liang M, Datti A, Pillai DR (2010) A repurposing strategy identifies novel synergistic inhibitors of *Plasmodium falciparum* heat shock protein 90. J Med Chem 53(9):3552–3557

Index

W. A. Houry (ed.), *The Molecular Chaperones Interaction Networks in Protein Folding and Degradation,* Interactomics and Systems Biology, DOI 10.1007/978-1-4939-1130-1,

MIX
Papier aus verantwortungsvollen Quellen
Paper from responsible sources
FSC® C105338

If you have any concerns about our products,
you can contact us on
ProductSafety@springernature.com

In case Publisher is established outside the EU,
the EU authorized representative is:
Springer Nature Customer Service Center GmbH
Europaplatz 3, 69115 Heidelberg, Germany

Printed by Libri Plureos GmbH
in Hamburg, Germany